T0358617

Principles of Occupational Health and Hygiene: An Introduction

Now in its fourth edition, this book allows for early career occupational hygienists and occupational health and safety professionals or students to develop their basic skills and knowledge to anticipate, recognise, evaluate and control workplace hazards that can result in injury, illness, impairment or affect the well-being of workers and members of the community. *Principles of Occupational Health and Hygiene: An Introduction* offers a comprehensive overview of occupational health risks and hazardous environments encountered in a range of industries and organizational settings.

This new edition offers information on the current techniques and equipment used in assessing workplace hazards. Methods of assessment are developing at a rapid rate due to the new technologies now available. Featuring new chapters on occupational hygiene statistics and psychosocial hazards and fully updated throughout, leading industry professionals and educators explain how to identify key workplace hazards including chemical agents such as dusts, metals and gases; physical agents such as noise, radiation and extremes of heat and cold; and microbiological agents. The book highlights assessment procedures and processes for identifying exposure levels and explains how to evaluate risk and follow safety guidelines to control and manage these hazards effectively. Highly illustrated, up to date with current Workplace Health and Safety legislation and written in a jargon-free manner, this book will be a bible to any student or professional.

Principles of Occupational Health and Hygiene: An Introduction is an essential reference for students, early career Occupational Hygienists professionals and anyone in an Occupational Health and Safety role.

Principles of Occupational Health and Hygiene

An Introduction

Fourth Edition

Edited by

Sue Reed

CRC Press is an imprint of the
Taylor & Francis Group, an **informa** business

Designed cover image: custom designed by the editor team, from images provided by Dr Ross Di Corletto, Martha Maravi, Melanie Reed and courtesy of Nyrstar.

Fourth edition published 2025
by CRC Press
2385 NW Executive Center Drive, Suite 320, Boca Raton FL 33431

and by CRC Press
4 Park Square, Milton Park, Abingdon, Oxon, OX14 4RN

CRC Press is an imprint of Taylor & Francis Group, LLC

First edition published by Allen & Unwin 2007
Second edition published by Allen & Unwin 2013
Third edition published by Allen & Unwin 2019

Library of Congress Cataloging-in-Publication Data
Names: Reed, Susan (Occupational hygienist), editor.
Title: Principles of occupational hygiene : an introduction / edited by Sue Reed.
Description: Fourth edition. | Boca Raton, FL : CRC Press, 2025. | Includes bibliographical references
and index. | Contents: The hazardous work environment: the hygiene challenge / Charles Steer,
Melanie Windust and Kate Cole, OAM -- Occupational health, basic toxicology andepidemiology /
Martyn Cross and Geza Benke -- The concept of the exposure standard / Robert Golec and Ian Firth. |
Summary: "Now in its fourth edition, Principles of Occupational Health and Hygiene:An Introduction allows
for early career occupational hygienists and occupational health and safety professionals or students to develop
their basic skills and knowledge to anticipate, recognize, evaluate, and control workplace hazards that can result in
injury, illness or affect the well-being of workers and the community. It offers a comprehensive overview of
occupational health risks and hazardous environments encountered in varying industries. This book is an essential
reference for students, early career Occupational Hygienists professionals and anyone in an Occupational Health
and Safety role"-- Provided by publisher. Identifiers: LCCN 2024024503 (print) | LCCN 2024024504 (ebook) |
ISBN 9781032645858 (hbk) | ISBN 9781032590578 (pbk) | ISBN 9781032645841 (ebk)
Subjects: LCSH: Industrial hygiene. | Industrial safety.
Classification: LCC RC967 .P729 2025 (print) | LCC RC967 (ebook) |
DDC 613.6/2--dc23/eng/20240906
LC record available at https://lccn.loc.gov/2024024503
LC ebook record available at https://lccn.loc.gov/2024024504

ISBN: 978-1-032-64585-8 (hbk)
ISBN: 978-1-032-59057-8 (pbk)
ISBN: 978-1-032-64584-1 (ebk)

DOI: 10.1201/9781032645841

Typeset in Times
by SPi Technologies India Pvt Ltd (Straive)

Contents

Foreword

This book is for students and early career practitioners of occupational hygiene, whether they be full-time professional hygienists working for employers or working as consultants or as is commonly the case, workers for whom occupational hygiene is part of their responsibilities.

The Australian Institute of Occupational Hygienists (AIOH) defines occupational hygiene as the art and science of occupational disease prevention. More specifically, occupational hygiene addresses the anticipation, recognition, evaluation, communication and control of environmental hazards in or arising from, the workplace that can result in injury, illness, impairment, or affect the well-being of workers and members of the community.

The importance of occupational hygiene has been increasingly recognised over the past three centuries. In 1775, Percival Pott, a UK surgeon noted an increased incidence of scrotal cancer in chimney sweeps in the United Kingdom, documenting the first causal association between workplace exposure to hazards and the subsequent development of disease. Since this time, we have continued to see clusters of occupational diseases associated with particular industries and hazards, including the prevalence of asbestos related lung diseases in the mid-late 20th century and the recent recognition of accelerated silicosis in manufactured stone workers in 2018. No doubt, more hazards will emerge in the future. With technological change comes changes to workplaces and hazards and the continued need for the fundamental principles of occupational hygiene – hazard anticipation and recognition and risk evaluation and control.

To promote these principles, the first edition of *Principles of Occupational Health and Hygiene: An Introduction* was published in 2007. This book was based on the 1992 book *Occupational Health and Hygiene Guidebook for the WHSO* (Workplace Health and Safety Officer) written by Dr David Grantham who generously ceded the rights and royalties for this book to AIOH in 2002. The new book, aimed at a wider audience of health and safety professionals, includes issues and topics that had become important since the original 1992 book was published. Many respected authors, most of them AIOH members, freely contributed their time to review the text and deliver updated chapters on their areas of expertise.

In 2013, a second edition of the *Principles of Occupational Health and Hygiene: An Introduction* was published. By this time, the book had been adopted as a valued reference source by many occupational health and hygiene professionals and as a standard textbook by a growing number of educational institutions. A third edition was published in 2019.

The fourth edition retains the original focus, aimed at a wide audience of health and safety professionals; it includes issues whose significance to workers, employers and PCBUs has risen appreciably since the third edition was published. The authors, again most of them AIOH members with demanding professional responsibilities, have freely contributed their valuable time. This select expert group, along with Sue Reed who has taken on the herculean task of editing the book, representing some of the best practitioners of our profession, deserves our deepest gratitude for producing the most significant body of work underpinning the practice of occupational hygiene in Australia today.

The AIOH sees this as a significant contribution to the Australian Work Health and Safety Strategy 2023–2033, promoted by Safe Work Australia, and specifically action areas 1 and 4.

Occupational hygienists' principal responsibility is to worker health. In meeting this, we must be skilled in the anticipation, recognition, evaluation, communication and control of physical, chemical and biological hazards in a diverse range of work environments. This is true for hygienists as a group but, as with most professions, no individual can expect to know everything. Happily, the *Principles of Occupational Health and Hygiene: An Introduction*, distils the knowledge of

Australia's best, most respected and experienced hygienists, across a comprehensive range of issues, providing a robust professional reference for the practice of our science. It is an ideal resource for occupational hygiene and occupational health and safety students, professionals and anyone working in an occupational health and safety role. The increased use of this book as a textbook and reference source will promote the correct and consistent application of the principles of occupational hygiene to maintain and intensify the focus on occupational disease prevention in Australia and overseas.

Jeremy Trotman
AIOH President, 2024

Preface

In 2003, an ambitious project was commenced by the AIOH to revise and expand David Grantham's (AOM) *Occupational Health and Hygiene Guidebook for the WHSO* under the guidance of Dr Cherilyn Tillman. The result, published in 2007, was the first edition of this book and covered a broad range of topics, which have continued to grow in importance. The second edition was published in 2013, the third edition was published in 2019 and at the beginning of 2023, the AIOH initiated work on this fourth edition.

The editor and the Council of the AIOH thank all the authors and reviewers for their professionalism and attention to detail and for their generous collaboration with us during this project. The majority of the 39 authors of this fourth edition are members of the AIOH. All authors have written the chapters in their own time, without payment or favour, to support the AIOH in its role of furthering the practice and recognition of occupational hygiene in the Australasian region, and in support of the Australian Work Health and Safety Strategy 2023–2033.

I would like to thank Dr David Grantham for his interest, encouragement and assistance in updating and expanding his original material. In addition, I would like to thank the Council of the AIOH for inviting me to serve as editor of the fourth edition. It was an honour to be entrusted with delivery of this important publication. I would also like to thank Prof Dino Pisaniello and Dr Geza Benke for their editorial advice early in the project.

Where appropriate, the book emphasises issues as seen in Australian workplaces such as the coverage of thermal stress, focussing on heat rather than cold. The book refers to Australian legislation where it is appropriate but in the main legislation, it is not covered in depth due to differences between states. Australian standards are only referred to where appropriate and, in some areas, such as indoor air quality, international standards are discussed. Occupational exposure standards are referred to as OES, and not WES, due to the broad readership of the book and similarly OHS or H&S is used in preference to WHS.

One of the greatest changes to the management of workplace airborne contaminant exposures seen by a generation of Occupational Hygienists in Australia will be heralded by the implementation of Workplace Exposure Limits (WELs), which will come into force on 1st December 2026. The WELs will replace the current Workplace Exposure Standards (WESs), the majority of which have not changed since their initial introduction by the National Occupational Safety and Health Commission (NOHSC) over 30-years ago in 1995. The WELs will signify the culmination of a more than eight-year process of major revision of OEL setting methodology in Australia and the review of the limits for over 750 individual substances and consequent changes to 286 of these limits.

Finally, I would like to thank the families of all the contributors for their patience, understanding and support during this project, which consumed far more of everyone's lives and free time than it was envisaged.

Honorary Associate Professor Sue Reed
Perth, 2024

Editor Biography

Sue Reed is a Fellow member of the AIOH (FAIOH) and a Certified Occupational Hygienist (COH)®, a retired CIH and a past President of the Australian Institute of Occupational Hygienists. Sue has a PhD, a Master of Science (Occupational Hygiene), a Master of Engineering Science and Bachelor of Science (Textile Engineering). Sue has over 45 years' experience in occupational hygiene as a consultant, in government roles, and 35 years as an academic. She is a Honorary Associate Professor at Edith Cowan University, Perth, and a director of a small occupational hygiene consultancy. Sue has published over 50 scientific papers and technical reports on a range of topics including chemical exposure assessment and control, noise, indoor air quality and OHS education.

Author Biographies

Linda Apthorpe FAIOH COH (*https://orcid.org/0000-0003-3025-6325*) has a master's degree in occupational hygiene practice. She has been working in occupational hygiene since 1994, and currently lectures in occupational hygiene at University of Wollongong, provides consultancy and laboratory services to a wide variety of workplaces and facilitates training for occupational hygiene students and worker groups. With her analytical background in asbestos, dust and quartz, Linda contributes volunteer technical assistance to the AIOH, National Association of Testing Authorities (NATA) and Proficiency Testing Australia. She has also authored or co-authored various papers in the occupational hygiene field.

Anthony Bamford MAIOH, COH (*https://orcid.org/0009-0006-6012-2255*) has qualifications in Environmental Health and Mine Ventilation Engineering. He started his career in occupational hygiene in 2007 at Anglo American Thermal Coal, where he managed a team of occupational hygienists and ventilation officers. Anthony moved to Australia and has worked as a consultant with focus on control selection, implementation and effectiveness with a keen interest in control technology and real-time sensing. Currently Anthony is the Group Principal Hygienist at South32 overseeing multi-national operations. Anthony contributes towards the AIOH, ICMM and WASMA on various working groups and mentoring programmes.

Geza Benke FAIOH COH (*https://orcid.org/0000-0001-9019-9364*) has a BSc (Physics), MAppSc (Environmental Engineering) and PhD (Epidemiology). He worked with the Victorian Environment Protection Authority for five years before joining the Victorian state government's Occupational Hygiene branch in 1985. His work in occupational hygiene mainly involved noise assessment and control, radiation safety and asbestos work. Since 1994, Geza has undertaken research in a range of occupational and environmental epidemiology studies. He is currently Senior Research Fellow with the Centre for Occupational and Environmental Health, Department of Epidemiology and Preventive Medicine, Monash University. Geza is a past President of the Australian Institute of Occupational Hygienists.

Claire Bird (*https://orcid.org/0009-0009-9373-695X*) holds an Honours degree in Environmental Science (environmental toxicology and air quality) and a doctoral degree in the ecology of bioaerosols. She has conducted academic research and private consultancy in the microbiological and chemical aspects of indoor air quality for 20 years in the United Kingdom and Australia. She is a highly active advocate for indoor air quality, helping drive several national and international not-for-profit organisations, leading technical groups and providing technical input across occupational exposure, pathogen transmission, air handling and indoor air quality. She is the Director of LITMAS Pty Ltd, a laboratory facility conducting microbial and molecular analytical services associated with building contamination.

Jodie Britton (*https://orcid.org/0009-0002-6018-5265*) MAIOH has a MSc Occupational Hygiene Practice and a Graduate Diploma in Occupational Health and Safety. Jodie is a Certified Occupational Hygienist with over 32 years of experience working in heavy industry. Jodie has extensive practical experience which compliments her current role within the Rio Tinto Global Health team as a Senior Advisor – Occupational Hygiene. Her areas of interest include the thermal work environment and coal tar pitch volatiles (CTPV).

Kerrie Burton FAIOH, COH (*https://orcid.org/0009-0002-9796-0720*) has a PhD and Masters of Science by research in Occupational Hygiene, focussed on the effectiveness of respiratory protection against diesel particulate matter. She has a Graduate Diploma of Occupational Hygiene, as well as an undergraduate degree majoring in Chemistry. Kerrie is Senior Occupational Hygiene Specialist for Coal Services and has over 20 years of experience in manufacturing and 5 years of experience in mining. She is experienced at facilitating the Basic Principles of Occupational Hygiene Course.

Ron Capil FAIOH COH Retired has a background in Chemistry and a Graduate Diploma in occupational hygiene. He has over 45 years of experience including a decade in environmental monitoring and the remainder in occupational hygiene. His occupational hygiene work has involved workplace environment assessments in aluminium smelters and alumina refineries with a major focus on asbestos removal and management. He also provided strategic guidance as principal advisor health and hygiene to a leading Australasian aluminium industry. Ron continued to provide occupational hygiene consulting services to the aluminium industry, other resources companies as well as mentoring until his retirement in 2022.

Kate Cole OAM FAIOH COH (*https://orcid.org/0000-0002-4612-4565*) has a BSc (Biotechnology) and master's degrees in engineering and occupational hygiene. She is a Winston Churchill Fellow and Top 100 Women of Influence (Australian Financial Review). Kate's contributions to workplace health and safety earned her a Medal (OAM) of the Order of Australia and inclusion in the COVID-19 Honour Roll in the 2022 Australia Day Honours. Kate works as an independent consultant. Kate is a PhD candidate at the University of Sydney where she is researching respirable crystalline silica exposures to tunnel construction workers. Kate is a past President of the Australian Institute of Occupational Hygienists.

Martyn Cross MAIOH Retired began his career as a Toxicologist working in the UK chemical industry. He has an honours degree in Toxicology, a Master of Public Health and a PhD. Martyn has considerable experience in Occupational Hygiene, as Occupational Hygienist for WorkSafe WA, the WA Department of Health, and Principal Occupational Hygienist for Minara Resources. Martyn has 40+ years as a safety professional and Occupational Hygienist in a variety of industries. He is currently Honorary Senior Lecturer at Edith Cowan University and provides occupational hygiene consulting services.

Margaret Davidson MAIOH (*https://orcid.org/0000-0002-8426-1843*) is a senior lecturer and researcher in the field of occupational hygiene and environmental health at Western Sydney University. Margarets' PhD was on the impact of smoking bans on indoor and outdoor air quality in NSW licensed clubs. Research interests include the fields of biological hazards and bioaerosols, having completed post-doctoral research on respiratory hazards in Colorado dairies at the NIOSH High Plains Intermountain Centre for Health and Safety. Current research projects include promoting health and safety in the Australian industrial hemp industry and worker exposure to respirable crystalline silica and solvents in the use of artificial stone products.

Ross Di Corleto FAIOH COH (*https://orcid.org/0000-0001-5130-298X*) has a BSc in applied chemistry, a graduate diploma in occupational hygiene, a research master's degree in heat stress and a PhD in occupational health in the area of polycyclic aromatic hydrocarbon exposure. He has worked for over 40 years in the areas of power generation, minerals, mining, refining and smelting. This has predominantly been across the chemical, health, safety and environment fields, with occupational hygiene the main emphasis. The thermal environment and its impact on the employees is a key area of Ross's interest and involvement. He is currently the Principal Consultant at Monitor Consulting Services. Ross is a past President of the Australian Institute of Occupational Hygienists.

Ian Firth FAIOH COH has an MSc in Applied Biology (toxicology) and over 40 years' work experience, including a decade in environmental sciences and the rest in occupational hygiene. His environmental work involved research on freshwater animal toxicology, environmental management and acid rock drainage management. His occupational hygiene work has involved workplace environment assessment and management in hard rock mining and smelting in zinc, lead and aluminium industries. Ian provided tactical and strategic advice on health management as the corporate principal adviser of a leading international resources company. He is currently an occupational hygiene consultant. Ian is a past President of the Australian Institute of Occupational Hygienists.

Robert Golec FAIOH COH (*https://orcid.org/0009-0006-9960-8543*) has a degree in Applied Chemistry, a Graduate Diploma in Analytical Chemistry, and a Master of Applied Science in Applied Toxicology. He has over 40 years' experience in occupational hygiene, initially with Victorian State government and then in consulting as Principal Occupational Hygienist/Director with AMCOSH Pty Ltd. He is a member of the AIOH Exposure Assessment Committee, Standards Australia Committee CH/31 Methods for Examination of Workplace Atmospheres, a Technical Assessor for Occupational Hygiene monitoring and analysis and a former member of the Life Sciences Accreditation Advisory Committee; and Occupational Hygiene Technical Advisor to NATA.

Terry Gorman FAIOH COH retired has a Master's degree in Safety Science, is a Certified Occupational Hygienist and has been involved in workplace safety and hygiene for over 30 years. He is a member of the Australian/New Zealand Standard Committees responsible for respiratory protection (AS/NZS 1715 & 1716) as well as the Eye/Face Protection Standards Committee (AS/NZS 1337 & 1338). He currently represents Standards Australia on the International Standards Organisation (ISO) Committee TC94/SC15, a team of international representatives creating a set of global respiratory standards.

Beno Groothoff FAIOH COH retired studied mechanical engineering in the Netherlands. In Australia, he gained a degree majoring in OHS. He worked in private practice and government positions in environmental protection, OHS and occupational hygiene, conducting compliance audits and assessments in a variety of industries. In 2015 he was course coordinator at QUT's Occupational Hygiene and Toxicology course. He is a part-time lecturer at UQ, lecturing post-graduate and masters' students. He is a member of the AV10 Committee of Standards Australia on noise and vibration. Beno is a past President of the Australian Institute of Occupational Hygienists.

Jennifer Hines FAIOH COH (*https://orcid.org/0000-0001-5429-6212*), an accomplished professional in the field of occupational hygiene, has completed her PhD, in emissions-based diesel engine maintenance and its workplace benefits. Her extensive industry experience includes roles in alumina refining, copper refining, smelting and underground coal mining. Through her consultancy company, EHS Solutions, she is committed to worker health protection and promoting and teaching occupational hygiene. Jennifer serves as a facilitator for the Australian Department of Defence Monitoring of Occupational Hygiene Course and the AIOH Basic Principles Course and lectures at the University of Wollongong. Her contributions extend to active participation on multiple committees within the AIOH.

Kelly Johnstone FAIOH COH (*https://orcid.org/0000-0002-4808-6435*) has a Bachelor of Applied Science with Honours, Master of Science (Occupational Hygiene Practice) and a PhD. She has over 25 years of occupational hygiene and OHS experience and has worked in academia as well as consultancy with experience in construction, oil and gas, transport, and agriculture.

Kelly is currently an Associate Professor and Director of the Occupational Hygiene program at The University of Queensland. She is a member of the AIOH Professional Development and Education Committee and has served on the AIOH Council of Management.

Ryan Kift MAIOH (*https://orcid.org/0000-0001-8282-3868*) has a PhD in Occupational Hygiene and undergraduate degrees in Environmental health and occupational health and environment. Ryan has over 20 years of occupational hygiene and safety experience in academic, industrial and local government settings. He has been involved in occupational hygiene management on large construction projects, and worked in the oil and gas and mining industries. Ryan is currently a Associate Professor at CQUniversity, Australia and a Senior Consultant for GCG Health, Safety, Hygiene. Ryan has published over 50 scientific papers in the areas of biological and dust monitoring, cognitive ergonomics and safety management.

Peter Knott FAIOH COH (*https://orcid.org/0000-0002-4775-7567*) holds a Master's Degree in Clinical Epidemiology and Bachelor of Applied Science in Environmental Health. Peter has over 35 years of experience in occupational hygiene in heavy industry, mining, manufacturing and consulting, where he is now Principal Hygienist with GCG Health Safety Hygiene. His current interests include the use of Bayesian statistical approaches for occupational hygiene data and risk communication and the effects of multipollutant exposures on worker health.

SoYoung Lee (*https://orcid.org/0000-0002-8215-9183*) has experience in the fields of academia expertise, occupational hygiene laboratory and industries for over 10 years. She has worked with industry and undertaking research in the field of Work and Vision with a particular specialization in exposure assessment of light sources used in the workplaces. Her research was about the photochemical damage from blue light exposure in the workplace and she is also interested in the health effects of workers from exposure to intense light sources. She is working in the mining industry and managing Health and Hygiene for workers based on hygiene monitoring.

Adélle Liebenberg FAIOH, COH (*https://orcid.org/0000-0003-0467-9412*) has a PhD in Environmental and Occupational Health and has been a lecturer in Occupational Hygiene for 10 years, currently serving as the Course Coordinator for the postgraduate Occupational Hygiene programs at Edith Cowan University. She started her career in occupational hygiene in 2003 at Anglo Ashanti, South Africa and moved to Australia in 2009 where she took up a consultant role as Occupational Hygienist with Simtars and then AECOM. Adèlle has experience in multiple industry sectors including gold mining, coal mining, manufacturing, and agriculture. Adèlle contributes to the AIOH in various committees including Chair of the COH Board.

Elaine Lindars FAIOH COH (*https://orcid.org/0009-0001-8401-691X*) has a geochemistry PhD, an Industrial Hygiene post-graduate and an environmental chemistry honours degree. She is Chartered Chemist (RACI), a Certified OHS Practitioner (AIHS) and an ECU adjunct senior lecturer. Elaine has volunteered with the AIOH for over 15 years and was the founder/first President of WHWB (Australia). Elaine has 35 years' experience in occupational, environmental and analytical chemistry, has worked in mining and engineering companies, taught at several universities, teaches the OHTA/BOHS hygiene courses, has more than 20 scientific papers and is co-owner of RED OHMS Group (consultancy specialising in hygiene, environment and training).

Dr Michael Logan is the Director Scientific Branch at the Queensland Fire and Emergency Services. His research interests focus on integrating science and risk management to drive strategic and operational performance enhancement n emergency management, research and training organisations. He is internationally recognised in HAZMAT/CBRN (Chemical Biological Radiological Nuclear) emergency management and has been adept at identifying emerging trends and opportunities across all spheres of emergency management.

David Lowry MAIOH COH (*https://orcid.org/0000-0002-9216-9350*) holds BSc degrees in Biological Science and Occupational Therapy, and a PhD in Public Health focussing on professional judgement and exposure assessment. David has published several peer-reviewed articles, has presented original research nationally and internationally at various scientific conferences and holds an adjunct appointment as Associate Professor at Edith Cowan University's School of Medical and Health Sciences. David has spent the last 12 years working for Rio Tinto, and in his current role as Principal Occupational Hygienist works across mining, supply chain, projects and exploration activities within Western Australia.

Ken Martinez CIH (*https://orcid.org/0000-0001-7651-8023*) is a Certified Industrial Hygienist and Environmental Engineer experienced in conducting large-scale research; managing programs in occupational safety and health, and emergency response; and teaching professional development courses. Brings 33 years of CDC expertise in the area of hazardous agent exposure characterisation and mitigation control practices in the manufacturing and healthcare industry. Subject matter expert in biological agents including infectious disease and bioterrorism agents. Bioaerosol research efforts have resulted in 38 peer-reviewed journal articles and book chapters and over 100 technical presentations. He currently serves as the Chief Science Officer for a non-profit (Integrated Bioscience and the Built Environment Consortium – IBEC).

Martin Mazereeuw MAIOH (*https://orcid.org/0009-0009-7960-570X*) gained a PhD in Analytical Chemistry at Leiden University in The Netherlands and has since worked in several analytical and managerial roles within biotechnology, research support and large industry. He has published over 15 scientific papers and book chapters, has been a GLP study director for food safety studies and implemented the first NATA accredited ISO17025 system for research within NSW. Martin joined TestSafe Australia in 2012 as the manager of the Chemical Analysis Branch of TestSafe Australia, which is part of SafeWork NSW.

Gregory E. O'Donnell MAIOH (*https://orcid.org/0000-0003-0276-9721*) has a PhD in chemistry, a degree in Applied Science (Chemistry), is a Chartered Chemist with the Royal Australian Chemical Institute and a member of the AIOH. He has worked in occupational exposure chemical analysis for over 27 years with SafeWork NSW, a member/chair of the SafeWork NSW BOEL committee; chairman of the Standards Australia committee CH-31 Methods for Examination of Workplace Atmospheres; member of the AIOH Exposure Standards Committee; a member of the Safe Work Australia's Health Monitoring Review Expert Working Group. Published 14 scientific papers and over 35 conference presentations in occupational health/hygiene and biological monitoring.

Bruno Piccoli (*https://orcid.org/0000-0003-0520-6924*), Occupational Physician graduated and specialised in OH&S at the Dept. of Occupational Health, Milan State University. Founder and first Chair of the ICOH SC on Work and Vision (2003) and coordinator of several editions of the Italian Guidelines on 'VDU work'. Currently Professor in OH&S (retired) at the Department of Biomedicine of the University of Tor Vergata in Rome, Italy, and Adjunct Professor at the University of Adelaide. Bruno is the author of 154 scientific articles and over 200 conference communications on long-term effect of near vision, indoor lighting assessment, operator's ocular surface disorders due to office indoor pollutants.

Dino Pisaniello FAIOH COH retired (*https://orcid.org/0000-0002-4156-0608*) has a PhD in chemistry and a master's degree in Public Health. He is a Fellow of the AIHS and the Royal Australian Chemical Institute. Dino has served as President of AIOH, Chairman of the Congress of Occupational Safety and Health Association Presidents (2001-5) and Australian Secretary for the International Commission on Occupational Health. He is currently Professor in Occupational and Environmental Hygiene in the School of Public Health at the University of Adelaide. Dino has published over 250 scientific papers and technical reports. His research interests include chemical exposure assessment and control, intervention research, occupational and environmental epidemiology, work and vision, heat and work injury, and health and safety education. Dino is a past President of the Australian Institute of Occupational Hygienists.

Martin Ralph (*https://orcid.org/0000-0002-5893-6886*) has a PhD and BSc (Honours) and is a Fellow of the AIHS and a CRadP. Martin's career started in the uranium industry before joining the Western Australian mine safety regulator. Later as the managing director of a not-for-profit organisation, he designed the regulator-approved courses for applied occupational hygiene for the WA mining industry; was the radiation safety officer for a mining company; and was an expert witness for the Victorian Civil Administrative Tribunal. Martin rejoined the WA mining regulator in 2017 and leads the team that identifies and responds to mine worker health hazards such as silica, noise, diesel particulate and radiation.

Sue Reed FAIOH COH (*https://orcid.org/0000-0003-1384-5208*) has a PhD and a Master of Science in Occupational Hygiene and is a retired CIH. Sue has over 45 years of experience in occupational hygiene as a consultant, in government roles, and as an academic. She is a Honorary Associate Professor at Edith Cowan University, Perth, and a director of a small occupational hygiene consultancy. Sue has published over 50 scientific papers and technical reports on a range of topics including chemical exposure assessment and control, noise, indoor air quality and OHS education. Sue is a past President of the Australian Institute of Occupational Hygienists.

Mark Reggers MAIOH COH (*https://orcid.org/0009-0001-6079-7926*) is a Certified Occupational Hygienist, has a master's degree in Science (Occupational Hygiene Practice) and currently works for the 3M Personal Safety Division as a Specialist Application Engineer Respiratory Protection Portfolio. He has over 20 years of experience in the safety equipment and training industry. Mark is the Chair of RESP-FIT, which is a respiratory fit testing training and accreditation program (AIOH). Mark is currently the chair of the Australian/New Zealand Standard Committee SF-010 Respiratory Protection and a member of SF-060 Eye/Face Protection. He represents Standards Australia on ISO TC94 SC15 WG6 Respiratory Protection. Mark co-hosts the '3M Science of Safety Podcast'.

Joshua Schaeffer (*https://orcid.org/0000-0003-3367-4696*) is an Associate Professor at Colorado State University in the Department of Environmental and Radiological Health Sciences and an adjunct Associate Professor at the Colorado School of Public Health (Department of Environmental and Occupational Health). He has over 20 years of experience in industrial hygiene that integrates a background in biochemistry and toxicology. Dr Schaeffer's research assesses occupational and environmental exposure and health outcomes. This broad focus includes measuring bioaerosols, characterising the aerobiome (including pathogens, antibiotic resistant genes and viruses), modelling exposure–response relationships in the context of the microbiome, inflammatory markers, and pulmonary function and evaluating control strategies.

Michael Shepherd FAIOH COH has a BSc (Chemistry), Graduate Diploma in OHS, and is a Chartered Chemist. Michael served as a AIOH Council member in 2010-2012 and 2015, Chair of the Membership and Qualifications Committee from 2010 to 2023 and is on the AIOH Certified

Occupational Hygiene Examiners panel. Michael is a NATA Technical Assessor and contributes to industry standards as a committee member of Standards Australia CH-31 Methods for Examination of Workplace Atmospheres. Michael has over 30 years of experience in multidisciplinary occupational health and hygiene assessments and advises government on health and safety policy and legislation. He is the Principal Occupational Hygienist for COHLABS Pty Ltd.

(John) Charles Steer FAIOH COH retired has a BAppSc (Applied Chemistry), a GDip (Environmental Studies) and completed the 13-week University of Sydney postgraduate occupational hygiene course. His experience includes occupational hygiene, health, safety, environment, risk management and chemical technology holding senior corporate positions in these areas in the electricity industry until mid-2000. His subsequent consultancy business included developing sustainable occupational health and hygiene management frameworks and mentoring as well as exposure assessment and hands on monitoring. He retired in mid-2020. Charles is a past President of the Australian Institute of Occupational Hygienists (AIOH) and inaugural Chair of the AIOH Foundation.

Aleks Todorovic MAIOH (*https://orcid.org/0009-0000-0293-3240*) has a Master of Science (Occupational Hygiene Practice). He has over 28 years of experience as a supplier of occupational hygiene monitoring equipment and 23 years has owned and managed Active Environmental Solutions (AES). He has been a guest lecturer and presenter at tertiary institutions and associations. Aleks has been involved in the development and testing of many new types of monitoring instrumentation, focusing on chemicals in the petrochemical, defence and emergency services industries. His specific interest is in the advancement of wearable connected real-time wireless monitoring solutions with an emphasis on total worker exposure from simultaneous multiple stressors. Aleks will be President of the Australian Institute of Occupational Hygienists (AIOH) in 2025.

Jane Whitelaw FAIOH COH (*https://orcid.org/0000-0003-2077-9640*) has a PhD in Medicine, Master of Applied Science, a Post Graduate Diploma in OHS and is a Certified Industrial Hygienist. She has been an active member of the AIOH since 1984. She is currently Academic Program Director for OHS at the University of Wollongong, a member of the AIOH PD&E committee, a RESP-FIT Board member and the AIHS representative on the Standards Australia SF-10 committee. Jane's research interests are in protecting worker health from chemical and physical hazards, and her major grants and research have been in evaluating respiratory health effects and the efficacy of protective equipment.

Melanie Windust MAIOH COH has a BSc in Chemistry, Post Graduate Diploma in Occupational Hygiene and Masters in Science (Health and Safety). Melanie has over 25 years of experience in a wide variety of industries, demonstrating a strong track record in ensuring workplace safety and health. Her expertise spans across diverse and complex sectors, including major hazard facilities, aviation, mining, dangerous goods, logistics, oil and gas, regulatory compliance and manufacturing. Melanie was the 2010 Drager Young Hygienist Awardee and is current Chair of the AIOH Awards Committee, Inaugural (2017) Director (Secretary) of the AIOH.

Acknowledgements

The AIOH editorial committee (Dr Sue Reed, Prof Dino Pisaniello and Geza Benke) gratefully acknowledges the pioneering work of Dr David Grantham (AM) on which this book was originally based.

We also acknowledge the invaluable efforts of Dr Cherilyn Tillman, editor of the first edition along with authors who decided not to be involved in the later editions but have allowed their work to be used, and this is gratefully acknowledged. The authors from the first edition who are not part of the later editions are: Dr Ian Grayson (concept of the exposure standard), Dr Steven Brown (indoor air quality), Dr Georgia Sinclair (ionising radiation) and Dr Denise Elson (biological hazards).

The authors from the second edition who are not part of the later editions are: Dr David Grantham AM (Chapter 1: The hazardous work environment: the hygiene challenge; Chapter 3: The concept of the exposure standard; and Chapter 4: Control of workplace health hazards), Dr John Edwards (Chapter 2: Occupational health, basic toxicology and epidemiology; and Chapter 10: Biological and biological-effect monitoring), Geoff Pickford (Chapter 7: Aerosols), Dr Brian Davies AM (Chapter 7: Aerosols), Gary Rhyder (Chapter 8: Metals in the workplace), Noel Tresider AM (Chapter 9: Gases and vapours), Kate Leahy (Chapter 9: Gases and vapours), Michelle Wakelam (Chapter 13: Radiation-ionising and non-ionising), Sarah Thornton (Chapter 16: Biological hazards), Dr Howard Morris (Chapter 17: Emerging and evolving issues) and Dr Bob Rajan OBE (Chapter 17: Emerging and evolving issues).

The authors from the third edition who are not part of the fourth edition are: Caroline Langley (Chapter 1: The hazardous work environment: the hygiene challenge), Garry Gately (Chapter 4: Control of workplace health hazards; and Chapter 6: Personal protective equipment); Wayne Powys (Chapter 4: Control of workplace health hazards); Dr David Bromwich (Chapter 5: Industrial Ventilation); Roy Schmid (Chapter 13: Radiation-ionising and non-ionising); Martin Jennings (Chapter 15: Lighting); and Prof Dooyong Park (Chapter 17: Emerging and evolving issues).

The AIOH is grateful to the AIOH members and non-members who volunteered to peer-review the latest chapters (4th Edition) of the book, namely Dr Julia Norris (Chapter 1), Dr Roger Drew (Chapter 2), Jackii Shepherd (Chapter 3), Dr Greg O'Donnell (Chapter 4), Dr Ryan Kift (Chapter 5), Andrew Orfanos and Simon Witts (Chapter 6), Jane Whitelaw (Chapter 7), Assoc. Prof. Kelly Johnstone (Chapter 8), Robert Golec (Chapter 9), Dr Sharann Johnson AM (Chapter 10), Assoc. Prof. Kennedy Osakwe (Chapter 11), Dr Peter McGarry (Chapter 12), Dr Edmore Masaka (Chapter 13), Dr Adélle Liebenberg (Chapter 14), Assoc. Prof. Ken Karipidis (Chapter 15), Dr Vinod Gopaldasani (Chapter 16) and Melanie Cox (Chapter 17).

We would also like to thank the many people who have supported this edition by providing images; they are acknowledged individually throughout the text. Images without source citation have been provided by the chapter author/s. In addition, we would like to thank Melanie Reed who assisted in drawing and upgrading of many of the images.

Finally, the editor would like to thank the AIOH council for supporting the project and the AIOH Office for administrative support.

Abbreviations and Definitions

A	ampere
A/m	ampere per metre
ABCB	Australian Building Codes Board
AC	asbestos cement
ACGIH®	American Conference of Governmental Industrial Hygienists
ACH	air changes per hour
ACM	asbestos-containing materials
ADI	acceptable daily intake
ADME	absorption, distribution, metabolism and excretion
ADWG	Australian Drinking Water Guidelines
AFOM	Australasian Faculty of Occupational Medicine
AICIS	Australian Industrial Chemicals Introduction Scheme
AIDS	acquired immunodeficiency syndrome
AIHA	American Industrial Hygiene Association
AIHW	Australian Institute of Health and Welfare
AIOH	Australian Institute of Occupational Hygienists
AIRAH	Australian Institute of Refrigeration Air Conditioning and Heating
ALARA	as low as reasonably achievable
ALARP	as low as reasonably practicable
ALI	annual limit of intake
AMR	Australian Mesothelioma Registry
AMR	antimicrobial resistance (Chapter 13)
ANOVA	analysis of variance
ANSTO	Australian Nuclear Science and Technology Organisation
ANZ	Australian New Zealand
ANZEx	Australian Certification/Recognition System for Hazardous Area Equipment and Services
ARPANSA	Australasian Radiation Protection and Nuclear Safety Agency
ARPS	Australian Radiation Protection Society
AS	Standards Australia
AS/NZS	Australian/New Zealand Standard
ASHRAE	American Society of Heating, Refrigerating and Air-conditioning Engineers
ASTM	American Society for Testing and Materials
AT	apparent temperature
ATT	aerosol taste test
A-weighting	how the human ear responds to sound
B	magnetic flux density
BAT values	biological tolerance values (*Biologische Arbeitsstoff Toleranz Werte*)
BCIRA	British Cast Iron Research Association
BDA	Bayesian decision analysis
BEI®	biological exposure index
BeLPT	beryllium lymphocyte proliferation testing
BET	basic effective temperature
BFv	Barmah forest virus
BLV	biological limit values

BMGV	biological monitoring guidance value
BMRC	British Medical Research Council
BOEL	Biological Occupational Exposure Limits
BOHS & NVvA	British Occupational Hygiene Society & Nederlandse Vereniging voor Arbeidshygiëne
BOSSA	Building Occupant Satisfaction Survey Australia
Bq	becquerel
Bq/g	becquerel per gram
Bq/m^3	becquerel per cubic metre
BSE	bovine spongiform encephalopathy
BTRA	basic thermal risk assessment
C/kg	coulomb per kilogram
CCA	copper-chrome-arsenic
CCT	correlated colour temperature (k)
cd	candela
CDC	Centers for Disease Control
CE mark	affirms the goods' conformity with European health, safety and environmental protection standards
CEC	Commission of European Communities
CEN	European Committee for Standardization
CFC	chlorofluorocarbons
CFL	compact fluorescent lamp
Ci	curie
CI	chemical intolerance
CIE	International Commission on Illumination
CJD	Creutzfeldt–Jakob disease
Cl_2	chlorine
cm	centimetre
cm^2	centimetre square
CMDLD	Coal Mine Dust Lung Disease
CNC	condensation nuclei counter
CNP	controlled negative pressure
CNS	central nervous system
CNT	carbon nanotubes
CO	carbon monoxide
CO_2	carbon dioxide
COH	Certified Occupational Hygienist (COH)® status is recognition of a professionally competent, independent and ethical practitioner and an industry leader in occupational hygiene
COHb	blood carboxyhaemoglobin
COPD	chronic obstructive pulmonary disease
COSHH	control of substances hazardous to health
CPC	chemical protective clothing (Chapter 7)
CPCs	condensation particle counters (Chapter 12)
CRI	colour rendering index
CS	chemical sensitivity
CSA	Canadian Standards Association
CSIRO	Commonwealth Scientific and Industrial Research Organisation
CTPV	coal tar pitch volatiles

CWHSP Coal Workers Health Surveillance Program
CWP coal workers' pneumoconiosis
d dose
D absorbed dose
d-ALA *d*-aminolaevulinic acid
DAWR Department of Agriculture and Water Resources
dB decibel
dB(A) decibel measured on the A-weighting scale
dB(C) decibel measured on the C-weighting scale
dB(Z) decibel measured on the Z-weighting scale, old SLMs refer to Lin weighting
DC direct current
DCCEEW Department of Climate Change, Energy, the Environment and Water
DECOS Dutch Expert Committee on Occupational Safety
df degrees of freedom
DFG Deutsche Forschungsgemeinschaft
DMEL derived minimal effect level
DNA deoxyribonucleic acid
DND daily noise dose
DNELS derived no-effect levels
DNPH Dinitrophenylhydrazine
DPIE Department of Planning and Environment NSW
DPM diesel particulate matter
E illuminance (Section 17.5.3)
E electric field strength (Section 15.3.6.3)
$E_{A,T}$ A-weighted noise exposure
EAV daily exposure action
EC elemental carbon
EC European Commission
ECESAI European Commission, Employment, Social Affairs and Inclusion
ED effective dose
EDTA ethylenediaminetetraacetic acid
E_{eff} effective irradiance
EF exceedance fraction
ELF extremely low frequency
ELV daily exposure limit
EMB eosin methylene blue
EMC electromagnetic compatibility
EMF electromagnetic frequency
EN European Standard
EN ISO European and International Standard
EPA Environmental Protection Agency
ES exposure standard
ETS environmental tobacco smoke
EU European Union
EU/m³ Endotoxin units per cubic metre
eV electron-volt
EWO emergency work only
f frequency
FCA flux-cored arc

FIDOL	NZ Ministry for the Environment, 2016
FTIR	Fourier transform infrared spectrometry
g	gram
G	Gauss; 10,000G = 1 T
g/mL	gram per millilitre
GC	gas chromatograph
GESTIS	Information system on hazardous substances of the German Social Accident Insurance
GHS	globally harmonized system
GHz	gigahertz
GM	geometric mean
GMO	genetically manipulated organisms
GPS	global positioning system
GSD	geometric standard deviation
Gy	gray
H	magnetic field strength
H&S	health and safety
H_2S	hydrogen sulfide
HAV	hand-arm vibration
HAVS	hand-arm vibration syndrome
HAZOP	hazard and operability
HCIS	hazardous chemical information system
HDM	house dust mite
HEPA	high-efficiency particulate air filter
HID	high-intensity discharge
HIV	human immunodeficiency virus
HoC	hierarchy of controls
HP	hypersensitivity pneumonitis
HPD	hearing protective devices
HPLC	high-performance liquid chromatography
Hr or h	hour
HSE	Health and Safety Executive (United Kingdom)
H_T	equivalent dose
HVAC	heating, ventilation and air-conditioning
Hz	Hertz
IAEA	International Atomic Energy Agency
IAQ	indoor air quality
IARC	International Agency for Research on Cancer
IATA	International Air Transport Association
ICNIRP	International Commission on Non-ionising Radiation Protection
ICOH	International Commission on Occupational Health
ICP	inductively coupled argon plasma
ICP-MS	inductively coupled plasma mass spectrometry
ICRP	International Commission on Radiological Protection
IDLH	immediately dangerous to life or health
IE	indoor environment
IEC	International Electrotechnical Commission

IECEx	International Electrotechnical Commission System for Certification to Standards Relating to Equipment for Use in Explosive Atmospheres
IEI	idiopathic environmental illness
IEQ	indoor environment quality
IHDA	IH Data Analyst
IHSTAT	Industrial Hygiene Statistical Package
IICRC	Institute of Inspection Cleaning and Restoration Certification
ILO	International Labor Office
IOHA	International Occupational Hygiene Association
IOM	Institute of Occupational Medicine
ipRGC	intrinsically photosensitive retinal ganglion cells
IR	infrared
IR-A	wavelengths 780–1400 nm and frequencies around 10^{14} Hz
IR-B	wavelengths 1400–3000 nm and frequencies around 10^{14} Hz
IR-C	wavelengths 3000 nm–1 mm and frequencies around 10^{11}–10^{14} Hz
IREQ	required clothing index
ISIAQ	International Society of Indoor Air Quality and Climate
ISO	International Organization for Standardization
IWBI	International WELL Building Institute
J/kgm^2	joule per kilogramcubic metre
JASP	a popular software program for conducting and teaching statistics developed by the University of Amsterdam
JIS	Japanese Industrial Standards
JSA	job safety analysis
K	Kelvin (6.6.2)
k	Atkinson friction factor (kg/m^3 or Ns^2/m^4)
kg	kilogram
kg/hr	kilogram per hour
kg/m^3	kilogram per cubic metre
kHz	kilohertz
kPa	kilopascals
kW	kilowatts
L/min	litres per minute
L/s	litres per second
$L_{Aeq,8h}$	equivalent sound pressure level, A-weighted, over 8 hours
$L_{Aeq,T}$	equivalent sound pressure level, A-weighted, over specified time period
LAL	limulus amoebocyte lysate
$L_{c,\ peak}$	sound pressure level, C-weighted, peak
LC_{50}	lethal concentration for 50 per cent of specific population
LCL	lower confidence limit
LD_{50}	lethal dose for 50 per cent of specific population
LED	light-emitting diode
LEL	lower explosive limit
L_{eq}	equivalent sound pressure level
LEV	local exhaust ventilation
LiOH	lithium hydroxide
lm	lumen
LOAEL	lowest observed adverse effect levels

LOD limit of detection
LOQ limit of quantitation
LOR limit of reporting
LPG liquid petroleum gas
LPS lipopolysaccharide
L_{Wa} A-weighted sound power level
lx lux
L_λ spectral radiance
m metre
m/s metre per second
$m/s^{1.75}$ metres per second to the power of 1.75
m/s^2 metres per second squared
m^2 square metres
MAC maximum allowable concentrations
MCS multiple chemical sensitivity
MDHS methods for the determination of hazardous substances
MDI methylene bisphenyl isocyanate
MEA malt extract agar
MEK methyl ethyl ketone
MERS Middle East Respiratory Syndrome
mg/m^3 milligrams per cubic metre
mGy milligray
MHz megahertz
mm millimetre
MMA manual metal arc
MMT methylcyclopentadienyl manganese tricarbonyl
MMVF man-made vitreous fibres
MRI magnetic resonance imaging
MSA Mine Safety Association
MSHA Mine Safety and Health Administration
mSv millisievert
mSv/h millisievert per hour
mT millitesla
MVE Murray Valley Encephalitis
MVOC microbial volatile organic compounds
MVUE estimate of the arithmetic mean (Est. AM)
mW milliwatt
n number of measurements
NABERS National Australian Built Environment Rating System
NADC National Air Duct Cleaners Association
NATA National Association of Testing Authorities
NCC National Construction Code (formerly the Building Code of Australia)
ND non-detect
NDIR non-dispersive infrared
NEAT Nanomaterials Emissions Assessment Technique
NEPC National Environment Protection Council
NEPM National Environment Protection Measures
NES National Environmental Standards
NGS next-generation sequencing

nGy	nanogray
NHANES	National Health and Nutrition Examination Survey
NHMRC	National Health and Medical Research Council
NiCAD	nickel–cadmium
NIHL	noise-induced hearing loss
NIOSH	US National Institute for Occupational Safety and Health
nm	nanometre (10^{-9} metre)
NO	nitric oxide
NO_2	nitrogen dioxide
NOA	naturally occurring asbestos
NOAEL	no observed adverse effect levels
NOHSC	National Occupational Health and Safety Commission
NORM	naturally occurring radioactive material
NOS	dusts not otherwise specified
NOx	oxides of nitrogen
NRR	American system of determining the expected sound attenuation of a hearing protector
NSW	New South Wales
NTDEs	new technology diesel engines
NTP	normal temperature and pressure
NZ	New Zealand
O_3	ozone
OAM	Medal of the Order of Australia
OARS	Occupational Alliance for Risk Science
OECD	Organisation for Economic Co-operation and Development
OEL	occupational exposure limits
OES	occupational exposure standard
OGTR	Office of the Gene Technology Regulator
OHS	occupational health and safety
ONIHL	Occupational noise-induced hearing loss
OPCs	Optical Particle Counters
ORCID	Open Researcher and Contributor ID
OSHA	Occupational Health and Safety Administration
OTDS	organic toxic dust syndrome
OVF	occupational visual field
P or p	pressure
P_0	reference sound pressure, normally 20 μPa
Pa	Pascal
Pa^2hr	Pascal squared hours
PAH	polycyclic aromatic hydrocarbon
PAPR	powered air-purifying respirator
PC	physical containment
PCBU	person in control of a business or undertaking
PCR	polymerase chain reaction
PD	personal dissatisfaction
PELs	permissible exposure limits
PF	protection factor
PFAS	per- and polyfluoroalkyl substances
PG	peptidoglycan

PHS	predicted heat strain
PID	photoionisation detector
PM	particulate matter
PM_1	particulate mass with a median aerodynamic equivalent diameter cut-point of 1 μm
PM_{10}	particulate mass with a median aerodynamic equivalent diameter cut-point of 10 μm
PM_{100}	particulate mass with a median aerodynamic equivalent diameter cut-point of 100 μm, also known as the inhalable fraction
$PM_{2.5}$	particulate mass with a median aerodynamic equivalent diameter cut-point of 2.5 μm, also known as the high-risk respirable fraction
PM_4	particulate mass with a median aerodynamic equivalent diameter cut-point of 4 μm, also known as the respirable fraction
PMF	progressive massive fibrosis
ppb	parts per billion
PPE	personal protective equipment
ppm	parts per million
PSMs	phenol-soluble modulinos
PTS	permanent threshold shift
P_v	velocity pressure
Pa	pressure (Pa)
PVC	polyvinyl chloride
Q	flow rate
QAP	quarantine-approved premises
QCM	quartz-crystal microbalance
Qld	Queensland
QUT	Queensland University of Technology
R	roentgen
RCD	respirable coal dust
RCF	refractory ceramic fibre
RCS	reuter centrifugal sampler
REACH	Registration, Evaluation, Authorization and Restriction of Chemicals
REGNET	School of Regulation and Global Governance, Australian National University
REL	recommended exposure limits
RESP-FIT	Respirator Fit Testing Training and Accreditation program
RF	radiofrequency radiation
RfD	reference doses
RG	risk grouping
RH	relative humidity
RHIBs	rigid hull inflatable boats
RMPF	required minimum protection factor
rms	root mean square
RNA	ribonucleic acid
RPE	respiratory protection equipment
Rpm or RPM	rotations/revs per minute
RR	relative risk
RRv	Ross River virus
RSH	mercaptan
RSV	respiratory syncytial virus

S	stopping distance (Section 6.6.2)
S	surface area
SAD	seasonal affective disorder
SARS	severe acute respiratory syndrome
SAS	surface air system
SBS	sick building syndrome
SCBA	self-contained breathing apparatus
SCOEL	Scientific Committee on Occupational Exposure Limits
SDA	Sabouraud dextrose agar
SDS	safety data sheet (previously called MSDS—material safety data sheet)
sec or s	second (time)
SEGs	similar exposed groups
SEM	scanning electron microscope
Sen	respiratory sensitiser
SI	International System of Units
SIMPED	Safety in Mines Research Establishment
Sk	skin
Sk:Sen	skin sensitiser
SKC	Scientific Kit Corporation (previously)
$\mathbf{SLC_{80}}$	sound level conversion, used for hearing protection selection
SMR	standardised mortality ratio
$\mathbf{SO_2}$	Sulfur dioxide
***S*-PMA**	*S*-phenylmercapturic
SSBAs	security-sensitive biological agents
STEL	short-term exposure limit
Sv	sievert
SVOC	semi-volatile organic gases
SWA	Safe Work Australia
T	time period (Chapter 14)
T	tesla (Chapter 15)
$\mathbf{T^{1/2}}$	radiological half-life
$t_{1/2}$	biological half-life or elimination half life
$\mathbf{T_a}$	air temperature
TC	total carbon
TDEs	traditional diesel engines
TDI	tolerable daily intakes
TEM	transmission electron microscope
TEOM	tapered element oscillating microbalance
$\mathbf{T_g}$	globe temperature
TI	tolerable intakes of chemicals
TILT	toxicant induced loss of tolerance
TLD	thermoluminescent dosimeter
TLV®	threshold limit value (ACGIH®)
TNT	2,4,6-Trinitrotoluene
$\mathbf{T_{nwb}}$	natural wet bulb temperature
TSA	tryptic soy agar
TTS	temporary threshold shift
TVOC	total volatile organic compounds
TWA	time-weighted average

TWL	thermal work limit
UCL	upper confidence limit
UEL	upper explosive limit
UK	United Kingdom
UKAEA	United Kingdom Atomic Energy Association
UNEP	United Nations Environment Programme
US EPA	United States Environmental Protection Agency
USA	United States of America
U_{SG}	specific gravity of urine
USSR	Union of Soviet Socialist Republics
UV	ultraviolet
UV-A	wavelengths 315–400 nm and frequencies around 10^{14} Hz
UV-B	wavelengths 280–315 nm and frequencies around 10^{15} Hz
UV-C	wavelengths 100–280 nm and frequencies around 10^{16} Hz
v	speed or velocity (m/s)
V/m	volts per metre
VBNC	viable but not culturable
vCJD	variant CJD
VDV	vibration dose value
VE	the ability of an air distribution system to remove internally generated pollutants
VEM	video exposure monitoring
VOC	volatile organic compound
VTV	vibration total value
VWF	vibration white finger
W	watt (equivalent to J/sec)
WBGT	wet bulb globe temperature
WCI	wind chill index
WEL	workplace exposure limit
WELL®	WELL Building Standard, published by International WELL Building Institute
WES	workplace exposure standard
W_h	frequency weighting
WHO	World Health Organization
WHS	workplace health and safety
WHSQ	Workplace Health and Safety Queensland
WPMN	Working Party on Manufactured Nanomaterials
w_R	radiation weighting factors
WRA	work-related asthma
W_T	tissue weighting factor
$\bar{x}$	arithmetic mean
XRF	x-ray fluorescence
ZPP	zinc protoporphyrins
λ	wavelength
π	pi
Φ	luminous flux
a	alpha
b	beta
g	gamma
%	per cent
s	standard deviation

q	angle
s^2	variance
ρ	density of air (kg/m^3)
°C	degrees Celsius
µg	microgram (10^{-6} g)
µg/dL	micrograms per decilitre
µg/g	micrograms per gram
µg/L	micrograms per litre
$µg/m^3$	micrograms (10^{-6} g) per cubic metre
µGy	microgray
µm	micrometre
µmol/L	micromoles per litre or micrograms per 100 mL
µPa	micropascal
µSv	microsievert
µSv/h	microsievert per hour
µT	microtesla
95% UCL	lands exact one-sided 95% upper confidence limit of the arithmetic

1 The Hazardous Work Environment
The Occupational Hygiene Challenge

Charles Steer, Melanie Windust and Kate Cole OAM

1.1 INTRODUCTION

Occupational hygiene is a profession grounded in science and engineering, requiring good communication skills and a dedication to protecting worker health. Its focus is to eliminate or, if that is not practical, to control chemical, physical, or biological hazards in workplaces. These hazards include hazardous chemicals, dusts, gases, vapours, mists, smokes, fumes, fibres, noise, vibration, light, heat and cold, ionising and non-ionising radiation, and biological hazards such as mould, fungi, bacteria, and viruses.

A key characteristic of many workplace health hazards is their slow and insidious effect on health, resulting in a toll of worker illness and death that greatly exceeds that of traumatic workplace injuries and fatalities. This is not to understate the importance of traumatic injuries and their impact. However, there is unrecognised and significant disease in our community, arising from exposure at work, to largely known hazards, that we can and should control.

While it would be ideal to eliminate rather than control these hazards, it is often not practical, for example, when mining an ore, using chemicals for which there is no safer substitute available, or working in sunlight or in hot or cold conditions. The examples are almost endless.

Nearly all activities with workplace health hazards involve some aspect of occupational hygiene practice. The preventative approach of occupational hygiene, namely interventions made to the work and work environment, rather than the worker, is responsible for a large proportion of improvements in worker health. To be effective, occupational hygienists need a detailed understanding of the nature of the hazards in the workplace, how these hazards arise, and the actual extent of exposure to them. They also require a sound knowledge of the risks to health which arise from exposure to each hazard and the means to mitigate the risks.

This chapter presents an overview of procedures for workplace health investigations using the occupational hygiene principles of anticipation, recognition, evaluation, and control of workplace hazards. It also lays out a framework for considering all the topics covered in this book.

This book is aimed as a general primer for occupational hygienists as well as those in other relevant disciplines who work together to achieve a healthy workplace. Figure 1.1 sets out the range of disciplines that may be involved.

Although the material in this book has its origins in Australia, it is also applicable to workplaces in the Asia Pacific region and other parts of the world.

DOI: 10.1201/9781032645841-1

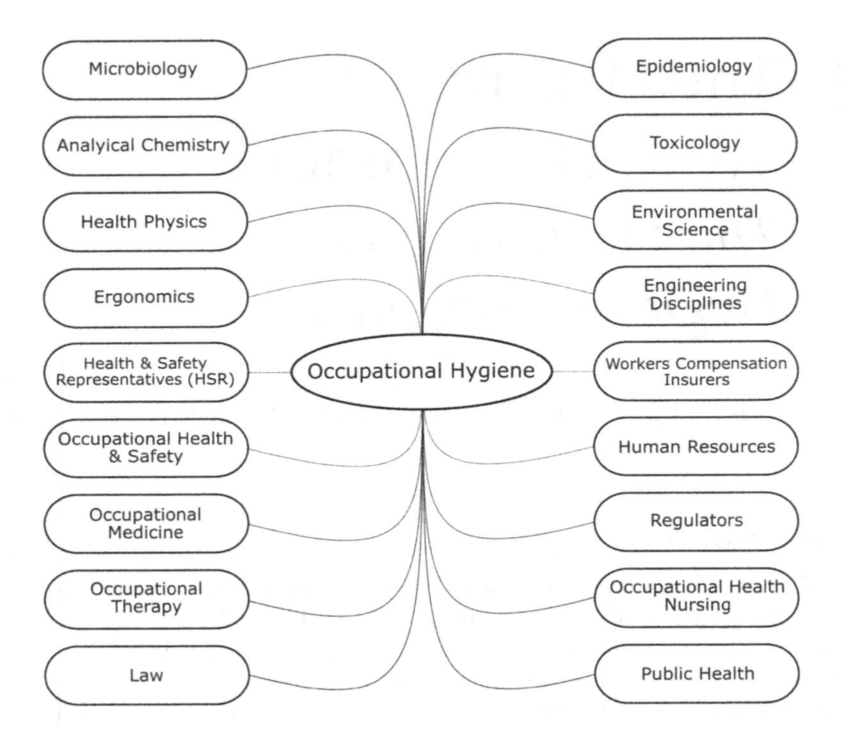

FIGURE 1.1 Occupational hygiene and allied professionals: collaboration for the protection of worker health. **Adapted from: Harden et al. 2015.**

1.2 HISTORICAL BACKGROUND

Human's efforts to create and build civilisations are intrinsically connected to health hazards. The first mention of occupational disease is credited to Hippocrates (c 460–370 BC) who documented lead poisoning in miners and metallurgists. Pliny the Elder may have been the first to document protective equipment when he noted in around 50 AD that animal bladders were used to prevent inhalation of dust and lead fume (Blunt et al. 2011).

Very little occupational disease was recorded until after the middle ages. Georgius Agricola (1556) described mining, smelting, and refining, including diseases and accidents, as well as the need for ventilation. Paracelsus, the medieval physician often described as the 'father of toxicology', described miner's respiratory diseases in 1567 and is said to have stated the words: 'All substances are poisons …the right dose differentiates a poison and a remedy'. In 1700, Bernadino Ramazzini, the 'father of occupational medicine', published a book on the diseases of workers and introduced the question – 'what trade are you?'.

From the late 1700s, the first Industrial Revolution accelerated workplace injuries and illness as new trades and industries such as mining and associated steam-powered factories emerged on an unprecedented scale. Awareness of hazards was low or non-existent. Many who survived physical injury or escaped death in the mines became ill from dust diseases. In mines, mills, and factories, inexperienced workers, including children as young as six, faced injury or death from machinery and work practices designed for output, not worker health and safety. Some factory and mill owners took account of the general safety, health, and welfare of their workers, but these were in the minority. Legislation in western countries became the driver for improvements to workplace health and safety but change was slow. It was not until 1833 that the first real labour laws and the Factory Inspectorate were established in the United Kingdom, and this was followed by the Coal Mines Act in 1842 (UK Parliament n.d.).

With the development of chemical-based industries in the late nineteenth and early twentieth centuries, many new occupational diseases emerged that are associated with exposures to chemicals, some of which continued unchecked until relatively recent times, despite readily available evidence of the hazards. Occupational diseases commonly associated with mining and quarrying (pneumoconiosis or dust diseases), tunnelling (silicosis), fur carroting to make felt (mercury poisoning), yarn and fabric manufacture and textile weaving (byssinosis, scrotal cancer), potteries (silicosis), metal casting and finishing (silicosis and metal fume fever), chimney sweeping (scrotal cancer), matchstick manufacture ("phossy" jaw or bone necrosis from oxide of phosphorus exposure), and bridge building (caisson disease) became accepted as ordinary risks of these jobs. While many of these industries and issues have disappeared, some remain such as silicosis, pneumoconiosis, and metal fume fever even though the causes and controls are well known (Petts et al. 2021).

While Georgius Agricola, Bernadino Ramazzini, and Charles Thackrah made astute observations about the importance of understanding a worker's job or trade to assess the impacts on health and occupational hygiene as we know it made no real advances until the First World War. Pioneering work done by Dr Alice Hamilton (Hamilton 1943) and others in the early twentieth century brought industrial disease, particularly in the war industries such as munitions, to the attention of legislators, employers, and workers in the United States and the United Kingdom.

In Australia, factory and mining work health and safety laws followed the UK example but were not enacted by the States until the late nineteenth century and early twentieth century, respectively. Australian social reform, including an emerging labour movement, and a federal arbitration system established in 1905 further improved protections for workers, including children (Damousi et al. 2014).

Like its UK predecessors, early Australian legislation was industry based with targeted prescriptive regulations to control specific hazards. These regulations set out complex requirements that were difficult to comply with and were, by nature, incomplete. Furthermore, with the rapid advancement of industry, including chemical manufacturing and defence, health and safety legislation was always lagging in terms of technical, medical, and epidemiological knowledge.

Health and safety legislation fundamentally changed in the 1970s and 1980s in Australia as the so-called Robens-style regulation took over, with its central tenets of overarching legislation featuring self-regulation and the statutory duty of employers to consult employees and to ensure their health and safety at work (Australian National University, REGNET 2017). This approach marked the beginning of Codes of Practice and specific sections in the regulations to provide guidance. At the time of publication, Australia has implemented consistent harmonised OHS legislation between all States and Territories with the exception of Victoria. The harmonised legislation sets out clear responsibilities and requirements for employers to eliminate risks to health and safety so far as is reasonably practicable and, if not reasonably practicable, to minimise those risks so far as is reasonably practicable. The legislation is backed up by extensive Codes of Practice and other guidance material that is freely available. This rather legalistic approach effectively enshrines the basic human right of being safe and healthy at work.

The profession we now know as occupational hygiene (also called industrial hygiene in the United States, Malaysia, Indonesia, etc.) emerged after the Second World War in the United States and the United Kingdom. Scientists and engineers from government and industry began working and collaborating in this field in Australia in the 1960s, with the Australian Institute of Occupational Hygienists, Inc. (AIOH) being formed in 1980.

The AIOH website (www.aioh.org.au) provides details of the professional grades, activities such as conferences as well as training courses including accredited university courses. The AIOH are also active in producing Position Papers and technical publications. Courses at various levels are also available online through www.OHLearning.com.

1.3 THE PRESENT

In Australia, there is a sound, although not fully consistent, legislative framework, with associated Federal and State/Territory regulators, a well-educated workforce, and a cohort of well-trained OHS professionals including occupational hygienists. However, there are continuing challenges as evidenced by the high rates of occupational disease as outlined in Section 1.4. Many of the older hazards have disappeared, but many remain along with new occupational health challenges. The following sets out the extensive range of health hazards that can be experienced in modern workplaces:

- *Hazardous chemicals* (Chapter 2) including:
  - carcinogens such as arsenic, chromium VI, and benzene;
  - substances that are toxic to organs or organ systems such as toluene, mercury, glutaraldehyde, or ethanol;
  - reproductive and teratogenic substances such as lead;
  - corrosive substances whether acidic or alkaline, such as acid gases;
  - sensitising agents such as nickel, isocyanates, and glutaraldehyde that can affect the skin, respiratory system, and gut;
  - irritants such as ammonia and ozone.
- *Aerosols* (Chapter 8) including airborne liquid and solid particulates (smokes, fumes, dusts, mists, and fibres) such as:
  - dusts from mining and quarrying, construction, and manufacturing including coal, metal, and wood dusts, respirable crystalline silica;
  - ultrafine particles including diesel particulate matter (DPM) from vehicle exhaust emissions;
  - biologically active particulates in pharmaceutical manufacturing, bioaerosols from cutting fluids;
  - nanoparticles in pharmaceutical and other manufacturing processes;
  - acid fumes from electroplating and anodising in manufacturing and defence-related industries;
  - metal fumes from welding, smelting, refining, and soldering (Chapter 8);
  - acid mists from electro-refining;
  - organic mists from spray painting;
  - aerosols generated from welding, grinding, and cutting;
  - fibres from manufacture and handling of man-made fibre products, asbestos removal, mining exploration, and cotton and paper manufacture;
  - pesticides, including insecticides and herbicides, and grain dusts found in agriculture;
  - airborne contaminants from chemical manufacture, process plants, mining, construction, and emergency services.
- *Gases and vapours* (Chapter 10) from almost all work settings including:
  - oxides of nitrogen, carbon dioxide, carbon monoxide, and cyanides from internal combustion engines in mining, logistics, and construction;
  - ozone from welding;
  - hydrocarbons from petroleum refining and storage, solvent degreasing;
  - anaesthetic gases and other biologically active vapours in medical settings;
  - hazardous and irritant chemicals such as formaldehyde from building materials in offices, glues, and manufacturing;
  - asphyxiants such as nitrogen, carbon monoxide, and refrigerants.

- *Noise and vibration* (Chapter 14) from almost all industries including manufacturing, agriculture, defence, construction, mining, tunnelling, and logistics.
- *Thermal (heat or cold) stress* (Chapter 16) arising from outdoor and cold store work.
- *Biological hazards* (Chapter 13) including:
  - fungi and mould in the built environment, particularly in warm, humid environments;
  - biological hazards, moulds, and fungi in agriculture;
  - bacteria such as *Legionella pneumophila* in excess concentrations in potable water and cooling towers, increasing disease risk;
  - zoonoses, including leptospirosis, anthrax, Q fever, bird flu, lyssavirus, and Hendra virus in animal husbandry;
  - biological hazards in health and medical settings, including accidental contact with blood and body fluids, and viruses including COVID-19, SARs, and influenza.
- *Potable water quality* where sites provide the drinking water supply, such as remote mine sites, this includes bacterial, chemical, and physical quality (such as turbidity) (NH&MRC & NRMMC 2011).
- *Ionising and non-ionising radiation* (Chapter 15) in a wide variety of industries including mining, power, manufacturing, medical diagnostics, research, and defence. Ultraviolet radiation exposure is a hazard for all outdoor workers as well as from welding processes and specialist equipment.

Occupational hazards such as those listed above can exist separately or together in a range of industries and w.orkplaces including construction, manufacturing, defence, research, health, mining, agriculture, and even office or retail environments. Although the body of knowledge on these hazards and their control is now extensive, some hazards are underestimated, some are ignored, and still others have re-emerged due to complacency. A salient example is the control of coal mine dust: the prevalence of coal workers' pneumoconiosis was as high as 27% in Australia before the Second World War (Moore & Badham 1931), and 16% in 1948 (Glick 1968). Unfortunately, there have been recent recurrences of pneumoconiosis in Queensland (Parliament of Australia 2017) as well as in the United States (Blackley et al. 2018) demonstrating the importance of maintaining controls and understanding the hazard.

The recent re-emergence of silicosis from working with engineered stone bench tops is an example of a long-standing hazard not being recognised in a new industry (The Thoracic Society of Australia and New Zealand 2017).

A further complication is the proliferation of small workplaces including the so-called 'gig economy', or piecemeal work where there is no union or organisational support or accumulated knowledge almost certainly resulting in vulnerable workers. Vulnerability increases where the workforce includes migrants, disabled workers, non-English speakers, and informal workers.

The challenge is significant for occupational hygienists and other professionals to provide good guidance, especially for hazards that cannot be seen, heard, felt, or smelt and can have health effects that may not be apparent for many years.

1.4 THE SERIOUS PROBLEM OF UNDERESTIMATION OF OCCUPATIONALLY RELATED DISEASE

Is work-related ill health important enough to deserve the attention it now receives? The last 35 years in Australia have witnessed an expansion of occupational health laws and regulations, with greater government administration and new inspectorates. There are many technical guides, numerous training programs, and expanded legal services. The number of OHS professionals and research programs is growing. But are workplace hazards – from chemicals to conditions – really detrimental to health? After all, there are relatively few ill people in any given workplace and our workers' compensation systems report relatively few cases. When a disaster like a rail crash or an explosion occurs, it is easy to count fatalities. In contrast, occupational disease resulting from exposure to physical, chemical, or biological agents may take years or decades to develop, so their causes are often not immediately apparent. Many similar illnesses are also the result of lifestyle or social conditions. Workers move away or change jobs or retire. A sick worker's doctor may simply be unaware of the kind of work that his/her patient does or has done in the past and asking about a patient's occupation is probably unusual in typical doctor–patient interactions unless the doctor has training in occupational health. Historical records of occupational exposures are rare, although new laws now mandate record-keeping in some cases.

Reliable data on the contributions of workplace conditions to ill health in the community have traditionally been difficult to assemble, and this is a worldwide problem. Data on compensation for work-related ill health is lacking, leading to a misleading impression of the true prevalence of occupational disease. Most work-related ill health is not the result of accidents (falls, high-energy impacts, crushing or piercing injuries, etc.) but rather, is the result of exposure to hazardous chemicals or environmental conditions. Consider the following examples where the findings of epidemiological studies, sometimes years after exposure first commenced, confirmed the need for the controls now widely demanded by law:

- The world's worst single-event industrial disaster (with the probable exception of Chernobyl): At Bhopal, India, in 1984, the inadvertent release of methyl isocyanate gas (an intermediate of manufacturing a pesticide) killed more than 3,000 people and injured some 17,000 more who lived in the chemical factory environs. Though the cause of the disaster was soon evident, its true scale was not so immediately obvious.
- The United States' worst individual industrial accident: Hawk's Nest Tunnel in West Virginia, built to divert a river in the early 1930s, required drilling through silica rock. As a result of inhaling the resulting dust, more than 600 men died from silicosis within two to six years. Neither the cause of the illness nor its true prevalence were obvious at the time of the work.
- Australia's worst industrial accident: Mining of blue asbestos at Wittenoom and surrounding areas, Western Australia, continued from 1937 to 1966. Workers, town residents, and visitors to Wittenoom developed asbestos-related diseases over the ensuing decades and deaths are still occurring. The death toll has exceeded 2,000. Worldwide, the death toll from asbestos exposure in the late twentieth and early twenty-first centuries could possibly reach into the millions. It is tragic to note that hints of the link between asbestos fibre exposure and asbestosis were first reported in 1906 increasing during the 1920 and 1930s (Murray 1990). Links with lung cancer emerged during the 1940s and were confirmed epidemiologically by Doll (1955). Wagner (1960) linked blue asbestos (similar to that mined in Wittenoom) to mesothelioma and took the scientific world by storm. Unfortunately, these scientific breakthroughs did not translate to the industrial world. The first Asbestos Regulations in Australia were not put in place until 1978.

A further disturbing aspect of asbestos-related disease is that the number of new mesothelioma cases continues to be high.

In 2012, the then Minister for Workplace Relations, the Hon Bill Shorten, stated that 642 people died from Mesothelioma in 2010 with many more with lung cancers related to asbestos exposure. He also noted that more Australians will die from asbestos-related diseases, than died in the whole of the First World War (Ireland & Willingham, 2012).

As the first wave of deaths from exposure during mining and processing prior to 1983 decline, a second wave of disease has emerged in tradespeople and workers in asbestos buildings (ASEA 2016).

Although these exposures were much lower, far more people were exposed, resulting in continuing high numbers of mesothelioma cases and deaths.

There were 722 mesothelioma cases reported in 2021, and 701 deaths recorded on the Australian Mesothelioma Registry in 2020 (AMR 2023). Asbestos is covered in more detail in Chapter 8.

These examples demonstrate that not only exposure to hazardous substances can cause severe ill health or death, but the true impact of this exposure can take years or decades to become apparent.

In the 1980s, the US National Institute for Occupational Safety and Health (NIOSH) identified the 'top ten' workplace injuries and illnesses (Millar 1984). The list included occupational cancers, dust diseases, musculoskeletal injuries and hearing loss, and a number of other diseases still with us today (Safe Work Australia 2022).

It is generally acknowledged (Schulte 2005) that the burden of occupational disease is significant but underestimated as the recording of long latency diseases is difficult and there may well be co-contributing causes.

The latest estimates of the global burden of occupational disease indicate a significant challenge. In 2017, the Singapore Workplace Safety and Health Institute (Hämäläinen et al. 2017) reported an estimated 2.78 million global deaths each year that can be attributed to work. This is approximately 5% of total global mortalities and higher than the 2.33 million occupational fatalities reported in 2014. Contrary to the focus of effort by many workplace regulators, the Institute states that 86% of these work-related deaths are caused by occupational disease compared with approximately 14% of fatalities that were directly related to accidents. Globally, the most significant occupational diseases are respiratory diseases including chronic obstructive pulmonary disease, circulatory diseases, and malignant neoplasms.

In the absence of any published Australian estimates (Steer & Bennett 2019) and using the ratio of around ten fatal illnesses to one fatal injury as has been derived in the World, the United States, and New Zealand estimates as shown in Table 1.1, this results in an estimate of 1690 Australian illness fatalities for the year 2021 (Safe Work Australia 2022). Although this result is an estimate, it serves to illustrate the size of the challenge.

TABLE 1.1
Summary of Estimated Fatalities by Illness and Accident in Various Jurisdictions

Jurisdiction	Estimated Annual Illness and Accident Fatalities	Ratio of Fatalities from Illness Compared to Injury
World wide	• 2.4 million deaths linked to work-related illnesses • 0.38 million fatal injuries (Hämäläinen et al. 2017)	6
United States	• 26,000–72,000 linked to work-related illnesses (Steenland 2003) • 5190 fatal injuries in the United States in 2021 (US Department of Labor 2022) • Total cost of work injuries and illness estimate in 2020 was 167 $US billion (National Safety Council 2023)	9 (Pre -COVID 19)
Great Britain	• 13,000 deaths related to work-related illness primarily linked to chemicals and dust (estimated 2021/22) • 135 fatal injuries (Health and Safety Executive, 2022/2023) • Estimated cost £11.2 billion for illness and £7.6 billion for injury (2019/2020)	96
New Zealand	• 750–900 deaths related to work-related illness (WorkSafe NZ 2019) • 59 fatal injuries (WorkSafe NZ 2023)	Approximately 15

An obvious question to consider is whether injury and illness rates will decrease in future as workplaces become better controlled.

On the positive side:

- Australia's injury fatality rates have been falling, from 259 in 2003 to 169 in 2021 suggesting improved management commitment, awareness, design, and operational improvements that are likely flowing into health hazard control.
- Legislation has improved (e.g. banning asbestos and quartz in grit blasting) along with more comprehensive Codes of Practice and consequent awareness.
- Smoking rates have declined decreasing the chance of synergistic adverse health effects.
- Engineering control technology has improved, for example, large-scale dust and vapour control, noise control, and hand-held tools with local exhaust ventilation.
- Environmental legislation has become more stringent coincidentally causing improvements to worker health and safety, for example, vehicle emission standards, environmental dust, and noise requirements.

On the negative side:

- The so-called 'gig' economy is increasing along with increasing numbers of small to very small businesses that have limited resources and workers in multiple transient jobs.
- Australia is 'exporting' more dangerous work, for example, in manufacturing and processing industries. This is a positive for injury and illness statistics in Australia but is likely to be compounding problems elsewhere and deeply misleading in terms of human suffering.
- New advances in technology such as nanomaterials may be outpacing health research.
- Reductions in the number of specialists employed by regulators including occupational hygienists, occupational physicians, and occupational health nurses.
- There are large areas of industry that have not experienced significant exposure to occupational hygiene principles and application, such as building, construction, manufacturing, agriculture, and health care. This is now changing in some sectors, for example, in tunnelling where occupational hygiene expertise is starting to be applied. However, it is by no means universal. In 2022, Australia had approximately 1.15M persons employed in construction, 895K in manufacturing, 309K in agriculture, forestry and fishing, and 1.91M in health care and social assistance. This compares to mining, which has a relatively mature occupational hygiene culture, and employed around 272K persons (Parliament of Australia, February 2022).
- It is estimated that around 40% (approximately 3.6 million) of the Australian workforce are still exposed to carcinogens in their current job role. Many of these are potentially exposed to multiple carcinogens (Carey et al., 2014; Carey et al., 2017; McKenzie et al., 2021).

Future challenges include:

- The need to keep industries up to date to reduce noise, chemical, and other exposures.
- The need for companies to ensure that lessons learnt from older facilities are applied when building new ones including ensuring retaining a 'corporate memory'. It is also important that company returns on investment considerations are made in the light of cost-effective health and safety options. Hygienists need to argue for the overall benefit of higher-order control measures – for example, design and engineering solutions for noise rather than supply of ear plugs.
- The need to consider emerging technologies, rapidly changing work environments and hazards including fatigue and psychosocial risks.

- The need to overcome the limitations of current workers compensation data systems for recording and estimating occupational illness. This recognition may then reduce the appreciable cost-shifting of long latency disease into the public health system in Australia and other countries.

The serious problem of the high levels of occupationally related disease demands that we ensure the basic, and accepted, human right of workers to have a healthy as well as safe workplace. Occupational hygienists therefore promote and advocate for embedded and structured occupational hygiene processes to address these current and future challenges.

1.5 THE OCCUPATIONAL EXPOSURE RISK MANAGEMENT PROCESS

The classic occupational hygiene process involves anticipation, recognition, evaluation, and control.

- Anticipation of problems – a vital skill, but while this usually requires considerable experience, assistance is now provided by abundant resources including Safety Data Sheets (SDS), various databases, and websites.
- Recognition – knowing the hazards and the processes by which they may affect health or identifying them through adverse health effects.
- Evaluation – assessing risk, measuring exposures, and comparing against standards.
- Control – providing contaminant or hazard control. The level of protection required is based on the knowledge of the toxic or other adverse effects produced by known quantitative exposures to the hazard.

Additional steps may need to be added to in more complex workplaces as depicted in Figure 1.2, where it is important to *characterise* the hazard and, if required, carry out *monitoring*. Once controls are in place, it is essential to carry out regular reviews of their continuing effectiveness.

Figure 1.3 provides more detail on the major risk assessment components. These are explained in the following sections. It is important to note that this section is a summary only. The reader is encouraged to refer to the references as well as the later chapters in this book to obtain a more detailed understanding.

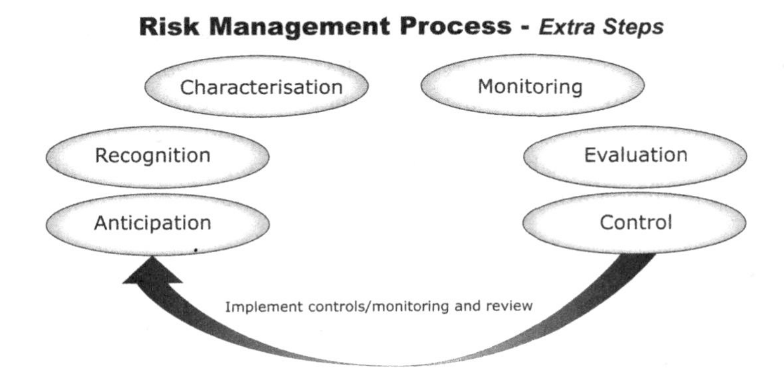

FIGURE 1.2 Additional steps in the risk management process.

Adapted from: Harden et al. 2015.

FIGURE 1.3 Detailed components of the occupational health risk management process.

Adapted from: Harden et al. 2015.

1.5.1 Anticipation

Anticipation involves identifying potential hazards before they are introduced into a workplace or before you visit the workplace. This can be undertaken using a Hazard and Operability Analysis (HAZOP) process, which would include collection and reviewing the following information:

- site and process information, including site plans and where specific chemicals are handled;
- employee numbers and organisation structure;
- site-specific legislation and regulations including exposure standards;
- any previous monitoring and analysis;
- injury and illness records taking into account the long-term nature of health effects of some chemicals and physical agents. This may include any patterns of adverse health effects;
- process plant incidents;
- company standards and procedures;
- SDS for hazardous chemicals. All SDS in Australia are now required to be in the Globally Harmonised System (GHS) format.

There are many work situations where an SDS is not available but the nature of the hazards is well known, for example, asbestos exposure in building maintenance, crystalline silica in sandstone, and wood dust exposure in small joinery workshops. In these cases, alternative information sources need to be used such as Codes of Practice, guidelines, and research publications.

There are other situations where health effects are not so well known and further research will be required possibly including using external specialist expertise.

1.5.2 Recognition

Recognition of health hazards in the workplace is fundamental to their proper control.

There are two major components of this step:

- Understanding the health effects and characteristics of the chemical, physical, or biological agent.

- Using that knowledge in a walk-through survey to assess exposure and how that agent may harm the worker, for example, by skin contact.

Key chemical, physical, or biological characteristics include the following:

- The inherent toxicity or other health effects of the various physical chemical and biological agents as set out in subsequent chapters.
- Potential exposure routes, for example, whole body, inhalation, ingestion, or absorption through the skin, eyes, or aurally. These are set out in subsequent chapters covering specific health-affecting agents.
- Whether the effects are acute or chronic.
- In regard to chemicals and some biological agents, it could include whether they are sensitisers, irritants, asphyxiants, target organ affecting, carcinogenic, corrosive to the skin or eyes, gut, or respiratory tract.
- The physical form of the agent is also important, for example, they are aerosols, solids, liquids, gases, or vapours.
- Other chemical considerations include whether they are lipophilic or lipophobic, acidic or alkaline, whether they may react with other chemicals to produce new hazards as well as their explosivity (heat, shock, friction, or a combination).

Recognition of hazards may be complicated in the following ways:

- Many health hazards cannot be seen, heard, smelt, or felt.
- Some hazards such as non-ionising radiation are not well understood by workers or employers.
- A chemical may have no adequate warning odour.
- Workers do not know what their exposure is.
- Workers and employers may accept exposures as an unavoidable part of the job.

A **site walk-through (or walk-through inspection or survey)** is a critical step in any assessment. It is the opportunity to integrate the knowledge of the hazard with how it is used or experienced in the workplace to gain insights into how workers can be affected.

Exposures can occur from handling, processing, transporting, packaging, and storage of raw materials, by-products, intermediates, waste, and finished materials. Exposure may be routine, occur intermittently (e.g. during a shutdown), be due to a non-standard operation, or occur only in an emergency.

Industries that typically give rise to occupational health hazards are widespread. They include major employment sectors in Australia and Asia including, but in no way limited to the following:

- Agriculture, fishing, and farming involve a wide variety of planting, harvesting, crop and livestock management practices that use or produce hazardous dusts, pesticides and herbicides, and microbiological hazards associated with moulds and fungi. Dangerous gases may arise in silage pits and enclosed feed sheds. Ultraviolet radiation, along with thermal stress, tool, and vehicle vibration, and noise may require control. In animal husbandry, parasites and other zoonotic agents may be present, as well as immunological sensitisers. Fin and shellfish processing involves exposure to aerosolised proteins that are immunological sensitisers. There is also extensive wet work and noise.
- Food processing and packaging industries may expose workers to zoonotic agents, wet work, as well as toxic, corrosive, irritant or sensitising chemicals, noise and vibration,

organic dusts, and thermal stress. Food flavours, colourants, and additives used in concentrated form in large quantities may be irritants or sensitising agents.

- Health industries including hospitals may produce a wide range of hazards including cytotoxic, sensitising and corrosive chemicals, equipment generating ionising radiation, unsealed radioactive materials, noise, and infectious disease. Anaesthetic administration in hospitals and veterinary clinics may pose a risk.
- Building and construction involve potential exposure to dusts including crystalline silica, wood dust, asbestos and man-made mineral fibres, metal fumes, as well as paints and coatings, solvents and adhesives, cements and fillers, and noise and vibration.
- Harmful or toxic gases, liquids, and solids form the basis of most chemical manufacturing, such as oil refining, plastics manufacturing, and cyanide production. Gas reactions and gassing procedures in the production of petrochemicals, the synthesis of plastics, rubbers and fuels, and in catalysis, fumigation and sterilising, have the potential to give rise to serious inhalation hazards.
- Heavy and light manufacturing and processes including welding, soldering, and thermal metal cutting, produce a range of potentially toxic metal fumes, and/or irritant, oxidising, and asphyxiating gases from welded material, welding rods, or fluxes. Industrial plating processes pose risks of skin damage, respiratory system damage from corrosive aerosols, and systemic poisoning from toxic gases and aerosols. Processes involving the manufacture, bagging, and pouring of powders and dry materials produce fine dust. Industrial processes using electromagnetic operations such as large direct current metal smelters, radiofrequency induction furnaces, microwave heaters, X-ray equipment, and radar signal generators can all produce biological effects unless equipment is properly shielded. Radio and microwave transmission towers produce significant risks in near-field radiation zones. Smelting and hot metal handling generate metal fumes, gases and vapours from decomposition of moulds, heat and radiant energy, noise, and light. Abrasive blasting typically uses steel shot, garnet, smelter (e.g. copper) slag, staurolite, and heavy mineral sands (ilmenite) and produces much fine dust and very high noise levels. Vapour degreasing is a widely used industrial process, posing hazards from vapour inhalation and skin and eye contact. Reinforced plastics usage may create exposures arising from various resins and man-made mineral fibres to potent skin and respiratory irritants and sensitising agents, corrosive, and toxic substances. Painting and coating cover a wide number of processes using organic and inorganic powders suspensions and solutions to spray, brush and roller-coat creating exposure to aerosols of organic materials as well as a range of solvent vapours, toxic metals, severe irritants, and respiratory sensitisers. Drying ovens are widely used in manufacturing, art and craft, and industrial surface coating. Ovens produce vapours from solvents, lacquers, paints, cleaning agents, and plastic resins as well as combustion gases from fuels. Nanotechnology engineering leads to exposure to superfine particles in a wide array of processes involved in the manufacturing of products from cosmetics, fabrics, to paints.
- Mining and quarrying may create exposures to dusts, radioactive gases, vehicle including diesel exhaust emissions, noise, vibration, risks of asphyxiation, poisoning by various gases, and injury or death from explosions.
- Transport and logistics operations including road and rail, shipping, aircraft, and warehousing (exposures from the materials and goods being handled, as well as dusts, vapours and solvents, vehicle exhaust, noise and vibration).
- Laboratories where exposure to toxic, irritant, sensitising, corrosive, or mutagenic chemicals may occur in novel operations. Physical hazards arising from equipment generating ionising and non-ionising radiation, and lasers are common. Unsealed radioactive materials and biological agents may also be in use.

- Enclosed buildings and the built environment including office and retail environments that may produce hazards from volatile organic compounds arising from office equipment such as printers and photocopiers to off-gassing by building products, carbon dioxide build-up from poorly ventilated buildings, and biological hazards (moulds and fungi).

These health hazards are covered in subsequent chapters and include aerosols, gases, vapours, biological agents, metals, noise and vibration, heat and cold, lighting, and radiation.

Information gathered could include the following:

- Detailed site and process information and any problems (e.g. symptoms) experienced including discussions with management, supervisors, and workers. Workers may often have specific knowledge of the idiosyncrasies of machinery (and operators), processes as well as how well the process works or doesn't work and non-routine operations.
- The materials used or handled and any difficulties experienced – again by specific questions.
- Worker training.
- The number of workers involved.
- Evidence of reactions, any material transformations (generated substances), by-products, intermediates, and wastes. The work process itself may generate a hazard from an apparently innocuous precursor material.
- Exposure times of directly involved employees as well as bystanders.
- Existing engineering controls in place and whether they work as expected.
- Housekeeping at the process site (e.g. spillages and cleaning methods).
- Visible conditions at the site (any dusts, mists, smoke, fumes, odours, and deposits of material).
- Possible routes of entry (inhalation, skin, ingestion, and injection).
- Personal protective equipment availability and use.
- Processes that expose workers to noise, vibration, heat, cold, or ionising or non-ionising radiation.
- Processes which may contact and affect the skin.

1.5.3 Immediate Controls

A walk-through survey or a preliminary medical examination may identify urgent issues that require immediate controls to protect worker health. In these circumstances, it is important not to wait until a formal characterisation process is completed before acting.

Examples of immediate controls include the need to use respiratory protection for workers where crystalline silica is being disturbed, the use of hearing protection where workers need to shout to be heard, or the use of specific personal protective equipment when handling hazardous chemicals.

1.5.4 Characterisation

This process involves characterising the workplace, workforce, and exposures to health-affecting agents in the workplace (Jahn et al. 2015). Steps include the following:

a) Using information gathered in the earlier stages to develop similar exposure groups (SEGs). A SEG is defined as a group (one or more) of employees similarly exposed to health-affecting agents. SEG development will require consultation with supervisory and workplace representatives. Note that workers may be exposed to more than one hazard at a time.

b) Next, develop a site-specific risk register (Firth et al. 2020) where the health risks of physical, chemical, or biological agents are assessed for individual SEGs. A risk register will include existing controls, calculated inherent and residual risks, and proposed controls. It provides a summary of the status of occupational health management at any workplace and can be used to further develop action plans for additional controls and provide input into determining any medical assessment requirements.
c) Members of the workforce, supervisory personnel, and subject matter experts need to be consulted during the development of the risk register.

1.5.5 Monitoring

Monitoring is a key aspect of the practice of occupational hygiene and is covered in some detail in subsequent chapters. The following has been developed as a summary, to demonstrate how monitoring fits into the occupational exposure risk management process.

- Once SEGs have been selected and a risk register developed, it is necessary to determine whether a more detailed exposure assessment is required. The key basis for this assessment is often workplace monitoring.
- The key criteria for monitoring is whether the exposure is likely to exceed an exposure standard or if there is uncertainty whether it is likely to affect health. Key considerations include practicality and whether an appropriate occupational exposure standard (OES) exists for the chemical, physical, or biological agent.
- An optimal occupational hygiene monitoring program will provide statistically valid monitoring, where appropriate, of health-affecting agents (Grantham & Firth 2019; Liedel et al. 1977). The general requirements are sufficient sample numbers, representative sampling times within a shift, and random sampling dates over a sufficient time period that takes account of seasonal variations, as well as day, night, and weekend shifts as appropriate. Some monitoring may be programmed, for example, campaign monitoring of specific agents during a plant shutdown or for batch processes. There are various options in relation to monitoring equipment and techniques including passive or active monitoring and direct reading instruments (Chapters 8 and 10).
- It is generally preferable to carry out personal monitoring but can include area, surface, biological and monitoring of specific physical and biological agents as set out in Chapters 8–17. There is a so-called hierarchy of exposure criteria (Jahn et al. 2015; Laszcz-Davis et al. 2014) with their foundation ranging from quantitative to more qualitative (Chapter 5).
- Exposure standards may need to be adjusted according to shift length and other factors (AIOH 2016; Standards Australia 2005). Attention should also be given to additive exposures, synergism, potentiation, and ototoxic effects (Chapters 2, 3, and 14).
- Monitoring must take account of the precision and accuracy of the sampling and analytical method if applicable. In all cases, monitoring equipment must be appropriately calibrated and maintained. Monitoring is of no value unless the results can be assured and potential errors are understood.
- Wherever possible, analysis of samples must be undertaken by a laboratory accredited for the test(s) being conducted and reported in accordance with recognised standards.
- Revert to original textExpertise will depend on the agents monitored and includes occupational hygienists for design and monitoring and may include analytical chemists, microbiologists, ergonomists, and acoustic engineers.

- All monitoring results should be reported and made available to the workforce except where inappropriate for privacy reasons (e.g. personal biological monitoring). These results should be reported to the individual only.
- A key document summarising the monitoring will be a SEG Matrix listing health-affecting agents; a partial example is set out in Table 1.2 for a mine that was also managing spontaneous combustion hazards:

In some circumstances, alternatives to workplace monitoring may be used including:

- Mathematical modelling (Keil et al. 2009) ranging from simple models to computational fluid dynamics. This can be useful for situations including:
  - Estimates during the design stage of a process or review of engineering design options for a control.
  - Scenario analysis.
  - Very large sites with many variations across processes resulting in excessive demands on monitoring and analytical resources.
  - Convenient methods for monitoring are not available.
- Control banding, which moves directly to the evaluation stage when monitoring data and or occupational exposure standards may not be available (Chapter 5).

1.5.6 Evaluation

Having measured an exposure, it is necessary to evaluate the risk associated with the hazard. This evaluation step is critical to answer the following questions:

- Is the particular risk from exposure acceptable?
- Does its existing level meet regulatory requirements?
- Will it need controlling to make it healthy and safe?
- Are there special controls for this hazard?
- How much control is needed?
- What is the most effective control mechanism for this risk?

There are, of course, many situations where evaluation will show that no further action is needed. However, a more formal and systematic approach is often required to evaluate a measured exposure.

Risks can be evaluated using a variety of methods depending on the circumstances. In the case of toxic chemicals, there are two components in the evaluation: the exposure profile and toxicity. The resulting health risk can be described as a combination of exposure and toxicity (Jahn et al. 2015):

This can be further developed to produce risk matrices to provide a qualitative risk estimate (Firth et al. 2020; ICMM 2016; Jahn et al. 2015). These matrices are generally based on likelihood (exposure) and consequence (from toxicity). It is this integration of the exposure profile with the toxicity or other health effects that provide the overall risk. The resultant risk can then be used to revise the risk developed in the Risk Register.

The following range of techniques can be used as part of the risk evaluation or legal compliance assessment processes:

- If an OES exists and statistically valid monitoring has been carried out, then formal exposure assessment methodologies can be used. These are most often used for some chemicals, aerosols, gases, vapours, and noise and are set out in various publications (BOHS & NVvA

TABLE 1.2
Example of Part of a SEG Matrix

SEG Number	SEG Name	Tasks	Employee Nos	Inhalable	Respirable Crystalline silica	Welding	Coal Tar Pitch Volatiles	H_2S, CO, SO2	Thermal Stress	Legionella/ Potable water
1	Dragline	Conduct dragline operations – operate auxiliary equipment, etc.	14	11	11	Not measured	7	Targeted – portable analysers	Targeted summer	Potable on plant
3	Field maintenance	Field maintenance of mobile plant	55	18	18	9 Welders opportunity basis	7	Targeted – portable analysers	Targeted summer	Potable on plant Legionella in workshop

2011; Grantham & Firth 2019; Jahn et al. 2015). Techniques used include Bayesian statistics (Hewett et al. 2006) for smaller sample sizes as well as tools such as the IHStat program (AIHA 2014). The measured exposure can then be used to provide guidance on compliance with legislation (Grantham & Firth 2019) as well as for further monitoring and controls (Jahn et al., 2015) and Chapters 3 and 4.

- Specific risk assessment methodologies may also be used for biological monitoring, heat, vibration, and radiation as set out in the relevant chapters of this book. More qualitative techniques will need to be used for monitoring such as surface wipes.
- For more straightforward workplaces, simplified risk management strategies may be used (Firth et al. 2020).
- Mathematical modelling (1.6.5(i)) can also be used to generate an exposure profile.
- Once the exposure assessment has been completed, it needs to be considered in the context of the likelihood of the consequence (i.e. health effect). The interplay between these two components using a risk matrix will result in an assessed risk rating ranging from low to high (or catastrophic). The interpretation of the implications of the risk rating will often require professional judgement in terms of appreciation of the challenge of managing that particular workplace exposure. For example, one scenario may be a likely exposure (at the OES) to an irritant such as ammonia compared to an exposure at less than 10% of the WES of a probable human carcinogen. The risk rating may be the same according to the matrix, but its management including risk communication will be quite different.

A typical 5x5 health risk matrix is set out in Table 1.3.

Likelihood or probability may be described in the context of health, as exposure to a ratio of the OES, for example, 'Almost Certain – frequent daily exposure at 10 x OES'; consequences may be described in the context of a health outcome, for example, 'Serious – severe reversible health effects of concern that would typically result in a lost time illness; can include acute/short-term effects associated with extreme temperature; or some infectious diseases'. Organisations may have varying thresholds and definitions for each of the categories of consequence and likelihood in the context of health.

Together, the descriptions of consequence and likelihood are used to describe the possible outcome if the event occurred. This in turn provides the risk rating from 'Low' risk through to 'Critical' risk. The risk rating will determine the action required, timelines for this action, management accountability, etc. For example, 'Low' risk may be managed in an organisation by adoption of routine procedures; whereas 'Critical' risk would require the attention of senior management along with implementing a detailed plan of action to reduce the risk to an acceptable level.

TABLE 1.3
Example of a Risk Matrix

	Consequence				
Likelihood	**1 – Minor**	**2 – Medium**	**3 – Serious**	**4 – Major**	**5 – Catastrophic**
A – Almost certain	Moderate	High	Critical	Critical	Critical
B – Likely	Moderate	High	High	Critical	Critical
C – Possible	Low	Moderate	High	High	Critical
D – Unlikely	Low	Low	Moderate	High	High
E – Rare	Low	Low	Moderate	Moderate	High

- The risk evaluation will inform the controls as set out below and will also signal other processes such as updating the risk register.
- The next tier down of exposure assessment is Control or Hazard Banding (NIOSH, 2023). This technique has a much higher degree of uncertainty and exhibits user and model variability (Van Tongeren 2014); however, it may be the only practical way to carry out an assessment. The outcome of this methodology is not a number, rather it moves to the next stage of guidance on controls.

1.5.7 Controls

Control remains the most important, yet least understood and often most poorly implemented OHS area. Establishing controls is, simply, how to protect the employee from that hazard. Controls are more fully covered in Chapters 5–7. Key aspects include the following:

a) The process of developing controls may require input from a range of employees and subject experts ranging from occupational physicians to engineers depending on the hazard (e.g. it may require an engineering control such as ventilation, or vaccination to induce immunity). While OHS Practitioners may not be required to design complex controls such as a complete ventilation system, a broad understanding of these processes is useful.
b) Controls need to follow the prioritising framework known as the 'hierarchy of control' (Chapter 5).
c) The recommended controls must be assessed against the regulatory requirement to control risk so far as is reasonably practicable (Section 1.2 of this chapter).
d) Selecting the correct type and level of control requires not only knowledge of the hazard, exposure risk and route of entry but also how much control is needed, comparative effectiveness of different control processes, maintenance and testing procedures, user preferences, and social impacts. The process also requires consideration of costs. Tools are available to assess the business value of implementing controls (AIHA 2014; Jahn et al. 2015).
e) There may be specific regulatory requirements. For example, mining inspectorates have specific regulatory workplace exposure standards for inhalable and respirable dust in mines and detailed medical surveillance requirements for lead workers.
f) The evaluation process can also be used to eliminate or control hazards in the design phase and also as part of change management. This is a powerful prevention tool when used.
g) The recommended controls can form part of any site planning process leading to management commitment to prioritise, resource, and implement controls.
h) The implementation of controls is fundamental to the elimination or control of health-affecting agents. There is no point in monitoring and evaluating hazards if controls are not implemented.
i) The management document for controls will be the health action plan which will set out responsibilities and key performance indicators.

1.5.8 Review

Controls will follow a continuous improvement process with reviews of all of the major elements of the health management framework including SEGs, the risk register, the monitoring program, monitoring standards, medical surveillance, and effectiveness of controls. The frequency of reviews will vary depending on the element of the health management framework. The system may also include audits.

It is beneficial for the overall health management framework to fit into an occupational health and safety management system such as those described in AS/NZS ISO 45001 (Standards Australia 2018). The AIOH's Occupational Hygiene Monitoring and Compliance Strategies (Grantham & Firth 2019) also provide a helpful Chart (11.1) laying out the process for compliance monitoring to meet Australian harmonised OHS legislation. However, as outlined above, exposure control is broader than just compliance monitoring.

There are many elements that make up a comprehensive health management framework. It can be a complex task bringing all these aspects together. An example occupational health management framework with key components is set out below. Table 1.4 can be used as a checklist of what documents and processes should exist in a well-managed operation. Review frequencies of each element can be set by a company's own risk assessment or by specific legislative requirements if they exist.

TABLE 1.4
Sample Occupational Health Management Framework

Key Health Management Documents	Components
Health Science Summary	This may be part of an annual OH report and include legislative updates, data reviews of health affecting agents, reviews of any carcinogens in the workplace, SEG reviews for appropriateness and quality assurance checks carried out on monitoring o health affecting agents
Medical Monitoring Summary	May include medical monitoring data and health awareness information.
Job Demands Manual	Sets out major physical job demands. *Used by Physician as an input into medicals.*
Risk Register	Sets out risks for major health affecting agents for each SEG, identifies gaps and sets out major controls using the hierarchy of controls.
Monitoring schedule	Sets out health affecting agents to be monitored, or assessed, frequency and sample numbers. *Drawn from Health Science Summary*
Site Specific Health Agent Management Plans	Sets out management of key health affecting agents as determined by company risk assessments or legislative requirements. Plans may include lead, radiation, asbestos, dusts, noise, health promotion, legionella and water quality.
Health Action Plan	Includes a review of progress against the Plan.
Standards & Procedures	
Contractor Management	A system to ensure that Contractors carry out their work in a healthy and safe way harmonised to company requirements.
Change Management	A system to ensure that any changes affecting plant, procedures or systems as well as procurement and design take health effects into account e.g. Buy Quiet.
Legal Register	Including legislative review.
First Aid	Included in safe work procedures This may include how first aid is organised, equipped, staffed, scenario analysis, and the level of training provided
Personal Protective Clothing & Equipment Standards	Includes management systems for respiratory protection, hearing protection etc.
Records System	Includes record types, retention policy and privacy considerations.
Hazard Identification & Risk Assessment Procedure	Includes the company Risk Assessment Matrix.
Hygiene Monitoring Quality Assurance	Sets out quality assurance requirements for providers and internal monitoring quality assurance requirements.

(Continued)

TABLE 1.4 (CONTINUED)
Sample Occupational Health Management Framework

Standards & Procedures	
Medical Surveillance Quality Assurance	Sets out quality assurance requirements for medical assessments, equipment maintenance and assessments including respiratory and audiometry.
Manual Handling Task Assessment	Includes task assessment methodology and risk assessment.
Hazardous Substance and Dangerous Goods system	Incorporates systems for bringing new chemicals on site, hazardous substance assessment procedures including storage and disposal. May also be major hazard facility legislative requirements
Thermal (heat and / or cold) Stress Assessment Procedure	Incorporates measurement, assessment and controls.
Fitness for Work	Includes fatigue and hydration management, drug and alcohol, employee assistance.
Legionella Monitoring Procedure	Legionella Standard Operating Procedure sets out monitoring strategy, schedules, and control strategies.
Potable Water Monitoring Procedure and action plan following exceedance	Sets out monitoring strategy, schedules and control strategies.
Exposure Standards	Integrates company, Australian and State exposure standards into context, taking working hours/shift and other adjustments into account.
Quality Assurance (QA)	
Occupational hygiene monitoring QA Procedures	May include particulates, vapours, gas, noise, vibration, biological, thermal alcohol and drugs testing.
Medical Systems QA procedures	Includes monitoring equipment register.
Internal or external audits of major occupational health systems	Including occupational hygiene and medical systems and any of the above management elements
OH Training	Training needs analysis; Review of effectiveness of training, records.
Respiratory & hearing protection FIT testing	May be quantitative or semi quantitative.
Non- Routine Monitoring	
Lighting Survey	Initial, Periodic and new plant.
Noise Contour Map of Fixed Plant	Must be made readily available to all workers and include designated mandatory hearing protection areas.
Electromagnetic Field Survey	Focus on major likely EMF sources such as substations and large electric motors.
New Plant assessments e.g. noise, vibration	As per Australian and International Standards
Ventilation	Includes local exhaust ventilation such as welding exhaust ventilation and its testing frequency, supplied air from compressors as per AS1715.
Registers	
Maintenance	Site specific maintenance relating to health affecting agents.
Signage	Includes mandatory Personal Protective Clothing & Equipment such as hearing protection areas, pacemakers and hearing aids (electric fields), ionising radiation, dangerous goods, confined spaces, authorised access, etc.
Monitoring Equipment Register	Includes both occupational hygiene and occupational medical equipment as well as calibration schedules
Training Register	All employees – training status reviewed as part of training needs analysis. Also includes FIT testing for respirators and hearing protection.

1.6 OTHER OCCUPATIONAL HYGIENE TOOLS

One very significant aspect of occupational hygiene is the extent of online tools and Apps available to help the occupational hygienist, health professional or workplace representative to assess the workplace and improve their skills.

Due to the dynamic nature of online tool development, it is suggested that the reader consult the web pages of major OH organisations in the first instance.

Major websites include the following:

- AIOH – position papers, exposure assessment spreadsheets.
- AIHA – exposure assessment strategies committee – excel-based tools including spreadsheet applications of statistical analysis, skin permeability, basic workplace exposure assessment, IH exposure scenario tool.
- BOHS – links to technical guidance.
- Breathe Freely Australia – links to 'Breathe Freely' for construction and manufacturing with significant material for OH risks in these industries.
- Occupational Hygiene Training Association (OHTA) – freely available access to a comprehensive range of occupational hygiene training courses.
- RESP-FIT – resources to support respiratory fit testing.
- UK Health and Safety Executive – Codes of Practice and numerous computer-based tools, for example, vibration and noise calculators as well as the COSHH control banding system.

There are numerous occupational hygiene Apps available in areas including noise exposure, octave band analysers, thermal risk, lux meters, and whole body vibration, as well as powerful Apps available from instrument manufacturers providing monitoring advice and instrument catalogues.

While many Apps are not validated, their application can be empowering to persons at the workface.

Two validated Apps that are likely to have longevity are:

- NIOSH Sound Level Meter App – backed up by quality NIOSH research, available for iPhone.
- The Predicted Heat Strain Mobile App developed by the University of Queensland – available for iPhone.

REFERENCES

Agricola, G. 1556, *De re metallica*, 1st edn. Basil: Hieronymus Froben & Nicolaus Episcopius.

American Industrial Hygiene Association (AIHA) 2014, Exposure Assessment Strategies Committee: '*Exposure Assessment Strategies Tools and Links New IHStat with multi-languages*', https://www.aiha.org/public-resources/consumer-resources/apps-and-tools-resource-center/aiha-risk-assessment-tools/ihstat-multi-language-version/ihstat-multi-language-version-version-tool-download [22 January 2024].

AIOH 2016, *Adjustment of Workplace Exposure Standards for Extended Workshifts Position Paper*, 2nd edn. Australian Institute of Occupational Hygienists Tullamarine Victoria, https://www.aioh.org.au/product/wes/ [22 January 2024].

Asbestos Safety and Eradication Agency (ASEA) 2016, *Future Projections of the Burden of Mesothelioma in Australia*, https://www.asbestossafety.gov.au/research-publications/future-projections-burden-mesothelioma-australia [22 January 2024].

Australian Mesothelioma Registry (AMR) 2023, *Mesothelioma in Australia 2021, Australian Institute of Health and Welfare, Australian Government*, https://www.aihw.gov.au/getmedia/034ebfb9-554f-4eb7-8f0f-ad56d9d3c5ae/aihw-can-152.pdf.aspx?inline=true [22 January 2024].

Australian National University (ANU) 2017, REGNET *Overview of work health and safety regulation in Australia* [ONLINE, 2017], http://regnet.anu.edu.au/research/centres/national-research-centre-ohs-regulation-nrcohsr/overview-work-health-and-safety-regulation-australia [26 July 2023].

Blackley, D., Halldin, C. & Laney, A. 2018, 'Continued Increase in Prevalence of Coal Workers' Pneumoconiosis in the United States, 1970–2017', *American Journal of Public Health*, vol. 108, no. 9, pp.1220–1222. https://doi.org/10.2105/AJPH.2018.304517

Blunt, L. A., Zey, J. N., Greife, A. L. & Rose, V. 2011, 'History and Philosophy of Industrial Hygiene', In D. H. Anna (Ed) *The Occupational Environment: Its Evaluation, Control, and Management*, 3rd edn. USA: American Industrial Hygiene Association Virginia.

BOHS & NVvA 2011, *Testing Compliance with Occupational Exposure Limits for Airborne Substances*, British Occupational Hygiene Society (BOHS) & Nederlandse Vereiniging voor Arbeidshygiëne (NVvA), https://www.arbeidshygiene.nl/-uploads/files/insite/2011-12-bohs-nvva-sampling-strategy-guidance.pdf [22 January 2024].

Carey, R., Driscoll, T., Peters, S., Glass, D., Reid, A., Benke, G. & Fritschi L. 2014, 'Estimated Prevalence of Exposure to Occupational Carcinogens in Australia 2011–2012', *Occupational and Environmental Medicine*, vol. 1, pp. 55–62. https://doi.org/10.1136/oemed-2013-101651

Carey, R., Hutchings, S., Rushton, L., Driscoll, T., Reid, A., Glass, D., Darcey, E., Si, S., Peters, S., Benke, G. & Fritschi, L. 2017, 'The Future Excess Fraction of Occupational Cancer among Those Exposed to Carcinogens at Work in Australia in 2012', *Cancer Epidemiology*, vol. 47, pp. 1–6. https://doi.org/10.1016/j.canep.2016.12.009

Damousi, J., Rubenstein, K. & Tomsic, M. (Eds) 2014, *Diversity in Leadership: Australian women, past and present*. Canberra, Australia: ANU Press.

Doll, R. 1955, 'Mortality from lung cancer in asbestos workers', *British Journal of Industrial Medicine*, vol. 12, no. 2, pp. 81–86. https://doi.org/10.1136/oem.12.2.81

Firth, I., Golec, R., Di Corleto, R., Ng, K. & Wilson, J. 2020, *Simplified Occupational Hygiene Risk Management Strategies*, Australian Institute of Occupational Hygienists, Inc (AIOH) Tullamarine Victoria, https://www.aioh.org.au/product/simplified-pdf/ [22 January 2024].

Glick, M. 1968, '*Pneumoconiosis in New South Wales coal mines*', First Australian Pneumoconiosis Conference, Joint Coal Board, Sydney, pp. 165–177.

Grantham, D. & Firth, I. 2019, *Occupational Hygiene Monitoring and Compliance Strategies* Australian Institute of Occupational Hygienists, Inc (AIOH) Tullamarine Victoria, https://www.aioh.org.au/product/monitoring-pdf/ [22 January 2024].

Hamilton, A. 1943, *Exploring the Dangerous Trades*, Boston: Little Brown and Company.

Hämäläinen, K., Takala, J. & Kiat, T. 2017, *Global Estimates of Occupational Accidents and Work-related Illnesses 2017*, Workplace Safety and Health Institute, Ministry of Manpower Services Centre, Singapore, https://www.icohweb.org/site/images/news/pdf/Report%20Global%20Estimates%20of%20Occupational%20Accidents%20and%20Work-related%20Illnesses%202017%20rev1.pdf [22 January 2024].

Harden, M., Steer, J., Harden, F., Butler, G., & Mengersen, K. 2015, 'Integrating Occupational Hygiene, Medicine and Engineering in a Mining Context', Abstract 0150, *10th International Occupational Hygiene Association (IOHA) London 2015* International Scientific Conference.

Hewett, P., Logan, P., Mulhausen, J., Ramachandran, G. & Banerjee, S. 2006, 'Rating Exposure Control Using Bayesian Decision Analysis', *Journal of Occupational and Environmental Hygiene*, vol. 3, no. 10, pp. 568–568.

International Council on Mining and Metals (ICMM) 2016, *Good Practice Guidance on Occupational Health Risk Assessment*, 2nd edn, https://www.icmm.com/website/publications/pdfs/health-and-safety/2016/guidance_health-risk-assessment-2016.pdf [22 January 2024].

Ireland, J. & Willingham, R. 2012, 'Shorten Warns over Asbestos Risk', *Sydney Morning Herald*, https://www.smh.com.au/national/shorten-warns-over-asbestos-risk-20120904-25ct3.html [22 January 2024]

Jahn S., Bullock, W. & Ignacio, J. 2015, *A Strategy for Assessing and Managing Occupational Exposures*, 4th edn. Virginia USA: American Industrial Hygiene Association (AIHA).

Keil, C.B., Simmons, C. E. & Anthony, T. R. (Eds) 2009, *Mathematical Models for Estimating Occupational Exposure to Chemicals*, 2nd edn. USA: American Industrial Hygiene Association (AIHA) Virginia.

Liedel, N., Busch, K. & Lynch, J. 1977, '*Occupational Exposure Sampling Strategy Manual: DHEW* (NIOSH) Publication No. 77-173' National Institute of Occupational Safety and Health Cincinnati, OH, USA. https://www.cdc.gov/niosh/docs/77-173/default.html [22 January 2024].

Laszcz-Davis, C., Maier, A. & Perkins, J. 2014, 'The Hierarchy of OELs: A New Organizing Principle for Occupational Risk Assessment' *The Synergist*, American Industrial Hygiene Association (AIHA) Virginia, USA.

Millar, J. D. 1984, 'The NIOSH-suggested List of the Ten Leading Work-Related Diseases and Injuries', *Journal of Occupational Medicine*, vol. 26, no. 5, pp. 340–361.

Moore, R. & Badham, C. 1931, 'Fibrosis in the Lungs of South Coast Coal Miners, New South Wales', *Commonwealth Department of Health*, vol. 9, no. 5, p. 33.

McKenzie, J. F., El-Zaemey, S., & Carey, R. N. 2021, 'Prevalence of Exposure to Multiple Occupational Carcinogens among Exposed Workers in Australia', *Occupational and Environmental Medicine*, vol. 78, no. 3, pp. 211–217. https://doi.org/10.1136/oemed-2020-106629

Murray, R. 1990, 'Asbestos: A Chronology of Its Origins and Health Effects', *British Journal of Industrial Medicine*, vol. 47, no.6, pp. 361–361. https://doi.org/10.1136/oem.47.6.361

NIOSH 2023, *Occupational Exposure Banding*, https://www.cdc.gov/niosh/topics/oeb/default.html [22 January 2024].

National Safety Council USA 2023, *Work Injury Costs 2021*, https://injuryfacts.nsc.org/work/costs/work-injury-costs/#:~:text=The%20total%20cost%20of%20work,administrative%20expenses%20of%20%2457.5%20billion [22 January 2024].

NH&MRC & NRMMC 2011, *Australian Drinking Water Guidelines Paper 6 National Water Quality Management Strategy*, National Health and Medical Research Council and National Resource Management Ministerial Council, Commonwealth of Australia, Canberra. https://www.nhmrc.gov.au/about-us/publications/australian-drinking-water-guidelines [22 January 2024].

Parliament of Australia 2017, '*Report No. 2, 55th Parliament Coal Workers*' Pneumoconiosis Select Committee.

Parliament of Australia 2022, *Snapshot of employment by industry*, https://www.aph.gov.au/About_Parliament/Parliamentary_departments/Parliamentary_Library/FlagPost/2022/November/Employment_by_industry_2022 [22 January 2024].

Petts, D., Wren, M., Nation, B. R., Guthrie, G., Kyle, B., Peters, L., Mortlock, S., Clarke, S. & Burt, C. 2021, 'A Short History of Occupational Disease: 2 Asbestos, Chemicals, Radium and Beyond', *The Ulster Medical Journal*, vol. 90, no. 1, pp. 32–34.

Safe Work Australia (SWA) 2022, *Key Work Health and Safety Statistics Australia 2022*, Safe Work Australia https://www.safeworkaustralia.gov.au/doc/key-work-health-and-safety-statistics-australia-2022 [22 January 2024].

Schulte, P. A. 2005, 'Characterizing the Burden of Occupational Injury and Disease,' *Journal of Occupational and Environmental Medicine*, vol. 47, no. 6, pp. 607–622. https://doi.org/10.1097/01.jom.0000165086.25595.9d

Standards Australia and New Zealand Standards 2005, *Occupational Noise Management, Part 1: Measurement and Assessment of Noise Emission and Exposure*, AS/NZS 1269.1:2005, SAI Global, Sydney. https://www.standards.org.au/access-standards

Standards Australia and New Zealand Standards 2018, *Occupational Health and Safety Management Systems – Requirements with Guidance for Use*, AS/NZS ISO 45001:2018. Sydney: SAI Global https://www.standards.org.au/access-standards

Steenland, K., Burnett, C., Lalich, N., Ward, E. & Hurrell, J. 2003, 'Dying for Work: The Magnitude of Us Mortality from Selected Causes of Death Associated with Occupation', *American Journal of Industrial Medicine*, vol. 43, no. 5, pp. 461–482. https://doi.org/10.1002/ajim.10216

Steer, C. & Bennett, A. 2019, *Australia's Workplace Illness Fatality Data Is Grossly Inadequate, and People Are Dying Because of It*, AIOH Foundation, https://aiohfoundation.org.au/media/2016/11/aioh-foundation-paper-2019.pdf [22 January 2024].

The Thoracic Society of Australia and New Zealand November 2017, Media Release '*Peak Body Calls for National Response Following Resurgence of Occupational Lung Diseases*', https://www.first5000.com.au/blog/lung-disease-makes-unwelcome-return-workplace/ [22 January 2024].

UK Parliament n.d., *The 1833, Factory Act*, https://www.parliament.uk/about/living-heritage/transforming society/livinglearning/19thcentury/overview/factoryact [22 January 2024].

US Department of Labor 2022, *News Release, National Census of Fatal Occupational Injuries in 2021*, https://www.dol.gov/newsroom/releases/osha/osha20221216-1#:~:text=%E2%80%9CIn%202021%2C%205%2C190%20workers%20suffered,their%20co%2Dworkers%20in%202021 [22 January 2024].

Van Tongeren, M. 2014, 'Use of Exposure Models for Regulatory Chemical Risk Assessment – Can We Rely on Them?' *Australian Institute of Occupational Hygienist, Inc (AIOH) 32nd Annual Conference*, December 2014.

Wagner, J. C., Sleggs, C. A. & Marchand, P. 1960, 'Diffuse Pleural Mesothelioma and Asbestos Exposure in the North Western Cape Province', *British Journal of Industrial Medicine*, vol. 17, no. 4, pp. 260–260. https://doi.org/10.1136/oem.17.4.260

WorkSafe New Zealand 2019, *Work Related Health Estimates and Burden of Harm*, https://www.worksafe.govt.nz/topic-and-industry/work-related-health/work-related-health-estimates-and-burden-of-harm/ [22 January 2024].

WorkSafe New Zealand 2023, *Fatalities*, https://data.worksafe.govt.nz/graph/summary/fatalities [22 January 2024].

2 Occupational Health, Basic Toxicology, and Epidemiology

Dr Martyn Cross and Dr Geza Benke

2.1 INTRODUCTION

This chapter is a basic introduction to the subjects of occupational health toxicology and occupational epidemiology as related to the workplace. The purpose of understanding toxicology and epidemiology is to assist in recognising hazards in the workplace with the intention of controlling them. While both these fields can require extensive expertise, most H&S practitioners will need an understanding of toxicology and epidemiology as they are the basis of many of our preventive actions. Practitioners needing in-depth toxicological or epidemiological information will need to consult further readings, some of which are identified at the end of this chapter, or experts in these areas. In addition, Chapter 11, which discusses biological monitoring and monitoring strategies, relies significantly on the toxicological principles of absorption, distribution, metabolism, and excretion (ADME) briefly presented in this chapter.

This chapter discusses why a systemic approach to occupational health is required, rather than merely a mechanistic outlook. The health impacts of hazardous substances and hazardous environments on the human body will be briefly considered. The concepts of chemical hazards, dose and risk, their relationships, and their role in workplace risk assessment will be explained, as will the way epidemiology in conjunction with toxicology can provide the link between exposure and disease.

2.2 THE HUMAN BODY IN THE WORKPLACE

To make any sense of occupational health and occupational hygiene, the H&S practitioner needs a basic knowledge of the worker and how this human machine interacts with the work environment. The body and its responses to exposures are highly complex. The workplace is a variable mix of different factors causing differing exposures, often acting in an uncontrolled way on the human body. Hence, worker–workplace interaction can be tremendously complicated.

The worker's body constantly interacts with the workplace environment and does so in a variety of ways. Some autonomic responses will be assessing the environment (such as sniffing the air), while others limiting exposures (e.g. breathing shallowly). Others may be defence mechanisms, either with short-term on a physiological level (e.g. coughing and tearing) or with longer-term exposure at a cellular level (e.g. the development of antibodies, increasing metabolism). The worker interacts with the workplace through:

- sensory input—sight, sound, smell, and feeling;
- ergonomic factors—posture, task, and energy demand;
- physical contact—pressure, vibrations, noise, heat, radiant energy, ionising radiation;
- chemical contact—irritation, inhalation or skin absorption;
- psychosocial factors—such as the way individuals interact with the demands of their job and their work environment, and past experience with such interactions;

DOI: 10.1201/9781032645841-2

Even moderately simple workplace tasks require the use of limbs, motor coordination, balance, vision, lungs, heart and circulatory system, the senses, the intellect, hearing, the voice, and body heating or cooling mechanisms. Some of the responses to these interactions are perfectly normal ones that the body is designed to withstand and respond to. But if we understand neither the factors in the workplace acting on the worker nor the way the body responds to or copes with these stressors, we will not be able to appropriately control any resulting strain on the body. Damage and/or illness may then occur.

Certain workplace substances may exert local effects directly at the point of contact, these range from sensory irritation of the eyes, nose, throat, and skin, to more damaging toxic effects. However, other substances may exert systemic effects elsewhere within the body after the chemical has been absorbed into the body via the bloodstream, distributed throughout the body, and possibly metabolised to a less or more toxic substance, before being excreted. Generally harmful toxic effects become manifest when damage to organs overwhelms their ability to repair.

In order to understand the potential for damaging effects from chemical hazards, two questions must be answered:

- What is the worker exposed to, how much, and for how long (i.e. the potential exposure)?
- What is the body's response?

Answers can be difficult to reliably establish. Aspects of measuring chemical exposure are dealt with in Chapters 8 to 11. However sometimes the factors affecting the worker will not be apparent. Historical examples are: in the 19th century, illness (lung damage and debilitating emphysema) among needle grinders was thought to be caused by inhalation of metal splinters. In fact, the damage was done by the crystalline silica from their grinding wheels. X-rays and radioactivity were initially seen as curious phenomena that could sometimes be industrially useful; their carcinogenic effects were understood too late for many.

Inevitably, two more questions arise. First, is the worker's illness linked to workplace chemical exposure? Second, if so, what can be done to prevent the damage to health? The first question will be examined in this chapter; the second question will be addressed throughout the remainder of the book.

2.3 ESTABLISHING CAUSAL RELATIONSHIPS

It is not always easy to recognise if a health effect is the result of a single workplace exposure. While we can identify individual hazards and measure them, exposure to workplace stressing agents doesn't occur in isolation; there may be many hazards acting on the worker simultaneously. Consider the foundry worker in Figure 2.1. Foundry workers' tasks include melting metals and pouring them into moulds, slag-skimming, and performing other operations on the top of the furnace, as well as cleaning and assembling the moulds. In these tasks, they might be subjected to heat, noise, carbon monoxide (CO), and metal fume particles, not to mention the stress of balancing atop a vat of molten metal. When tapping the furnace into a crucible for pouring metal into sand moulds, workers may be additionally exposed to additional chemical hazards, for example, amine fumes, gaseous cyanide, and/or isocyanates from mould constituents. The intensity, duration, and route of exposure (e.g. breathing, skin contact) determine the dose (the quantity of a substance absorbed during an exposure the worker receives from each hazard, which in this case, will differ at various times during a single work shift depending on the activity of the worker). Making evaluation even more complex, the amount of exposure varies from one work site and task to the next.

We need to know how the body responds to each and every one of these hazards in order to conduct a risk assessment to evaluate the potential health effects associated with these exposures

FIGURE 2.1 Foundry workers may be subjected to several health hazards, including various dusts, gases, and heat

Source: Courtesy of 3M.

for the foundry workers in this workplace. Failure to realise that CO is a chemical asphyxiant, which interferes with the transportation or absorption of oxygen in the body or, worse, failure to even recognise that it could be present, could make it difficult to understand, for example, why workers in this type of workplace could develop headaches.

The body exhibits different responses depending on the dose and the magnitude, duration, and frequency of exposure. For example, breathing in relatively low doses over a protracted period of time, such as the foundry workers exposed to CO over a work shift, may result in headache, loss of concentration, and lassitude. In contrast, exposure to high CO concentrations over a short period of time, such as from breathing in petrol engine exhaust in a confined space, may lead to severe mental confusion, unconsciousness, coma, and death. So, the relationship between the hazards which exist in the workplace, a worker's actual level of exposure, and the way the worker's body responds requires careful consideration in order to suggest appropriate ways to control exposures. Interactions between multiple hazards and their harmful effects also need to be considered. (This is especially true where interactions are additive; or possibly synergistic, meaning that the effect resulting from two hazards combined is greater than the added effects of the individual hazards.)

It is especially challenging to establish causal relationships between workplace exposures and ill health. Often the health effects of exposure can be delayed so that it is difficult for an individual or an organisation with poor record keeping to recall the exposure or its extent. Humans are seldom exposed to one hazard at a time and many workplace and lifestyle hazards may result in similar health effects or mask the effect of the other. Obviously, experimenting on humans is limited by ethical considerations, so science has been forced to develop alternative ways of eliciting information concerning the health effects due to workplace chemical exposures. One method uses experimental animals as test subjects in studies (toxicology), and another has been long-term studies of exposed human populations (epidemiology). Indeed, these approaches have provided much of our understanding of the potential health effects of metals, dusts, organic solvents, pesticides, and microbiological agents. In addition, New Approach Methodologies (NAMs) have been introduced to reduce the use of experimental animals. NAMs include in vitro studies using human or animal cells, in-chemico studies which simply evaluate how a substance chemically interacts/reacts with certain materials, and in silico computer-driven predictive tools which provide information on the mode of action and mechanisms by which chemicals exert their toxic effects. Data from these sources may assist in the H&S practitioner's assessment of risks in the workplace (US EPA 2023a), but usually, an expert will be needed for assistance.

However, even after all these efforts to study the hazards and the body's responses, individual response variability, and exposures required, within a given work group may result in different outcomes. Consequently, the significance of harmful exposure prevention for all workers becomes apparent.

2.3.1 Multiple Causes of Disease

Another problem to surmount in studying workplace illness is that the body may respond in similar ways to exposures from lifestyle factors (such as alcohol consumption, diet, and smoking) as well as workplace hazards. Foundry workers may develop asthma from exposure to workplace amines or isocyanates, or their asthma may be caused by pollens and other natural substances and not work-related at all. A headache may be the result of dehydration, or excessive noise, or carbon monoxide exposure.

2.3.2 Linking a Health Effect to a Particular Event

The observation of workers exposed to thousands of different hazardous substances such as asbestos or solvents, or processes such as welding, has provided the basis for making workplaces much healthier. For these well-acknowledged problems, the H&S practitioner can confidently use the available epidemiological and toxicological information and recommended exposure standards to facilitate control strategies. Occupational, or workplace, exposure standards are assigned by the regulatory authorities (Safe Work Australia). A workplace exposure standard (WES)/ workplace exposure limit (WEL) for an airborne contaminant is the maximum upper limit for exposure to that particular chemical that must not be exceeded. (The concept of the exposure standard is addressed in Chapter 3).

On the other hand, many hazardous-substance exposures, especially long-term, may lead to effects such as adverse mood or personality change, or loss of memory, which may be too subtle to be recognised easily and differentiated from non-workplace factors. It is helpful for H&S practitioners to understand such phenomena. Further, substances are constantly being investigated through toxicity testing to detect whether a chemical may cause harm to humans for possible effects such as carcinogenicity (the ability to cause cancer), mutagenicity (the ability to induce mutation in cells), developmental toxicity, or teratogenicity (the ability to cause defects in offspring). However, cancer, cell mutations, and birth defects occur in the general population for a variety of reasons. For epidemiological studies to identify occupational causes, one must show an increase in a particular disease rate following known occupational exposure. For example, exposure to radon gas, which occurs naturally in underground mines, can cause lung cancer, but so can smoking. To identify any increase in the incidence of lung cancer in a group of miners beyond that experienced in a comparable non-occupationally exposed group requires specific and sensitive epidemiological methods.

The study of the health of workers and the diseases caused by their work is hence very complex. The tools used by the epidemiologist to provide a better understanding of the relationship between exposure and health are presented later in this chapter.

2.4 OCCUPATIONAL HEALTH

Occupational health may be defined as the maintenance of the individual worker's state of well-being and freedom from occupationally related disease or injury. Occupational hygiene is the practice whereby this is achieved.

2.4.1 Occupational Disease Compared with Occupational Injuries

By now, the H&S practitioner will have realised that successful intervention in an occupational health issue requires a systemic approach rather than telling workers to minimise exposure. Indeed, every jurisdiction in Australia has a general duty of care requirements that a workplace is safe and without risk to health as far as reasonably practicable. However, there are few regulations that prescribe detailed courses of action for minimising or eliminating health impacts due to specific hazards. Any successful intervention in this self-regulatory environment, therefore, requires knowledge of:

- the workplace process and the hazards it produces;
- the equipment used in the processes;
- the toxicology of hazardous substances;
- the health effects of physical, chemical, and biological hazards in the workplace;
- the physical and organisational environment in which the task occurs;
- effective risk communication and consultation;
- the effectiveness and appropriateness of different control strategies;
- the relative costs of implementation.

The focus of occupational health management is to protect workers in their employment from health risks, where:

- The H&S practitioner has to intervene without the benefit of being physically able to see the hazard. This may require new insights in identifying problems and seeking solutions. In safety matters, one can define the hazards quite readily and also identify some of the outcomes without error. For example, unrestrained moving loads might be expected to fall, unguarded machinery can be expected to mangle limbs, and unsupported excavations can be expected to collapse.
- In dealing with hazardous substances, the H&S practitioner has to be able to identify possible outcomes, even where the worker does not show immediate effects. If workers are in the way when loads fall or trenches collapse, there is a high probability of resulting injury or death. In the case of overexposure to potentially hazardous substances or environments, the worker may appear not to be affected immediately afterwards yet fall ill or even die much later, including after retirement.
- In occupational health, there exists a clear possibility of intervention to change the course of events. In the case of accidental injury, intervention is aimed at preventing the event from occurring, since once an accident is underway, intervention is rarely possible. The use of personal protective equipment such as helmets or steel-capped boots does not prevent incidents but reduces the consequent harm.

Occupational health and safety practices are all based on risk assessment. That is, we aim to reduce exposure to the level where there is an acceptably small chance of harm occurring. The human body can tolerate some levels of exposure to some hazards without detriment. In general, occupational safety deals with accidents and their prevention, while occupational health deals with occupational diseases and their prevention.

A sudden injury at work can differ from a disease caused by work exposures in three ways. These are the dose factor, the time factor, and the damage factor.

2.4.1.1 The Dose Factor

Disease may be caused by a single large exposure or many small exposures to a workplace hazard. The likelihood of disease depends on the dose received. There may be some threshold dose below which there is no adverse effect, which will vary depending on the agent. It will be very small for highly hazardous agents (high potency) and large for those of lower hazard (low potency).

Throughout the following chapters, the significance of this dose factor will be discussed. Different doses can provoke different responses. The dose to which a worker is exposed affects the interpretation of risk and the kinds of controls we should use.

2.4.1.2 The Time Factor

An accident and the resulting injury occur at virtually the same point in time and from a single incident. This immediacy of injury means that the link between cause and effect is obvious. However, while occupational diseases may occasionally result from a single massive exposure they usually result from exposures to the causative agent over a period of time. The disease may take some time to develop, ranging from days to years.

Examples include:

- carbon monoxide poisoning—minutes;
- solvent intoxication—minutes–hours;
- metal poisoning—days;
- noise-induced hearing loss—years;
- asbestos-related diseases—up to 50 years.

Some health effects appear long after exposure has ceased. Consequently, it is important to recognise that workplace exposure to hazardous substances can be equated with multiple accidents occurring over a long time span. The typical time span for asbestos-related diseases to manifest, for example, may be 15 years for asbestosis, 20 years for lung cancer and 30 years for mesothelioma.

2.4.1.3 The Damage Factor

An accident may injure tissues but only at the point where the energy of the accident is applied—for example, the head or the arm. With disease, tissue damage may or may not occur where the causative agent is applied. For example, inhaled quartz dust has a direct effect on the lungs, but inhaled solvent vapours may produce effects on the liver, kidney or brain (e.g. headache or drowsiness

In addition, subtle changes, which are not obvious to the worker, may occur in bodily functions.

In summary health effects due to chemical exposure occur when either a single, short-term or long-term exposure results in a cumulative dose to the affected tissue that causes the rate of damage to exceed the rate of repair of the tissue to normality. Thus there are some exposures that are tolerated and do not cause harmful outcomes. While it is an axiom in toxicology that the dose makes the poison (see Section 2.5.4), in fact, harmful effects are a function of dose and time over (duration and frequency) which the dose occurs; the adverse effect (E) is thus a function of absorbed dose (D) and time (T), as shown in Equation (2.1).

$$E = \int (D X T) \tag{2.1}$$

For H&S practitioners, it is effectively a function of exposure and exposure time. Note also that the adverse effect is dependent on the hazard (toxicological) potency of the substance. For example, lower doses of carbon tetrachloride are needed to produce the same extent of liver damage than are required for toluene.

2.4.2 What the H&S Practitioner Needs to Know

Because of these three factors—time, damage, and dose—the link between the cause and the resulting occupational disease may not be immediately obvious. Therefore, to prevent disease, or to detect minor health changes early before the worker becomes ill, it is essential to have knowledge of the conditions in the workplace which pose a risk to health.

Detection and prevention of disease require an understanding of a wide array of work situations and knowledge of the effects that various hazardous agents can potentially have on workers. Therefore it is necessary for the H&S practitioner to seek good occupational hygiene knowledge of:

- how much exposure poses a risk;
- what procedures are necessary to monitor the workplace and worker exposure; and
- how exposure can be controlled to non-harmful levels.

Only when armed with this information can the H&S practitioner take steps to prevent work-related illness. The succeeding chapters provide details of workplace investigation, toxicological information on some regular workplace hazards, guidelines on how to assess the hazards and, lastly, a wide range of control techniques and methods. This information will not make the H&S practitioner an expert but should provide a foundation on which to build. Only practical experience will permit mastery over real work situations.

2.5 BASIC TOXICOLOGY

Toxicology may be defined as the description and study of the effects of a substance on living organisms. In terms of the workplace, we are interested in human toxicology. Few animals, even the canary used historically in mines to warn of poisonous gas exposures, are used nowadays in workplaces. Nonetheless, animal toxicology has provided valuable insight into the potential effects of hazardous substances on humans. Experiments on animals involving prolonged and/or high exposure to various hazards, which can be carried through the majority of the animal's life span, obviously cannot be conducted on human populations for ethical reasons. That is why toxicology took on a predictive role where new substances could be screened to determine their toxicity prior to the potential exposure of humans to prevent harmful effects from their use.

The H&S practitioner needs to understand basic toxicological concepts in order to:

- understand recommended or regulatory exposure standards;
- recognise and prioritise chemical hazards in the workplace; and
- determine control measures for a particular hazard.

A foundry, for example, might provide respiratory protection such as a dust mask but fail to appreciate that polyurethane moulds release isocyanate vapours into the workplace atmosphere—vapours that are not captured by a dust mask. The H&S practitioner's knowledge of all the risk factors will be needed to provide correct protection.

2.5.1 Hazard and Risk

Hazard and risk are two quite different concepts, although the terms are commonly and inappropriately used interchangeably in discussions of hazardous substances. Hazard represents the potential for a substance to cause adverse effects. A hazardous substance is one that can cause harm given the appropriate conditions of exposure or absorption (dose). Risk is a function of the likelihood of an adverse effect occurring in a particular exposure situation and of the magnitude of the adverse effect (the severity of the outcome). If a hazardous substance is to cause harm the worker must be exposed to a sufficiently large amount of the substance.

In short, a hazard becomes a risk when there is exposure to the hazard. Potential for exposure depends on the hazard itself, how the hazardous substance is used with particular focus on exposure pathways, the equipment involved in the task, the people involved and their susceptibilities, and the physical and organisational environment in which the task takes place. The likelihood and consequence of the risk is reduced by any controls put in place to reduce the potential for exposure. The hazardous property (toxicity) of a substance cannot be changed as this is determined by its chemistry.

All substances found in workplaces have the potential to cause harm. However, it is important to differentiate between hazards of no concern and those hazards that present a risk of harm. Asbestos in an asbestos-cement sheet is a hazard but may not present a serious risk because the likelihood of inhalation exposure occurring is very low if the material is not machined or abraded. However, the release of asbestos fibres into the workplace air during the removal of asbestos lagging will almost certainly result in inhalation exposure leading to a higher risk of toxic effects. Benzene is a potent leukaemia hazard, but while contained in a closed reaction vessel it presents a low risk; xylene used in open tanks or in painted coatings may pose a greater risk to health than when used in a fume hood.

In mathematical terms, risk is a function of consequence or severity and likelihood. Likelihood is related to the extent of exposure, with greater exposures more likely to result in injury or disease.

Expressed as a conceptual formula (Equation 2.2):

$$\text{RISK is proportional to HAZARD} \times \text{EXPOSURE}$$

or

$$R \propto H \times E \tag{2.2}$$

where

R = risk
H = hazard
E = exposure

If exposure is zero (exposure is controlled), then risk will also be zero and risk is controlled. (Chapter 5, 'Control of workplace health hazards', examines a number of ways to control exposure.) In the same manner, if the hazard is removed by replacing a hazardous substance with a substance of lower toxicological potency, or better with a non-hazardous substance (e.g. by replacing a solvent process with a water-based one), then both hazard and risk are reduced or become zero, and exposure may not need additional control.

'All substances are poisons; there is none which is not a poison. The right dose differentiates a poison and a remedy'. This comment, by the 16th-century physician and philosopher Paracelsus, summarises one of the fundamental principles of modern toxicology—this is the concept of threshold dose or No Observed Adverse Effect Level (NOAEL). That is, the greatest concentration or amount of a substance at which no detectable adverse effects occur in an exposed population, usually established in experimental animal populations (rats, mice, and less often rabbits). Doses of substances below the threshold for adverse effects may produce a temporary functional effect or a biochemical effect which are not harmful, and, in fact, some chemicals (notably medicines and vitamins) at low doses are beneficial. Essential elements such as cobalt, iron, sodium and calcium testify to the power of this observation. Doses below the threshold dose are needed to maintain health. Doses exceeding the threshold dose will give rise to adverse effects. The threshold information is useful when considering safe levels of a toxic substance and establishing workplace exposure standards.

2.5.2 Absorption of Hazardous Substances

For hazardous substances to be able to exert a toxic effect, they must have a route of entry into the body. In the work situation, we are generally concerned with three potential routes of entry:

- inhalation;
- skin contact;
- ingestion.

Inhalation of a dust, fume, vapour, mist, or gas is the principal route of entry in the work environment. The membranes lining the lungs provide a large area (approximately 140 m^2) which may react directly with the hazardous substance (local effect), or the substance may be absorbed into the circulatory system and exert its effects within the body (systemic effect). The lung has a large surface area that efficiently allows the absorption of substances, especially gases and vapours, into the blood.

Many substances in workplaces readily evaporate or become airborne. Similarly, work process such as welding, grinding or packaging dusty items may contaminate the air. For this reason, a large part of workplace hazardous-substance control is based on keeping the work atmosphere clean. This is why workplace exposure standards, discussed in Chapter 3, relate specifically to inhalation of airborne substances. It is also why occupational health and hygiene investigations often focus on material that can be inhaled.

Skin contact is the second most important route of absorption for workers. Entry through the skin applies most frequently to solvents, and some substances (e.g. oils and some pesticides) that enter predominantly via the skin. Sometimes substances that are present as vapours or gases also enter through the skin, this may result in significant absorption and consequent distribution through the body. Occasionally, substances will enter the body through both the lungs and the skin (e.g. mercury and hydrogen cyanide), where their effects will be additive (Semple 2004). Of course, there are also substances that have direct effects on the skin. These include corrosives including acids and alkalis, substances such as sodium hydroxide, sulfuric acid, and hydrogen peroxide. The eye is also at risk from direct contact, particularly with biohazards and corrosives.

Skin is a very effective barrier to prevent environmental substances from entering the body, however, individual workers vary in their susceptibility to skin penetration. Those with thicker skin will be at lower risk than those with thin skin. Oily skin may provide better protection than dry skin. Some areas of skin are very thin (e.g. on the scrotum or around the eyes) and provide less of a barrier than thicker skin (e.g. on the palms or soles of the feet). Damaged skin may often present a ready route of entry for a hazardous substance, particularly a microbiological hazard. Skin hydration can enhance skin permeation which is often dependent on the type of work and the extent of perspiration (Park et al. 2008).

Ingestion (swallowing), often the result of hand-to-mouth contact, is usually a minor route of chemical absorption in workplaces. Good personal hygiene, particularly attention to washing hands, and avoidance of eating, drinking, and smoking in the workplace can help reduce this route of entry. Some substances may be swallowed by virtue of internal respiratory clearance mechanisms depositing mucus at the oesophagus or following deposition of substances in the mouth or pharynx during mouth breathing.

Other, rarely encountered, absorption route includes direct injection of hazards into the bloodstream or body tissues. These may be relevant to specific workplaces and processes, such as the accidental injection of infective agents from contaminated syringe needles, transdermal injection of fluids from high-pressure hoses, or puncture and laceration injuries from chemically contaminated laboratory glassware.

Once absorbed into the bloodstream a chemical is distributed around the body to organs and tissue throughout the body. The rate of distribution is largely dependent on the blood flow through each organ or tissue. Many chemicals go through biotransformation (metabolism), a metabolic process that takes place mainly in the liver to facilitate the excretion of the parent chemical or its metabolites. The principal route of excretion is the urine and faeces (Slitt 2019). Biological and biological effect monitoring (see Chapter 11) is dependent on the toxicological principles of absorption, distribution, metabolism, and excretion (ADME).

2.5.3 Toxic Effects of Hazardous Substances

In the human body, the effects of hazardous substances vary greatly, as do the severities of those effects. Some are examined specifically in later chapters. In simple terms, effects can be described in terms of where and when they occur in the body.

Where:

- **Local effects** are adverse effects on the particular tissue that has been directly exposed to the hazardous substance. Examples include:
  - damage to the eyes or skin by corrosives;
  - dermatitis caused by contact with organic solvents;
  - respiratory and skin sensitisation, due to an adverse reaction of the immune system;
  - intense irritation of the respiratory tract by gases such as chlorine or ammonia.
- **Systemic effects** are adverse effects on one or more body systems after absorption of the hazardous substance into the blood. In these instances, the hazardous substance travels through the body to a distant, susceptible organ. For example:
  - the nervous system, blood, kidneys, and reproductive functions can be harmed by lead;
  - the nervous system can be damaged by organophosphate insecticides.

When:

Toxic effects may be immediate or delayed.

- **Immediate effects** develop immediately or soon after exposure. These may range from subtle (possibly reversible/short-term effects) to harsh (irreversible/long-term) adverse effects. These can include irritation, acute poisoning, or reproductive effects. For example:
  - the eyes and respiratory tract immediately may become irritated by gases such as ammonia;

- narcotic effects (depression of the central nervous system) (e.g., headache, dizziness, incoordination, and unconsciousness) can rapidly follow excessive exposure to toxic organic solvents;
- metal fume fever from high exposure to some metal fumes may result in flu-like symptoms within one to two days.

- **Delayed** adverse effects occur after a lapse in time. Delayed adverse effects generally occur when cell or tissue damage has not been appropriately repaired. For example, pulmonary adverse effects associated with particulates is due to sustained inflammatory response from high exposures that clearance mechanisms cannot cope with. Even a brief exposure to some type of toxicants, for example, genotoxic agents, may be sufficient to induce delayed adverse effects that develop years later. Some well-known delayed adverse health effects are:
  - asbestosis and silicosis following excessive exposure to asbestos and silica dust, respectively (Wagner 1997);
  - lung cancer following exposure to dusts containing arsenic (Wei et al. 2019);
  - chronic dermatitis from exposure to chromium-containing cements (Cahill et al. 2004);
  - lung cancer and mesothelioma from asbestos exposure (Huh et al. 2022);
  - damage to the DNA structure of sperm and ova, and reduced fertility (suppression of sperm production), for example, from lead exposure (Massányi et al. 2020).

Toxic substances may exhibit a wide range of longer-term adverse health effects depending on the chemical and include:

- cancer (genetic and non-genetic carcinogenicity);
- organ damage;
- weakening of the immune system;
- development of allergies, sensitisation, or asthma;
- neurological and behavioural effects;
- reproductive and developmental problems and birth defects;
- mental, intellectual, or physical development of children (Gilmour et al. 2019; Health Canada 2023; Klaassen 2019).

Toxicological information about possible short-term and long-term effects should appear on the safety data sheet (SDS) for the substance.

Toxicologists use the terms 'acute' and 'chronic' to denote duration and frequency of exposure in toxicity tests. Often referred to as short-term (acute) toxicity studies or long-term (chronic) toxicity studies. This usage occurs in animal toxicity-testing guidelines such as those specified by the Organisation for Economic Co-operation and Development (OECD 2023). The guidelines, as applied to laboratory-bred rats, identify acute studies (single exposures), short-term repeated-dose studies (14–28 days), sub-chronic studies (90 days) and chronic studies (six–30 months). The generally accepted usage of these terms in occupational toxicology relates to short exposures (acute) and exposures of several years (chronic).

2.5.4 The Importance of Dose

Earlier we quoted Paracelsus' observation that the dose makes the poison. A vital point to remember is that most hazardous substances are foreign to the body: they are not supposed to be there. In order to link the absorption of hazardous substances and the effects they produce, we have to link the observed effects with the amount absorbed (i.e. the bioavailability of the substance), stated in precise and measurable terms. It is preferable not to describe toxic doses using relative terms

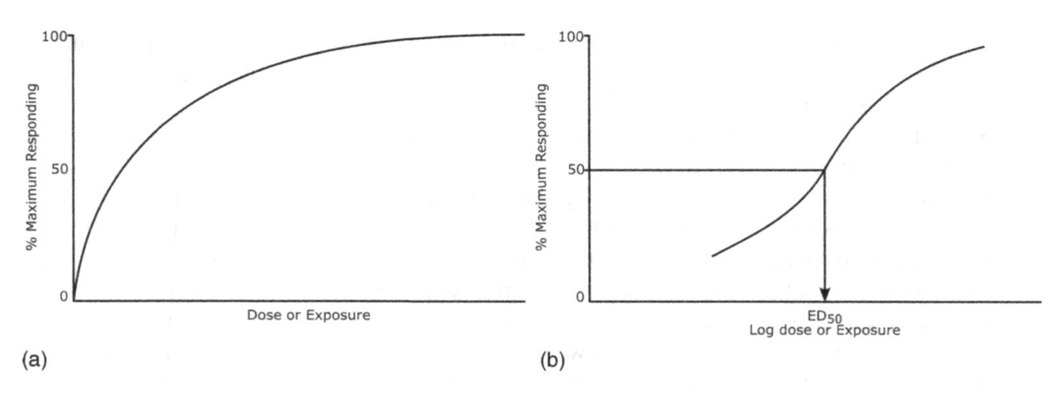

FIGURE 2.2 Graded dose–response relationships in an individual for a specific effect (e.g. a defined level of liver toxicity): (a) linear plot and (b) logarithmic plot.

Source: M. Cross.

such as low, moderate, or high exposure, but to obtain numerical data on the exact dose. All the exposure limits given in Chapter 3 are based on measurable quantities.

Toxicologists use two approaches to examine the relationships between dose and response. The typical dose–response relationship indicates that an adverse response or severity of an adverse effect increases as the dose increases. In the first, the extent of the effect in an individual (e.g. a human or an animal) is related to dose—for example, the degree of enzyme inhibition arising from chemical exposure, the amount of skin redness (erythema) from UV exposure, or a specific level of liver toxicity. When the data are plotted, this is referred to as a dose–effect, or dose–response curve. The expectation is that increasing the dose causes increasing effect.

Figure 2.2a shows the graded dose–response relationship for an individual for a specific effect (e.g. a defined level of liver toxicity) with arithmetic axes, and Figure 2.2b shows the same data plotted on a logarithmic dose axis. A dose–effect curve represented on a log axis virtually always results in a sigmoid curve (S-shaped), indicating that there are low doses that do not cause a response (effect), while high doses generate noticeable, marked, effects for an individual ultimately reaching the maximum effect. The log axes enable the linear data between 20 and 80 per cent of maximal effect to be accurately plotted, and the dose causing a 50 per cent maximal effect, the ED_{50}, can be estimated. Figure 2.2a and b also shows that as dose declines, a smaller effect is observed, until we see a dose at which no effect is seen—the region of the dose–response curve that indicates the transition from 'no toxicity' to 'toxicity' is known as the 'threshold' (Boston University 2019).

The second approach is the quantal dose–response relationship which depicts the distribution of individual responses for a population to a range of discrete doses (i.e. in animal toxicity tests). When epidemiologic studies are unavailable or not relevant, risk assessment is often based on studies of laboratory animals. This approach describes the per cent (%) of a population displaying a given response at a given dose, providing a dose–response curve. A defined response is examined in groups of rats exposed to a range of doses of a chemical. The response may be the presence or absence of a measurable outcome—for example, a defined level of liver toxicity, or at least 10 per cent inhibition of liver activity, or unconsciousness, or death. The proportion of the tested group which shows this response is plotted against the dose (or more usually log dose, as seen in Figure 2.3). Knowledge of the dose–response curve, the No Observable Adverse Level (NOAEL) and the Lowest Observable Adverse Effect Level (LOAEL) (shown in Figure 2.3) enables extrapolation to the 'threshold' and facilitates extrapolation of the animal toxicity data

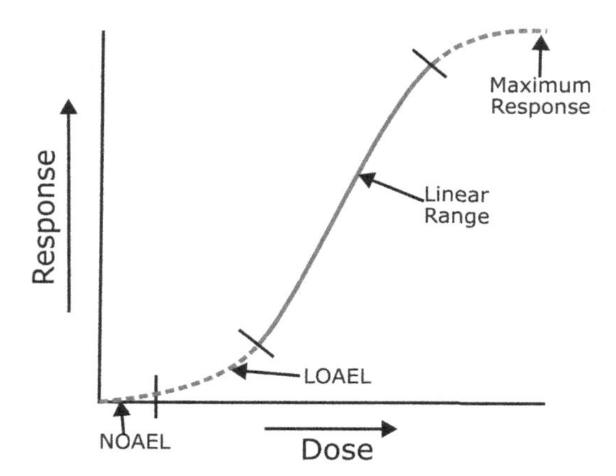

FIGURE 2.3 Quantal dose–response relationship for a defined effect in an experimental animal population as different groups get a higher dose of the substance.

Source: M. Cross.

to humans; with the aim of establishing an estimate of an exposure that is thought to 'have a reasonable certainty of no harm' (National Research Council 1994). Such information is used to help determine Workplace Exposure Standards (WESs) (AIOH 2021). These data provide the basis for making extrapolations needed for assessing the risks of non-carcinogenic substances (Kwon & Lee 2017). They provide a 'Point of Departure' to enable a quantitative estimate of the toxicity of each specific chemical substance (Sand et al. 2017).

2.5.5 Threshold Versus Non-threshold Effects

So far, we have considered the proposition that as dose declines, the adverse effect seen or the proportion of the population affected also declines and that there is a threshold dose below which no adversity is observable (No Observed Adverse Effect Level (NOAEL)). This threshold, even if derived from animal experiments, can be used to estimate an amount of the substance to which a human population can be exposed, and which is unlikely to cause harm. From the animal experiments, the NOAEL is divided by uncertainty (also called safety) factors to derive the so-called 'safe' dose for humans. Such doses or exposures may be termed tolerable daily intakes (TDIs), acceptable daily intakes (ADIs), reference doses (RfDs), or for airborne contaminants workplace exposure standards (WES). Workplace exposure standards (WES) are used to assist control of exposure to hazardous substances in the workplace. The safety factors are based on interspecies differences (extrapolating from animals to humans), inter-individual differences (accounting for different sensitivities of individuals of the same species), the extent of the toxicological data (how much we know about the toxicology of the substance), and the severity of the effect observed. This method of estimating acceptable/safe doses/exposures is usually sufficient for risk assessments of most chemical exposures.

For certain substances an alternative approach which suggests that at every dose above zero, there is a probability (or risk) of an adverse effect, a linear dose–response is assumed. This non-threshold approach is often applied to carcinogens which act through direct alteration of DNA (i.e. genotoxic carcinogens). For example, cancer may occur in 1 of every 1000 individuals with a given exposure. Reducing exposure by a factor of 100 may reduce the cancer rate to 1/100,000.

Reducing exposure a further 10-fold may result in one person developing cancer per million exposed, and so on. In these cases, a very low probability of an appreciable risk to worker health is normally defined by regulators (e.g. one per 100 000) which may be used as the basis for exposure standards for carcinogens, radiation and other hazards thought not to have a threshold. A more detailed examination of occupational carcinogens has been published by the Australian Faculty of Occupational Medicine (2003). The two major international agencies that have cancer classification systems are the US EPA (2023b) and the International Agency for Research on Cancer (IARC 2023) who evaluate the carcinogenicity of hazardous substances to humans but do not comment on risk.

2.5.6 Further Examination of Dose

The dose of a hazardous substance is the amount absorbed, taking into account both the concentration, duration of exposure and physiological factors that affect exposure (e.g. breathing rate), as shown in Equation 2.3.

$$D = C \times A \times T \times R \quad (2.3)$$

Where

D = dose
C = concentration of exposure
A = amount absorbed (bioavailability)
T = duration of exposure
R = breathing rate

This equation disguises a few oversimplifications. For example:

- The default assumption is 100% of gases are absorbed when breathed in; however, the dose may be less than the amount inhaled if most is exhaled without any absorption (e.g. many gases). The amount exhaled may vary with physical exertion.
- Workers with heavy workloads breathe more air than those with light workloads and so have larger doses.
- Dose may depend on whether the worker is a mouth or nose breather.
- Additional exposure may come from non-occupational sources (e.g. CO from cigarette smoking).

Clearly, if we can reduce either the concentration or the duration of exposure, we will reduce the dose. Exposure standards therefore take both of these factors into account. Our attempts to develop acceptable (safe) levels for occupational exposure standards are often based on dose studies in the workplace, volunteer studies, or animal studies in which various effects (even disease and death) have been identified. Acceptable or safe levels of exposure have been proposed from the results of epidemiology studies of exposures where injurious effects were observed in exposed workers and compared with groups with no exposures and where no effects were observed (see Section. 2.7.2.2, Case–control studies).

The dose–response curve shows the biological variation, or the range of susceptibility between individuals. There are some susceptible individuals, the ones who are affected at low concentrations. There are also some highly resistant individuals who aren't affected even at high exposure, and **average** individuals somewhere in the middle.

Figure 2.4 indicates the population variability of these susceptible, average, and resistant individuals in their increasing responses to differing doses. The y-axis shows relative effect, in

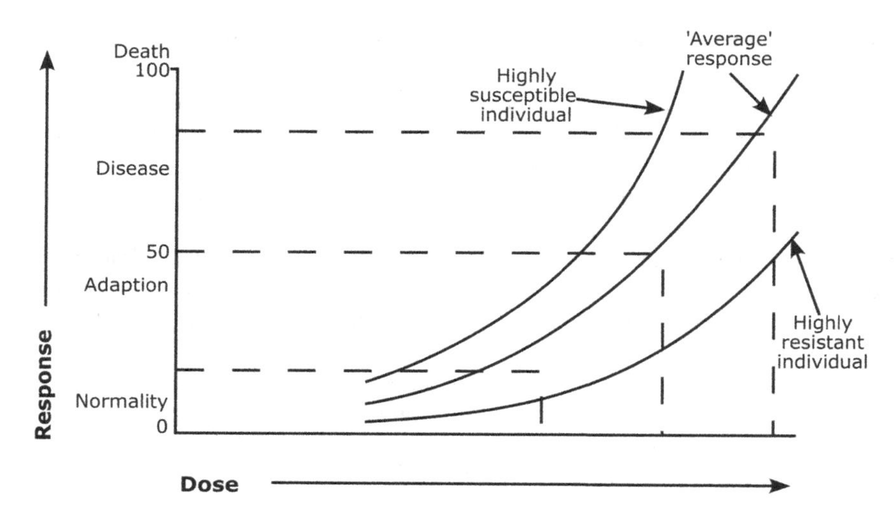

FIGURE 2.4 Variability of human exposure to dose.

Source: M. Cross.

increasing severity. The zone marked 'normality' indicates no response, 'adaptation' indicates that there are cellular, biochemical, or physiological changes but these are not considered adverse, while 'disease' indicates that changes in function are adverse effects compromising the health of the individual. Susceptible individuals exhibit more severe responses, such as disease or possibly death at doses lower than average.

Resistant individuals exhibit no response or may adapt with higher doses and may also tolerate much higher doses without disease or death.

Our determination of safe dose in general must be based on the responses of the most susceptible individuals in the working community. Exposure standards thus tend to be based on doses tolerated by 'susceptible' subjects without ill effect. That is, they are designed to protect nearly all workers. There may, however, be hypersensitive individuals—for example, those with a pre-existing disease or a missing metabolic enzyme—who may suffer adverse effects at doses lower than the rest of the population.

2.5.7 The ALARP Principle

H&S practitioners should always endeavour to reduce the exposure of hazardous substances to the worker to as low as reasonably practicable (ALARP). This means that if it is practicable to reduce the exposure (i.e. the dose) of the worker, you should do so. Practicability takes into account the technological means of control at your disposal, physical limitations, any added health benefits and economic cost factors.

For example, hospitals use the anaesthetic nitrous oxide. Operating staff may be exposed to anaesthetic gas in their breathing zone which exceeds the Workplace Exposure Standard (WES) at a level about 70 ppm when it is exhaled by the patient. However, the workplace exposure standard for nitrous oxide (N_2O) is 25 ppm, concentrations in these exposure situations can be reduced to 20 ppm by increasing cross-flow ventilation in the workplace, or to less than 1 ppm by using a proper scavenging mask on the patient. The choice of scavenging masks is good practice, easily achievable, economical, and consistent with the ALARP principle.

At work, the H&S practitioner will be required to make decisions on three types of exposure:

- Regular exposures that lie below the recommended exposure standards. Here the dose (exposure × time) is likely to be within acceptable limits and the risk is low. Nonetheless, you should still aim to reduce exposure to ALARP.
- Known operations producing significant exposures which could exceed exposure standards and hence exceed a safe dose—for example, applying toluene diisocyanate-based polyurethane cork floor sealant for 30 minutes each day without proper respiratory protection. Even though the exposure is relatively short, the risk to becoming sensitised and developing allergic asthma is potentially great, so control needs to be based on ALARP (Huuskonen, et al., 2023).
- Uncontrolled, gross overexposures that can result from an accident, from the absence or failure of control procedures, or from exceedingly poor hygiene practice—for example, the bursting of an organic-solvent delivery line, or entry into a confined space such as a vapour degreaser. Where practicable, even in emergency situations, controls must be available to limit exposure to ALARP.

You should always consider the ALARP principle to minimise risk even further. Good basic controls to prevent exposure embody the ALARP concept because practical technical measures are readily available to control most work exposures. ALARP does not imply relentless pursuit of the reduction of exposure, although some special high-risk industries demand this approach.

2.5.8 The Concept of Hazard, Risk, and Dose for Physical Agents

Not all workplace health hazards involve toxic substances. Significant health hazards are also posed by physical agents. These produce energy that is absorbed by various sense organs and by the body itself. Noise, heat, and radiation, for example, are all very commonly encountered physical agents. The concept of risk as a function of exposure and consequence is equally applicable to the effects of energy absorption from physical agents. In addition, there may be interactions between physical hazards and exposure to chemical hazards; for example, excessive long-term noise increases the ototoxicity of solvents such as toluene.

2.6 TYPES OF WORKPLACE HEALTH HAZARDS

This book presents to the H&S practitioner some basics in occupational health and hygiene in the following areas only:

- chemical hazards (Chapters 8–10);
- principles of biological monitoring (Chapter 11);
- indoor air quality (Chapter 12);
- biological hazards (Chapter 13);
- physical agents (Chapters 14–17).

The book does not attempt to deal with workplace ergonomics, stress-related disorders, or psychosocial effects related to work. These are large areas of study, each sufficiently so to require separate treatment. The H&S practitioner should consult other texts for assistance in these specialist areas.

2.6.1 Chemical Hazards

Chemicals give rise to a great number of health hazards in the workplace. The category 'chemicals' includes many naturally occurring substances, such as minerals and cotton, as well as both simple and complex manufactured chemical products. Chemical exposure can arise through direct use or from by-products.

Exposure to chemical hazards occurs through:

- inhalation including:
  - dusts—silica, coal, asbestos, lead, cotton, wood, and cement;
  - mists—acid mists and chrome plating;
  - gases—chlorine, sulfur dioxide, ethylene oxide, and ozone;
  - fumes—smoke and metal fumes from welding;
  - vapours—chlorinated and aromatic solvents, amines, ethers, and alcohols.
- skin contact including:
  - pesticides and phenol, absorbed directly through the skin;
  - acids and vapours, irritating eyes and mucous membranes;
  - acids, alkalis, and phenols, corroding the skin;
  - liquid toluene, methylene chloride, plus other agents removing lipids in the skin (defatting);
  - creosote and bitumen, photosensitising the skin;
  - nickel and chromium, which have an allergenic action on the skin.

2.6.2 Physical Agents

All workplaces may expose people to physical agents that are potentially harmful, including heat, noise, vibration, and light. Excess exposure to physical agents that may affect health includes:

- noise—processed by the ear; with some very low frequency (infrasonic) and ultrasonic sounds affecting the whole body;
- vibration—received by the body in contact with a vibrating object;
- light—visible, ultraviolet, and infrared are received by both eyes and skin; the eye is susceptible to laser energy; poor lighting may also be a workplace health hazard;
- heat—absorbed by all parts of the body;
- cold—experienced by the whole body; extremities in contact with cold;
- pressure—extremes affect body tissues that contain air spaces: lungs, teeth, sinuses, and inner ear;
- electromagnetic non-ionising radiation—microwaves, radiofrequency, and very low electromagnetic radiation are received by biological tissues;
- ionising radiation—X-rays and radioactive decay products—α (alpha) particles, β (beta) particles, and γ (gamma) rays—received directly by the body.

Physical agents and their health effects in the workplace are discussed in detail in Chapters 14–17.

2.6.3 Biological Hazards

Some workers are subject to potential health hazards when they work with biological materials or work in environments where microorganisms may abound. These hazards may arise from animal or plant materials, or sometimes from the handling or treatment of sick persons. A few biological hazards (e.g. *Legionella* and COVID-19) exist more widely and also affect members of the general working community. Chapter 13 looks at some of the biological or microbiological hazards of workplaces under the following classes:

- bacterial—*Legionella*, *Brucella*, tuberculosis, Q fever, etc;
- fungal—infective agents (e.g. tinea); allergenic agents (e.g. *Aspergillus*);
- viral—COVID-19, hepatitis B, and human immunodeficiency virus (HIV), which causes AIDS.

2.7 OCCUPATIONAL EPIDEMIOLOGY

Not all the activities of the H&S practitioner will be preventative. You may need to investigate how ill health has arisen. Sometimes the cause will be easy to recognise, particularly for hazards with well-known acute effects. If you find that older workers in a silk-screening department seemed extraordinarily clumsy, when their craft would suggest they were capable of great care, you might investigate the extent of their exposure to solvents. Other investigations may be more difficult to carry out. If you find, for example, that none of the 15 men engaged in common work activities has any children, you may need to investigate their occupational exposures. Such problem situations may be the starting point for epidemiological investigations.

We have imperfect knowledge of the health outcomes associated with our industrial use of hazardous substances, though new information on the health risks of substances and processes is continuously coming to light. The scientific study of disease distributions and their determinants in worker populations is known as occupational epidemiology, it is where occupational illnesses are correlated with occupational exposure information. Epidemiological studies tend to be medically and statistically complex. They often require large administrative resources to conduct and large sample populations to be meaningfully investigated. However, while these occupational epidemiological studies have often been disparagingly referred to as ‘counting dead bodies’ or ‘cases’, reflection indicates that they are far more than this. In conjunction with toxicology, they underpin much preventive health in the workplace, and many exposure standards are based on their findings.

H&S practitioners should be familiar with some occupational epidemiological concepts, which may help them draw conclusions, albeit guarded ones, from meagre or unrelated observations. These concepts may also guide the H&S practitioner to facilitate epidemiology investigations by putting together the information vital for valid studies in an accessible form. Such information should include, but not necessarily limited to, demographic information about the population and information about potential exposure histories, such as hazardous substance exposure, years and duration of employment, nature of the job, and any adverse health outcomes. Specifically, the data may include:

- identification of the population at risk, numbers, ages, gender, etc.;
- when each worker commenced and ceased employment;
- complete employment histories including with previous employers;
- nature and detailed description of jobs;
- hazardous-agent exposure;
- medical or accident records held by employers;
- details of workers’ compensation claims;
- smoking and possibly alcohol history for each worker and other possible confounding exposures.

One of the shortcomings of the majority of epidemiological studies conducted in the workplace is that data on the exposures of the population under study are either inadequate or absent.

2.7.1 Why Carry Out Epidemiological Studies?

The purpose of epidemiological studies is to find out whether there is an association between exposure and ill health so that appropriate action can be taken. For example, the association between mesothelioma and blue-asbestos exposure has been well demonstrated through epidemiology studies. The association is so strong and the outcome so unequivocal that the use of asbestos is now prohibited by law. Likewise, the increased rates of aplastic anaemia and leukaemia among workers exposed to benzene, compared with members of the general population, led to restrictive legislation and reduction of benzene concentrations in petrol and substitution of benzene by far less toxic solvents in other uses.

But such dramatic links between occurrence and cause are not always evident. Many, and probably most, studies suffer from deficiencies and limitations and while showing possible association does not demonstrate causation. Usually many studies of the same association are required to conclude an unambiguous causation between agent exposure and disease. Confounding factors such as drug or alcohol use, bias in the selection of the study population or control group, and poor or missing data often prevent firm conclusions from being drawn. Old, unverifiable clinical observations may be incorrect, workplace histories may be incomplete, or processes may change over time resulting in different exposure patterns. These limitations may seriously limit the validity of the findings.

Despite these shortcomings, occupational epidemiology often provides the only valid way to establish the human health hazards associated with many substances and processes (Beaglehole et al. 2006; Checkoway et al. 2004).

2.7.2 How Are These Studies Conducted?

Two epidemiological approaches commonly employed are:

- observational studies—for example, these are ecological, cross-sectional, case–control, longitudinal or cohort studies;
- experimental studies—for example, randomised controlled trials, field trials, or community trials.

Since nearly all studies in occupational epidemiology are observational studies, we will not deal further here with experimental studies. The observational studies can be listed as follows, in the order of methodological strength.

2.7.2.1 Cohort Studies

Cohort studies are used to ascertain if groups with an exposure to a hazardous agent have more disease than a non-exposed group. Studies of cohorts, or peer groups, involve:

- a known group of workers;
- a fixed time period of study;
- some knowledge of exposure to the agent under investigation.

The group may all be involved in a process at a particular factory (e.g. viscose rayon workers handling carbon disulfide), or in a particular industry (e.g. all workers involved in vinyl-chloride manufacture in a country). For statistical significance, these groups usually need to be large in number, several hundred or more. The assumption is that everyone within the cohort has a similar exposure to the agent that varies only with the length of time of exposure (employment).

The time period may be from some point in the past up to the present, in which case the cohort under investigation is called a historical cohort. Studies of such cohorts are called retrospective studies. If the study starts now and runs into the future, it is called a prospective study.

Knowledge of exposure to the agent requires hygiene measurement data. If this is not available, the surrogate 'years of exposure' is often used, in which the periods are split up into several classes: 1–5 years, 6–10 years, 11–15 years, and so on. It may or not be assumed that exposure was the same in these years.

Retrospective studies are comparatively cheap to carry out; however, there is little or no control over the quality of data available. Complete identification of the study population and obtaining all subjects' job and smoking histories is often problematic. The linking of subjects identified from company files to state and national cancer registry records is difficult and likely to be incomplete. Prospective studies are normally more expensive and may take 10–20 years to complete, but data on the cohort are easier to assemble and are likely to be more accurate and representative.

These studies have one other vital element: a group to use as a yardstick. This comparison group is called the control population. The control group should differ from the population under study only in not having experienced exposure to the agent of interest. For example, in a study of the health effects of foundry work, the population under investigation should not be compared with other foundry workers. The control population should be similar to the exposed population in age and gender, smoking history, and socioeconomic status. General population statistics are often used as the control, but special groups can be constructed when required. Investigators often have considerable difficulty matching the demographics of the control group to that of the study group.

These studies express findings as measured morbidity (e.g. incidence of a particular illness) or mortality rates compared to the control population. The **relative rate** (the rate experienced in the study group divided by the rate in the control group) is expressed as a ratio, such as Equation 2.4.

$$\text{Standardised mortality ratio}\left(\text{SMR}\right)=\frac{\text{observed number of deaths in the study group}\times 100}{\text{expected number of deaths in a control population}} \tag{2.4}$$

An SMR of 100 shows that the population under investigation is dying at the same rate as the comparison population. A higher SMR shows that the population under investigation has a higher mortality rate than the comparison population.

Industrial populations often have a low SMR. This is known as the **healthy worker effect**. The healthy worker effect arises because, on average, workers are in better health than the general population for two reasons: first, because the general population includes people unable to work because of ill health; and second, because workers in ill health tend to leave work.

To express risk in cohort studies, we need to establish which subjects have experienced the exposure of interest and then identify the proportion of the exposed and unexposed subjects who have experienced the health outcome of interest. A simple example involving a number of persons suffering from renal failure after exposure to cadmium (Cd) is given in Table 2.1.

Equation 2.5 expresses the relative risk (RR) of someone in a cohort study (such as that in Table 2.1) developing renal failure (column 1) associated with cadmium exposure (row 1) compared with the probability of a non-exposed person (row 2) developing renal failure (column 2). If the RR is greater than 1, there is a risk that the exposure causes that disease; statistical analysis is required.

$$\text{Relative risk}=\frac{\left[{}^{a}\!/\!_{(a+b)}\right]}{\left[{}^{c}\!/\!_{(c+d)}\right]} \tag{2.5}$$

TABLE 2.1
Schematic Showing Proportions of a Cohort Study Displaying Presence or Absence of Renal Disease Associated with Cadmium Exposure

	Renal Failure	No Renal Failure	Total
Cadmium exposure	a	b	a + b
No cadmium exposure	c	d	c + d
Total	a + c	b + d	

2.7.2.2 Case–Control Studies

A case–control study takes a different approach; it assesses if people with a disease of interest (a case) have more exposure to the agent of interest than a non-exposed population. The study populations comprise:

- a group of patients or workers with a particular disease; and
- a control group whose members do not have that particular disease.

The exposure histories of the cases and controls are then compared. It is possible that some cases and some controls will have had exposure to the agent under study. Being able to reach valid inferences depends on how well the exposure/non-exposure histories are recorded. This method is useful in the study of rare diseases because investigators can work with small numbers. To investigate rare diseases with a cohort study would require large numbers of participants and consequently be very expensive.

In case–control studies, epidemiologists use the measure called the **odds ratio**, which is given by Equation 2.6, where a, b, c, and d are defined as in Table 2.1.

$$\text{Odds ratio} = \frac{a \times c}{b \times d} \tag{2.6}$$

This is the odds comparing the probability of having the disease and exposure versus the probability of having the disease without exposure. If the number is larger than 1, the odds of having disease and exposure are higher than having the disease and no exposure. For relatively rare diseases, the relative risk calculated in a cohort study and the odds ratio in a case–control study are approximately the same.

2.7.2.3 Cross-Sectional Studies

These studies are generally undertaken at a 'point in time' or over a relatively short period of time. They comprise a snapshot of the health and exposure of a particular group. As a result, causation is difficult to determine, because it is not possible to establish whether the exposure preceded or followed the disease. A major limitation of these studies is the possible influence of a survivor effect, where affected individuals have left employment and are therefore not included in the study population. For instance, a cross-sectional study investigating the prevalence of asthma in a dusty industry may result in a 'low prevalence' being reported, suggesting that the dust exposure is not related to asthma. However, the heavily exposed workers who suffered from asthma may have left the industry before the study was conducted. Only those not so heavily exposed thus take part in the study, and these workers are predominantly non-diseased.

Cross-sectional studies are inexpensive and may be done quickly and easily. They are useful as a first step in identifying a suspected cause if exposure to many agents occurs in a workplace.

2.7.2.4 Ecological Studies

Ecological or correlational studies are considered 'hypothesis-generating' studies. They involve the analysis of the health and exposures of groups or populations, not individuals, so the potential link between exposure and disease cannot be made. An example of an ecological study would be the observation that population A smokes more than population B and has higher rates of heart disease. We do not know, but we might suspect, that the smokers in population A are the ones with the higher rate of heart disease. Ecological studies can be inexpensive and simple to conduct, as they often rely on banks of data collected for other purposes, such as compensation databases, cancer registers, or hospital records. These studies can easily be biased owing to various socioeconomic factors not being controlled. They are considered the weakest in terms of study design but are often initiated because of their low cost and quick results.

2.7.3 A Few Relevant Epidemiological Terms

Epidemiological studies express disease burden in terms of rates known as **incidence** or **prevalence**. Prevalence is the total number of cases in the population at any point in time while incidence is the rate of new cases appearing in the population during a certain time period. Incidence is expressed as new cases per 1000 people at risk per year. Determining incidence is crucial to studying changes in the rate of appearance of a disease. Prevalence relates to whether the pool of cases is growing or shrinking.

Figure 2.5 allows us to appreciate prevalence as the result of both incidence and loss. A cross-sectional study at a particular point in time will yield a point prevalence. When a study is conducted over a particular time span, it can produce a period prevalence. Cross-sectional studies are often called 'prevalence studies' because they can measure only the prevalence and not the incidence of a disease.

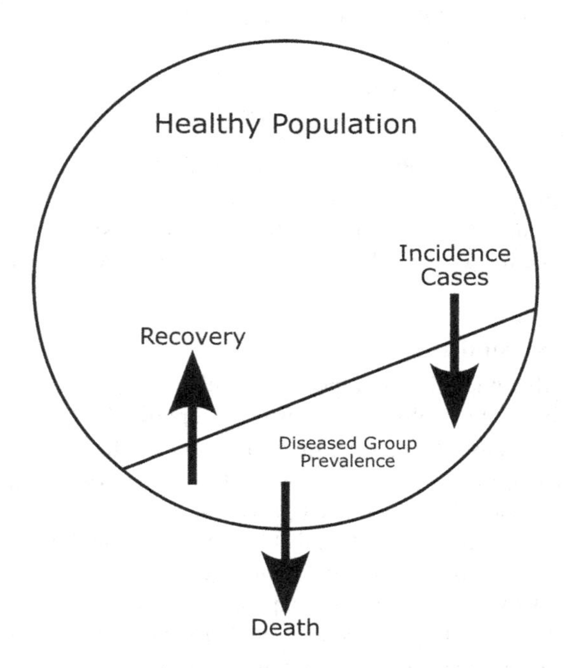

FIGURE 2.5 Diagram indicating the difference between incidence and prevalence.

Source: G. Benke.

All studies suffer from **bias**. This can occur in the measurement of health outcomes or in the selection of participants for the study. A selection bias means that the group chosen for study is not a fair sample of all individuals who could be available for study. In other words, the mix of individuals chosen to be observed differs from all those available in regard to characteristics such as age, state of health, smoking habits, and socioeconomic class. Selection bias is minimised if those chosen for observation have a similar range of characteristics to all of those who could be available. Selection bias can apply either to those being chosen for the group being assessed for the effects of exposure to an alleged harmful agent or to those being chosen for a comparison or control group.

However, even if the members of each group are fairly chosen, that is, if selection bias is minimised, the exposed group may happen to differ from the other group in regard to its mix of personal characteristics such as age, state of health, smoking habits, and socioeconomic class. Suppose that after a period of study, the incidence of a disease is higher in one group than the other. The cause of this may be because of one group's exposure to the alleged harmful agent or indeed it may be because of one of the other differences between the groups. The presence of other possible causes is said to confound the attribution of cause and effect.

Epidemiological studies always refer to **statistical significance** in their findings. This is the probability that the association between the exposure and the outcome occurs by chance. For example, $p < 0.05$ indicates that the finding has less than a 5% probability of occurring by chance.

For example, a study might look at bladder cancer among rubber workers. The epidemiologist naturally wants to attach some degree of confidence to the findings. A whole range of statistical methods are available for numerical testing of various hypotheses, but the tests generally seek one common goal—that is, a probability of 95 per cent or more that the observations did *not* arise by chance; in other words, a probability of 5 per cent or less that this *is* a random association. This is sometimes described as 'statistically significant at $p < 0.05$'.

If the results of a study are described as statistically significant, and all biases and confounders are dealt with, this means that the variation in disease rates between the study and control populations is unlikely to result from chance alone. For example, a study that shows an SMR of 300 (18 deaths found when only six were expected) might be able to quote its findings as statistically significant if the lower 95 per cent confidence limit for the SMR is greater than 100.

2.7.4 Practical Significance

Despite the obvious difficulties in occupational epidemiology, this field of study provides the surest basis for long-range improvements in occupational health. Such studies have provided much of the information we have to date on rare disease occurrence and causative exposure. Both toxicology and epidemiology are 'complementary methods of detecting whether a chemical causes harm to humans' (Adami et al. 2011). Toxicology may be used to evaluate the weight of evidence particularly when there is evidence of a specific effect in humans and evidence of exposure (Oosthuizen & Cross 2018). As a result, the H&S practitioner should always be receptive to the possibility that workplace factors may be implicated in ill health, while resisting the temptation to blame them for every illness.

2.8 CONCLUSIONS

Section 2.3 introduced some of the issues associated with establishing cause with respect to workplace chemical exposures and disease. In particular, if a worker has symptoms, how can they be confidently attributed to the workplace? With an understanding of basic toxicology and epidemiology, you may be in a position to suggest a cause for the person's ill health if there are data, observations or published evidence to support the following:

- Gradients exist in space. That is, those workers remote from the source of a pollutant responsible for disease have low rates of symptoms, those near the source have high rates of symptoms and those between have intermediate rates of symptoms.
- Distribution of cause reflects distribution of response. For example, workers at similar industrial plants without the proposed cause do not display symptoms.
- Time sequences are observed—for example, symptoms occur after exposure or abate at weekends when not working.
- Results are consistent with the findings of other published studies.
- The proposed mechanism for the effect is plausible.
- Experimental evidence (e.g. animal studies and human exposure studies) supports the results.
- Preventive trials or interventions reduce the effects.

In addition, to critically evaluate the extent of exposures and the veracity of symptom reports, we may ask:

- Was there a potential for exposure?
- Was the potential exposure level likely to cause symptoms?
- Are symptoms consistent with known effects of the suspected agent?
- Are there independent measurable disease/symptom indices?
- Are there independent measures of exposure?
- Are there other possible causes for the person's illness?

In some cases, there may be a clear association between exposure and reported symptoms. In other cases, some of the circumstances above may be met and others not, and the H&S practitioner will need to make a judgement based on the balance of evidence. Experienced hygienists may effectively meet these challenges. However, it is also important to recognise when specialist toxicological, epidemiological, or other assistance is required to reach appropriate conclusions.

REFERENCES

Adami, H-O, Berry, C.L., Breckenridge, C.B., Smith, L.L., Swenberg, J.A., Trichopoulos, D., Weiss, N.S. & Pastoor, T.P. 2011, 'Toxicology and Epidemiology: Improving the Science with a Framework for Combining Toxicological and Epidemiological Evidence to Establish Causal Inference', *Toxicological Sciences: An Official Journal of the Society of Toxicology*, vol. 122, no. 2, pp. 223–34. https://doi.org/10.1093/toxsci/kfr113

AIOH 2021, *Virtual Chapter Meeting 10 March, 2021: Reviewing the Australian Workplace Exposure Standards* https://www.youtube.com/watch?v=67q5rJhCdAA [28 December 2023].

Australasian Faculty of Occupational Medicine (AFOM) 2003, *Occupational Cancer: A Guide to Prevention, Assessment and Investigation.* AFOM Working Party on Occupational Cancer. Sydney: AFOM.

Beaglehole, R., Bonita, R. & Kjellström, T. 2006, *Basic Epidemiology*, 2nd edn. Geneva: World Health Organization.

Boston University, School of Public Health 2019, *Toxicology/Dose-Response Assessment*, https://sph.bumc.bu.edu/OTLT/swnas/info_site/health_assessment_tools/Dose-Response_Assessment.html> [29 December 2023].

Cahill, J., Keegel, T. & Nixon, R. 2004, 'The Prognosis of Occupational Contact Dermatitis in 2004', *Contact Dermatitis*, vol. 51, no 5–6, pp. 219–26. https://doi.org/10.1111/j.0105-1873.2004.00472.x

Health Canada 2023, *Exposure and Health Effects of Chemicals. Potential Health Effects*, https://www.canada.ca/en/health-canada/services/health-effects-chemical-exposure.html [29 December 2023].

Checkoway, H., Pearce, N. & Kriebel, D. 2004, *Research Methods in Occupational Epidemiology* 2nd edn. New York: Oxford University Press.

Gilmour, N., Kimber, I., Williams, J. & Maxwell. G. 2019, 'Skin Sensitization: Uncertainties, Challenges, and Opportunities for Improved Risk Assessment,' *Contact Dermatitis.* vol. 80, no. 3, pp. 195–200. https://doi.org/10.1111/cod.13167

Huh, D.A., Chae, W.R., Choi, Y.H., Kang, M.S., Lee Y.J. & Moon, K.W. 2022, 'Disease Latency according to Asbestos Exposure Characteristics among Malignant Mesothelioma and Asbestos-Related Lung Cancer Cases in South Korea', *International Journal of Environmental Research and Public Health*, vol. 19, no. 23, pp. 15934. https://pubmed.ncbi.nlm.nih.gov/36498008/

Huuskonen, P., Porras, S.P., Scholten, B., Portengen, L., Uuksulainen, S., Ylinen, K. & Santonen, T. 2023, 'Occupational Exposure and Health Impact Assessment of Diisocyanates in Finland', *Toxics*, vol. 11, no. 3, pp. 229. doi: https://doi.org/10.3390/toxics11030229

International Agency for Research on Cancer (IARC) 2023, IARC *Monographs on the Identification of Carcinogenic Hazards to Humans*: Volumes 1–135, https://monographs.iarc.who.int/agents-classified-by-the-iarc/ [29 December 2023].

Klaassen, C.D. (Ed.) 2019, *Casarett and Doull's Toxicology: The Basic Science of Poisons* 9th edn. New York: McGraw-Hill Education.

Kwon, S. & Lee, B-M. 2017, 'Chapter 30: Risk Assessment and Regulatory Toxicology, Appendix 30.1. Terminology Use in Risk Assessment' (pp. 616–618). In B.-M. Lee, S. Kacew & H. S. Kim (Eds), *Lu's Basic Toxicology: Fundamentals, Target Organs, and Risk Assessment*, 7th edn. Boca Raton, FL: CRC Press.

Massányi, P., Massányi, M., Madeddu, R., Stawarz, R. & Lukáč, N. 2020, 'Effects of Cadmium, Lead, and Mercury on the Structure and Function of Reproductive Organs', *Toxics*, vol. 8, no. 4, p. 94. https://doi.org/10.3390/toxics8040094

National Research Council (US) 1994, *Science and Judgment in Risk Assessment. 4, Assessment of Toxicity.* Committee on Risk Assessment of Hazardous Air Pollutants. Washington, DC: National Academies Press (US), https://www.ncbi.nlm.nih.gov/books/NBK208246/ [29 December 2023].

Oosthuizen, J. & Cross, M. 2018, 'Establishing Cause, What Does That Mean from an Epidemiological and Legal Perspective?' *Environmental and Planning Law Journal*, vol. 35, pp. 426–429.

Organisation for Economic Co-operation and Development (OECD) 2023, *OECD Guidelines for the Testing of Chemicals, Section 4.* Paris: OECD https://www.oecd-ilibrary.org/environment/oecd-guidelines-for-the-testing-of-chemicals-section-4-health-effects_20745788 [29 December 2023].

Park, J.H., Lee, J.W., Kim, Y.C. & Prausnitz, M.R., 2008, 'The Effect of Heat on Skin Permeability', *International Journal of Pharmaceutics*, vol. 359, no. 1–2, pp. 94–103. https://doi.org/10.1016/j.ijpharm.2008.03.032

Sand, S., Parham, F., Portier, C.J., Tice, R.R., and Krewski, D. 2017, 'Comparison of Points of Departure for Health Risk Assessment Based on High-Throughput Screening Data', *Environmental Health Perspectives*, vol. 125, no. 4, pp. 623–633.

Semple, S. 2004, 'Dermal Exposure to Chemicals in the Workplace: Just How Important Is Skin Absorption?' *Occupational Environmental Medicine*, vol. 61, no. 4, pp. 376–376. https://doi.org/10.1136/oem.2003.010645

Slitt, A. 2019, 'Chapter 5. *Absorption, Distribution, and Excretion of Toxicants'*. In: C.D. Klaassen (Ed.), *Casarett and Doull's Toxicology: The Basic Science of Poisons*, 9th edn. New York: McGraw-Hill.

United States Environmental Protection Agency (US EPA) 2023a, *EPA New Approach Methods: Efforts to Reduce Use of Vertebrate Animals in Chemical Testing*, https://www.epa.gov/chemical-research/epa-new-approach-methods-efforts-reduce-use-vertebrate-animals-chemical-testing [28 December 2023].

United States Environmental Protection Agency (US EPA) 2023b, *Risk Assessment for Carcinogenic Effects*, https://www.epa.gov/fera/risk-assessment-carcinogenic-effects [28 December 2023].

Wagner, G.R. 1997, 'Asbestosis and Silicosis', *The Lancet*, vol 349, no. 9061, pp. 1311–1315. https://doi.org/10.1016/S0140-6736(96)07336-9

Wei, S., Zhang, H. & Tao, S. 2019, 'A Review of Arsenic Exposure and Lung Cancer'. *Toxicology Research*, vol. 8, no.3. pp. 319–327. https://doi.org/10.1039/c8tx00298c

3 The Concept of the Exposure Limit for Workplace Health Hazards

Robert Golec and Ian Firth

3.1 INTRODUCTION

The occupational exposure limit (OEL) has been, and continues to be, the cornerstone of occupational hygiene risk assessment and risk management and forms an important element in the control of occupational disease and the setting of policies on occupational health (Henschler 1984; Howard 2005; Vincent 1999). In Australia, Safe Work Australia (SWA) has set OELs (termed Workplace Exposure Standards for Airborne Contaminants or WESs) for over 700 workplace chemicals (SWA 2021, 2024). It is estimated that there are OELs for more than 6000 specific chemicals globally (Brandys & Brandys 2008). By way of comparison, the number of chemicals in commercial use exceeds 40 000, although approximately 6000 industrial chemicals account for more than 99% of the total volume of industrial chemicals in commerce globally (UN Environment & International Council of Chemical Associations 2019).

To understand the concept of the OEL, it is important to understand the history of its development, the thinking behind the setting and use of OELs, and the strengths and weaknesses of the OEL approach, including criticisms and challenges.

3.2 A BRIEF HISTORY OF OCCUPATIONAL EXPOSURE LIMITS

The OEL has its basis in the toxicological concepts of a threshold of effect for exposure to chemicals and a relationship between dose and response. This concept was first theorised in the 16th century by the Swiss physician and chemist Phillippus Paracelsus, who declared that it is the dose of a chemical which determines whether it exhibits toxic properties or not (Gallo 2013). Hippocrates in the 4th century BC first recorded cases of occupational disease from exposure to chemicals among mine workers with lead poisoning (Dawson et al. 2021), and Paracelsus, who had also worked for 10 years as a labourer in mining and smelting, later wrote an important treatise that described various miners' diseases (Dawson et al. 2021). However, the systematic study of industrial diseases caused by exposure to chemical contaminants in the workplace did not begin until the late 17th century, when the Italian pioneer of occupational medicine Bernadino Ramazzini observed, for example, that in the dusty trades 'ill-health was caused by the inhalation of subtle particles that were offensive to human nature' (Oliver 1902, p. 28). Ramazzini urged physicians to ask their patients for the details of their occupations in order to make the correct diagnoses (Harbison et al. 2015). At that time, meaningful quantification of workplace exposures to chemicals was not possible owing to a lack of suitable sampling and analytical techniques (Paustenbach & Cyrs 2021). Researchers could make only qualitative assessments based on visual observations of the workplace, medical examination of workers, and post-mortem findings.

 DOI: 10.1201/9781032645841-3

In the late 19th century, the German hygienist and toxicologist Karl Bernhard Lehmann and his pupils conducted a series of quantitative studies on 35 gases and fumes involving animal and human experiments and observations in industry. These laid the foundations for the quantitative assessment of chemical exposures at which various health impacts occurred (Kober et al. 1924). Lehmann also wrote about the quantitative analysis of workplace air for chemical and biological contaminants (Lehmann 1893).

By the start of the 20th century, advances in analytical chemistry and the development of occupational hygiene sampling techniques made it possible to reliably quantify exposure (dose) of chemicals to workers in the workplace. This led the way to estimating levels of exposure that were relatively 'safe', and those at which serious effects might be expected. One of the first attempts to formulate a set of acceptable acute OELs was that of the German researcher Rudolf Kobert. In 1912, he published a list of 20 chemicals, and the doses and durations of exposure required for them to bring about effects from minimal disturbance through to rapid death (Paustenbach & Cyrs 2021). Kobert's list is reproduced in Table 3.1.

In 1921, the US Bureau of Mines published a technical paper listing thresholds for odour, irritation of the eyes and throat, and coughing, maximum one-hour exposure levels without symptoms, and lethal exposure doses for some 33 chemicals commonly found in industry at the time (Henderson 1921). The International Critical Tables of 1927 listed 'maximum safe concentrations' for 27 gases and vapours, a list partially based on Kobert's original 1912 data combined with data from other sources (Sayers 1927). The USSR was the first country to establish a set of compulsory statutory OELs for 12 chemicals in 1930 (ILO 2011).

In 1941, The National Conference of Governmental Industrial Hygienists (NCGIH), since renamed the American Conference of Governmental Industrial Hygienists (ACGIH®), published a list of maximum allowable concentrations (MACs) for 63 chemicals (Baetjer 1980). The MAC list was intended to be reviewed and published annually from 1946 onwards. The ACGIH® stated from the outset that these MACs were not to be taken as being recommended safe concentrations (Ziem & Castleman 1989). The ACGIH® MACs of 1946 drew heavily on a 1945 list used by various US states and the US Public Health Service and recommended by the American Standards Association compiled by the industrial hygienist Warren Cook. Some of Cook's values were based on extensive 'animal experiments and experience with workers under actual industrial conditions', while others were based on limited animal experiments or on professional judgement (Cook 1945). Cook noted that his list was intended as a 'handy yardstick to be used as guidance for the routine industrial control of these health hazards – not that compliance with these figures listed would guarantee protection against ill-health on the part of exposed workers' (Cook 1945). The term Threshold Limit Values (TLVs®) was introduced in 1956, and the first Documentation of Threshold Limit Values for 257 chemicals was published in 1962 (ACGIH® 2023). Despite the limitations and warnings by the ACGIH® that TLVs® should not be used for compliance purposes, the 1968 TLV® list was adopted by the OSHA as US federal law.

The ACGIH® list of TLVs® grew over time and was used as the basis for many other nations' occupational exposure limits, including the German MAKs, British WELs, and Finnish OELs (Henschler 1984). Reliance on the ACGIH® list has declined in Europe since the European Union developed its own set of guidelines for setting OELs (Adkins et al. 2009a), although they still rely on similar data cited by the ACGIH®. The European Scientific Committee on Occupational Exposure Limits (SCOEL) has defined OELs as

> 'limits for exposure via the airborne route such that exposure, even when repeated on a regular basis throughout a working life, will not lead to adverse effects on the health of the exposed person and/or their progeny at any time (as far as can be predicted from the contemporary state of knowledge)' (European Commission, Employment, Social Affairs and Inclusion [ECESAI] 2010).

TABLE 3.1
Kobert's 1912 List of Acute Exposure Limits Compared to 2024 Safe Work Australia WESs

	Kobert's 1912 List of Acute Exposure Limits				
Chemical	**For Human and Animal Rapid Death**	**0.5–1 hour Exposure Serious Threat to Life**	**0.5–1 hour without Serious Health Effects**	**Repeated Exposure, Minimal Symptoms**	**Safe Work Australia 2024 WES's TWA**
Ammonia		240–450 ppm	300 ppm	100 ppm	17 mg/m^3 (25 ppm)
Aniline			400–600 mg/m^3	100–250 mg/m^3	7.6 mg/m^3
Benzene			10,000–15 000 mg/m^3	~5,000 mg/m^3	3.2 mg/m^3
Bromine	~10,000 ppm	400–600 ppm	40 ppm	10 ppm	0.1 ppm (0.66 mg/m^3)
Carbon dioxide	30 per cent	60,000–80,000 ppm	40,000–60,000 ppm	20,000–30,000 ppm	5000 ppm (9000 mg/m^3)
Carbon disulfide		10,000–12,000 mg/m^3	2,000–3,000 mg/m^3	1,000–1,200 mg/m^3	31 mg/m^3 (10 ppm)
Carbon monoxide		20,000–30,000 ppm	5000–10 000 ppm	2,000 ppm	30 ppm (34 mg/m^3)
Carbon tetrachloride	300,000–400,000 mg/m^3	~150,000–200,000 mg/m^3	25,000–40,000 mg/m^3	~10,000 mg/m^3	0.63 mg/m^3 (0.1 ppm)
Chlorine	~10,000 ppm	400–600 ppm	40 ppm	10 ppm	1 ppm *peak* (3 mg/m^3 *peak*)
Chloroform	300 000–400 000 mg/m^3	70 000 mg/m^3	25 000–30 000 mg/m^3	~10 000 mg/m^3	10 mg/m^3 (2 ppm)
Gasoline			15 000–25 000 mg/m^3	5000–10 000 mg/m^3	No WES
Hydrogen chloride		1500–2000 ppm	500–1000 ppm	100 ppm	10 ppm (14 mg/m^3)
Hydrogen cyanide	~3000 ppm	1200–1500 ppm	500–600 ppm	200–400 ppm	10 ppm *peak* (11 mg/m^3 *peak*)
Hydrogen sulfide	10 000–20 000 ppm	5000–7000 ppm	2000–3000 ppm	1000–1500 ppm	10 ppm (14 mg/m^3)
Iodine			30 ppm	5–10 ppm	0.1 ppm *peak* (1 mg/m^3 *peak*)
Nitrobenzene			1000 mg/m^3	200–400 mg/m^3	5 mg/m^3 (1 ppm)
Phosphine (phosphane)		400–600 ppm	100–200 ppm		0.3 ppm (0.42 mg/m^3)
Phosphorus trichloride	3500 mg/m^3	300–500 mg/m^3	10–20 mg/m^3	4 mg/m^3	1.1 mg/m^3 (0.2 ppm)
Sulfur dioxide		4000–5000 ppm	500–2000 ppm	200–300 ppm	2 ppm (5.2 mg/m^3)
Toluidine			400–600 mg/m^3	100–250 mg/m^3	8.8 mg/m^3 (2 ppm)

Adapted from: Ripple 2010; Safe Work Australia 2024.

This definition implies that the objective of European OELs is to protect all workers, including potentially susceptible individuals such as unborn offspring. By contrast, the ACGIH® approach is to set limits that protect 'nearly all workers'. SCOEL's methodology discusses setting two types of OELs: 'health-based' OELs, which are established based on a clear threshold dose below which exposure is not expected to cause adverse effects; and 'risk-based' OELs, where it is not possible to define a threshold (e.g. genotoxic carcinogens and respiratory sensitisers) and where any exposure may carry some risk (SCOEL 2013).

In Australia, the influence of the ACGIH®'s TLVs® was evident by the early 1970s, when the Occupational Hygiene subcommittee of the National Health and Medical Research Council (NHMRC) published a schedule of recommended concentrations for occupational exposure which were largely based on the ACGIH® limits. It should, however, be noted that the setting of OELs in Australia significantly predates the NHMRC schedule, as 'permissible standards for toxic gases, vapours and fumes in Industry' had been published by the NHMRC Committee on Industrial Hygiene in Munitions Establishments in 1944, and a list of maximum allowable concentrations of chemicals was gazetted in 1945 in the Victorian Health (Harmful Gases Vapours Fumes Mists Smokes and Dusts) Regulations (De Silva 2000). The NHMRC schedule was initially a more or less direct adoption of the ACGIH®'s TLV® list, but the Occupational Hygiene subcommittee later developed some of its own limits based on Australian experience and approaches (De Silva 2000).

The formation of the National Occupational Health and Safety Commission (NOHSC) in 1985 saw the transfer of responsibility for the development of OELs away from the NHMRC to the NOHSC Exposure Standards Working Group, under the auspices of the Standards Development Standing Committee (SDSC). The terms of reference for the Exposure Standards Working Group were to 'Consider and recommend options to the Standards Development Standing Committee on occupational exposure standards for atmospheric chemical contaminants based on consideration of the best available technical data from Australian and overseas sources' (National Occupational Health and Safety Commission [NOHSC] 1995a). Initially, a list of exposure standards was adopted based on the NHMRC's schedule, but in 1995, the NOHSC published a list of OELs based on the recommendations of the working group (NOHSC 1995b). The NOHSC had no statutory authority over the states and territories, which have ultimate responsibility for occupational health and safety. However, the NOHSC's exposure standards became de facto legal compliance standards for workplace chemical exposure as they were called up in the various OHS regulations in each jurisdiction. The role of OELs in Australian legislation has been reinforced with the introduction of harmonised work health and safety (WHS) regulations in most states and territories, in which the workplace exposure standard has been prescribed as 'the airborne concentration of a particular substance or mixture that must not be exceeded' (SWA 2023a).

Between 1995 and 2021, there were some 48 changes (amendments and additions) to the NOHSC's list of approximately 700 exposure standards. These included lowering the limits for crystalline silica, benzene, and chrysotile asbestos. In 2018, the methodology for setting Workplace Exposure Standards in Australia was revised, drawing from a body of knowledge developed by OEL setting bodies in other countries including the ACGIH®, Deutsche Forschungsgemeinschaft (DFG), EU Scientific Committee on Occupational Exposure Limits (SCOEL), American Industrial Hygiene Association/Occupational Alliance for Risk Science (AIHA/OARS), and Health Council of the Netherlands (Dutch Expert Committee on Occupational Safety). More information about the changes to the Safe Work Australia WES setting process is available from the Safe Work Australia website (SWA 2023b). In April 2024. Safe Work Australia published a set of OELs derived using the 2018 methodology which will replace the Workplace Exposure Standards. These new OELs are to be known as Workplace Exposure Limits (WELs) and are to come into force in Australia on 1st December 2026 (SWA 2024).

3.3 DEFINITION OF OCCUPATIONAL EXPOSURE LIMITS

Safe Work Australia's version of an OEL is referred to as a **Workplace Exposure Standard** (WES). Safe Work Australia defines a WES as 'the airborne concentration of a particular substance or mixture that must not be exceeded' (SWA 2024). The WESs are available in a searchable database on the Safe Work Australia's Hazardous Chemical Information System website (SWA 2021).

While past SWA guidance has stated that WESs are not dividing lines between safe and hazardous exposures and not measures of relative toxicity, they now state that it is important the airborne concentration of a chemical substance or mixture hazardous to health is kept as low as is reasonably practicable to minimise the risk to health (refer to Section 3.12.1).

An OEL (or WES) can take one or all of three forms:

- ***Time-weighted average (TWA)*** – an eight-hour TWA exposure standard is the maximum average airborne concentration of a particular substance permitted over an eight-hour working day and a five-day working week. The TWA is the most common of the exposure standards and, except where a peak limitation has been assigned, virtually all substances are listed with a TWA.
- ***Short-term exposure limit (STEL)*** – a short-term exposure limit is the time-weighted maximum average airborne concentration of a particular substance permitted over a 15-minute period. STELs are established to minimise the risk of:
    - intolerable irritation;
    - irreversible tissue change;
    - narcosis to an extent that could precipitate workplace incidents.

  Exposures at the STEL should not be longer than 15 minutes and should not be repeated more than four times per day. There should be at least 60 minutes between successive exposures at the STEL.
- ***Peak*** – a maximum or peak airborne concentration of a particular substance determined over the shortest analytically practicable period that does not exceed 15 minutes.

Chemical substances that have workplace OELs are also classified as known, presumed, or suspected carcinogens or are known respiratory or skin sensitisers according to the criteria of the Globally Harmonized System of Classification and Labelling of Chemicals (GHS). Important notations include the following:

- *Skin (Sk)* – chemicals where skin, mucous membrane, or eye absorption may be a significant route of exposure (ACGIH® 2023).
- *Respiratory sensitiser (Sen)* – a chemical substance that leads to hypersensitivity of the airways after being inhaled.
- *Skin sensitiser (Sk:Sen)* – a chemical substance that leads to an allergic response after skin contact.

3.3.1 Carcinogen Categories

Carcinogenicity Category 1A – Known to have carcinogenic potential for humans. The classification of a chemical into this category is based largely on human evidence from studies that have established a causal relationship between human exposure and the development of cancer.

TABLE 3.2
Examples of Workplace Exposure Standards in Australia

Contaminant	TWA Exposure Limit	STEL Exposure Limit	Carcinogen Category	Notation
Toluene	50 ppm	150 ppm	–	Sk
Isocyanates, all (as –NCO)	0.02 mg/m³	0.07 mg/m³	–	Sen
Chromium (VI) compounds (as Cr), certain water insoluble	0.05 mg/m³	–	1A	Sen
Glutaraldehyde	0.1 ppm peak limitation	–	–	Sen
Crocidolite asbestos	0.1 f/mL	–	1A	–
Phenylhydrazine	0.1 ppm		1B	Sk, Sen

Adapted from: GESTIS 2023

Carcinogenicity Category 1B – Presumed to have a carcinogenic potential for humans. The classification of a substance into this category is based largely on animal evidence where there is sufficient evidence to demonstrate carcinogenicity in animals or where there is limited evidence of carcinogenicity in humans and animals.

Carcinogenicity Category 2 – Suspected human carcinogen. The classification of a chemical into this category is based on evidence from human and animal studies, where the evidence is not sufficiently convincing to place the chemical into Category 1 or from limited evidence of carcinogenicity in human or animal studies.

3.4 UNITS OF CONCENTRATION

OELs are concentrations in air which can be expressed as a mass per unit volume, such as milligrams of substance per cubic metre of air (mg/m³) or parts per million (ppm). For gases and vapours, the units can be converted from one to the other using Equation (3.1).

$$\text{Concentration}\left(\text{ppm}\right) = \text{Concentration}\left(\text{mg/m}^3\right) \times \left(\frac{24.45}{\text{Molecular weight of contaminant}}\right) \tag{3.1}$$

where 24.45 is the molar volume at 25°C and 1 atmosphere (101.325 kPa).

For example, the TWA-OEL for formaldehyde is 1 ppm. The average concentration of formaldehyde over eight hours measured in the breathing zone is found to be 2 mg/m³. The molecular weight of formaldehyde is 30.03. The concentration of formaldehyde in the breathing zone in ppm is:

$$2\,\text{mg/m}^3 \times \left(\frac{24.45}{30.03}\right) = 1.6\,\text{ppm}\ \left(\text{which is above the OEL}\right)$$

For mineral fibres such as asbestos and ceramic fibres, the concentration is expressed as the number of respirable sized fibres (f) per millilitre of air (mL). This is discussed further in Chapter 8, 'Aerosols'.

FIGURE 3.1 Worker's breathing zone.

Source: Safe Work Australia, 2023c, Occupational Lung Disease, Figure 2.

3.5 INTERPRETATION OF EXPOSURE STANDARDS

It is important to understand that OELs:

- refer to airborne concentration of chemical contaminants in the worker's breathing zone;
- are based on current knowledge; and
- apply to nearly all workers.

3.5.1 Worker's Breathing Zone

Exposure standards relate to personal exposure to airborne contaminants in the worker's breathing zone. The breathing zone is defined as a hemisphere with a radius of 300 mm extending in front of the face, measured from the midpoint of an imaginary line joining the ears (Figure 3.1). OEL values do not account for other routes of exposure such as skin absorption or ingestion. It follows that valid quantitative assessment of exposure for comparison against the OEL can *only* be made via personal monitoring where the sampling device is worn within the breathing zone. Static (or area) monitoring, although useful for providing general information on contaminant release and for assessing engineering controls, does not measure the concentration of a contaminant in the worker's individual breathing zone and therefore should not be used to assess compliance with a WES (SWA 2013).

3.5.2 According to Current Knowledge

The OEL setting process involves the rigorous scientific evaluation of information from a number of sources, including the following:

- physicochemical data and structure–activity relationship for the contaminant;
- threshold toxicological data such as relevant no observed adverse effect levels (NOAEL) and lowest observed adverse effect levels (LOAEL), and non-threshold data from acute, chronic, and sub-chronic animal experiments;

- human studies and/or literature on accidental poisonings or adverse reactions from exposure to the contaminant;
- epidemiological data on exposed work populations and experience from industry;
- analogy with homologues, or similar substances.

Over time, new toxicological and epidemiological data or re-evaluation of existing data may alter the basis upon which the original OEL is set. For example, some substances that were originally thought to have primarily irritant effects have later been found to cause chronic disease. Inevitably, such new information will result in a review and amendment (usually lowering) of the OEL.

Figure 3.2 shows how OELs have reduced over time (Schenk 2011). For example, the ACGIH® TLV® for benzene in 1976 was 25 ppm as a ceiling value. This was reduced to a TWA–TLV® of 10 ppm in 1977 and to a TWA–TLV® of 0.5 ppm in 1996 based on updated toxicological and epidemiological data.

3.5.3 Nearly All Workers

OELs should not be considered as a fine dividing line between exposures that are safe and those that are hazardous – a fact often overlooked by those who use them. The ACGIH® states in its 'Introduction to the Chemical Substances of the TLVs®' that 'TLVs® will not adequately protect all workers. Some individuals experience discomfort or even more serious adverse health effects when exposed to a chemical substance at the TLV® or even at concentrations below the TLV®' (ACGIH® 2023).

Factors that may make some individuals more susceptible to chemicals include genetic predisposition, hypersensitivity/sensitisation, pre-existing disease and medication, lifestyle (tobacco, alcohol, and recreational drug use), age, gender, and reproductive status (De Silva 1986).

In some instances, the OEL may not be 'health-based'. Not all workers may be adequately protected when the OEL has been set as a 'regulatory adjusted' OEL, which involves the additional consideration of technical and economic feasibility. Technical feasibility can relate to measurability (due to instrumentation and/or analysis constraints) and economic impact may relate to one industry, many industries or return on investment for a required change.

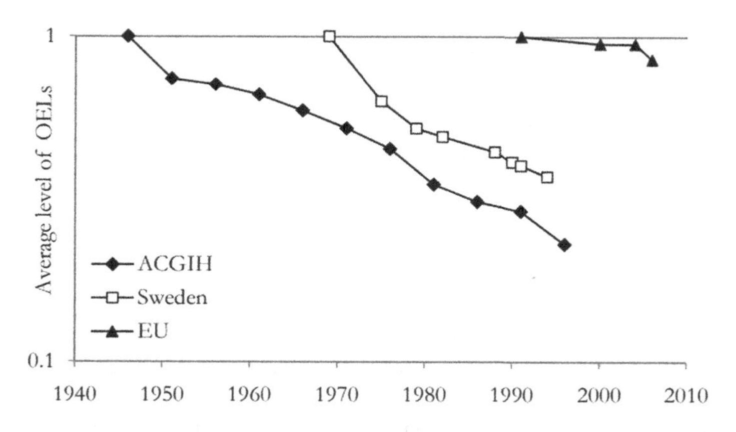

FIGURE 3.2 Reduction of occupational exposure limits over time.

Source: Schenk 2011, p. 17, Figure 4.

3.6 ADJUSTMENTS FOR EXTENDED WORK SHIFTS

TWA-OELs assume that individuals are exposed for 8 hours per day, 5 days a week (a 40-hour work week) for an entire working lifetime. The design of work or changes in work patterns means that some workers may be exposed for periods longer than the assumed standard working hours. In these cases, to provide an equivalent level of protection, it may be appropriate to adjust the TWA-OEL to a lower value in consideration of the longer exposure period and shorter recovery times. It should be noted that such adjustments are not appropriate in the case of OELs that are based on effects such as acute irritation and cyanosis, nor should they be used to adjust STEL and peak exposure limits.

Several methods for adjusting TWA exposure limits have been proposed, including mathematical models such as the Brief and Scala and OSHA models, and complex pharmacokinetic models.

The Brief and Scala method (Brief & Scala 1975) is regarded as the most conservative model. It considers the impact of the number of increased hours worked and the recovery time between exposure periods, but not the contaminant's activity in the body. Using either the daily or weekly equation detailed below, a reduction factor is determined and then applied to the TWA-OEL.

The Brief and Scala model reduces exposure limits according to a reduction factor calculated by Equations 3.2 and 3.3.

Daily exposure:

$$RF_{daily} = \frac{8}{h} * \frac{24-h}{16} \tag{3.2}$$

Where

RF_{daily} = daily reduction factor

h = hours worked per shift

Note that 24-h represents exposure-free hours per day.

Weekly exposure: for the special case of a seven-day work week is thus:

$$RF_{weekly} = \frac{40}{h} * \frac{168-h}{128} \tag{3.3}$$

where

RF_{weekly} = weekly reduction factor

h = average hours per week over full roster cycle

Note that 168-h represents exposure-free hours per week

The adjusted TWA-OEL is calculated by applying the reduction factor to the eight-hour TWA-OEL, as shown in Equation 3.4:

$$\text{Adjusted TWA-OEL} = \text{8h-TWA-OEL} \times \text{RF} \tag{3.4}$$

The OSHA method assumes that the magnitude of the toxic response of a contaminant is a function of the concentration that reaches the site of action for that contaminant. The model was designed to be applied to most systemic toxic substances, but not irritants, sensitisers, or carcinogens.

Under the OSHA method, each contaminant that has a TWA-OEL is categorised based on its toxic effect. The assigned category (these should not be confused with the carcinogen categories

TABLE 3.3
Reduction Factors Based on Adjustment Categories (Adj Cat)

Adj Cat.	Adjustment Classification	Type of Adjustment
1A	Substances regulated by a ceiling value	**No** adjustment
1B	Irritating or malodorous substances	
1C	Simple asphyxiants, substances presenting a safety risk or a very low health risk, whose half-life is less than four hours. Technological limitations	
2	Substances that produce effects following *short-term* exposure	**Daily** adjustment (8/daily exposure hours)
3	Substances that produce effects following *long-term* exposure	**Weekly** adjustment (40/weekly exposure hours)
4	Substances that produce effects following *short-* or *long-term* exposure	Daily or weekly adjustment **the most conservative of the two**

Adapted from: Drolet (2015).

assigned according to the GHS) is then used to determine whether any adjustment is required, and if so, what equation is to be used. The categories used are shown in Table 3.3.

The Quebec model (Drolet 2015) is essentially based on the OSHA model and uses the most recent information on toxicological effects, including sensitisation, irritation, organ toxicity, reproductive system toxicity, and teratogenicity (Verma 2000), as the critical health effect used to derive the Quebec regulatory exposure limits.

Depending on the category assigned, a recommendation of either:

- no adjustment is made to the TWA-OEL;
- a daily or weekly adjustment is made; or
- where both apply, the most conservative of the daily or weekly adjustments is made.

The Quebec model is supported by a comprehensive technical guide and a selection tool to assist in determining the most appropriate adjustment category. Caution must be applied when using the Quebec model to ensure that TWA-OELs appropriate to the jurisdiction are used (e.g. SWA's WES for use in Australia).

To address this situation, Firth and Drolet (2014) documented the adaptation of the Quebec method to be used with Safe Work Australia WESs. They also developed an Australian version of the downloadable tool, available from the AIOH website, to easily make decisions about adjustments of the WESs using the Quebec model or the Brief and Scala method. This Excel spreadsheet has a drop-down list of assessed substances that provides the adjustment category or code and computes the RF (called the adjustment factor in the spreadsheet) based on the daily and weekly average working hours.

Several different pharmacokinetic models are also available to adjust TWA-OELs for non-traditional working patterns. These are suitable for application to TWA-OELs that are based on accumulated body burden. They consider the expected behaviour of the contaminant in the body based on knowledge of its properties, using information such as the biological half-life and exposure time to predict body burden. The use of pharmacokinetic models can be complicated by the lack of biological half-lives for many chemicals. The most widely used pharmacokinetic model is the Hickey and Reist model, which requires knowledge of the chemical's biological half-life and hours worked per day and per week. The Hickey and Reist (1977) model, like other pharmacokinetic models, views the body as one compartment – that is, a homogeneous mass.

Pharmacokinetic models are less conservative than the Brief and Scala and OSHA/Quebec models, usually recommending smaller reductions of the TWA-OEL. While pharmacokinetic models are theoretically more exact than other models, their lack of conservatism may not allow adequately for the unknown additional effects on the body from night work or extended shifts, both of which may influence how well the body metabolises and eliminates the contaminant.

3.7 MIXTURES OF CHEMICALS

OELs are normally applied in situations where it is assumed that there is a single chemical present. In many working environments, workers are commonly exposed to a mixture of chemicals which could present an increased risk to health. Applying OELs in circumstances where multiple chemical contaminants are present is a complex task and requires care. Interactions between different chemicals should be assessed by appropriately qualified professionals such as occupational hygienists, occupational physicians, and toxicologists to ensure that specific toxicological consideration is made of all chemicals involved.

3.7.1 Independent Effects

In cases where there is clear toxicological evidence that two or more chemical contaminants have separate and different effects on the body, then each chemical may be separately evaluated against its appropriate OEL. For example, wood dust affects the respiratory system, while cyclohexane vapour acts upon the central nervous system. Each of these substances may therefore be assessed individually against its appropriate OEL.

3.7.2 Additive Effects

When the body is exposed to two or more chemical contaminants, an additive effect results when the contaminants have the same target organ or the same mechanism of action. For example, ethyl acetate and methyl ethyl ketone both affect the central nervous system. In this situation, the total effect upon the body equals the sum of effects from the individual chemicals (Equation 3.5). For chemicals whose effects are purely additive, conformity with the standard results when:

$$\left(\frac{C_1}{L_1}\right)+\left(\frac{C_2}{L_2}\right)+\ldots+\left(\frac{C_1}{L_n}\right)\leq 1 \tag{3.5}$$

where

C_1, $C_2 = C_n$ are the average measured airborne concentrations of the particular chemicals $1, 2 \ldots n$.

L_1, $L_2 = L_n$ are the appropriate OELs for the individual chemicals.

An example of an additive effect is the general effect of organic solvents on the central nervous system (narcotic or anaesthetic effect). However, the OELs for a few solvents, such as benzene and carbon tetrachloride, have been assigned on the basis of effects other than those on the central nervous system. Therefore, it is essential to refer to the documentation for the specific chemicals to ascertain the basis of the standard and any potential interactions.

TABLE 3.4
Differences between Additive, Synergistic, and Potentiation Effects

Type of Interaction	Effect of Substance X (%)	Effect of Substance Y (%)	Combined Effects of Exposure to Substances X and Y (%)
Additive	25	15	40
Synergistic	10	30	60
Potentiation	0	15	100

3.7.3 Synergism and Potentiation

Synergism and potentiation occur when exposure to two or more chemical contaminants results in an adverse health effect more severe than would be expected from the sum of the individual exposures.

Synergism occurs when two or more chemicals or mixtures have an effect individually, but their total effect exceeds the additive effect. For example, exposure to carbon tetrachloride and consuming alcohol (ethanol) presents a risk of liver damage much greater than exposure to those substances individually.

Potentiation occurs when one chemical or mixture, which is sometimes of no or low toxicity, enhances the effect of another chemical or mixture. For example, exposure to ototoxins such as white spirits can worsen the impact of noise on the hearing or balance functions of the inner ear.

Table 3.4 illustrates the differences between additive, synergistic, and potentiation effects for exposure to substances X and Y.

3.8 EXCURSION LIMITS

The concentration of an airborne contaminant arising from a particular process may fluctuate significantly over time. Even where the TWA concentration does not exceed the OEL over the work shift, excursions above the eight-hour TWA-OEL should be controlled.

Where there is no STEL limit assigned to a substance, a process is not considered to be under reasonable control if short-term exposures exceed three times the TWA-OEL for more than a total of 30 minutes per eight-hour working day, or if a single short-term value exceeds five times the eight-hour TWA exposure standard.

3.9 OCCUPATIONAL EXPOSURE LIMITS FOR PHYSICAL HAZARDS

In addition to OELs for chemical substances, the ACGIH® sets TLVs® for physical hazards, including noise and vibration, ionising and non-ionising radiation, heat and cold, and musculoskeletal hazards.

In Australia, OELs for physical hazards are set by a number of organisations. SWA has an OEL for noise exposure of 85 dB(A) as an eight-hour equivalent continuous A-weighted sound pressure level ($L_{Aeq,8h}$) and 140 dB(C) C-weighed peak limit ($L_{C,peak}$) as a part of its *Model Code of Practice – managing noise and preventing hearing loss at work* (SWA 2020a). This will be discussed in more depth in Chapter 14.

The responsibility for setting OELs for ionising and non-ionising radiation lies with the Australian Radiation Protection and Nuclear Safety Agency (ARPANSA). The agency published the *Fundamentals for Protection Against Ionising Radiation* in 2014, and *Standard for Limiting Exposure to Radiofrequency Fields – 100 kHz to 300 GHz (Rev. 1)* in 2021. This is discussed in Chapter 15, 'Ionising and non-ionising radiation'.

Exposure limits have not been set for musculoskeletal hazards or thermal stress in Australia (see Chapter 16 for more detail).

3.10 BIOLOGICAL EXPOSURE INDICES

OELs are an invaluable tool for assessing the risk posed by exposure to inhaled chemical contaminants in the workplace. However, since many chemicals are significantly absorbed through the skin, focusing on the inhalation route of entry alone provides only part of the risk estimate. For example, many organophosphate pesticides are lipid soluble and hence readily absorbed through the skin, mucous membranes, and gastrointestinal tract, yet the OELs for organophosphate pesticides are based exclusively on exposure by inhalation.

Biological monitoring measures the amount of a chemical, its metabolites, or its biochemical effect in a suitable biological sample such as urine, blood, or exhaled breath to determine exposure by all routes. It is considered complementary to atmospheric personal monitoring, which does not account for protection due to the use of personal protective equipment. Biological monitoring can be used to assess both the toxicological burden on the body and the adequacy of personal protective equipment. Biological monitoring may include testing for lead in blood, mandelic acid in urine as a measure of styrene exposure, carboxyhaemoglobin in blood as a measure of carbon monoxide exposure, and tetrachloroethylene in exhaled breath.

Biological exposure indices (BEIs®) are guidance values for assessing biological monitoring results. They represent the levels of the analyte (chemical, metabolite, or biochemical effect) in the biological sample that are likely to be observed if a person is exposed to the OEL. There are some exceptions for chemicals that might cause non-systemic effects, or where the OEL is directed towards preventing responses such as irritation or the skin is a significant route of entry.

The ACGIH® establishes BEIs® and publishes them in its annual TLVs® and BEIs® booklet, as well as in documentation which supports the BEIs®.

Safe Work Australia has established several Guidelines for Health Monitoring for a range of hazardous chemicals (SWA 2020b), which include biological monitoring action and removal levels for a number of chemicals including cadmium, chromium, mercury, and organophosphate pesticides. Chapter 11 discusses biological monitoring in more depth.

3.11 THE FUTURE OF OCCUPATIONAL EXPOSURE LIMITS

Formal processes of setting OELs have been in place for over 60 years, yet of the tens of thousands of potentially hazardous chemicals commonly used in industry today, only about 800 have been developed by organisations such as the ACGIH® and Safe Work Australia. The GESTIS (2023) international limit values for chemical agents database has OEL values gathered from 35 lists from 29 countries and contain OELs for 2332 substances. Indeed, for many new chemicals, there may be little or no toxicological data available, making it impossible to develop and apply OELs in the risk assessment process. Additionally, supposedly health-based OELs developed in different countries for the same chemicals can vary significantly, a fact that may reflect divergent philosophies and variations in the scientific robustness of the OEL setting processes. Some researchers have questioned the OEL development process and the adequacy of OELs to protect against adverse health effects (Borak & Brosseau 2015; Castleman & Ziem 1988; Ziem & Castleman 1989) The GESTIS (2023) database and Table 3.5 illustrate how OELs can vary between different countries and organisations.

These limitations have prompted some to rethink the concept of the OEL and explore alternative approaches to OEL setting and chemical risk assessment, such as control banding, the development of derived no-effect levels (DNELs), and setting global OELs.

TABLE 3.5
Examples of OELs in Different Countries and Organisations

Chemical Substance	ACGIH	Australia	European Commission	Finland	Germany	Sweden	Quebec, Canada	US OSHA
Carbon tetrachloride	5 ppm	0.1 ppm	1 ppm	1 ppm	0.5 ppm	1 ppm	5 ppm (15-minute average)	10 ppm (Ceiling Value)
Halothane	5 ppm	0.5 ppm	–	1 ppm	5 ppm	5 ppm	50 ppm	–
Toluene	20 ppm	50 ppm	50 ppm	25 ppm	50 ppm	50 ppm	20 ppm	200 ppm

Adapted from: GESTIS (2023).

3.11.1 Control Banding

Control banding is a technique that utilises the occupational hygiene approach of anticipation, recognition, evaluation, and control in cases where there is a lack of data on quantitative exposure and level of hazard for a chemical in a workplace. Processes for control banding are discussed more fully in Chapter 5, but in general it is used to aid selection of appropriate measures for controlling a risk (e.g. ventilation, containment, and substitution) according to the extent to which their hazards are banded (i.e. based on their GHS hazard phrases) and on a subjective estimate of exposure (high to low). Control banding has been effectively used for many chemical substances that have not had OELs assigned, such as pharmaceuticals and nanomaterials. It is a particularly useful tool for small to medium enterprises where there is often a lack of occupational hygiene expertise and expensive exposure monitoring. A number of digital tools have been developed by organisations such as the UK COSHH Essentials, the German REACH-CLP Helpdesk, and the Dutch Ministry of Social Affairs and Employment to assist in conducting control banding assessments.

3.11.2 Derived No-Effect Levels (DNELS)

The European Commission (EC) brought into force the REACH (Registration, Evaluation, Authorization and Restriction of Chemicals) regulations in June 2007. Under the REACH legislation, chemical substances subject to registration that are manufactured, imported, or used in European markets in quantities of greater than 10 tonnes a year must have chemical safety assessments undertaken by a responsible party (normally the manufacturer or supplier), including the establishment of DNELs, or levels beyond which humans should not be exposed.

These DNELs must be established based on a number of exposure scenarios for workers, consumers/general populations and in some cases specific vulnerable subpopulations (European Chemicals Agency 2012). Thus, DNELs are to be established for all relevant combinations of:

- population groups (workers, general population, human via the environment);
- exposure routes (oral, dermal, and inhalation);
- duration of exposure (long-term and short-term); and
- type of effects (systemic and local).

Each exposure scenario and health effect for a substance requires the establishment of a DNEL, which is calculated by dividing the dose descriptor from experimental toxicological data (e.g. NOAEL, LOAEL, LD_{50}, and LC_{50}) for the health effect by an appropriate assessment factor allowing the data to be extrapolated for human exposure situations. For non-threshold effects (e.g. genotoxicity and carcinogenicity), the derived minimal effect level (DMEL) is the level of risk below which exposure is of no concern. The lowest DNEL for a given health effect is documented for each exposure scenario and included in safety data sheets (SDSs).

Unlike OELs, DNELs consider exposure from all routes, not just inhalation, therefore the basis upon which they are established differs from that of OELs. OELs are generally developed by scientific specialists in occupational hygiene, occupational medicine, and toxicology and/or government agencies using published data and are measures of 'acceptable risk'. DNELs, on the other hand, are 'no-effect levels' developed by manufacturers, importers, or suppliers, sometimes using proprietary data. Some have speculated that DNELs may replace OELs (Paustenbach & Cyrs 2021), while others regard DNELs as guidance values that can coexist with OELs and believe both should be used as part of a chemical risk assessment and management process (ECESAI 2010).

3.11.3 Global Occupational Exposure Limits

Extensive efforts have been made to harmonise the classification and labelling of chemicals under the Globally Harmonized System of Classification and Labelling of Chemicals (GHS) adopted by the United Nations in 2003. However, harmonising OELs across the globe remains a major challenge. The establishment of the Scientific Committee on Occupational Exposure Limits (SCOEL) and development of a methodology for setting European OELs (SCOEL 2013) are important steps towards unifying these, at least among EU members, and signals a possible diminution of the influence of the ACGIH® TLVs®. However, it remains to be seen whether the regulatory, political, cultural/philosophical, and economic factors currently impeding the progress of developing globally harmonised OELs can be overcome. Some have argued that as OELs are not absolute thresholds, but allow for 'acceptable risk' levels, 'the acceptability of the risk should ultimately be determined by the cultural body politic of the society and thus could be different for different groups' (Adkins et al. 2009b). Nonetheless, many experienced hygienists still consider OELs to be the most effective risk assessment tools or guides for developing strategies to protect workers from exposure to chemicals (Adkins et al. 2009b).

But the OEL has no utility in itself unless there are reliable complementary tools for measuring workplace exposures and interpreting them against the OEL in a consistent fashion across workplaces, industries, and countries. In Australia, the challenge is how to consistently use them within the regulatory framework. Enshrining OELs in law or state regulation has created another major challenge for regulators and users alike, particularly when the accuracy of measurement is an issue. Though the OEL has considerable appeal as a number against which workplace performance or compliance could be judged, in reality the myriad workplace factors, the individual workplace tasks and work design, the consideration of 'reasonably practicable', limitations on monitoring, the differences or errors in measurement, and the variety of ideas about what constitutes 'compliance' all militate against consistent outcomes. Some workplace monitoring issues, the tools that have evolved, and the requirements for decision-making about compliance with exposure regulations are all covered in the following section.

3.12 MAKING COMPLIANCE DECISIONS BASED ON REGULATORY USE OF EXPOSURE STANDARDS

Interpreting exposure standards, deciding which type to use, allowing for mixed exposures, and so on are all relatively straightforward compared with making reliable decisions about compliance with regulations.

There are several reasons for this. First, the use of a simple numerical limit in regulation does not facilitate easy decision-making for exposures which are known to vary widely depending on processes, environmental conditions, work practices, control efficacy, and individual worker behaviour. Second, to accommodate all this variability, measurements need to be subject to statistical examination before informed judgements about exposure can be made. Third, even when fairly robust statistical measures of exposure can be determined, there is no common understanding of which measure should be used for making decisions on what is an exceedance of the OEL/WES. Finally, undertaking a valid sampling and monitoring exercise requires a competent person, such as an occupational hygienist, to develop protocols for identifying differently exposed groups in a given workplace and to establish a monitoring regime with adequate numbers of samples and an appropriate frequency of sampling.

One of the indeterminacies with which regulation must deal is that 'compliance' is not a numerical condition, but the yardstick by which it is assessed is a numerical one. Further, the tools for assessing compliance are mostly numerical measurement ones and, when used, must be able to provide a sufficiently robust outcome for making a non-parametric compliance decision. Despite these obstacles, the National Occupational Health and Safety Compliance and Enforcement Policy prepared by Safe Work Australia and the state and territory Workplace Health and Safety Regulators (SWA 2020c) has identified a number of aims including ensuring compliance with the WES by:

- providing advice and information to duty holders and the community;
- monitoring and enforcing compliance with WHS laws.

3.12.1 Understanding Regulatory Compliance

Within the Australian context, regulations for airborne hazardous chemicals are based on the principle that 'no one is exposed to a substance or mixture in an airborne concentration that exceeds the exposure standard'. Here it is clear that compliance applies not only to groups of workers but to every individual within a given group and others who may enter the workplace while work is being undertaken. To encompass all the variability imposed on exposure by process, work environment, control, and so on, and to manage coverage of all workers without embarking on the impossible task of sampling for every individual worker, statistical procedures of some kind are needed for sampling and analysis. For utility, compliance decisions for WESs must be statistically based, and compliance is hinged around the most widely accepted interpretation of 'exposed'. Thus, it is usually an exposure profile that is compared with a WES, not individual measurements. As a result, an individual measurement that exceeds the TWA-WES cannot be unequivocally interpreted as evidence of non-compliance (though it may be such), other than for application of OELs utilising either peak limitations or STELs.

The base problem is that measurements made in most exposure situations are known to vary widely for no immediately apparent reason. Therefore, no individual reading represents a common value that can be compared with the WES: it is the whole exposure profile that should be compared with the standard. Even so, this profile is only a probability distribution of a relatively small number of measurements, and so has a good deal of uncertainty about it. Significantly reducing the uncertainty requires a very large number of measurements to be made (200–300), which is generally not feasible.

So, 'compliance' has evolved into a process of comparing probability estimates about exposure for groups and individuals with an exposure standard, and decisions on compliance are based on as few measurements as is practicable.

To promote flexibility in approaching this problem, the Australian WHS regulations invite the use of statistical interpretation by using the term 'not certain on reasonable grounds'. This encourages

H&S practitioners to employ the concept of statistical probability rather than demanding absolute certainty of compliance, since even rudimentary monitoring exercises demonstrate that certainty can rarely be achieved. Safe Work Australia (2013) provides the general guidance that interpreting compliance will require a sound understanding of the nature of contaminant concentrations, together with the statistics relevant to their measurement and interpretation of measurement results. Determining that exposure of individual workers or groups of workers complies with the regulations by being below a WES requires a degree of certainty, although that degree might be as low as 50 per cent or as high as 99 per cent.

It remains to be ascertained, however, what is an accepted degree of certainty for making a compliance decision, and what tools are available to assist in making those decisions. A common convention in the social sciences is to assess statistical significance. At a 95 per cent level (usually expressed as $p < 0.05$), this indicates that an event has probably occurred for reasons other than chance. Some argue that a similar convention should be employed in occupational hygiene to aid compliance decision-making with respect to OELs.

Three factors are needed by H&S practitioners to make compliance decisions about WESs:

- a sound knowledge of how to conduct atmospheric personal monitoring at the workplace;
- an understanding of and access to appropriate tools for statistical analysis of monitoring data; and
- a compliance decision-making strategy.

3.12.2 Air Monitoring in the Workplace

Monitoring the workplace for airborne contaminants is a critical step in the evaluation of risks from airborne contaminants and comprises a significant part of the work of H&S practitioners. Monitoring provides the bulk of the data about exposure that are to be compared with an OEL.

Air monitoring is separated into:

a. identifying the contaminants that require monitoring;
b. devising the most appropriate monitoring strategy for measuring exposures and making compliance decisions;
c. selecting the most appropriate sampling method for the contaminant in question in conjunction with an analytical laboratory that can provide the required analysis; and
d. making calculations about exposures from each individual field measurement.

Recognising which contaminants require monitoring may seem obvious, but this is not always the case. Failure to identify airborne contaminants with significant toxicity will result in incorrect or incomplete monitoring, measurement, and management regimes and ultimately result in adverse health effects in workers. For example, failing to recognise pneumotoxic oxidant gases among the fumes arising from an arc welding process will limit the value of a gravimetric monitoring of welding fumes. Discussions in later chapters on dust, gases and vapours will assist the H&S practitioner to ensure that monitoring is directed to measuring the important contaminants.

A monitoring strategy must be tailored to a specific workplace. The monitoring strategy addresses questions of where and when to collect samples, from whom to collect samples, for how long to sample and how many samples need to be collected. H&S practitioners need to be familiar with the individual tasks performed by workers to identify which are at risk and also to determine whether there are different similarly exposed groups (SEGs) in the workplace. For example, fettlers or metal dressers in a metal foundry would be classed as an individual SEG, as would moulders or shakeout and knockout operators. The temporal pattern of possible exposure needs to be determined to accommodate different exposures on night shifts, maintenance procedures, and seasonal variations. If there are contaminants with a combination of WESs, such as those with TWAs and STELs or peak limitations, then both long-term and short-term monitoring may be needed.

A good monitoring strategy observes the number of sources of contaminant release, the rate of release, the period of exposure, the type and level of maintenance of controls in place, local environmental factors, and worker idiosyncrasies, which will all contribute to the overall exposure of the worker that needs to be measured. Monitoring strategies should also consider the possibility of mixed exposures, and other active routes of exposure, such as skin absorption. A monitoring strategy may also have to determine an appropriate re-sampling period, depending on the toxicity of the airborne contaminant, the reliance on lower-level control measures and how close to the WES the result is. A monitoring strategy may also be devised for determining the most suitable protective factors of personal protective equipment where control of the working environment is not practical.

Selecting the most appropriate sampling technique requires time and attention and as previously noted, only personal sampling should be used. Personal sampling is often carried out in a sensitive environment, subject to some invalidation of measurement through delinquency, sample mishandling, transport loss and, occasionally, equipment failure. There are very few sampling methods that do not require laboratory analysis. Both the way the air is sampled and the method of analysis are interlinked.

To analyse for the thousands of possible workplace atmospheric contaminants, some hundreds of standardised and validated methods are now available (HSE 2023; NIOSH 2018; OSHA 2022). It is important to match the analytical measurement range of the method selected to the expected concentration range of the workplace contaminant. Sampling and analytical constraints such as sample breakthrough, adsorption and desorption efficiency, limits of detection, interferences, and measurement error should be discussed with the laboratory carrying out the analysis.

Because a field measurement ultimately has to reflect the health risk, specialised sampling protocols may be mandatory. For example, dusts that damage the lungs, such as crystalline silica, are required to be sampled with an elutriator meeting a penetration criterion matched to the human respiratory system (SWA 2013). Dusts that have systemic health effects (e.g. lead), or effects that are dependent on fibre morphology (e.g. asbestos) are sampled and assessed differently. Sampling techniques may differ when measuring over short periods, for example, for STELs.

Commercial suppliers of air sampling equipment can provide a vast array of sampling pumps, different kinds of plain and treated filters, absorption tubes, adsorption badges, as well as direct reading instrumentation for specialised purposes for gases, vapours, and some aerosols.

It should be noted that the SCOEL (2017) states that ‘Measurement techniques should be able to assess exposure at: 0.1 times the OEL for 8-hour TWA’ when assessing whether accurate sampling and analytical methods are available to measure exposure to compare with or assess compliance against a recommended exposure standard.

Making calculations about exposure from collected samples requires correct measurement of sampling volumes, a thorough appreciation of units of measurement for conversion between gravimetric and volumetric expressions of concentration and making corrections for temperature and pressure (if applicable). A laboratory may report a metal concentration when its OEL is expressed as an oxide, so familiarity with the periodic table and chemical formulae for making the necessary conversions are useful skills for some applications.

3.13 CONCLUDING REMARKS ON OELs AND THEIR USE

OELs or exposure standards, despite criticisms, have provided valuable service to occupational hygienists and the health and safety community for several generations. Their expression in simple numbers often belies the complexity behind their derivation, their continual refinement, their occasional differences, and the myriad factors required in their correct application. Their use and decisions against the measurement of them must always be accompanied by competent best practice in occupational hygiene. However, the enshrinement of these values in regulations has bestowed on them both legality and an aura of greater authoritativeness than was ever intended.

Despite this, even good exposure standards have very limited value in law without a robust framework into which they can be inserted. Decisions in law require some acceptable degree of consistency and certainty if they are to be upheld and to achieve an intended outcome (protection of workers). The monitoring of exposure and the process for making compliance decisions based on that monitoring has rarely been attended by either complete consistency or certainty. The regulation of hazardous chemicals has established both that compliance with an exposure standard needs to be achieved and that airborne monitoring may need to be undertaken where there is uncertainty about compliance, but without identifying what is meant by compliance. Despite their necessary resort to some statistics, the monitoring programs described in Section 3.12.2 and the decision-making protocols for determining compliance in Chapter 4 have shown that it is possible to make consistent, understandable, and acceptable decisions about compliance. These decisions should satisfy both the protective intent of the regulation and withstand legal scrutiny, to the extent possible given their basis in predictions about probability.

Properly structured monitoring programs combined with the best decision-making processes will suit the needs of occupational hygienists, H&S practitioners, and regulators alike. The framework of statistically based monitoring and decision-making about compliance is essential if OELs are to serve any regulatory function. When monitoring programs and decision-making are well integrated, OELs established as guides for professional use can also operate effectively as legal regulatory benchmarks.

REFERENCES

Adkins, C., Booher, L., Culver, D., et al. 2009a, *Occupational Exposure Limits—Do They Have a Future?* Derby, UK: International Occupational Hygiene Association, ftp.cdc.gov/pub/Documents/OEL/12.%20Niemeier/References/Adkins_2009_OEL%20future%20%2009.18.09.pdf [2 January 2024].

Adkins, C., Booher, L., Culver, D., et al. 2009b, 'The Future of Occupational Exposure Limits—Can OELs Be Saved?', *The Synergist*, vol. 20, no. 9, pp. 46–8.

American Conference of Governmental Industrial Hygienists (ACGIH) 2023, *Introduction to the Chemical Substances TLVs®*, www.acgih.org/science/tlv-bei-guidelines/tlv-chemical-substances-introduction [2 January 2024].

Baetjer, A.M. 1980, 'The Early Days of Industrial Hygiene—Their Contribution to Current Problems', *American Industrial Hygiene Association Journal*, vol. 41, no.11, pp. 773–7

Borak, J. & Brosseau, L.M. 2015, 'The Past and Future of Occupational Exposure Limits', *Journal of Occupational and Environmental Hygiene*, vol. 12, no. 1, S1–S3, oeh.tandfonline.com/doi/pdf/10.1080/15459624.2015.10.1263 [1 January 2024].

Brandys, R.C. & Brandys G.M. 2008, *Global Occupational Exposure Limits for Over 6000 Specific Chemicals* 2nd edn. Hinsdale, IL: Occupational and Environmental Health Consulting Services https://www.oehcs.com/global_exposure_standards.htm [2 January 2024].

Brief, R.S. & Scala, R.A. 1975, 'Occupational Exposure Limits for Novel Work Schedules', *American Industrial Hygiene Association Journal*, vol. 36, no. 6, pp. 467–9.

Castleman, B.I. & Ziem, G.E. 1988, 'Corporate Influence on Threshold Limit Values', *American Journal of Industrial Medicine*, vol. 13, no. 5, pp. 531–59, doi: 10.1002/ajim.4700130503

Cook, W.A. 1945, 'Maximum Allowable Concentrations of Industrial Atmospheric Contaminants', *Industrial Medicine*, vol. 11, pp. 936–46.

Dawson, B.J., Dotson, K.B., Grimsley, F., Grumbles, T., Mansdorf, Z., Roskelley, D., Sahmel, J., Tresider, N. & Tsai C. 2021, 'Occupational and Industrial Hygiene as a Profession: Yesterday, Today, and Tomorrow', in B. Cohrssen (Ed), *Patty's Industrial Hygiene, Volume 1: Hazard Recognition*, 7th ed, pp. 3–18, Hoboken, NJ: John Wiley & Sons.

De Silva, P. 1986, 'TLVs to Protect "nearly all workers"', *Applied Industrial Hygiene*, vol. 1, no. 1, pp. 49–53.

De Silva, P.E. 2000, *Science at Work: A history of occupational health in Victoria*. Blackburn, Victoria: PenFolk.

Drolet, D. 2015, *Technical Guide T–22: Guide for the adjustment of permissible exposure values (PEVs) for unusual work schedules*, 4th ed, Montreal: IRSST. www.irsst.qc.ca/en/publications-tools/publication/i/100349/n/guide-for-the-adjustment-of-permissible-exposure-values-pevs-for-unusual-work-schedules [2 January 2024].

European Chemicals Agency 2012, *Guidance on Information Requirements and Chemical Safety Assessment Chapter R.8: Characterisation of dose [concentration]-response for human health*, ECHA, Helsinki, echa.europa.eu/documents/10162/13632/information_requirements_r8_en.pdf/e153243a-03f0-44c5-8808-88af66223258 [30 May 2023].

European Commission, Employment, Social Affairs and Inclusion (ECESAI) 2010, *Guidance for Employers on Controlling Risks from Chemicals: Interface between Chemicals Agents Directive and REACH at the workplace*, European Commission, Brussels, osha.europa.eu/en/legislation/guidelines/guidance-employers-controlling-risks-chemicals-interface-between-chemicals-agents-directive-and-reach-workplace [2 January 2024].

Firth, I. & Drolet, D. 2014, 'Development of a Tool for the Adjustment of Workplace Exposure Standards for Atmospheric Contaminants Due to Extended Work Shifts', *Journal of Health & Safety Research & Practice*, vol. 6, no. 2, pp 6–10.

Gallo, M.A. 2013, 'History and Scope of Toxicology', in C.D. Klaassen, L.J. Casarett & J. Doull (Eds) *Casarett & Doull's Toxicology: The Basic Science of Poisons*, 8th ed, pp. 3–10. New York: McGraw Hill Medical.

GESTIS 2023, *International Limit Values for Chemical Agents (Occupational exposure limits, OELs)*, www.dguv.de/ifa/gestis/gestis-internationale-grenzwerte-fuer-chemische-substanzen-limit-values-for-chemical-agents/index-2.jsp [2 January 2024].

Harbison, R.D., Bourgeois, M.M. & Johnson, G.T. 2015, *Hamilton & Hardy's Industrial Toxicology*, 6th edn. Hoboken, NJ: Wiley.

Health and Safety Executive (HSE) 2023, *Methods for Determination of Hazardous Substances (MDHS) Guidance*. London: HSE, www.hse.gov.uk/pubns/mdhs [2 January 2024].

Henderson, Y. 1921, 'Effects of Gases on Men and the Treatment of Various Forms of Gas Poisoning', in A.C. Fieldner, S.H. Katz and S.P. Kinney, *Gas Masks for Gases Met in Fighting Fire*. Washington, DC: Department of Interior, US Bureau of Mines.

Henschler, D. 1984, 'Exposure Limits: History, Philosophy, Future Developments', *Annals of Occupational Hygiene*, vol. 28, no.1, pp. 79–92.

Hickey, J.L. & Reist, P.C. 1977, 'Application of Occupational Exposure Limits to Unusual Work Schedules', *American Industrial Hygiene Association Journal*, vol. 38, no. 11, pp. 613–21.

Howard, J. 2005, 'Setting Occupational Exposure Limits: Are We Living in a Post-OEL World?' *University of Pennsylvania Journal of Labor and Employment Law*, vol. 7, no. 3, pp. 513–28.

ILO 2011, 'Occupational Exposure Limits: The History of Occupational Exposure Limits', in *Encyclopaedia of Occupational Health & Safety Part IV Occupational Hygiene*, https://www.iloencyclopaedia.org/part-iv-66769/occupational-hygiene-47504/item/575-occupational-exposure-limits [2 January 2024].

Kober, G.M., Hayhurst, E.R. & Rober, G.M. 1924, *Industrial Health*. Philadelphia, PA" P. Blakiston's.

Lehmann, K. B. 1893, *Methods of Practical Hygiene*, vol. 1. London: Kegan Paul, Trench, Trübner & Co.

National Institute for Occupational Safety and Health (NIOSH) 2018, *NIOSH Manual of Analytical Methods (NMAM)*, 5th edn. Cincinnati, OH: NIOSH, www.cdc.gov/niosh/nmam/ [2 January 2024].

National Occupational Health and Safety Commission (NOHSC) 1995a, *Guidance Note on the Interpretation of Exposure Standards for Atmospheric Contaminants in the Occupational Environment*, 3rd edn, NOHSC 3008. Canberra: Australian Government Publishing Service.

National Occupational Health and Safety Commission (NOHSC) 1995b, *Adopted National Exposure Standards for Atmospheric Contaminants in the Occupational Environment*, NOHSC:1003(1995). Canberra: Australian Government Publishing Service, www.safeworkaustralia.gov.au/system/files/documents/1702/adoptednationalexposurestandardsatmosphericcontaminants_nohsc1003-1995_pdf.pdf [2 January 2024].

Oliver, T. 1902, 'Dusts as a Cause of Occupational Disease,' in T. Oliver (Ed), *Dangerous Trades: The Historical, Social and Legal Aspects of Industrial Occupations as Affecting Health*, pp. 267–276. London: John Murray.

OSHA 2022, *Sampling and Analytical Methods*. Washington, DC: OSHA, www.osha.gov/dts/sltc/methods [2 January 2024].

Paustenbach, D.J. & Cyrs, W.D. 2021, 'The History and Biological Basis of Occupational Exposure Limits for Chemical Agents', in B. Cohrssen (Ed), *Patty's Industrial Hygiene, Volume 1: Hazard Recognition*, 7th edn, pp. 161–211. Hoboken, NJ: John Wiley & Sons.

Ripple S.D. 2010, 'History of Occupational Exposure Limits', *8th International Occupational Hygiene Association Conference*, 28 September–2 October, IOHA, Rome, ftp.cdc.gov/pub/Documents/OEL/12.%20Niemeier/References/Ripple_2010_History%20oel.pdf [2 January 2024].

Safe Work Australia (SWA) 2013, *Guidance on the Interpretation of Workplace Exposure Standards for Airborne Contaminants*. Canberra: SWA, www.safeworkaustralia.gov.au/doc/guidance-interpretation-workplace-exposure-standards-airborne-contaminants [2 January 2024].

Safe Work Australia (SWA) 2020a, *Model Code of Practice—Managing Noise and Preventing Hearing Loss at Work*. Canberra: SWA, www.safeworkaustralia.gov.au/doc/model-code-practice-managing-noise-and-preventing-hearing-loss-work [2 January 2024].

Safe Work Australia (SWA) 2020b, *Health Monitoring Guidelines*. Canberra: SWA, www.safeworkaustralia.gov.au/safety-topic/managing-health-and-safety/health-monitoring/resources [30 May 2023].

Safe Work Australia (SWA) 2020c, *National Compliance and Enforcement Policy*. Canberra: SWA, www.safeworkaustralia.gov.au/resources-and-publications/legislation/national-compliance-and-enforcement-policy [2 January 2024].

Safe Work Australia (SWA) 2021, *Hazardous Chemical Information System (HCIS)*. Canberra: SWA, hcis.safeworkaustralia.gov.au/ [2 January 2024].

Safe Work Australia (SWA) 2023a, *Model WHS Regulations*. Canberra: SWA, www.safeworkaustralia.gov.au/doc/model-whs-regulations [2 January 2024].

Safe Work Australia (SWA) 2023b, *Workplace Exposure Standards Review*. Canberra: SWA, www.safeworkaustralia.gov.au/safety-topic/managing-health-and-safety/workplace-exposure-standards-chemicals/workplace-exposure-standards-review [2 January 2024].

Safe Work Australia (SWA) 2023c, *Occupational Lung Diseases*. Canberra: SWA www.safeworkaustralia.gov.au/safety-topic/hazards/occupational-lung-diseases/air-monitoring [25 March 2024].

Safe Work Australia (SWA) 2024, *Workplace exposure limits for airborne contaminants*. Canberra: SWA, https://www.safeworkaustralia.gov.au/doc/workplace-exposure-limits-airborne-contaminants [25 August 2024].

Sayers, R.R. 1927, 'Toxicology of Gases and Vapors', in *International Critical Tables of Numerical Data, Physics, Chemistry, and Toxicology*, vol. 2, pp. 318–321. New York: McGraw-Hill.

Schenk, L. 2011, *Setting Occupational Exposure Limits: Practices and outcomes of toxicological risk assessment*, PhD Thesis, Södertörn University, Stockholm, http://ftp.cdc.gov/pub/Documents/OEL/12.%20Niemeier/References/Schenk_2011_Thesis.pdf [2 January 2024].

Scientific Committee on Occupational Exposure Limits (SCOEL) 2013, *Methodology for the Derivation of Occupational Exposure Limits*, Key documentation, Version 7, European Commission, Employment, Social Affairs and Inclusion, Brussels, ec.europa.eu/social/BlobServlet?docId=4526&langId=en [2 January 2024].

Scientific Committee on Occupational Exposure Limits (SCOEL) 2017, *Methodology for Derivation of Occupational Exposure Limits of Chemical Agents – The General Decision-Making Framework of the Scientific Committee on Occupational Exposure Limits (SCOEL)*, European Commission, Luxembourg: Scientific Committee on Occupational Exposure Limits, op.europa.eu/en/publication-detail/-/publication/3c8ef3e0-48fc-11e8-be1d-01aa75ed71a1 [2 January 2024].

UN Environment and the International Council of Chemical Associations 2019, *Knowledge Management and Information Sharing for the Sound Management of Industrial Chemicals*, http://www.saicm.org/Portals/12/Documents/EPI/Knowledge_Information_Sharing_Study_UNEP_ICCA.pdf [2 January 2024].

Verma, D.K. 2000, 'Adjustment of Occupational Exposure Limits for Unusual Work Schedules', *American Industrial Hygiene Association Journal*, vol. 61, no. 3, pp. 367–74.

Vincent, J.H. 1999, 'Occupational Hygiene Science and Its Application in Occupational Health Policy, at Home and Abroad', *Occupational Medicine*, vol. 49, no. 1, pp. 27–35.

Ziem, G.E. & Castleman, B.I. 1989, 'Threshold Limit Values: Historical Perspectives and Current Practice', *Journal of Occupational Medicine*, vol. 31, no. 11, pp. 910–18.

4 Occupational Hygiene Statistics

Dr Peter Knott and Dr David Lowry

4.1 INTRODUCTION

The definition of occupational hygiene, centred around the tenets of anticipation, recognition, evaluation and control, speaks to the complexities and diversity within the profession. The practising hygienist will be expected to do many things; however, the 'evaluation' component of a hygienist's role directly relates to the concept of exposure assessment (Jahn et al. 2015).

An exposure assessment can be carried out for a variety of reasons, and the design of the assessment strategy should be dependent on the context (Stewart & Stenzel, 2000). The most common reason is the measurement of worker exposures to chemical and physical hazards and the comparison of these exposures to an occupational exposure limit (OEL) (Jahn et al. 2015; Ramachandran 2008; Stewart & Stenzel 2000). This can be done by occupational hygienists employed or engaged by a company or by regulatory enforcement agencies to determine whether exposure levels meet legal standards.

Sampling may also be conducted to verify the impact of a control measure implemented in the workplace. Here a sampling strategy would be to obtain sufficient power to correctly gauge the impact of the control.

Another reason might be to determine a relationship between exposure and health outcome in an epidemiological study (Banerjee et al. 2014; Kauppinen 1994; Nieuwenhuijsen 2003). In which case, the sampling strategy would be to randomise the collection as much as possible to obtain an unbiased estimate of the true exposure.

The purpose of conducting an exposure assessment also drives the choice of the decision statistic in the analysis. For example, if the exposure assessment is done in the context of an epidemiological study, some measure of central tendency such as the arithmetic mean is appropriate (Bullock & Ignacio 2006; Ramachandran 2008; Stewart & Stenzel 2000). In contrast, if exposure assessment is done for routine risk management, that is, to ensure that most of the workers have acceptable exposure levels, then some upper percentile of the exposure distribution (e.g. the 95th percentile) may be a better decision statistic (Banerjee et al. 2014; Nieuwenhuijsen 2003).

But what do we mean by these terms? Occupational hygiene results are usually reported in statistical language which can be confusing to newcomers, as basic statistics taught in undergraduate courses does not cover the complexity often encountered in exposure data. This chapter is written to explain key terms and to assist in interpreting results in a meaningful way. Oftentimes as hygienists, we do not need mathematical precision, just enough information so that we understand what the results are telling us and an indication of uncertainty around the results.

DOI: 10.1201/9781032645841-4

4.2 DESCRIPTIVE STATISTICS

Regardless of the strategy, the collection of individual exposure measurements leads the occupational hygienist to obtaining a sample from the population of all possible exposures to workers (which is unknown), a relationship is depicted in Figure 4.1. In this chapter, the term *sample* means a set of measurements obtained representing the overall population of exposures.

Descriptive statistics are used to summarise the individual measurements in a sample. They describe the aggregate qualities of a sample obtained from a population. Occupational hygiene measurements are typically continuous variables which are bounded at the lower end by zero, although in practice they will be censored at some point above zero, for example, <LOQ. Descriptive statistics for sample data typically will describe their central tendency (arithmetic mean, median and geometric mean) and their spread (range, minimum and maximum, standard deviation and variance). The following descriptive statistics are generally automatically generated by most occupational hygiene statistical analysis tools.

- Number of measurements (n): a count of how many measurements in the sample;
- Maximum: the maximum value measured;
- Minimum: the minimum value measured;
- Range: the magnitude of the difference between the max and the min;
- Arithmetic mean ($\overline{x}$): the sum of all measurements divided by n;
- Median: the middle value in a set of measurements ranked from; min to max
- Sample variance (s^2): a measure of dispersion of a sample data set (how far values are from the mean) and is reported in the units of the data squared;
- Sample standard deviation (s): a measure of the spread (how far apart values are) of a sample data set described in the units of the data;
- Geometric mean (GM): antilog of the arithmetic mean of the log-transformed values;
- Geometric standard deviation (GSD): antilog of the standard deviation of the log-transformed values;

In occupational hygiene data, the distribution of measurements in a sample is commonly approximated to fit a lognormal distribution (Huynh et al. 2014). This distribution has been shown to fit multiple data sets over decades of research and is the accepted rule of thumb. However, unless there are a sizeable number ($n > 30$) of measurements, formally testing the lognormality of a

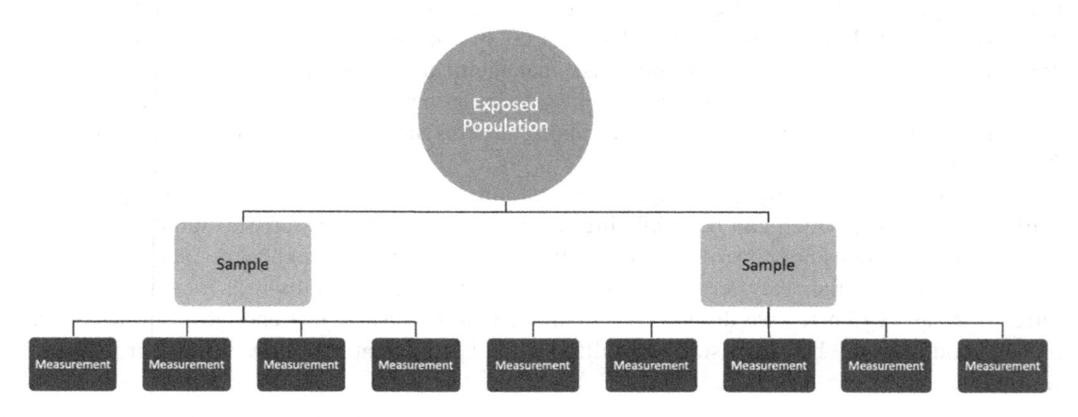

FIGURE 4.1 Relationship between population exposures, samples and individual measurements.

Source: P. Knott 2023.

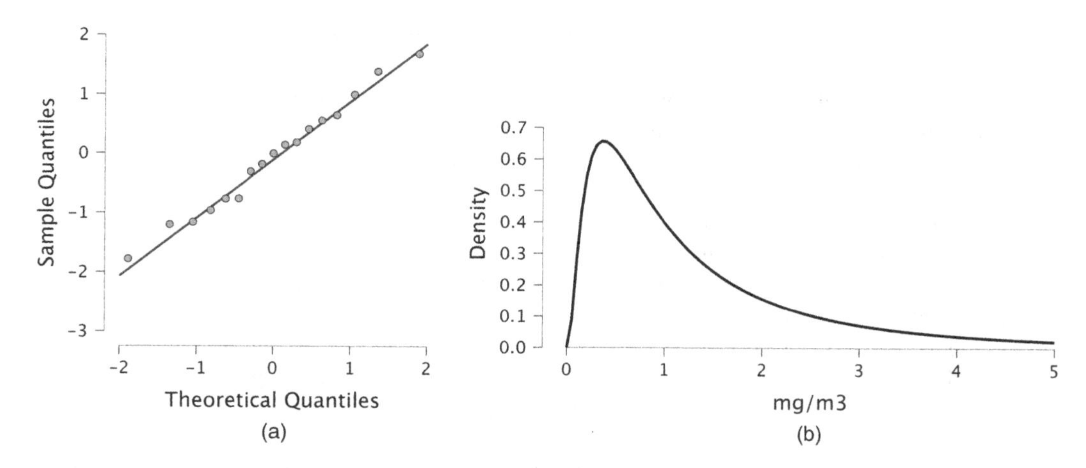

FIGURE 4.2 Q-Q Plot of a sample of exposure measurements (a) and its corresponding lognormal distribution (b).

Source: P. Knott 2023.

data set may not show significance; however, plotting a probability plot of the log of individual samples and examining if these fit a straight line will establish whether the data is lognormally distributed and if there are deviations from it. The scatterplot in Figure 4.2a represents a sample obtained from the lognormal distribution as shown in Figure 4.2b.

In lognormal distributed data, the GM is usually equal to the distribution median. Because the GM is lower than the arithmetic mean in a data set that is lognormally distributed, using the GM will underestimate the average exposure. The difference between the two grows as variance in the distribution increases. Thus, as the GSD gets larger, the GM further underestimates the average exposure, and if used may not adequately estimate actual workplace exposures.

When most of the data are clustered well below or well above the OEL, one can generally decide on workplace acceptability by using descriptive statistics and professional judgment. However, when the range of data approaches or includes the OEL, inferential statistics, as described in the next section, must be used for decision-making.

4.3 INFERENTIAL STATISTICS

Samples collected from a population cannot describe accurately the population itself because it is generally not possible to measure the entire population. It therefore goes that the sample is used to infer the most likely estimates of the population's exposure. Using that sample, we can calculate the corresponding sample characteristic which we then use to infer information about the unknown population. The population characteristic of interest is called a *parameter* and the corresponding sample characteristic is the sample *statistic*.

If the exposure distribution is approximated by a lognormal distribution, as is usually the case, there are several methods for estimating the arithmetic mean of the lognormal distribution and for calculating the confidence limits. *Remember that the best predictor of dose is the exposure distribution's arithmetic mean, not the geometric mean.* The general technique is to:

1. Calculate an estimate of the exposure distribution's arithmetic mean.
2. Characterise the uncertainty in the arithmetic mean's point estimate by calculating confidence limits for the true mean.

3. Examine the arithmetic mean's point estimate and true mean confidence limit(s) in the light of an OEL or other information to make a judgement on the exposure profile. Because of the inherent variability of workplace concentrations, statistically guaranteeing that all exposures are below an OEL is usually impossible; however, demonstrating statistically that no more than a given percentage of exposures are >OEL with some confidence is possible.

If the data do not seem to fit either the normal or lognormal distribution, the data should be examined to determine whether the similar exposure group (SEG) has been properly defined or whether there has been some systematic change to the underlying exposure distribution while the monitoring data were being gathered. Many statistical methods rely on the assumption that the measurements are drawn randomly from a distribution that is identical each time the measurements are drawn. If no reason can be found for splitting the data so that it represents two or more SEG exposure profiles that fit either the lognormal or the normal distribution, other options such as Bayesian methods may be needed, if the data confidently represents a single SEG.

To commence, say we have a set of 8-hour time weighted average results for diesel particulate monitoring for a maintenance similar exposure group (SEG) where the occupational exposure limit (OEL) is 0.1 mg/m^3. We have 16 monitoring results as shown in Table 4.1.

First, we need to determine the single value that represents average exposure of the group, the middle or centre of the exposures. This will most commonly be the geometric mean or arithmetic mean.

The **geometric mean** is always the lower estimate of the central tendency in a lognormal distribution and in our case equals 0.025 mg/m^3, which is not at the top point of the distribution.

In a lognormal distribution, the size of the highest results (in our case 0.088 mg/m^3) has a greater effect on the **arithmetic mean** and in our case results in a value of 0.034 mg/m^3. The calculation of the arithmetic mean assumes that several conditions are met, one of which is that the data is normally distributed, which in the case of exposure data is generally acknowledged not to be the case.

TABLE 4.1
Elemental Carbon Exposure Measurements

Elemental Carbon (mg/m^3)	
0.059	0.013
0.083	0.037
0.024	0.015
0.088	0.022
0.029	0.027
0.005	0.061
0.015	0.044
0.009	0.008

Note that this set of results has most of the numbers low with two values (0.088, 0.083 mg/m^3) being quite a distance from the next lowest value (0.061 mg/m^3). This creates a distribution of results which is skewed to the right and follows a lognormal shape. Our data is represented in Figure 4.3, where most of the results are likely to be less than 0.02 mg/m^3, but there is a decreasing probability that we might experience a result >0.02 mg/m^3, and a small (but not zero) chance of experiencing a result >0.1 mg/m^3.

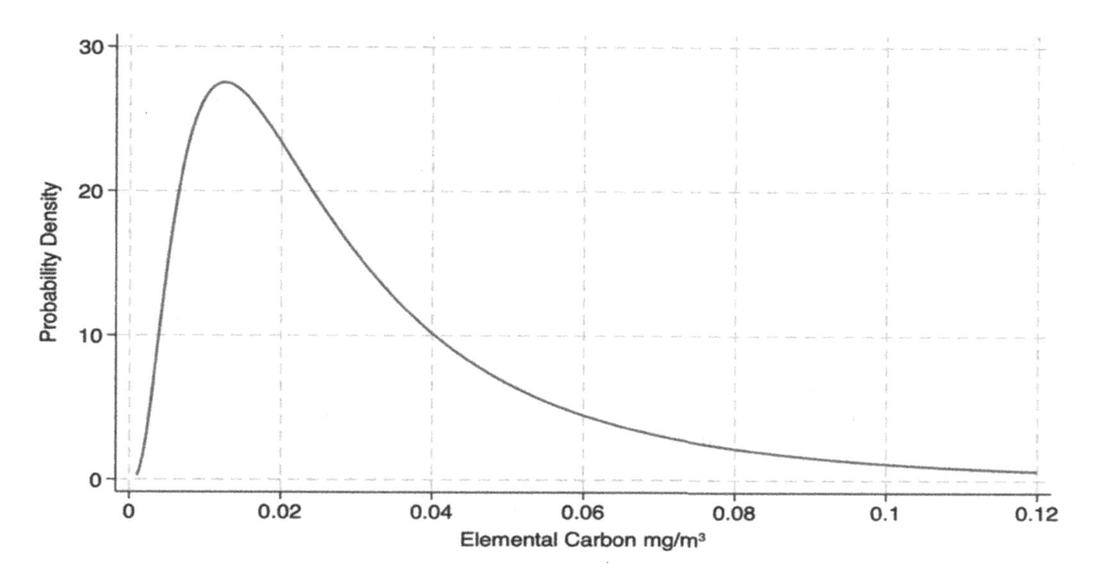

FIGURE 4.3 Distribution of elemental carbon exposures from Table 4.1.

Source: P. Knott 2023.

Because the arithmetic mean can be biased when used on a skewed distribution, there are alternative methods which can calculate an *unbiased* estimator, in our case, it is the *minimum variance unbiased estimator of the arithmetic mean*, which is commonly termed MVUE AM. It represents the estimate of the population mean from which our sample is drawn:

$$\textbf{MVUE AM} = 0.034\ \text{mg}/\text{m}^3$$

The **MVUE AM** is not the same as the arithmetic mean and uses a complex iterative formula to calculate. In this example, the MVUE AM and the sample arithmetic mean are the same but can be considerably different in other data sets.

This represents our best estimate, but we should also acknowledge the uncertainty in our estimate and do this by calculating the confidence limits around our estimate. In this case, the upper estimate is:

$$\textbf{95\% UCL} = 0.061\ \text{mg}/\text{m}^3$$

'**95% UCL**' is an abbreviation of *Lands exact one-sided 95% upper confidence limit of the arithmetic mean* (Land 1971) sometimes written as '$\mathbf{UCL_{1,95\%}}$' where the 1 refers to one-sided and 95% is the 95th percentile of the distribution. If we look at our example above, this indicates that if we collected 100 sets of samples, the mean of 95 of these sets would be less than 0.061 mg/m^3, but 5 out of 100 sets would have a mean > 0.061mg/m^3. Lands exact is dependent on the data closely fitting a lognormal distribution and having a moderate amount of variability. If these assumptions are not met and *n* is small, the 95% UCL can be unrealistically high given the range of our measurements.

When we come to quantifying the spread of our exposures, several measures can describe this parameter.

Range is the difference between the highest and the lowest readings in the monitoring results, so we know that all the other results fit between these two numbers; in our case, it is 0.083 mg/m^3.

The **Geometric Standard Deviation** (GSD) is a dimensionless value which describes the spread of a lognormal distribution; when the GSD = 1, all values are equal; when the GSD = 1.4, the distribution resembles a normal distribution; when the GSD = 2, it is slightly skewed. The higher the GSD, the more the data is spread or variable. Historically, a GSD > 3 is considered highly variable and warrants closer examination. The sources of variability can come from the process, workers, location or even the methods of sampling and analysis themselves. In our case, the GSD is 2.35, which indicates a moderate amount of variability.

95th Percentile – this is the value in the distribution above which 5% of all exposures are likely to be. In our case, it is 0.101 mg/m^3.

The **exceedance fraction (EF)** is the proportion of the exposure distribution which is greater than the OEL. It can be described in terms of % of all possible exposures, or more simply the probability of any individual sample being > OEL. In our data set, the EF is 5.3%.

The GSD, 95th percentile and exceedance fraction (EF) also have a level of uncertainly associated with them, which can be calculated to varying degrees of confidence commonly 95% or 70%.

So, what is being said here is:

> For our work, 16 samples were taken and the estimated arithmetic mean of all exposures in the population was 0.034 mg/m^3. We are also confident that 95% of all exposures in this group would return a result less than 0.101 mg/m^3, which is just above the OEL of 0.1 mg/m^3. Correspondingly 5.3% of all exposures are likely to be >OEL. We also know that our highest result was 0.088 and our lowest 0.005, and as our GSD is 2.35, the results were not too variable.

In occupational hygiene, chemical exposure measurements are commonly continuous values, which are compatible with the statistical approaches described above. The measurement of noise exposures expressed as decibels may appear to be a continuous value and is in fact a relative value, being the ratio of logs of pressure (Pa) between a reference level and the measured value. Because of this characteristic, when conducting any statistical assessment of noise exposure measurement, the measured pressure value should be used and then the summary statistic or parameter of the pressure value is converted back to a decibel and reported. In practice, this involves obtaining the Pascal squared hour value (Pa^{2}hr) of the noise exposure, calculating the statistic from the sample of Pa^{2}hr values and then converting the relevant value to a decibel (dB).

4.3.1 Between/within Worker Variance

As previously noted, the distribution of worker exposures is typically assumed to be lognormal and the aggregation of workers into similar exposure groups leads to the overall distribution of results in a SEG being our evaluation of compliance or calculation of exposure parameters for the SEG. Since the SEG is a compilation of individual workers, so too is the overall distribution of results in a SEG being a compilation of the distributions of all the individuals in the SEG. For each of these workers, factors which create variation in every exposure instance are the product of many sources including the type and the amount of contaminant, generation processes, location, wind speed and direction, temperature, effectiveness of controls, tasks and individual work practices.

In our sample of 16 elemental carbon exposures, the MVUE AM was calculated to be 0.034 mg/m^3, with a sample geometric standard deviation of 2.35, which by most compliance measures would indicate that the SEG is compliant with an OEL of 0.1 mg/m^3.

After adding the workers associated with these individual measurements (Table 4.2), the overall compliance measures do not provide any indication of the individual workers average exposure

TABLE 4.2
Elemental Carbon Exposures by Worker

Elemental Carbon (mg/m³)	Worker
0.059	1
0.083	1
0.024	1
0.088	1
0.029	2
0.005	2
0.015	2
0.009	2
0.013	2
0.037	3
0.015	3
0.022	3
0.027	3
0.061	4
0.044	4
0.008	4

or variation. Since we are dealing with the health of workers and not SEGs, it would be useful to understand if any worker is potentially higher exposed than others in the SEG.

This point was observed by Oldham and Roach who noted *'It was found that significant variation was occurring in the dust concentrations from one collier's experience to another's, and from one day to another in the same collier's experience'* (Oldham & Roach, 1952, p. 117). In their final summary, the importance of the effects of variability of exposure both in duration and in intensity as well as using exposure estimates from the actual workers at risk were recommended for studies of pneumoconiosis.

Exposure variability was noted by Liedel (1977) in relation to the probability of individual worker overexposure following a TWA measurement on a worker as a fraction of the OEL. Three types of variability are described:

1. inter-operator (between workers in the same job category);
2. intraday (between samples taken on one shift on one worker); and
3. inter-day (between daily exposures averages on the same worker).

Sources of variability were then revisited by Kromhout et al. (1993) in a series of analyses and the concept of between-worker and within-worker variation became firmly established in the occupational hygiene field (Rappaport et al. 1993).

In relation to our diesel particulate samples, Table 4.3 displays the results of a simple ANOVA of the log-transformed results indicating that there is a significant difference ($p = 0.038$) between the workers exposures. Here the between groups and within groups are between workers and within workers, respectively.

Where the difference exists is not shown by the ANOVA, additional analysis using a post hoc comparison shown in Table 4.4 can be used to identify the workers whose exposure distributions are different. In this case, the source of the difference is between the exposure profile of worker 1

TABLE 4.3
ANOVA of EC Measurements by Worker

Source	Sum of Squares	df	Mean Square	F	p
Between groups	5.382	3	1.794	3.863	0.038*
Within groups	5.573	12	0.464		
Total	10.955	15	0.730		

* $p < .05$, ** $p < .01$
Note: Results based on uncorrected means.

TABLE 4.4
Multiple Comparison of EC Results by Worker

Dunnett Post Hoc Comparisons – Worker

Comparison (Worker – Worker)	Mean Difference	SE	t	$p_{dunnett}$	
2 – 1	−0.049	0.013	−3.747	0.008	**
3 – 1	−0.038	0.014	−2.758	0.044	*
4 – 1	−0.026	0.015	−1.725	0.250	

* $p < .05$, ** $p < .01$
Note: Results based on uncorrected means.

and worker 2 ($p = 0.008$), and worker 1 and worker 3 ($p = 0.044$), with the direction of the mean difference showing worker 1 exposures are higher than the others.

In Figure 4.4, it is visually shown that the distribution of each workers exposure shows worker 1 having a much higher probability of experiencing individual exposures >OEL than the other workers in the SEG.

It would then be the focus of the occupational hygienist to ascertain the factors which contribute to the propensity of worker 1 to be more highly exposed than others in the same SEG.

4.3.2 Censored Data

Where an exposure measurement is less than the limit of detection (LOD) of the instrument used to measure a contaminant, the result is typically reported to be < LOD. Statistically, these values represent a non-measurable exposure with an uncertain value between zero and the LOD. The real value of the measurement could be just a fraction below the LOD, or it could be a true zero. How do we address these uncertainties in our data?

The term limit of detection is commonly used but sometimes in practice refers to the limit of quantitation (LOQ) or limit of reporting (LOR), which represent amounts of analyte which can be quantified to a predetermined level of uncertainty (ISO 1997). As this is a property of the analytical method, variations in sampling volume for air contaminants may then lead to multiple LOD in a final air concentration data set.

Less commonly, results may be obtained which are greater than the capacity of the analytical method to quantify an exposure. This may be due to sample overloading, for example, thermal desorption tubes, or colorimetric detector tubes or results greater than the calibration ranges of direct reading instruments, for example, gas detectors.

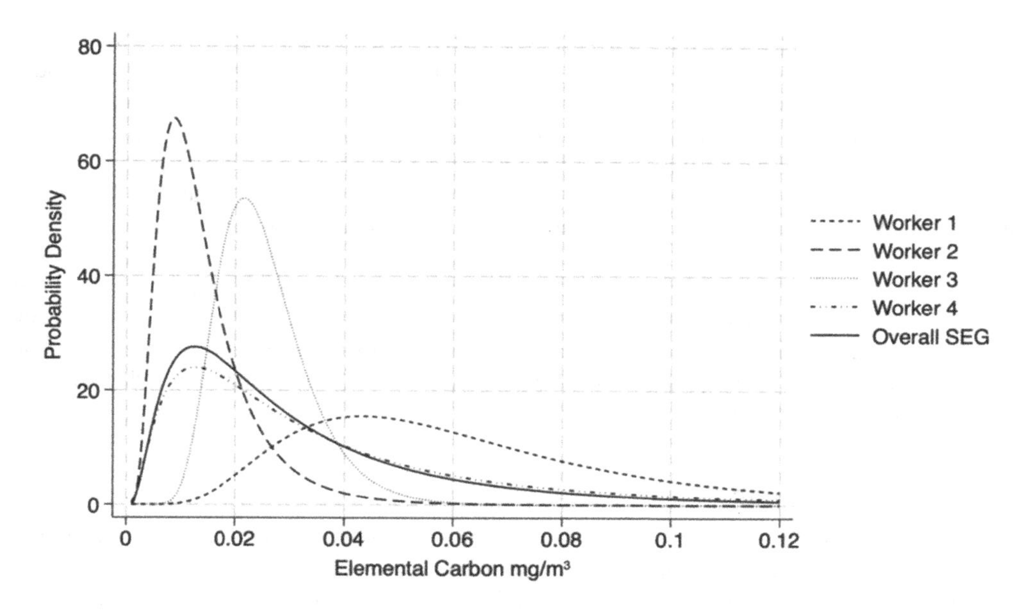

FIGURE 4.4 Distributions of EC exposures by individual worker and overall SEG.

Source: P. Knott 2023.

Results reported as being below a quantifiable limit are termed *left-censored*, and conversely results which may be reported as being above method limits are termed *right-censored* (Shoari & Dubé 2018). In rare situations, results may be *interval censored*, which refers to a result where the lower limit and the upper limit are known but the actual value between these is not.

How these types of results are dealt with in occupational hygiene is dependent on their abundance in a data set and the aim of the statistical analysis. Censored results should be taken into consideration in any analysis and should not be removed or ignored (Hewett & Ganser 2007).

The simplest method and one most prone to bias is replacing censored values with a fixed value, commonly applied to left censored values and replaced with LOD/2. So, in the case of a result of <0.1, a value of 0.05 is substituted in its place. Where there is a small proportion of left censored values in a data set (<30%), some authors (Hewett & Ganser 2007) suggest that this approach will not lead to excessive bias. However, a replacement method that substitutes a censored value with a uniform single value can lead to overestimates of the mean and underestimates of spread, which can lead to bias in compliance decisions.

Several methods are available, using free of charge software, to address censored data in a manner which reduces bias. The accessibility and ease of use of these tools eliminate any reasons to use substitution methods in the future. Even in the absence of software, manual plotting methods exist which interpolate censored data in a distribution (Hewett & Ganser 2007).

4.4 PROFESSIONAL JUDGEMENT

Professional judgement plays a crucial role in any field in which decisions must be made in the absence of a complete data set (O'Hagan 2006). Medical professionals, pilots, financial analysts and occupational hygienists all use professional judgement to facilitate their decision-making.

For occupational hygienists, professional judgment is commonly used to assess exposures where monitoring data is limited or not yet available, and accurate exposure judgments are the foundation of efficient and effective exposure management (Lowry et al. 2022). Research to date has indicated that this 'art' of making exposure judgments is some combination of professional experiences, educational background and other unknown factors. A key factor relating to accuracy is cognitive biases which may present when a hygienist is trying to interpret lognormal distributions (Vadali et al. 2012). When reviewing these distributions, mental shortcuts, known as heuristics, are often used which can lead to errors in judgment and introduce bias. Using heuristics leads to a pattern that assigns weights to decisions that differ from the true probabilities of these outcomes. Improbable outcomes are over-weighted, while outcomes that are almost certain are under-weighted, leading to a pattern of inaccurate thinking that may 'under' or 'over' estimate true exposures. Occupational hygienists should be aware of such biases that may influence their day-to-day exposure decision-making.

4.4.1 Bayesian Methods

Consider a situation where an experienced occupational hygienist conducts an exposure assessment of a workplace they have not previously attended; however, the process conducted in the workplace is familiar and they have conducted many quantitative assessments in other locations with the same process. Under a conventional statistical framework, we would assume the distribution of exposures in the workplace is completely unknown and a set of measurements collected from which sample statistics are calculated and estimates of the population parameters obtained with 95% confidence limits.

If our hygienist employed a Bayesian approach to this scenario, they could use their prior knowledge to add additional information to their assessment, supporting the measurements they collect and calculate the probability of the most likely exposure distribution for the workplace. How this is done is by using Bayes Rule, a calculation based on conditional probabilities (probability of an event (A) occurring, given something else (B) has already occurred), which can consider our previous experience and judgement (Banerjee et al. 2014; Lavoué et al. 2019). It can be simply explained in this context as:

1. There is a new exposure situation to be explained/described.
2. Before collecting measurements, we have some prior idea of what the exposure situation is likely to be.
3. We collect some new data (measurements) and then shift our understanding towards the exposure situation which better accounts for the data.

Fundamental to Bayesian methods are the concepts of the Prior, Likelihood, and Posterior probabilities. **Prior** distributions are probability distributions of exposure parameters (GM, GSD, etc.) based on some a *priori* knowledge about the parameters. Prior distributions are independent of measurements.

These can be from:

- Same SEG/same site/different years – past years results.
- Same SEG/different site/different years – past years results from a comparable site.
- Similar SEG/different site/different years – past years results from similar operations.
- Walk thru survey observations.
- Published literature.

The **likelihood** is the distribution of collected measurements. The **posterior** is the updated probability exposure distribution of our parameters after applying Bayes rule. The outputs from Bayesian inference are probabilities of events occurring such as the probability of the arithmetic mean >OEL or the probability of an individual exposure > OEL, which are capable of being communicated in simple terms, that is, there is a 7% chance of an overexposure in this SEG. Bayesian methods in occupational hygiene have been adopted by the American Industrial Hygiene Association (AIHA) with several tools available for the occupational hygienist.

4.4.2 Communication of Results and Associated Risk

The pursuit of taking measurements according to scientific methods and the statistical evaluation of the results to obtain a conclusion in relation to worker health risks is diminished if practitioners are unable to effectively communicate the results of their efforts to those with the ultimate responsibility and accountability for protecting worker health. Managers of work sites are confronted with a multitude of competing priorities on a day-to-day basis and may often be looking for a simple binary response against risk (good/bad, compliant/non-compliant). Experts and non-experts (i.e., the public, members of the workforce) will often make decisions about risk in very different ways, as shown in Figure 4.5.

Risk communication is a discipline in and of itself, and although not covered at length in this chapter, the link between statistical evaluation of exposure and assessment of risk to communicating the implications of the assessment to the workforce is an important one. The role of the occupational hygienist as a risk communicator is challenging and should not be understated because most of the public are apathetic about most risks, and it is extremely difficult to generate concern (Sandman 1993).

However, when people are concerned about a risk, it is also extremely difficult to calm them down and provide reassurance.

What does this mean for the practising occupational hygienist? First, in any high-outrage risk issue, the chief task of communication is to address the outrage, not to state or debate assessments of the hazard itself. The best antidote for outrage is to build sustainable trust through listening to

FIGURE 4.5 Factors underpinning 'expert' and 'public' views of risk.

Source: D. Lowry 2023.

concerns, echoing what has been heard, asking questions, finding common ground and finding areas to respectfully voice reservations to the argument.

Second, occupational health risk is imprecise and probabilistic in nature because of variation between workers and changes in day-to-day exposure. The intrinsic variability of workplace exposures means that a single day of monitoring is unlikely to provide a clear answer. This leans heavily on the occupational hygienist to communicate the uncertainty of risk as well as the most probable estimate in a manner to which the receiver is most likely to understand.

Considerable work has gone into understanding the characteristics of risk communications (Lipkus 2007). This has led to the use of graphical representations of adverse outcomes which have been shown to be effective in communicating risk in public health and applied to workplace applications. Graphics showing the extent of a population affected such as icon arrays provide a clear concise representation of parameters such as the proportion > OEL (Stege et al. 2022).

4.4.3 Tools

4.4.3.1 American Industrial Hygiene Association IH_STAT

IH_STAT is an Excel-based spreadsheet which both tests underlying assumptions about an OH data set and calculates sample statistics and parameter estimates. Developed in conjunction with the AIHA reference text 'A Strategy for Assessing and Managing Exposure Occupational Exposures' (Jahn et al. 2015) IH_STAT is a widely used free tool for producing statistical assessment of single SEG, Single Contaminant, Single OEL data sets. The package calculates both normal and lognormal parameter estimates, there is no capacity for handling of censored results and needs greater than six samples for reliable parameter estimates.

IH_STAT_Bayes is a recently developed Excel-based version of IH_STAT which makes use of the Bayesian priors developed for the Expostats web package. The package provides parameter estimates, graphical outputs such as box and whisker plots as well as a bar graph showing the probability of mean exposures within the AIHA exposure risk bands.

4.4.3.2 Expostats

A web-based package designed specifically for industrial hygiene statistical data analysis (Expo Stats 2023) (https://expostats.ca/site/en/index.html). Exposure data is either uploaded or typed into fields for analysis. The core of Expostats is a Bayesian calculation engine making use of historical prior data sets accessed from a graphical interface. Expostats contains three tools which conduct analysis of exposure measurements including censored results to determine compliance to an OEL, determine compliance to an OEL incorporating within and between worker variation and assessment of the effects of determinants of exposure.

4.4.3.3 IH Data Analyst (IHDA)

A dedicated analysis package which conducts both descriptive and compliance statistics as well as Bayesian decision analysis (BDA) incorporating treatment of censored data (https://www.easinc.co). The content follows AIHA analysis strategies, with the BDA either customisable or using default priors. Available in a free-ware and fully featured version.

4.4.3.4 BWStat

This web-based application developed by the Belgian Society for Occupational Hygiene executes the BOHS/NVvA guidance on 'Testing Compliance to Occupational Exposure Limits' (Ogden et al. 2011) and EN689:2018 (CEN 2018). Data is uploaded as a file and then analysis is executed. A wide range of graphical and statistical outputs are generated with specific compliance/non-compliance statements made relevant to the guidance and EN standard.

4.4.3.5 R package 'STAND'

An open-source package which conducts statistical analysis of IH data according to AIHA conventions with a variety of procedures for dealing with censored data. The STAND package (Frome & Frome 2015) is accessible through R, a free statistical and computing language and environment which runs on a wide variety of systems including Linux, Windows and MacOS.

4.4.3.6 NDExpo

A free web-based application on the Expostats website (https://expostats.ca/site/en/index.html) (Expo Stats 2023) which conducts regression on order statistics for a censored data set and calculates an alternative interpolated data set suitable for further statistical analysis.

4.4.3.7 ProUCL 5.2

A free software package from the United States Environmental Protection Agency (USEPA 2022) for analysis of environmental data sets with and without censored observations (https://www.epa.gov/land-research/proucl-software). The package contains graphical and statistical methods applicable for handling lognormal exposure data with a range of estimators commonly used in occupational hygiene. The software runs only in Windows environments.

4.4.3.8 JASP

Not strictly occupational hygiene software, JASP (JASP Team 2023) is an open-source statistics platform from the University of Amsterdam which uses a graphical interface to provide end users with a choice of either Bayesian or frequentist statistical analysis using an underlying R framework. The output from JASP is annotated to provide additional understanding of results and provides end users with access to high-quality graphics and statistical tables in an easy to use drag and drop interface.

4.5 A COMPLIANCE DECISION-MAKING STRATEGY – WHICH APPROACH TO ADOPT?

The regular statistical tools discussed in this chapter and used by occupational hygienists and other H&S practitioners for examining monitoring survey data provide a number of measures that might be considered useful for comparing with an OEL to make compliance decisions. The measures of interest are the MVUE AM, the GM, the 95% UCL of the GM, the GSD and the 95th percentile of a lognormal probability distribution. However, knowing how to calculate these parameters is not the same as knowing how to use them in making a compliance decision.

4.5.1 What Should Constitute Compliance?

Figure 4.6 demonstrates what a compliance decision will infer about the (stylised) probability distribution of monitoring results with respect to the OEL when each of the five different criteria for making a compliance decision is employed. In this example, the monitoring results exhibit a GSD of about 2. Note that, at present, different organisations and regulators are at liberty to choose whichever one of these compliance parameters they might consider most appropriate. The five, corresponding to Examples 1 to 5 on the following page, are:

1. all measurements to be below the level of the regulatory OEL;
2. the GM of measurements to be below the level of the regulatory OEL;
3. the MVUE AM of measurements to be below the level of the regulatory OEL;
4. the 95% UCL of the MVUE AM to be below the level of the regulatory OEL;
5. the 95th percentile of the probability distribution (represented by the few measurements that one can afford to make) to be below the regulatory OEL.

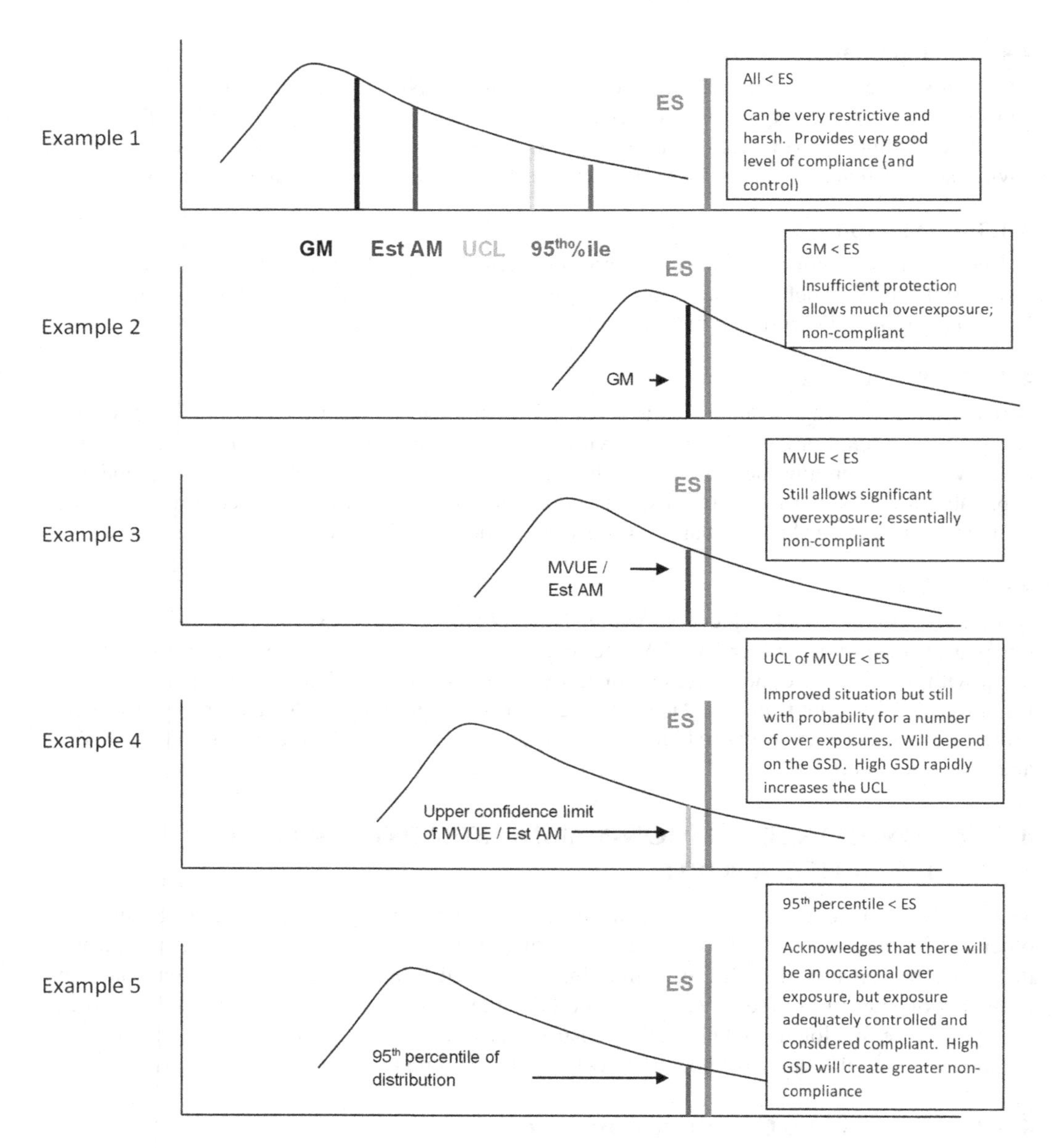

FIGURE 4.6 Outcomes arising for making compliance decisions based on different interpretations from monitoring data.

Source: Grantham & Firth 2014, p. 84.

Some of the outcomes in terms of the practicability of meeting compliance with each option are given. A number of the options will permit excessive exposures to occur.

From these options for compliance, Examples 2 and 3 must be rejected, since they permit the possibility of far too many instances of overexposure when applied to group measurements. An example of averaging is found in the Queensland Coal Mining Safety and Health Regulation (2017) to indicate compliance, though it is not specified whether the average is a GM or an Est.

AM, and these can be quite different depending on the GSD. This averaging system can confirm compliance only where all measurements relate specifically to one individual.

Example 4, using the 95% UCL of the AM to be <OEL, is employed by many organisations but will still permit a proportion of exposure measurements to be greater than the OEL. That problem is exacerbated where the probability distribution relates to a group of workers but some of those workers may consistently constitute the non-compliant tail. Bad performers within a group cannot be traded off against good performers. Nevertheless, the 95% UCL is applicable when applied to OELs for chronic toxicants (Grantham & Firth 2014).

A compromise between the stringency of Examples 1 and 4 is to have the 95th percentile to be <OEL (as in Example 5). Adoption of the 95th percentile being <OEL is a balance between the uncertainty inherent in taking a small number of measurements and the need for a practical and acceptable yardstick for making correct compliance decisions. If 95 per cent of the exposure profile represented by the probability distribution lies <OEL, then there is a high probability that the exposure profile will not exceed the exposure standard. This decision must necessarily be conservative, since the derivation of the OEL itself will also have some uncertainty associated with it. As multi-tasking has increased in the Australian workforce, so the degree of exposure variability (GSD) has increased such that when using the 95th percentile, most exposures appear to not be in compliance.

4.5.2 What Other Strategies Are There for Judging Compliance Against an OEL?

When making decisions about compliance where a series of measurements have been made on a group of workers, there are two considerations:

- Is the whole group compliant?
- Are all individuals in the group also compliant?

These two considerations have led to the development of several strategies to answer both questions. Each strategy aims to achieve this by taking a minimum of measurements. The two simple methods that follow have some serious limitations but can be used without the need for any complex statistical calculation. A modern robust approach is also outlined in EN 689: 2018 (CEN 2018), though its complete methodology cannot be fully reproduced here.

4.5.2.1 Simple Methods

One of the simplest methods proposed for evaluating compliance with regulatory standards for a group of workers all involved in a similar task is to use the NIOSH compliance strategy as follows: typically, the maximally exposed worker in a group is identified and monitored for exposure to environmental agents. If that personal exposure is below the standard, all worker exposures are also presumed to be compliant with the regulatory exposure standard (Jahn et al. 2015).

This approach requires prior advice so this ‘worst case’ or maximally exposed worker can be differentiated from all other workers. Professional input, perhaps aided by measurements made with a direct reading instrument, may assist in identifying that worker.

Similar to the NIOSH strategy, the HSE (2022) shortcut method (using measurements <OEL/3) is an approach that includes more of the exposed workers.

Simple methods such as these are applicable only in situations where compliance is highly probable. Where it is doubtful, more advanced decision-making methods are needed.

4.5.2.2 Advanced Methods – Group and Individual Compliance

In practice, making measurements to determine compliance faces two main constraints. To keep costs low, monitoring programs need to be as economical as possible in terms of the number of samples taken. But the sampling regime must be designed to take into account all the factors which typically impart variability to its measurements. Most sampling exercises limit themselves to a cross-sectional study over one or two days, thus missing or discounting many of the variables (e.g. environment and variations in production schedule) that contribute to an exposure profile. Further, unless workers are individually sampled repeatedly, most group sampling assists only in making broad compliance decisions about the group as a whole.

EN 689: 2018 meets these challenges. The process can be economical in terms of the number of measurements required, and it accommodates the variability issues through a program of structured and timed campaigns. It differs from the conventional compliance decision, which uses the 95th percentile of the probability distribution for a group, in that it can address differences both within and between workers, and it is based on the concepts of group and individual compliance with an OEL.

The use of the method depends on valid measurements and a correctly defined SEG. It is a sequential process, in which its simple first steps are intended to provide:

- an indication of compliance if all results are low (<0.1 OEL); or
- immediate implementation of controls if there is evidence of possible non-compliance (one result >OEL); or
- the undertaking of two further monitoring campaigns if uncertainty remains (any results > 0.1 OEL but <OEL).

Group compliance is determined by a simple calculation based on a minimum of around nine structured samples from three campaigns six samples using the technique set out in the French law on regulatory compliance EN 689:2018. However, group compliance will not necessarily imply individual compliance and may disguise individual non-compliance if there are idiosyncratic worker effects operating, even though in the same SEG. Individual compliance can be examined by using an analysis of variance (ANOVA) of the means of measurements within and between workers.

The two measures for specifying compliance in EN 689:2018 are:

- **Group compliance**: the group complies if, with 70 per cent confidence, <5 per cent of the exposures in the SEG exceed the OEL.
- **Individual compliance**: the SEG complies in terms of individual exposure if there is <20 per cent probability that >5 per cent of the exposures of any individual exceed the OEL.

The methodology is not reproduced here, but all analysis can be conveniently carried out using Microsoft Excel software (refer Section 4.4.3.4). Calculations are based on the use of the GM, the GSD and the OEL.

It is recommended that occupational hygienists and others required to undertake compliance analysis on sampling data acquaint themselves with this method and the ANOVA tools. Procedures are also provided for making the calculations if a limited amount of data is missing because of work changes, worker absences and the like. Benefits include:

- simple protocols;
- asingle compliance decision-making criterion for groups;
- suitability for regulatory applications;

- the ability to examine for the likelihood of individual non-compliance without having to sample each worker;
- considerably minimised sample set (at the expense of an increased number of campaigns);
- an immediate feedback mechanism indicating when control is required;
- an approach that can also be applied when using STELs.

4.6 CHAPTER SUMMARY

Understanding statistics comes fundamentally from the practical application of statistics as they apply to exposure assessment. This chapter serves as a brief overview on what is currently on offer to the practising occupational hygienist; however, this topic could serve as a standalone book given the different test cases that hygienists may find themselves in where they need to apply a level of statistical treatment to a dataset. The statistical parameters explained within this chapter can and should be thought of as a minimum level of understanding required to be able to analyse sample results statistically. An important point to consider is the following – occupational hygienists are not statisticians. Hygienists use statistics only as a tool in the toolkit of protecting and optimising worker health. Statistics are generally used first to understand the nature of measured results, and, second, to predict what might happen if the measurements were to be repeated in the same or similar circumstances (Dewell 1989). But prior to statistical treatment of data, it is also important to ascertain the objectives of the sampling in order to identify potential biases in the data. For example, if the objective was to sample only well-controlled facilities, then the results would probably not represent the exposure in the industry as a whole. If the monitoring resulted from worker complaints, then exposures may not represent typical exposures. If the monitoring was conducted to evaluate engineering controls or as a preliminary screening of exposure, the results may not represent actual employee exposure. It is important that all potential variables be identified and evaluated, and this is balanced against the application of any statistical testing to the data (US EPA 1994).

REFERENCES

Banerjee, S., Ramachandran, G., Vadali, M. & Sahmel, J. 2014, 'Bayesian hierarchical framework for occupational hygiene decision making', *Annals of Occupational Hygiene*, vol. 58, no. 9, pp. 1079–1093. doi: 10.1093/annhyg/meu060

Bullock, W. H. & Ignacio, J. S. 2006, *A Strategy for Assessing and Managing Occupational Exposures*, Fairfax, VA: AIHA Press.

Dewell, P. 1989, *Technical Handbook Series No. 1: Some Application of Statistics in Occupational Hygiene*, BOHS, British Occupational Hygiene Society Science Reviews Ltd with H & H, Scientific Consultants, Leeds, UK.

CEN 2018, *European Standard EN689: Workplace Exposure – Measurement of Exposure by Inhalation to Chemical Agents – Strategy for Testing Compliance with Occupational Exposure Limit Values*, Comite Europeen de Normalisation, Brussels.

Expo Stats 2023, *Expostats Bayesian Calculator, Statistical Tools for the interpretation of Industrial hygiene data*, https://expostats.ca/site/en/index.html [6 January 2024].

Frome, E.L. & Frome, D. P. 2015, *STAND: Statistical Analysis of Non-Detects*: R package version 2.0, https://CRAN.R-project.org/package=STAND [15 January 2024].

Grantham, D. & Firth, I. 2014, *Occupational Hygiene Monitoring and Compliance Strategies*, Australian Institute of Occupational Hygienists (AIOH) guidebook, www.aioh.org.au/product/monitoring-pdf/ [15 January 2024].

Hewett, P. & Ganser, G. H. 2007, 'A comparison of several methods for analyzing censored data" *Annals of Occupational hygiene*, vol. 51, no. 7, pp. 611–632.

Health and Safety Executive (HSE) 2022, *Exposure Measurement: Air Sampling, COSHH Essentials General Guidance G409*, HSE, London, www.hse.gov.uk/pubns/guidance/g409.pdf [6 January 2024].

Huynh, T., Ramachandran, G., Banerjee, S., Monteiro, J., Stenzel, M., Sandler, D. P., Engel, L. S., Kwok, R. K., Blair, A. & Stewart, P. A. 2014, 'Comparison of methods for analyzing left-censored occupational exposure data', *Annals of Occupational hygiene*, vol. 58, no. 9, pp. 1126–1142. doi: 10.1093/annhyg/meu067

ISO 1997, *Capability of Detection-Part 1: Terms and definitions, ISO 11843: 1997*, International Organization for Standardization. https://www.standards.org.au/access-standards

Jahn, S. D., Bullock, W. H. & Ignacio, J. S. 2015, *A Strategy for Assessing and Managing Occupational Exposures* (4th ed.). Fairfax, VA: AIHA Press.

JASP TEAM 2023. *JASP* (Version 0.17.3) [Computer software]. University of Amsterdam, https://jasp-stats.org [6 January 2024].

Kauppinen, T. P. 1994, 'Assessment of exposure in occupational epidemiology', *Scandinavian Journal of Work, Environment & Health*, vol. 20, pp. 19–29.

Kromhout, H., Symanski, E. & Rappaport, S. M. 1993, 'A comprehensive evaluation of within-and between-worker components of occupational exposure to chemical agents', *Annals of Occupational hygiene*, vol. 37, no. 3, pp. 253–270.

Land, C. E. 1971, 'Confidence intervals for linear functions of the normal mean and variance', *The Annals of Mathematical Statistics*, vol. 42, no. 4, pp. 1187–1205.

Lavoué, J., Joseph, L., Knott, P., Davies, H., Labrèche, F., Clerc, F., Mater, G. & Kirkham, T. 2019, 'Expostats: A Bayesian toolkit to aid the interpretation of occupational exposure measurements', *Annals of Work Exposures and Health*, vol. 63, no. 3, pp. 267–279. doi: 10.1093/annweh/wxy100.

Liedel, N. A. 1977, *Occupational exposure sampling strategy manual*, US Department of Health, Education, and Welfare, Public Health Service. https://www.cdc.gov/niosh/docs/77-173/pdfs/77-173.pdf?id=10.26616/NIOSHPUB77173 [6 January 2024].

Lipkus, I. M. 2007, 'Numeric, verbal, and visual formats of conveying health risks: Suggested best practices and future recommendations', *Medical Decision Making*, vol. 27, no. 5, pp. 696–713.

Lowry, D. M., Fritschi, L., Mullins, B. J. & O'Leary, R. A. 2022, 'Use of expert elicitation in the field of occupational hygiene: Comparison of expert and observed data distributions'. *PLoS One*, vol. 17, no.6. doi: 10.1371/journal.pone.0269704.

Nieuwenhuijsen, M. J. 2003, *Exposure Assessment in Occupational and Environmental Epidemiology*, OUP, Oxford.

O'Hagan, A., Buck, C. E. & Daneshkhah, A. 2006, *Uncertain Judgements: Eliciting Experts' Probabilities*, John Wiley & Son, Chichester.

Oldham, P. D. & Roach, S. A. 1952, 'A sampling procedure for measuring industrial dust exposure', *British Journal of Industrial Medicine*, vol. 9, no. 2, pp.112–119.

Ogden, T., Kromhout, H., Hirst, A., Honnes, K. Ingle, J., Rooij, J., Kennedy, A., Scheffers, T., van Tongeren, M. & Tielemans, E. 2011, *Testing Compliance with Occupational Exposure Limits for Airborne Substances. BOHS& NVvA*, https://www.arbeidshygiene.nl/-uploads/files/insite/2011-12-bohs-nvva-sampling-strategy-guidance.pdf [15 January 2024].

Ramachandran, G. 2008, 'Toward better exposure assessment strategies—The new NIOSH initiative', *Annals of Occupational Hygiene*, vol. 52, no. 5, pp. 297–301. doi: 10.1093/annhyg/men025.

Rappaport, S., Kromhout, H. & Symanski, E. 1993, 'Variation of exposure between workers in homogeneous exposure groups', *American Industrial Hygiene Association Journal*, vol. 54, no. 11, pp. 654–662.

Sandman, P. M. 1993, *Responding to Community Outrage: Strategies for Effective Risk Communication*, American Industrial Hygiene Association, Fairfax, VA.

Shoari, N. & Dubé, J. S. 2018, 'Toward improved analysis of concentration data: Embracing nondetects', *Environmental Toxicology and Chemistry*, vol. 37, no. 3, pp. 643–656. doi: 10.1002/etc.4046.

Stege, T., Bolte, J., Claassen, L. & Timmermans, D. 2022, 'Risk communication about particulate matter in the workplace: A digital experiment', *Safety Science*, vol. 151. doi: 10.1016/j.ssci.2022.105721.

Stewart, P. & Stenzel, M. 2000. 'Exposure assessment in the occupational setting', *Applied Occupational and Environmental Hygiene*, vol. 15, no. 5, pp. 435–444. doi: 10.1080/104732200301395.

U.S. Environmental Protection Agency 1994, *Guidelines for Statistical Analysis of Occupational Exposure Data*, USEPA, Washington, DC, https://www.epa.gov/sites/default/files/2015-09/documents/stat_guide_occ.pdf [6 January 2024].

U.S. Environmental Protection Agency 2022, *ProUCL: Statistical Software for Environmental Applications for Data Sets with and without Nondetect Observations*. Version 5.2, https://www.epa.gov/land-research/proucl-software [14 January 2024].

Vadali, M., Ramachandran, G. & Banerjee, S. 2012, 'Effect of training, education, professional experience, and need for cognition on accuracy of exposure assessment decision-making', *Annals of Occupational Hygiene*, vol. 56 no. 3, pp. 292–304. doi: 10.1093/annhyg/mer112.

5 Control Strategies for Workplace Health Hazards

Dr Kerrie Burton and Associate Professor Kelly Johnstone

5.1 INTRODUCTION

As outlined in Chapter 1, there are four fundamental principles in occupational hygiene—**anticipation, recognition, evaluation**, and **control** of health hazards. *Control* is a key objective and is extremely important as it involves the actions required to eliminate or reduce exposure to workers. Control involves the application of technological, engineering, and operational measures directed at the health hazard, work environments, or workers to eliminate any adverse outcomes or reduce risk to acceptable levels. This chapter aims to explain each element of the hierarchy of controls (HoCs) and outline its application. More specific examples of control solutions are also provided in Chapter 6: Ventilation and Chapter 7: Personal Protective Equipment.

5.2 LEGAL REQUIREMENTS

Occupational Health and Safety (OHS) legislation in Australian jurisdictions and many other countries contains requirements for a general duty of care, together with some specific directives to manage chemical and other hazardous agents, such as asbestos, silica, lead, carcinogens, noise, and ionising radiation sources. There is a framework in this legislation, within which the obligations of various parties, including both employers and workers, are established. Persons conducting a business or undertaking (PCBU) have obligations to ensure the health and safety of workers by controlling hazards at the source. Workers also have obligations to cooperate with employers to maintain their own health and safety. The law imposes obligations regarding control of traditional safety hazards (guarding, electrical, prevention of falls, etc.) as well as to the more difficult aspects of occupational health and hygiene, such as preventing chemical exposures.

It is not acceptable to allow hazards to persist simply because it will cost too much to control them. It is equally not acceptable to leave an identified hazard in an uncontrolled state simply because workers seem prepared to tolerate it. In almost all workplaces, there is still room for improvement in the control of health hazards. Besides the regulatory imperative, it has been shown that good work health and safety improves long-term business productivity (Safe Work Australia 2022).

The expectation of OHS legislation in Australia and many other countries is that risk is controlled 'as far as reasonably practicable' or 'so far as is reasonably practicable'. Therefore, the highest level of protection that is reasonably practicable, commensurate with the risk posed by the hazard, should be used.

DOI: 10.1201/9781032645841-5

5.3 THE HIERARCHY OF CONTROL

Some control options are more effective than others, leading to an order or 'hierarchy', commonly referred to as the 'Hierarchy of Controls' [HoC], generally set out in the following order:

1. elimination;
2. substitution;
3. engineering controls;
4. administrative controls; and
5. personal protective equipment (PPE).

The elements of the HoC are shown in Figure 5.1, as well as the control effectiveness.

The most effective control strategy should be implemented in accordance with the HoCs. Control strategies will often involve more than one level of the HoC. Controls can be directed at

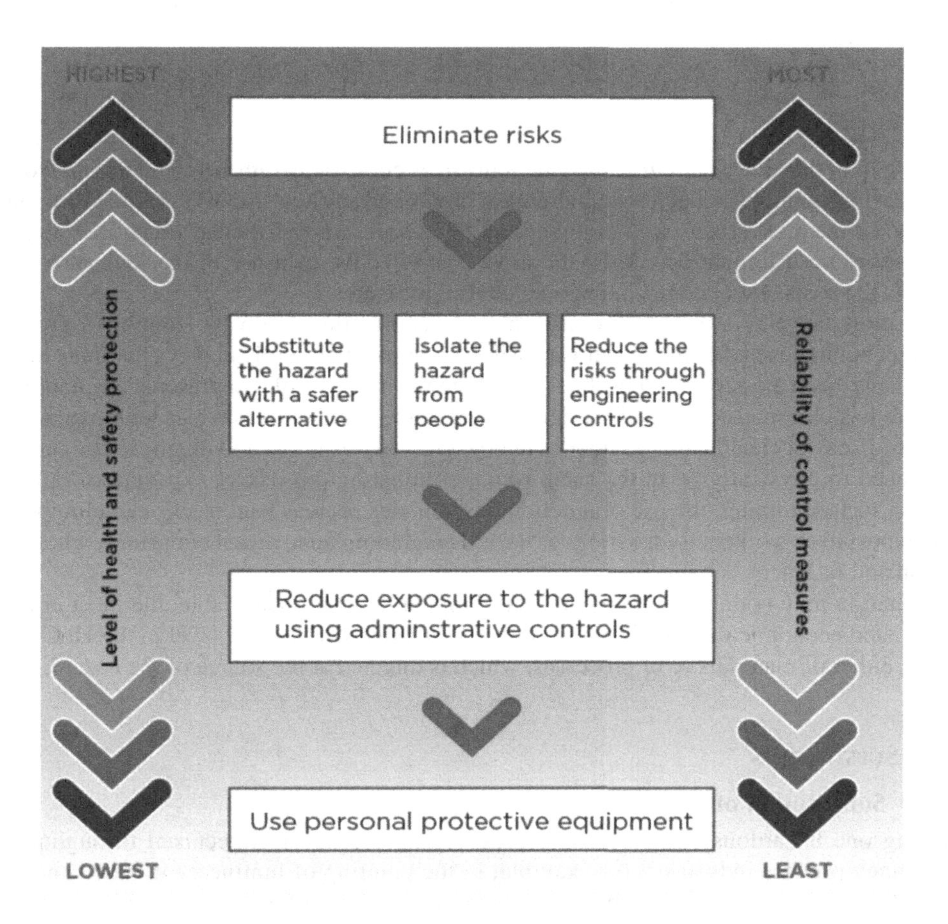

FIGURE 5.1 A representation of the hierarchy of control.

Source: Safe Work Australia 2018, p. 19, Figure 2.

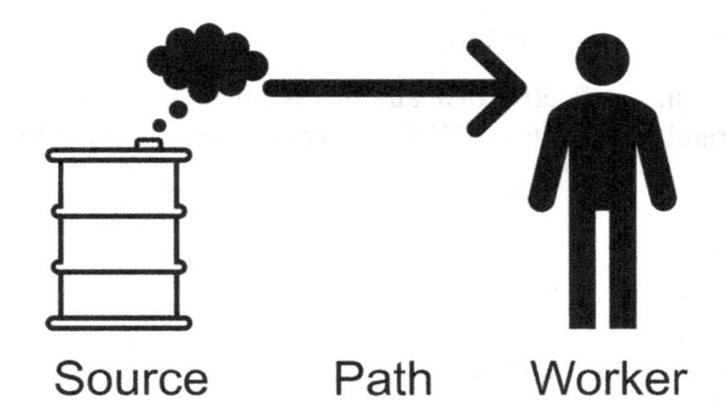

FIGURE 5.2 Source, transmission path, receiver model.

Source: K. Johnstone.

the source of the hazard, in the path from the source to the receiver, or at the receiver, a concept illustrated in Figure 5.2.

5.3.1 Elimination

The first step in the HoC is elimination, which involves completely removing the hazard from the workplace. For example, many jurisdictions have prohibited any new uses of asbestos. Elimination is considered the most effective control measure because it removes the hazard entirely thus preventing exposure. Elimination can be the most cost-effective measure in the longer term as it eliminates the costs associated with ongoing control measures.

Innovation can play a critical role in achieving elimination. With the continued growth in modern technologies and processes, it is possible to eliminate hazards in ways that may not previously have been possible or feasible. For example, the use of telepresence technology in healthcare has eliminated the need for doctors and nurses to directly interact with patients with infectious diseases. Healthcare providers can now remotely examine and diagnose patients without the need to physically be in the same room, eliminating the risk of exposure to infectious agents. In surface mining, the use of autonomous vehicles such as haul trucks can eliminate the risk of exposure to workers for a variety of hazards including dust, diesel emissions, whole body vibration and fatigue.

Elimination may not always be possible. If elimination is not achievable, the most effective, practical, and economic means of control must be established. The next level in the HoC is substitution, either of materials or of processes, which is directed at the source of the hazard.

5.3.2 Substitution

5.3.2.1 Substitution of Materials

Replacing one hazardous chemical with a less hazardous one has occurred throughout history in many process industries; for example, in the painting of luminescent watch and clock dials, radium was replaced with phosphorescent zinc sulfides after it was established that the radioactivity of radium paint caused bone and tissue cancer in the painters (particularly those who licked their brush to form a fine tip) (Martland 1929). In the manufacture of matches, white and yellow phosphorus, responsible for, a disfiguring and potentially fatal necrosis of the jawbone (termed 'phossy jaw'), was replaced by less dangerous red phosphorus (Felton 1982). Modern safety matches incorporate a safer form of phosphorus on the striking-friction side of the box.

In fact, many of the most important developments in occupational health have come from the search for less hazardous substitutes for dangerous materials. Some well-known examples include the following:

- Asbestos, whose fibres cause mesothelioma and other cancers, has been replaced by safer synthetic substitutes (glass foam, rockwool, and glass wool). Notwithstanding the removal of asbestos continues to result in new exposures.
- Benzene, which causes leukaemia, has been replaced as an industrial solvent by less hazardous solvents (e.g., Xylene); changed refining methods have reduced the level of benzene in gasoline.
- Beach and river sands, which have a high quartz content, have been replaced as abrasive blasting agents with low-quartz materials such as ilmenite, zircon, and copper slag.
- Mercury in the extraction of gold from ore has been replaced by cyanides and 'carbon-in-pulp' leaching.

These examples represent some classic advances but there are still many workplaces where there are opportunities for substitution of less hazardous materials. Lead can be phased out, for example, as can mercury and hexavalent chromium salts. Other, far less toxic aliphatic hydrocarbon solvents are replacing the neurotoxic aliphatic solvent n-hexane used in printing. Care must always be taken, however, to ensure that the substitute does not itself pose a hazard.

5.3.2.2 Substitution of Processes

In some industrial processes, where substitution of a less hazardous chemical is not possible, the risk associated with handling hazardous chemicals can be reduced by a change in the process. For example:

- To limit dust, a wet process can replace a dry process (e.g., Damp sawdust), or sweeping can be replaced by an Australian standard compliant industrial vacuum cleaner of the correct class (Figure 5.3).
- The working temperature of a process can be lowered to reduce the evaporation of volatile materials.
- A pelletised form of a chemical can be used instead of a dusty powder (e.g., to control lead exposure when handling stabilisers in PVC product production).
- Organic solvents can be used in gelled form to reduce the rate of vapour emissions (e.g., gelled styrene and gelled paint strippers).
- Changing the process to reduce fugitive dust or vapour hazards (e.g., conduct cold acid pickling processes rather than hot ones to reduce evaporation of hazardous substances and use floating ping-pong balls on the pickle liquor to reduce airborne droplets; use liquid catalysis of chemical reactions, such as in foundry moulding, rather than gaseous catalysis).

Periodically reviewing work processes is crucial to ensure that they are designed in a way that minimises workers' exposure to health hazards. A task may have been done in a traditional way for years without any questions arising about its efficiency or safety. For example, it may be

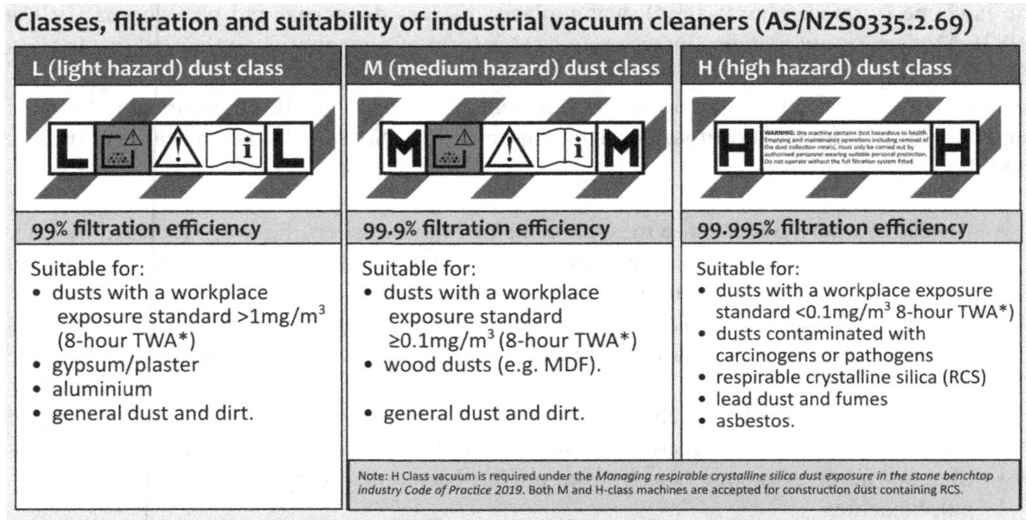

FIGURE 5.3 Examples of hazardous dusts and the required class of industrial vacuum cleaner.

Source: Workplace Health and Safety Queensland 2019, p. 2.

possible to substitute a product obtained from a purpose-designed workshop rather than operate a hazardous process in-house. The periodic review should consider:

- Reducing the number of times a worker handles a hazardous material (e.g., the lead used in manufacturing a battery may be handled in 20 or more separate operations).
- Changing processes that produce hazards, particularly airborne hazards (e.g., using larger bulk bags instead of multiple smaller bags that require extra handling).

As well as chemical hazards, the strategy of substitution of processes can be applied successfully to other occupational hygiene hazards, for example:

- Reducing noisy operations by replacing rivets with welds.
- Replacing noisy rollers with less noisy conveyor belts.
- Replacing high vibration hand and other tools with lower vibration models.
- Replacing vehicle driving seats with vibration-dampening seats.

The challenge to the H&S practitioner is to be alert to the possibility of substitution to reduce hazards. If the hazardous materials cannot be substituted because they are integral to a process, or the processing cannot be improved, then one or more of the following control strategies should be implemented.

5.3.3 Engineering Controls

When the use of hazardous materials or processes is unavoidable, the next best control measure is often to engineer out the hazard. A wide range of engineering controls are possible depending on the hazard in question, including various types of isolation/segregation, containment, and ventilation systems for hazardous chemicals, physical, and biological hazards. Engineering controls are used to modify the work environment to isolate workers from an established or perceived hazard or contain the hazard. Figure 5.2 shows how engineering controls can be directed at the hazard source, the transmission path, and/or the receiver (worker).

5.3.3.1 Isolation

If the worker can be isolated completely from the hazard, the risk to health is removed. Isolation may be achieved by a physical barrier or by distance. The following examples illustrate the principle:

- Installing noisy compressor units well away from worker-frequented areas within process plants.
- Using interlocked doors or barriers to prevent entry into an area while toxic substances are present.
- Relocating workers not directly engaged with processes that may give rise to exposure to a low-risk/hazard area.
- Separating materials that could create hazards if they come into contact with each other by accident (e.g., keeping oxidants and fuels in separate buildings or compartments).
- Locating rest areas away from hazard sources (heat, noise, dust, etc.).
- Installing a sound-proof control room in noisy environments.
- Using a remotely controlled laboratory to handle radioactive isotopes.

Time (e.g., timing of activities) is also a barrier, although time may equally be considered an *administrative* control. For example, burning-off of plastic extrusion dyes might be restricted to evening hours, allowing several hours for the air to clear before workers re-enter. Painting of a workplace could be conducted outside normal work hours to prevent unnecessary exposure to solvent vapours. Work on air-conditioning plants that may involve their shutdown should be restricted to periods outside normal working hours, particularly when duct cleaning is required. Fumigating restaurants with pesticides cannot be conducted when staff or patrons are on the premises.

5.3.3.2 Containment

Once an agent (be it chemical, physical, or biological in nature) has escaped from its source, it becomes far more difficult to capture or control. A better strategy is therefore to maximise the containment by engineering means. For example:

- A whole process can be enclosed and coupled with an exhaust extraction system (e.g., a spray booth where only the spray painter remains exposed to the hazard).
- A glove box or biological safety cabinet can be installed for handling infectious agents.
- Noisy machinery can be enclosed in sound-proof structures.
- Gas-tight systems can be used in chemical processing or in many sterilising or fumigation procedures.
- Using bunding to contain hazardous chemical spills.
- Using control options like skirts or shrouds to reduce the spread of contaminants (e.g., on conveyors or transfer points).

The containment approach has some drawbacks. For example, installing enclosed or contained systems may carry a high initial cost and may introduce the issue of workers coming into contact with the hazard if the system fails.

FIGURE 5.4 An example of a totally enclosed process: an ozone-generating corona discharge supplemented with an extraction fan.

Source: G. Gately.

The complete enclosure of a hazard is usually for high hazards where the escape of the hazardous substance could have serious health consequences or may be immediately life-threatening. Containment is normally supplemented by a ventilation system to ensure complete containment. Figure 5.4 is an example of an enclosed process; here a Plexiglas enclosure is built around a corona discharge unit to contain the generated ozone. A small exhaust ventilation fan supplements the containment. In Chapter 6, 'Ventilation', there are examples of the use of partial enclosures, a common and widespread method for control.

While a totally enclosed process is in operation, operator exposure will be limited. However, when the process is disrupted or maintenance is needed, it may be necessary for workers to enter the enclosure. This event must be treated with extreme care and utmost caution. For example, entering a toxic pesticide melting oven for maintenance or entering a fumigation process may require the operator to use high-integrity personal protective equipment, according to the specific risk. Such cases are often managed via a work permit, which is an *Administrative* control, discussed later.

Even where hazardous processes are totally enclosed, their location is important. They should not be situated where users or bystanders could be harmed if the enclosure system fails. Moreover, they may need to be interlocked so the process cannot be operated unless the isolation system is operating. Finally, primary enclosed systems should be alarmed and have secondary back-up controls. Planning for the siting and operation of totally enclosed systems is crucial to their continued safety. An extreme example of an incorrect location would be an ethylene oxide fumigation chamber, including cylinders of fumigant, in the centre of a library. A far better location would be to have the cylinders outside the normal workspace, alarmed for leakage and fitted with a continuous exhaust system to handle any accidental discharge of sterilising gas.

In any workplace, the H&S practitioner should be aware of the locations of all potentially hazardous operations and ensure that those locations will not make control difficult if the isolation system breaks down. For example, workers should not be able to stray into situations where, without warning, they could be at unnecessary risk. If new processes are being installed, particularly enclosed processes, they should not be sited in areas of constant or high worker access, and they must not block or impede emergency exit routes.

5.3.3.3 Ventilation

Industrial ventilation—the control of contaminants by dilution or local exhaust ventilation—is one of the main methods of controlling airborne chemical and biological hazards. The topic has already been touched on in this chapter and is covered in more detail in Chapter 6, 'Industrial ventilation'.

5.3.4 Administrative Controls

The exposure controls examined so far work by eliminating or substituting the material or process, or by engineering out the risk. Frequently, these controls are insufficient by themselves, and it becomes necessary to change work methods or systems to achieve the desired level of control. Such measures have traditionally been considered 'administrative controls'. It would be both simplistic and optimistic to suppose that a single control strategy (except for complete elimination) can result in satisfactory control of exposure. Sometimes higher-level control mechanisms are found to be impractical in use and are by-passed by workers or they don't work well enough to reduce the risk to an acceptable level. Consider the following workplace situations:

- Working inside a deep freezer;
- Working inside a hot oven; or
- Working underwater at a depth of 100 metres.

In the freezer and the oven, it is rarely practicable to introduce a warm (or cool) micro-environment to compensate for the cold (or heat) in the work environment. Keeping divers safe while working at great depths for long periods—and returning to the surface—requires elaborate equipment.

The use of administrative controls in managing workplace hazards is a strategy that concentrates on work processes and systems, and worker behaviours, rather than workplace hardware. Administrative controls do not alter the hazard at the source, or along the pathway, they target the worker. While preference must be given to higher-order controls, special attention must also be given to worker education, behaviour, or work practices because higher-order methods alone may not be sufficient to effectively control the risk. Administrative controls should not be confused with management functions, such as responsibility, audit, and review. Administrative controls should mainly be employed to supplement higher-order controls or in emergency situations (e.g., clean-up of chemical spills) and include measures like documented work procedures, provision of information, training and education, job rotation, job scheduling, housekeeping, supervision, labelling, and signage.

All OHS legislation, codes of practice, and industry standards incorporate forms of administrative control. The way in which administrative controls are implemented will depend on the workplace. Company OHS policies, procedures, government regulations, and some industrial relations arrangements all have a role.

5.3.4.1 Documented Policy and Procedures

Documented work procedures or work permits that are based on a risk assessment can be an effective administrative control process for some high-risk tasks. For example, work in confined spaces is controlled by performing a risk assessment and using work permits (with detailed control measures) as described the SafeWork Australia Confined Spaces Code of Practice (SafeWork Australia 2020a).

5.3.4.2 Housekeeping

Housekeeping is an administrative control measure that helps limit inadvertent exposure to workplace hazards. The importance of maintaining high standards of housekeeping cannot be overstated. Dirty and untidy workplaces not only increase the likelihood of secondary exposures (e.g., by inhaling dust raised by draughts and wind, or by inadvertent skin contact with dirty surfaces and equipment) but may also send a message to workers that poor work habits are acceptable.

5.3.4.3 OHS Information

The provision of information, such as the mandatory availability of a safety data sheet (SDS) and labelling for hazardous chemicals, are administrative control mechanisms. Signage is another mechanism to provide information to the worker. Information systems are powerful administrative control mechanisms that operate unobtrusively in most workplaces and are often taken for granted.

5.3.4.4 Education and Training

To use administrative controls properly, workers must be adequately trained so they know:

- The full nature of the hazard and potential health impacts.
- Why the administrative control is being used.
- The exact procedures and guidelines to be followed.
- The limitations of administrative control procedures.
- The consequences of ignoring administrative control.

In other words, worker involvement, participation, training, and education are critical to the success of administrative control programs.

OHS legislation in Australia and many other countries imposes specific requirements for training workers and others involved in health and safety activities. Where administrative controls are instituted (such as may occur under the various health and safety or hazardous chemical regulations), the law generally requires training and induction of workers.

Training programs should be formalised and administered throughout the length of employment. Training should always incorporate the practical aspects of a job and include some form of competency assessment. If workers potentially exposed to a hazard are made fully aware of the consequences of over-exposure, and the routes and mechanisms of exposure, they are more likely to identify other exposure situations and act to reduce exposures in new situations. In this regard, it is useful in a training setting to visually represent exposure. For example, fluorescein dye (used by plumbers to trace drains) can be added to aqueous solutions to simulate toxic liquids. To show the degree of containment, or the spread of contamination, the traces of fluorescein can be made visible by using a UV lamp. Smoke generators can be used to test ventilation systems and can show the potential movement of contaminants within the workplace. Intense lighting can also be used to show dust generation, as shown in Figure 5.5.

5.3.4.5 Worker Rotation and Removal from Exposure

A form of exposure control by changing a work schedule would be to rotate tasks within the work group to spread the exposure across a larger number of workers. For this to be a viable strategy, the H&S practitioner must have a reliable system to measure the exposures of all members of the work group.

FIGURE 5.5 Intense light illumination being used to highlight secondary dust from work clothes as a source of exposure.

Source: HSE 2024a.

In many circumstances, exposure to hazardous chemicals or a hazardous environment cannot be avoided. If any workers are exposed to the maximum permissible level, then they may need to be removed from exposure. For example:

- In the lead industry, if worker blood lead levels exceed the prescribed limit, they may be removed from further exposure until their levels decrease to an acceptable level.
- In industries where ionising radiation is involved, workers are permitted a maximum radiation dose over a specified time period. When this is exceeded they are removed until such time as ' effective dose' is reduced. (ARPANSA 2020).
- Workers in excessively noisy industries, who cannot be adequately protected by hearing protection, should have noise exposure reduced by reassignment so that their daily equivalent noise exposure does not exceed the legal limit of 85 dB(A) L_{eq}.

Where workers have developed sensitivity to a substance, a common administrative control is to prevent any further exposure. In other instances, workers may be predisposed to experiencing effects at lower thresholds than the average worker, or they may be medically diagnosed as showing effects of exposure without being symptomatic. For example:

- Isocyanate-sensitised workers should be prevented from any further exposure.
- Workers with radiologically confirmed dust disease should be precluded from further work in dusty underground mining.

- Workers with certain genetic dysfunction should not be occupationally exposed to TNT or chemicals causing haemolytic anaemia.
- Pregnant workers should not be exposed to known foetal toxins (e.g., lead, methyl ethyl ketone, and other solvents).
- Asthmatics should not work with strong irritant gases.

There are examples from history where workers' exposures were controlled by compulsorily withdrawing them from exposure rather than by limiting the actual routes of exposure. This method of exposure control is suggestive of the inability of other control measures to limit exposures. In the current-day context, it would be considered unethical to continue exposing a new group of workers without rectifying the root causes of the exposure. Therefore, if such a situation arises, the H&S practitioner would be advised to re-evaluate all aspects of the control systems and strategies.

5.3.5 Personal Protective Equipment (PPE)

PPE takes its place at the bottom of the HoC. PPE represents the absolute last resort; beyond it is the unprotected worker and inevitable exposure if the PPE is not correctly selected, maintained, and used appropriately. Even though PPE is at the bottom of the hierarchy, it is still widely used and accepted as a backup and supplement to other controls. There will also be situations where higher-level controls cannot be used, and PPE will be the only practicable solution, for example, during an emergency. A detailed discussion of various aspects of the selection, care, and use of PPE is covered in Chapter 7.

5.4 APPLICATION OF THE HIERARCHY OF CONTROLS

The HoCs are applied to eliminate or reduce exposure to as low as reasonably practicable based on an evaluation of the risk posed by the hazard. In this way, the process of applying the HoC forms part of the broader risk management process. Figure 5.6 provides an overview of the key steps involved in the application of the HoC.

5.4.1 Anticipation and Recognition of Health Hazards

As we have seen in Chapter 1, when investigating a potential exposure, the H&S practitioner may first conduct a walkthrough survey of the workplace, during which observations are made of factors such as the work processes used, personnel at work, hazards to which they are potentially exposed, and control measures in use. This may include collecting information and data relating to workers' exposure; this could be in the form of actual exposure measurements and/or health-impact data (e.g., reports of incidence of symptoms or illness). The observations and data gathered will help the H&S practitioner establish the current adequacy of controls or whether additional controls are required.

Often materials and processes find their way into the workplace without any effort being made to investigate their hazards or alternatives. In many cases, materials are introduced, usually to make the task easier, without sufficient thought being given to potential new hazards. For example:

- New solvents to clean residues from parts may be toxic.
- Compressed air to blow components clean may generate a dust hazard.
- UV lamps to cure resins may expose workers to UV radiation.

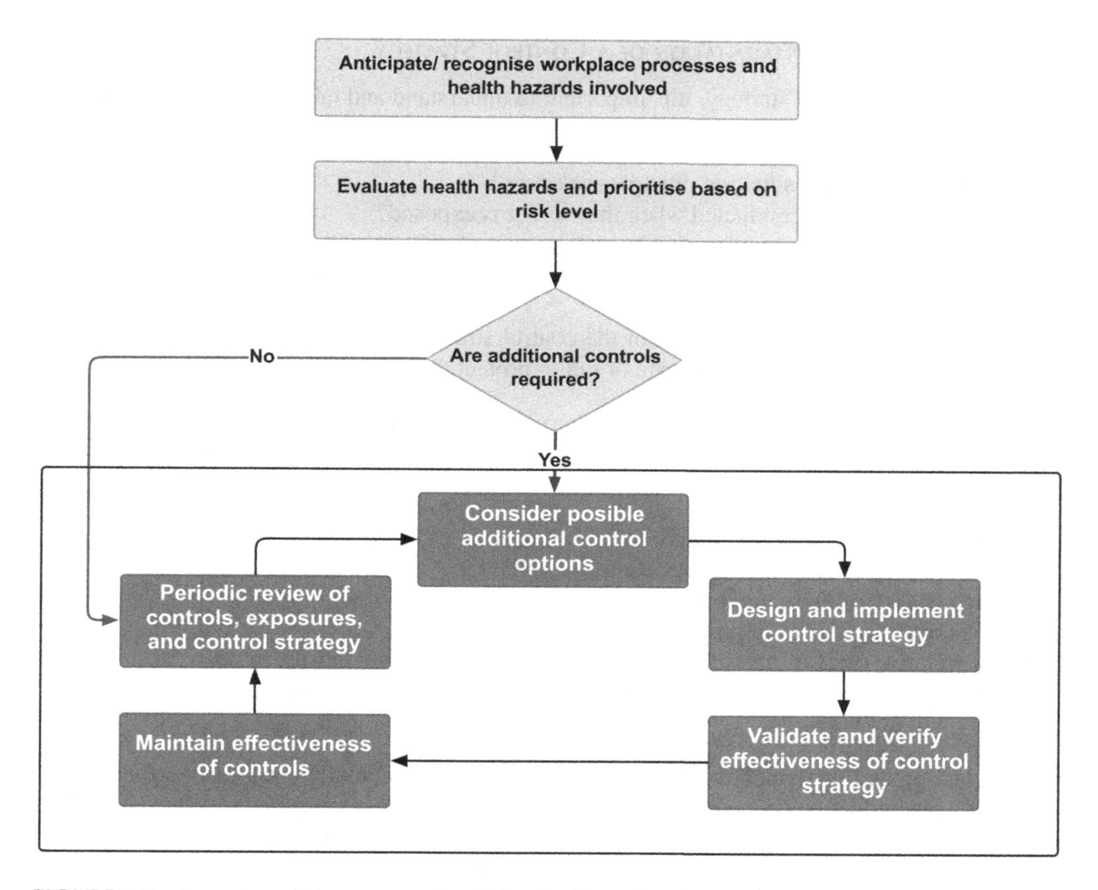

FIGURE 5.6 Overview of the process of applying the hierarchy of controls.

Source: K. Johnstone & K. Burton.

A change management system should be in place to ensure that risk assessments are conducted each time a new chemical or process is introduced, or an existing one is changed. Change management controls are required by most OHS regulations and should involve representatives from all affected work groups.

5.4.2 Evaluation and Prioritisation of Health Hazards

Invariably in any workplace, there are many and varied OHS issues that require some form of action to reduce the risks to those exposed. Consequently, there will be competing projects for the inevitably limited funds available to manage risks. It is therefore necessary to prioritise all the identified hazards and issues, be they chemical, physical, safety, biological, or environmental in nature, and it is common practice to undertake a health risk assessment for this purpose. There are many tools available to facilitate such assessments; typically, these tools include a risk matrix to determine the consequence of exposure and the likelihood of the consequence occurring to rank a hazardous scenario (AIOH 2020).

5.4.3 Design and Implementation of a Control Strategy

When designing a control strategy, it is important to understand and take into consideration matters such as:

- How the hazardous situation or exposure arises?
- What task is being conducted when the worker is exposed?
- Why the task is being undertaken?
- Whether there are any other people in the area being exposed and what are the impacts of that exposure?
- What the intended outcomes are of the control strategy and what the likely consequences are if the controls are not adequate or fail to protect the worker?

It is generally found that a combination of controls results in a more robust and reliable system of control than would be achieved using a single method. Deciding which control solutions to recommend will require consideration of:

- The hazard and the magnitude of the risk it poses.
- The practicability of the various controls available.
- The efficacy of those controls.
- The consequences of failure of controls.
- The relative costs of providing, operating, and maintaining controls.
- The likely acceptance of the controls by the workforce (if they are viewed as impractical, they will not be used).

When selecting and implementing the final control strategy, eventually a point is reached where any further risk reduction becomes out of proportion to the resources required to reduce it further. This point is referred to as 'as low as reasonably practicable' (ALARP) and represents the level to which we would expect to see workplace risks controlled.

5.4.3.1 Considerations of Cost Versus Benefit

When considering the cost versus benefit of possible control solutions, the following points are important guidance for selecting controls:

- Controlling a hazard by elimination and substitution may in some cases be simple and effective and should always be considered first.
- Higher-order controls should be implemented at the design stage as they can be expensive to retrofit. For example, a common control measure for noise is to 'buy quiet', which is incorporated in a company purchasing policy to specify maximum noise levels new equipment needs to achieve.
- Administrative controls and PPE can involve complex decision-making and rely heavily on worker compliance and acceptance, which can be unpredictable and uncertain. Making decisions on respiratory protection requires detailed information about exposure levels, equipment performance, worker training, and maintenance. If respiratory protection is used as the primary control mechanism and fails, there is no backup system.
- During each step of developing the controls, all the stakeholders (e.g., process managers, exposed work group, engineers and H&S practitioners) should be involved to ensure that they all take ownership of the solutions and that all practical solutions are considered.
- When considering the trade-offs of *control* effectiveness against the consequences of failure and cost, where there is a high consequence hazard (e.g., a carcinogen), a more robust and effective control must be used.

5.4.3.2 Application of the HoC at the Design Stage

Despite the HoC being a well-known, understood, and accepted construct, it is frequently found that the lower-order control strategies of Administration and PPE are those most likely to have been implemented in the workplace. The best time to implement the higher control strategies is during the design of workplaces or processes, since the cost is much less and technically easier to undertake than during the operations.

It is important for H&S practitioners to be involved in the design of new workplaces or processes. In this way, their specific expertise, knowledge, and best practices in design can be applied to eliminate hazards or at least reduce the associated risk, to generally acceptable levels. Retrofitting higher-order controls is expensive and generally less effective.

There are several resources to support 'designing out' health and safety hazards during developments, including:

- The Safe Work Australia 'principles of good work design', which addresses health and safety issues at the design stage (Safe Work Australia 2020b).
- The US National Institute for Occupational Safety and Health (NIOSH) 'prevention through design' (ptd) initiative, the mission of which is to 'prevent or reduce occupational injuries, illnesses, and fatalities through the inclusion of prevention considerations in all designs that impact workers' (National Institute for Occupational Safety and Health 2023).

5.4.3.3 Implementation and the Management of Change

The development of a control strategy should follow change management processes to ensure that the implementation of the strategy is well considered and does not introduce any unintended risks. For example, this will need to incorporate consultation, education and training, and a thorough understanding of the requirements for control effectiveness. Consideration should be given to any interactions with other controls or systems. Implementation should include processes for the maintenance and monitoring of the ongoing effectiveness of the controls (e.g., alarms, visual displays, inspection regimes, and triggers for maintenance or production actions).

5.4.4 Validate and Verify Effectiveness of Control Strategy

Any control strategy needs to be validated to verify that it is working as intended and is effective at reducing exposures. There are a number of tools available to assist H&S professionals such as the use of real-time monitors with or without video exposure monitoring in fixed locations and/or as personal monitoring. This approach provides useful real-time information about control effectiveness and highlights areas of high and lower exposure as well as any change in expected levels. Biological monitoring can also be useful for verifying the effectiveness of controls. For example, metals analysis of a welder's exhaled breath condensate can be used to evaluate the effectiveness of workshop ventilation and respiratory protection (Hulo et al. 2014).

5.4.5 Periodic Review of the Control Strategy

As the risk associated with a particular process may change over time (e.g., the output of hazardous gas may increase), periodic reviews of control strategies will be needed. This will ensure that controls remain effective, as well as being efficient and economical. Simplicity is important; complicated control systems require ongoing attention (e.g., training, checking, inspections and maintenance), which increases the likelihood that they become ineffective and possibly their failure could go undetected, thus endangering worker health.

5.5 CONTROL OF CHEMICAL HAZARDS

5.5.1 Regulatory Requirements for the Management of Hazardous Chemicals

Within Australia, hazardous chemical regulations are based on the Safe Work Australia model regulations (Safe Work Australia 2023). These set out the basic obligations of manufacturers, importers, and suppliers to provide information to the workplaces in which their products are used. Employers also have obligations to identify hazards, mainly based on the hazardous chemicals risk phrases used in SDS, provide relevant information on hazardous chemicals in their workplaces, to assess the extent of the exposure and control risks, for exposure routes to be defined, to train staff, to undertake health surveillance, and to keep records where necessary. These regulations incorporate all the HoC principles described in the various sections above.

Many published guidelines on the management of hazardous chemicals contain useful information on the HoC, but the optimal control strategy for any workplace will depend on its unique set of circumstances. Some additional regulations on asbestos, lead, and carcinogens detail specific actions to be taken for their management.

5.5.2 Control Banding

The traditional structured approach of occupational hygiene is built around quantitative risk assessment, typically by measurement of exposures to a health hazard (e.g., noise, biological hazard, toxic dust, gas or vapour, and radiation). The measured exposures can then be compared against an agreed exposure standard to determine the level of control needed to be applied. However, not all hazards encountered have exposure standards, thus making it difficult to do the risk assessment. In the 1990s, the pharmaceutical industry developed a control approach based on the comparison of the toxicological data combined with physical form and properties to apply one of a prescribed set of controls to a material in particular handling scenarios. This semi-quantitative approach was further refined by the UK HSE and published as Control of Substances Hazardous to Health (COSHH) Essentials (Health and Safety Executive) and referred to as Control Banding (CB). The initial focus was on assisting small and medium enterprises in complying with chemical control regulations. The process steps are:remove "the"

- Identify the 'hazard band' based on:
  - The toxicity (e.g., hazard statements).
  - The ease of becoming exposed (e.g., volatility or dustiness).
  - The work processes (e.g., container filling, grinding, and bag emptying).
  - The duration of the potential exposure (e.g., minutes, hours, and/or continuous or intermittent).
  - The quantity of material handled (e.g., grams, kilograms, and tonnes).
- Based on the above, a generic and conservative control approach can be identified. These are typically:
  - Ventilation
  - Engineering controls
  - Containment
  - Additional administrative and PPE controls
  - Seek expert advice

Resources are available for the application of the CB approach, see, for example, the UK HSE COSHH Essentials (Health and Safety Executive 2024b) and WorkSafe QLD Nanomaterial Control Banding Risk Assessment (Workplace Health and Safety Queensland 2017).

REFERENCES

AIOH 2020, *Simplified Occupational Hygiene Risk Management Strategies*, Australian Institute of Occupational Hygienists Inc, Gladstone Park, Victoria.

Australian Radiation Protection and Nuclear Safety Agency (ARPANSA) 2020, *Code for Radiation Protection in Planned Exposure Situations*, Australian Radiation Protection and Nuclear Safety Agency, Commonwealth of Australia, https://www.arpansa.gov.au/sites/default/files/20220404-rps_c-1_rev_1.pdf [30 December 2023].

Felton, J.S. 1982, 'Classical syndromes in occupational medicine: Phosphorus necrosis—A classical occupational disease', *American Journal of Industrial Medicine*, vol. 3, no. 1, pp. 77–120.

Health and Safety Executive (HSE) 2024a, *How Does Skin Come into Contact with Chemicals?*, https://www.hse.gov.uk/skin/chemicals.htm [30 December 2023].

Health and Safety Executive (HSE) 2024b, *COSHH Essentials*, http://www.hse.gov.uk/coshh/essentials/index.htm [30 December 2023].

Hulo, S., Chérot-Kornobis, N., Howsam, M., Crucq, S., de Broucker, V., Sobaszek, A. & Edme, J-L. 2014, 'Manganese in exhaled breath condensate: A new marker of exposure to welding fumes', *Toxicology letters*, vol. 226, no. 1, pp. 63–69.

Martland, H.S. 1929, 'Radium poisoning', *Monthly Labor Review*, vol. 28, no. 6, pp. 20–95.

National Institute for Occupational Safety and Health (NIOSH) 2023, *Prevention Through Design Program*, National Institute for Occupational Safety and Health, https://www.cdc.gov/niosh/topics/PTD/ [30 December 2023].

Safe Work Australia (SWA) 2018, *How to Manage Work Health and Safety Risks Code of Practice*, Safe Work Australia, Commonwealth of Australia, https://www.safeworkaustralia.gov.au/doc/model-code-practice-how-manage-work-health-and-safety-risks [30 December 2023].

Safe Work Australia (SWA) 2020a, *Confined Spaces Code of Practice* Safe Work Australia, Commonwealth of Australia, https://www.safeworkaustralia.gov.au/doc/model-code-practice-confined-spaces [30 December 2023].

Safe Work Australia (SWA) 2020b, *Principles of Good Work Design*. Safe Work Australia, Commonwealth of Australia, https://www.safeworkaustralia.gov.au/resources-and-publications/guidance-materials/principles-good-work-design [30 December 2023].

Safe Work Australia (SWA) 2022, *Safer, Healthier, Wealthier: The Economic Value of Reducing Work-Related Injuries and illnesses – Technical Report*, Safe Work Australia, Commonwealth of Australia, https://www.safeworkaustralia.gov.au/doc/safer-healthier-wealthier-economic-value-reducing-work-related-injuries-and-illnesses-technical-report [30 December 2023].

Safe Work Australia (SWA) 2023, *Model Work Health and Safety Regulations*, Safe Work Australia, Commonwealth of Australia, https://www.safeworkaustralia.gov.au/law-and-regulation [30 December 2023].

Workplace Health and Safety Queensland (WHSQ) 2017, *Nanomaterial Control Banding Risk Assessment*, WorkSafe Queensland, https://www.worksafe.qld.gov.au/safety-and-prevention/hazards/hazardous-exposures/nanotechnology/nanomaterial-control-banding-risk-assessment [30 December 2023].

Workplace Health and Safety Queensland (WHSQ) 2019, *Selecting the Right Portable Extractor or Industrial Vacuum Cleaner for Hazardous Dusts*, Queensland Government, https://www.google.com/url?sa=t&rct=j&q=&esrc=s&source=web&cd=&ved=2ahUKEwir86--_qSAAxX2amwGHdfCCtEQFnoECA4QAQ&url=https%3A%2F%2Fwww.worksafe.qld.gov.au%2F__data%2Fassets%2Fpdf_file%2F0024%2F23649%2Findustrial-vacuum-cleaners-and-hazardous-dust.pdf&usg=AOvVaw3Bx440vBpbNq4cbcBH9iui&opi=89978449 [30 December 2023].

6 Industrial Ventilation

Dr Elaine Lindars and Anthony Bamford

6.1 INTRODUCTION

Ventilation is one of the most effective ways to both reduce toxic airborne exposures in the workplace and maintain the thermal comfort of workers. However, to remain effective, a ventilation system must be well designed and maintained. As an occupational hygienist, you are likely to encounter various forms of ventilation, including:

- natural ventilation;
- primary and secondary ventilation systems (mining and tunnelling);
- local exhaust ventilation (LEV), including on tool extraction;
- indoor ventilation systems (heating, ventilation, and air conditioning);
- cabin pressurisation and filtration ventilation systems.

This chapter outlines some basic principles of industrial ventilation to help aid the reader in identifying poor and inefficient designs, and to identify strategies to improve the effectiveness of a poorly designed ventilation system. While minor modifications to a system can be made through this process, significant changes or new ventilation designs will require specialist knowledge.

6.2 BASIC SCIENCE OF VENTILATION

Air will only move from one point to another if a pressure difference exists between the two points. The volume of air that will move in unit time depends on the magnitude of the pressure difference and the resistance offered by the duct to the flow of air. Just as unassisted water will only flow from a higher to a lower elevation, airflows only from an area of higher to a lower atmospheric pressure unless assisted by a fan, other appliance, or density differences.

Under still conditions, we will not notice the atmospheric pressure. A strong wind not only ruffles our hair and clothes but it is also difficult to walk against. Moving air can exert considerable pressure. For example, very strong winds of gale force can blow people off their feet and topple giant trees. This pressure is called velocity pressure (P_v), which increases as the square of the velocity (v) of the air (Equation (6.1)):

$$P_v = \rho \frac{v^2}{2} \tag{6.1}$$

where

ρ = density of air (kg/m^3)
v = velocity (m/s)
P_v = velocity pressure (Pa)

DOI: 10.1201/9781032645841-6

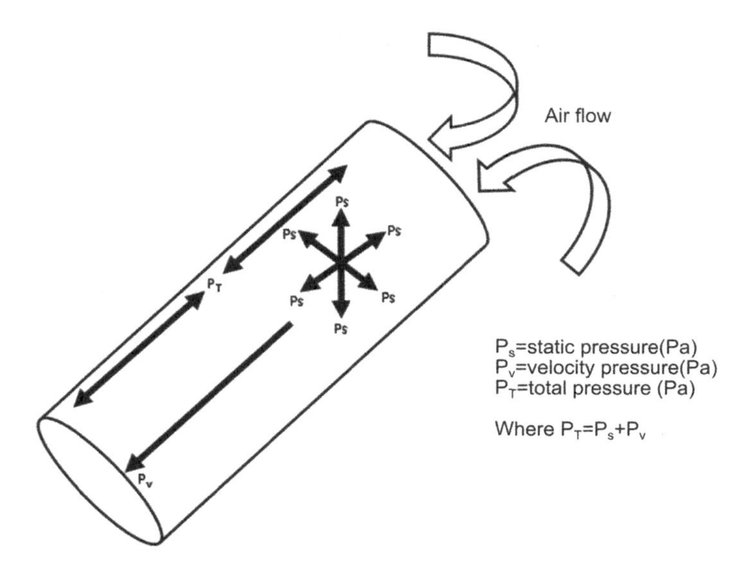

FIGURE 6.1 Pressure within a ventilation column.

Source: A. Bamford.

Static pressure (P_s) is defined as the pressure exerted in all directions by a fluid that is stationary; it represents the potential energy of the air while velocity pressure (also known as dynamic pressure) represents kinetic energy of the air. The algebraic sum of the two is the total pressure of the air as represented in Figure 6.1.

Occupational hygienists should be able to perform basic ventilation calculations to determine the amount of air needed to dilute or extract contaminants and provide guidance on how an extraction system should be designed. The following are the key equations in describing a ventilation system.

6.2.1 Volume of Air within a System

When air moves through a system, it is a result of the area of the system and the velocity of the air going through it; it is represented by Equation (6.2):

$$Q = v \times A \tag{6.2}$$

where

Q = volumetric flowrate (m^3/s)
v = velocity (m/s)
A = area of duct (m^2)

6.2.2 Flow of Air within Ducts

Flow within air ducts is influenced by resistance of the duct, and other fittings, to airflow. This resistance can be characterised by the Atkinson friction factor (k). The friction factor can be taken into account when calculating a pressure drop in a ventilation system using the Atkinson equation (Equation (6.3)):

$$P = \frac{k \times c \times L \times v^2}{A} \quad (6.3)$$

where

P = pressure loss (Pa)
k = friction factor (kg/m³ or Ns²/m⁴) obtained from manufacturer, estimated, or calculated
c = perimeter (m)
L = length (m)

6.2.3 Factors Affecting Airflow

Duct design: the geometry of the duct, including its size, shape, and presence of bends and fittings, can significantly impact airflow and thermal distribution.
Airflow rate: the velocity of the air flowing through the duct affects both pressure losses and heat transfer rates.
Air temperature and properties: the temperature and humidity of the air, as well as its density and viscosity, influence its flow behaviour and heat transfer characteristics.
External factors: factors like ambient temperature, solar radiation, and building envelope characteristics can also impact the thermal performance of ducts.

6.3 PRINCIPLES AND MYTHS

The study of industrial ventilation and its application to hazard control begins with an understanding of how air moves in addition to several key concepts described herein.

6.3.1 Moving Contaminated Air Away from the Breathing Zone

If contaminated air is made to move away from a person's breathing zone (Figure 6.2), then it is likely that the person will be exposed to less of the contaminant. While this may be obvious, it is not uncommon to observe industrial ventilation systems that fail to achieve this goal due to inadequacies in the design, maintenance, or incorrect use of the system.

In Figure 6.2, the worker is grinding a casting with a small wheel. Clean air is supplied through a duct and contaminated air is removed by an extraction system under the grate at his feet. The air is drawn away from his face and away from the source of contamination. In ideal conditions, the flow would be laminar with minimal eddies.

One of the most common errors in ventilation system design is to place a canopy hood directly above a worker or require the worker to operate beneath the hood, as shown in Figure 6.3. Such designs result in contaminated air being drawn past the worker's breathing zone. This is very common with welding hoods and if the welding booth is near a wall, replacement air can swirl in front of the welder as it rises, making the fume concentration inside the welding helmet higher than that outside the helmet.

FIGURE 6.2 Contaminated air moving away from worker's face.

Source: Great Britain Home Office Committee on Ventilation of Factories and Workshops 1907, p. 29.

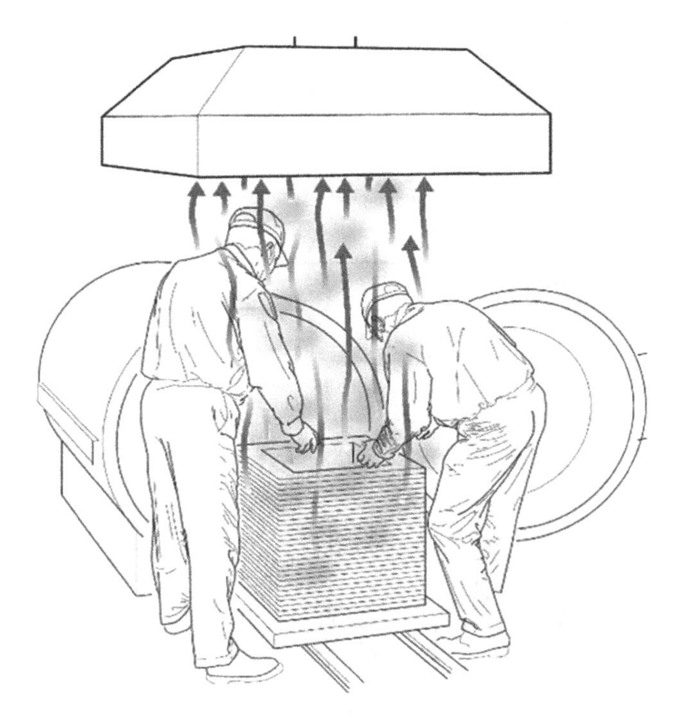

FIGURE 6.3 Poor canopy hood placement.

Source: Health and Safety Executive [HSE] 2017, p. 41.

6.3.2 Eddies around Downwind Objects

When cooking on a barbecue, most people know that if they stand downwind the smoke will blow into their face. However, novice cooks may be surprised to find that when they move upwind, the smoke seems to follow them. This happens because air swirls, or eddies, downwind of the body, trap some smoke. This phenomenon is shown in Figure 6.4 with eddies forming in front of the person at the canopy face: similar effects occur in front of a fume cupboard, with spray-painting booth or with welding (Figure 6.18).

In many workshops, a fume extraction system is placed above the worker to capture fumes from processes such as welding. The air thus tends to flow towards and past the worker's face. This is exacerbated because the extraction system produces a local flow of replacement air that swirls around the worker, so that instead of moving directly upwards, the contaminated airflow tends to curve towards the worker's face. The concentration of fume inside a welding helmet can thus be higher than that outside the helmet.

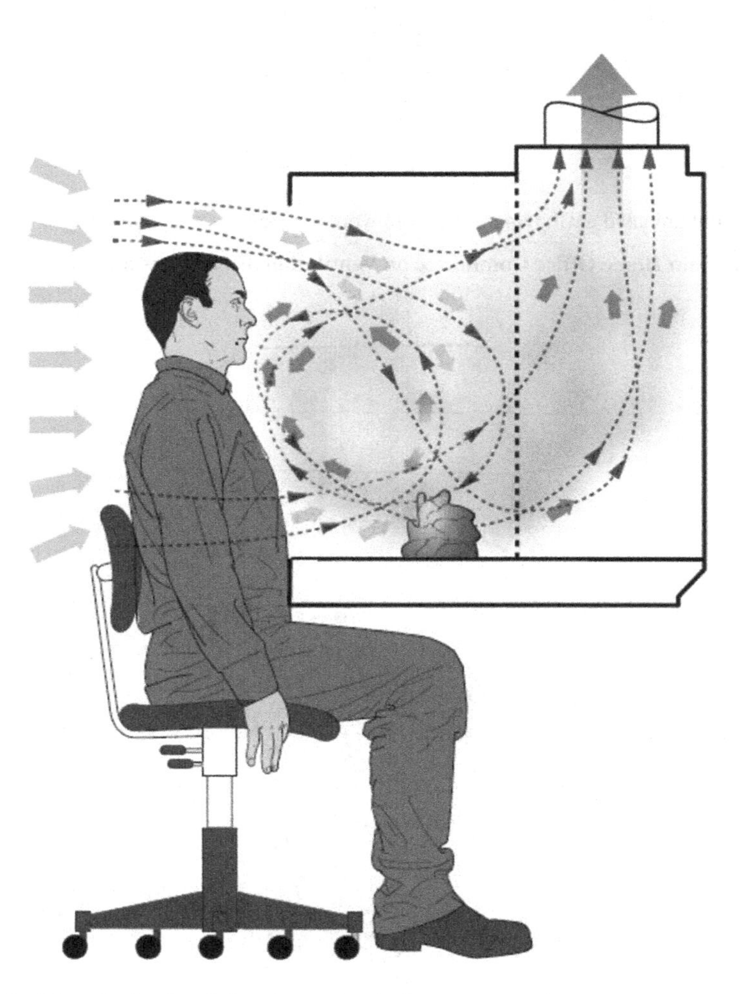

FIGURE 6.4 Partial enclosing hood showing eddies in front of worker's body.

Source: Health and Safety Executive [HSE] 2017, p. 37.

On a larger scale, when air swirls around buildings, contaminants released from a short stack downwind of a building, but still in the building's wake, can be drawn back inside through windows and ventilation inlets.

6.3.3 Extracting vs. Blowing

An understanding of the difference between extracting (sucking) and blowing air is fundamental to industrial ventilation design. The difference can easily be demonstrated.

Hold a lit birthday candle a distance from your mouth and *blow* towards it. The air movement easily extinguishes the candle because the airflows in a jet in the direction of the candle. Re-light the candle and now try to extinguish the candle at the same distance by inhaling. Now you are *sucking* (exhausting) air in an attempt to put it out. In this latter scenario, the airflow into your mouth is from all directions, much of it over the surface of your face. So, while you can blow out a candle at arm's length, if you try to "suck out" the candle, your lips will need to be only millimetres from the flame before you can extinguish it by inhaling (see Figure 6.5). This fundamental difference between sucking and blowing can be used to explain why placing a hood too far from the source of contaminated air is ineffective. It also explains why a good seal between the face and a respirator is so important—air is inhaled from all directions and so if any part of the seal is poor, the airflow will take the path of least resistance under the seal, bypassing the respirator filtration.

When a fan (or open duct) blows air, the velocity drops to roughly 10 per cent of the maximum in a distance equal to about 30–60 times the diameter of the fan/duct if no obstruction is placed at the inlet. However, when air is sucked, this velocity drop occurs within a distance equal to the diameter of the fan or duct. This can be observed with a small fan such as a desk fan. If a piece of paper is held some distance in front of the fan, the airflow ruffles it easily. If the paper is held behind the fan, it will not be disturbed until it is very close to the blades. Similarly, with a vacuum cleaner, the suction operates only close to the nozzle, but the exhaust plume can be felt at some distance.

Industrial ventilation systems are often designed and installed by engineers who are more familiar with HVAC (heating, ventilation, and air-conditioning) systems, which condition the air and moves *(blow)* it into buildings. The fans, ducts, and some air-cleaning systems are much the same in industrial ventilation, but now the air is being sucked rather than blown, and that difference and its consequences may not be understood. As a result, industrial ventilation systems often fail to protect workers adequately from contaminated air.

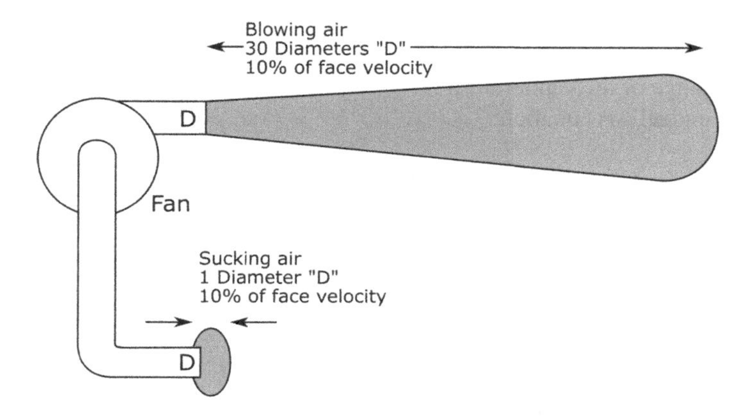

FIGURE 6.5 Extracting versus blowing air.

Source: D. Bromwich.

6.3.4 The 'Heavier than Air' Myth

It is not uncommon to find industrial ventilation systems designed to remove 'heavier than air' (density greater than air) vapours and gases from near the floor. This premise that heavier than air vapours and gases will settle to the floor is not always correct. Vapours and gases usually concentrate near the floor in the following circumstances.

- Where there are very high concentrations, this can be as result of catastrophic release and/or concentrations orders of magnitude above workplace exposure standards.
- Where there are temperature gradients (this can result in reduced mixing).
- There are no minimal air disturbances from moving people or moving plant to facilitate the mixing of air. It should be noted that even in a closed room with no external air movement, Brownian motion will still promote the mixing of air.

As indicated, these conditions may occur in spaces such as the confined bilge of a boat or during a catastrophic large volume release such as a tank rupture, but these are exceptions, because while the gas or vapour can be heavier than air, imperceptible temperature gradients can be significant in creating slow convection currents that mix the air.

Imagine a million air molecules, among which are 1000 molecules of a solvent (10 times the workplace exposure standard for many solvents) whose vapour density is twice that of air (do not confuse vapour density with the density of the liquid). Some simple calculations show the impact of molecule density vs. thermal changes in a room.

The density change when there is a solvent vapour/air mix:
density change = $\Delta\rho$

$$\Delta\rho = \frac{(\text{molecules of air} \times \text{density of air}) + (\text{molecules of solvent} \times \text{density of solvent})}{(\text{original molecules of air} \times \text{density of air})} \quad (6.4)$$

$$\Delta\rho = \frac{(999{,}000 \times 1) + (1000 \times 2)}{(1{,}000{,}000 \times 1)}$$

$$\Delta\rho = 1.001 \equiv 0.1\%$$

The thermal change of the vapour/air mix:
So consider the ideal gas equation:

$$\text{PV} = \text{nRT} \quad (6.5)$$

where
P = pressure
V = volume
R = gas constant
T = temperature in Kelvin

$$\text{n}(\text{number of moles}) = \frac{\text{mass of substance}}{\text{MW}(\text{molecular weight})}$$

Note: ignore the weight of the air column in the room air pressure is effectively the same at floor (P1) and ceiling (P2)volume of air is constant

$$PV = \frac{\text{mass of substance}}{MW} \times RT \tag{6.6}$$

Rearrange:

$$(P \times MW)/RT = \text{mass}/V = \rho(\text{density}) \tag{6.7}$$

If the air temperatures is 20°C at the floor and just 1°C warmer or 21°C at the ceiling, the change in density ($\Delta\rho$) is calculated by:

$$\Delta\rho = \frac{\rho_1}{\rho_2} = \frac{P_1 \times MW}{RT_1} \times \frac{RT_2}{P_2 \times MW} \equiv \frac{T_2}{T_1} \tag{6.8}$$

$$\Delta\rho = \frac{294}{293} = 1.0034 = 0.34\%$$

where
T_1 = 20°C = 273 + 20 = 293°K at floor
T_2 = 21°C = 273 + 21 = 294°K at ceiling

In this case, the difference in density from a temperature change is 3.4 times that of the vapour density change. A temperature difference of just 1°C from floor to ceiling will produce greater convective forces to mix the air than the density of a vapour at 1000 ppm. This means that in most workplaces, where there will likely be a temperature gradient of at least few degrees between the floor and ceiling of a room, normal convective air currents should mix the air and no blanket of 'heavier than air' vapour should form on the floor. However, if there is significant stratification, for example, with hot air in the rafters of a workshop and cooler air at the slab, then this may have the impact of reducing mixing. Situations need to be reviewed by a competent hygienist on a case-by-case basis.

6.4 GENERAL VENTILATION DESIGN

The design of ventilation systems is well covered in texts such as the ACGIH®'s *Industrial Ventilation: A Manual of Recommended Practice* (2023), and the 2017 booklet from the UK Health and Safety Executive, *Controlling Airborne Contaminants at Work—a guide to local exhaust ventilation (LEV)*. Section 6.1 of the latter text covers the basics to determine pressure and volume for a system which is essential for a hygienist to help with design specifications and assessment of a system for control effectiveness.

6.4.1 Key Considerations

For LEV, design considerations include the following:

- The toxicity and pattern of release of contaminated air. Is LEV needed or is there a better solution?
- Hood design and capture efficiency. Factors to consider include the specific work task, the movement of people past the hood, and the potential impact of cross draughts.

- Transport velocity of the contaminated air within ducts. Factors to consider include the size of ducts to ensure that the hood can operate as designed, and the roughness of the ducts' surfaces, which will impact both the airflow and the noise generated within the ductwork due to air movement.
- The type of air filtration required: this can be researched through the use of standards (e.g. ASHRAE) or from commercial suppliers of the filters.
- Fan selection. Factors to consider include efficiency, power, size, noise, and cost.
- Fan location: extraction fans should ideally be placed outside of the building so that the low-pressure part of the system is internal and the high pressure on the external side of the building. That way any holes in the ductwork on the inside of the building will continue to extract contaminated air, but any holes on the outside of the building, after the fan will exhaust contaminated air out of the hole into the ambient air. Ductwork should have minimal (smooth) direction changes, be sufficiently strong, be well supported, and be capable of withstanding normal wear and tear that may be brought about by the contaminant.
- The ease of maintenance (accessibility) and monitoring of ventilation systems.
- Noise prediction and reduction.
- Design and location of stacks. Factors to consider include the risk of re-entrainment due to air inlets within close proximity and impact to other external populated areas such as courtyards, and break areas. AS/NZS 1668.1:2015 discusses the issues in the design and location of stacks (Standards Australia, 2015).
- Explosion and/or flame prevention design principles when dealing with explosive gases, vapours, and flammable dusts.

6.4.2 Pressure Drops

In Australia, the United Kingdom, and Europe, pressure drops in LEV systems are estimated with the 'velocity pressure' method, where the pressure drop is estimated for each straight run, bend, or join. In the United States, the trend has been towards using the 'equivalent foot' method, where the pressure drops for bends and joins are estimated for a specific length of ducting. Outside of the United States, this is known as 'static regain design'. For complex systems, the flows are balanced by means of air regulators at the design stage so that the designed flow rates at each hood are achieved by making the pressure drop to the same extent along each 'leg'.

6.5 GENERAL VENTILATION

General ventilation is aimed at reducing the concentration of a contaminant by adding fresh air to the workplace. There are two categories: dilution ventilation and displacement ventilation.

6.5.1 Dilution Ventilation

In dilution ventilation, the contaminant is mixed with fresh air, diluting it. The air in the whole room may be mixed with the fresh air or a stream of air from a fan or an open window may perform the task more locally. If natural airflows are used to dilute contaminants, wind direction, wind speed, and air temperatures are likely to have significant impacts on the effectiveness of this approach.

6.5.1.1 Use

Dilution ventilation may be appropriate when:

- the air contaminant has low toxicity;
- there are multiple sources, that is, not a single point source;
- the emission is continuous;
- the concentrations are close to or lower than the workplace exposure standard;
- the volume of air needed is manageable;
- the contaminants can be sufficiently diluted before inhalation;
- comfort (or odour) is the issue, in the absence of other contaminant toxic effects;
- a spill has occurred, and extended airing of the workspace is needed.

6.5.1.2 Limitations

Extraction fans are sometimes mounted above a workbench to remove contaminated air. Since, in reality, the air moves towards the fan from all directions in a collapsing hemisphere (Figure 6.6), only a small amount of the contaminated air from the bench is actually extracted. In the case of Figure 6.6, the fan is extracting from the whole room, the table happens to be placed closest to the extraction point; however, the premise shown in Figure 6.6 is true regardless of where the extraction point is set.

In many cases, poorly designed ventilation systems that aim to remove contaminated air from around a worker (Figure 6.6) only dilute the contaminant, giving little local protection to the worker, and only slowly cleaning the air in the workplace as a whole.

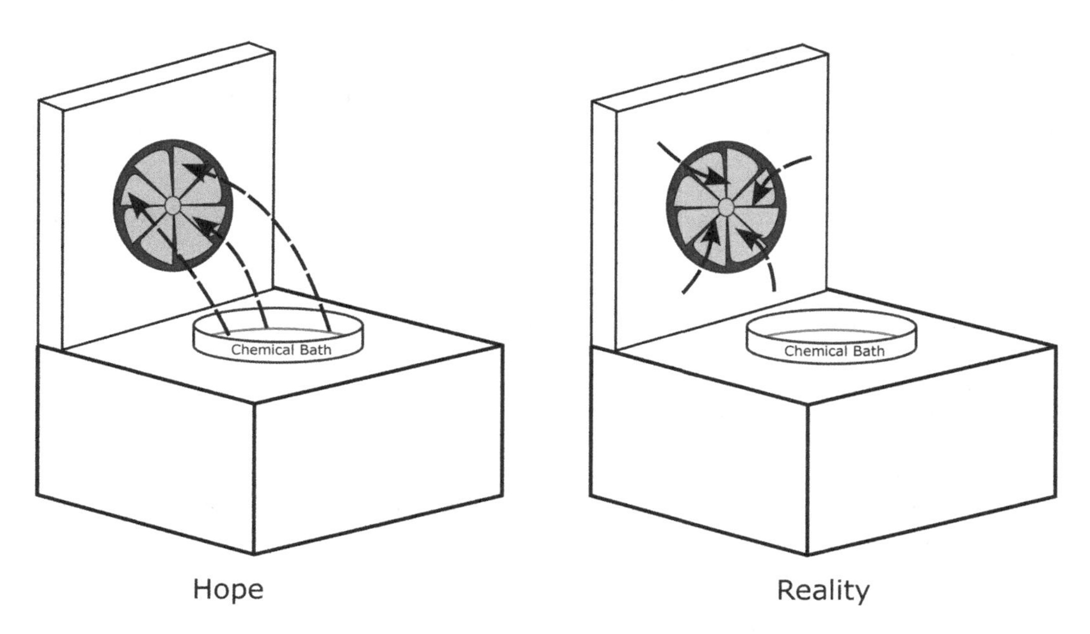

FIGURE 6.6 A fan in the wall: hope for the extraction zone (left) and reality of the extraction zone (right).

Source: M. Reed.

6.5.1.3 Air Exchanges and Mixing

Dilution ventilation of contaminated air works in much the same way that pouring clean water into a jug of coloured water dilutes the colour. If the added water forms a layer on the top and overflows the jug without mixing with the coloured water, then the colour (contaminant) persists. If the fresh water and the coloured water are well mixed, the colour gradually fades as it is diluted. However, traces of the chemical colourant remain long after all visible colour has been removed.

Air changes per hour or ACH is a calculation that describes how many times per hour the entire volume of air in a room is replaced. Calculation of the ACH allows for the air exchanges for rooms of different sizes to be normalised so that they can easily be compared.

Calculating air exchanges per hour in a room:

$$\text{ACH} = \frac{\text{time}\left[3600\,\text{seconds}\right]\times\text{volume of air exhausted}\left[\text{or supplied in m}^3/\text{s}\right]}{\text{volume of the room}\left[\text{m}^3\right]} \tag{6.9}$$

$$\text{ACH} = \frac{3600\times Q}{V}$$

where

V = volume of room in m^3

Q = quantity of air supplied in m^3/s

ACH = air exchanges per hour

Note that the 3600 seconds converts Q from m^3/s to m^3/hour.

6.5.1.4 Decay Rate Calculations

In dilution ventilation, the basic formula to describe the exponential reduction in contaminant concentration (C) over time (after removal of the source) is given by Equation (6.10), if we assume complete mixing:

$$C = C_0\,e^{-Rt} \tag{6.10}$$

where

C_0 = the initial contaminant concentration in air by volume (ppm)

V = volume of ventilated space (m^3)

T = time in seconds (s)

R = air exchange rate (s) [Q/v = t], where Q is the airflow rate into the space (m^3/s).

As an example, in a room of 10 m^3, with 1000 ppm contaminant in the air and a diluting airflow of 0.1 m^3/s, the concentration in the room 10 minutes after the source has stopped is calculated using Equation (6.10), as follows:

Initial contaminant concentration: C_0 = 1000 ppm

Volume of room: V = 10 m^3

Time since ventilation stopped: t = 10 minutes or 600 s

Airflow rate into the space: Q = 0.1 m^3/s

Therefore,

air exchange rate: Q/V = 0.1/10 = 0.01 s

and using Equation (6.10)

$$C = C_o \times e^{-(R \times t)} = 1000 \times e^{-(0.01 \times 600)} = 1000 \times e^{-6} = 1000 \times 0.002479 = 2.479 = 2.5\,\text{ppm}$$

This is a small fraction of the original concentration, but it is never zero.

The exponential decay of the contaminant over time is shown in Figure 6.7 on the linear plot, with the same data, plotted on a logarithmic scale, producing a straight line. The slope of the line represents the rate of contaminant decrease, which in turn is indicative of air changes, usually as air changes per hour.

This calculation would underestimate the final concentration, as mixing is always incomplete; hence, the flow should be increased by a safety factor, K, of between 3 and 10 to calculate the expected concentration at a given time (ACGIH 2023). The K factor accounts for potential uncertainties and deviations from ideal conditions to ensure adequate worker protection.

Circumstances:

- well-designed ventilation systems with consistent contaminant release: K factor closer to 3;
- poorly designed systems or highly variable contaminant sources: K factor closer to 10.

The American Conference of Governmental Industrial Hygienists (ACGIH, 2023) also suggests that proximity, location, and number of sources should be considered when selecting the 'K' factor'. K can be incorporated into the dilution ventilation equation (Equation (6.11)) as:

$$C = C_0 e^{-Rt/K} \qquad (6.11)$$

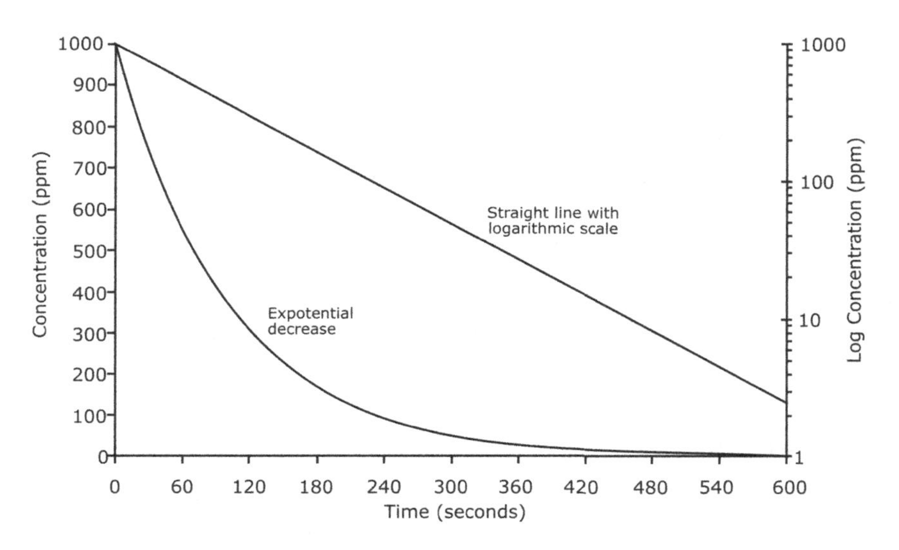

FIGURE 6.7 Impact of dilution ventilation represented on linear and logarithmic scales.

Source: D. Bromwich.

where
C = final contaminant concentration in air by volume (ppm)
C_0 = initial contaminant concentration in air by volume (ppm)
R = air exchange rate (s) [Q/v = t]
t = time in seconds (s)
K = safety factor.

Many texts give tables of K, terming it a 'mixing and safety factor', but this does not take into account either the variability of local concentrations or the degree of mixing (Feigley et al. 2002).

The degree of mixing of air in a room largely determines how well dilution ventilation works. For example, it is not uncommon for fresh air in an air-conditioned office to be introduced through slots in the ceiling, only to travel along the ceiling, with little effect on the air quality for the occupants. If the air inlet is directed downwards to achieve better mixing, however, some workers might complain of draughts. By ensuring good mixing of fresh and contaminated air while limiting draughts, the contaminated air is diluted and, so long as no more contaminant is introduced, the contaminant concentration will reduce exponentially over time.

To dilute the air after a (volatile) contaminant spill, the required flow, Q, is defined by Equation (6.12):

$$Q = \frac{r}{\rho \times ES} \tag{6.12}$$

where
Q = flow rate (m^3/s)
r = rate of evaporation (mg/s)
ρ = density of liquid (kg/m^3)
ES = exposure standard (ppm).

The required flow rate would need to be multiplied by the safety factor K (3–10) to account for incomplete mixing.

6.5.1.5 Key Considerations

Some things to watch for when considering ventilation problems:

- is there make-up air to replace the exhausted air;
- is the incoming or make-up air always clean;
- where does the contaminated air go; and
- if the ventilation is natural, what happens on still days or when the wind blows from another direction?

6.5.2 Confined Spaces

Confined spaces pose unique ventilation challenges due to potential toxins, flammable gases, and lack of oxygen. Entry requires specialised training and certification. Here is the gist.

- **Safety first**: prioritise contaminant (toxic, flammable) and oxygen monitoring throughout entry and occupancy.

- **Ventilation strategies**:
  - air exchange: provide fresh air to displace contaminants and maintain a breathable atmosphere;
  - source extraction: capture contaminants at their source for efficient removal;
  - air supply units: use as a last resort with redundancy for emergencies.
- **Planning considerations**:
  - space dimensions: size and entry points influence ventilation needs;
  - contaminants: identify types and quantities for effective removal;
  - supply air source: ensure compatibility with confined space atmosphere.

Remember, confined space ventilation demands careful planning and specialised expertise. Prioritise safety and follow established guidelines for a successful operation. It is advised that additional resources are consulted to fully understand this topic, such as the Safe Work Australia Model Code of Practice on Confined Spaces and the Australian Standard on Confined spaces as well as similar documents by the HSE and NIOSH (HSE 2013; Safe Work Australia 2021; Standards Australia 2009).

6.5.3 Thermal Displacement Ventilation (TDV)

In TDV, contaminated air is removed from the work area by the plume of warmer, more buoyant air that is generated from people and work processes. Figure 6.8 compares TDV with dilution ventilation.

Air that is around 2–3 °C cooler is gently introduced near floor level to fill the area to a height of 2–3 metres, in a process similar to filling a swimming pool with water. Special diffusers (often long fabric ducts) are used to ensure that the airflow does not create draughts or settle. Unlike conditioned air, which is usually cool and fed in from vents in the ceiling, displacement ventilation works with natural convection patterns and can be much more efficient than dilution ventilation because only the occupied depth of the room is cooled. Without such a temperature gradient, the whole room would have to fill to remove contaminated air.

In Figure 6.8, the contaminant concentration in the standing person's breathing zone is more effectively reduced by TDV than by traditional dilution ventilation. This method not only works better but also saves energy.

6.5.3.1 Use

TDV works best when:

- the contaminants are warmer than the surrounding air;
- the supply air is slightly cooler than the surrounding air;
- the ceiling is relatively high (more than 3 metres); and
- there is limited movement in the room and there are no major openings.

6.5.3.2 Limitations

TDV:

- is ineffective for lighter-than-air contaminants: gaseous pollutants like VOCs and lighter-than-air gases can rise with the warm air plume, potentially exposing occupants;
- has limited mixing at high heat loads or obstructions (strong heat sources or furniture placement can disrupt the thermal gradient and lead to pockets of higher contaminant concentrations);

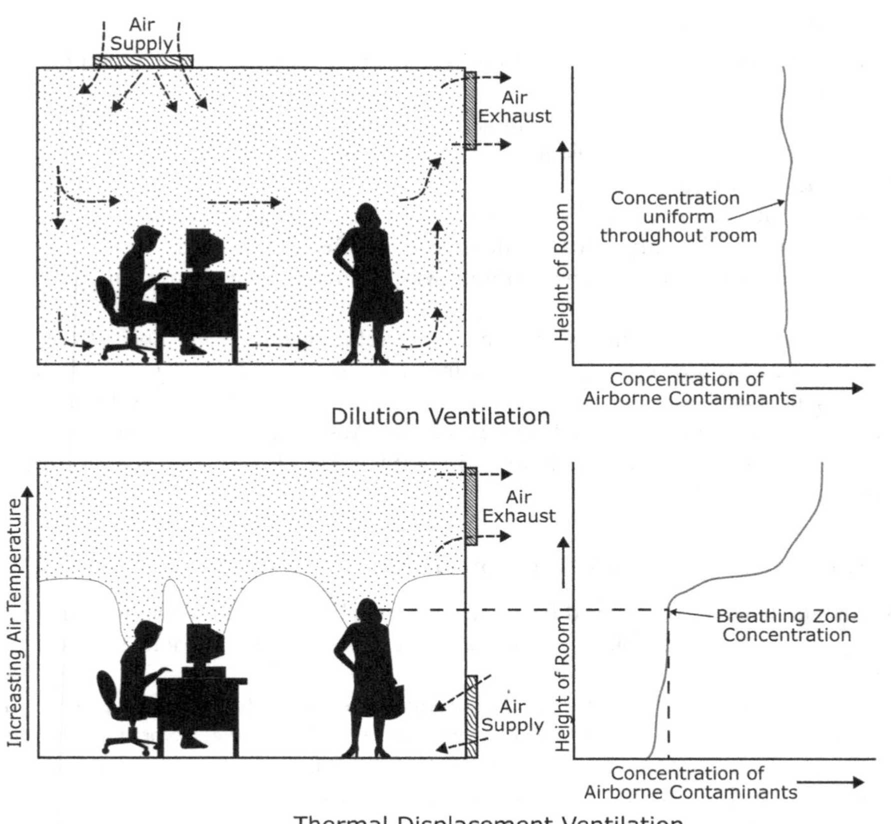

FIGURE 6.8 Dilution ventilation and thermal displacement ventilation.

Adapted from: Skistad 1994, pp. 7–9.

- has slower removal of contaminants compared to dilution ventilation: TDV may take longer to remove airborne contaminants, especially in large spaces or with low supply airflow rates; and
- has draft concerns: depending on the design and airflow rates, TDV can create cold drafts near the floor, impacting occupant comfort.

If the ceiling height is less than 2.3 metres, the contaminated air is cooler than the incoming air, or there are draughts, then displacement ventilation does not work as well. Cold windows or walls can create downdraughts, and hot spots in the sun can create local updraughts.

6.6 LOCAL EXHAUST VENTILATION

LEV aims to remove air contaminants at the source before they have a chance to be inhaled. Although poorly understood at the time, the principles of LEV have been known for over a century. A simple LEV system (Figure 6.9) most commonly comprises a hood to capture and remove contaminated air near the point of release, ducting to connect the hood to an air-cleaning system, a fan to move the air through the system, and an exhaust stack outside the building to disperse the cleaned air.

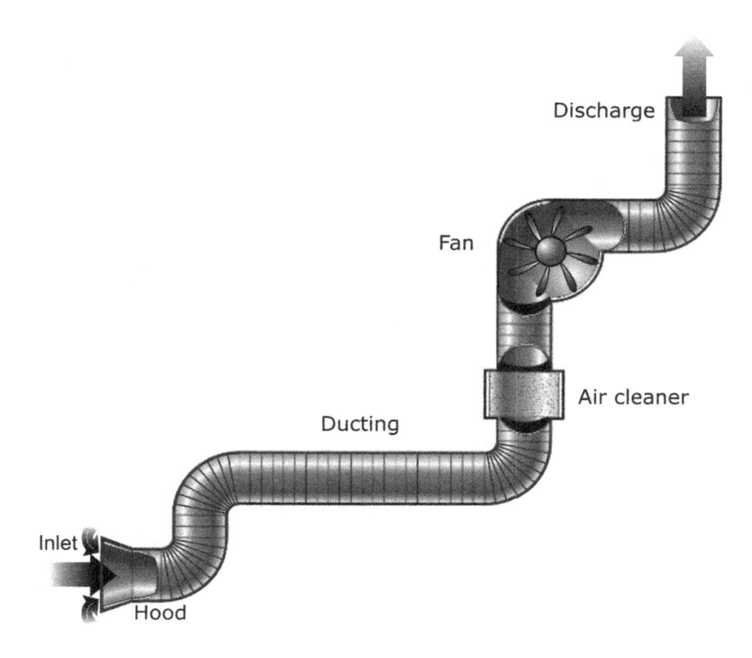

FIGURE 6.9 Components of a simple LEV system.

Adapted from: HSE 2017, p. 7.

6.6.1 Good Design

The weak point in LEV design tends to be in the hood, the function of which is to shape the flow of contaminated air and ensure its efficient capture. To work as designed, a hood requires a predetermined air inflow rate, which in turn influences the design of the rest of the LEV system. The maximum distance of the hood from the contaminant source can generally be no more than 25 cm. If this is not considered during the design phase and the hood is placed as an afterthought, air capture efficiency is likely to be compromised.

In Figure 6.10, workers are finishing fired earthenware with emery paper, producing dust containing silica. This example shows that modern technology is not necessary where there is good design. The hoods are ventilated enclosures with a glass top to let in light. Each hood is joined to a duct that is attached to a centrifugal fan. The exhaust from the (unguarded) belt-powered fan is cleaned by a cyclone. Although good for its time, this design could be improved by cleaning the air before it reached the fan to lessen the wear and tear on the fan blades and limit the deposition of particulates in the ducting.

Before a LEV system is designed, the potential for exposure to toxic materials should be examined so that the required degree of protection can be estimated. This assumes that a LEV system is the best approach and that the hazard cannot be eliminated by substitution of the hazardous material or through isolation of the process.

If LEV is the preferred approach, then an appropriate hood is designed to capture the contaminated air, and the airflow rate needed for the hood to function is calculated. When considering the diameter of any duct, it should be noted that a large duct costs more and may result in settling of particulates within the duct. In contrast, a small duct may result in large pressure drops or unacceptable noise, which depending on its intensity, may result in the system being turned off. Consideration must also be given to the presence of duct bends and joins. As with merging traffic on a highway, a duct's size should increase gradually as the airflow (traffic) increases so the

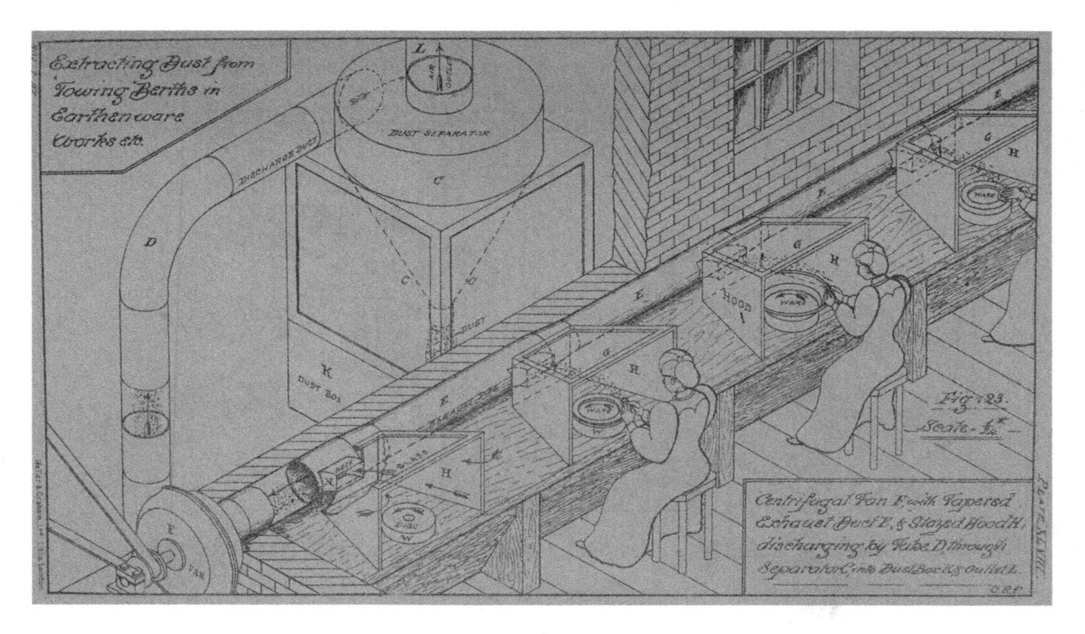

FIGURE 6.10 LEV system in a pottery.

Source: Great Britain Home Office Committee on Ventilation of Factories and Workshops 1907 p. 58.

velocity remains the same. The contaminated air may be filtered before being released into the atmosphere, with the air-cleaning device placed ahead of the fan that sucks the air through the system, so that the fan is protected.

Once the air has passed through the fan, it is pushed through a stack under pressure. In an exhausting LEV system, the ductwork in front of the fan is under negative pressure. Therefore, any leaks will result in lower capture velocities (Standards Australia 2012). The ductwork located after the fan is under positive pressure and any leaks will result in contaminated air leaking outwards. It is therefore optimal to place the fan outside of the building. Where this is not practical a routine inspection program should be implemented to identify any leaks. Alternatively, real-time pressure sensors can be installed to identify a sudden or gradual drop in pressure that could be brought about by a leak. In determining the height of the stack, planners should take into consideration buildings and structures both upwind and downwind to ensure that the air released does not enter other inhabited buildings. As a rule of thumb, the stack should be at least 2.5 times the height of surrounding buildings or countryside to minimise turbulence and re-entrainment of exhaust into surrounding buildings.

Lastly, the type and size of the fan should be chosen to produce the desired flow into the hood and overcome any pressure drops within the system. The components of a LEV system will now be considered in their design order.

6.6.2 Hoods

Hoods are the most important part of a LEV system, as a poorly designed hood limits the performance of the whole system. It is not uncommon to see a complex system installed with little regard to how effectively a hood will capture the contaminated air from the process or machinery to which it is attached.

6.6.2.1 Key Requirements

The most efficient hoods smoothly accelerate air from near stationary to the proper duct velocity with few eddies or changes of direction. It is difficult to capture billowing contaminants without first enclosing them in a larger volume. Similarly, high-velocity air contaminants from operations such as grinding are usually intercepted, and the residual contaminants then captured.

Simple principles that can make hoods more effective include:

- reduction of the source of emissions—such as closing the lid on a vessel, making the process wet to reduce dust, or using premixed formulations to limit the dustiness of a toxic component;
- placing the hood as close as possible to the source, preferably enclosing it;
- if the source includes fast-moving particles, positioning the hood to receive those particles (such as on a grinder or cut-off saw);
- specifying a 'capture velocity' greater than the particle velocity;
- locating the hood such that an imaginary line drawn from the operator's face to the contaminant source leads directly towards the hood.

6.6.2.2 Capture Zone and Capture Velocity

When installing a hood, it is first necessary to determine the zone where the air contaminants are generated and need to be captured from. This in turn requires estimates of the air velocity needed at the edges of the zone to ensure that capture occurs. In Figure 6.11, the trajectories of air into a slot hood are visualised in the laboratory using a smoke tube, the chalk lines represent the lines of airflow into the hood, and a capture zone for an air speed of 0.25 m/s (measured with a hot wire anemometer) is shown by the line intersecting the air trajectory lines.

Air is drawn from all directions into the hood and almost as much uncontaminated air enters this unflanged hood from behind as from in front.

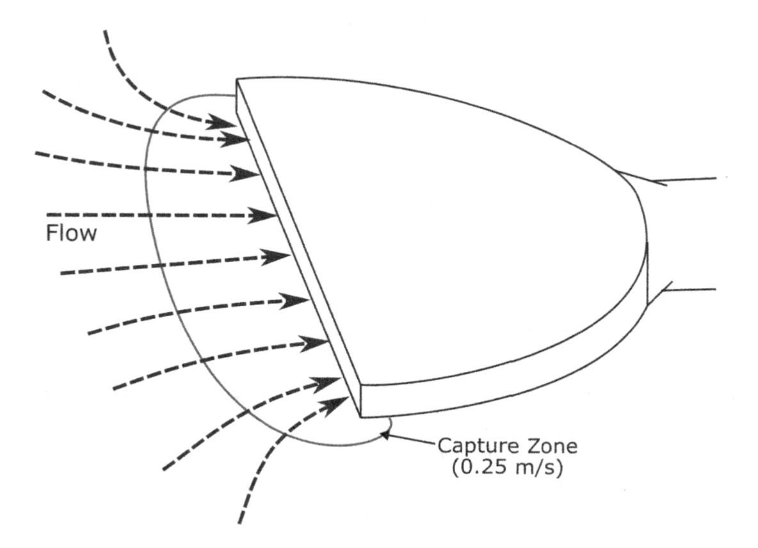

FIGURE 6.11 Airflow lines and 0.25 m/s capture zone for a slot hood.

Source: D. Bromwich.

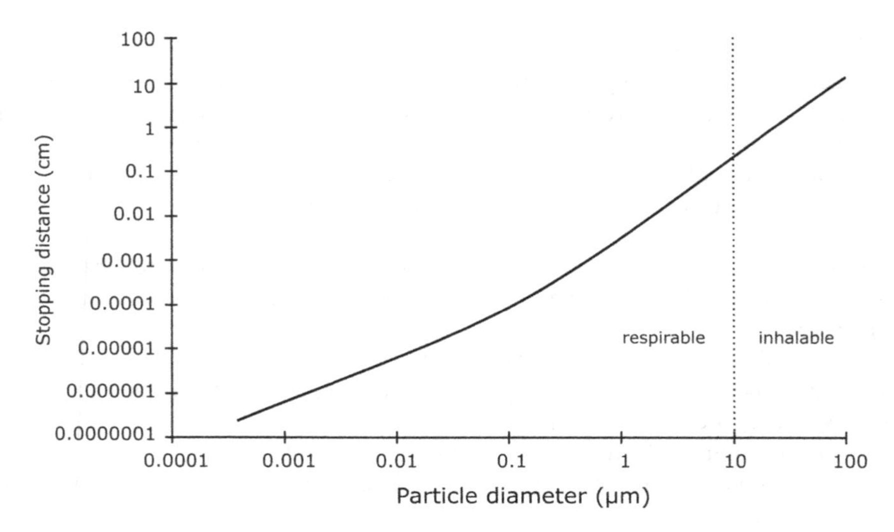

FIGURE 6.12 Stopping distance of unit-density particles: it is a useful concept for estimating how far a particle will travel before it can be affected by airflow into a hood.

Source: D Bromwich.

The 'stopping distance', S, is a useful concept for estimating how far a particle will travel before it can be affected by airflow into a hood. For particles with the density of water (unit density) that can be inhaled (<100 µm diameter) and travelling at 10 m/s at normal temperature and pressure (NTP), the stopping distance varies from 13 cm at 100 µm diameter to 88 µm at 1 µm diameter. The stopping distance can be visualised in Figure 6.12, which shows the stopping distance for given particle diameters and a density. (NTP is defined as air at 20°C [293.15K] and 1 atmosphere [101.325 kPa].)

Particles greater than 100 µm diameter tend not to be inhaled and can move some distance before the viscous drag from the air slows them. They may be more of a housekeeping problem than an inhalation issue. The stopping distance of very heavy (lead) or very light particles (glass micro-balloons or expanded styrofoam beads) has to be scaled by their density, since light particles penetrate deeper and more easily into the lungs.

6.6.2.3 Hood Types

There are three main types of hoods, capturing, receiving, and enclosing, these hood types are shown in Figure 6.13. Flanges (mainly on receiving hoods) and baffles (mainly on enclosing hoods) can help shape the airflow and make the hoods more efficient.

Flanges inside a hood help shape the airflow and reduce the amount of uncontaminated air entering the hood. Many hoods are poorly shaped, making them very inefficient at capturing air contaminants. The most efficient hoods avoid eddies, so that most of the suction accelerates the contaminated air to the duct velocity smoothly and with few eddies. However, a hood that is highly efficient at collecting contaminated air may obstruct the work process and get removed. Close observation and investigation of work practices are therefore needed to design a hood that is both acceptable to the workers and efficient at collecting contaminated air before it is inhaled.

In Figure 6.14, the flange prevents most clean air coming from behind the hood from being captured and so extends the range of the hood to capture more contaminated air with the same airflow. It also decreases pressure drops associated with the hood, making it about 25 per cent more efficient.

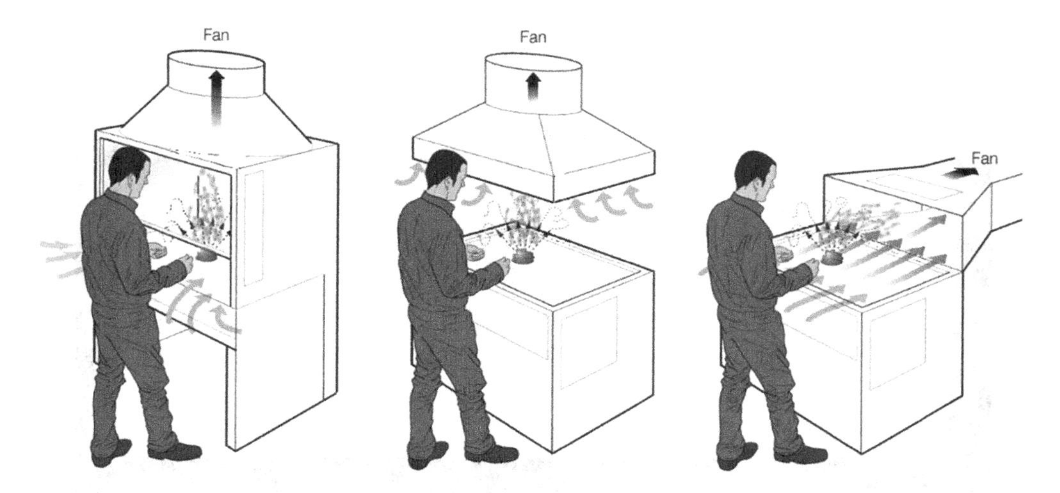

FIGURE 6.13 Enclosing, receiving, and capturing hoods (left to right).

Source: HSE 2017, p. 30, Figures 11, 12 & 13.

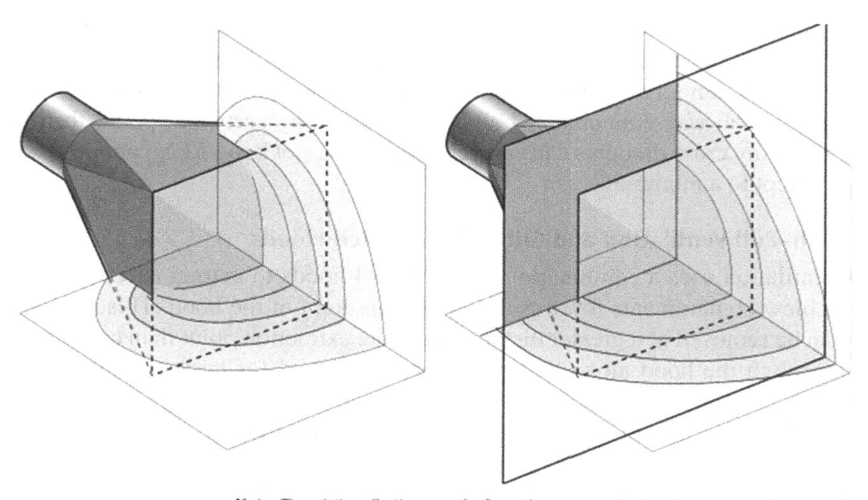

FIGURE 6.14 Velocity contours: flanged duct and plain duct end.

Source: HSE 2017, p. 47, Figure 32.

A hood may have internal baffles to better distribute the airflow and improve its performance. Almost all fume cupboards contain a rear baffle.

6.6.2.4 Slots

Where the contaminant is not a point source—such as in a drum-filling operation or a dipping tank—long hoods with slots can make capture more efficient (see Figure 6.11). The hood may be straight or curved but, most importantly, it should be very close to the source of emission, as effectiveness drops rapidly with distance. It is very common for the capture range of a slot hood to be grossly overestimated, particularly in the case of drums, tanks and large work surfaces.

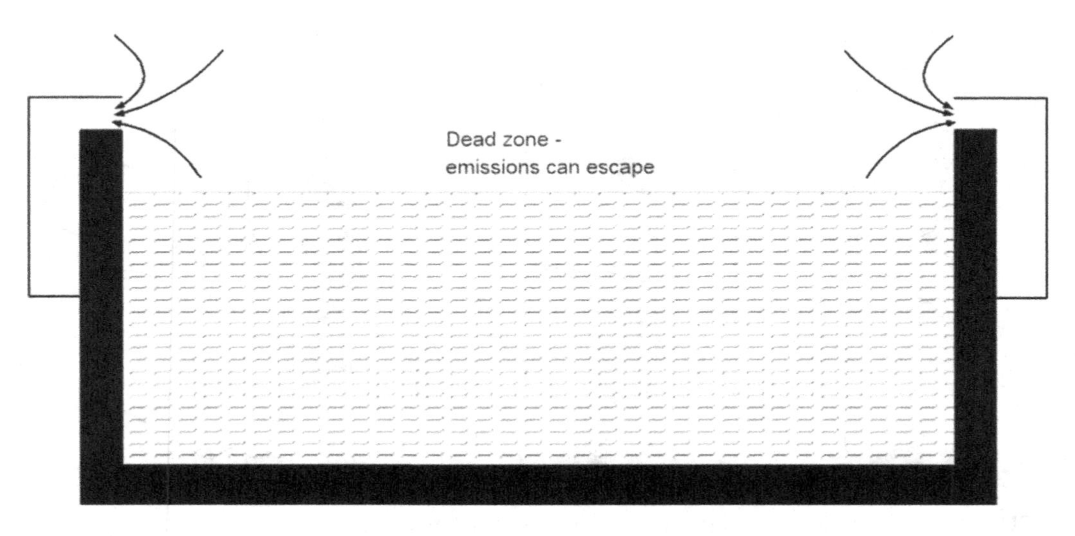

FIGURE 6.15 Slot hoods along the length of a tank (Adapted from Alden & Kane 1982).

For a long tank, a slot hood along each side can be effective (Figure 6.15), but this may create a dead zone down the middle of the tank where along the centre line of the tank, the hazardous emissions are drawn in both directions equally and can escape into the workplace. If it is impossible to predict which way contaminated air will move at any particular point, then there is a design problem. Any cross draughts can severely impact the slot hood LEV capture effectiveness and should be kept to a minimum.

6.6.2.5 Push–Pull Ventilation and Other Air-Assisted Hoods

Push–pull ventilation uses a combination of traditional hoods to extract air contaminants, and jets of air to blow contaminants towards the extraction point of the hood. This means that contaminants can be removed at a greater distance from the extraction point than could be managed by suction through the hood alone. It can be very successful for large rectangular areas like tanks, where there is slot ventilation at one end and a jet of air is blown across the tank from the opposite side. It can also be used to avoid much of the swirling of air that occurs in front of a worker.

In Figure 6.16, on the left-hand side, it can be seen that the tank is too wide for capture slots to be effective, with the vapour not being extracted, but instead entering the ambient environment. However, when push–pull ventilation is used (see the right-hand side of Figure 6.16), the air inlet on the right of the tank blows the vapour across the tank towards the hood, carrying and entraining the contaminant cloud. In push–pull systems, flows of extracted air must significantly exceed those of supply air, if they do not, then the blown air will be forced out of the tank through the open top.

The major drawback of push–pull systems is that when the air jet is interrupted (e.g. as objects are added to or removed from the tank or there are drafts to interfere with the flow of air), contaminants can be deflected into the workplace.

Various attempts have been made to achieve directional airflow with hoods to extend their range. There has been some success with air-assisted Aaberg-type hoods (Cao et al. 2021). In Figure 6.17, the increase in capture range from a simple flanged circular hood (bottom) and Aaberg circular hood (top) is shown. The circumferential jet of air used by the Aaberg circular hood (right) entrains clean air that would otherwise be captured, leaving a tunnel in front of the

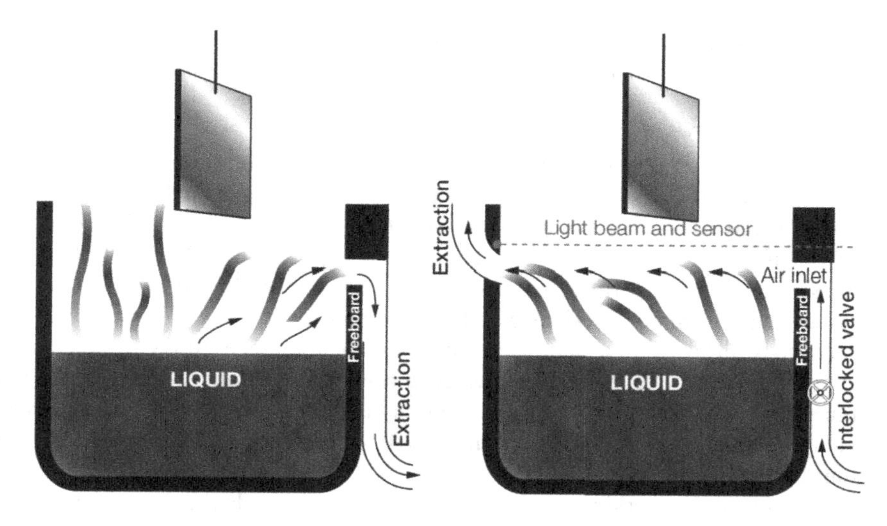

FIGURE 6.16 Push–pull applied to an open-surface tank.

Source: HSE UK, 2017, p. 41, Figure 29.

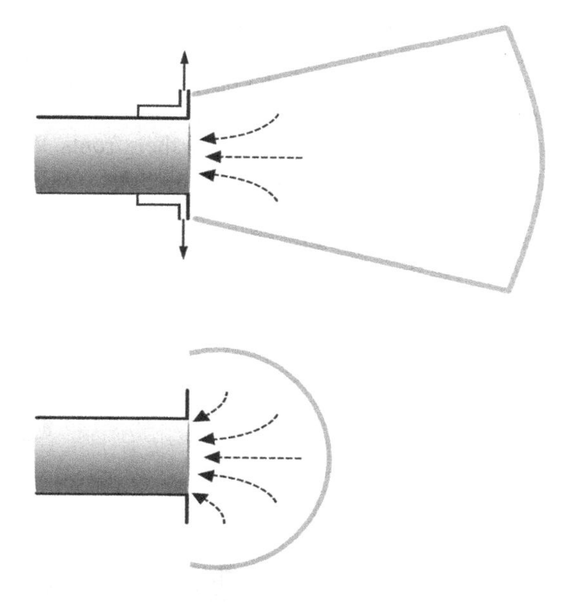

FIGURE 6.17 Aaberg-type hoods (top) can greatly extend the capture range of a simple flanged hood (bottom).

Source E. Lindars.

hood to capture contaminated air from a greater distance. A linear version of the air jet can be used to shape the airflow into a slot hood.

Air-assisted hoods largely overcome the problem that arises with push–pull ventilation when objects are raised from dipping tanks, as there is no jet of air directed at the contaminated object. However, if the air jets designed to entrain clean air also entrain some contaminated air, then they will spread this contaminated air about the workplace, as such assurances of optimal working conditions are essential. Air-assisted hoods may also be unacceptably noisy.

6.6.2.6 Fume Cupboards

Fume cupboards are a special type of enclosure often found in laboratories. Good designs have been available for over 50 years, but design flaws can still limit a fume cupboard's effectiveness. While a respirator may reduce air contaminants by a factor of between 10 and several hundred, even a poorly designed fume cupboard may reduce toxic exposure by a factor of 1000 or more. However, the slightest reverse airflow inside the fume cupboard will limit its effectiveness. The best designs can outperform poor designs by a factor of 10–100.

It is easy to track airflow into a fume cupboard with a smoke tube. Many technicians testing fume-cupboards release smoke outside and watch it flow into the fume-cupboard and wrongly report the fume cupboard as working well. In assessing the effectiveness of a fume cupboard the question that needs to be answered is 'how much smoke (representing contaminated air) released inside the fume cupboard escapes?'.

Smoke released near the working surface will often move towards the front of the fume cupboard to take the place of air entrained in the flow near the front sill (see Figure 6.18 a) or just inside the sash when the incoming air cannot follow the contours of the sill, which is often raised. This smoke that has moved toward the sill (or the sash), like the smoke of a barbecue—becomes entrained in the air swirling in front of the worker. The problem is compounded by the air wake left by passers-by and the thermal plume of warm air generated by the fume cupboard operator (see Figure 6.18b). This greatly lowers the degree of protection offered by the fume cupboard.

If the fume hood has a sash, then it must be at the operational height when the hood is in use. Many older fume cupboards currently in use in schools, industry, and tertiary institutions do not have aerodynamic surfaces, and reverse airflows occur near their fronts. Working close to the sash to see inside greatly increases exposure to contaminated air.

Airflow in fume cupboard ducts should not generally be more than 7 m/s to mitigate excessive noise; however, it is recognised that for the extraction of heavy particulates up to 20 m/s may be required, along with specific duct design to stop these solids settling out within the ductwork (AS/NZS 2243.8).

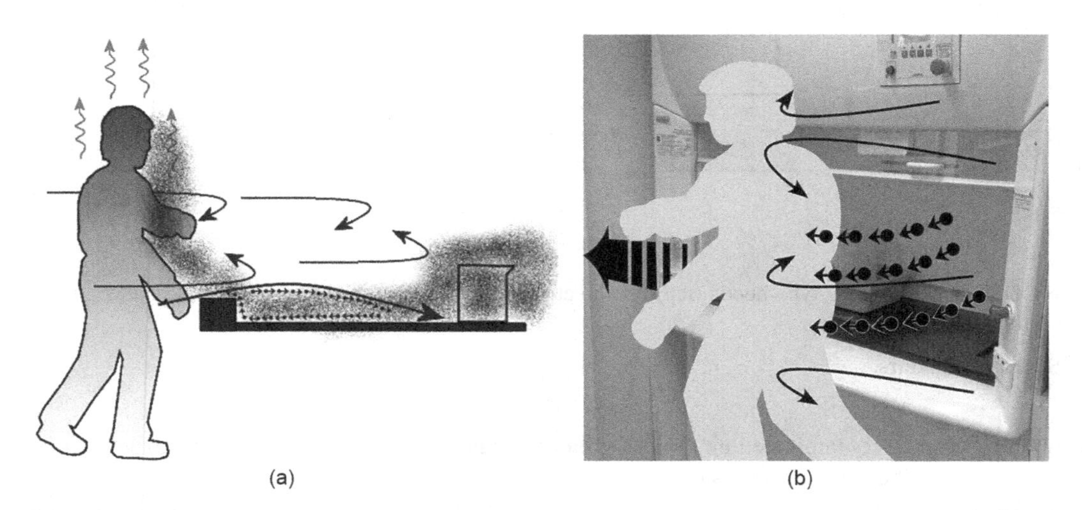

FIGURE 6.18 Complex airflows in front of a fume cupboard increase the exposure of the user.

Source: D. Bromwich.

Potentially contaminated air is not permitted to be used for fume cupboard make-up air. As such, the storage of chemicals in cabinets or benches ventilated via the fume cupboard system is not allowed under AS/NZS 2243.8 nor is the storage of chemicals and other equipment within a fume cupboard, as this will likely reduce their effectiveness (Standards Australia 2014).

6.6.3 Hot Processes

Designing ventilation for hot processes must take into account how to manage the buoyant air generated from the heat source (Figure 6.19). In such instances, the canopy must be large enough to capture fume from the hot process as it rises and spreads laterally. Rising and expanding fume is complicated by the following.

- Air extraction will likely cool the process and result in greater energy demands to maintain the required hot process temperatures.
- Air extraction may also take cool air-conditioned air out of the area, resulting in greater energy demands to replace the cool air.

This additional energy required to maintain equilibrium will result in increased costs. As such, the most efficient method of extracting contaminants is likely to be directly from the breathing zone of workers.

Where hot processes such as welding are undertaken, the extraction ventilation may extract the warm, contaminated air towards and across the worker's face. Replacement air then swirls past the worker, so contaminated air moves in exactly the wrong direction (as in Figure 6.18, left).

For ventilation of moveable processes, the canopies must be sufficient to mitigate the contaminants released at all stages of the process. For example, in smelting, the canopies are often

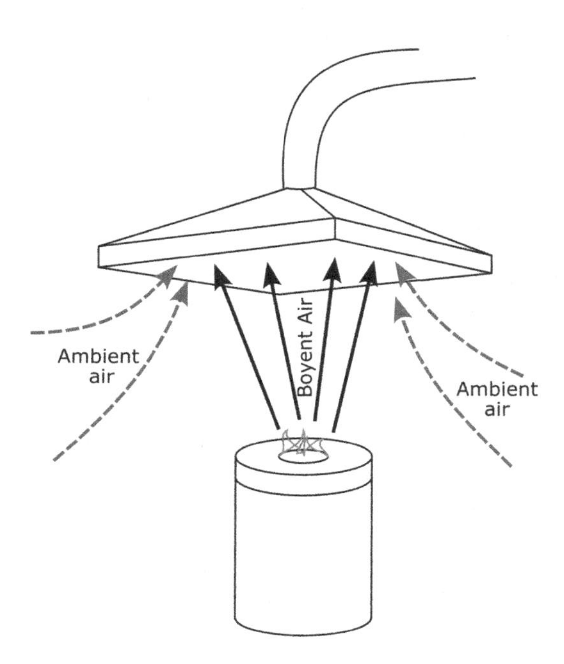

FIGURE 6.19 The arrows indicate the spread of the fume from the source as the fume rises and expands.

Source: M. Reed.

FIGURE 6.20 (a) Furnace and canopy with the arrows indicating the direction of the fume once pouring commences. (b) The pour of molten metal and its associated fume shows how ineffective the canopy hood is once the pour commences.

Source: E. Lindars.

designed to extract the fume during the smelting process (Figure 6.20); however, when pouring commences, the furnace tips, causing the open face of the furnace and the molten material being poured into the mould to protrude outside of the canopy extraction zone (Figure 6.20b). Hence, it is critical that canopies for hot processes are designed, managed, and maintained by someone with training and experience in this type of ventilation. Some guidance for hot work is outlined by the ACGIH (2023).

Figure 6.20a shows a canopy placed over the furnace to extract the buoyant fume from the furnace during the smelting process. The arrows indicate the fume extraction zone falling within the canopy.

In Figure 6.20b, the arrows show the fume entering the workplace because the canopy cannot extract the fume during the pour.

6.6.3.1 Key Considerations

The key considerations for ventilating hot processes are:

- ensuring contaminated airflows away from the worker's face;
- minimising cross-drafts;
- defining the release area (area of the hot source) and the calculated destination area (hot air column diameter at the canopy face);
- the potential for a sudden release of a lot of contaminated air that overwhelms the collecting hood;
- setting the canopy at the correct height to ensure that all buoyant air is captured;
- consideration of an enclosure to aid with extraction; and
- the use of baffles for rectangular hoods.

6.6.3.2 Limitations

The following can result in reduced capture of contaminated air.

- Nearby structural walls can cause the flow of contaminated air to become attached to the walls and not be extracted into the ventilation system. This is known as the Coandă effect.
- High canopy hoods allow small cross-drafts to significantly affect the canopy extraction. They may also place a worker under the hood (see Figure 6.3).
- A lack of sidewalls or enclosures will reduce extraction efficiency.
- Hood shapes not matching the source shape will result in inefficient capture.

6.6.4 Hand Tools with LEV

Pneumatically powered and electrically powered hand tools are commonly designed with provision for dust extraction. This is particularly important in the construction industry, where concrete and manufactured stone benchtops produce toxic silica dust, and in the wood-working industry, where dusts from both hardwoods and softwoods have been associated with an increase in nasal cancers, occupational asthma, and dermatitis. Even with dust-extracting tools, a worker may still require properly selected and fitted respiratory protection, particularly with concrete and brick 'chasing' and other dusty grinding and cutting operations. In addition, other controls such as wet works could be undertaken.

Often all that is required to reduce the dust is for an appropriate vacuum to be connected to the tool. Although this simple method can reduce inhalable dust by up to 98 per cent, it is less effective with the very fine respirable particles. Respirable dust is invisible to the naked eye and tends to deposit not in the nose or throat but deep within the lungs. Some domestic vacuum cleaners may have high-efficiency particulate air (HEPA) filters which can remove 99.97 per cent of particulates 0.3 μm or larger. However they are not designed to meet industrial standards and have been found to pass up to 50 per cent of particles 0.35 μm in diameter (Trakumas et al. 2001). Toxic dusts collected by a vacuum cleaner can also become a potential source of air contamination when they are re-suspended inside the cleaner, particularly with cyclonic and wet vacuum cleaners.

The guns of some MIG and TIG welding torches are equipped with LEV designed to capture fumes but not interfere with the shielding gas.

6.6.5 Fans

Poor LEV design may require more powerful fans to achieve the required extraction efficiency, which costs more to buy and run, and may do little to protect the worker. As such, fan design and selection is a specialised area.

The relationship between the flow rate through a fan (Q), the pressure drop (P) it can produce, the fan speed (RPM), and size (d) are related and given in terms of fan laws (or fan affinity laws), as shown in Table 6.1. An understanding of the affinity laws is important to accurately match new design points of operation between fan and system curves.

The formulas in Table 6.1 shows how small increases in flow (Q) to compensate for poor hood design can result in large and costly increases in fan motor size—and power consumption. A design that requires a doubling of airflow means an eightfold increase in fan wattage, larger ducts, and higher running costs, as well as more noise.

TABLE 6.1
Fan Laws

Fan Law	Formula	Comment
Volume flow (Q)	$Q \propto RPM$ $Q_2/Q_1 = RPM_2/RPM_1$	Volumetric flow rate varies directly with fan speed (RPM)
Fan pressure (P)	$P_2/P_1 = (Q_2/Q_1)^2\ (\rho_2/\rho_1)$ $P_2/P_1 = (RPM_2/RPM_1)^2\ (\rho_2/\rho_1)$	Since fans are always in turbulent flow, pressure varies directly with the change in air density (ρ) and the square of the change in either flow rate (Q) or fan speed (RPM)
Fan power (PWR)	$PWR_2/PWR_1 = (Q_2/Q_1)^3\ (\rho_2/\rho_1)$ $PWR_2/PWR_1 = (RPM_2/RPM_1)^3\ (\rho_2/\rho_1)$	Since fans are always in turbulent flow, power (PWR) varies directly with the change in air stream density and the cube of the change in either flow rate (Q) or fan speed (RPM)

Adapted from: ACGIH (2023).

6.6.5.1 Types

Many types of fans are used for industrial ventilation, but in small industry, three types dominate: axial fans, forward-bladed centrifugal fans, and radial-bladed centrifugal fans (Figure 6.21a and b). Large axial fans are often used to move large volumes of air against a small resistance. Some axial fans are also used in LEV systems, being hidden inside a duct. In many systems, good flows

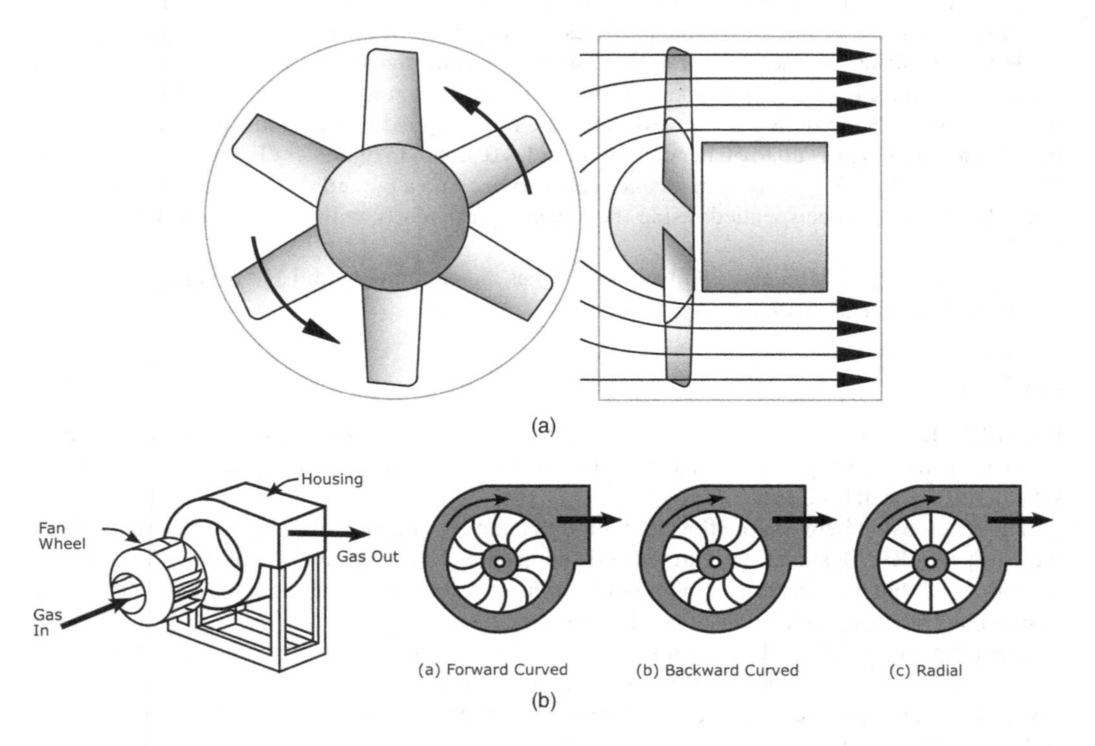

FIGURE 6.21 (a) A diagram of an axial fan **Source: HSE UK, 2017, page 55**. (b) Diagrams of a centrifugal fan and the types of fan blades that can be present: demonstrating the curvature of the fan blades.

Source: Wikimedia Commons.

can be obtained for long runs of ducts and air-cleaning devices with a forward-bladed centrifugal fan. If the air is contaminated, then a less efficient but more robust radial-bladed centrifugal fan may be required.

Regardless of the fan type, fans can be either single-phase or three-phase, where three-phase power is used for the higher loads usually used in commercial settings.

6.6.5.2 Which Fan Size to Choose

Once the airflow (Q in m^3/s) needed to make a LEV system work is estimated, and the total pressure drop (P in Pa) in the system is calculated, the wattage of the fan motor required can be calculated using Equation (6.13):

$$W = P \times Q \tag{6.13}$$

where

W = power of fan motor (kW)
P = total pressure drop (Pascal [Pa])
Q = flow rate (m^3/s)
E = fan efficiency.

However, allowance should also be made for the actual overall efficiency of the fan and the fan motor, which is generally about 60 per cent:

$$\text{power of fan motor}\left(W\right) = \frac{PQ}{E} \tag{6.14}$$

For example, to move 10 m^3/s of air with a total pressure drop of 250 Pa for the whole system with a 60% efficiency, the wattage of the fan motor would have to be:

$$\text{Power of fan motor} = \frac{250\left(\text{Pa}\right) \times 10\left(\text{m}^3/\text{s}\right)}{0.6\left(\text{overall efficiency}\right)} = 4\ \text{kW}$$

In practice, 4.5–5 kW would be chosen to allow for declines in efficiency over time.

6.6.5.3 Fan Curves

The ‘Fan laws’ are used by designers and show how the flow, pressure, and power usage are influenced by factors like fan speed (n) and fan diameter (d). A hood will need a design flow calculated to make it work properly, because an inefficient hood could make a huge difference to the size of the fan (initial cost and running cost) and the size of the ducting. Fan curves provide the optimal design pressure and volume flow for a ventilation system, this is the duty point.

A ventilation system can be represented by a curve (Figure 6.22) relating the pressure drop (p) and the airflow (Q), the system curve. The system curve is parabolic, with the pressure drop increasing with the square of the flow. The fan's performance can also be represented by a similar curve: this plot is known airflow at a given speed against the resistance of the air-cleaning devices, ducts, and hoods on the suction side and the ducts and stack on the exhaust pressure side. Each type of fan will have a different set of fan curves for different fan speeds—these fan curves are published by fan manufacturers. It is usual to select a duty point on the fan curve where the mechanical efficiency is high.

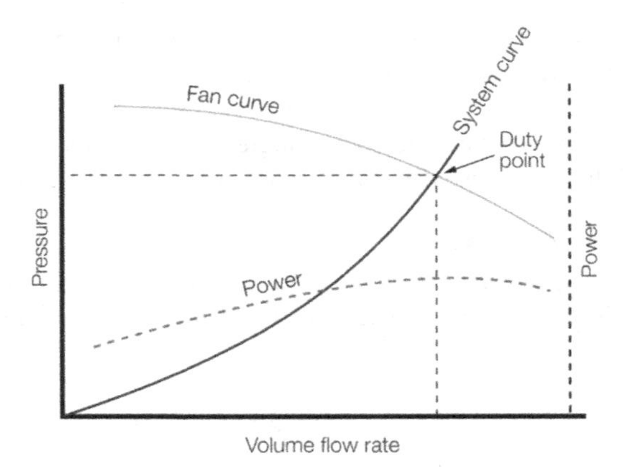

FIGURE 6.22 Fan and system curves

Source: HSE 2017, p. 57, Figure 35.

6.6.5.4 Possible Problem

After maintenance, it is possible for three-phase fan motors to be reconnected backwards, that is, two of the three wires can be changed causing the motor to change direction.

- With radial fans, the airflow is reversed and hence the miswiring is obvious.
- With centrifugal fans, the airflow continues in the same direction and the power consumption is much the same; however, the reversal of the centrifugal fan can greatly reduce the fan efficiency. When the wiring is corrected, it is not uncommon for a large amount of material that has settled in horizontal ducts to noisily hit the fan, so internal inspection and cleaning of ducts may be warranted before turning the system back on.

6.6.6 Ducting

6.6.6.1 Key Requirements

Ducts in LEV systems may be likened to highways, with merging lanes and sweeping bends. If there are unnecessary or sharp bends, greater pressure drops occur, and less suction is available to power the hoods, meaning that energy is wasted, and hoods underperform in capturing contaminated air. It is important to consider the following when designing LEV ductwork.

- If sections of duct bend at 90° angles, there will be increased turbulence where the airflow changes direction. It is common for an otherwise well-designed system to be impaired when the installers introduce additional sharp bends and tortuous paths around beams that greatly increase pressure drops and which encourages the deposition of particulates. Airflow is preferably smooth, with gradual corners and no abrupt changes in direction and speed. Badly installed ducting is usually easy to identify.
- Sections of duct should not join at 90^0 angles; the airflows should intersect at shallow angles and increase in duct diameter after the join.
- Avoid flexible ducting. Flexible ducting is sometimes supported by light spiral wire (and sometimes left coiled in roof spaces), but it is usually inappropriate through poor installation and multiple bends in the ducting.
- Duct diameter changes should be smooth and tapered, not abrupt.

- Ducts should be provided with 'cleanout ports' at strategic intervals to enable the removal of deposited particles and condensed vapours, and to permit inspection of the duct for clogging and corrosion.
- Duct material should be sturdy enough to prevent the material from being slowly eroded away during normal operation. Many materials may be used to form ducts, but steel sheeting and PVC pipe predominate. Abrasive or corrosive agents and clean operations may require special ducts, particularly on bends. For example, snorkel hoods in welding shops have tough, flexible ducts.

6.6.6.2 Duct Transport Velocity

If the air velocity in a duct is too low, then particulates will settle in straight sections of the duct and clog them. If the air velocity is too high, large pressure drops may develop, requiring a larger fan. High air velocities lead to excessive noise and ventilation systems will sometimes get turned off to avoid the noise. The minimum design velocities set out in Table 6.2 are based on the ACGIH (2023) recommendations.

6.6.6.3 Resistance

Hood and duct resistance are critical in maximising efficiency; hence, the design phase is important to ensure the energy used in the system is optimal for extraction of the contaminants. While calculations of pressure drops in ducts are beyond the scope of this book, their reduction will reduce resistance, which will improve efficiency. For example (ACGIH 2023), while a smaller duct will be cheaper at the design stage, it will have a larger pressure drop and need a larger fan (more expense), which will result in higher operating costs; however, a larger duct will provide a lower pressure drop and need a lower power fan (i.e. lower cost) and have lower ongoing energy consumption.

6.6.7 Air Treatment

There are many treatment technologies available; for some industrial plants, like power stations, the installations can be huge. In principle, exhaust contaminants can be considered as either particulates (particulates, fumes, and mist) or gases.

Although beyond the scope of this book, most texts on industrial ventilation will give a good introduction to this topic. Suppliers can be good sources of information on selection of the appropriate technology; however, their advice can be potentially conflicted.

TABLE 6.2
Duct Transport Velocities Needed to Transport Various Contaminants through Ducts

Type of Contaminant	Examples	Recommended Duct Velocity (m/s)
Gases and non-condensing vapours		5
Condensing vapours, fume, and smoke	Welding fumes	10
Low or medium density, low moisture content dusts, fine dusts, and mists	Plastic dust, sawdust, and flour	15
Process dust	Cement dust, brick dust, wood shavings, grinding dust, soap dust, brick cutting, and clay dust	Approx. 20
Large particles, aggregating and damp dusts	Metal turnings, moist cement dust, and compost	Approx. 25

Adapted from: HSE, 2017, p. 54, Table 12. Note the velocities in m/s are rounded from imperial conversions.

6.6.7.1 Organics

Organic vapour scrubbers that use molecular processes such as absorption, adsorption, condensation, and incineration are available to remove toxic gases and vapours. However, they will have a limited life-span as they become saturated over time requiring replacing. Scrubbers present in high humidity environments can become saturated and need replacing before the expected volume of contaminant has been absorbed due to the active sites being taken by water vapour and any other organic compounds extraneous to the contaminant.

6.6.7.2 Particulates

To remove particulates, a number of filtration and inertial separation technologies are used, including cyclones and inertial separators, electrostatic precipitators, and fabric filters (sometimes located in large 'bag houses'). Understanding the optimal method for particulate removal and the management requirements necessary to maintain its effectiveness requires specialist knowledge (ACGIH 2023; HSE 2017).

6.6.8 Stacks

6.6.8.1 Key Requirements

A stack serves to disperse the air emitted from a ventilation system outside a workplace. Good design limits the amount of exhausted air that re-enters buildings. As shown in Figure 6.23, a stack should not incorporate structures that cap to reduce ingress of rain, as that will limit dispersion or even direct foul air back into a building. Good alternatives are shown in Figure 6.24.

A rule of thumb is that the stack should be 2.5 times higher than the tallest building (Hughes, 1989). An alternative rule is that the stack height (H_s) should be:

$$H_s = H + 1.5D \tag{6.15}$$

where

H = the height of the building

D = the lesser of the building height and the maximum length of the building across the prevailing wind.

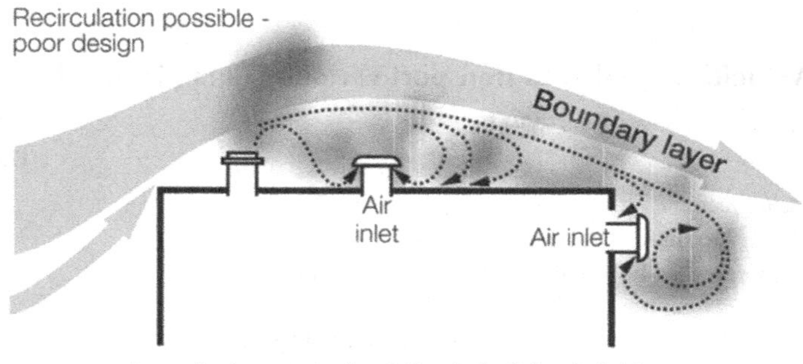

FIGURE 6.23 Stack height and location are important, position here shows bad design.

Source: HSE 2017, p. 62, Figure 41.

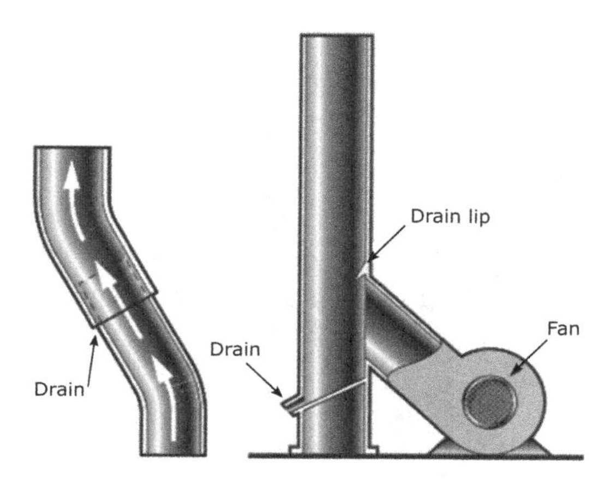

FIGURE 6.24 Good stack design alternatives.

Source: HSE 2017, p. 64, Figure 42.

Sometimes scaled wind-tunnel studies are needed, to understand the dispersion routes for exhausts, particularly if the building has a courtyard. Releasing smoke into the system from a cheap theatrical smoke generator will often reveal inadequacies in stack height and location. Additionally, stacks need to be:

- open at the exhaust point to use the inherent air velocity to continue taking the exhaust up above the building for local removal;
- have annular opening around the base, which allows the upper section of the stack to sit outside of the inner section so that rain will strike the outer sleeve and run off to the roof.

6.6.8.2 Use of Wind Roses

A wind rose is a diagram (Figure 6.25) showing the relative frequency of wind directions at a place in a graphical form: it can show the strength, direction, and frequency in a single diagram. They are used to look at potential dispersal patterns to identify air inlets that may be:

- at risk of re-entrainment; or
- of concern as a result of stack placement though airflows around building produce reverse airflow (see Figure 6.23).

Note that local airflows around buildings can negate the effect of a prevailing wind.

6.6.8.3 Common Problems

Common issues usually revolve around stack exhaust being re-entrained into the building or other buildings. This is often a result of:

- the stack being too short;
- incorrect placement of the stack;
- an exhaust cover being used on top of the stack to stop rain entering the ducting (see the left side of Figure 6.24).

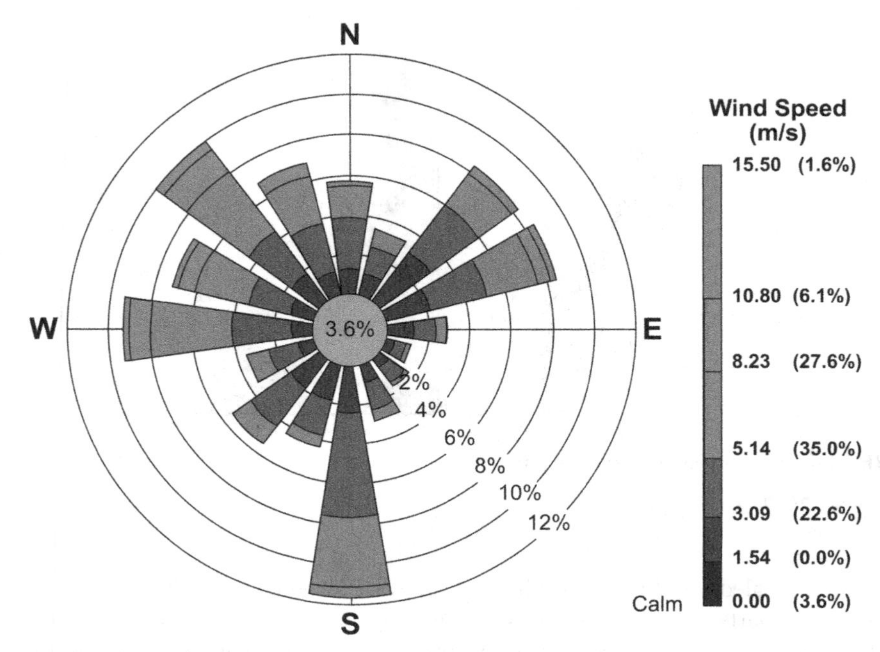

FIGURE 6.25 Example of a wind rose.

Source: BREEZE Software, Wikimedia Commons.

6.7 MEASUREMENTS

To comprehensively assess the effectiveness of an extraction unit (or canopy), measurements on the edges of the canopy are more important in assessing its overall performance than measurements obtained from the centre of the face of an extraction unit (or canopy). Forward flows on the edges place contaminated air at a point where it may be entrained in the swirling air (eddies, or turbulence) at the canopy face. For fume cupboards, eddies may form in front of a person using the fume cupboard (see Figure 6.18). The front of the fume cupboard is also more exposed to room draughts, and people moving past it drag contaminated air from the fume cupboard in their wake.

Small reverse airflows near the face of the fume cupboard (at 'X' in Figure 6.26) lead to significant loss of containment in almost all fume cupboards, a fact easily demonstrated with a smoke tube. These fugitive emissions are important in the case of highly toxic chemicals.

6.7.1 Face Velocity

Quantitative measurement of airflows can be performed with either vane or hot-wire anemometers (Figure 6.27). Vane anemometers are mechanical and have a start-up speed around 0.2 m s^{-1}; they are often used for measuring the face velocity of fume cupboards and for room outlets in air-conditioning systems. They can give an air speed averaged over a few seconds, and over the area of the vane.

Thermo-anemometers, or hot-wire anemometers, as shown in Figure 6.26, are generally considered unsafe in explosive atmospheres, unless certified otherwise, but are useful for general

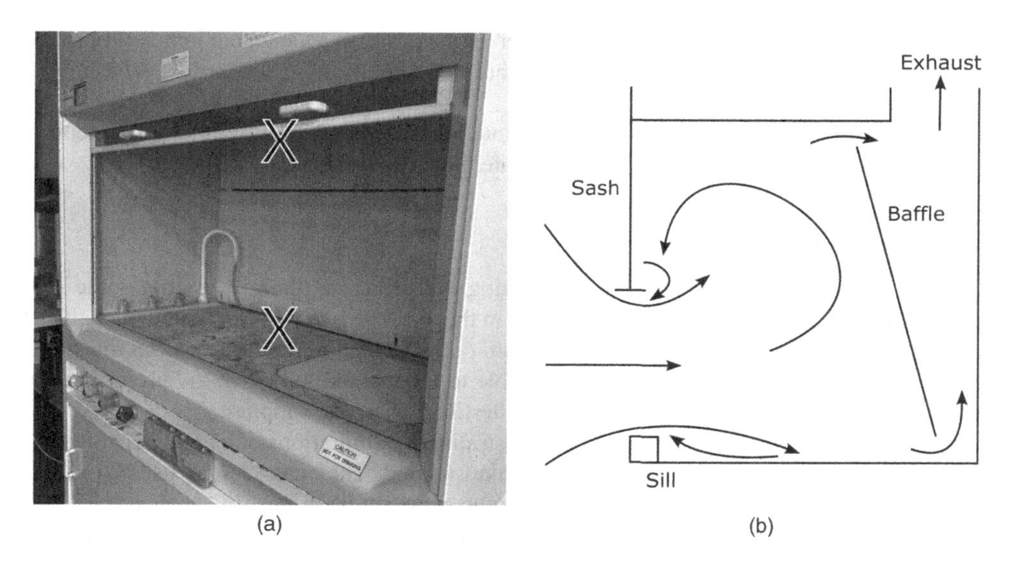

FIGURE 6.26 Potential loss of containment of a fume cupboard at 'X'.

Source: D. Bromwich.

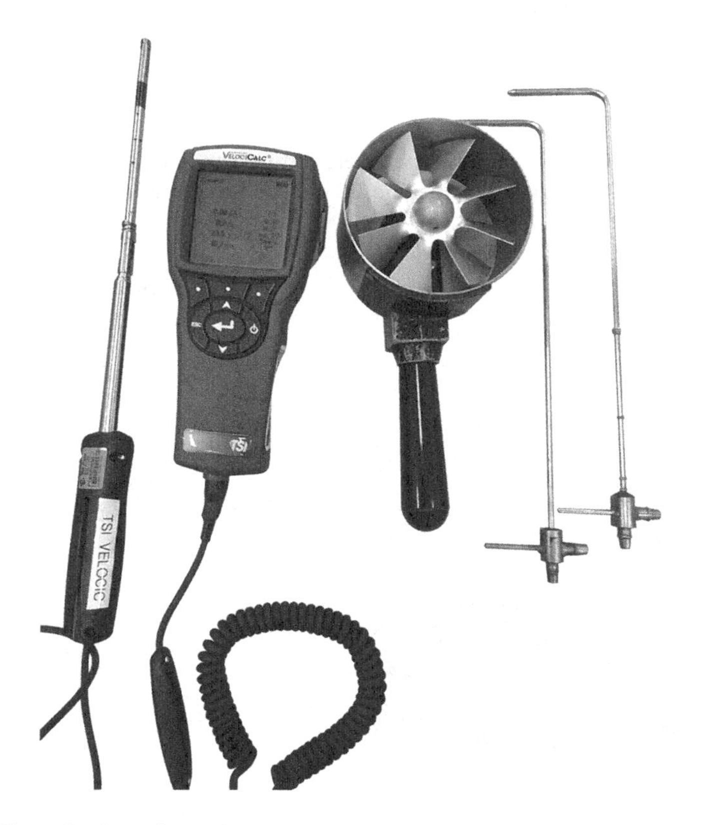

FIGURE 6.27 Pitot tube, hot-wire, and vane anemometers.

Source: S. Reed.

industrial ventilation work and can measure lower air speeds than a vane anemometer. In recently manufactured hot-wire anemometers, hot wires are not actually used for airflow measurements and as such present a low ignition hazard.

Although cumbersome, pitot tubes give absolute measurements of air velocity by measuring 'velocity pressure' and are usually used to estimate the airflow in ducts.

6.7.2 Smoke Tubes

The cheapest and most useful tool for investigating industrial ventilation systems is the smoke tube. A chemical in the tube reacts with moisture in the air to form a white smoke that enables the visualisation of airflows and the estimation of both direction and magnitude of even the smallest air movements (Figure 6.28). You may not be able to take smoke tubes on an aircraft and may need to have a packet of smoke tubes sent to the destination by the supplier.

Smoke tubes are particularly useful for demonstrating the limitations of a hood in capturing air contaminants and for showing airflows from behind a hood. They can also be used to visualise airflows through windows and doors. Sometimes the flow can be in one direction at the top of a door and the opposite direction at the bottom. It is also possible to make a good estimate of air speeds inside a room by estimating the time a puff of smoke takes to travel 1 metre. This makes smoke tubes particularly useful in indoor air-quality investigations.

As shown in Figure 6.26, it is usually possible to demonstrate reverse airflows in a fume cupboard by puffing smoke just inside the sash and on the working surface near the front of the fume cupboard.

6.7.3 Hot Processes

When measuring the flow for hot processes, where temperatures may exceed 1000 °C, the anemometer needs to be high-temperature specific. At elevated temperatures, standard anemometers

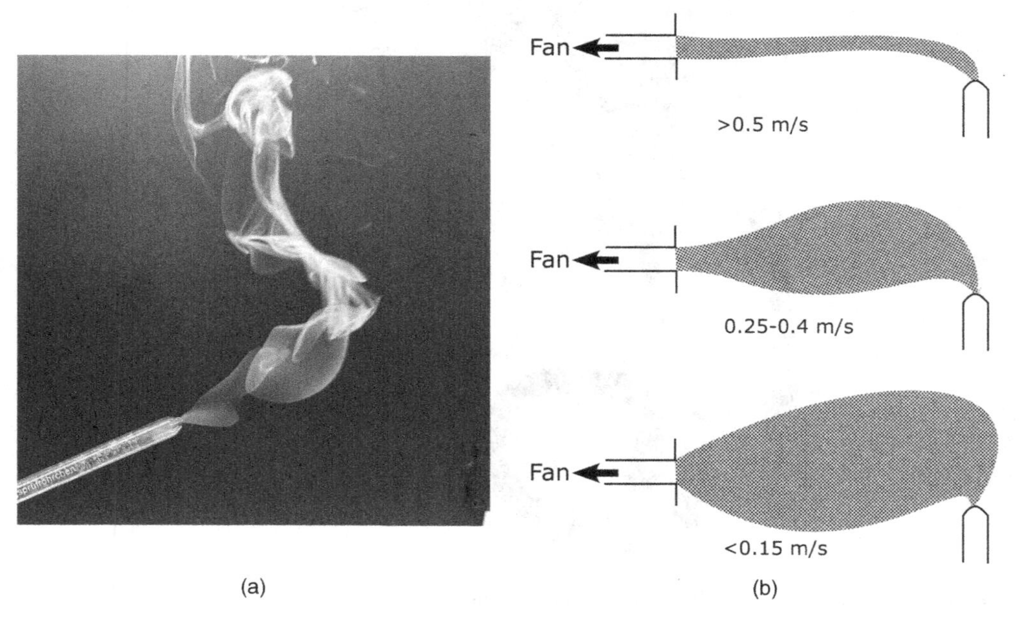

FIGURE 6.28 Smoke tube (a) and its use in estimating air velocity (b).

Source: D. Bromwich

may not give the correct readings, and so an anemometer capable of air velocity measurements in the region of at least 500 °C must be used.

6.8 NOISE CONTROL IN VENTILATION SYSTEMS

A common reason for ventilation systems failing to provide the expected protection is that they are too noisy and get switched off. Often the fan is the source of the noise, which travels along the ducting like sound through a large trumpet. It is possible to predict the levels of noise generated by industrial ventilation systems, but the calculations are complex, with the degree of turbulence, stiffness, reflections, and openings all affecting the amount of noise generated throughout the system.

6.8.1 Common Problems

- The noise associated with the 'passing frequency' of the fan blades is a common noise source. This is the noise generated as a result of the number of revolutions a fan makes per second multiplied by the number of blades in the fan.
- If a fan is isolated from a duct using a flexible join or some other sort of suspension that is too floppy it will vibrate and increase the noise level.
- A fan placed too close to a bend, or wrongly sized for the job, can easily produce 10 dB of additional noise.
- Low-frequency noise generated by the vibration of air-handling systems is usually the hardest to attenuate.
- In a straight duct, the noise level varies with at least the fifth power of the air velocity (Sharland 1972), so that for a given airflow, very small changes in duct diameter can have a large effect on noise levels.
- Small increases in duct roughness can also add 5 dB or more to the noise.

6.8.1 Fixes

- Passing frequency noise can be greatly reduced by mechanically isolating the fan from the duct using a flexible join located at least one duct diameter away from the fan.
- If ducting is suspended from a roof or other large surface, where vibrations in the duct make the whole roof vibrate like a giant loudspeaker, a duct suspension system can reduce this 'flanking' and can have a marked effect on the noise levels in the workplace.
- Variable speed fans can reduce the sound level as the fan speed decreases.
- The use of smooth ducting and transitions can reduce sound levels.
- Reducing the number and sharpness of bends.
- Installation of absorptive or dissipative silencers that are lined with sound absorbing material.
- Regular maintenance of the system to ensure the system is working optimally.

REFERENCES

ACGIH 2023, *Industrial Ventilation: A Manual of Recommended Practice* (31st edn), American Conference of Governmental Industrial Hygienists, Cincinnati, OH.

Alden, J.L. & Kane, J.M. 1982, *Design of Industrial Ventilation Systems* (5th edn), Industrial Press, New York

Cao Z., Zhou Y., Cao S. & Wang, Y. 2021, 'Chapter 2. Local Ventilation', in D. Goodfellow and Y. Wang (Eds), *Industrial Ventilation Design Guidebook, Volume 2, Engineering Design Applications* (2nd edn), Cambridge, MA: Academic Press.

Feigley, C.E., Bennett, J.S., Lee, E. & Khan, J. 2002, 'Improving the use of mixing factors for dilution ventilation design', *Applied Occupational and Environmental Hygiene*, vol. 17, no. 5, pp. 333–44.
Great Britain, Home Office Committee on Ventilation of Factories and Workshops 1907, *Second Report of the Departmental Committee Appointed to Inquire into the Ventilation of Factories and Workshops*, His Majesty's Stationery Office, London.
Health and Safety Executive (HSE) 2013, *Confined Spaces: A Brief Guide to Working Safely*, HSE, London, UK, https://www.hse.gov.uk/pubns/indg258.pdf [12 January 24].
Health and Safety Executive (HSE) 2017, *Controlling Airborne Contaminants at Work: A Guide to Local Exhaust Ventilation (LEV)* (3rd edn), HSE, London, UK, https://www.hse.gov.uk/pubns/books/hsg258.htm [12 January 24].
Hughes, D. 1989, *Discharging to Atmosphere from Laboratory-scale Processes*, H&H Scientific Consultants, Leeds, UK.
Sharland, I. 1972, *Woods Practical Guide to Noise Control*, Woods Acoustics, Colchester, UK.
Skistad, H. 1994, *Displacement Ventilation*, Research Studies Press, Taunton, UK.
Standards Australia 2009, *Confined Spaces*. AS 2865-1995, SAI Global, Sydney, https://www.standards.org.au/access-standards
Standards Australia 2012, *Ductwork for air-handling systems in buildings, Part 2: Rigid duct*, AS 4254.2-2012, SAI Global, Sydney, https://www.standards.org.au/access-standards
Standards Australia 2014, *Safety in Laboratories—fume cupboards*, AS/NZS 2243.8, SAI Global, Sydney, https://www.standards.org.au/access-standards
Standards Australia 2015, *The Use of Ventilation and Air Conditioning in Buildings, Part 1: Fire and Smoke Control in Buildings*, AS 1668.1:2015, SAI Global, Sydney, https://www.standards.org.au/access-standards
Standards Australia 2021, *Ductwork for air-handling systems in buildings, Part 1: Flexible duct*, AS 4254.1:2021, SAI Global, Sydney, https://www.standards.org.au/access-standards
Trakumas, S., Willeke, K., Reponen, T., Grinshpun, S.A. & Friedman, W. 2001, 'Comparison of filter bag, cyclonic, and wet dust collection methods in vacuum cleaners', *AIHA Journal*, vol. 62, no. 5, pp. 573–83.

7 Personal Protective Equipment

Mark Reggers and Terry Gorman

7.1 INTRODUCTION

In Chapter 5, the hierarchy of hazard controls was discussed. At the bottom of the hierarchy is personal protective equipment (PPE). PPE, as a control strategy, represents the last resort (though no less important if required); and beyond it is the unprotected worker, who will inevitably face exposure to a toxic agent or hazard if the PPE is not correctly selected, used, and maintained. Despite being the lowest place in the control hierarchy, PPE is still widely used and accepted as a backup and as a supplement for other controls. There will also be situations where higher-level controls cannot be used, and PPE is the only practicable solution or a promptly available temporary solution.

This chapter will discuss general safety apparel, gloves, and respiratory protective equipment. Refer to Chapter 14, 'Noise and vibration', for a detailed discussion of hearing protection.

7.2 OVER PROTECTION

It must be borne in mind that excessive use of PPE can sometimes interfere with the worker's ability to accurately perceive the work environment and may therefore compromise personal safety; consequently, PPE should always be used wisely. For example, thick or double gloves will reduce dexterity. In some instances, workers may complain of being treated like 'Christmas trees', with pieces of PPE hung on them like ornaments. Considerable care should be taken when choosing basic PPE to ensure the correct level of protection. Overprotection, for example, using heavy duty style equipment when lightweight is suitable is likely to result in the PPE not being used at all or removed frequently, leaving the worker totally unprotected.

7.3 PPE SELECTION STRATEGY

A thorough understanding of the hazard is essential when selecting the appropriate PPE. The H&S practitioner needs to understand both the physical and chemical properties of the agent(s), the relevant routes of exposure and the circumstances of exposure, the what, when, where, and why. The physical, chemical, and relevant toxicological properties are normally gleaned from the Safety Data Sheet (SDS), but it is imperative that the H&S practitioner observes the practices and behaviours of the workplace in question before selecting the PPE.

7.4 SAFETY EQUIPMENT

Basic safety equipment is a fundamental means of protecting the body from physical hazards in the workplace, it includes items like hard hats, safety boots, overalls, safety spectacles or goggles, hearing protection, and gloves. Each protective device is designed to provide a barrier between the worker and the hazard. It does not remove the hazard from the workplace but

DOI: 10.1201/9781032645841-7

reduces the risk of injury if the worker comes into contact with the hazard. For the most part, immediate protection of the eyes, hands, head, and feet is relatively uncomplicated. The hazards are usually obvious and programs to use PPE can be successfully implemented with appropriate attention and resources. Measures of this basic type of PPE will not be described extensively in this text.

7.5 GLOVES

Most people use their hands a lot at work, so hand injuries are very common. They may take the form of physical trauma or chemically induced damage, as shown in Figure 7.1. Gloves are widely used in industry to protect hands against many different hazards. The selection of the correct glove type is neither straightforward nor simple, particularly when considering chemical protection. For example, gloves that protect against methylated spirit solvents may not protect against turpentine. Knowledge of the chemical resistance and permeability of different glove materials is required before making a choice. Glove manufacturers and others have published data on the resistance of glove material to permeation by the common solvents used in industry. However, permeation resistance data on a given material from one manufacturer may not hold true for the same type of glove material from another manufacturer. There are international standards, such as EN ISO 374-1:2016+A1:2018 2018, Protective gloves against dangerous chemicals and micro-organisms (CEN), and the ASTM's Method F739–12E1, Standard Test Method for Permeation of Liquids and Gases Through Protective Clothing Materials Under Conditions of Continuous Contact (ASTM International 2020), that describe test methods and terminology associated with chemical-resistant gloves. There are also independent online resources, such as ProtecPo (Institut National de Recherche et de Sécurité and Institut de recherche Robert-Sauvé en santé et en sécurité du travail 2011), that the H&S practitioner can use to select the category of polymer most likely to provide protection against individual chemicals or classes of chemical. Given the rate of research in this field, it is recommended that the most current information be consulted.

Other aspects to consider when choosing gloves include the dexterity required by the worker, thermal resistance, abrasion, and puncture resistance. The widely favoured leather 'rigger's' glove

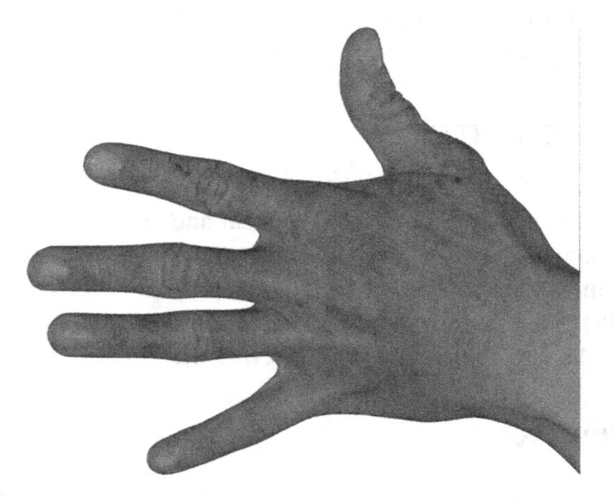

FIGURE 7.1 An example of severe contact dermatitis caused by direct chemical contact (e.g. MDI).

Source: R. Nixon.

offers virtually no resistance to chemical hazards and in fact can act as a reservoir, allowing the chemical to accumulate, then slowly penetrate the glove to the user's skin.

The re-use and laundering of gloves can be fraught with problems and any decision to do either should be approached with caution. The risks include:

- gloves may contain small holes that are not detected;
- contaminants can be translocated to the inside of the glove during washing;
- the washing process can physically damage the glove;
- the washing process may not remove chemicals that have started to migrate through the glove material.

A successful skin protection program includes correct choice of gloves as well as training in their correct use, that is, in the limitations to their protection, and in their correct removal without causing contamination of the skin.

Some skin lotions and hand cleaners contain petroleum products that can cause swelling and degradation of the glove material. Specialist advice should be sought from the glove supplier if there are any concerns about this. Users of the gloves need to be alert and if any deterioration is noticed, the gloves should be removed and their suitability reassessed.

Most users need to be shown how to remove gloves correctly without contaminating the unprotected skin. Loose-fitting gloves can be shaken off quite easily. Tight-fitting gloves in general should not be reused because of the risk of introducing contamination onto their internal surfaces. If contamination does occur, the enclosed nature of the glove material may increase the rate of absorption of the very agent the gloves were intended to protect against.

7.6 CHEMICAL PROTECTIVE CLOTHING

All clothing provides a degree of protection, even cotton overalls, but fabrics can act as a reservoir of contaminants, leading to continued contact with the embedded chemicals long after the wearer has left the workplace. Some form of chemical protective clothing (CPC) is required wherever hazardous chemicals are used that can be absorbed through the skin. Contaminated clothing can also act as a mechanism to transport the contaminant away from the workplace in harmful quantities. For example, there are many documented examples where toxic agents like lead, asbestos, or mercury have been taken home with toxic consequences to family members.

Again, a full understanding of what, when, where, and why is required to ensure the correct protection is achieved by the CPC. Examples of the types of CPC materials and their uses and limits are listed below:

- plastic-coated woven polyolefin overalls with hood—hazardous dusts and low toxicity liquid chemical splashes;
- impervious plastic pants and coat set—moderate-hazard materials, simple splash protection;
- impervious plastic fully enclosed suits—high-hazard materials, such as chemical warfare agents, emergency response, spills cleanup, and radiological hazards.

The selection of the correct chemically resistant protective material is only part of the protection of skin against toxic hazards. Additional measures that need to be taken include:

- sealing of gaps between boot or gloves and the suit material, commonly with adhesive tape;
- protecting the skin where respiratory protection integrates into the suit construction;

- sealing zippers or other garment closures;
- providing a 'dresser' or assistant given the potential difficulty and complexity of donning and achieving the seals;
- providing effective decontamination facilities so the exterior of the suit can be cleaned before it is removed and then suitably cleaned for reuse if possible;
- recognising and managing heat stress during use (see Chapter 16, 'Thermal stress').
- communication between workers wearing fully enclosed suits;
- difficulties in manoeuvrability or limits to the range of vision when wearing enclosed suits;
- operating effectively within the time constraints of the suit and/or breathing equipment.

Any situation where CPC is needed requires a detailed assessment and consideration of all the above measures. For an excellent guide to the selection and management of CPCs, refer to the Occupational Safety and Health Administration's Technical Manual (Section VIII Chapter 1).

7.7 RESPIRATORY PROTECTIVE EQUIPMENT

Of the three principal routes of exposure, inhalation is considered the most significant (everyone must breathe, and the inhalation and absorption of airborne agents via the lungs can be very efficient). Consequently, the provision of clean air to breathe is paramount. There are many situations in which the use of respiratory protective equipment (RPE) is an established method of protection, such as where:

- other control methods are too costly or impracticable (e.g. electrical power may not be available, or ventilation controls cannot be arranged around a large open formwork metal structure);
- the tasks may be carried out at a range of locations (e.g. a pesticide applicator providing termite treatments to buildings);
- the tasks may involve only short-term exposures (e.g. one job taking two hours a month).
- exposure may be low risk, not requiring elaborate controls (e.g. nuisance dust exposure);
- the agents used may have poor, or no, warning properties and overexposure can readily occur;
- RPE is required for emergency procedures, including emergency escapes (e.g. firefighting, escape from chemical leaks or spills);
- oxygen-deficient atmospheres (e.g. confined-space entries);
- RPE is necessary to supplement other control procedures (e.g. air-supplied RPE is still required for some underground tunnelling tasks or abrasive blasting even when water suppression is also used).

RPE should not immediately be the first choice nor should it be used merely to quell concerns or complaints of overexposure to contaminants. Before any RPE is issued, a full investigation of the task in question should be carried out to ensure that RPE is an appropriate control measure. See Chapter 5 for a full discussion on the application of the Hierarchy of Controls.

RPE offers relatively simple, low-cost protection from hazards that arise in many workplaces. However, it must be stressed that effective use of RPE programs is often difficult (and sometimes impossible) to achieve. Any H&S practitioner embarking upon an RPE program needs to be aware that this protective strategy requires 100% ongoing commitment from both management and the worker. If there is any lack of enforcement on the part of management or lack of compliance on the part of the worker, then it is possible that the protective measure will be ineffective or completely fail. RPE use requires much greater organisational effort than other hazard-control measures. If RPE is adopted, it should only be after all other options have been examined.

7.7.1 International Standards

Across the globe, there are a number of national respiratory protection standards created for use in various jurisdictions, for example, EN standards in Europe, CSA Group Standards in Canada, JIS in Japan, and NIOSH standards that apply in the United States and also used by some other countries. The equipment selected for use will depend on the relevant standards applicable in the jurisdiction.

In Australia and New Zealand, the current standards relating to respiratory protection are AS/NZS 1715 and 1716. AS/NZS1716 is the standard that outlines the performance requirements and associated testing protocols for the various types of respirators used in ANZ workplaces; while AS/NZS1715 provides guidance on selection, use, and maintenance of respirators.

AS/NZS1716 recognises three levels of particle filter performance:

P1—for mechanically generated particulates, for example, from sanding, crushing, sawing, drilling, etc.
P2—for mechanically and thermally generated particulates, that is, as for P1 plus those particulates produced from thermal effects, for example, welding fume, molten metal fume, and laser cutting.
P3—for all particulates especially where high protection levels are required.

At the time of writing (early 2024), the relevant Australian and New Zealand Standards Committee has adopted the full suite of ISO respiratory protective device standards, for example, those dealing with product test methods (AS/NZS ISO 16900 Series), Classification (AS/NZS ISO 17420 Series), Human Factors (AS/NZS ISO 16976 Series), the selection and use standards (AS/NZS ISO 16975 Series), and others. These standards are intended to supersede AS/NZS 1715 and 1716 on 1/1/2030 and products from then should be rated and used according to the new AS/NZS ISO standards. Until this time, both sets of AS/NZS 1715 & 1716 and the new AS/NZS ISO standards are current during this transition period.

The EN respiratory protection standards use similar classes to AS/NZS 1716 for particle filters, for example, FFP1, FFP2, and FFP3.

The USA's NIOSH standard classifies particle filters into three types that indicate the minimum per cent efficiency of the filter against the required test challenge aerosol. These are 95, 99, and 100. Another filter characteristic tested by NIOSH is to rank performance against an oil-based challenge. There are three classes here also:

N—not resistant to oil
R—somewhat resistant to oil
P—strongly resistant to oil

From the testing data for these two characteristics, a particle filter is rated, for example, N95, P95, R100 or P100, etc. Other national standards have a broadly similar approach to particle filter classification.

For gas/vapour filters on tight-fitting facepieces, AS/NZS1716 uses the same class identifiers as Europe, for example, A1, B2, E1, K1, and Hg indicating the relevant chemical target groups (the letter) and capacity (the number). NIOSH has abbreviations for their specified classes, for example, OV (organic vapours), Cl (Chlorine), AG (Acid Gas), Amm (Ammonia), etc.

Almost all national respirator standards promote the use of a comprehensive respiratory protection program to deliver the protection required to the workers.

7.7.2 Australian Regulatory Requirements

Any RPE program should comply with the below to help demonstrate compliance to PPE regulatory requirements and duty of care:

- any applicable advisory standard or code of practice on PPE, including AS/NZS 1715:2009, Selection, Use and Maintenance of Respiratory Protective Devices (Standards Australia 2009);
- the selection of RPE that meets the relevant performance requirements, for example, AS/NZS 1716:2012, Respiratory Protective Devices (Standards Australia 2012).

The hazardous chemical regulations mention the use of PPE, including RPE, but only as a last resort in the control of hazardous chemicals.

7.7.3 Respiratory Protection Programs

The implementation of a respiratory protection program needs to be based on a detailed assessment of each workplace hazard and the extent of worker exposure to these respiratory hazards. For example, if the workplace uses lead, and other controls are not possible, it will be necessary to determine which class and type of RPE is appropriate for lead dust. The wrong choice of RPE could expose workers to excessive lead levels, resulting in possible long-term illness, as well as regulatory actions or fines, civil action by employees, and other unnecessary economic costs to the business. Therefore, the procedures for selecting and implementing RPE require considerable attention. Where doubt exists, specific needs should be discussed with informed specialists, for example, occupational hygienists or representatives from the RPE supply company.

There are some excellent respiratory protection programs operating throughout Australia. However, many purchasers of respiratory protection fail to get satisfactory performance from their RPE, some of the possible reasons include:

- a lack of knowledge about the nature of the airborne contaminant and its concentration levels;
- incorrect selection of air-purifying devices including wrong filter class or protection factor;
- poor fitting of tight-fitting respirators;
- a lack of scheduled maintenance;
- ill-treatment and contamination of RPE in the workplace;
- lack of knowledge of filter life;
- workers failing to use the RPE provided at the appropriate time.

There are many different types of respirators, and it is not the intention of this text to provide a detailed evaluation of each type. As a high-level guide, the main types of air-supplied and air-purifying respirators are summarised in Figure 7.2 and described in Table 7.1. For more detail, consult AS/NZS 1715:2009 (Standards Australia 2009).

Great attention to the technical detail and administration of RPE programs is necessary. There are many anecdotal examples that illustrate the incorrect use of respirators, such as the dusty respirator hanging on a nail in a workshop, fitted with a dust filter, and used during spray painting with a paint containing solvents and toxic pigments, by a bearded operator who could not recall

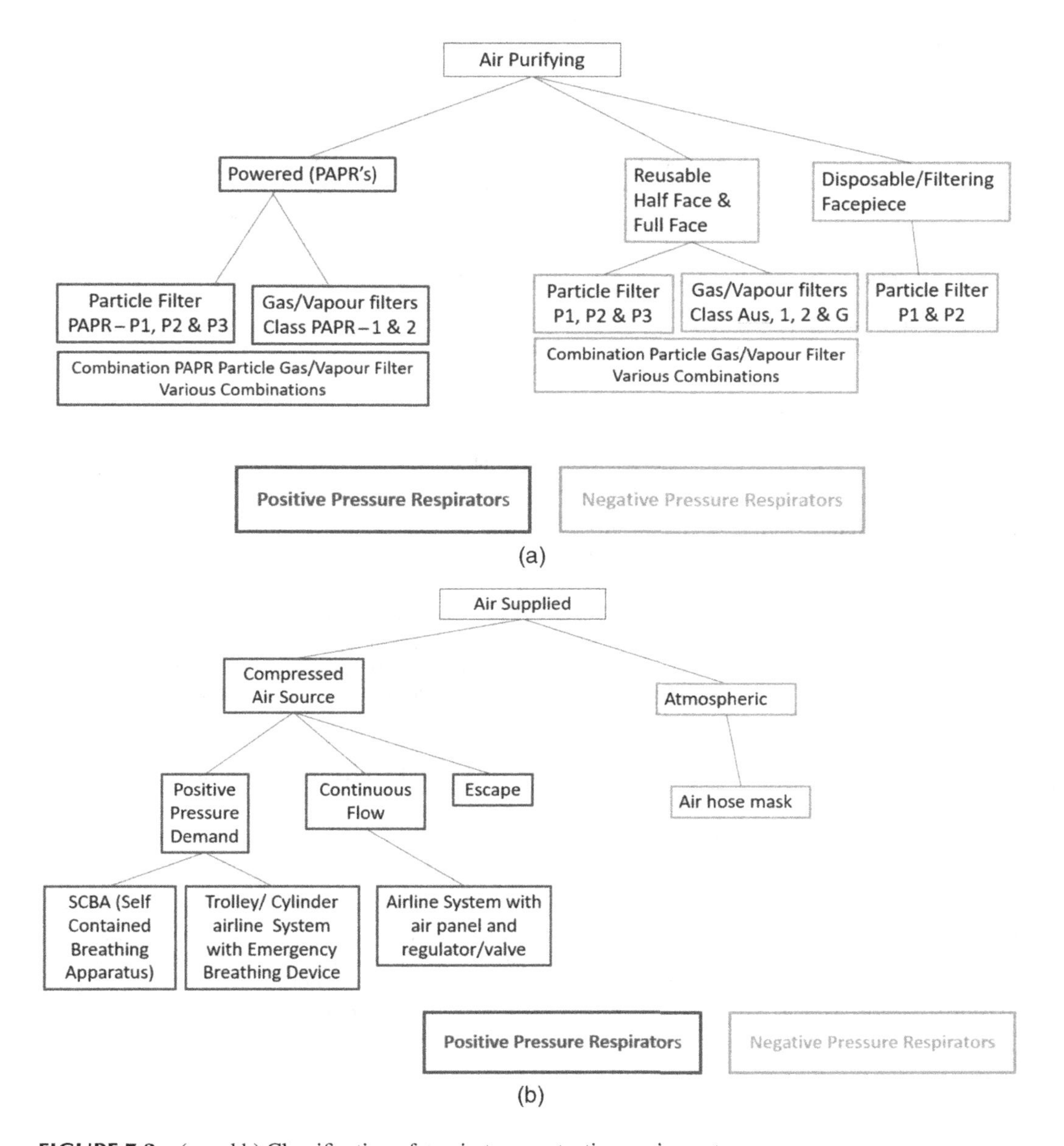

FIGURE 7.2 (a and b) Classification of respiratory protective equipment.

Source: M. Reggers and T. Gorman.

when the filter was last changed but says the man who sold it to him assured him it was the correct one for spray painting.

The key requirements of an RPE program include:

- managerial capacity to administer and resource an RPE program (it is part of the H&S practitioner's task to guide the managers); and
- knowledge of respiratory hazards.

TABLE 7.1
Main Types of Respirators and Filters

Particulate Filter Types/Classes	Function or Purpose
P1	Protection against mechanically generated particulates
P2	Protection against mechanically and thermally generated particulates
P3	Protection against all particulates including highly toxic materials

[*Note*: 'handyman' dust masks that do not carry the Australian Standards claim or mark do not provide protection against toxic or hazardous dusts. Extreme caution should be exercised when selecting this type of protection.]

Gas/vapour filter classes

The larger the class number, the larger the capacity of the gas filter—that is, it will last longer for a given gas or vapour concentration.

Class AUS	Low absorption capacity filters
Class 1	Low to medium absorption capacity filters
Class 2	Medium absorption capacity filters
Class 3	High absorption capacity filters

Gas/vapour filter types

For specific chemicals, confirm suitability of filter types with the manufacturer

Type A	Organic gases and vapours
Type B	Acid gases and vapours as specified, excluding carbon monoxide
Type E	Sulfur dioxide and other inorganic gases and vapours as specified
Type G	Low vapour pressure organic compounds for vapour pressures less than 1.3 Pa (0.01 mm Hg) at 25°C as specified. These filters incorporate a particulate filter at least equivalent to P1
Type K	Ammonia and organic ammonia derivatives
Type AX	Low-boiling-point organic compounds (less than 65°C), for example, methanol
Type NO	Oxides of nitrogen
Type Hg	Metallic mercury
Type MB	Methyl bromide

Types of respirators	**Function or purpose**
Air purifying, powered type	Powered air purifying respirator (PAPR)—uses a battery-driven fan to force air through a filter assembly and deliver cleaned air to a helmet, hood or face mask. May be continuous flow or pressure demand positive pressure
Air purifying, replaceable filter type	A facepiece (full face or half face) to which a replaceable filter assembly is connected. The user's lung power is used to draw the air through the filter. Negative pressure
Air purifying, disposable	A respirator with the filter as an integral part of the facepiece and the filter is not replaceable. When exhausted, the whole assembly is discarded. Negative pressure
Air hose—natural breathing	A wide diameter hose is located outside the contaminated zone and connected to the respirator facepiece. The user's lung power draws air to the facepiece (can be low-pressure fan assisted). Air hose systems supply air at or near atmospheric pressure
Supplied air, compressed air line	A small-bore line connected to a compressed air source supplies clean air for respiration. This may use a compressor, or compressed air cylinders located a distance from the work location. Positive pressure
Supplied air, self-contained	The respirator facepiece is connected by a breathing tube to a cylinder of breathable gas that is carried by the wearer. Often referred to as SCBA. Positive pressure

Note: AS/NZS 1715 Appendix A specifies air purity for air-supplied systems including SCBA systems.

Source: Reproduced from p. 21 & 22 from AS/NZS 1716:2012.

- selection and purchase of the appropriate type of RPE, including the appropriate protection factor;
- acceptance and appreciation of RPE by workers so it will be worn when required;
- medical assessment of respirator use for RPE users;
- training in use, including correct fitting of respirator;
- fit testing of tight-fitting respirators;
- inspection, maintenance, and repair of RPE; and
- audit and review.

The person selecting and supervising the RPE program may also require training. Written guidelines should be prepared and strictly followed. The use of RPE must be strictly monitored to ensure that the equipment is not used for purposes beyond those designated. Workers must receive adequate instruction, training, and supervision to ensure they are using RPE correctly and do not inadvertently use the RPE in life-threatening situations.

7.7.4 Knowledge of Respiratory Hazards

It is not possible to proceed with an RPE program unless there is a clear understanding of why RPE is being used. Many factors need to be considered.

To provide proper respiratory protection, the following steps must be taken:

1. Determine the identity and nature of the hazardous contaminants to be controlled (i.e. dust, mist, fume, gas, vapour, asphyxiant, or any mixture of these) and the expected range of concentrations in various work situations/areas.
2. Determine whether the workplace situation is or can become immediately dangerous to life or health (IDLH), and the potential consequences of failure of the RPE.
3. Determine which workers require RPE. All workers involved in a particular process may not be at equal risk.
4. Distinguish whether RPE is the major control method or a backup or secondary control.
5. Determine whether there are any additional routes of entry other than inhalation, or other effects. Skin absorption and effects on the eyes are the most common considerations.

After the above factors have been considered, then the suitability aspects should also be evaluated. Many respirator types may be adequate for the contaminant and contaminant level, but not all will be suitable. Things to consider include:

- Compatibility with other PPE.
- Communication requirements.
- Thermal impact on the worker.
- Fit and facial hair, and
- Wear time

It is key any respirator is worn correctly and for the whole time it is required.

7.7.5 Protection Factors

Selection of appropriate RPE should not be attempted by the inexperienced or the untrained. Some regulations or codes of practice, for example, Safe Work Australia's Model Code of Practice—How to Safely Remove Asbestos (2020) spell out the minimum RPE for situations where it is required. On occasions, trials of different types and brands will be necessary to determine the RPE that is most suited to a particular workplace and/or wearer. Suppliers of RPE can be an excellent source of information and guidance.

Two pieces of numerical information are crucial to applying RPE successfully are:

- the concentration of contaminant present or potentially present in the workplace; and
- the target concentration inside the respirator.

Based on this data, it is possible to calculate the required minimum protection factor (RMPF) needed to reduce exposures to the minimum acceptable level. Ideally, the exposure inside the respirator is well below the WES rather than just below it. The following formula should be used to determine the RMPF:

$$\text{Required minimum PF} = \frac{\text{concentration in workplace}}{\text{WES} * (\text{or other target concentration})} \qquad (7.1)$$

Note: Both the concentration and the *WES (workplace exposure standard) should be in the same units.

The RMPF is a crucial piece of information when selecting RPE. It relates to the capacity of the respirator to adequately reduce concentrations of the contaminant between the outside and the inside of the respirator. No respirator prevents all the contaminants from entering the breathing zone. AS/NZS 1715:2009 assigns maximum use limits for different types of respirators, summarised in Table 7.2.

It is notable that the Australian Standard uses the term 'acceptable exposure level/standard' to refer to the concentration inside the respirator. It is the position of the authors that simply applying the WES may not provide enough margin of safety to accommodate errors in the various estimations. As an example, let us say the cyclohexane concentration in a workplace is measured at 1800 ppm. Cyclohexane can be removed with Type A gas/vapour filtering-type respirators with Class AUS, 1, 2, or 3 filters. The required protection factor to reduce the 1800 ppm concentration to below, say, a practically achievable target amount well under the WES of 50 per cent of the WES target concentration of 100 ppm will be 1800/50 = 36.

Table 7.2 indicates that only full facepiece respirators with Class 2 or 3 filters will have the required minimum protection factor to be used above 1000 ppm. (There are, however, other types available and suitable, as outlined in Table 4.5 of AS/NZS 1715: 2009—half-face, air-line, continuous flow; Class PAPR-2 in full-face, etc.) AS/NZS 1715:2009 contains a flowchart/decision

TABLE 7.2
Required Protection Factors for Selected Respiratory Devices

Required Minimum Protection Factor	Respirator Type
Up to 10	Disposable facepiece respirator, half-face respirator for particulates and gas/vapour concentrations <1000 ppm
Up to 50	PAPR or full facepiece P2 for particulates, and Class AUS or 1 gas filters filter for vapour concentrations <1000 ppm
Up to 100	Full facepiece P3 for particulates or Class 2 gas filters up to 5000 ppm
100 +	Full facepiece positive pressure airline or SCBA positive pressure demand

Adapted from: Tables 4.2, 4.3, and 4.5 from AS/NZS 1715:2009.

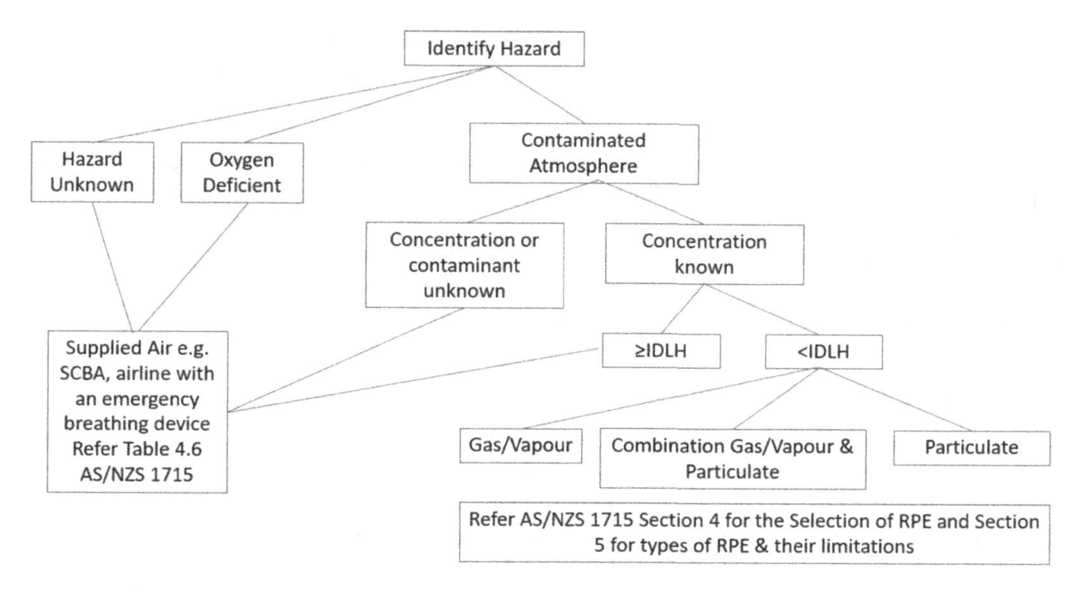

FIGURE 7.3 Simplified RPE selection decision tree.

Source: M. Reggers and T. Gorman.

tree (Figure 7.3). This standard deals with respirator selection under the simplified headings of contaminant, task, and operation. A simple decision tree is included in Figure 7.3 to help identify adequate RPE for most situations.

7.7.6 Filter Service Life and Breakthrough

When introducing air-purifying RPE to control exposure, attention must be directed to usage patterns and exposures in order to estimate the service life of protective filters. This is a complex issue and depends on many factors, including the filter capacity and construction, the concentration of the contaminant, temperature and humidity, the worker's breathing rate, the physical and chemical composition of the contaminant, and general respiratory competence. Service life estimates must include some unexpended reserve capacity as a safety margin. Importantly, filters should not be used beyond the expiry of their shelf life—this date should be marked on the filter itself or its packaging.

A recurring question is: how long will the filter in the respirator last? There is no simple answer, so it is better to rely on outside experience or advice from manufacturers rather than chance unsafe practices. For any type of respirator, wearer acceptance and comfort are also important factors in adoption of RPE. For particulate filters, filtration efficiency usually also increases with use as dust particles gradually block the filter. This also causes increased inhalation resistance. Similarly, the higher filter efficiency of P3 filters comes with increased initial inhalation resistance. This may have adverse effects for the wearer when RPE is used for continuous work or if the wearer has some respiratory ailment that makes the use of a respirator difficult. The service life of such particle filters is usually over when the wearer has difficulty breathing through the filter.

For gas and vapour filters, expected service life can be calculated only if there is reliable data on exposure conditions. Otherwise, scheduled maintenance and replacement programs with a reasonable margin for safety must be scrupulously followed. Several RPE manufacturers have software to perform service-life calculations. It is essential to check the compatibility of the software with parameters such as the WESs, the RPE specifications on flow or breathing rates and relative humidity in the situation under consideration. In short, the calculation of service life is fraught and should be approached with caution.

Breakthrough of the contaminant, as indicated by odour, taste or irritation, is an unreliable means of determining the end of service life (i.e. exhaustion of capacity) of a respirator gas/vapour filter. Some contaminants have no odour (carbon monoxide); others have workplace exposure standards below their odour threshold (isocyanates). Some individuals become insensitive to odour or develop nasal or lung irritation. Some individuals have no sense of smell. Some chemicals rapidly deaden the sense of smell (carbon disulfide). A calculated gas/vapour filter service life and change schedule from the RPE manufacturer can assist in resolving some of these issues.

7.7.7 Respirator Fit Testing

Tests for the correct fitting of tight-fitting RPE can only be achieved using the proper equipment and established protocols. To qualitatively test for facial leaks, the wearer typically is subjected to an aerosol (Bitrex or saccharin) as illustrated in Figure 7.4. It is also known as the Aerosol Taste Test (ATT). The respirator user has a hood over their head and an aerosol of saccharin or Bitrex is aspirated through a nebuliser into the hood without a respirator to confirm they can taste the agent and at what threshold. If they are unable to taste the agent within 30 sprays, this method is not suitable for them and a quantitative test method is required. If they can detect the aerosol, the user

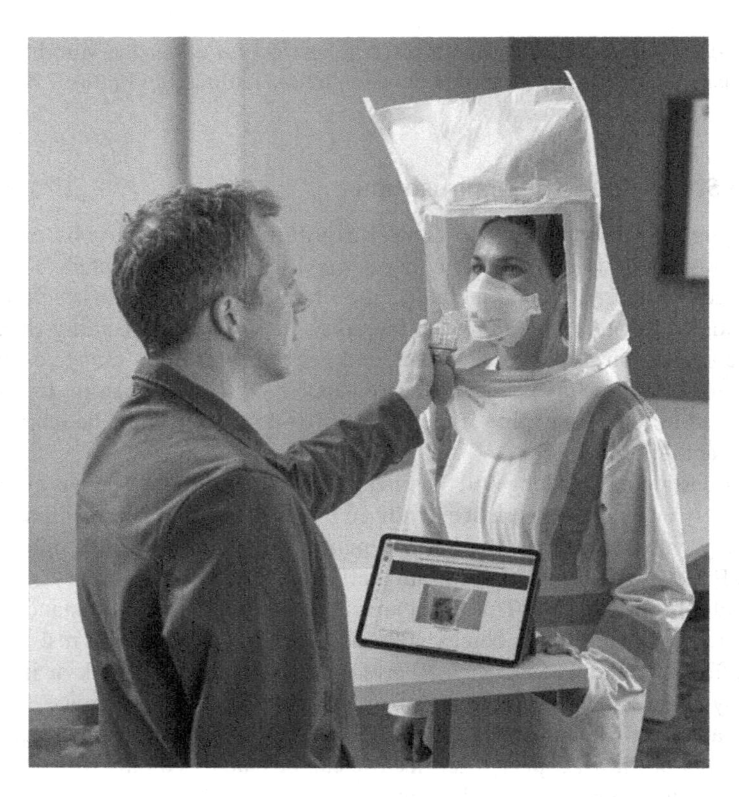

FIGURE 7.4 Qualitative aerosol taste tvest facial fit testing using a saccharine aerosol.

Source: 3M.

is asked to perform a series of exercises involving talking and various head movements with the respirator worn. Any significant failure of the face seal will be detected by the taste of saccharin or Bitrex by the user. This method is suitable only to test face seals on half-face respirators fitted with particulate filters, as the saccharin or Bitrex particles can pass through gas and vapour filters. A similar test can be conducted for organic-vapour air-purifying respirators using isoamyl acetate vapour as the challenge agent. Fit testing is described broadly in AS/NZS 1715:2009 Section 8 (Standards Australia 2009) and in greater detail in AS/NZS ISO 16975-3:2023. If the wrong test agent is used it will penetrate the filter, producing misleading results. RPE suppliers and fit testing service providers can provide assistance with fit testing.

The more complex quantitative face-fit tests use a modified respirator facepiece and a particle detector to measure the inward leakage of an aerosol or pressure change. Some equipment of this type uses ambient aerosols naturally occurring in the environment (condensation nuclei counter, CNC), while others use a created negative pressure inside the facepiece to measure air leakage (controlled negative pressure). These tests require both specialist equipment and training. There are pros and cons for each fit testing method that workplaces should be aware of when determining which method to use.

Also of particular importance is the competence of person conducting the fit testing to ensure the results are reliable and administered correctly. AS/NZS ISO 16975.3:2023 Section 5 outlines the areas a fit tester should have demonstrated knowledge and ability such as:

- Knowledge of the respirator used for the fit test.
- Knowledge of the fit test method.
- Ability to set up and monitor the function of fit test equipment.
- Ability to conduct the fit test.
- Ability to identify likely causes of fit-test failure.

RESP-FIT is a Respirator Fit Testing Training and Accreditation program developed and managed by the Australian Institute of Occupational Hygienists (AIOH). Refer to www.respfit.org.au for further information on respirator fit testing and fit test service providers who employ RESP-FIT accredited fit testers. Note there are similar organisations in other countries.

Both types of tests (qualitative and quantitative) help ensure that the selected respirator can effectively fit the wearer's face and therefore delivers adequate protection when worn correctly. The results indicate satisfactory fit, *not* applicable workplace hazard protection levels. The level of protection is highly dependent on how the device is used in the workplace. Results of such fit tests should not be used to increase the protection factors above those recommended in codes of practice or standards.

The fit test should be performed when the respirator is first issued and whenever the respirator model or type is changed. They serve the added purpose of reminding users of the need to always put on their respirator with care and can also demonstrate the effect of damaged parts or facial hair, including moustaches, stubble, and beard growth, on the completeness of the face seal. All wearers of tight-fitting respirators should be clean-shaven in the area of the face seal and any facial hair inside the face piece should not interfere with exhalation valves.

Each time the mask is put on, a negative-pressure user seal check should be done to check for leaks by blocking off the filter inlets using the palms and inhaling (see Figure 7.5). Inward leakage can be detected, though this test is not very sensitive because the negative pressure within the facepiece causes the respirator to be drawn onto the face by the external atmospheric pressure, improving the seal on the face. A similar positive pressure user seal check can be done by exhaling while covering the filters or exhalation valve and trying to detect leakage. These seal checks are *not* substitutes for fit testing but are used each time after donning respirators to ensure there are no gross seal failures.

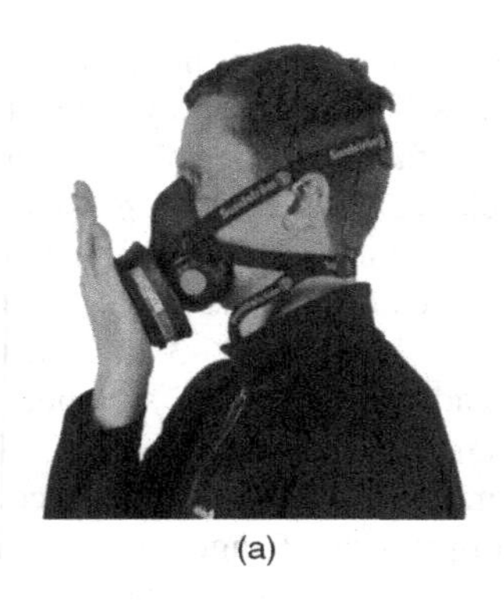

(a)

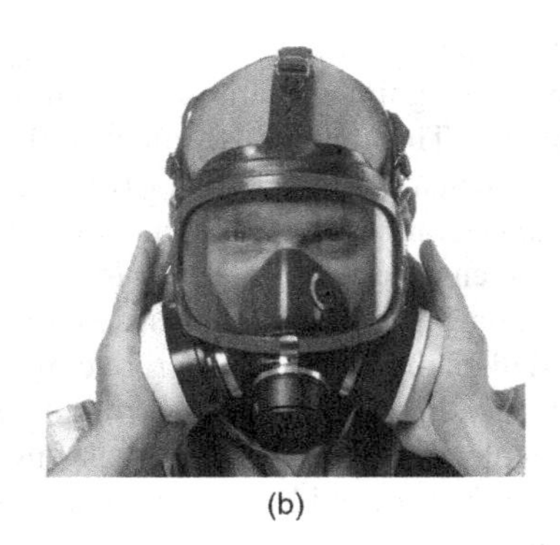

(b)

FIGURE 7.5 (a and b) Negative pressure user seal check.

Sources: SEA & Draeger.

7.7.8 Inspection, Maintenance, and Repair

With the exception of disposable type respirators, the use of RPE in the workplace requires a constant program of inspection, maintenance, and repair. Maintenance includes washing, cleaning, disinfecting where necessary, inspecting for wear and tear, checking for leaks, and replacement of worn components and filters. It is essential to store RPE properly between uses. Gas and vapour filters can continue to absorb contaminants from the air when they are not in use, further exhausting their capacity. Plastic sealable storage containers or zip-lock plastic bags are ideal for between-use storage. Suppliers will be able to provide details of suitable maintenance programs for their equipment. Where possible, each wearer should be provided with individual RPE. Where air compressors are used for air-supplied respirators, there should be a maintenance, inspection, and testing program in place for the compressors as well, to ensure the air quality. See AS/NZS 1715:2009 Appendices A and F (Standards Australia 2009) for guidance on compressed air quality testing.

7.7.9 Medical Assessment of RPE Users

Wearing RPE is not without its physiological and psychological demands. There are several medical conditions, including diabetes, asthma, emphysema, skin sensitivity, or chronic airway disease that can prevent a worker from using RPE. Some workers also feel claustrophobic wearing a normal filtering respirator but may find a powered air purifying respirator (PAPR) more acceptable. Always seek expert medical advice from an occupational physician if prospective wearers have any problems when wearing respirators.

In some cases, biological testing (e.g. blood lead testing) can be used as a supplementary means to assess the overall effectiveness of exposure control programs. While respirators provide respiratory protection, other, uncontrolled routes of intake of the toxic agent may remain, for example, ingestion or skin absorption, leaving the worker less protected than expected.

7.8 EYE PROTECTION

Eye protection is a common requirement in many working environments. The hazards are mostly associated with flying particles or sparks, but chemical exposure is also significant in certain industries and protection from splashes, sprays, gases, and vapours is needed. Workplace experience is that many eye injuries occur to workers who are not wearing eye protection at all, but there are also a significant number of incidents where eye injuries occur to those wearing some form of eye protection. This largely occurs to those wearing protective eyeglasses with no side shields or those who are wearing a model that leaves large gaps between the spectacles and the face. Style can also be a factor to gain acceptance from the workers so a range of products should be offered to allow for this as well as getting a suitable fit outcome on the wearer's own facial contours.

The crucial factors to prevent eye/face injuries are:

- Always wear effective eye protection that complies with the local standards. To be effective, the eyewear or face shield must be of appropriate type and rating for the eye/face hazards encountered and should be properly fitted to reduce any gaps and fit securely in position.
- Potential interference with other PPE: the eye or face protection needs to be compatible with any other PPE being used. Safety spectacles with wide or stiff arms may cause a significant loss of the attenuation provided by a pair of earmuffs, or the frame may interfere with the correct fit of a half-face mask respirator. This issue needs to be assessed for each individual and their selected suite of PPE.
- Training and education. Some workers injured while not wearing protective eyewear have indicated they did not believe it was required for their tasks. Many employers furnish eye protection at no cost to employees, but commonly many of these workers receive no information on the kind of eyewear they should use and for what tasks.
- Maintenance. Eye protection devices (spectacles, goggles, and face shields) must be properly maintained. Scratched and dirty devices reduce vision, can cause glare, and may contribute to accidents. Scratched or damaged eyewear should be replaced immediately.

7.9 A FINAL WORD ON PPE

If PPE is used as a primary method (or any level) of protection, then it is essential that robust systems be in place to ensure that the correct level of protection continues to be achieved. Such systems may include workplace behavioural safety programs, in-house inspections and audits, routine training or 'toolbox talks', or biological monitoring (mentioned above). The H&S practitioner should resort to PPE only when other means of exposure control are impracticable. The use of any protective equipment places restrictions upon workers, reduces the flexibility of their operation, and can affect their performance. Others have suggested that if an H&S practitioner wants a worker to work permanently in PPE, they should try wearing the same protection for a week to judge its real suitability as a long-term measure.

Implementing and management of a robust PPE program requires great attention to detail and careful record-keeping, for example, to track who has received what training and fit testing for which types of respirators and for which specific tasks.

REFERENCES

ASTM International 2020, *Standard Test Method for Permeation of Liquids and Gases Through Protective Clothing Materials under Conditions of Continuous Contact*, ASTM F739–20, ASTM International, West Conshohocken, PA, www.astm.org [26 December 2023].

EN ISO 374-1:2016+A1:2018 2018, *Protective Gloves Against Dangerous Chemicals and Micro-organisms*, European Committee for Standardisation, Brussels.

Institut National de Recherche et de Sécurité and Institut de recherche Robert-Sauvé en santé et en sécurité du travail 2011, ProtecPo Software for the Selection of Protective Materials, INRS and IRSST, http://protecpo.inrs.fr/ProtecPo/jsp/Accueil.jsp [26 December 2023].

Occupational Safety and Health Administration n.d., *OSHA Technical Manual*, Section VIII: Chapter 1 chemical protective clothing, US Department of Labor, Washington, DC, https://www.osha.gov/otm/section-8-ppe/chapter-1 [26 December 2023].

Safe Work Australia (SWA) 2020, *Model Code of Practice—How to Safely Remove Asbestos*, SWA, Canberra, https://www.safeworkaustralia.gov.au/doc/model-code-practice-how-safely-remove-asbestos [26 December 2023].

Standards Australia 2009, *Selection, Use and Maintenance of Respiratory Protective Devices*, AS/NZS 1715:2009, Sydney, https://www.standards.org.au/access-standards

Standards Australia 2012, *Respiratory Protective Devices*, AS/NZS 1716:2012, Sydney

Standards Australia and Standards New Zealand 2023, *Respiratory Protective Devices – Selection, Use and Maintenance Fit-testing Procedures, AS/NZS ISO 16975.3:2023*, Sydney, https://www.standards.org.au/access-standards

8 Aerosols

Linda Apthorpe and Dr Jennifer Hines

8.1 INTRODUCTION

Thousands of workplaces have many hazardous materials in them – the risk materialises only when workers are exposed to the hazard. This chapter considers aerosols from the perspectives important to the H&S practitioner and covers:

- the definition and types of aerosols;
- work situations where aerosols are commonly encountered;
- what happens to inhaled aerosols in the human respiratory system;
- how to assess various kinds of workplace aerosol hazards;
- an historical review, to help ensure lessons from the past are not lost.

8.2 WHAT ARE AEROSOLS?

The term 'aerosol' applies to a group of liquid or solid particles usually in the range of 0.001–100 µm in size (100 µm = 0.1 mm), suspended in a gaseous medium (Kulkarni et al. 2011). Naturally and artificially produced aerosols are found in ambient and industrial air environments and vary greatly in size, density, particle shape and chemical composition. For example, the shapes of particles include spheres (water or oil droplets and welding fume), cylinders (asbestos and glass fibres), crystals (crystalline silica) and both regular and irregular particles (fly-ash and road dust). The term 'particle' implies a small, discrete object and 'particulate' indicates that a material is made up of particles or has particle-like characteristics.

There is no simple system for classifying aerosols found in the workplace based on the nature of the aerosol, its toxic effect or the particle size, although all of these are important for different reasons. Some aerosols exert their toxic effects in the nose, throat and upper airways and others in the lung, while for some, the lung is merely the route of entry, not the ultimate target organ. Further, some inhaled aerosols can cause more than one effect, depending on the amount to which a worker is exposed (i.e., the dose).

8.2.1 Dust

Dust usually comprises solid particles, generally greater than 0.5 µm in size, formed when a parent material is crushed or subjected to other mechanical forces. Dust is found everywhere and remains one of the most difficult of workplace problems. Operations carried out in today's industrial workplaces, that is, crushing, sieving, milling, grinding, sawing, sanding, machining and pouring, all contribute to dust generation. Particles generated in one workplace can become airborne and be carried into other locations. While many of these dusts are relatively harmless, causing only transient irritation, some give rise to lung fibrosis, others to carcinoma, bronchitis, asthma or other lung disorders.

DOI: 10.1201/9781032645841-8

8.2.2 Fibres

There are many different types of fibres. They can be naturally occurring, that is, asbestos and cotton, or man-made such as man-made vitreous fibres (MMVF) (formally known as synthetic mineral fibres), Kevlar and carbon fibres. Fibres are generally defined by their aspect ratio, that is, the ratio of their length to width. The aspect ratio is important as it defines the ability to become airborne and where it deposits in the human respiratory system. Fibre composition is also important in considering the impact on health from exposure.

8.2.3 Fume

Fume is produced when vaporised materials, usually metal, condense to solid particles. These are initially less than 0.05 µm in size, and generally agglomerate. Smelting, thermal cutting and welding operations all produce fume. 'Diesel fume', as it is commonly and incorrectly known, is a particulate and is discussed in Section 8.7.

8.2.4 Mist

Mist or fog is an aerosol composed of liquid droplets. The droplets may be generated mechanically, such as by spraying, or by condensation of vapour, such as in fog. The particles of mists and fogs are commonly greater than 10 µm in size. Examples of mists are oil mists, paint spray, chemical mists and sprays.

8.2.5 Nanoparticles

Nanoparticles can be naturally occurring as by-products of heating/combustion such as diesel emissions, fume and smoke or as by-products of processes such as laser and 3D printing. They can also be present in the form of engineered (i.e., man-made) nano-sized particles. The International Organization for Standardisation (ISO) (2023) indicates that the size range for nanoparticles or sub-micrometre particles is 1 nm–100 nm (i.e., 0.001 µm–0.1 µm).

8.2.6 Smoke

Smoke is an aerosol, which is a suspension of fine solid particles and/or liquid droplets in a gas. Smoke is generated from the combustion of products and contains gases (carbon monoxide, carbon dioxide), particulate matter, usually carbon-based, and other chemicals, depending on the parent material. Smoke can be generated from industrial processes such as smelting and refining metals; and agricultural practices such as burning crop residues or clearing land for farming and burning fuel. Smoke particles can range from coarse (2.5–10 µm), to fine (2.5 µm or smaller), to ultrafine, (less than 0.1 µm), with most being less than 1 µm in size. The size of the smoke particle affects where it is deposited in the lungs and the subsequent health effects from inhalation.

8.3 AEROSOLS AND THE WORKPLACE

Sources of aerosol include naturally occurring and synthetic manufactured materials. Coal, quartz-bearing rock and metal dusts (e.g., lead) commonly associated with mineral extraction and processing industries are all significant sources of inorganic workplace dust. Many metal manufacturing and heavy industries give rise to dusts and fumes of metals (e.g. lead, zinc, copper and arsenic). Construction and demolition activities potentially expose many workers to dust.

Asbestos can still be present in some 'asbestos-free' friction materials manufactured overseas (e.g., brake linings), despite the ban on importation of all types of asbestos. Workers are also potentially exposed to asbestos fibres in the asbestos removal industry. Man-made vitreous fibres (MMVF) used in insulation and fire protection provide another source of workplace dust exposure; however, they are not toxic like asbestos, with the exception of some types that are classified as 'possibly carcinogenic to humans'. Refer to Section 8.9.1 for more detailed information.

Naturally occurring organic dusts are common in some workplaces. Rural workers are exposed to natural dusts of grain. Sugar-mill workers may be exposed to dust from bagasse, the waste cane that remains after the sugar has been extracted. Downstream processing industries expose workers to dusts containing wood, flour, cotton, paper, felt, fur, feathers and pharmacologically active plant materials.

Within industrial manufacturing processes, the H&S practitioner can encounter manufactured dusts from a range of plastic polymers, including epoxies, polyvinyl chloride, acrylates, polyesters and polyamides. They are found in workplaces as diverse as foundries, plastic pipe manufacture, packaging, surface coating industries, manufacture of composite materials and dental laboratories. The wide spectrum of dusts that occur in different workplaces provides a constant challenge to the H&S practitioner to devise various strategies for their control.

From the point of view of the H&S practitioner, besides dose, the two factors important for assessing the impact of inhaled aerosols in the workplace are:

- Particle size.
- Chemical composition of the aerosol.

Particle size and composition are important in terms of how an inhaled aerosol affects the worker because together they govern how much of a material actually enters the body, where it is finally deposited and what sort of toxic effect it can exert.

8.3.1 Chemical Composition of the Aerosol

It is known that different kinds of aerosol can have different effects on health. Thus, the composition of the aerosol, principally its chemical composition, is important. In some cases, the toxic effect caused by the inhaled particles occurs quickly (e.g., within a few minutes) and this acute effect is easy to associate with exposure to the inhaled aerosol.

In other cases, the effect of inhaling the aerosol may not appear for many years following exposure. It then becomes difficult to make an association between the inhaled aerosol and its chronic (or long-term) effect. This long period, or latency, between exposure and disease has often made it difficult to establish causal links between particular aerosol exposures and disease. This can also lead to a false sense of safety in dealing with such aerosols. To assess any likely health impact of the aerosol in workplace air, the H&S practitioner first needs to know the composition of the aerosol.

The following examples highlight the importance of correct identification of the aerosol. Some particulates are acute respiratory hazards, and some are chronic respiratory hazards; some are sensitisers which can cause increasing symptoms on further contact; and some are not particularly hazardous at all.

- Welding fume may consist mostly of relatively non-toxic iron oxide or it may be combined with a small proportion of highly toxic cadmium or other toxic metals, which have acute effects (refer to Section 8.10 and Chapter 9, 'Metals in the workplace').

- Airborne asbestos fibres present significant chronic and life-threatening respiratory hazards; airborne MMVF do not. Both types of fibre can be present in some workplaces.
- Airborne quartz-containing dusts are far more hazardous than limestone dusts because quartz (i.e., crystalline silica) causes silicosis, a chronic lung disease, whereas limestone is more a 'nuisance' dust, with little effect on the body.
- Many wood dusts, sap, latex and lichens associated with wood can lead to skin irritation, sensitisation, dermatitis and respiratory effects such as rhinitis, nose bleeds, asthma and other allergic reactions. However, toxic activity is determined by the species of tree, with some more toxic than others, even leading to nasal cancer in the case of some hardwood dusts.

In some cases, the identity of an aerosol can be obtained directly from a safety data sheet (SDS), particularly where no chemical transformation occurs during processing. For example, in manufacturing lead-acid accumulator (rechargeable) batteries, it would be reasonable to expect to find lead-containing dust and sulfuric acid mist in the workplace atmosphere. In cases of uncertainty where the identity of particular aerosols is unknown, expert advice and analysis should be sought from a suitable laboratory. Workplace aerosols can contain more than one hazardous component (e.g., refractory ceramic fibre (RCF) and cristobalite, or coal dust and crystalline silica). Where this is the case, each component may need to be monitored. Effective control solutions can be arrived at only if the identities of hazardous materials are known, otherwise, strategies may be based on wrong information and hence ineffective.

Inhaled particles can sometimes be toxic to organs other than the lungs. Uptake may be directly via the bloodstream in the lungs, or by secondary absorption in the gut. Solubility in the gut may be low, while absorption in the lungs is very high. For example:

- Soluble salts such as nicotine enter via the lungs and gut and target the brain.
- Metals such as arsenic, zinc, cadmium or lead may enter primarily via the lung and target various other organs.

8.3.2 Particle Size of Aerosols

Aerosols found in the workplace vary widely in size, as shown in Figure 8.1. Workplace particulates of interest range in diameter from around the width of a human hair (approximately 30–100 μm), to less than 0.1 μm (i.e., nanoparticles). Particle size is important for two reasons. First, particle size determines how long a particle remains airborne and, hence, how far an aerosol cloud will disperse in a workplace before settling which can influence the selection and effectiveness of control strategies.

Second, the effects on health caused by many aerosols depend on their site of deposition in the respiratory tract. The site of deposition depends largely on the size of a particle or, more correctly, its aerodynamic settling velocity. Large particles (up to several hundred micrometres in diameter) settle in the nose and throat and may exert their effects at these sites. Smaller particles are deposited in the upper airways (the bronchi and bronchioles), from where they can be cleared by the mucociliary escalator (a mucus layer moved by cilia, or fine hairs, beating in an upward direction) to be ultimately swallowed and eliminated through the gastrointestinal tract. The very small particles, termed respirable and generally smaller than around 3–5 μm in size, penetrate to the alveolar gas-exchange region of the lung, from where they are cleared very slowly. Particles in the sub-micrometre range (e.g., nanoparticles) may also be transported through the alveolar cell walls directly into the bloodstream.

Consequently, any assessment made in a workplace of an aerosol that presents a disease risk must consider the particle size of interest based on the location of deposition in the respiratory

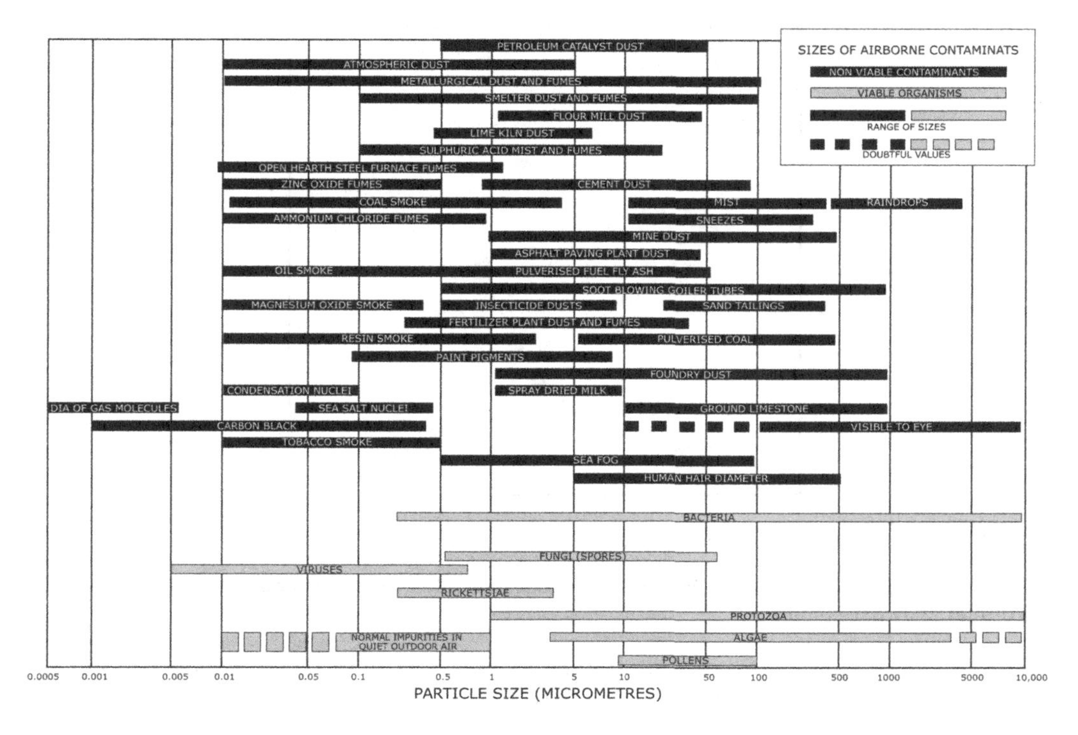

FIGURE 8.1 Sizes of various particles

Adapted from: Daly 1978, p. 65, Figure 5.3.

tract and associated health effects. Different sampling methods are required to achieve suitable assessment.

Figure 8.2 shows, in simplified form, the basic structure of the human respiratory system and each of its components. It also illustrates a number of concepts used in the study and measurement of dust deposition as follows:

1. **inhalable mass** – for materials that are hazardous when deposited anywhere in the respiratory tract.
2. **thoracic mass** – for materials that are hazardous when deposited anywhere in the lung or gas-exchange region.
3. **respirable mass** – for materials that are hazardous when deposited in the gas-exchange region.

The primary agent measured for the thoracic area is sulfuric acid. In Australia, inhalable and respirable fractions are of major significance, and both of these fractions are examined in more detail in Section 8.6. Note that the Australian Standard AS 3640 (Standards Australia 2009a) uses the term 'inhalable dust' (formerly 'inspirable dust').

8.4 AEROSOL MEASUREMENTS IN THE WORKPLACE

To obtain reliable and valid aerosol sampling results, significant attention to detail, appropriate skills and the use of relatively costly and calibrated equipment are required.

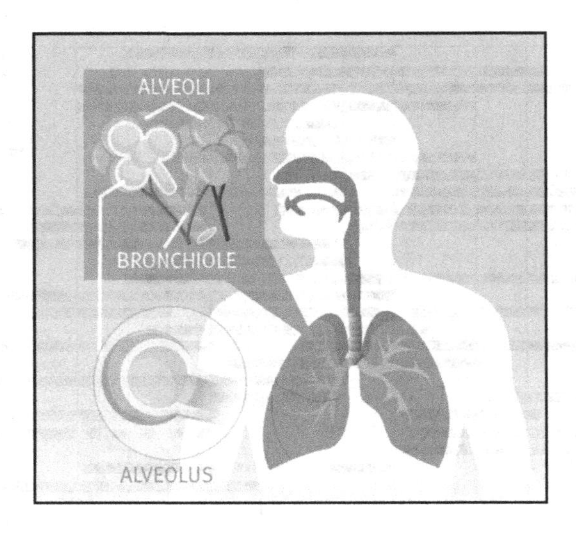

FIGURE 8.2 Basic structure of the respiratory system.

Adapted from: Resources Safety & Health Queensland 2024, https://www.rshq.qld.gov.au/miners-health-matters/what-is-mdld.

Four different classes of aerosol measurements important to the H&S practitioner are discussed here. They involve the sampling and analysis of:

- inhalable particles;
- respirable particles;
- fibrous dust; and
- diesel particulate.

Different devices are used to sample each of these. Sampling with the wrong device can lead to an over- or **underestimation** of the risk, depending on the nature of the contaminant. This can lead to inadequate protection of workers.

Differences between inhalable and respirable particulate sampling are outlined below, together with typical applications and sampling procedures. The laboratory analysis for some applications is also described.

8.4.1 General Approach

The sampling of aerosols to estimate disease risk has taken centuries to understand properly, and just as long to develop into a practical form. The types of aerosol clouds found in the workplace have complex characteristics in terms of particle size and composition. Numerous different sampling procedures and extensive chemical analysis may be necessary. However, if H&S practitioners understand the basic principles of aerosol measurement, then with adequate equipment and training, they will be able to perform some workplace monitoring for aerosols.

For example, with dusts, the H&S practitioner will first need to determine:

- whether a dust should be assessed as a respirable or an inhalable dust;
- whether a dust produces acute or long-term chronic health effects.

Such information provides the basis for proper assessment and control. It will also allow the H&S practitioner to give specific guidance to the employer or a consultant engaged to undertake workplace dust monitoring and to interpret the results of the monitoring. Not all dusty situations will require monitoring. If a competent assessment is made based on knowledge of the dust, it is possible to recommend control procedures without the need for dust monitoring.

8.4.2 Development of Workplace Aerosol Monitoring

Before the current methods of aerosol sampling are examined, it is worth looking at the development of dust sampling over the last century. With the very earliest of dust-sampling equipment, those interested in examining the dust conditions in a workplace, typically a mine or a tunnel, reasoned that the ill health arising from inhaling the dust was probably correlated in some way with the amount the worker inhaled. Dust disease in miners had been known since the days of King Solomon, yet it did not attract much compassionate concern in the days when most mine workers were slaves or prisoners. In fact, it was not until the 15th and 16th centuries, when there was an unacceptable toll amongst the silver miners in Europe (who were responsible for maintaining the coffers of the kings) that some interest (though not much assistance) was taken in their respiratory health.

It was not until the early 1900s that methods of quantifying dust levels became available, and although they were crude, they were ingenious. One of the earliest methods of quantifying mine dust involved pumping a large quantity of air through a tube containing sugar, dissolving the sugar and weighing the trapped dust that remained (Lanza and Higgins 1915).

Such methods were not very sensitive, and only one sample could be collected per shift. In the United States in the 1920s, air samplers were developed that trapped dust particles in a liquid impinger and impacted the dust onto a cold glass slide (by inertial or thermal force) (Marple 2004). These portable samplers allowed an inspector to take many samples per shift. Impingement samples were assessed using a microscope, giving rise to the term 'dust count'. In Australia, these and similar methods were employed by the NSW Joint Coal Board until as late as 1984, when they were replaced by gravimetric sampling methods.

Particle-counting methods, while far more sensitive than the early methods of weighing dust, had numerous drawbacks. Sensitivity is important because some inhaled particles cause respiratory disease even at very low levels. One of the fundamental problems found by medical researchers was that there was no relationship between the number of particles inhaled and the severity of the disease they caused. This was because the size-selecting characteristics of the human respiratory system were largely ignored.

Over the 25 years following the Second World War, the largest ever health study of a group of workers was undertaken in the British coal mining industry to establish the relationship between inhaled dust and dust disease. By examining the dusts deposited in the lungs of deceased miners, the British Medical Research Council was able to propose a size distribution as one of the fundamentally important parameters (i.e., 'respirable' dust) (BMRC 1952). Instruments have since been devised to capture particles fitting this size profile.

Current methods used for aerosol sampling consider the size-selective nature of the human respiratory system, and measurements are expressed in terms of the mass of particles, rather than the number of particles, entering different parts of the respiratory system.

8.5 TYPES OF AEROSOL-SAMPLING DEVICES

Filtration samplers and direct reading devices are common methods used in workplace monitoring. Both have advantages and disadvantages, but for most applications, the filtration types are valid for exposure monitoring. Any H&S practitioner involved in workplace monitoring should seek the advice of an experienced occupational hygienist.

8.5.1 Filtration Sampling Instruments

The majority of aerosol-sampling devices use filtration techniques which involve 'filtering' the air to capture the airborne contaminant (i.e., aerosol) on a filter. Advantages include relatively low cost, robustness and the ability to sample different types of airborne particles (e.g., wood, crystalline silica, coal, metals, grains, organic dusts, powders, acid mists and fibres).

Collecting particles on a filter and measuring subsequent concentration occurs via the following steps:

- The filter media is positioned within a sampling head or sampling cassette, to collect the desired aerosol.
- A known volume of aerosol-laden air is drawn by a sampling pump through the filter media.
- The collected deposit is weighed on a microbalance or analysed using contaminant-specific techniques in a laboratory.
- Equipment required to undertake filtration sampling for aerosols (e.g., in Figure 8.3) includes:
    - Filter.
    - Sampling head which is specific to the type and size of aerosol to be collected.
    - Air sampling pump.
    - Connecting tubing.
    - Flow meter – used to set the required flow rate through the sampling head.

A belt, harness or backpack can be used to conveniently attach the sampling equipment to the person wearing it.

8.5.1.1 Filters

Many different types of filters are useful for workplace monitoring. Filters are chosen depending on the method and aerosol being sampled, the analysis to be carried out and the environment being sampled. Some filters have grids, some are acid-resistant, and some are transparent in certain light wavelengths. Filter media may include polyvinyl chloride (PVC), polycarbonate, glass

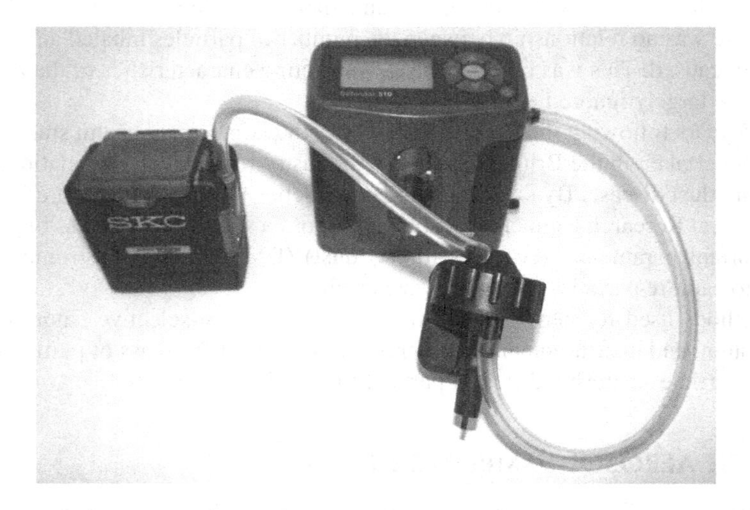

FIGURE 8.3 Typical personal monitor showing pump, sampling head and electronic flow meter in a calibration train.

Source: L. Apthorpe.

fibres, polytetrafluoroethylene (PTFE), mixed cellulose esters etc. and have pore sizes ranging from 0.3 to 5 µm. Specialist advice on appropriate filter type and pore size should be sought for any new sampling task.

8.5.1.2 Sampling Pumps

Different types of sampling pumps are available for aerosol monitoring in the workplace. Some operate on mains power, most are small, convenient, battery-powered pumps that can be worn by the worker. These pumps are designed to provide flow rates of between 0.5 and 5 litres per minute (L/min), with most aerosol sampling carried out between 2.0 and 3.0 L/min.

Good dust-sampling pumps have four important features:

- They are controlled so that the volume of air can be accurately measured.
- They are pulsation dampened so that any size-selective device connected may operate correctly (i.e., constant flow).
- They have the ability to set flow rates over a wide range.
- They have the capacity to operate at a reasonable suction pressure (e.g., up to 10 kPa).

A fully charged pump in good operating condition should be able to operate for 12–16 hours. Intrinsically safe pumps, designed to be incapable of generating sufficient energy to ignite a flammable atmosphere, are mandatory for use in workplaces with potentially explosive atmospheres (e.g., coal mines, petroleum refineries and grain-handling facilities).

Flow meters built into sampling pumps should not be relied upon because they can indicate incorrect and variable flow rates. Any flow meter used to measure the flow rate of a pump must be calibrated with a primary flow meter: this is a flow meter whose key properties are traceable to national measurement standards of length, mass and/or time. One example of a primary flow meter is the 'soap film flow meter' used in a calibration train similar to that in Figure 8.4. The soap film flow meter uses the principle of timing the movement of a soap bubble along a transparent tube of known volume as a result of air flow produced by a pump. There are several types of portable flow meters (e.g., dry piston) that have been shown to provide consistent and reliable results. Although the makers of some of these instruments claim that they are primary meters, they do not meet the Australian National Measurement standard in this respect and should be checked against a bona fide primary flow meter.

8.5.1.3 Affixing Sampling Equipment to a Worker

Tubing used to connect the pump to the filter holder needs to be made of high-quality PVC (e.g., Tygon) or polyethylene so that it does not readily crush, retains its elasticity and does not absorb chemicals.

Pumps typically are worn on the belt or may be carried in a backpack to secure the pump to the worker. For exposure monitoring, the sampling head must be placed within the worker's breathing zone.

8.6 AEROSOL SAMPLING DEVICES

There are many commercially available sampling devices for aerosols and it is important that the correct one is selected for each sampling task. Details of the recommended samplers are given on the following pages. Only samplers with the correct performance can be used to make measurements that are in accordance with the exposure standards (ES).

While most sampling commonly involves dusts, similar principles underpin sampling for other aerosols such as fibres (see Section 8.8.8).

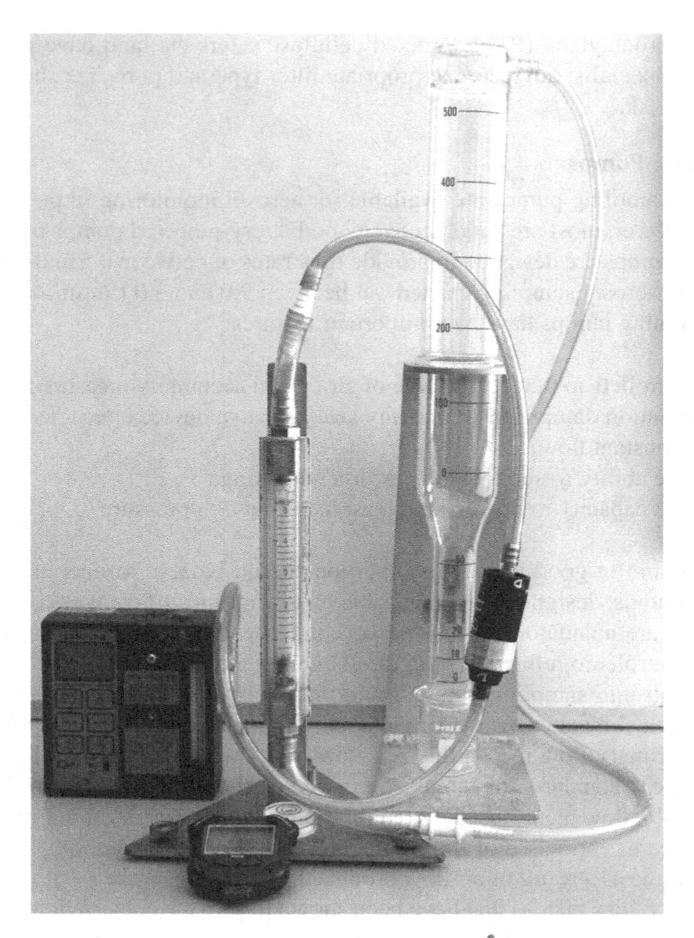

FIGURE 8.4 Flow rate calibration train and soap film flow meter.

Source: L. Apthorpe.

8.6.1 Inhalable Dust Sampling

Dust hazards will mostly be assessed based on the dust a worker can inhale from the workplace air. This is commonly known as inhalable dust and is similar to, but does not produce the same sampling results as, 'total dust' (the total-dust sampler was developed in North America and is not used in Australia or Europe). Many particles in a visible dust cloud are aerodynamically too heavy to be captured by the respiratory system, which means that only a fraction of them are inhaled. It is important to bear in mind that inhalable dusts represent a wide size range and include respirable particles (respirable dust is discussed in Section 8.6.2).

Some of these inhalable dusts cause their effects at the site of deposition in the upper airways. They include the following examples:

- wood dusts causing nasal cancer (e.g., oak, beech, birch and mahogany);
- cement dusts causing airway irritation;
- sensitising wood dusts causing asthma (e.g., western red cedar);
- proteolytic enzymes, which attack cell structure.

Inhalable mists can also contain oil, acids or alkalis, which require special analysis and expert advice for sampling.

There is no universally accepted size criterion for inhalable dusts, because inhalability varies with the density of the dust. It may also depend on a number of workplace and human factors, such as wind speed and whether the worker is a nose or mouth breather. For practical purposes, inhalable dust is defined as the fraction of a dust cloud collected by certain types of dust samplers. Australian Standard AS 3640 (Standards Australia 2009b) provides definitions for two different samplers based on the common standard defined by the International Organization for Standardization (ISO 1995), published as ISO 7708 and harmonised with the definition used by the American Conference of Governmental Industrial Hygienists (2024). The ISO criteria are seen in Figure 8.5, which shows that the collection efficiency of any inhalable dust sampler is above 90 per cent for particles below 4 µm in diameter and drops to around 50 per cent efficiency for particles of diameter greater than 100 µm. In other words, as with the human nose and mouth, not all particles are caught by the sampler.

The main inhalable dust sampler recommended by AS 3640 which is used in conjunction with the ES is shown in Figure 8.6. The IOM dust sampler has been designed to overcome problems of different wind speeds past the wearer and sampler orientation (i.e., sampler facing horizontal or downwards).

Although not commonly used, the United Kingdom Atomic Energy Association (UKAEA) seven-hole sampler may also be used. This device has been shown to be less precise in both workplace and laboratory-based sampling for numerous contaminants and requires a correction factor (Hanlon et al., 2021, 2023).

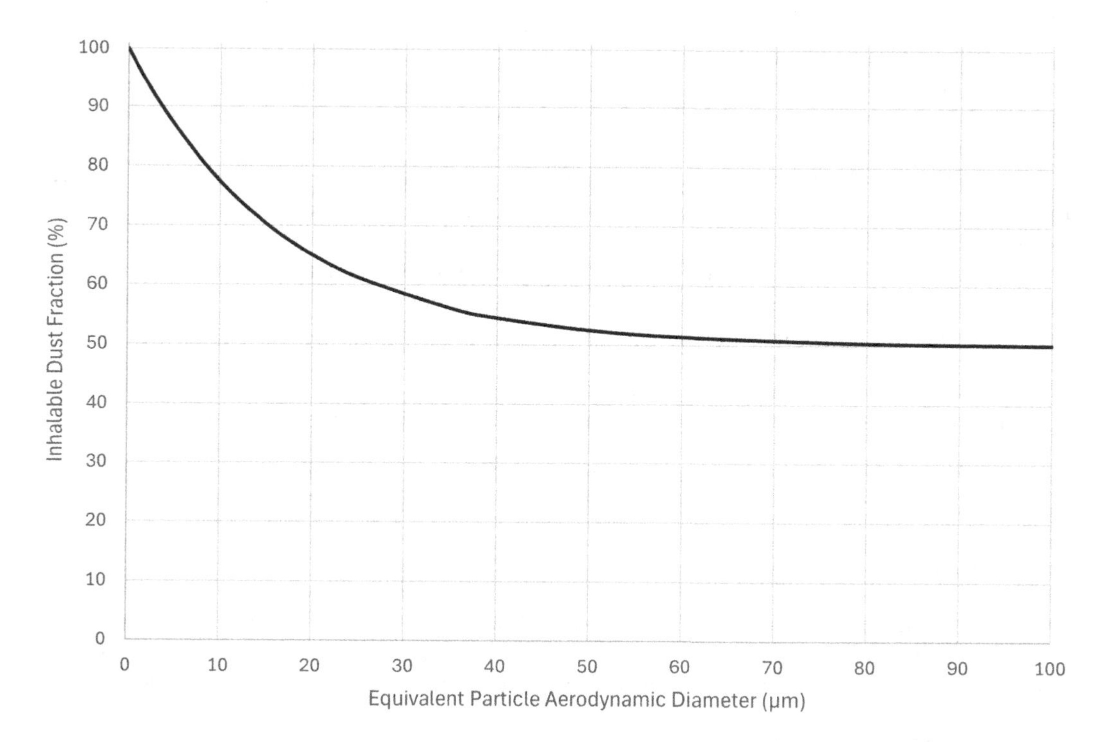

FIGURE 8.5 Inhalable dust sampler performance.

Adapted from: International Organization for Standardization 1995, ISO 7708, Table B.2, p. 7.

FIGURE 8.6 IOM inhalable dust sampler.

Source: L. Apthorpe.

A detailed practical procedure for measuring inhalable dust in the workplace is given in AS 3640. The field sampling procedure specified in this standard can be conducted by H&S practitioners with appropriate sampling equipment and training. The laboratory utilises analytical equipment and techniques (e.g., microbalance and static eliminating equipment) to analyse samples. Analysis should be conducted by laboratories accredited by the National Association of Testing Authorities (NATA) for the specific test.

On completion of sampling, the concentration in mg/m^3 of inhalable dust is calculated by dividing the net weight gain (after taking into account weight change of a blank filter) by the total volume of air sampled.

Gravimetric weighing of the dust is not the only method of evaluating concentration in the workplace. When there is a mixed dust, other types of laboratory analysis will be needed to determine the concentration of the contaminant of interest. Examples include:

- measuring metal contaminants in fumes or dusts;
- measuring polycyclic aromatic hydrocarbons (PAHs) in bitumen or smoke from fires.

For additional information on the interpretation of inhalable dust sampling results, refer to AIOH (2014).

8.6.2 Respirable Dust Sampling

The respirable fraction is a subset of the inhalable fraction of aerosols. Evaluation of respirable aerosols is important when assessing certain workplace hazards. Prior to assessing respirable dusts, consideration is required for the chemical composition and shape (e.g., irregular or fibrous).

8.6.2.1 Dust Respirability

Typical aerosols in a mine, quarry or factory contain particles of irregular shapes ranging in size from 100 µm aerodynamic diameter to as little as 0.1 µm. The particles are usually generated whenever a solid material undergoes a mechanical process that breaks up or disturbs the material to create smaller particles. When inhaled, the smaller particles can travel to the lower parts of the lung where damage to lung tissue can occur.

Historically, workers in some dusty occupations have developed specific diseases based on the type of dust they were exposed to. Diseases such as silicosis (fibrosis of the lungs) have been known to occur in miners, quarry workers, tunnellers and stonemasons; and coal workers pneumoconiosis (also known as 'dusty or black lung' which is a permanent alteration of the lung structure) occurring in coal miners.

To assess health risks from exposure to common respirable dusts such as those containing crystalline silica or coal, the sampling technique must measure the concentration of dust particles within the size range reaching the critical parts of the lung, the alveolar, non-ciliated regions (i.e., those without the hair-like structures that help to transport secretions). These particles are very small (less than around 5–10 μm).

8.6.2.2 Respirable Dust Measurements

Fibrous dusts (not to be confused with dusts that produce fibrosis of the lung) are examined in Sections 8.8 and 8.9.

The British Medical Research Council (BMRC) original definition of 'respirable dust' was recommended at the Pneumoconiosis Conference held in Johannesburg in 1959. In 1995, the International Organization for Standardization's technical report ISO 7708 *Air Quality—particle size fraction definitions for health-related sampling* (ISO, 1995) modified the BMRC definition of respirable dust so that the same definition was used worldwide (now often referred to as the CEN-ISO-ACGIH respirable fraction curve).

When measuring respirable dust, the dust is commonly measured by mass, with compositional analysis to be undertaken as required, for example, for crystalline silica.

The practical sampling device meeting the respirable dust size-selection criterion is the miniature cyclone elutriator, a small, portable device well suited to personal sampling. The cyclone elutriator operates by utilising a rapid cyclonic circular action to separate out particles according to their aerodynamic diameter.

The original Higgins and Dewell miniature cyclone was developed by the British Cast Iron Research Association with a lighter cyclone developed by the Safety in Mines Research Establishment at Casella as a personal dust sampler. This modified version was known as the SIMPEDS cyclone. Other manufacturers, for example, SKC, also produced their versions of the Higgins and Dewell miniature cyclones. The Casella and SKC versions are now known as 'modified' Higgins and Dewell cyclones. The aluminium cyclone, commonly used in North America, was originally designed to meet the ACGIH® sampling curve.

Figure 8.7a shows the miniature cyclone sampler, with its filter cassette and sampling filter. Figure 8.7b displays the aluminium cyclone, along with the SKC and Casella modified Higgins and Dewell cyclone sampling devices.

AS 2985:2009 Workplace Atmospheres – Method for Sampling and Gravimetric Determination of Respirable Dust (Standards Australia 2009b) provides the methodology for respirable dust sampling. Based on AS2985 and information from the manufacturers of the cyclones, examples of flow rates used are provided in Table 8.1.

If set at the correct flow rate, the samplers in Table 8.1 meet the requirements of the sampling efficiency curve for respirable dust as per the ISO 7708 respirable dust convention (ISO 1995), as seen in Figure 8.8. These flow rates satisfy the criteria for size-selective sampling whereby unit-density particles less than 2 μm in diameter are collected at greater than 97 per cent efficiency, 5 μm particles are collected at 34 per cent efficiency and particles larger than 18 μm are not collected at all. As the specific gravity (density) of particles increases, smaller particles become

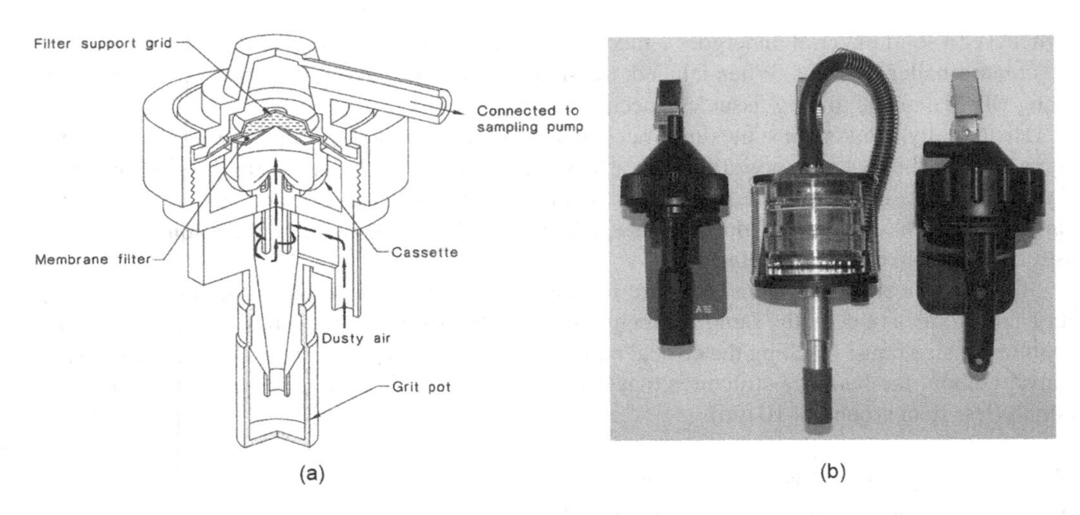

FIGURE 8.7 (a) Sectional view of cyclone sampler for respirable dust sampling (Reproduced Figure 1 from AS 2985-2009, p. 8). (b) Casella, Aluminium and SKC cyclone for respirable dust sampling.

Source: L. Apthorpe.

TABLE 8.1
Flow Rates for Size-Selective Cyclone Samplers

Size-Selective Sampler	Flow Rate (L/min)
Casella modified Higgins and Dewell cyclone	2.2
Aluminium cyclone	2.5
SKC modified Higgins and Dewell cyclone	3.0

aerodynamically equivalent to larger 'unit-density' particles. For example, 3 μm quartz particles are equivalent to 5 μm unit-density particles.

The H&S practitioner involved in respirable dust monitoring must ensure that sampling is conducted using appropriate equipment and with the correct flow rate to sample respirable dust that meets the ISO 7708 (1995) sampling efficiency curve. A detailed practical procedure for measuring respirable dust in the workplace is given in AS 2985 (Standards Australia 2009b). Monitoring procedures for respirable dust permit very little latitude and flow control is critical for correct adherence to size-selective criteria. The field sampling procedure specified in this standard can be conducted by H&S practitioners who have appropriate sampling equipment and training.

The laboratory analysis requires equipment such as micro-balance and static eliminator with various analytical instrumentation required for secondary analysis for analytes such as crystalline silica. This usually makes it necessary to utilise specialist laboratory services accredited by NATA for the analysis.

On completion of sampling, the concentration (mg/m^3) of respirable dust is calculated by dividing the net weight gain (after taking the weight gain or loss of a blank filter into account) by the total volume of air sampled. Where speciation (i.e., secondary analysis) has occurred (e.g., for quartz), the amount per filter is utilised in the concentration calculation.

It is strongly recommended that the H&S practitioner requiring dust measurements should contact an experienced occupational hygienist or a laboratory experienced in this field.

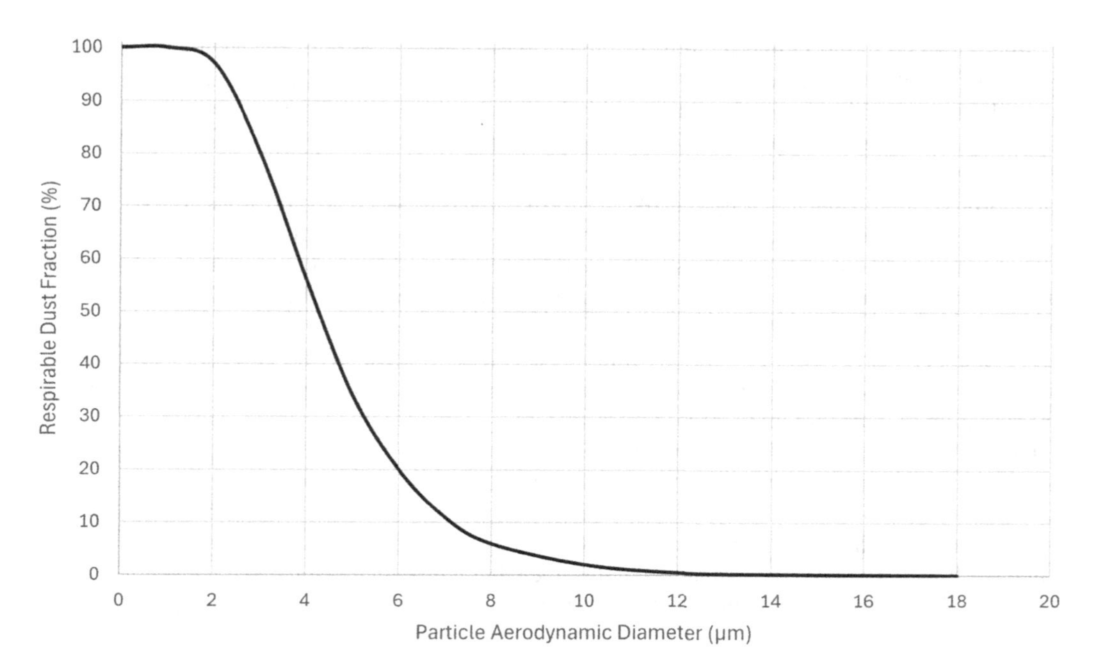

FIGURE 8.8 Respirable dust sampler performance.

Adapted from: International Organization for Standardization 1995, ISO 7708, Table B.2, p. 7.

8.6.2.3 Assessing Workplaces for Common Respirable Dusts

Most dusts producing pneumoconiosis are assessed by monitoring respirable dust. Following is a list of the most common of these:

- Respirable crystalline silica:
  - alpha quartz
  - cristobalite
  - tridymite
- Tripoli
- Fused silica
- Fumed silica
- Microcrystalline silica
- Coal dust
- Graphite (natural).

Alpha quartz and coal dust are the most prevalent and depending on the workplace, there can be large numbers of workers exposed.

8.6.3 Silica Dusts

8.6.3.1 Silica Sources and Exposures

Silicon is the most abundant element in the earth's crust, and a number of geological silicates are toxic to humans. It is ubiquitously distributed, occurs in most rocks and soils and is widely found in many workplaces including mining, construction and manufacturing industries.

Silicon can be found in two main forms, that is, crystalline or amorphous, each with the same chemical formula. Both crystalline and amorphous forms can be found as a naturally occurring mineral. The crystalline form of quartz (α or alpha) is often known by the general term, crystalline silica. The most common type of respirable dust found in workplaces is crystalline silica in the form of α-quartz. α-quartz in its original state may be transformed by heat to two of its other forms (i.e., polymorphs), cristobalite and tridymite. All forms of crystalline silica are fibrogenic to lung tissues.

Typically, the fibrotic lung disease of silicosis is a chronic disease that may progress from a relatively benign form where there is little impairment to lung function to a severe form if exposure continues. Permanent disability or death may ensue as the parts of the lung affected by fibrosis are rendered incapable of transferring oxygen into the bloodstream. In some cases of very heavy exposure over short periods of time, an accelerated form of the silicosis can develop. Furthermore, heavy exposure to silica can also increase the risk of lung cancer. Early diagnosis and cessation of exposure is therefore imperative.

Crystalline silica in the lungs initiates a reaction involving macrophages (scavenging cells), which die, triggering an inflammatory reaction. This reaction leads to the development of fibrotic nodules, which grow and coalesce as the disease progresses even after the exposure ceases. The reason that crystalline silica promotes fibrogenic disease of the lung is not entirely clear, although it may be linked to the surface structure of the free silicon dioxide crystals.

Silicon dioxide which is bound up in complex silicates (e.g., basalt, ilmenite) produces little or no lung fibrosis. The presence of other minerals (e.g., aluminium) on crystalline silica particles has been shown to reduce toxic effects. Fresh cleavage of respirable quartz causes the formation of reactive radical species which increases toxicity to lung tissues. Where the particles have been 'aged' by the process of time (i.e., in air over time) or by water (which ages the particles quickly), the free radical action, that is, toxicity, decreases (AIOH, 2018a).

Figure 8.9a shows the radiograph of the lungs of a miner with no dust disease; Figure 8.9b shows a case of advanced silicosis. The normal air space of the lung shows up as dark shadow.

Table 8.2 provides examples of various workplaces and processes which can lead to respirable crystalline silica exposure for workers. In these workplaces, H&S practitioners may need to undertake respirable dust measurement and respirable crystalline silica (i.e., quartz) analysis.

When considering potential exposures to respirable crystalline silica, it is important to consider both primary sources, that is, via processing a solid material to generate airborne dust; and secondary sources, where airborne dust which contains silica has settled on surfaces that can become airborne via disturbance (e.g., dry sweeping or blowing).

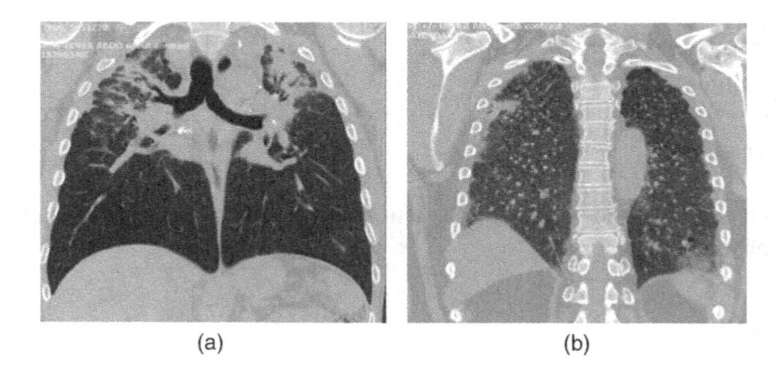

FIGURE 8.9 (a) Radiograph (CT Scan) showing lungs with early stage silicosis (Courtesy of Prof Deborah Yates) (b) Radiograph (CT Scan) showing lungs with late stage silicosis.

Courtesy of Prof Deborah Yates.

TABLE 8.2
Workplaces and Processes Which Can Lead to Silica Exposures

Abrasive blasting	Abrasive blasting on large steel structural components, processing plants and exposed aggregate concrete products produces extremely high concentrations of respirable dust. Australian abrasive blasting regulations prohibit the use of crystalline silica for dry abrasive blasting, and all sands are excluded from use. Only abrasives which meet the stringent standards can be used for abrasive blasting purposes which usually occur inside an enclosed cabinet or space.
Beverage production	A common material used in the filtering and clarification of beverages is diatomaceous earth (amorphous silica). In its calcined (roasted) form, this material may contain 40–50 per cent cristobalite and dusts from this material pose a potential hazard unless well controlled.
Concrete and masonry work	Chasing is the process where concrete or masonry is cut to allow the laying of services such as electrical conduit into a wall, floor or ceiling. Masonry is the skill of building with brick and stone (i.e., natural and man-made) and involves processes such as cutting and finishing tasks. Mechanical removal of grout from brickwork is also undertaken. Traditional sculptors and craftspeople can be exposed to crystalline silica from working with sandstone and granite. The use of newer mechanical tools (chipping hammer and air bottle), which creates more dust from grinding and sanding than occurred with traditional hand methods, can lead to greatly increased risks unless properly controlled.
Construction and demolition	Workers who undertake construction and demolition work may be at risk of exposure to many hazards, including silica. Buildings and structures can contain silica in concrete, bricks, tiles, natural and engineered stone and any process which disturbs the solid materials can lead to respirable crystalline silica exposures.
Pottery and brick-making industries	In these traditional industries, the handling of clay dusts containing up to 20 per cent quartz has produced many cases of respiratory disease. A secondary hazard exists because some of the quartz can be converted to cristobalite in the furnace or kiln. On a much smaller scale than occurs in industrial processes, art studios and pottery schools also produce respirable dusts which can contain crystalline silica. Often smaller scale industries and artisan workers have limited access to suitable control strategies.
Quarrying and mining	Many quarries for building materials and road-base rock produce respirable dusts containing crystalline silica in their crushing plants and screening operations. Depending on the mineral being extracted, processes at some metalliferous and non-metalliferous mines can lead to crystalline silica (i.e., quartz) exposure for workers.
Stonemasonry	Stonemasons work with various natural (e.g., sandstone) and man-made (e.g., engineered) stone materials. Their work involves processes including chipping, cutting, grinding and polishing using traditional hand tools or mechanically operated equipment. Techniques involving the use of mechanical tools on materials such as engineered stone without effective controls in place have led to significant exposures and development of accelerated silicosis for workers involved in fabricating and installation of kitchen bench tops.
Tunnelling	With increasing population, there is a need for large infrastructure projects for services, road and rail. Construction works include extensive earthworks and tunnelling and these activities can expose workers to respirable crystalline silica (i.e., quartz). Historically, silicosis was common amongst tunnellers, however, chronic exposure may also lead to chronic obstructive lung disease and lung cancer.

8.6.3.2 Laboratory Analysis of Respirable Crystalline Silica

Most particulate samples obtained by air sampling will contain a mixture of materials, and laboratory analysis may be required to determine their composition.

In practice, laboratory analysis for crystalline silica (i.e., quartz) is conducted on most respirable dust samples. This analysis should be conducted by specialist laboratories accredited by NATA using one of the two National Health and Medical Research Council methods adopted and published by NOHSC: infrared spectrometry and X-ray diffractometry (National Health and Medical Research Council [NHMRC] 1984).

Infrared spectrometry or X-ray diffraction analysis can be conducted on the sample filters; however, prior to selecting an analytical method, it is important to understand the strengths and limitations of the methods as it may bias the results. If interfering minerals are present, for example, kaolinite, amorphous silica or cristobalite, then X-ray analysis will be necessary. Cristobalite can be present naturally or where quartz has undergone heat treatment above 450°C. To select the most appropriate analytical technique for the workplace where the measurements were taken, the H&S practitioner should contact the analysing laboratory.

The respirable concentration of alpha quartz in the air can then be calculated once the volume of air sampled is known.

8.6.4 Coal Mining Dust

Coal is a combustible mineral, formed predominantly of carbon, with variable quantities of other elements such as hydrogen, sulfur, oxygen and nitrogen and other impurities. Dust in coal mines includes carbon, stone dusts, quartz and silicate particulates which originate primarily from cutting adjacent or inseam rock strata. Respirable coal dust (RCD) is formed during coal mining by the action of coal cutting machinery, from transfer and handling (e.g., conveyors and transfer points), stockpile building and reclamation, train loading and unloading processes.

Coal mine dust lung disease (CMDLD) is defined as an occupational lung disease attributed to the exposure and cumulative inhalation of RCD. The three most common health outcomes are:

- coal workers pneumoconiosis (CWP) (also known as 'black lung');
- progressive massive fibrosis (PMF);
- chronic obstructive pulmonary disease (COPD).

Each of these affects the lungs in different ways; however, common themes include impaired lung function, emphysema and reduced life expectancy.

CWP is caused by the inhalation of coal dust, and possibly crystalline silica. In its simpler form, CWP will usually be asymptomatic, even though detectable on lung X-rays. It can progress following further exposure to show extensive but discrete lesions on an X-ray. Advanced simple CWP is significant because it may lead to the most severe form, PMF. In PMF, the lesions coalesce to show extensive large opacities on the X-ray. The lung in these areas becomes a hard, black mass, severely reducing breathing capacity, which leads to disability and likely premature death. COPD is a term that encompasses long-term lung conditions such as emphysema, chronic bronchitis and chronic asthma.

In addition to RCD, coal mines and coal handling facilities may also have exposure to other forms of dust such as:

- inhalable fraction of coal dust;
- respirable crystalline silica (RCS); and
- dusts not otherwise specified (NOS) (AIOH 2018b).

The United States led the way in reducing CWP through the enactment of the 1969 Federal Coal Mines Health and Safety Act (Coal Act) and implementation of the Coal Workers Health Surveillance Program (CWHSP) that was administered by the National Institute of Occupational Safety and Health (NIOSH). Along with identifying CWP amongst workers, a permissible exposure limit for respirable coal was introduced and resultant preventative measures were adopted by coal mines as mandated by the Coal Act to comply with this exposure limit. In the mid-1970s, more than 30% of underground coal miners who had worked for greater than 25 years had evidence of CWP with 3.5% having PMF. This dropped to 5% and 0.5%, respectively, by the late 1990s; however, by 2000, these trends reversed and, by 2015, the prevalence of PMF reached the highest level ever recorded (Attfield et al. 2011). Australia followed a similar downward trend due to the relentless application of improved controls in response to more stringent respirable dust ES. However, Australia also detected a change in 2015–2017, when more than 20 cases of coal workers pneumoconiosis (CWP) were diagnosed in Queensland, and one in NSW in 2017 (AIOH 2018b). This was a disturbing revelation as the latency period is long, and the indications are that more cases may be found. Investigations determined that the increases are due to controls, including health surveillance not being appropriately maintained, longer duration of exposures (longer work hours), thinner seams and the resulting increase in exposure to rock and other intrusions with the overall result being a resurgence in dust diseases.

Control technologies have been developed that can, and have been, successfully implemented across parts of the coal mining industry. The hierarchy of controls must be applied when determining the appropriate controls to be utilised. OHS in the Australian mining industry is regulated by individual States and Territories. They have their own Legislation which includes specific information on dust control. A useful resource regarding controls for dust in underground Mines is 'Recognised Standard 15 – Underground respirable dust control' (Department of Natural Resources and Mines, 2017) which provides descriptions and images of control strategies including minimising dust exposure by considering the cutting sequence and method; automating the process wherever possible; applying water via sprays with recognition given to droplet size, pressure and interaction with dust particles; ventilation; maintenance and implementation of a comprehensive respiratory protection program if dust exposure remains high (including fit testing and being clean shaven). AS/NZS ISO 23875:2003 (Standards Australia, 2023b) refers to the best practice design, testing, operation and maintenance of air quality control systems for heavy machinery cabs and other operator enclosures, particularly in the mining industry.

It should be understood that there are multiple sources (and types) of dust and every workplace is different; hence, a number of control strategies will be required to reduce worker exposures to acceptable levels. Whatever strategy is adopted, it should be supported by an effective maintenance program on coal cutting and handling equipment and ventilation systems, so that dust control effectiveness is sustained (AIOH, 2018b).

Apart from underground mining, the H&S practitioner will also find coal dust associated with coal-fired furnaces, at ports during transportation (loading and unloading), warehousing, foundry work and in laboratory facilities that analyse coal.

8.6.5 Direct Reading Devices

Accurate and precise data for inhalation hazards can be provided by integrated air sampling and laboratory-based analysis of field-collected samples as described in Section 8.5.1. The limitation of this approach includes a time delay between sample collection and laboratory analysis; the need for additional sampling if results are inconclusive, the source cannot be identified, or speciation is required. Acute short exposures are generally not well determined with time-weighted average concentration.

Portable or wearable continuous monitors can provide immediate concentration data which has the advantages of temporal resolution of concentration variations providing information on peaks and average exposures; allowing an understanding of specific activities that create elevated exposures; and immediate identification of dust sources and/or exposures of concern which can be used to justify actions to promptly reduce exposure.

However, obtaining useable data from this instrumentation requires reliable and cost-effective instruments. Although real-time aerosol monitors have been available for many years, they have generally been limited to light scattering and estimates of mass concentration, which are not suitable in all instances. Recent advances in sensor development, electronics within the instruments and information processing technologies are providing more sensitive, continuous monitoring in a field that is rapidly changing. It is important that users of real-time monitoring equipment understand the technology and methodology used behind the data and results obtained, along with the advantages and limitations of such equipment; and how information provided from this equipment can be used to the best advantage to protect worker health for the specific workplace context.

With an ever-increasing variety of commercial instruments available, it is necessary for the H&S practitioner to understand the operating principles of the instruments and the type of aerosol to be detected. Some of the criteria for selecting an appropriate instrument include consideration of the particle size range of interest; the system parameters of interest (e.g., mass, surface area concentration or number concentration) versus particle size distribution or chemical composition measurement, etc., as well as the calibration, cost and convenience (e.g., fixed location or portability) of the instrument.

The technologies utilised in the instruments include light scattering devices, optical particle counter (OPC), piezoelectric mass sensors, tapered element oscillating microbalance (TEOM) and condensation particle counters (CPC). Often the size of the particle to be measured helps determine which is the most appropriate instrument to use. However, it should be noted that the size of the particle measured by real-time instruments is often termed particle matter (or PM) (e.g., $PM_{1.0}$, $PM_{2.5}$, PM_4 or PM_{10}) which does not relate to the particle sizes measured by traditional inhalable or respirable dust sampling equipment and the respective Australian Standards AS2985 and AS3640.

Light scattering devices have traditionally been the most common type of aerosol monitor. These devices employ a light source, typically a laser or light-emitting diode (LED), that illuminates aerosol particles as they pass through the instrument. The particles scatter the incident light, and the scattered light is detected at specific angles. The intensity of the scattered light is proportional to the size and concentration of the aerosol particles, which results in real-time aerosol measurements. OPCs also utilise the light scattering principle to measure aerosol particle concentration and size distribution. OPCs differ widely in their design and performance characteristics. These devices are user-friendly and applicable to various work settings, including where immediate detection of changes in particle concentration is needed. However, their accuracy can be compromised by variations in particle refractive indexes and difficulty differentiating between single large particles and clusters of smaller ones. They are affected by moisture (including fog or mist), which can lead to inaccurate readings and misinterpretation of data. OPCs are widely used for size distribution measurements in both the indoor and outdoor environments.

Piezoelectric mass sensors also known as quartz-crystal microbalance (QCM) provide near real-time mass concentrations. Particles are deposited on a quartz crystal; the natural vibrating frequency of the crystal is shifted and is used as a measure of the deposited particle mass. The deposited particle mass is proportional to the frequency shift. Particle deposition is achieved either by electrostatic precipitation or by inertial impaction. Impactors such as respirable impactors can be incorporated to remove non-respirable particles allowing respirable particles (e.g., $PM_{4.25}$) to be deposited by electrostatic precipitation onto the quartz crystal for measurement.

TEOMs use the same operation principle of the QCMs; however, instead of relying on the natural frequency of the quartz crystal, the vibration of the hollow tapered element is initiated and maintained by an electronic feedback system. A LED and phototransistor monitor the oscillation of the taped element. The collected particle is proportional to the difference of the inverse square of frequencies before and after sampling.

CPCs are direct reading instruments that detect and count aerosol particles based on their ability to condense and grow on a working fluid (such as water). To condense water vapour on particles, the instrument cools the sampled aerosol and delivers the aerosol stream into a moisture-saturated heated condenser. Within the condenser, the water vapour diffuses into the core region of a condenser tube. The water vapour condenses on the particles causing the particles to grow, producing supersaturation. Particles are then detected optically by light scattering. This method is effective for measuring ultrafine particles with diameters down to a few nanometres.

The accuracy of the different types of direct reading instruments can be affected by environmental factors, such as humidity, temperature and the presence of other gases.

One of the main advantages of direct reading instruments is that they generally have data-logging capabilities, allowing them to record and store data over time. It is important when using real-time instrumentation to understand what the displayed data means. For example, some instruments calculate the average dust concentrations over a 15-minute period and assume zero exposure for the remainder of the day, that is, a 15-minute result of 12 mg/m^3 assumes 0 mg/m^3 for the remainder of the 8-hour day. The 8 hr TWA for this result is actually 0.3 mg/m^3.

Other instruments provide a time constant or rolling average over a set time, or the results may be averaged over a time period (e.g., 1 hour) and the result is displayed at the end of that time period, therefore, it is not exactly an instantaneous result. Other factors to consider include the logging intervals for data to be recorded.

Some instruments also offer wireless connectivity for remote monitoring. Both features can be valuable for tracking trends and identifying high- and low-exposure patterns. It is also possible to pair direct reading instrumentation with video footage to gain a visual representation of events occurring at the time of elevated results. This is a very powerful combination of techniques to determine where controls are most required, and to assist in educating workers where and when high dust levels occur.

Many direct reading instruments come with alarm systems that trigger when dust concentrations exceed pre-defined thresholds. These alerts can help prompt immediate action when hazardous conditions are detected.

The cost of the instrument, including maintenance and calibration expenses, should be taken into consideration when selecting appropriate monitoring equipment for the workplace. Some devices are not able to be calibrated, and many are not calibrated to the specific type of aerosol being monitored.

Real-time monitoring equipment is very beneficial when used as a screening tool or to assess engineering controls; however, they should not be used as a decision-making tool or replacement for exposure monitoring when undertaking health risk assessments. The equipment and technologies utilised are rapidly evolving, and it is imperative that the H&S Practitioner understands the advantages and limitations of the individual pieces of equipment prior to using them in the workplace. Examples of real-time monitoring equipment are shown in Figure 8.10a and b.

8.7 DIESEL PARTICULATE MATTER

Diesel machinery is used widely in industry, due to its reliability, fuel efficiency and power. It was purposefully designed to handle strenuous loads in challenging environments. Unfortunately, it emits diesel engine exhaust which is carcinogenic and may cause other health-related issues to the workers utilising this technology. Diesel exhaust is an ever changing and complex mixture of gases, adsorbed organics and particulate components. Engine type and mode of use (load state, maintenance condition, driving style, etc.) result in a wide variation of emitted compounds and

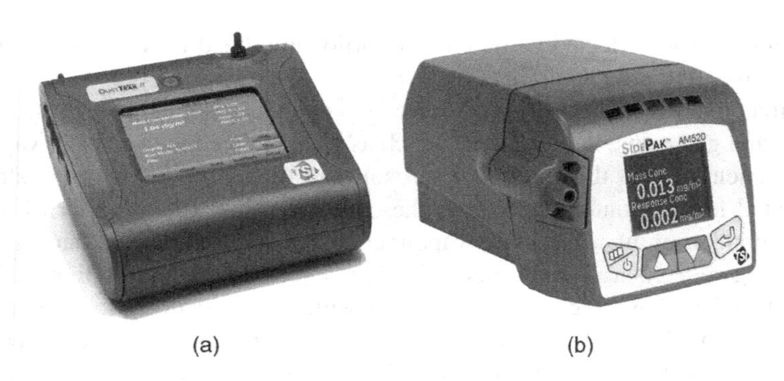

(a) (b)

FIGURE 8.10 (a) DustTrak™ direct-reading aerosol monitor (Photo courtesy of TSI Incorporated). (b) SidePak ™ AM520 personal direct-reading aerosol monitor.

Photo courtesy of TSI Incorporated.

the mixture is of interest to the occupational hygienist with respect to exposure to workers. The particulate phase of the emissions [or exhaust] is the focus of this section.

8.7.1 Composition of Diesel Particulate Matter

Amman and Siegla (1982) summarised the early research into the composition of diesel particulate matter (DPM) defining it as 'consisting principally of combustion-generated carbonaceous soot with which some unburned hydrocarbons have become associated'. Using photomicrographs of the exhaust from a diesel passenger car, they demonstrated that diesel particulates were made up of a collection of spherical primary particles, termed 'spherules', which formed aggregates resembling in appearance a range of forms from a cluster of grapes to a chain of beads. Subsequent researchers have confirmed that the spherules vary in diameter from 10 to 80 nm, with most in the 15–30 nm range.

High-resolution electron microscopy (Figure 8.11a) indicated that the basic spherule consists of an irregular stacked graphitic structure, referred to as elemental carbon, shown schematically in Figure 8.11b (Rogers & Whelan 1996; World Health Organization 1996).

The graphitic nature and high surface area of these very fine particles (typically <100 nm in diameter) means they can absorb significant quantities of hydrocarbons (the organic carbon fraction) originating from unburnt fuel, lubricating oils and compounds formed in complex chemical reactions during the combustion cycle. Traces of inorganic compounds (e.g., sulfur, zinc, phosphorus, calcium, iron, silicon and chromium) have also been found in the particulates. These are believed to have arisen from the fuel, additives and lubricating oil used in diesel engines.

Improvements in fuel quality and emissions controls have led to a change in the chemical fingerprint of diesel exhaust. This change is commonly referred to as pre- and post-2007, as this is when regulations regarding improving the quality of emissions began in the United States. The terms traditional diesel engines (TDEs) and new technology diesel engines (NTDEs) are commonly used, with TDEs being pre-2007 and NTDE for those manufactured post-2007. The key features of NTDEs include a focus on 'in cylinder' combustion processes and associated electronic control management of engine parameters.

These and other advancements in engine combustion and improvements in fuel types have altered the compounds commonly found in diesel exhaust from engines pre-2007 to post-2007. Advances in diesel engine technology and fuel have resulted in reductions in emissions, chiefly particulate matter, oxides of nitrogen and hydrocarbons. Therefore, the emission profile of diesel

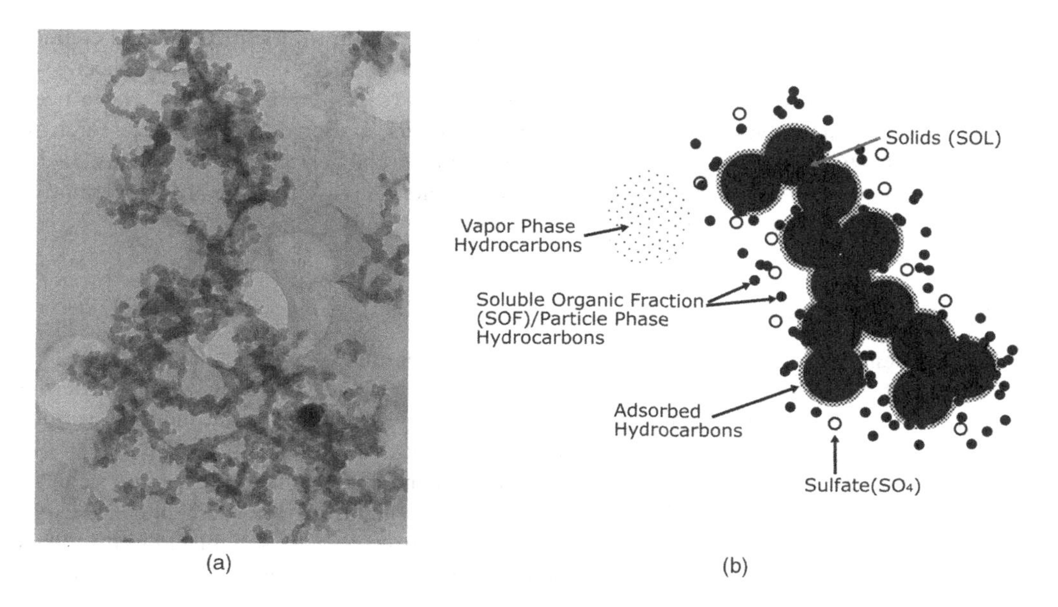

FIGURE 8.11 (a) Electron micrograph – mine diesel particulate showing spherules, chains and agglomerates (Source: A Rogers). (b) Schematic – mine diesel particulate showing spherules, chains and agglomerates **Adapted with permission.** © 1994 SAE International.

exhaust differs between NTDE and TDE (McClellan et al. 2012). Particle mass of DPM in new engines (i.e., post-2007) has been reduced by an order of magnitude, although the number of smaller particles has increased with NTDE (Hesterberg et al. 2011; Kittelson 1998; Matti Maricq 2007; McClellan et al. 2012). It remains uncertain how a shift to smaller particles size will affect the toxicity of NTDE compared to TDE and more research is required (Landwehr et al. 2019). Substantial reductions have also been reported for carbon monoxide, non-methane hydrocarbons, formaldehyde, benzene, acetaldehyde and PAHs (McClellan et al. 2012).

8.7.2 Health Effects

The potential for adverse health effects arising from occupational exposure to diesel particulate has been the subject of intense scientific debate for many years. Part of this debate is attributed to the way diesel fuel and its subsequent exhaust composition have changed over the years. It is challenging to predict the health effects of a substance that has changed so intensely over time. A detailed critical review of this research is provided in the Australian Institute of Occupational Hygienists' Position Paper *Diesel Particulate Matter and Occupational Health Issues* (AIOH 2017). This document should be consulted for detailed information on the health effects of diesel particulate. However, the following general statements on adverse health effects can be made:

- Diesel particulate has the potential to irritate the mucous membranes of the eyes and respiratory system and cause neurophysiological symptoms such as headaches, light headedness, nausea and vomiting.
- Non-malignant respiratory disease including increased levels of coughing and phlegm, and some evidence of altered pulmonary function.

- Animal and epidemiological studies link exposure to DPM from traditional diesel engines (pre-2007) with an increased risk of lung cancer, and to a lesser extent bladder cancer.
- The combined gaseous and particulate emissions, as 'diesel engine exhaust', have been classified as carcinogenic to humans (Group 1) by the International Agency for Research on Cancer (IARC Working Group on the Evaluation of Carcinogenic Risks to Humans 2013).
- The potency of DPM has been subject to intense scientific discussion without a definitive answer currently available.

8.7.3 Exposure Standards

The development of workplace ES for diesel particulate continues to be in a state of flux. This is a result of the paucity of data on dose–response with regard to NTDE emissions, differing approaches to sample collection and analysis methodology and differing attitudes of various industry segments, advisory groups and regulatory authorities.

In 2020 and 2021, respectively, NSW and WA gazetted and implemented an exposure standard of 0.1 mg/m^3 (8-hour TWA measured as EC) which was adopted from the AIOH guidance exposure value of the same concentration. This standard comes from research completed in the 1990s (Pratt et al. 1997; Rogers & Davies 2001). It was determined that if EC was kept below 0.1 mg/m^3, the level of irritation in the eye and upper respiratory tract was significantly reduced. The AIOH also recommends an action level of 0.05 mg/m^3 for DPM which is to trigger an investigation into the source of exposure and implementation of suitable control strategies. As with all worker exposures, the AIOH supports the principle that diesel emissions should be controlled to levels that are as low as reasonably practicable (ALARP) (AIOH 2017). Safe Work Australia is introducing an WEL of 0.01 mg/m^3 for DPM on 1 December 2026.

New Zealand, Switzerland and Austria have legislation in place to control DPM exposures in all workplaces including mining and tunnelling to 0.1 mg/m^3 EC. Mine Safety and Health Administration (MSHA) enforces an ES of 0.16 mg/m^3 total carbon (TC).

Germany, Sweden, France, Belgium and the European Union amongst others have moved or are moving to a diesel engine emissions exposure limit in workplace air of 0.05 mg/m^3, measured as EC, based on what can be achieved using technically available measures (GESTIS, 2023).

In summary, the promulgation of a dose–response workplace ES linked to sound epidemiological or dose–response evidence continues to be researched, and a national nor indeed international consensus has not been established.

8.7.4 Monitoring Methods

EC is used as a surrogate for DPM as it provides the best fingerprint of particulate in diesel emissions. It is relatively free of interferences and is chemically stable, unlike the adsorbed organic carbon fraction.

In the early 2000s, there was a shift from measuring the mass of DPM to EC concentration (by thermal analysis).

The internationally accepted monitoring and analytical method for DPM as submicron EC is NIOSH Method 5040 *Diesel Particulate Matter as Elemental Carbon* (NIOSH 2016). Diesel particulate is collected using either a single use impactor cassette or a 3-piece cassette with a specialised quartz filter. Where size selection is required, such as in situations that are inherently dusty (i.e., mines), or where contamination from other carbonaceous materials is likely (i.e., breakdown of conveyor material), an impactor and cyclone sampler should be used to remove the

FIGURE 8.12 DPM SKC modified size-selective sampler (left) and 3-piece cassette (right).

Source: L. Apthorpe.

larger particles. Refer to Figure 8.12 for two types of DPM samplers. Analysis involves carbon speciation for EC using thermal-optical instrumentation techniques.

At this stage, there is no evidence that methods that rely on measuring the number of ultrafine droplets of semi-volatile organic compounds provide a suitable alternative or better method of defining exposures for health assessment purposes.

Real-time monitoring instruments currently on the market can be useful indicative instruments to help identify DPM sources and manage and reduce overall DPM concentrations. However, these need to be adequately calibrated against traceable primary standards such as for TC or EC (AIOH 2017). Research is continuing in the quest to develop a reliable real-time monitor, focusing on the development of a mid-infrared spectrometry real-time field-portable device (Parks et al. 2019) and a particle number counter calibrated to site-specific correlation factors (Habibi et al. 2021).

8.7.5 Control Technologies

Experience has shown that no single simple solution exists to control diesel exhaust exposure, with the exception of changing to electric vehicle. Individual organisations need to explore which of the following control technologies best fit their circumstances. Often more than one control strategy may be required to reduce worker exposures to ALARP levels. It must be remembered that strategies that work at one location may not necessarily work at another due to differing ventilation, equipment, etc. Whatever strategy is adopted, it should be underpinned by an effective maintenance program so that emission reductions are sustained (AIOH 2017; Swanepoel et al. 2023).

Although improvements are mandated by regulatory authorities in both the United States and Europe requiring engine manufacturers to produce cleaner engines, many older diesel engines remain in service and will be around for some time to come. The control of emissions from these engines presents unique challenges, and experience has shown that while there is no single simple method of controlling particulate levels, a range of options in one or more configurations can be effective. In some cases, this may be as simple as redirecting an exhaust away from personnel; in others, it may involve retrofitting the workplace with one or more sophisticated control technologies.

Control options for DPM include:

- the use of vehicles powered by natural gas, low-sulfur diesel, or other cleaner fuels delivered to the engine in a clean state (i.e., not contaminated with dirt);
- tailpipe ventilation in workshops where vehicles are required to idle constantly to maintain a state of readiness or being serviced while operating;
- ventilation – taking into consideration the number and size of vehicles operating;
- vehicle restrictions – limiting the number of diesel engines operating in an area;
- exhaust treatment devices such as wet scrubber systems, regenerative ceramic filters, disposable exhaust filters or exhaust dilution/dispersal systems. Exhaust filters can be permanently or temporarily fitted. Filters can be added to the exhaust to operate either while the vehicle is idling or operating under load. Ensuring the effectiveness of the filter on its own, as well as within the filter system is vital to the success of this strategy;
- new-generation low-emission engines, such as Euro V, VI or VII;
- maintenance programs targeted at minimising exhaust emissions (Hines 2019);
- well-sealed, positive pressure, filtered and maintained air-conditioned operator cabins driver (AS/NZS ISO 23875: 2023) and workforce education;
- ensure leased vehicles meet the same criteria as site-based vehicles;
- respiratory protective equipment (RPE). In all situations, other control technologies should always be explored in preference to respiratory protection. Care should be exercised when selecting respirators to ensure they have been validated against challenge particles of comparable size and makeup to that of DPM (Burton 2023).

Comprehensive diesel exhaust management plans assist the site to reduce:

- diesel engine emissions at the source;
- diesel exhaust transmission throughout the mine; and
- importantly, worker exposures.

8.8 FIBROUS DUSTS – ASBESTOS

Of all the hazards, asbestos probably has the highest profile and involves the most debate, emotion and worker concern. Asbestos is one of the most widely evaluated workplace health hazards, and there is a significant amount of literature published regarding this fibrous mineral silicate. 'Indestructible' asbestos has been used since antiquity in lamps, pottery and woven garments; however, only recently has its less fortunate legacy been widely recognised. Asbestos was prized for its resistance to heat and other important qualities of strength and friction resistance. Sadly, these properties today are overshadowed by the grim toll of deaths caused by inhalation of the fine fibres. In the past, commercial mining and the use of asbestos in manufacturing have often occurred under conditions that could only be described as horrific.

8.8.1 Public Concern about Asbestos

There is concern regarding asbestos as the fibres are a human carcinogen and any issues raised regarding asbestos exposure or potential exposure are highlighted which has led on many occasions to exaggerated public concerns. Assessment of the actual risks asbestos poses can be affected by emotion rather than reliance on scientific information and data.

8.8.2 Physical Characteristics of Asbestos

For the H&S practitioner, the term asbestos is limited to the commercially used fibrous minerals from one serpentine rock and the amphibole series. The fibrous forms of these minerals have qualities of flexibility and good tensile strength, and some are able to be woven. They show resistance to heat and are non-conductive, and the amphiboles especially show good acid resistance. Many other minerals occur in fibrous form (e.g., wollastonite, brucite and erionite); however, they are not asbestos. Fibrous minerals may also be present in some mineral deposits worked for their metals (e.g., nickel, iron ore) or in quarried minerals which can lead to increased risks for workers.

8.8.3 Sources and Types of Asbestos

Many years ago, asbestos was mined in Australia at locations in New South Wales and Western Australia. The mineral continues to be mined and/or extensively used in other countries because of its useful properties. Unlike Australia, these countries still have significant numbers of workers involved in the manufacture of asbestos-containing products.

The main types of asbestos commonly found in workplaces are:

- chrysotile, belonging to the serpentinite family (also commonly known as 'white' asbestos).
- amphiboles, including:
  - amosite, commonly known as 'brown' asbestos; and
  - crocidolite, or 'blue' asbestos.

Other members of the fibrous amphibole series include actinolite, tremolite and anthophyllite; however, these forms were not commonly used in commercial products.

It is useful to know about the different types of asbestos fibre in order to understand the various diseases they can cause and the risks associated with each fibre type in the workplace.

8.8.4 Health Effects

Where asbestos is present and a process generates airborne fibres, the risk arises if these fibres are inhaled.

Today, asbestos may be present in the workplace as part of the fabric of a building (e.g., insulant and construction materials). Asbestos poses a risk solely when fibres become airborne, with ingestion suspected to cause disease in a few sites (liver, prostate) in cases of heavy ingestion of fibres. Occupational disease is not related to skin contact. As with crystalline silica, the occupational diseases asbestos causes are primarily respiratory diseases, related to fibres of respirable size.

Fibres of respirable size, known as 'respirable fibres', are those longer than 5 μm and < 3 μm in diameter, with length-to-width ratios of more than 3:1.

The three major occupational lung diseases caused by asbestos are:

- asbestosis;
- lung cancer; and
- mesothelioma.

Not all the types of asbestos listed above have strong associations with these diseases.

Section 8.8.6 examines the risk factors and the likelihood of any of these diseases occurring from present occupational exposure to asbestos.

8.8.4.1 Asbestosis

Pulmonary and pleural asbestosis is found only in asbestos workers who have been exposed to high fibre concentrations over a long period. It is the classic disease of asbestos miners, millers, weavers and those involved in processing fibre in large quantities (e.g., manufacturing brake linings or asbestos-cement products). Chrysotile and the amphiboles (amosite and crocidolite) have been shown to cause asbestosis. In pulmonary asbestosis, the inhaled fibres penetrate the alveolar region of the lung, where a fibrotic (scarring) reaction takes place. There are no well-defined nodules as seen in silicosis. Pleural asbestosis, also known as pleural plaques, presents in the form of calcification of the outer and generally top surface of the lung. It does not in itself progress to pulmonary asbestosis and is usually not debilitating.

Mild cases of pulmonary asbestosis are usually asymptomatic; however, they may progress, particularly with further exposure, to cause increasing breathlessness. The onset of asbestosis typically occurs after 15–40 years of substantial exposure to airborne asbestos fibres. As fibrosis progresses, the oxygen-exchange capacity of the lungs can decrease drastically, leading to associated heart failure.

8.8.4.2 Lung Cancer

Historically, an increased incidence of lung cancer was observed amongst workers heavily exposed to any type of asbestos, and a greatly increased risk of lung cancer occurred when these workers were also heavy smokers.

Research has shown that where cigarette smokers with lung cancer have had only brief exposure to airborne asbestos, their cancer is completely or almost completely caused by smoking (Hodgson & Darnton 2000). Swiatkowska et al. (2015) report that where there is increased asbestos exposure, workers are at increased risk of lung cancer and the effect of smoking is somewhere between additive and multiplicative. Latency periods for lung cancer are around 20 years or more, and Swiatkowska et al (2015) state that smoking cessation is beneficial.

8.8.4.3 Mesothelioma

Mesothelioma is a malignancy of the cells in the mesothelium, or lining, of the pleura surrounding the lung (pleural mesothelioma) or abdomen (peritoneal mesothelioma). Radiologically, a mesothelioma presents as a large mass of tumour protruding into the lungs. The disease is usually fatal, usually within one to three years after diagnosis. Mesothelioma is a rare disease, occurring most often amongst people exposed to asbestos but also in some unexposed people. Its latency period is usually 30–40 years and it typically follows substantial exposures. In some cases, however, it is thought to have been caused by brief yet intense exposures over a few months or less. Adults are reported to have developed the disease in their twenties after being exposed as young children to dust carried home on a parent's clothing.

Where there is no history of asbestos exposure, it is believed that the development of mesothelioma in these cases may be caused by other factors such as genetic predisposition, exposure to other biopersistent fibres or radiation exposure (Attanoos et al. 2018).

It is well established that exposure to crocidolite (blue asbestos) causes this malignancy, with some cases also attributable to amosite (brown asbestos) or to mixtures of these fibres with chrysotile (white asbestos). There is a weak association (some orders of magnitude lower than that of crocidolite) between exposure to chrysotile asbestos and mesothelioma. Crocidolite asbestos was mined and milled at Wittenoom, Western Australia until 1966, and was later found to have caused mesothelioma and other asbestos diseases in many of the workers.

The greatest risk of mesothelioma may be associated with the ability of an asbestos mineral or product to produce biodurable and long (>10 μm) fibres. The evidence for mesothelioma amongst asbestos workers being caused by exposure to chrysotile alone is less convincing. However, the extensive industrial use of chrysotile means that more cases of chrysotile-only related mesothelioma may appear in the future.

8.8.5 Occurrence of Asbestos in the Workplace

Asbestos still exists in insulation in older buildings and equipment and in some asbestos-cement (AC) materials manufactured before 1966. Industrial use of asbestos diminished rapidly during the 1980s throughout Australia. In the building materials industry, its use in the manufacture of asbestos cement (AC) sheeting and piping was phased out completely by 1984. Any use of crocidolite or amosite in new applications in Australia was banned in the 1990s. Chrysotile asbestos continued to be used in manufacturing in Australia until the early 2000s; however, since December 2003, the use (i.e., new use and reuse) of any type of asbestos is prohibited.

Asbestos can still be encountered during activities involving maintenance, refurbishment and demolition.

Examples of where asbestos can be found in Australian workplaces include:

- insulation on boilers and pipes used for steam in buildings, structures and ships. The asbestos can occur as raw fibre lagging, or in a cementitious form combined with magnesite (magnesium silicate) as a trowelled plaster on steam pipes, calorifiers, outer furnace skins, etc.
- fire-retarding insulation on steel-framed supports and fire stoppings between floors in buildings, particularly high-rise buildings;
- decorative finishes and acoustic attenuation on ceilings and walls in auditoria, public halls, schools and hospitals;
- space insulation in buildings, particularly beneath metal-sheeted roofs and in the ceiling spaces of homes (e.g., sprayed 'Mr Fluffy' amphibole insulation) and other buildings;
- AC building products in flat or corrugated sheets, pipes and moulded products (e.g., gutters and cisterns);
- friction brake and clutch products;
- gaskets and valve packing;
- millboard in air-conditioning heater banks;
- asbestos fabric as fire-proof rope, gloves, mats and hoses;
- older-style vinyl flooring materials;
- bituminous felt used on roofs and around oil and petrol pipelines;
- asbestos-containing materials (mainly AC) in soil as the result of demolition or dumping;
- mines and quarries, as a naturally occurring mineral.

Asbestos will still be present in walls, ceilings, floors, roofs, pipes, etc. of many buildings and structures for many decades to come. The ubiquitous use of AC in domestic, industrial and commercial premises causes significant problems when previously used land is contaminated, and remediation is necessary for redevelopment.

8.8.6 Asbestos Exposures in the Workplace

The current Safe Work Australia exposure standard for all forms of asbestos is 0.1 fibres per millilitre of air (fibres/mL) in the breathing zone of an exposed person, averaged over a full work shift. Past industrial procedures involving the handling of asbestos fibre on a large-scale generated relatively large risks compared with those occurring today.

The largest use of asbestos in Australia was in the production of AC building products. Since the early 1980s, asbestos use has been superseded by an asbestos-free cellulose technology. Imported products labelled 'asbestos free' sometimes contain a small proportion of asbestos fibre (<1–5 per cent). The following processes that generate fibres in workplace air may still be encountered, even though the presence of asbestos may be unknown to the worker:

- grinding, drilling, sanding and sawing (using power tools) of building materials containing asbestos (2–50 fibres/mL), including high-pressure water blasting (up to 1 fibre/mL);
- exposure can still occur in the friction material industry, particularly in workplaces that strip and replace brake linings blowing down brake drums with compressed air during repair (believed to be up to 10 fibres/mL for less than one minute); and
- asbestos-stripping and removal operations (up to 100 fibres/mL).

The last category, the asbestos removal industry, can be a potential source of exposure to significant amounts of asbestos fibres. Asbestos-removal programs are discussed in Section 8.8.6.3 and 8.8.6.4.

8.8.6.1 Asbestos in Soil

In Australia, the major asbestos 'problem' (Pickford et al. 2004) is not the removal of asbestos from buildings or structures, it is the treatment of land contaminated with asbestos containing materials (ACM), of which the majority is AC. This material has come from illegal dumping, inappropriate demolition of AC structures in the past and inadequate remediation of contaminated land sites.

Strategies for remediation and validation of ACM-contaminated soils must incorporate risk-management approaches rather than ad hoc procedures.

Many public, environmental and government stakeholders have been unduly concerned that disease can arise from casual and brief contact with a small amount of non-friable AC in soil. Together with substantial and often ill-informed press coverage, the sight of workers in suits and respirators performing trivial tasks associated with minor amounts of AC in good condition has created a public perception that these tasks are dangerous to workers and communities.

There may be a perception that buried AC can become friable, and it should be noted the common definition of friable asbestos material is any material that contains asbestos and is in the form of a powder or can be crumbled, pulverised or reduced to powder by hand pressure when dry (SWA 2020a). Current Safe Work Australia publications on asbestos (SWA 2020a, 2020b) indicate a risk-assessment approach should be made to determine control and remediation strategies. It is important a risk assessment considers appropriate techniques whenever any disturbance of asbestos-contaminated soil is required. Assessment techniques and criteria for asbestos in soil are provided in the National Environment Protection Measure, Schedules B1 and B2 (NEPC 2023).

There is a widespread perception AC becomes friable if it is being processed or tracked over by heavy plant equipment and trucks. While some pieces may fracture or become slightly abraded, essentially none become 'friable' by the broad definition of that term (SWA 2020a). Laboratory analysis of thousands of soil samples taken from AC-contaminated sites, often in the immediate vicinity of the AC itself, indicates no respirable asbestos fibres are released into the soil (Pickford et al. 2004).

The public image of asbestos as inherently dangerous has impelled a push for 'zero' tolerance in soils that is impossible to achieve, scientifically or practically. Consequently, soil which is contaminated with AC is removed to an approved waste-management depot or depending on the jurisdictional requirements may be used as fill material with suitable capping. Small amounts of AC contamination on the surface may be also removed by hand, with the remaining soil inspected and sometimes retested for the presence of asbestos before being validated as satisfactory for the intended use.

8.8.6.2 Naturally Occurring Asbestos

Naturally occurring asbestos (NOA) includes fibrous and semi-fibrous asbestos (including elongated mineral particles) which can be found in nature. In certain workplace locations, NOA can be found in rocks or soils where agriculture, mining, quarrying, construction and road building occur. There may be veins within the earth and certain rock formations which, when disturbed during excavation or drilling, may release fibres into the air. It is important for workers that this potential risk be considered by H&S practitioners in locations where geological information indicates that NOA may be present. Where NOA is to be disturbed, an asbestos management plan is required, and specialised work practices and control strategies must be employed to protect workers and the community (SWA 2020a).

8.8.6.3 Asbestos Management

It is essential wherever asbestos is present (e.g., buildings, structures, ships or where natural materials are disturbed), and there are interactions with people, that the asbestos is managed to prevent exposure.

Detailed guidelines for management, control and removal of asbestos are provided in these Australian documents:

- *Code of Practice: How to manage and control asbestos in the workplace* (SWA 2020a) applies to persons conducting a business or undertaking. The document includes:
  - managing and assessing risks, including formulating an asbestos management plan;
  - requirements for an asbestos register;
  - safe practices for ACM.
- *Code of Practice: How to safely remove asbestos* (SWA 2020b) applies to those involved in the task of safely removing asbestos. Important topics covered include:
  - removal techniques and equipment for different kinds of ACM;
  - respiratory protective equipment for asbestos removal work;
  - purpose of air monitoring and criteria for air monitoring results.

Of particular interest in the removal document is the section on choosing respiratory protection. It includes estimates of the expected asbestos fibre concentrations in air for different activities and the appropriate type of respirator to be worn using guidance from Australian/New Zealand Standard AS/NZS ISO 16975.1:2023 (2023a) Respiratory protective devices - Selection, Use and Maintenance, Establishing and implementing a respiratory protective device programme (Standards Australia 2023a). Some activities (e.g., dry stripping) involve fibre concentrations of several hundred fibres/mL, justifying the necessity to check the code for respiratory protection requirements.

Guidance Note on the Membrane Filter Method for Estimating Airborne Asbestos Fibres (NOHSC 2005) deals with the sampling and analysis of airborne asbestos fibres, discussed in Section 8.8.7. For additional information, refer to the AIOH Position Paper: Asbestos and its potential for occupational health issues (AIOH 2016a).

8.8.6.4 Asbestos Removal

Where removal is required, there are specific Legislation and Codes of Practice in each jurisdiction that cover arrangements for asbestos removal, including the use of licensed personnel and special isolation and sealing procedures to prevent the spread of asbestos fibres during the work.

The merits of removing ACM which is in good condition before the end of its normal service life is not examined here. In most cases, there is no health-related necessity for such removal, since fibre monitoring where ACM is in good condition and remains in situ returns results of <0.01 fibres/mL. Sometimes employee demands, or a proactive approach, drive removal works to occur. Generally, the planned removal of ACM (particularly friable ACM) is a good principle, which can avoid accidental exposure in the future. Removal is usually necessary before building alterations or refurbishment commences.

When an asbestos-lagged installation requires maintenance, or where an insulating product fails in service, asbestos removal is necessary. An example of a commonly failing product is insulation on fire-rated structural beams that have lost sections of insulation or been damaged by water penetration, and these must be targeted for remediation involving asbestos removal and reinstatement with an asbestos-free insulation. Where sprayed asbestos exists as insulation on the inside surfaces of roof spaces or air-conditioning systems, removal is a high priority.

When a building or structure is to be demolished or extensively refurbished, all ACM should always be removed beforehand. For the H&S practitioner, the greatest priority is to ensure the removal is conducted safely and in accordance with local legislation. Both the asbestos removal worker and the bystander require suitable exposure protection. Poor work practices in the lucrative business of removing asbestos insulation from commercial, high-rise and domestic dwellings in the 1980s and 1990s resulted in poorly conducted removals with remnants of asbestos insulation left in place. There may be legacy issues in some buildings due to these previous practices.

8.8.7 Asbestos Exposure Risks

8.8.7.1 Asbestos Dose

All three asbestos-related diseases discussed in Section 8.8.3 are dose related. The larger the inhaled dose of asbestos fibre, the greater the risk of developing disease. Dose (otherwise known as cumulative exposure) is a function of the amount of asbestos fibres in the air and the duration of exposure. This is usually expressed in terms of fibre/mL years – that is, the *time-weighted average* airborne-fibre concentration inhaled by the worker in fibres/mL of air multiplied by the exposure duration. In this instance, *time-weighted average* airborne-fibre concentration must be estimated for the entire period of exposure, not just over a single day. Periods of non-exposure and variable exposure must therefore be taken into account:

$$\text{Dose} = \text{average airborne concentration of inhaled fibre x years of work} \tag{8.1}$$

Using historical industrial exposures and response data in attempting to extrapolate today's risk (with its low exposures) presents considerable problems. It cannot be assumed that any of the asbestos diseases follow a linear dose–response model (i.e., there may be a threshold level below which asbestos exposure has no effect). Research on the fibre contents of the lungs of people without asbestos disease has quashed the fallacy of the 'one-fibre' theory (i.e., that a single asbestos fibre in a person's lungs can cause an asbestos-related disease). It has been shown that non-exposed people who die in old age from non-asbestos-related causes can have significant quantities of asbestos fibres present in their lungs, presumably from environmental exposure (Berry 2002; Rogers et al. 1994).

The risk of developing asbestosis in any Australian workplace is now remote as previous poor industrial conditions have been eradicated and asbestos is a prohibited substance. While the Safe

Work Australia exposure standard for asbestos in air is set at a level that will preclude any occurrence of asbestosis, it is intended primarily to prevent the rarer lung cancer and mesothelioma, which can occur after considerably smaller exposures.

Regarding the ES, it is important to know that the current ES of 0.1 fibres/mL is a very low level compared with fibre concentrations before awareness of asbestos health issues, where an asbestos worker could be exposed to tens or hundreds of fibres/mL each working day. In addition, as for any ES, the recommended level is not intended to separate safe and unsafe conditions. These standards have been arrived at after extrapolating data from historical occupational hygiene surveys and epidemiological studies. Further, the improved sensitivity of today's measurement techniques (at least 10 times more sensitive than 60 years ago) means that conditions of the 1950s and earlier may have been much worse than the few measurements available from that time suggest. In other words, workers may have had exposures of hundreds to thousands of fibres/mL in terms of modern measurement methods.

The risk of lung cancer associated with today's levels of airborne asbestos fibres also appears to be negligible, since the high levels of exposure which led to lung cancer no longer exist. Since it is clear that smoking far exceeds asbestos as a cause of lung cancer, H&S practitioners will be far more effective in the task of promoting health in the workplace if they focus on altering lifestyle habits (e.g., promoting quit-smoking programs) in conjunction with controlling asbestos exposure.

Today, the risk of developing mesothelioma from the current low levels of exposure is also unlikely. The ES has been set to cater for the two fibre types strongly implicated in mesothelioma (i.e., crocidolite and amosite) and also for chrysotile. Exposure data for those who developed mesothelioma outside industry or in situations not directly associated with asbestos work have not been recorded. Further, the methods of extrapolating backwards linearly from high to very low doses are unreliable.

8.8.7.2 Practical Assessment of the Risk

The risk posed by asbestos arises from inhalation of respirable asbestos fibres. A visual assessment of the workplace is therefore the first important step when assessing risk.

The following steps indicate the basic procedures for risk assessments for asbestos. Accurate fibre identification and fibre counting may also be required as part of the risk assessment process. H&S practitioners will not be able to conduct the identification or counting processes without extensive training; however, they must be able to assess the information about the types of asbestos fibre present, airborne asbestos fibre concentrations and risk factors.

8.8.7.2.1 Ascertain that Asbestos is Present

Asbestos is an established constituent of many products and asbestos is only one of many materials used; others include:

- MMVF – fibrous glass, rock wool and RCF;
- vermiculite;
- shredded cellulose (using newsprint).

The use of optical microscopy is appropriate to distinguish these other materials from asbestos fibres. If reliable information is not available, collecting and analysing a sample of the material may be necessary to confirm or eliminate the presence of asbestos.

8.8.7.2.2 Collect An Appropriate Sample

A representative sample is needed for laboratory analysis, and there may be more than one fibre type in the sample. Sampling for ACM is a destructive process, and there are certain procedures required when collecting the sample to minimise disturbance and to protect the sampler. The sampling process is described in the Code of Practice: How to manage and control asbestos in the workplace (SWA 2020a).

Enough sample should be collected to include all types/layers of material present to ensure the sample is representative. Usually, a sample of 10–50 g of insulation materials will be sufficient; for differentiating older building products from newer asbestos-free materials, a piece approx. 5 cm square should be submitted; for difficult to analyse vinyl materials floor tiles, a minimum 10 cm square is needed; and for other materials such as gaskets and friction blocks, they can be submitted whole. All samples should be packed carefully in a labelled and sealed container to prevent contamination during transport or in the testing laboratory.

8.8.7.2.3 Have the Asbestos Fibre Type Identified Positively

Crocidolite and amosite are more hazardous and generate higher airborne asbestos concentrations than chrysotile, so it is important to know which type of fibre is present, especially for risk-assessment purposes. Analysis should only be carried out by a specialist laboratory accredited in this field by NATA, using AS 5370 (Standards Australia 2024). Capital costs of the equipment used in identification are significant, and a high degree of skill is necessary.

Fibre analysis requires observation of various optical properties using complex diagnostic criteria to identify different kinds of fibres.

The analytical techniques used include low and high-power stereomicroscopy, polarised light microscopy (PLM) and dispersion staining microscopy. Sometimes, for confirmatory purposes, additional techniques can be used such as:

- infrared spectroscopy;
- X-ray diffractometry;
- electron microscopy (scanning (SEM) or transmission (TEM) electron microscopy incorporating X-ray analysis).

Examples of different types of fibres commonly submitted for identification are shown in Figure 8.13. Note the wavy shape of chrysotile fibres compared with straight amosite and crocidolite fibres. MMVF are commonly very large in diameter compared with asbestos fibres or show long filaments of uniform diameter.

PLM and dispersion staining microscopy analysis using AS 5370 for bulk materials and non-homogeneous samples such as soils and aggregates has a detection limit of 0.01%. When a mixture of chrysotile and other asbestos fibres is detected, the risk associated with handling the material is usually assessed based on the more hazardous type of fibre. X-ray and infrared spectrophotometric techniques are not able to differentiate between the non-fibrous form of asbestos minerals and the fibrous form. However, when combined with a technique to ensure that the fibrous form is present, X-ray and infrared spectrophotometric techniques with spectral subtraction facilities can differentiate between types of asbestos in mixtures, although they are not suitable for identifying low and 'trace' levels of fibres in a mixture. SEM or TEM, although specialised and expensive and requiring skilled interpretation, can provide information if fibre identification by PLM using AS 5370 results in an inconclusive finding.

8.8.7.2.4 What is the Type of Asbestos-containing Material?

Knowledge of how the fibres are contained or in what form they are present in a material can assist when assessing risk. Is the material friable or is it non-friable (i.e., all other ACMs), including those materials where the fibres are bonded or locked into the matrix (e.g., AC) (SWA 2020a).

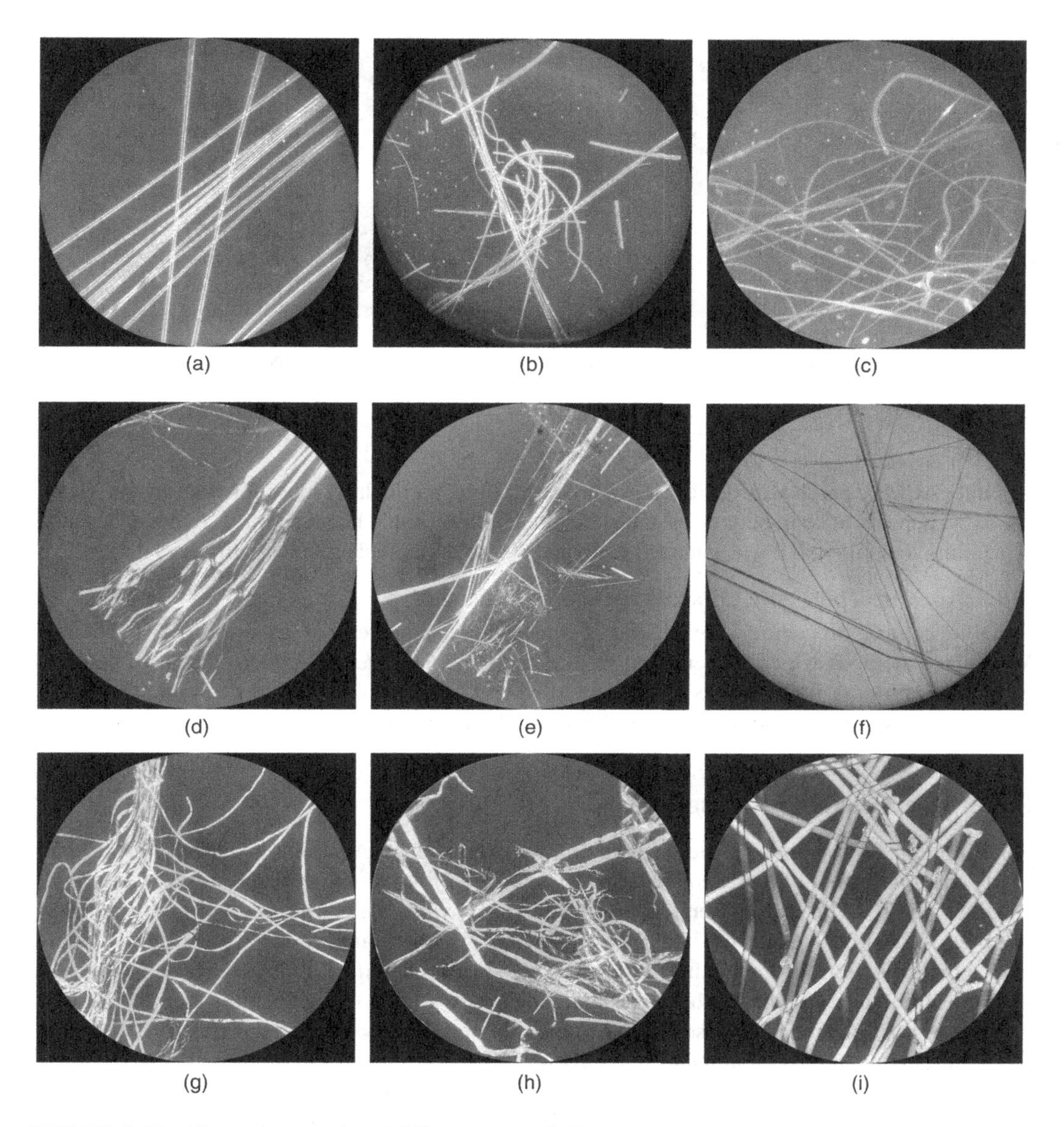

FIGURE 8.13 Photomicrographs of different types of fibre commonly found in workplaces. (a) Fibrous glass filaments. (b) Rock or glass wool. (c) Refractory ceramic. (d) Chrysotile. (e) Amosite. (f) Crocidolite. (g) Cotton. (h) Cellulose (paper). (i) Wool.

Source: L. Apthorpe.

8.8.7.2.5 What is the Condition of the Asbestos?

Exposure risk depends on the amounts of fibre released into the workplace. Ascertaining the state or condition of the ACM is therefore crucial to its control. Some ACMs release no fibres to the workplace (e.g., AC in situ, sealed gaskets, vinyl-asbestos floor tiles, encapsulated fireproofing and resin-bonded friction materials).

FIGURE 8.14 Asbestos and MMVF sampling cowl, assembled and disassembled.

Source: L. Apthorpe.

Other ACMs can release considerable amounts of fibre, particularly loose forms of insulation. If the ACM appears to be in poor or unstable condition, plans should be made for its removal because it is failing to do the job it was intended to do and is also more liable to release fibres as it ages.

8.8.7.2.6 Are There Any Control Procedures In Place?

Suitable control procedures are crucial to the safe management of asbestos (SWA 2020a). Workplace methods which prevent dust from being generated are the primary means of control. Suitable respiratory protection is also required when dust-control procedures cannot control the release of fibres. Refer also to Section 8.8.6.4 on asbestos removal.

8.8.8 Air Sampling for Asbestos

Air sampling for asbestos in the occupational environment in Australia is carried out according to the *Guidance note on the membrane filter method for estimating airborne asbestos fibres* (NOHSC 2005). This method employs light microscopy, with electron microscopy sometimes required for sensitive environmental investigations. It is important the H&S practitioner understands the limitations of the method and the results that are obtained. Laboratories accredited by NATA or by state regulatory authorities should always be sought to conduct this work. Asbestos-removal regulations in most jurisdictions require monitoring be conducted and that it conforms to the requirements of the *Guidance Note* (NOHSC 2005).

Airborne dust is collected on a 25 mm membrane filter with a pore size of 0.8 μm, usually housed in a three-piece conductive cowl as shown in Figure 8.14. After sampling, the filter is rendered transparent and mounted on a microscope slide together with a cover slip. A phase-contrast optical microscope is used to count the number of respirable fibres, geometrically defined by their size criteria.

The membrane filter method was initially developed for measuring fibre levels in workplaces using asbestos. In these work environments, there was usually a mixture of dusts, of which the largest component was asbestos fibres. Disadvantages of the optical counting method include analysis difficulties when there are both dust and asbestos fibres, as it cannot distinguish between true asbestos fibres and other fibres such as fine cellulose or ceramic fibres which leads to conservative estimates (i.e., overly high concentrations) for environments containing mainly non-asbestos fibres. False positives will almost certainly be obtained even if the air sample derives from an asbestos-free environment, because of the presence of fibres from plant matter, carpets, fabric and clothes.

The membrane filter method can be deployed for sampling airborne respirable asbestos fibres circumstances, such as:

- Control monitoring. The results are used to ensure controls such as isolation enclosures and negative-pressure air fans are effective during removal work. When friable removal work is completed, air monitoring is required as a 'clearance' to validate that the removal area is safe for reoccupation.
- Environmental monitoring. While there are no formal standards in Australia for environmental asbestos in air, where there is the potential for very low fibre concentrations, electron microscopy (i.e., SEM or TEM) can be used. These methods have the added advantage that fibres can be positively identified by energy dispersive X-ray analysis. They also permit the differentiation of other fibres from asbestos and detection of fibres which are too small to be seen by standard optical phase-contrast microscopy. For these techniques, certain sampling and preparation techniques are needed.
- Exposure monitoring – undertaken in the worker's breathing zone to determine exposure compliance with the ES is not normally conducted as persons involved in asbestos disturbance will already be wearing appropriate respiratory protection. If it is required, then advice from an occupational hygienist with significant experience in the methodology should be sought.

It should be noted that SEM and TEM are not used routinely in Australia for exposure or control monitoring because of the high cost of both equipment and analysis.

8.9 FIBROUS DUSTS – MAN-MADE VITREOUS FIBRES

8.9.1 Types of Man-Made Vitreous Fibres

MMVF have been used as a replacement for most asbestos-based insulation materials for a mixture of health-related, technical and economic reasons. The properties of many MMVF allow them to be used in applications where asbestos could not (e.g., glass-fibre reinforcing). MMVF have been around for more than 90 years. Some initial concern that they might turn out to have effects similar to those of asbestos has proved largely unfounded. MMVF present some problems in the workplace because they can cause contact irritation, mainly due to the coarse fibres. The major commercial types of MMVF are listed in Table 8.3.

Primarily, most MMVF have large fibre diameters compared with asbestos fibres. *Continuous-filament glass fibres* are used in textiles, and to reinforce plastics and concretes in typical applications such as swimming pools, boats, surfboards and plumbing materials. Typical diameters range from 5 to 30 μm, depending on the product, with very few or no respirable fibres present. There is generally a narrow range of fibre diameters in any single product.

TABLE 8.3
Major Commercial Types of Man-Made Vitreous Fibres

Type	Raw Material	End Product Examples
Filaments	Borosilicate glass	Continuous-filament reinforcing fibres, woven cloth and electrical insulators
Wools	Basalt + fluxes Borosilicate glass	Insulation and acoustic batts, tiles and preformed sections
Ceramic fibre	Alumina + silica	Blanket, boards, modules, plasters, textiles, etc.

Glass fibre or glass wool mainly takes the form of insulation mats or blankets, with a significant percentage of fibres of respirable size (less than 3 μm diameter), even though many of the fibres are in the range of 5–15 μm diameter. Rock wools (or slag wools) contain fibres in a range of sizes similar to that of glass wool and, however, have a larger percentage of respirable-sized fibres.

RCF are aluminosilicates and found mainly in the form of high-temperature insulation blankets. Common trade names are Kaowool and Fiberfrax. They range in diameter from sub-micrometre to around 6 μm, with a large proportion of respirable fibres. RCF are required for applications involving temperatures >900°C.

Airborne fibre concentrations for MMVF are usually low unless hygiene practices are poor. Personal dust and respirable fibre exposures for workers installing building insulation (e.g., mineral wool or fibreglass batts) in attic spaces and walls are generally low. However, exposures can be much higher during installation of loose-fill materials. Removal of ceramic fibres can result in significantly higher exposures.

8.9.2 Health Effects

The major acute health effect from exposure to MMVF is irritation of the skin, eyes and upper respiratory tract. The use of eye protection is required and garments with close-fitting collars and cuffs are useful for reducing skin contact and, hence, skin irritation. Exposure to high airborne dust concentrations can be prevented by using P1 particulate respiratory protection.

With regard to lung diseases, factors such as fibre dose (cumulative exposure), dimension (diameter and length) and durability (biopersistence) are important. Other factors to consider include chemical composition of the fibre type (either filament, wool or ceramic fibre) and whether the MMVF is new or old. Older-style MMVF have health effects different from those of newer ones, which are specifically designed to have low biopersistence in the lungs (AIOH 2016b).

MMVF do not remain in the lung very long because they dissolve there. The new fibre types generally dissolve in the lung even more quickly (i.e., biosoluble fibres). While this development is an attempt to further distance MMVF from asbestos fibres in terms of physical properties and health effects, the clearance rates of any type of MMVF are very rapid compared with those of asbestos.

The current workplace Exposure Standard for all forms of MMVF (except RCFs) is 0.2 mg/m^3 (as inhalable dust) due to its dusty nature, and for RCF, there is a dual standard of 0.5 fibres/mL and 2 mg/m^3 (SWA, 2022).

If a workplace is involved in activities that use MMVF, refer to the NOHSC *Code of Practice for the Safe Use of Synthetic Mineral Fibres* (NOHSC 1990). If sampling and analysis of respirable airborne MMVF is required, the NOHSC's *Technical Report and Guidance Note* (1989) should be consulted. This publication provides guidance on the membrane filter method, which has many similarities to the asbestos-monitoring method (NOHSC 2005).

In 2002, the IARC (2002) reviewed available epidemiological data on MMVF and determined that:

- special-purpose glass fibres such as E-glass and '475' glass fibres are possibly carcinogenic to humans (Group 2B).
- RCF are possibly carcinogenic to humans (Group 2B).
- insulation glass wool, continuous filament, rock (stone) wool and slag wool are not classifiable as to their carcinogenicity to humans (Group 3).

Further information regarding MMVF can be found in the AIOH position paper on synthetic mineral fibres (AIOH 2016b).

Where MMVF (e.g., RCF) is subjected to elevated temperatures over long periods of time, the fibres can devitrify, and various polymorphs of crystalline silica can be formed.

8.10 FUME

The H&S practitioner may be required to undertake risk assessments for workers who are exposed to fumes from thermal processes. The most common exposure occurs for welders who generate fumes during welding processes. Depending on the metals to be welded (i.e., surfaces joined together), the fumes generated will contain a variety of elements as a result of the different welding techniques and consumables used (AIOH 2022).

Health effects from exposure to welding fume include metal fume fever, and IARC has classified welding fume as a Group 1 carcinogen (IARC 2018).

Measurement of welding fume is conducted using AS 3853.1 (Standards Australia 2006a). The Australian Standard describes the use of sampling devices suitable for insertion behind protective face shields normally worn by welders. The companion standard AS 3853.2 (Standards Australia 2006b) deals with the measurement of gases relating to welding.

Exposure monitoring can be undertaken for welders as part of the risk assessment, with results compared to the welding fume (not otherwise classified) ES of 1 mg/m^3 (SWA 2022) and to the individual elements within the fume.

8.11 MISTS

The presence of mists in workplaces typically leads to chemical exposure via inhalation, skin contact or in some cases, ingestion. Mist droplets contain liquids, and they may also hold suspended solids. Risk assessments must consider the chemical nature and toxicity of the mist components, and the potential deposition site, that is, section of the respiratory tract; or part of the body where contact may occur, that is, mouth, face, eyes, hands, arms, torso, etc. Depending on the type of mist and route of entry, exposure evaluation techniques may include airborne sampling via a certain filter or sorbent tube, and/or dermal sampling techniques. Due to the complexity involved, advice from an experienced occupational hygienist is recommended.

8.12 NANOTECHNOLOGY

Nanotechnology involves the precision-engineering of materials at the nanoscale (10^{-9}–10^{-8} metres), at which point unique and enhanced properties can be utilised. These properties have led to the development of new products, procedures and processes.

The field of nanotechnology is growing rapidly with input from multiple disciplines (e.g., science, medicine and engineering) with advances in nanotechnology driven by research, industry and customer demands.

Despite these advancements, the properties associated with engineered and manufactured nanomaterials arising from, for example, high surface area per unit mass, may give rise to health and safety concerns, and nanomaterials generally are more toxic than the corresponding macrosize substance (Toxikos 2013). The following example assists in understanding the concept of how much more surface area is available when considering nanomaterials: If a cube with an edge length of 1 cm and a surface of 6 cm^2 is divided into cubes with an edge length of 1 nm, the result is 1021 cubes with a total surface area of 6 000 m^2. The total mass and volume of the cubes remain unchanged.

There is considerable knowledge about the work health impacts of fine and ultrafine particulate air pollution that can be applied when considering the potential health effects of manufactured or engineered nanomaterials. While there is potential for both ingestion of, and dermal exposure to, nanomaterials, the main concern in the workplace is potential inhalation exposure.

8.12.1 Key Definitions

The International Organization for Standardization (ISO) has published a number of definitions relevant to nanotechnologies in ISO 80004-1:2023, including:

- Nanoscale – length range approximately from 1 nm to 100 nm.
- Nanomaterial – material with any external dimension in the nanoscale or having internal structure or surface structure in the nanoscale.
- Nanotechnology – application of scientific knowledge to manipulate and control matter predominantly in the nanoscale to make use of size- and structure-dependent properties and phenomena distinct from those associated with individual atoms or molecules, or extrapolation from larger sizes of the same material (ISO 2023).

8.12.2 Nanotechnology and the WHS Regulatory Framework

The Australian model work health and safety (WHS) legislation includes both general care duties and more specific obligations for managing the risks of hazardous chemicals. These provisions apply to nanotechnologies and nanomaterials. However, there are nanomaterial-specific issues that impact the application of these provisions, notably in the areas of:

- uncertainty about the hazardous properties of engineered nanomaterials and
- capability in measuring the emissions and exposures of nanomaterials in workplaces, including personal exposure assessment.

As a general approach, where there is limited understanding of hazards, Safe Work Australia advocates taking a precautionary approach to prevent or minimise workplace exposures.

8.12.3 Safety Data Sheets and Labelling

To comply with hazardous chemical regulations, manufacturers and importers must classify these chemicals correctly and provide appropriate SDS and workplace labels based on that classification.

Information is provided regarding nanomaterials in the model codes of practice for workplace labelling (SWA 2023a) and SDSs (SWA 2023b). The codes note that SDSs and labels should be provided for engineered nanomaterials unless there is evidence that they are not hazardous.

Extra parameters have been added to Section 9 ('Physical and Chemical Properties') of the model code of practice for SDSs that are particularly relevant for nanomaterials and also relevant for chemicals more generally. Amongst many parameters, key elements include particle size (average and range) and distribution, aggregation and/or agglomeration, shape and aspect ratio, crystallinity, surface area and coating and dustiness.

The ISO Technical Report (13329) on the preparation of SDSs for manufactured nanomaterials contains advice that can help manufacturers and importers provide accurate and relevant information in SDS for nanomaterials (ISO, 2012).

8.12.4 Workplace Exposure Standards

There are currently only a limited number of ES for nanoscale materials. Exposure limits vary according to factors such as type of nanomaterial (composition), the size of particle and crystallinity, all of which can influence toxicity.

To increase the guidance available for decision-making on control measures, other values such as NIOSH's recommended exposure limits (RELs) may be used as guidance (QUT 2012). As an example, the NIOSH proposed REL for carbon nanotubes (CNT) is based on a measurement detection limit. NIOSH recommends that efforts should be made to reduce airborne concentrations of CNT as low as possible below the REL (NIOSH 2013).

8.12.5 Hazardous Properties of Nanomaterials

8.12.5.1 Health Hazards

There has been a significant amount of research conducted on nanomaterial toxicity and potential health hazards. Consistent with the research findings, there are many factors that impact on toxicity which can lead to a range of hazard severities, and generally, nanomaterials are more toxic than larger particles.

As an example, CNT can exist in both fibre-like and non-fibre-like structures, and both are potentially hazardous. The Australian Industrial Chemicals Introduction Scheme (AICIS) has recommended that these are classified as hazardous (AICIS 2020). Based on a number of studies (Orsi et al. 2020), there is potential for CNT to cause mesothelioma.

As this area of science is constantly changing, it is important for the H&S practitioner to remain up to date with the latest health-related information on nanoparticles from organisations such as Safe Work Australia.

8.12.5.2 Physicochemical Hazards

The physicochemical properties of nanoscale materials, such as high surface area per unit mass, make them widely useful for many applications – for example, as catalysts. There are also potential safety hazards arising directly from the physicochemical properties, such as risk of fire or explosion, or unexpected catalytic properties. Physicochemical hazards of engineered nanomaterials must also be considered as they have propensity for causing fires and explosions as they can be readily combustible due to powdered or granular forms (SWA 2013).

8.12.6 Measuring Nanomaterials

Measurement of airborne nanoparticles and subsequent interpretation of results are highly complex and require considerable experience, as the particles exhibit different behaviour and toxicity characteristics.

There are several challenges associated with the measurement of nanomaterials in air. Besides many different types, shapes and sizes; a high tendency to agglomerate, to aggregate or to stick to other particles and surfaces; there are significant amounts of background nanoparticles in air from natural or incidental sources – for example, combustion products such as diesel exhaust emissions. The background levels can vary significantly, which can make it difficult to detect and quantify emissions of nanoparticles from processes.

In relation to hazards, a number of parameters provide relevant information, including mass concentration, number concentration, size distribution, shape and chemistry and surface area. There are different instruments that potentially can be used to measure these parameters, examples include the scanning mobility particle sizer, electrical low-pressure impactor and fast mobility particle sizer.

Work has been undertaken with a focus on practical emissions and exposure measurement in the workplace. In 2009, the Organisation for Economic Co-operation and Development (OECD) Working Party on Manufactured Nanomaterials (WPMN) published *Emission Assessment for the Identification of Sources and Release of Airborne Manufactured Nanomaterials in the Workplace: Compilation of existing guidance* (OECD WPMN 2009). This is based on the US NIOSH document: Nanomaterials Emissions Assessment Technique (NEAT) (Methner et al. 2010). The approaches to measurement recommended in these documents were validated in research undertaken by Queensland University of Technology (QUT) and Workplace Health and Safety Queensland (WHSQ) (2013), investigating the operations of six nanotechnology processes with a number of different engineered nanomaterials. The research confirms that a three-tiered approach is effective in assessing worker exposure to emissions (QUT 2012).

- **Tier 1**: The tier-1 assessment involves a standard occupational hygiene survey of the process area, plus measurement of aerosols, to identify likely points of particle emission.
- **Tier 2**: Tier-2 assessment involves measuring particle number and mass concentration to evaluate emission sources, worker breathing zone exposures and the effectiveness of workplace controls. A combination of instruments such as a portable CPC, OPC and photometer can be used effectively.
- **Tier 3**: If further information is required, a tier-3 assessment can be undertaken. This involves repeating tier-2 measurements together with simultaneous collection of particles for off-line analysis of particle size, shape and structure, mass and fibre concentration, and chemical composition. Offline particle analysis can be compared with real-time measurement results, and with ES or other limit values.

It may not be necessary to undertake all three tiers of assessment. The findings of tier 1 and/or tier 2 may be sufficient to identify that controls are effective, or that work needs to be done to improve controls and prevent exposure.

Therefore, in practical terms, to assess whether workplace controls are effective, the parameters that need to be measured are number concentration and mass concentration. This can be achieved using a combination of handheld instruments, such as a CPC, OPC and photometer (refer to Section 8.6.5), in conjunction with conventional sampling techniques.

8.12.7 Controlling Exposure to Nanomaterials

As is the case for substances with larger particles, nanomaterial exposure levels will be process and material dependent, with the highest exposures likely when handling 'free' or uncontained nanomaterials. Regarding exposures to hazardous chemicals, the hierarchy of control should also be applied for nanomaterials as discussed in Chapter 5.

A range of substitution and modification options may be applicable to make nanomaterials less hazardous; for example, making the nanomaterials more hydrophilic, more soluble or less biopersistent. However, as with any substitution control strategy, it is important that product properties can be maintained, and no new hazards are introduced.

When appropriately designed and maintained, conventional engineering controls such as local exhaust ventilation can effectively reduce exposures to nanomaterials.

Nanoparticles can move through filter media by diffusion, and there is a probability that they will impact on the filter fibres and be captured. This means that air-purifying respirators with P2 and P3 filters may reduce exposure to nanomaterials. Air supply respirators with higher protection factors are likely to be more effective. Respirator manufacturers may be able to assist in determining which products are most effective to capture the size range (and shape) of the particles of interest.

8.12.8 Risk Management for Nanoparticles

Safe Work Australia has published *Safe Handling and Use of Carbon Nanotubes* (SWA 2012), which provides two different CNT risk management approaches – risk management with detailed hazard analysis and exposure assessment, and risk management by control banding – either or both of which may be used.

Safe Work Australia (2010) has developed an assessment tool for handling engineered nanomaterials. This is a useful tool which can be used to identify hazards and develop relevant policies and procedures for nanomaterials in the workplace.

Further information regarding managing risks for manufactured nanomaterials can also be found in the ISO/TR 13121:2011 (2011) document: Nanotechnologies – nanomaterial risk evaluation. This document includes health and safety information for identification, evaluation, decision-making and communicating risks for public, consumers, workers and the environment.

REFERENCES

American Conference of Governmental Industrial Hygienists (ACGIH) 2024, *Threshold Limit Values and Biological Exposure Indices*, ACGIH®, Cincinnati, OH.

Amman, C.A. & Siegla, D.C. 1982, 'Diesel Particulates: What Are They and why?', *Aerosol Science and Technology*, vol. 1, no. 1, pp. 73–101.

Attanoos, R.L., Churg, A., Galateau-Salle, F., Gibbs, A.R. & Roggli, V.L. 2018, 'Malignant Mesothelioma and Its Non-Asbestos Causes', *Archives of Pathology & Laboratory Medicine*, vol. 142, no. 6, pp. 753–760. https://doi.org/10.5858/arpa.2017-0365-RA

Attfield, M., Castranova, V., Hale, J.M., Suarthana, E., Thomas, K.C. & Wang, M.L. 2011. *Coal Mine Dust Exposures and Associated Health Outcomes; A Review of Information Published since 1995*, National Institute for Occupational Safety and Health, https://www.cdc.gov/niosh/docs/2011-172.pdfs/2011-172.pdf?id=10.26616/NIOSHPUB2011172 [27 December 2023].

Standards Australia 2024. Sampling and qualitative identification of asbestos in bulk materials (ISO 22262-1:2012, MOD), AS 5370:2024, SAI Global, Sydney, https://www.standards.org.au/access-standards

Australian Industrial Chemicals Introduction Scheme (AICIS) 2020, Public Report, Carbon Nanotubes File No.: STD/1724, https://www.industrialchemicals.gov.au/sites/default/files/2021-04/STD1724%20public%20report%20%5B472%20KB%5D.pdf [11 January 2024].

Australian Institute of Occupational Hygienists 2014, *Position Paper: Dusts Not Otherwise Specified (Dust NOS) & Occupational Health Issues*, AIOH, Melbourne, https://www.aioh.org.au/product/dust-nos/ [27 December 2023].

Australian Institute of Occupational Hygienists 2016a, *Position Paper: Asbestos and Its Potential for Occupational Health Issues*, AIOH, Melbourne, https://www.aioh.org.au/product/asbestos/ [27 December 2023].

Australian Institute of Occupational Hygienists 2016b, *Position Paper: Synthetic Mineral Fibres (SMF) and Occupational Health Issues*, AIOH, Melbourne, https://www.aioh.org.au/product/synthetic-mineral/ [27 December 2023].

Australian Institute of Occupational Hygienists 2017, *Position Paper: Diesel Particulate Matter and Occupational Health Issues*, AIOH, Melbourne, https://www.aioh.org.au/product/diesel/ [27 December 2023].

Australian Institute of Occupational Hygienists 2018a, *Position Paper: Respirable Crystalline Silica and Occupational Health Issues*, AIOH, Melbourne, https://www.aioh.org.au/product/respirable-crystalline-silica/ [27 December 2023].

Australian Institute of Occupational Hygienists 2018b, *Position Paper: Respirable Coal Dust and Its Potential for Occupational Health Issues*, AIOH, Melbourne, https://www.aioh.org.au/product/respirable-coal-dust/ [27 December 2023].

Australian Institute of Occupational Hygienists 2022, *Position Paper: Welding and Thermal Cutting Fume – Potential for Occupational Health Effects*, AIOH, Melbourne, https://www.aioh.org.au/product/welding-and-thermal-cutting-fume-potential-for-occupational-health-effects-2022/ [27 December 2023].

Berry, G. 2002, 'Asbestos Lung Fibre Analysis in the United Kingdom, 1976–96', *Annals of Occupational Hygiene*, vol. 46, no. 6, pp. 523–6.

British Medical Research Council (BMRC) 1952. *Recommendations of the BMRC panels relating to selective sampling, from the minutes of a joint meeting of Panels 1, 2 and 3 held on March 4th 1952*, MRC, London.

Burton, K.A. 2023, *Do AS/NZS Respiratory Protection Standards for Filter Penetration Ensure that Worker Health is Protected Against Nanoparticle Sized Diesel Particulate Matter?*, Doctor of Philosophy thesis, School of Medicine, University of Wollongong, https://ro.uow.edu.au/theses1/1563 [27 December 2023].

Daly, B.B. 1978, 'Chapter 5: Pollution Control', *Woods Practical Guide to Fan Engineering*, Woods of Colchester Ltd.

Department of Natural Resources and Mines 2017, Recognised standard 15: Underground respirable dust control—Coal Mining Safety and Health Act 1999. Brisbane, Australia.

GESTIS 2023, *GESTIS International Limit Values*, https://limitvalue.ifa.dguv.de/ [25 August 2023].

Habibi, A., Bugarski, A.D., Loring, D., Cable, A., Ingalls, L. & Rutter, C. 2021, 'Evaluation of Methodology for Realtime Monitoring of Diesel Particulate Matter in Underground Mines', In P. Tukkaraja, *Mine Ventilation* (pp. 115–123), Boca Raton, FL: CRC Press.

Hanlon, J., Galea, K.S. & Verpaele, S. 2021, 'Review of Workplace Based Aerosol Sampler Comparison Studies, 2004–2020', *International Journal of Environmental Research and Public Health*, vol. 18, no. 13, pp. 6819. https://doi.org/10.3390/ ijerph18136819

Hanlon, J., Galea, K.S. & Verpaele, S. 2023, 'Review of Published Laboratory-Based Aerosol Sampler Efficiency, Performance and Comparison Studies (1994–2021)', *International Journal of Environmental Research and Public Health*, vol. 20, no. 1, pp. 267. https://doi.org/10.3390/ ijerph20010267

Hesterberg, T.W., Long, C.M., Sax, S.N., Lapin, C.A., McClellan, R.O., Bunn, W.B. & Valberg, P.A. 2011, 'Particulate Matter in New Technology Diesel Exhaust (NTDE) Is Quantitatively and Qualitatively Very Different from that Found in Traditional Diesel Exhaust (TDE)', *Journal of the Air & Waste Management Association*, vol. 61, no. 9, pp. 894–913.

Hines, J. 2019, *The Role of Emissions Based Maintenance to Reduce Diesel Exhaust Emissions, Worker Exposure and Fuel Consumption*, Doctor of Philosophy thesis, School of Health and Society, University of Wollongong, 2019, https://ro.uow.edu.au/theses1/854 [27 December 2023].

Hodgson, J.T. & Darnton, A. 2000, 'The Quantitative Risks of Mesothelioma and Lung Cancer in Relation to Asbestos Exposure', *Annals of Occupational Hygiene*, vol. 44, no. 8, pp. 565–601.

International Agency for Research on Cancer (IARC) 2002, *IARC Monographs on the Evaluation of Carcinogenic Risks to Humans Volume 11: Man-made Vitreous Fibres*, IARC, Lyon, France, https://publications.iarc.fr/Book-And-Report-Series/Iarc-Monographs-On-The-Identification-Of-Carcinogenic-Hazards-To-Humans/Man-made-Vitreous-Fibres-2002 [27 December 2023].

International Agency for Research on Cancer (IARC) 2013, *IARC Monographs on the Evaluation of Carcinogenic Risks to Humans Volume 105: Diesel and Gasoline Engine Exhausts and Some Nitroarenes*, IARC, Lyon, France, https://publications.iarc.fr/_publications/media/download/3181/e6bd0692f1a9bb46589d3ca2d8178fa8dcd05ba5.pdf [27 December 2023].

International Agency for Research on Cancer (IARC) 2018, *IARC Monographs on the Evaluation of Carcinogenic Risks to Humans Volume 118: Welding, molybdenum trioxide, and indium tin oxide*, IARC, Lyon, France, https://publications.iarc.fr/Book-And-Report-Series/Iarc-Monographs-On-The-Identification-Of-Carcinogenic-Hazards-To-Humans/Welding-Molybdenum-Trioxide-And-Indium-Tin-Oxide-2018 [27 December 2023].

International Organization for Standardization (ISO) 1995, *Air Quality—Particle Size Definitions for Health-Related Sampling*, ISO 7708:1995, ISO, Lyon, France, https://www.standards.org.au/access-standards

International Organization for Standardization (ISO) 2011, *Technical Report; Nanotechnologies—Nanomaterial Risk Evaluation*, ISO/TR 13121:2011, ISO, Geneva, https://www.standards.org.au/access-standards

International Organization for Standardization (ISO) 2012, *Technical Report; Nanomaterials—Preparation of Material Safety Data Sheet (MSDS)*, ISO/TR 13329:2012, ISO, Geneva, https://www.iso.org/standard/53705.html [27 December 2023].

International Organization for Standardization (ISO) 2021, *Mining—Air Quality Control Systems for Operator Enclosures—Performance Requirements and Test Methods* ISO 23875:2021, ISO, Geneva, https://www.standards.org.au/access-standards

International Organization for Standardization (ISO) 2023, *Nanotechnologies—Vocabulary, Part 1: Core vocabulary*, ISO 80004-1:2023, ISO, Geneva, https://www.standards.org.au/access-standards

Kittelson, D. 1998, 'Engines and Nanoparticles: A Review', *Journal of Aerosol Science*, vol. 29, no. 5, pp. 575–88.

Kulkarni, P., Baron, P.A. & Willeke, K. (eds.) 2011, *Aerosol Measurement: Principles, Techniques, and Applications*. New York, NY: John Wiley & Sons.

Landwehr, K., Larcombe, A., Reid, A. & Mullins, B. 2019, *Critical Review of Recent Diesel Exhaust Exposure Health Impact Research Relevant to the Underground Hardrock Mining Industry*, Department of Mines, Industry Regulation and Safety, Western Australia, https://www.dmp.wa.gov.au/Documents/Safety/MSH_nPDM_Study_LitReview.pdf [27 December 2023].

Lanza, A.J. & Higgins, E. 1915, *Pulmonary Disease among Miners in the Joplin District, Missouri, and Its Relation to Rock Dust in the Mines: A Preliminary Report* (Vol. 105). US Government Printing Office.

Marple, V. A. 2004, 'History of Impactors-The First 110 Years', *Aerosol Science and Technology*, vol. 38, no. 3, pp. 247–292. https://doi.org/10.1080/02786820490424347

Matti Maricq, M. 2007, 'Chemical Characterization of Particulate Emissions from Diesel Engines: A Review', *Journal of Aerosol Science*, vol. 38, no. 11, pp. 1079–1118. https://doi.org/10.1016/j.jaerosci.2007.08.001

McClellan, R.O., Hesterberg, T. & Wall, J. 2012, 'Evaluation of Carcinogenic Hazard of Diesel Engine Exhaust Needs to Consider Revolutionary Changes in Diesel Technology', *Regulatory Toxicology and Pharmacology*, vol. 63, no. 2, pp. 225–58.

Methner, M., Hodson, L., Dames, A., & Geraci, C. 2010, 'Nanoparticle Emission Assessment Technique (neat) for the Identification and Measurement of Potential Inhalation Exposure to Engineered Nanomaterials—Part b: Results from 12 Field Studies', *Journal of Occupational and Environmental Hygiene*, vol. 7, no. 3, pp. 163–176. https://doi.org/10.1080/15459620903508066

National Environment Protection Council (NEPC) 2023, National environment protection measure, Schedules B1 and B2, Australian Government, https://www.nepc.gov.au/publications/archive/ephc-archive/ephc-archive-assessment-site-contamination-nepm [27 August 2023].

National Health and Medical Research Council 1984, *Methods for Measurement of Quartz in Respirable Airborne Dust by Infrared Spectroscopy and X-ray Diffractometry*, NHMRC, Canberra.

National Institute for Occupational Safety and Health (NIOSH) 2013, *Current Intelligence Bulletin 65: Occupational Exposure to Carbon Nanotubes and Nanofibers*, Department of Health and Human Services, Washington, DC, https://www.cdc.gov/niosh/docs/2013-145/pdfs/2013-145.pdf [6 August 2023].

National Institute for Occupational Safety and Health (NIOSH) 2016, 'Diesel particulate matter (as elemental carbon), Analytical Method 5040', in *NIOSH Manual of Analytical Methods (NMAM)*, 5th edn, NIOSH, Atlanta, GA, www.cdc.gov/niosh/docs/2014-151/pdfs/methods/5040.pdf [6 August 2023].

National Occupational Health and Safety Commission (NOHSC) 1989, *Technical Report on Synthetic Mineral Fibres and Guidance Note on the Membrane Filter Method for Estimation of Airborne Synthetic Mineral Fibres* [NOHSC: 3006 (1989)], Australian Government Publishing Service, Canberra https://www.safeworkaustralia.gov.au/system/files/documents/1702/guidancenote_membranefiltermethod_estimationofairbornesyntheticmineralfibres_nohsc3006-1989.pdf [27 December 2023].

National Occupational Health and Safety Commission (NOHSC) 1990, *National Code of Practice for the Safe use of Synthetic Mineral Fibres*, NOHSC:2006, Australian Government Publishing Service, Canberra, https://www.safeworkaustralia.gov.au/system/files/documents/1702/nationalstandard_syntheticmineralfibres_nohsc1004-1990_pdf.pdf [28 August 2023].

National Occupational Health and Safety Commission (NOHSC) 2005, *Guidance Note on the Membrane Filter Method for Estimating Airborne Asbestos Fibres*, 2nd edn, NOHSC:3003 (2005), Australian Government Publishing Service, Canberra, www.safeworkaustralia.gov.au/doc/guidance-note-membrane-filter-method-estimating-airborne-asbestos-fibres-2nd-edition [6 August 2023].

Organisation for Economic Co-operation and Development Working Party on Manufactured Nanomaterials (OECD WPMN) 2009, *Emission Assessment for Identification of Sources and Release of Airborne Manufactured Nanomaterials in the Workplace: Compilation of Existing Guidance*, OECD, Paris, www.oecd.org/dataoecd/15/60/43289645.pdf [6 August 2023].

Orsi, M., Hatem, C.A., Leinardi, R. & Huaux, F. 2020, 'Carbon Nanotubes Under Scrutiny: Their Toxicity and Utility in Mesothelioma Research', *Applied Sciences*, vol. 10, no. 13, pp. 4513.

Parks, D.A., Raj, K.V., Berry, C.A., Weakley, A.T., Griffiths, P.R. & Miller, A.L. 2019, 'Towards a Field-Portable Real-Time Organic and Elemental Carbon Monitor', *Mining, Metallurgy & Exploration*, vol. 36, no. 4, pp. 765–772.

Pickford, G., Apthorpe, L., Alamango, K., Conaty, G. & Rhyder, G. 2004, 'Remediation of Asbestos in Soils: A Ground Breaking Study', *Proceedings of the Australian Institute of Occupational Hygienists Annual Conference*, 4–8 December, Fremantle, WA.

Pratt, S., Granger, A., Todd, J., Meena, G.G., Rogers, A. & Davies, B. 1997, 'Evaluation and Control of Employee Exposure to Diesel Exhaust Particulates at Several Australian Coal Mines', *Applied Occupational and Environmental Hygiene*, vol. 12, no.12, pp. 1032–1037.

Queensland University of Technology (QUT) 2012, *Measurements of Particle Emissions from Nanotechnology Processes, with Assessment of Measuring Techniques and Workplace Controls*, Safe Work Australia, Canberra, https://www.safeworkaustralia.gov.au/system/files/documents/1702/measurements_particle_emissions_nanotechnology_processes.pdf [6 August 2023].

Rogers, A. & Davies, B. 2001, 'Diesel Particulate (soot) Exposures and Methods of Control in Some Australian Underground Metalliferous Mines', *Proceedings of the Queensland Mining Industry Health and Safety Conference*, 26–29 August 2001, Townsville, Qld. https://www.qmihsconference.org.au/wp-content/uploads/qmihsc-2001-writtenpaper-rogers_davies.pdf [27 December 2023].

Rogers, A., Leigh J., Ferguson, D., Mulder, H., Ackad, M. & Morgan, G. 1994, 'Dose–response Relationship between Airborne and Lung Asbestos Fibre Type, Length and Concentration, and the Relative Risk of Mesothelioma', *Annals of Occupational Hygiene*, vol. 38, suppl. 1, pp. 631–8

Rogers, A. & Whelan, W. 1996, 'Elemental Carbon as a Means of Measuring Diesel Particulate Matter Emitted from Diesel Engines in Underground Mines', *Proceedings of the 15th Annual Conference of the Australian Institute of Occupational Hygienists*, 1–4 December 1996, Perth, WA.

Safe Work Australia (SWA) 2010, *Work Health and Safety Assessment Tool for Handling Engineered Nanomaterials*, Safe Work Australia, Canberra, https://www.safeworkaustralia.gov.au/system/files/documents/1702/work_health_safety_tool_handling_engineered_nanomaterials.pdf [27 December 2023].

Safe Work Australia (SWA) 2012, *Safe Handling and Use of Carbon Nanotubes*, Safe Work Australia, Canberra, https://www.safeworkaustralia.gov.au/system/files/documents/1702/safe_handling_and_use_of_carbon_nanotubes.pdf [27 December 2023].

Safe Work Australia (SWA) 2013, *Information Sheet: Safety Hazards of Engineered Nanomaterials*, Safe Work Australia, Canberra https://www.safeworkaustralia.gov.au/system/files/documents/1702/safety-hazards-engineered-nanomaterials.pdf [27 December 2023].

Safe Work Australia (SWA) 2020a, *Code of Practice: How to Manage and Control Asbestos in the Workplace*, Safe Work Australia, Canberra https://www.safeworkaustralia.gov.au/sites/default/files/2020-07/model_code_of_practice_how_to_manage_and_control_asbestos_in_the_workplace.pdf [11 January 2024].

Safe Work Australia (SWA) 2020b, *Code of Practice: How to Safely Remove Asbestos*, Safe Work Australia, Canberra, https://www.safeworkaustralia.gov.au/doc/model-code-practice-how-safely-remove-asbestos [27 December 2023].

Safe Work Australia (SWA) 2022, *Workplace Exposure Standards for Airborne Contaminants*, Safe Work Australia, Canberra, https://www.safeworkaustralia.gov.au/doc/workplace-exposure-standards-airborne-contaminants-2022 [27 December 2023].

Safe Work Australia (SWA) 2023a, *Code of Practice: Labelling of Workplace Hazardous Chemicals*, Safe Work Australia, Canberra, https://www.safeworkaustralia.gov.au/sites/default/files/2023-06/model_code_of_practice_labelling_of_workplace_hazardous_chemicals.pdf [27 December 2023].

Safe Work Australia (SWA) 2023b, *Code of Practice: Preparation of Safety Data Sheets for Hazardous Chemicals*, Safe Work Australia, Canberra, https://www.safeworkaustralia.gov.au/sites/default/files/2023-06/model_code_of_practice_preparation_safety_data_sheets_for_hazardous_chemicals.pdf [27 December 2023].

Standards Australia 2006a, *Fume from Welding and Allied Processes—Guide to methods for the sampling and analysis of particulate matter*, AS 3853.1:2006, SAI Global, Sydney, https://www.standards.org.au/access-standards

Standards Australia 2006b, *Fume from Welding and Allied Processes—guide to methods for the sampling and analysis of gases*, AS 3853.2:2006 (R2016), SAI Global, Sydney, https://www.standards.org.au/access-standards

Standards Australia 2009a, *Workplace Atmospheres—method for sampling and gravimetric determination of respirable dust*, AS 2985:2009, SAI Global, Sydney, https://www.standards.org.au/access-standards

Standards Australia 2009b, *Workplace Atmospheres—method for sampling and gravimetric determination of inhalable dusts*, AS 3640:2009, SAI Global, Sydney, https://www.standards.org.au/access-standards

Standards Australia and Standards New Zealand 2023a, *Respiratory protective devices – Selection, use and maintenance Fit-testing procedures. AS/NZS ISO 16975.3:2023*, SAI Global, Sydney, https://www.standards.org.au/access-standards

Standards Australia and Standards New Zealand 2023b, *Mining—Air Quality Control Systems for Operator Enclosures—Performance Requirements and Test Methods* AS/NZS ISO 23875:2023, SAI Global, Sydney, https://www.standards.org.au/access-standards

Swanepoel, J.D., Hines, J., Gopaldasani, V. & Davies, B. 2023. 'Hitting Two Birds with One Emissions-Based Maintenance Stone–A Literature Review on Improving Overall Productivity of Underground Diesel Fleets', *Journal of Sustainable Mining*, vol. 22, no. 1, pp. 55–64. https://doi.org/10.46873/2300-3960.1376

Swiątkowska, B., Szubert, Z., Sobala, W., & Szeszenia-Dąbrowska, N. 2015, 'Predictors of Lung Cancer among Former Asbestos-Exposed Workers', *Lung Cancer*, vol. 89, no. 3, pp. 243–248. https://doi.org/10.1016/j.lungcan.2015.06.013

Toxikos 2013, *Evaluation of Potential Safety (physicochemical) Hazards Associated with the Use of Engineered Nanomaterials*, Safe Work Australia Canberra, https://www.google.com/url?sa=t&rct=j&q=&esrc=s&source=web&cd=&ved=2ahUKEwiWzLystK6DAxXrVmwGHcFCA58QFnoECBYQAQ&url=https%3A%2F%2Fwww.safeworkaustralia.gov.au%2Fsystem%2Ffiles%2Fdocuments%2F1702%2Fsafety-hazards-engineered-nanomaterials.pdf&usg=AOvVaw3O-qpi3_ftsfYd3Jl481r_&opi=89978449 [27 December 2023].

Workplace Health and Safety Queensland 2013, *Silica and the Lung*, https://gards.org/asbestos/wp-content/uploads/2019/11/What-is-silica-lung-factsheet.pdf [1 January 2024].

World Health Organization 1996, 'Diesel Fuel and Exhaust Emissions', *Environmental Health Criteria 171*, WHO, Geneva, www.inchem.org/documents/ehc/ehc/ehc171.htm [6 August 2023].

9 Metals in the Workplace

Ian Firth and Ron Capil

9.1 INTRODUCTION

The world's industrial and pre-industrial civilisations have depended in numerous ways on metal-ore extraction and metal fabrication. Coinage, precious metals, the implements of war and industry – they have all been linked with occupational health hazards since the Bronze and Iron Ages. During the Industrial Revolution, and more recently in the technological age, metals have been implicated in occupational disease in many industries.

The toxic nature of metals and metal salts has also long been recognised, with lead and arsenic compounds often favoured by poisoners. Most people today are aware of the possibility of lead poisoning in children who may eat or chew the sweet-tasting flakes of lead paint in old houses. The Mad Hatter in Lewis Carroll's *Alice in Wonderland* may have been sent 'mad' by mercury poisoning; psychotic symptoms were common among workers in the fur and hat-making industries in the early 19th century, owing to excessive mercury exposure (see Section 9.11).

The current challenge is to monitor and measure those metals that are increasingly being used in various nanomaterials. Research in this area may raise concerns, since some metals are more toxic in nanoparticle form; this is discussed briefly in Chapter 8.

9.2 MAJOR METALS OF CONCERN IN THE WORKPLACE

This chapter examines the more toxic of the most commonly encountered metals, namely lead, cadmium, chromium, mercury, nickel and zinc, and metalloid arsenic. (Metalloids have properties of both metals and non-metals.) The following aspects are covered for each of these materials:

- typical occurrence and use;
- basic toxicology;
- assessment in the workplace; and
- typically used control procedures.

A few less occupationally significant metals, such as aluminium, beryllium, cobalt, copper, lithium, manganese, selenium and thallium, and the metalloids antimony and boron, are examined in less detail. This is not an exhaustive list, and the H&S practitioner should seek more authoritative references for metals not covered here.

9.3 METAL TOXICITY

The forms in which the metals exist are important. They may exist as the native material (e.g. chromium metal) or as various salts (e.g. chromium oxide) and, depending on these forms, their ions may exert a range of toxic effects, from dermatitis through neurotoxic effects to cancer. Some exposure standards have different exposure limits, depending on the form of the metal, its

DOI: 10.1201/9781032645841-9

chemical valency or whether it is in the form of an inorganic salt or organometallic compound (e.g. chromium and nickel).

Assessing the toxic dose of various metals is often more complicated than with other hazardous substances. Indeed, some metallic elements are essential to human life because of their role in cellular functioning, bone structure, or blood and enzyme systems. Fourteen metals, including sodium, potassium, calcium and magnesium, are involved in the body's basic building blocks. Trace elements, including zinc, selenium, iron, cobalt, arsenic and copper, are all essential in the right amounts – they have a narrow 'window of life' range of concentrations, with higher and lower concentrations being detrimental. A number of metal-based compounds, themselves potentially toxic to humans, have found great service in pharmacologically active drugs, including the early anti-syphilitic drugs, Mercurochrome, and platinum-containing cytotoxic (anti-cancer) drugs.

9.4 NATURE OF METAL CONTAMINANTS AND ROUTES OF EXPOSURE

The extraction, processing, refining, fabrication and widespread use of metals, their compounds and salts produce hundreds of situations in which hazardous exposures can occur. Because most metals and their salts are solids, most exposure to metals and metal salts in the workplace occurs through inhalation of their particulate (or aerosol) forms (i.e. dust, fume or mist). However, the contribution of ingestion should not be overlooked, as it is possible to transfer significant amounts of metals into the mouth during smoking and eating when personal hygiene is poor.

Most metals are solid at room temperature, though mercury, a few metal hydrides (e.g. arsine and stibine) and some organometallic compounds (e.g. tetraethyl lead) are common exceptions to this rule. These are either gases or can exert enough vapour pressure at room temperature to be present in the vapour state. In such cases, the metal can be inhaled as a vapour.

Some significant exposures also occur via the skin. Mercury salts, thallium and organometallic liquids can penetrate the skin, and metals and metallic salts can enter the body through damaged skin, cuts and abrasions. In some cases (e.g. nickel and other skin-sensitising metals), the skin is the target organ, and direct skin contact is a route of exposure.

Processes giving rise to metals in a form that can be absorbed are:

- metal-ore extraction (e.g. mining of iron ore, manganese, lead, zinc, copper and uranium, and their subsequent processing prior to smelting);
- metal smelting (e.g. arsenic, cadmium and selenium are liberated in lead and zinc smelting, and mercury is liberated in gold refining and in alumina refining);
- metal founding (e.g. lead and brass);
- metal machining (e.g. beryllium drilling, grinding or polishing, and cobalt in dental and hip prostheses);
- hot metal processing (e.g. hot zinc galvanising and metal recycling);
- welding, soldering, brazing and thermal cutting of metals (producing potentially hazardous metal fumes of aluminium, cadmium, chromium, copper, iron, lead, magnesium, manganese, molybdenum, nickel, titanium, vanadium and zinc);
- handling powders of metal salts (e.g. lead battery manufacture, zinc and copper oxide manufacture, lead stearate used in PVC pipe manufacture).

9.5 ASSESSING EXPOSURE TO METALS IN THE WORKPLACE

Most monitoring for metals in the workplace requires sampling for dusts and fumes. In the case of electroplating, some metals, such as chromium and nickel, become airborne as mists. These mists are monitored in much the same way as dusts and fumes containing metals. However, monitoring for some metals or their compounds (e.g. mercury vapour, arsine and stibine) requires special techniques.

Because similar sorts of air-monitoring processes are used for most metals and metallic compounds, a procedure is detailed here for only one metal, lead. As the toxic effects of metals often result from a combination of absorption from the lungs and ingestion after deposition in the nasopharyngeal (nose and throat) region, the inhalable fraction of the particulate is most often appropriately sampled (although in some cases that will be explained later, the respirable fraction should also be sampled). See Chapter 8, 'Aerosols', for other practical details or, for more complete procedures, AS 2985 (Standards Australia 2009a) on gravimetric determination of respirable dusts, AS 3640 (Standards Australia 2009b) on gravimetric determination of inhalable dusts and MDHS 14/4 (Health and Safety Executive 2014a) on gravimetric analysis of respirable, thoracic and inhalable aerosols.

Air monitoring may indicate compliance with the relevant exposure standard (refer to Chapter 3, 'The concept of the exposure standard'). To assess the exposure from all routes (inhalation, ingestion and skin absorption), however, biological monitoring of exposure may be necessary in particular circumstances to evaluate the accumulated dose experienced by individual workers. The general principles of biological monitoring are discussed in Chapter 11, 'Biological monitoring for chemical exposures'. Throughout this chapter, reference is made to:

- Safe Work Australia (SWA 2021) workplace exposure standards (WESs) for airborne contaminants, available from the web-based Hazardous Chemical Information System (HCIS).
- American Conference of Governmental Industrial Hygienists' (ACGIH® 2023) threshold limit values (TLVs®).
- US National Institute for Occupational Safety and Health (NIOSH) recommended exposure limits (RELs).
- Occupational Health and Safety Administration's (OSHA 2021) permissible exposure limits (PELs).
- UK Health and Safety Executive's (HSE 2020) workplace exposure limits (WELs).
- The GESTIS (2023) international limit values for chemical agents database of occupational exposure limits (OELs) (refer to Chapter 3).

9.6 METHODS OF CONTROL

Although metal contaminants are often present as dust, fumes or mists, control procedures vary greatly depending on how the contaminant is generated. Further, some of the more toxic metals take greater effort to control (e.g. lead dusts require more stringent control procedures than iron dusts). The example of lead (Section 9.10) provides the detail that may typically be required for an H&S practitioner involved in the control of hazardous metals in the workplace.

Workers in industries where toxic metals such as lead, cadmium, mercury, arsenic, chromium, nickel and zinc are handled must be fully informed of the routes of exposure, the nature of the health hazards and the measures required to prevent hazardous exposure, including respiratory protective equipment (RPE) and its use and maintenance. Appropriate RPE must be selected based on the assessed exposure and it should be fit tested for each worker (see Chapter 7, 'Personal protective equipment', Section 7.7). Hand-washing and other personal hygiene measures and separate eating facilities must be provided, and their use must be enforced to prevent any possibility of accidental ingestion in the workplace. The need to prohibit smoking as a further guard against ingestion must be stressed.

9.7 ARSENIC

9.7.1 Use and Occurrence

Arsenic, particularly its trioxide, finds a curious place in history as a poison favoured by murderers. Metallic arsenic has a few industrial uses in alloys with lead for bearings, cable sheaths and battery grids. In contrast, compounds of arsenic (oxides and complex salts with other metals) are widely used in the manufacture of weed killers, insecticides, wood preservatives, anti-fouling paints and fungicides. Tobacco crops were once widely sprayed with arsenicals (arsenic compounds). A well-known copper–chrome–arsenic (CCA) preparation has been widely used in the production of logs and timbers for outdoor and garden use, although in Australia and other countries, CCA use has been restricted in residential situations. CCA should not be used on high-contact timber structures: these include garden furniture, picnic tables, exterior seating, children's play equipment, patio and domestic decking, and handrails.

Arsenic compounds are no longer used as paint pigments for obvious public health reasons, but they are still used in some fireworks. Arsenic has found new applications in the manufacture of semiconductors and radiation detectors. Workers in these industries potentially risk exposure to arsenic or its compounds. Smelting of ores containing arsenic impurities (e.g. copper, lead and zinc ores) is also an important source of potential exposure.

Users of arsenic-containing agricultural chemicals are also at risk. End-users of CCA-treated timber should not be at undue risk, provided that timber offcuts are not burned (e.g. in barbecues). Burning can convert the arsenic bound into timber to the volatile and hazardous arsenic trioxide. The health implications of frequent contact with CCA-treated timber structures, particularly for children, are uncertain (Australian Pesticides and Veterinary Medicines Authority 2005).

9.7.2 Toxicology

The toxicity and action of arsenic and its compounds are varied, depending on the metal's chemical form. Arsenic and its compounds can be both ingested and inhaled as dusts, and arsenicals can cause corrosive or ulcerative effects in the skin and mucous membranes. The major routes of entry into the workplace are by inhalation. Acute effects include haemorrhagic gastritis, muscular cramps, facial oedema, peripheral neuropathy, corrosive actions and skin lesions. Chronic effects include irritation to the nasal mucosa, with penetration of the nasal septum in some workers exposed over long periods to low levels of arsenic dusts. Dermatitis may be observed, with heavy skin pigmentation and peripheral vascular disorders in some people who ingest arsenic-contaminated water.

The International Agency for Research on Cancer (IARC 2023) and SWA (2021) consider that arsenic and arsenic compounds are carcinogenic to humans and give arsenic a Group 1 or category 1A classification, respectively. Arsenic causes cancers of the lung (e.g. in smelter populations) and respiratory tract (e.g. in workers making arsenical pesticides). There is evidence that ingestion of inorganic arsenic is associated with skin and perhaps liver cancer. Paradoxically, arsenic is also a therapeutic agent, still used in the treatment of cancer (acute promyelocytic leukaemia) and African trypanosomiasis (sleeping sickness). Some skin cancers and hyperkeratosis are attributable to medicinal arsenic exposure.

Arsenic is one of a group of metals/metalloids that form a volatile hydride by reacting with nascent hydrogen (e.g. from contact with acids). Arsenical contamination can generate this hydride – arsine – accidentally. A gas with a garlic-like odour, it presents as an extremely acute inhalation hazard. For example, in a Queensland case, several people – including children – were affected when ground contaminated with old arsenical cattle dip came into contact with an acid source, probably superphosphate, and arsine was liberated. The onset of symptoms occurs within a few hours of exposure, with headache, giddiness, abdominal pain and vomiting. The urine is stained by haemolysed, excreted blood cells. Anaemia and jaundice follow, which may result in kidney failure.

9.7.3 Standards and Monitoring

Most of the Western world has set an exposure limit for inorganic arsenic and its compounds, as well as for gaseous arsine, as per the GESTIS (2023) database for international limit values. For inorganic arsenic and its compounds, most are set at a time-weighted average (TWA) value of 0.01 mg/m^3, (range of 0.00083–0.1 mg/m^3), with some having a STEL value, which ranges from 0.002 to 0.4 mg/m^3. The TWA value is set to protect for excess skin, lung and liver cancers in exposed workers (ACGIH® 2023). For arsine, most are set at a TWA value of 0.05 ppm (equivalent to 0.16 mg/m^3) but range from 0.003 to 0.1 ppm. Some countries also have a STEL for arsine, which ranges from 0.02 to 0.25 ppm.

9.7.3.1 Air Monitoring

Monitoring methods depend on the form of the arsenic. The ones most widely used in the workplace involve personal air sampling, with a sampling pump connected to filter(s) or a solid sorbent tube.

Arsenic trioxide vapours (a major source of exposure in smelting) must be collected on filter papers treated with sodium hydroxide or sodium carbonate. Any sampling head suitable for use with treated filters will suffice.

Arsenic particulates are collected on mixed cellulose ester filters using the IOM sampler or equivalent. Refer to AS 3640 (Standards Australia 2009b) and MDHS 14/4 (HSE 2014a) for sampling details.

Arsine is collected using a charcoal sorbent tube followed by analysis using atomic absorption spectrophotometry with graphite furnace, as detailed in NIOSH method 6001 (NIOSH 2020).

Most H&S practitioners should be able to conduct the sampling with appropriate training, but all the methods require laboratory analysis. Arsenic can be analysed according to MDHS 91/2 (HSE 2015) or ISO 15202 (2020).

Simple direct-reading indicator stain tubes (see Chapter 10, 'Gases and vapours', Section 10.11) are available for both arsenic trioxide vapour and arsine. These indicator tubes may require a very large number of pump strokes to be readable at concentrations much lower than the TWA OEL. The gaseous-arsine indicator stain tubes are more manageable, requiring only 20 strokes, with a measuring range of 0.05–3 ppm. The use of direct-reading indicator tubes is not appropriate for assessing TWA exposures.

9.7.3.2 Biological Monitoring

Inorganic arsenic is a scheduled hazardous substance and may require health monitoring if exposure is significant (SWA 2020). Though air monitoring remains the most appropriate means of workplace surveillance, monitoring of urine in workers manufacturing arsenicals is recommended so recent exposures can be reviewed. The preferred biological indicator of the absorption of elemental arsenic and soluble inorganic-arsenic compounds is the metal's concentration in urine. ACGIH® (2023) recommends a BEI® of 35 µg/L, taken at the end of the working week.

Normal, unexposed workers have arsenic in urine levels below 10 µg/L in European countries, slightly higher in the United States, but around 50 µg/L in Japan (ACGIH 2001). Seafood can be a prime source of non-occupational exposure to arsenic (mainly as organic compounds). A laboratory experienced in trace-level arsenic determinations can advise on sampling procedure. Urinary testing requires a 200 mL sample preserved with 0.5 gm EDTA and refrigerated until analysis. Exposure to 0.01 mg/m^3 of arsenic in air for eight hours will most likely result in a urinary concentration of about 35 µg/L. Workers should be asked how much seafood they have eaten in the 48 hours before the test, and high quantities should be noted. Caution is necessary when interpreting single high values in urine. In these cases, a second sample should be taken, either after asking the worker to reduce seafood consumption or to allow testing for both inorganic and organic forms of arsenic.

9.7.4 Controls

Several control procedures can be required when working with arsenic-containing compounds. Substitution of arsenic preparations in agriculture by organochlorine and organophosphate pesticides has markedly reduced the potential for occupational exposure. All processes involving the handling of powders should be totally enclosed. When this is impractical, a high standard of local exhaust ventilation (LEV) must be employed to remove dust from the workplace.

Processes utilising arsine must be fully enclosed, with extraction systems and reliable leak detectors. Accidental arsine generation in smelters and metal shops must be prevented by keeping dross dry and forbidding all hazardous reaction ingredients (acids, alkalis, zinc and aluminium).

Skin protection with impervious gloves is mandatory for workers handling liquids (e.g. in the CCA wood treatment industry), where splashes to the hands and forearms are common. Face shields may also be required when working with solutions of arsenic salts.

Where RPE is required (i.e. where engineering controls have not reduced air concentrations to required levels), careful attention is needed in its selection and fit. The H&S practitioner should take account of:

- the nature of the arsenic hazard (dust, vapour or gas); and
- the required minimum protection factor.

For example:

- Arsenic-containing dusts require particulate-filter respirators with filtration efficiency (P1, P2 or P3) according to the concentration of dust in the workplace.
- Arsenic trioxide vapour requires a filter treated with the appropriate absorbent, soda lime, which can be used up to 1000 ppm only.
- Arsine requires supplied air with a full-face mask. Normally, this would be for emergency use only, in the event that arsine is generated accidentally.

In addition to being fully instructed about health effects and control techniques, workers should regard health monitoring, which includes medical assessments, arsenic biological monitoring and skin examination, as central to their well-being when working with arsenic or any of its compounds.

9.8 CADMIUM

9.8.1 Use and Occurrence

The white, ductile metal cadmium finds several industrial uses because of its low melting point, conductivity and resistance to corrosion. It is used in the manufacture of nickel–cadmium (NiCAD) batteries, in cadmium electroplating to apply a protective coating to steel, in welding rods, brazing solders, low-melting-point safety valves and metal alloys. Cadmium salts are also widely used in pigments, rubbers, paints, inks, plastic stabilisers, fireworks, rectifiers, solar cells and television phosphors.

Exposure occurs principally by inhalation, usually from processes that involve handling the material or its salts as powders, or where thermally generated fumes occur in the workplace. Recovery of cadmium from NiCAD scrap batteries and welding of cadmium-plated metals are also potential sources of exposure. Accidental ingestion is rare.

9.8.2 Toxicology

Cadmium shows both acute and chronic toxic effects. Cough, headache, eye irritation, chill and fever, with chest pain, may follow acute inhalation of a cadmium fume, with possible delayed lung damage (pulmonary oedema and pneumonitis). A metal worker overexposed to cadmium fume may develop a typical metal-fume fever in the evening (or even days later) and not relate it to work carried out the previous day.

Chronic effects of exposure include damage to the kidneys, sometimes with the formation of kidney stones, as well as to the respiratory system (fibrosis). IARC classifies cadmium and cadmium compounds as being carcinogenic to humans, Group 1 (IARC 2023), while SWA (2021) classifies it as being a category 1B carcinogen (presumed human carcinogen).

9.8.3 Standards and Monitoring

Most of the Western world has set an exposure limit for cadmium and its compounds, as per the GESTIS (2023) database for international limit values. For inhalable cadmium and its compounds, most are set at a TWA value of 0.01 mg/m^3 (range of 0.001 to 0.05 mg/m^3), with some having a STEL value, which ranges from 0.004 to 0.05 mg/m^3. The TWA value is set to protect from effects on the kidneys in exposed workers (ACGIH® 2023). A few jurisdictions quote an exposure standard for respirable cadmium and its compounds. These range from 0.002 to 0.02 mg/m^3 and are set to minimise the potential for cadmium accumulation in the lower respiratory tract, which could induce lung cancer.

9.8.3.1 Air Monitoring

Workplace monitoring requires sampling the personal breathing zone with an IOM inhalable dust sampler or equivalent and a respirable cyclone dust sampler. Other practical details are similar to those outlined for lead sampling. Laboratory analysis is required and described in method MDHS 91/2 (HSE 2015) on detection and measurement of inorganic cadmium and its compounds in air.

9.8.3.2 Biological Monitoring

Cadmium is a scheduled hazardous substance for which health monitoring must be provided if the risk of exposure is found to be significant. The monitoring includes testing of respiratory function, questionnaires and urinary and blood testing (SWA 2020).

For workplaces that use cadmium regularly, blood testing for cadmium may be necessary. The biological exposure index (BEI®) for blood cadmium is 5 µg/L (ACGIH® 2023). Alternatively, urinary cadmium can be measured, for which the BEI® is 5 µg/g creatinine. Urinary excretion of cadmium is related to body burden, recent exposure, and renal damage, so the interpretation of urinary cadmium levels is not simple.

9.8.4 Controls

The highly toxic nature of cadmium requires that its use in the workplace is extremely well controlled. Elimination and substitution are rarely feasible, so prevention of airborne dust and fume production is mandatory. This is achieved by measures such as minimising temperatures in welding and soldering and using mechanical cutting instead of thermal cutting of cadmium-coated products.

Where LEV is employed, it will need to be of a high standard to control the dust or fume hazard. The filtration and recovery of cadmium or its salts from air discharged from the extraction system also require consideration.

RPE for use with cadmium or its salts may be required in certain work operations where higher-level controls (i.e. engineering controls, ventilation) are impractical or cannot adequately control the hazard. Respirators with medium particulate filtration efficiency (P2) may be used for airborne concentrations up to 10 times the TWA OEL. However, high-temperature soldering/brazing or thermal cutting with cadmium-containing materials can generate concentrations up to 50 mg/m^3 (cadmium evaporates significantly at its melting point). Mistaken reliance on filtration RPE could have disastrous consequences. For thermally generated fumes, one of the following is needed:

- primary fume/dust control and medium-efficiency particulate filtration (P2); or
- an air-supplied system with backup high-efficiency particulate (P3) protection, while ensuring there is no subsequent exposure of bystanders.

All workers involved with cadmium in the workplace require thorough instruction in the hazards and routes of exposure, and methods of safe handling and use. Particular attention must be paid to the use, fit and maintenance of RPE whenever it is required and to the need for biological monitoring.

9.9 CHROMIUM

9.9.1 Use and Occurrence

The hard, grey metal chromium, obtained from chromite ore, and many chromium salts find wide uses in a variety of industrial applications. Chromium metal is extensively used as an alloy in stainless steel, special tooling metals, welding rods and electrical resistance wires. Chromium exists in several valency states, as II (chromous, basic), III (chromic, trivalent chromium, amphoteric) and VI (hexavalent chromium or chromate, acidic). The chromium II salts are relatively unstable and not widely used. The chromium III and VI salts find wide use, the chromium VI salts because of their strong acid and oxidative properties. Chromium III salts are used most widely, including in the production of pure chromium metal and chromium VI compounds.

Typical workplace processes where chromium exposure can occur include:

- chromium electroplating;
- manual metal arc (MMA) and flux-cored arc (FCA) welding of stainless steels;
- aluminium anodising;
- chromium-based timber treatments (CCA);
- tanning of leather hides;
- manufacture and use of spray paints containing chromium salts as pigments or zinc chromate as a rust inhibitor;
- chromium bichromates of ammonia, sodium and potassium used as mordants in dyeing;
- photography and photo-engraving;
- manufacture and use of high-temperature chromium-containing cements for aggressive environments (e.g. refractory products).

9.9.2 Toxicology

While the metal chromium is inert, its salts are irritating and destructive to human tissue. Uptake of chromium in the workplace occurs mainly by inhalation. Chromium VI salts in particular are an irritant and may cause dermatitis and skin ulcers, and in extreme cases ulceration and perforation of the nasal septum. Chrome 'holes' around fingernails, finger joints, eyelids or sometimes on the forearms may occur, though they are not proliferative and may be caused more by the strong oxidative power of these materials than by the chromium itself.

Studies of workers producing chromates, bichromates and chromic acid, however, have established that prolonged inhalation of chromium VI dust causes lung cancer. Exposure to roasted chromite ore may likewise cause cancer. IARC (2023) and SWA (2021) consider that chromium VI compounds are carcinogenic to humans and give chromium VI a Group 1 or category 1A classification, respectively. For this reason, chromium VI salts are considered environmentally hazardous; many local government authorities require them to be converted to a reduced form prior to disposal.

Chromium metal, raw chromite ore and chromium III salts may cause respiratory tract and skin irritation and dermatitis but are not considered carcinogenic. IARC (2023) considers that metallic chromium and chromium III compounds are not classifiable as to their carcinogenicity to humans and gives these compounds a Group 3 classification. In fact, some chromium III salts are essential nutrients.

9.9.3 Standards and Monitoring

Most of the Western world has set an exposure limit for chromium and its compounds, as per the GESTIS (2023) database for international limit values. Because the toxic effects of chromium salts depend on oxidation state, different TWA OEL values are provided for different salts, as shown in Table 9.1. A few jurisdictions also recommended a STEL value, which ranges from 0.0005 to 0.08 mg/m^3 for chromium VI soluble compounds. The ACGIH® (2023) TWA values for the metal and the salts in lower oxidation states are believed to be sufficient to minimise irritation and lung effects in exposed workers. For chromium VI salts, the lower TWA OEL in most cases is primarily to protect long-term workers against increased risk of cancer in addition to respiratory tract and skin irritation and dermatitis, although the ACGIH® value is to also minimise respiratory sensitisation and the likelihood of asthmatic responses in already sensitised individuals. It should be noted that NIOSH (OSHA 2021) and the Australian Institute of Occupational Hygienists (AIOH 2023) have also recommended a REL/Trigger-TWA of 0.0002 mg/m^3 for chromium VI compounds.

TABLE 9.1
Typical Exposure Standards for Chromium and its Salts

Chromium Compound	Common International TWA OELs with (range)	ACGIH® TLV®-TWA
Chromium metal	0.5 mg/m^3 (0.5–2 mg/m^3)	0.5 mg/m^3
Chromium II compounds	0.5 mg/m^3 (0.5–2 mg/m^3)	-
Chromium III compounds	0.5 mg/m^3 (0.5–2 mg/m^3)	0.003 mg/m^3
Chromium VI compounds	0.005 mg/m^3 (0.00002–0.025 mg/m^3)	0.0002 mg/m^3

Source: GESTIS 2023; ACGIH® 2023.

9.9.3.1 Air Monitoring

Air monitoring for chromium particulates requires personal sampling with an IOM inhalable dust sampler or equivalent. Refer to AS 3640 (Standards Australia 2009b) or MDHS 14/4 (HSE 2014a) for details. Other technical details are fully covered in methods such as MDHS 91/2, *Metals and metalloids in air by X-ray fluorescence spectrometry* (HSE 2015), MDHS 52/4, *Hexavalent Chromium in Chromium Plating Mists* (HSE 2014b), *NIOSH Methods* 7605, 7302 and 7304 (NIOSH 2020), and ISO 16740 (2005).

Laboratory advice should be sought before attempting to monitor chromium, as a number of technical problems are present. Sampling chromic acid mists requires a non-metallic open-faced filter holder, and a filter material (e.g. PVC) that will withstand the action of acid and not affect the stability of the chromium. Sampling of chromium in welding fume is time-critical; the chromium oxidation state will depend on the welding technology (inert-gas shielded or manual metal arc, etc.) and the time after thermal generation of the fume.

9.9.3.2 Biological Monitoring

As chromium is a scheduled hazardous substance, health monitoring for chromium workers may be required, depending on the significance of risk (SWA 2020). Urine testing is required for those working with water-soluble chromium VI salts. ACGIH® (2023) recommends a BEI® of 0.7 µg/L, with a Pop notation (i.e. based on the levels in the environmentally exposed population), for total chromium in urine, taken at the end of the last shift of the working week. The AIOH (2023) recommend a guidance value of 1 µg/L for total chromium in urine, taken at the end of the last shift of the working week, to check the efficacy of controls. Annual physical examination with emphasis on the respiratory system and skin and weekly inspection of hands and forearms are advisable if there has been significant exposure (e.g. greater than half the OEL). Air monitoring provides a better indication of the level of exposure than blood or urinary monitoring.

9.9.4 Controls

Control of chromium dusts in the workplace is of paramount importance in preventing exposure by inhalation. Enclosed systems and high-quality LEV are priority control procedures.

In chromium electroplating, the chromium or chromic acid mists generated over air-agitated tanks require several levels of control:

- surface active additives to prevent the formation of stable bubbles;
- floating surface balls to prevent a large surface-to-air interface;
- push–pull ventilation to capture any escaped mists (refer also to Chapter 6, 'Industrial Ventilation', Figure 6.16).

Skin protection is necessary where splashes of liquid may occur, or where dried salts may be picked up by sweat on the skin (e.g. in the leather-tanning industry). A paraffin and lanolin barrier cream should be used as an added protection. A 10 per cent CaNa EDTA ointment should be applied to any cuts or abrasions contaminated by chromium salts (this converts all chromium VI to chromium III, which can be safely bound or chelated). Full impermeable protective equipment against accidental exposure should be used where appropriate in electroplating.

Because of chromium's carcinogenic potential, RPE should be a last resort, to be used only when all other methods of controlling exposure have been exhausted (e.g. when stripping chromium cements inside a warm furnace flue). The choice of RPE depends on the following considerations:

- valency state of chromium (III or VI);
- airborne concentration of chromium;
- protection factor required; and
- physical demands of tasks being undertaken.

Workers involved in handling chromium salts, chromium electroplating or welding of stainless steels should be fully instructed in the health effects and specific hazards of their particular task. Attention should be paid to safety measures and controls, and their correct implementation.

9.10 LEAD

9.10.1 Use and Occurrence

The soft, bluish-white to dull greyish-coloured, malleable metal lead is obtained by smelting ores containing lead sulfide (galena), lead sulfate or lead carbonate. Lead ores often contain zinc and other toxic metals such as cadmium in minor concentrations. Chile, Australia and the United States are the largest producers and exporters of lead. However, about half of all lead produced each year comes from recycled material (Bell 2020). Lead's main industrial use is in lead-battery manufacture, but it also finds applications in automotive paints, solders, ceramic glazes, metal alloys (e.g. gun metal), bearings and lead shot.

Industrial workplaces where there is **potential exposure to inorganic lead** include:

- the lead mining and refining industry (see Figure 9.1);
- the battery industry, both manufacture and reclamation;
- the radiator repair industry;

FIGURE 9.1 Smelting and refining can result in a risk of exposure to fumes and gases.

Source: Photo from the Nyrstar Collection, taken by Karen Seindanis.

- propeller grinding;
- lead lighting;
- non-ferrous metal foundries manufacturing gun metal or leaded bronzes;
- the spraying of lead-based paints;
- the sanding or torch cutting of lead-painted metals (e.g. bridge painters and demolition workers);
- the assaying of gold and silver;
- indoor shooting galleries and rifle ranges;
- house painting, in the sanding of some houses, painted prior to the 1950s;
- ceramic glazing.

Organic lead exposure occurs in petroleum workers potentially exposed to tetramethyl or tetraethyl lead. These are being phased out of use and are no longer used in most countries, but they are an issue in some developing countries. Additionally, PVC pipe manufacturers may use lead stearate as a stabiliser, although such use has been phased out in Australia.

9.10.2 Toxicology

Lead is the metal most likely to harm both workers and members of the public. More is known about the toxic effects of lead than of any other metal. Toxicity depends mainly on particle solubility and size, since these determine how easily the metal is absorbed. The greatest hazard in the workplace has typically been inhaled lead, either as particulate (dust) or as very fine lead fume, but workers may also accidentally ingest lead if there is lead dust on their hands or face when they smoke or eat. Some inhaled dust and fume will also ultimately be swallowed after coughing or clearing the throat. Soluble lead salts are very toxic if swallowed. The smaller the particles, the more rapid their absorption is, and the more acute and severe their toxic effect. Thermally generated lead fumes are often involved in lead poisoning: fumes inhaled into the lungs can pass easily through the alveolar walls directly into the bloodstream. These fumes contain the easily soluble lead suboxide, common in the grey fume that occurs in and around lead smelters and brass foundries. For inorganic lead, absorption through the skin is not a significant route of exposure.

Once in the body, lead is transported in the bloodstream to all tissues and is predominantly stored in the bones, where it replaces calcium. Mobilisation of lead from bone to blood is slow and can lead to slight elevations of blood lead levels for many years after exposure ceases. Absorbed lead is excreted from the body primarily via the kidneys, in urine. Blood lead levels are expressed in micromoles per litre (µmol/L) or micrograms per 100 mL, or decilitre, (µg/dL) and are a good reflection of absorption of inorganic lead into the body.

Lead is a neurotoxin that can slow the transmission of impulses along the nerves. It has been implicated in affecting intellectual development in the young (exposed to lead during gestation and early childhood) and is also associated with kidney dysfunction, elevated blood pressure and sperm abnormalities. Other serious effects can accompany acute and chronic lead intoxication. Historically, the major toxic effect of lead has been on the haemopoietic (blood generation) system, resulting in anaemia. At very high levels, which are no longer typical, lead absorption can cause constipation, abdominal pain, blue lines along the gums, convulsions, hallucinations, coma, weakness, fatigue, tremors and wrist drop.

IARC classifies inorganic lead compounds as being probably carcinogenic to humans, Group 2A (IARC 2023), while SWA (2021) classifies it as a category 2 carcinogen (suspected human carcinogen). Most importantly, lead is considered a category 1A (known human) reproductive toxin (SWA 2021).

9.10.3 Standards and Monitoring

Regulations require both:

- monitoring of air in the workplace for lead;
- biological monitoring of the worker for lead in the blood.

Blood lead levels should be used as the primary indicator of both inhalation of airborne lead and ingestion via eating and smoking. Air monitoring in the workplace should also be considered as a complementary measure to evaluate the effectiveness of controls for airborne lead (AIOH 2018a).

9.10.3.1 Air Monitoring

Adequate control of exposure to airborne lead should be employed to maintain workplaces within a TWA OEL of 0.05 mg/m^3. The ACGIH®, NIOSH and SWA use this limit value for lead and its inorganic compounds, although some jurisdictions still use 0.15 mg/m^3 (e.g. UK HSE). The TLV®-TWA of 0.05 mg/m^3 is intended to maintain worker blood lead levels below 30 µg/dL (1.45 µmol/L). Monitoring provides information about:

- effectiveness of control measures;
- reasons for some high blood–lead levels;
- specific work factors that may be hazardous;
- the correct level of intervention required for control (e.g. LEV or RPE).

Personal air monitoring must be undertaken using a standard method (e.g. AS 3640; Standards Australia 2009b). The basic principles are:

- Workplace air is sampled using an IOM sampler or equivalent (Figure 8.6) fitted in the worker's breathing zone, using a portable monitoring pump running at 2 L/min attached to the worker's belt or a pump harness.
- The particulate or fume is trapped on an appropriate membrane filter.
- The sampled material is analysed directly without further sample preparation by X-ray fluorescence spectrometry according to MDHS 91/2 (HSE 2015).
- Or, the sampled material is dissolved in nitric acid.
- Then the amount of lead is measured by atomic absorption spectrometry (AAS) or inductively coupled plasma mass spectrometry (ICP-MS) (ISO 15202:2020, NIOSH 2020).
- The resultant concentration of lead in the air is calculated, taking into account the sample volume.

Analysis for lead requires laboratory facilities. Most H&S practitioners will probably need to limit their involvement to sampling and the subsequent calculations and reporting.

9.10.3.2 Biological Monitoring

Air monitoring may indicate compliance with the OEL. However, monitoring the worker's blood lead levels may still be necessary where:

- it is required by regulation;
- there may be accidental uptake (e.g. via smoking or eating);
- control of the lead hazard has failed;
- the worker's tasks (e.g. irregular duct cleaning) might still result in increased exposure to lead;

- workers have a history of excessive exposure to lead;
- primary control processes are not used and RPE is the only defence;
- a health and safety inspector requests blood–lead monitoring.

Biological monitoring for lead represents more than just checking hygiene in the workplace. It is an active intervention necessary for maintaining the health of exposed workers and ensuring that lead is not absorbed in deleterious quantities. Wherever lead is used, some lead exposure is inevitable, but symptoms do not appear in most workers if the blood lead level can be kept below the action level chosen (e.g. 0.97 μmol/L or 20 μg/dL of blood). Specific action levels may not apply to all workers, however. Lead is toxic to the foetus, and lower levels of exposure will be necessary for most women. SWA has model regulations (Part 7.2) relating to lead exposure for women of childbearing age (SWA 2023). For females of reproductive capacity, a 'lead risk job' is one in which blood lead is likely to exceed 5 μg/dL (0.24 μmol/L).

Measurement of pre-existing blood lead levels may be required by regulation before a worker begins a job where lead is a potential hazard. There are specific requirements for the sampling and analysis of blood for lead, such as those provided by the World Health Organization (WHO 2020). Some other test indicators of the biological effect of lead, such as zinc protoporphyrin (ZPP), can be used to supplement the basic blood–lead measurement, but they are not generally used in routine health monitoring.

Regulations can have a complex regime of actions to control workers' blood lead levels. For example, in the SWA (2023) amended model regulations, monitoring frequency can vary from every six months when less than 10 μg/dL, to every six weeks if greater than 20 μg/dL (0.97 μmol/L) for all but females of reproductive capacity, where 5 μg/dL (0.24 μmol/L) will trigger six-weekly monitoring. Removal from exposure is mandatory at greater than 30 μg/dL (1.45 μmol/L) for males, with return to work allowed only when blood lead returns to less than 20 μg/dL. Some companies may use different blood–lead action levels to these.

Testing of urine for lead level is not a good indicator of exposure or body burden in the case of inorganic lead, because it is cleared from the blood and bone at quite different rates. However, urine testing is the method of choice for assessing exposure to organic lead compounds such as tetraethyl lead.

9.10.4 Controls

Where levels of lead in air or blood indicate that exposure is deleterious, the cause of such exposure needs to be determined and appropriate controls instituted. Control of both the inhalation and ingestion routes will usually be required.

Lead is essential in many industries, but substitution with another metal or a change in the form of the lead should always be considered. Though the possibility of elimination or substitution is limited, it has occurred in a number of cases, including the manufacture of fishing weights and some PVC stabilisers. In addition, changing the form of the material, such as using pelletised forms of lead stabilisers instead of fine powders in PVC pipe production, reduces the probability of lead becoming airborne. As fumes from molten metallic lead are generated at more than 450 °C, reducing the temperature of the molten metal will also reduce airborne lead exposure.

The most common secondary control procedures, with examples, are:

- Enclosing processes to control dust or fume:
  - lead alloying plants and lead oxide furnaces are completely enclosed to prevent any escape of lead dust (Figure 9.2a);
  - using dust minimising techniques;
  - keeping process materials wet;

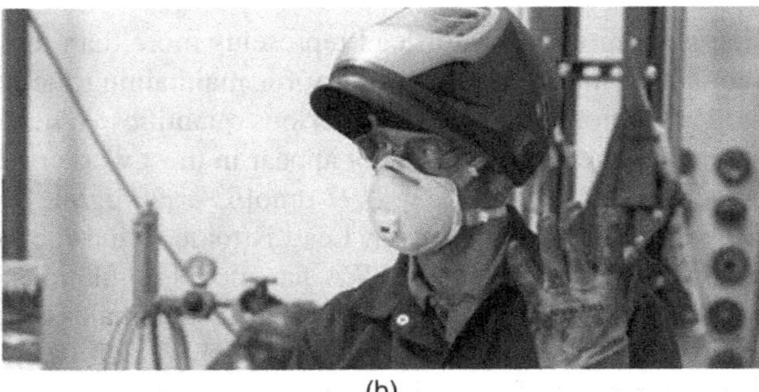

(a) (b)

FIGURE 9.2 Examples of fume control and RPE against thermally generated lead fume. (a) full extraction on a lead oxide production process. **Courtesy of Penox.** (b) P2 particulate filtration respiratory protection. Courtesy **of 3M.**

  - mixing spilled material with wet sawdust as a dust suppressant;
  - using a vacuum cleaner (with a HEPA filter) in place of sweeping.
- Use of dust or fume extraction equipment:
  - fume extraction hoods or LEV over high-temperature furnaces (>500 °C);
  - portable hand tools fitted with dust extractors for sanding/grinding;
  - exhausted enclosures in plate stacking in battery manufacture.
- Administrative controls:
  - permitting blood levels to increase above a background level but limiting the increase to a specified level;
  - removing lead-affected workers from lead work;
  - ensuring that eating, drinking and smoking do not occur in workplace areas where lead is handled;
  - maintaining good housekeeping;
  - ensuring that facilities for good personal hygiene are provided and used.
- Use of RPE, the type of which will depend on the concentration of lead in the workplace atmosphere, how it is generated, the protection factor required and the physical demands of tasks being undertaken. For example:
  - Thermally produced fumes will require medium efficiency particulate (P2) filtration for air concentrations up to 10 times the TWA OEL (Figure 9.2b).
  - Higher concentrations will require the use of powered air-purifying respirators with P2 or P3 filters.
  - Airline respirators will be necessary for filling batteries in submarines.

Instruction, training and maintenance of monitoring and health-monitoring programs are all very necessary in workplaces where lead is handled. Workers must also know and understand the hazards of handling lead. In addition to its toxic qualities, lead is heavy, and lifting and moving it safely by hand can require ergonomic considerations.

9.10.5 Lead Regulations

Lead is subject to regulation in most countries. In Australia as an example, regulations in each state are to be based on the *Model Work Health and Safety Regulations* (SWA 2023).

These embody the concepts of a lead process (certain conditions must be met before the task is considered a lead hazard) and a lead risk job (a job involving lead exposure which results in various blood–lead levels – hence, categories – depending on individual circumstances). There are also special requirements regarding equal-employment opportunities for women and men. The requirements emphasise biological and air monitoring, followed up with a strict control regime. H&S practitioners involved in the assessment or management of lead risk jobs will need to know:

- how the operations produce lead dust or fume contamination;
- the exposure to airborne lead in each job or similar exposure group;
- the categories of job involving a risk of lead exposure and the health-monitoring requirements;
- the blood–lead levels of individual operators (subject to medical confidentiality);
- what control processes are in place and their relative effectiveness; and
- procedures to undertake when action levels in blood or air are exceeded.

For much of this work, the H&S practitioner will have to collaborate with a medical practitioner experienced in the interpretation of lead exposure. The assessment of control measures may be facilitated by knowledge of the blood–lead levels of individual workers. To preserve medical confidentiality, it may be advisable to ask workers to sign a release form for their blood–lead results. The H&S practitioner must ensure that this information is used solely to help the workers control their lead absorption rates.

9.11 MERCURY

9.11.1 Use and Occurrence

Metallic mercury is a heavy, silvery liquid obtained from roasting cinnabar ore. The use of mercury-containing compounds has a long history, dating at least from Roman times. In the 17th century, the use of mercury nitrate in the hat trade for carrotting of fur (raising the scales on fur shafts) was widespread. Mercury has found more recent use in submarine ballast, mercury-fulminate explosive detonators, barometers, thermometers, pressure pumps, electric lamps and mercury rectifiers, chlor-alkali cell electrodes, dry-cell batteries, electrical switches, chemical catalysis, dental amalgams, pesticides and in the extraction of gold from ores. Cyanide leaching of low-grade gold and silver ores also collects any mercury contained in the ore, which is volatilised off by heating during further processing for the gold. It also finds use in mould and fungus inhibitors for wood, paper and grain, some medicinal preparations, and paints for inhibiting marine growth on ships' hulls. It can still be found providing a frictionless 'float' bearing for the lens assembly in a few lighthouses. There have been calls to severely limit the use of mercury, owing mainly to its tendency to accumulate in brain and foetal tissues and in breast milk, with possible adverse consequences for foetal and child growth.

Mercury may occur as a contaminant in ore (e.g. bauxite) and coal, to be concentrated to environmentally significant levels by processing.

9.11.2 Toxicology

Liquid mercury vaporises readily at room temperature, so inhalation of mercury vapour is the primary route of entry, although, to a lesser extent, both the metal and its compounds can be absorbed through the skin.

Cases of gross poisoning with skin ulceration and gastrointestinal symptoms are now a thing of the past. Accidental poisonings of children with mercury and its compounds still occur but are rare.

Mercury accumulates mainly in the brain and kidneys. The main target organ is the central nervous system (including the brain), although kidney damage may also occur with some mercury salts. The toxic action of mercury compounds occurs by precipitation of protein and inhibition of sulfhydryl enzymes. These damage the central nervous system (CNS), causing headache, tremors, weakness and psychotic disorders that may present as shyness, irritability and excitability. The classical mad hatter's disease, with its spidery writing and withdrawn behaviour, is typical of mercury poisoning but rarely seen these days. Some people may develop sensitivity to mercury whereby their skin reacts even to the vapour, causing contact dermatitis. Ingestion of organomercury compounds has resulted in a particular type of CNS debilitation, which is often irreversible. This has occurred in Japanese people who have eaten fish containing high levels of methylmercury (the so-called Minamata disease), and in Middle Easterners who have consumed seed wheat treated with mercury.

Recent data indicate that mercury can affect both male and female reproduction and lead to children with abnormal cognitive and physical functioning; hence, its category 1B reproductive toxicity classification (SWA 2021). IARC (2023) considers that mercury and inorganic mercury compounds are not classifiable as to their carcinogenicity to humans and gives these compounds a Group 3 classification.

9.11.3 Standards and Monitoring

Most of the Western world has set an exposure limit for mercury vapour and inorganic mercury compounds as per the GESTIS (2023) database for international limit values. Most are set as a TWA value of 0.025 mg/m^3 (range of 0.02 to 0.05 mg/m^3), to minimise the potential for damage to the CNS and kidneys (ACGIH® 2023). Note that mercury and its compounds can have a notation for skin absorption. A few jurisdictions also list a STEL value, which ranges from 0.4 to 0.04 mg/m^3.

9.11.3.1 Air Monitoring

Several methods are available for monitoring mercury in the workplace, all of which rely on the fact that mercury can be trapped and measured as a vapour. They are:

- indicator stain tubes (see Chapter 10, 'Gases and vapours', Figure 10.2); this method is generally useful for exposures greater than the TWA OEL;
- the direct monitor (Figure 9.3a) using ultraviolet detection or a gold film amalgam; this method is useful both above and below the TWA OEL;
- impingement into an acid permanganate solution (Figure 9.3b) followed by cold-vapour atomic absorption spectrometry (the most sensitive method), although ICP-MS is now more often used;
- passive or active sampling (see Chapter 10, 'Gases and vapours', Section 10.15) on to a solid sorbent device (using Hydrar or hopcalite as the sorbent), followed by cold-vapour atomic absorption spectrometry. This method is also useful both above and below the TWA OEL and is often employed, although ICP-MS is more commonly used. Refer to the ISO 17733 (2015) method on determination of mercury and its inorganic divalent compounds in the air by cold-vapour atomic absorption spectrometry.

9.11.3.2 Biological Monitoring

Inorganic mercury is a scheduled hazardous substance. Health monitoring (SWA 2020) may be required if the risk from exposure to mercury is found to be significant, and any affected workers

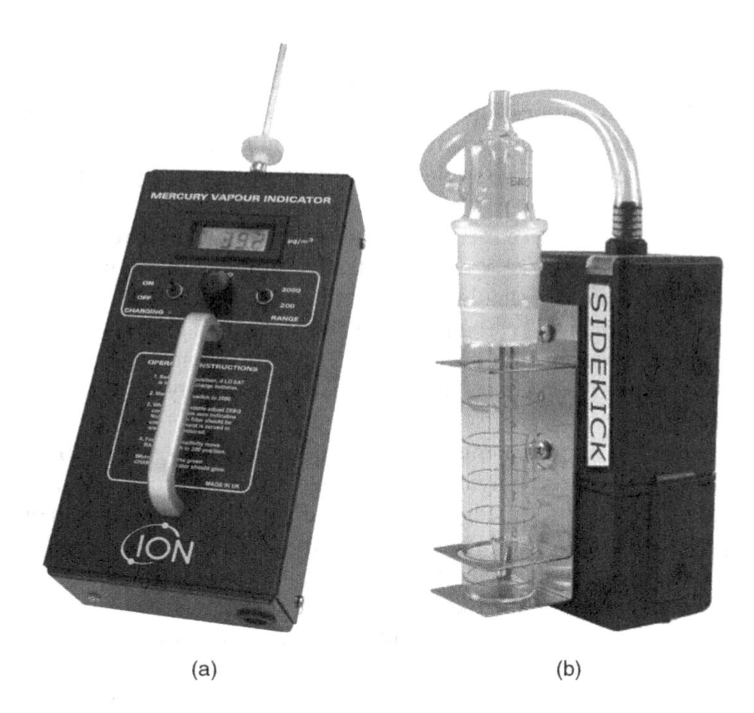

FIGURE 9.3 Mercury monitoring methods (a) portable mercury vapour indicator using ultraviolet detection **(Source: Airmet)**; (b) impinger method **(Source: SKC Ltd, www.skcltd.com)**.

should immediately be removed from further exposure. The preferred indicator of elemental mercury absorption is its concentration in urine. ACGIH® recommends a BEI® of 20 μg/g creatinine, from a sample taken pre-shift.

Some mercury is normally contributed from the diet (e.g. from eating fish). Those exposed occupationally to mercury show elevated levels of mercury in urine; background mercury levels of unexposed persons are generally less than 5 μg/g creatinine, while levels above 35 μg/g creatinine are considered significant. Urinary levels of about 50 μg/g creatinine are seen after occupational exposure to about 0.04 mg/m^3 mercury in air. Blood levels of mercury can be monitored, but sample collection is much more complicated than simple collection of urine samples and interpreting the significance of the results can be difficult.

9.11.4 Controls

The Minamata Convention on Mercury (2023), a United Nations multilateral environmental agreement, addresses the adverse effects of mercury through practical actions to protect human health and the environment from emissions and releases of mercury and mercury compounds. The primary methods used to prevent mercury uptake in the workplace are as follows:

- **Replacement of mercury-containing equipment** such as thermometers with non-mercury-containing alternatives such as alcohol thermometers or thermocouples. Mercury from mercury-in-glass thermometers that break in hot ovens will rapidly evaporate.
- **Substitution of mercury-containing processes** wherever possible (e.g. use of composite restoration material in place of mercury dental amalgams; use of organic chemicals for fur

carrotting; use of carbon-in-pulp leaching of gold rather than the older-style mercury retorting; replacement of neon/fluorescent tubes that contain mercury).

- **Prevention of vapour generation**. Mercury evaporates readily, so it should be stored in closed, water-sealed containers. Where mercury metal is heated, the hazard is enormously increased, so operations such as welding or brazing on mercury-in-metal thermometers must be strictly controlled.
- **Enclosure of processes where mercury vapours may be generated** to prevent escape of vapours. Where not possible, LEV to a high standard is required. All mercury-contaminated air exhausted from a process should be routed through a mercury scrubbing device using a sulfur- or iodine-impregnated carbon pack, or bubbled into a tank that contains a mercury-complexing agent in conjunction with a de-mister. Mercury tends to condense in ventilation ducts. Having smooth-walled ducts that slope towards a gravity collection trap should control this condensation.
- **Designing areas for handling mercury to cope adequately with spills**, with impervious flooring (e.g. use epoxy, polyurethane or vinyl sheeting; wood or carpeting should be avoided) and bunded (built-up) edges to prevent dispersion of the small mercury droplets. Proprietary products (HgX®) or a paste of slaked lime and flowers of sulfur (1:1) can be used to assist in removing the metal from inaccessible and otherwise hard-to-clean cracks and crevices, equipment and the like. Contaminated materials such as carpet are best discarded, but their disposal must comply with local hazardous-waste requirements.
- **Preventing mercury from spraying into small droplets** that are difficult to coalesce and collect. A given volume of mercury has a much larger surface area when broken into small droplets and evaporates much more rapidly. Mercury should not be collected with a vacuum cleaner unless it is specifically designed for mercury (e.g. Tiger-Vac® or Nilfisk®). Any vacuum pick-up system must have an adequate mercury-vapour trap.
- **Using personal protective equipment for mercury vapour** when significant mercury spills have to be cleaned up. Respirators impregnated with a mercury-vapour absorbent can be used with relatively small spills.
- **Using airline breathing apparatus for major spills**, particularly if the ambient temperature is elevated. The use of overalls and gloves will be necessary if any skin contact is likely.
- **Instructing workers handling mercury about the hazards of the material**, as well as in adequate methods of containment and control, and in decontamination procedures should a major spill occur.

9.12 NICKEL

9.12.1 Use and Occurrence

Nickel is obtained from a mixture of ores including sulfides, silicates and laterites. One industrial process (Mond) converts the sulfide to nickel oxide by roasting, followed by conversion to nickel carbonyl by reaction with hydrogen and carbon monoxide, and subsequent reduction to pure nickel. Nickel carbonyl may also occur in the chemical, glass and metal plating industries, where it is used as a catalyst in various chemical reactions, in glass plating and in the forming of nickel films and coatings.

Nickel finds use mainly in stainless steels and in a huge number of alloys, in which nickel confers high resistance to corrosion and to extremes of temperature. It is used in nickel–cadmium batteries, nickel plating, Monel alloy, nickel chemical-reaction catalysts, magnetic tapes, coins, jewellery and electronic, electrical and engine parts. Workers are exposed to nickel-containing dusts and fumes during plating and grinding, mining and nickel refining, and in steel plants, foundries and other metal industries. Exposure can also arise from welding of metals containing nickel.

9.12.2 Toxicology

Nickel exerts two major workplace-related health effects. The first is nickel itch or nickel eczema on the arms of those whose skin is exposed to nickel (e.g. platers). The itch may spread to the face and other parts of the body, constituting an allergic response. Workers so sensitised may not be able to return to this kind of nickel-related work. Nickel-plated watches, jewellery and earrings may likewise cause an allergic reaction on the skin, produced when sweat comes in contact with the metal. Nickel can sensitise the skin by direct contact, and there is good evidence that it can also do so after being ingested.

Inhalation of nickel carbonyl can result in acute symptoms of headache, dizziness and vomiting, followed by chest pain, dry cough, shortness of breath and extreme weakness, depending on the degree of exposure. Nickel carbonyl causes haemorrhagic pneumonia. Most inhaled nickel ends up in the brain and lung. The process of nickel roasting produces nickel fumes that are considered to be carcinogenic. IARC (2023) considers that nickel compounds are carcinogenic to humans and gives these compounds a Group 1 classification, while SWA (2021) considers them to be a category 2 (suspected) carcinogen. Nickel metals and alloys are classified as Group 2B by IARC (possibly carcinogenic to humans). The compounds principally implicated in causing respiratory cancer are sulfidic nickel, particularly nickel sub-sulfide and oxidic nickel, which includes a range of insoluble nickel compounds. There is debate about whether soluble nickel compounds are carcinogenic (AIOH 2016).

9.12.3 Standards and Monitoring

Most of the Western world has set an exposure limit for nickel compounds as per the Gestis (2023) database for international limit values. There are a number of exposure limits for nickel compounds, depending on the form of the nickel, as shown in Table 9.2. Nickel carbonate, nickel sulfide and nickel oxide are insoluble compounds of nickel, whereas nickel chloride, nickel sulfate and nickel nitrate are soluble compounds. The AIOH (2016) recommend an OEL-TWA of 0.1 mg/m^3 for all forms of nickel, as the inhalable fraction.

Indicator stain tubes are available for direct measurement of nickel carbonyl in air, though the method is not particularly sensitive. Better sensitivity and precision in the assessment of inorganic nickel in workplace air are obtained by methods such as MDHS 91/2 (HSE 2015) or NIOSH method 7302 (NIOSH 2020). As with other metals, the dust or fumes are collected on a filter, dissolved by acid and measured by atomic absorption spectroscopy, inductively coupled argon plasma (ICP) or X-ray fluorescence (XRF).

TABLE 9.2
Typical Exposure Standards for Nickel and its Compounds

Nickel Compound	Common OEL-TWA	OEL Range
Nickel metal	1 mg/m^3	0.15–1.5 mg/m^3
Soluble compounds of nickel	0.1 mg/m^3	0.015–1 mg/m^3
Nickel carbonyl	0.007 mg/m^3 (peak)	0.01–0.35 mg/m^3
Insoluble compounds as Ni	0.2 mg/m^3	0.015–1 mg/m^3

Source: GESTIS, 2023.

9.12.4 Controls

Because of the irritation and carcinogenic effects of some nickel dusts and aerosols, control of processes involving nickel by enclosure or LEV is often mandated. Workers in nickel electroplating should have skin protection against nickel salts and access to adequate washing facilities in case of skin splashes.

Refining processes involving nickel carbonyl must be completely enclosed, preventing any exposure. Only air-supplied RPE is suitable for entry into any area of a plant where nickel carbonyl is processed.

Workers handling nickel metal, its salts or nickel-containing compounds should be instructed about the hazards, particularly the need to be aware of skin conditions. Workers also require instruction in the use of controls appropriate to the potential exposure.

9.13 ZINC

9.13.1 Use and Occurrence

Metallic zinc is obtained by treatment (froth flotation and roasting) of zinc sulfide ore. Smelting produces an impure grade of zinc suitable for galvanising, and electrolytically refined, higher-grade zinc is suitable for metal die-casting and moulding. Zinc reacts rapidly with oxygen to form a protective layer of zinc oxide that prevents any further reaction. Zinc finds use in the protective galvanising of steel and roofing iron, in lightweight metal castings for toys and fancy goods, in zinc oxide for paints and vehicle tyre fillers, in zinc dry-battery cases and in alloys such as brass.

Workers most likely to be affected by zinc exposures are welders working on galvanised iron in poorly ventilated or unventilated conditions, hot-dip galvanisers and workers in zinc smelters and zinc oxide and granulated zinc production plants. Zinc powder also presents a combustion and explosion hazard.

9.13.2 Toxicology

Because zinc is an essential metal, occupationally related poisonings are not numerous. While some poisonings do occur with zinc metal, most arise from zinc compounds. The major route of intake in the workplace is inhalation, with possible subsequent ingestion after clearance from the lung. Zinc and its compounds are readily absorbed owing to their solubility. The major health consequence of zinc inhalation is metal-fume fever, sometimes known as 'brass founder's ague' or just 'zinc chills'. This fever occurs only because of exposure to freshly generated fumes of either the metal or zinc oxide. The particles must be sufficiently small to enter into the lung space. Fever usually begins some hours after exposure and is accompanied by increased leukocyte count. A continuously elevated leukocyte count can confer some resistance to a recurrence of fever. There is no indication of chronic health effects from zinc exposure, but other substances in the fume (e.g. lead, arsenic, cadmium or oxidant gases from welding) may cause chronic illness.

Zinc chloride fumes (from smoke bombs or soldering fluxes) are known to be toxic, causing inflammation and corrosion of lung tissue and subsequent rapid lung fibrosis. Zinc chloride solutions have a caustic action on the skin. Zinc phosphide used in rat poison is toxic when inhaled or ingested due to the release of phosphine gas.

9.13.3 Standards and Monitoring

Most of the Western world has set exposure limits for zinc chloride and zinc chromates only as per the GESTIS (2023) database for international limit values. For zinc chloride, most are set at a TWA value of 1 mg/m^3 (range of 0.5–2 mg/m^3), with a STEL of 2 mg/m^3 (range of 1–4 mg/m^3). For zinc chromate, most are set at a TWA value of 0.01 mg/m^3 (range of 0.001–0.05 mg/m^3), with some jurisdictions also having a STEL value (range of 0.01–0.2 mg/m^3). The recommended ACGIH® (2023) TLV®-TWA limit for zinc oxide as respirable dust is 2 mg/m^3, with a STEL of 10 mg/m^3, to prevent metal-fume fever.

It is therefore necessary to know whether an exposure was to zinc fume or dust. In cases where the zinc is thermally generated (e.g. welding, burning of zinc in a candle furnace and smelting or hot-dip galvanising), it is appropriate to assume that it will be finely divided fume. Concentrations up to 10 times the TWA OEL may be reached during unventilated welding of galvanised steel.

There are no simple, direct methods for measuring zinc fume in the field. Monitoring for zinc fume or zinc oxide particulate requires personal sampling with a respirable dust sampler. Refer to AS 2985 (Standards Australia 2009a) or MDHS 14/4 (HSE 2014a) for details. Monitoring for zinc dust and zinc chloride requires personal sampling with the IOM inhalable dust sampler or equivalent. Consult AS 3640 (Standards Australia 2009b) for details. Measurement requires laboratory assessment, usually by atomic absorption spectroscopy. Sampling in a hot-dip galvanising works requires special efforts to protect the sampling head from direct splashes of hot metal, which will burn holes in the filter. In this case, the use of a 7-hole inhalable dust sampler may be more beneficial.

9.13.4 Controls

Processes using zinc should be controlled by enclosure, adequate LEV or other engineering measures. The metal and oxide fumes are not life threatening, and RPE with a protection factor suitable for metal fumes can often be considered satisfactory when other controls are impractical. For welders working regularly on galvanised steels, airline RPE rather than filter RPE should be considered if work practices consistently place them in the plume of metal fume originating from the welding task. Special RPE products for welding are available from several manufacturers. The Welding Technology Institute of Australia supplies a good publication titled *Fume Minimisation Guidelines* (Weld Australia 2021).

Those who work with zinc and zinc products should be fully instructed in the nature of the associated health hazards and the control procedures required to prevent exposure.

9.14 MINOR METALS OF WORKPLACE CONCERN

The preceding sections have dealt with the more common hazardous metals the H&S practitioner is likely to encounter, but there are some others. They are classed as metals of minor significance because of their relatively rare use and occurrence, not because of their toxicology, which may be both complex and fascinating. Their classification as 'minor' also stems from the fact that they are seldom the subjects of extensive workplace health or hygiene investigations.

For example, boron is a toxic metalloid, but its use as a high-energy neutron shield in nuclear reactors is of little consequence to the average H&S practitioner, as there are few reactors in most countries. On the other hand, compounds of boron find wide use in ceramic glazes, fireproofing materials, low-grade insecticides, organic chemical synthesis, printing and painting. So there may well be good reasons to consider these materials as other than 'minor'.

H&S practitioners seeking more extensive information on metals of minor significance should consult some of the references at the end of this chapter.

Details on sampling procedures for these metals are not presented here, but the NIOSH *Manual of Analytical Methods* (2020) covers most of them. The H&S practitioner should contact a specialist hygiene laboratory if measurement of these metals is required. There are no simple techniques available to the H&S practitioner for assessing any of them in the field.

9.14.1 Aluminium

Aluminium is a relative newcomer to the family of commercial metals. In its mineral form, bauxite, aluminium is the most abundant metal in the earth's crust. Mined bauxite is refined into alumina (aluminium oxide) using a chemical refining process (Bayer), whereby finely ground bauxite is digested in a hot caustic soda solution, then clarified, precipitated and calcined. The alumina is then smelted into aluminium using electrochemical reduction, whereby the alumina is dissolved in molten cryolite in cells through which a direct electrical current is passed via carbon cathodes and anodes. The molten metal formed in each cell is drawn off at regular intervals for casting into various shapes. Aluminium is often alloyed with small amounts of copper, magnesium, silicon, manganese and other elements to impart a variety of useful properties.

Aluminium oxide is used as an abrasive, in refractory material and for electronic applications. The uses of aluminium are numerous: in automobile engines, transmissions, bodies and suspension components, shipbuilding, aircraft manufacture, home products such as doors, window frames, roofing and insulation, electrical wire and transmission cable, packaging material such as aluminium foil, drink cans and wrap, aluminium coatings for telescope mirrors, and decorative paper, packages and toys. Aluminium compounds and materials also have a wide variety of uses in the production of glass, ceramics, rubber, wood preservatives, pharmaceuticals and waterproofing textiles. Salts of aluminium include alum, which is used in water treatment, and natural aluminium minerals (e.g. bentonite and zeolite), which are used in water purification, sugar refining, brewing and the paper industry.

Aluminium metal readily oxidises to aluminium oxide, forming a protective layer against further ordinary corrosion. Powder and flake aluminium are flammable and can form explosive mixtures in air, especially when treated to reduce surface oxidation (e.g. pyro powders). The toxicity of aluminium is still controversial. The most common OEL for aluminium is 10 mg/m^3, ranging from 0.5 to 10 mg/m^3 (GESTIS 2023). ACGIH® (2023) has set a TLV®-TWA of 1 mg/m^3 (as a respirable fraction) for aluminium metal and insoluble compounds to protect from lower respiratory tract irritation, pneumoconiosis and neurotoxicity. There appears to be no serious adverse effect on respiratory health associated with exposure to bauxite in open-cut bauxite mines. There is a belief among the general public that aluminium plays some role in Alzheimer's disease, although research findings so far do not support this notion. Of more importance for those working in the aluminium electrolytic smelting industry is the potential for a condition known as 'pot-room asthma', characterised by cough with dyspnoea, wheezing and chest tightness, usually occurring within the first year of exposure to fumes from smelting cells or pots. While peak concentrations of fluoride in air have been implicated as a causative agent, the actual cause of this asthma is yet to be clearly determined. For those working in aluminium smelters, the production of carbon cathodes and anodes can cause exposure to polycyclic aromatic hydrocarbons (PAHs) from coal-tar pitch volatiles (CTPVs). PAHs are carcinogenic, and thus of primary concern.

General control procedures for smelting aluminium should principally be aimed at preventing pot-room fumes and CTPVs from entering the workplace atmosphere. Expert opinion and advice are needed to accurately assess exposures in an aluminium smelter.

9.14.2 Antimony

Metallic antimony finds use in alloys with lead (e.g. battery grids), in cable sheaths, pewter, ammunitions and some solders. Salts of antimony are used in paints, rubber, glass and ceramics.

One regular application of antimony trioxide is as a fire retardant in weatherproofing/insulation membranes for housing construction. The toxic effects of antimony resemble those of arsenic and include irritation of mucous membranes, gastrointestinal symptoms, sores in the mouth and skin lesions. The hydride of antimony, stibine, is an extremely toxic haemolytic agent. Antimony trisulfide is also very toxic and is reported to cause heart failure. This effect on cardiac muscle may be shared by other antimony compounds as well.

The use of antimony in lead-acid battery grids leads to a small possibility of stibine production during battery charging. Good ventilation procedures should be followed, particularly in large battery facilities.

Generally, however, antimony does not find significant use in industry. The OEL-TWA is commonly 0.5 mg/m^3 for all forms, but it should be noted that the production process for antimony trioxide is suspected of being carcinogenic.

General control procedures for working with antimony should be aimed at preventing antimony compounds from entering the workplace atmosphere. Expert opinion and advice should be sought if antimony exposure is a possibility.

9.14.3 Beryllium

Beryllium finds its principal use as an alloying agent in steel, copper, magnesium and aluminium. Beryllium oxide is also a component of ceramics. Its use as a phosphor in fluorescent tubes has long since been abandoned. Its major application today is in the nuclear industry, aerospace products and high-stress, low-strain metals and alloys.

Beryllium is cited by SWA (2021) as being a category 1B carcinogen (presumed human carcinogen), while the ACGIH® (2023) classify it as a confirmed human carcinogen, causing lung cancer. IARC (2023) considers that beryllium and beryllium compounds are carcinogenic to humans and gives these compounds a Group 1 classification. Mining of the beryl ore is not associated with lung disease. Beryllium poisoning is related to the very toxic effects of beryllium dust, which affects a number of organs. Delayed effects from material deposited in the body can occur. Sensitisation or development of a beryllium-specific, cell-mediated immune response arises in 2–19 per cent of exposed individuals. Sensitisation usually precedes the development of chronic beryllium disease, or berylliosis, characterised by a Type IV, delayed-hypersensitivity, cell-mediated immunity. Relatively small amounts of inhaled material can induce acute effects. Expert diagnostic opinion, with particular emphasis on complete occupational history, is extremely important in establishing beryllium-induced disease. Medical control in beryllium-using industries rests mostly on regular health monitoring (SWA 2020), including chest X-rays and beryllium lymphocyte proliferation testing (BeLPT).

Most of the Western world has set an exposure limit for beryllium and inorganic compounds as per the GESTIS (2023) database for international limit values. The typical inhalable TWA for beryllium is 0.0002 mg/m^3 (range of 0.00002–0.002 mg/m^3). Some jurisdictions have a typical STEL value of 0.01 mg/m^3 (range of 0.00014–0.01 mg/m^3). The TWA value is set to prevent beryllium sensitisation and chronic beryllium disease (berylliosis). Thus any grinding, polishing or machining process on beryllium-containing metals or ceramics should be considered very carefully. Good occupational hygiene practice will be required to keep the inhalation risks of working with beryllium to acceptable levels. Industrial or research operations using beryllium should be assessed carefully.

9.14.4 Boron

Boron is an essential element for plants and animals, including humans. Boron compounds are widely used in applications from household cleaning products (e.g. detergents and bleaches) to boron-fibre technology. Borax and boric acid are used in smelting, glazes for ceramic ware, and production of glass and glass-related products (e.g. insulation and textile fibreglass, Pyrex), and borax is still used in fireproofing of pulped-cellulose insulation, low-activity insecticide mixtures and leather tanning. These compounds do not pose significant hazards in normal use, and consequently the OEL-TWA ranges from 1 to 10 mg/m^3, depending on the type of boron compound (GESTIS 2023). The OEL is generally set to protect for irritation of the mucous membranes in exposed workers. H&S practitioners need to ensure that neither inhalation nor ingestion of dusts can occur.

Several other boron compounds, including boron trifluoride and boron trichloride, find use in chemical synthesis and organic chemical reactions.

There are several boron hydrides (e.g. diborane, pentaborane, decaborane and others), which are highly toxic. They are used mainly in high-energy rocket 'zip' fuels and in pharmaceutical and rubber vulcanising. They are very irritating to skin and mucous membranes and induce headache, chills, dizziness and weakness. These compounds also present extreme fire and explosion risks. Good occupational-hygiene practice (engineering control, personal protective equipment) is required to provide adequate control of these substances.

9.14.5 Cobalt

Cobalt, being a relatively rare element, is also relatively rare in the workplace. The metal finds use in Al–Ni–Co permanent magnets, as a binder in hard-metal cutting tools, in some electrical alloys and in some dental metal alloys. Some cobalt salts, such as cobalt blue and cobaltous chloride, have been used in glass and ceramic enamels for colouring. Cobalt is an essential element in the formation of vitamin B12.

Cobalt metal workers have developed a pulmonary disease often known as 'hard metal disease', though exposure to other materials might be implicated also. The effects of cobalt poisoning include diarrhoea, loss of appetite, hypothermia and possible death. Inhalation of cobalt dust can cause pulmonary oedema in animals and so should be avoided. An allergic dermatitis resulting from exposure to minute amounts has been seen in some workers in the building, metalwork, pottery, leather and textile industries.

Exposure to cobalt by inhalation should be limited, preferably by isolation, LEV, or other engineering controls. This is particularly important in grinding operations that produce fine dust. Control of airborne dusts should also reduce the need for skin protection. Additional skin protection, including barrier creams, may prove necessary in situations where such dusts are difficult to control. RPE must be of a high standard for mechanically generated dusts, preferably an airline respirator or high-efficiency filter respirator, depending on the protection factor required.

Most of the Western world has set an exposure limit for cobalt and its compounds and cobalt carbonyl as per the GESTIS (2023) database for international limit values. For cobalt and its compounds, most are set at a TWA value of 0.02 mg/m^3 (range of 0.01–0.1 mg/m^3). For cobalt carbonyl, all are set at a TWA value of 0.01 mg/m^3.

9.14.6 Copper

Another essential element, copper finds common use in electrical wiring, alloys of brass, bronze or Monel, plumbing services and cookware. Copper is also used in insecticides and as an algaecide and a bactericide and in electroplating. Industrial exposure to copper fume is not common – copper melts at about 2350°C – though the fume can give rise to a metal-fume fever like that caused by zinc. Copper salts, copper carbonate and copper sulfate are all relatively safe to use, provided proper precautions are taken to prevent inhalation or accidental ingestion of dusts.

Ingestion of copper has occurred accidentally when acidic foodstuffs have been in contact with copper vessels (e.g. fruit juice, moonshine liquor from copper stills). Copper exposure presents no significant problem in industry, other than for workers with Wilson's disease, a rare genetic inability to excrete excess copper.

The OEL-TWAs for copper fume and copper dusts are typically 0.2 mg/m^3 (range of 0.01–1 mg/m^3) and 1 mg/m^3 (range of 0.01–2 mg/m^3), respectively (GESTIS 2023).

9.14.7 Lithium

Lithium is another relative newcomer to the family of commercial metals. Commonly known uses are in pharmaceuticals as lithium carbonate or lithium citrate, used to treat manic-depressive illness (bipolar disorder), and in rechargeable batteries as elemental lithium. Lithium compounds are also used: in polymerisation catalysts for the polyolefin plastics industry; in manufacturing of high-strength glass and glass-ceramics; in production of organometallic alkyl and aryl lithium compounds; in production of high-strength, low-density aluminium alloys for the aircraft industry; in extremely tough, low-density alloys with aluminium and magnesium used for armour plate and aerospace components; and as a chemical intermediate in organic syntheses. Mining and processing of lithium ore (e.g. spodumene) create the greatest exposure hazards.

Lithium does not have a known biological use and does not appear to be an essential element for life, but it may influence metabolism. Lithium in dust may result in serious injury to the eyes, nasal passages, upper airways and lungs due to the formation of LiOH, a strong alkali that is highly corrosive. Due to the use of lithium salts in pharmaceuticals, there is good information on the toxicity of these salts when ingested; in fact, most information on lithium toxicity is derived from its pharmaceutical use. Ingested in excessive amounts, lithium primarily affects the gastrointestinal tract, the CNS and the kidneys (AIOH 2018b).

International studies have shown that systemic adverse effects due to lithium (e.g. increased urination, shakiness of the hands and increased thirst), as sometimes seen when used therapeutically to treat psychiatric illnesses, are unlikely to occur at lower exposure levels of lithium and lithium compounds in the workplace. The key effect of breathing in lithium dust or alkaline lithium compounds is irritation of the respiratory tract and mucous membranes (AIOH 2018b).

Prevention of the effects of lithium depends mostly on preventing inhalation or ingestion of lithium-containing dusts. LEV and other engineering controls should be adequate to maintain exposures within safe levels. Typical OELs for lithium hydride are a STEL of 0.02 mg/m^3 (as inhalable fraction), while the ACGIH® (2023) recommend a ceiling TLV® of 0.05 mg/m^3 (as inhalable fraction) to prevent irritation of the eyes and respiratory tract. The AIOH (2018b) recommend a STEL guidance value of 0.02 mg/m^3 (as inhalable fraction) to prevent worker adverse exposure to lithium and its hydride and hydroxide compounds, as well as lithium carbonate. RPE with fit testing and training should be used where other controls are inadequate.

9.14.8 Manganese

Manganese finds its major use in steel making, but it is also used in the production of aluminium alloys and cast iron, in alkaline manganese dry batteries, as manganese dioxide for brick colouring, as an oxidising agent in dyes and chemicals, as a chemical catalyst, as manganates and permanganates for disinfecting and bleaching, and as a fungicide. Mining and smelting of manganese ore (pyrolusite) create the greatest exposure hazards. Subsequent environmental contamination around smelters can also occur. Other small exposures occur in welding with manganese-containing rods and in (dark) paint pigments. The organic manganese compound methylcyclopentadienyl manganese tricarbonyl (MMT) has been used as an additive to increase the octane rating in petrol.

Manganese is an essential trace element, but in excessive amounts, it can cause neurological disorders involving the central nervous system. Symptoms of manganism range from apathy, anorexia and mental excitement to speech disturbance, clumsiness and a stone-faced appearance. In the established phase of the disease, staggering gait, muscular rigidity (e.g. finger or hand deformation), spasmodic laughter or tremors may occur. Manganese-poisoning victims may be cripplingly debilitated but otherwise well. There was concern by some that MMT might contribute to neurological effects like those caused by tetraethyl lead in petrol, but this has not been eventuated.

Metal-fume fever can follow inhalation of fine manganese fume, and a manganese-induced pneumonia can occur. Manganese poisoning has also been associated with a decline in lung function and reduced fertility in male workers.

Prevention of manganese-related disease depends mostly on preventing inhalation of manganese-containing dusts and fumes. Manganese can also be accidentally ingested. LEV and other engineering controls should be adequate to maintain exposures within safe levels. The typical OEL-TWA for manganese fume, dust and compounds is 0.2 mg/m^3 (range of 0.02 to 1 mg/m^3) (GESTIS 2023). The ACGIH® (2023) recommend a TLV-TWA of 0.1 mg/m^3 (as inhalable fraction) and 0.02 mg/m^3 (as respirable fraction), levels set to protect against central nervous system effects. The use of RPE, subject to achievement of appropriate protection factors, could be appropriate, if conditions permit wearing it.

Medical supervision of those exposed occupationally to manganese dust is advisable. Manganism exhibits three distinct phases, so detection of any slight neurological abnormality may allow workers to be removed from exposure before symptoms worsen.

9.14.9 Selenium

Selenium is a non-metallic element that can occur in a metal-like form. This is obtained as a by-product of refining copper (and sometimes zinc) ore, so exposure to selenium fume can occur during the smelting and refining of these ores. Some of its forms are volatile and so more difficult to contain. Occupational exposure occurs primarily via inhalation, although in some cases it can be by direct skin contact. Selenium is used in the manufacture of glass, pigments, ceramics and semiconductors; in photocopiers and cameras; in photoelectric cells, steel, rectifiers in electronic equipment; as catalysts; and in the vulcanising of rubber. Selenium compounds are also used in the treatment of several animal and human diseases – it is a common ingredient in anti-dandruff shampoos. It is an essential trace element for animals and humans but can concentrate to toxic levels in the body, causing selenosis.

The acute toxicities of inorganic selenium compounds vary greatly. Hydrogen selenide, selenium oxychloride, selenium dioxide and selenium hexafluoride are highly poisonous, while the element and sulfides are much less toxic. Selenium dust, selenium dioxide and hydrogen selenide are more likely to be encountered in the workplace. The chronic effects of nearly all forms of selenium in humans appear to be similar: depression, languor, nervousness, dermatitis, upset stomach, giddiness, a garlic odour of breath and sweat, excessive dental caries and in extreme cases loss of fingernails and hair. Most knowledge of selenium's toxic effects is derived from clinical toxicology and the incidence of selenium poisoning because of ingestion of seleniferous grains and other food; there have been no reports of disabling chronic disease or death from industrial exposures. The typical OEL-TWA value is 0.1 mg/m^3 for selenium compounds (excluding hydrogen selenide), which is half that of the ACGIH® (2023) and OSHA (2021) 0.2 mg/m^3 limit, set to minimise irritation of the eye and upper respiratory tract as well as systemic effects such as headache, garlic breath and skin rashes. The OEL-TWA standard for hydrogen selenide is typically 0.05 ppm or 0.16 mg/m^3 (GESTIS 2023).

General control procedures for working with selenium should be aimed at preventing selenium compounds from entering the workplace atmosphere. Expert opinion and advice should be sought if selenium is encountered in a workplace.

9.14.10 Thallium

This metal was once widely used in the preparation of salts for rodenticides because it is undetectable by taste or smell, but it is no longer readily available in developed countries. Thallium is also used in special optical lenses for scientific equipment (infrared cells), glass tinting, semiconductors and photoelectric cells. It is also alloyed with lead, zinc, silver and antimony to enhance their resistance to corrosion. A rare application of thallium is in Clerici's solution, thallium malonate, which has a high specific gravity and is used in heavy-mineral test separations in the mineral sands industry. Thallium may also be encountered in the production of cement and in the handling of pyrites and flue dusts.

Most deaths due to thallium are from poisoning, either accidental (when poisoned grain is mixed in with grain to be milled) or deliberate (suicidal or homicidal). Thallium is also absorbed through the skin. Symptoms of severe poisoning include swelling, joint pain, vomiting, mental confusion and loss of hair.

The effects of exposure during the preparation of thallium salts have been milder, thanks mostly to the care taken in handling the salts, but they are usually more gradual than acute poisonings. A worker's occupational history is extremely important in diagnosis.

The typical OEL-TWA for thallium and compounds is 0.1 mg/m^3 (as inhalable fraction) (range 0.02–0.1 mg/m^3). The ACGIH® (2023) TLV®-TWA is 0.02 mg/m^3 (as inhalable fraction), to protect for gastrointestinal and neurological disturbances, peripheral neuropathy and alopecia in exposed workers. Monitoring of urine may be a useful measure for those regularly involved in the preparation of thallium compounds.

If thallium must be used, processes must be isolated and strict precautions taken to prevent the dispersal of airborne dusts in the workplace (LEV) and skin contact (personal protective equipment). Hand-washing facilities and separate eating facilities are mandatory and eating and smoking in the workplace and the wearing of work clothes to and from home should be forbidden.

REFERENCES

Australian Institute of Occupational Hygienists 2018a, *AIOH Position Paper: Inorganic Lead – Potential for Occupational Health Issues*, AIOH, Melbourne, www.aioh.org.au/education/publications/#position [26 December 2023].

Australian Institute of Occupational Hygienists 2016, *AIOH Position Paper: Nickel and Its Compounds and Their Potential for Occupational Health Issues*, AIOH, Melbourne, www.aioh.org.au/education/publications/#position [26 December 2023].

Australian Institute of Occupational Hygienists 2018b, *AIOH Position Paper: Lithium and Its Hydride and Hydroxide Compounds – Potential for Occupational Health Issues*, AIOH, Melbourne, www.aioh.org.au/education/publications/#position [26 December 2023].

Australian Institute of Occupational Hygienists 2023, *AIOH Position Paper: Chromium VI – Potential for Occupational Health Issues*, AIOH, Melbourne, www.aioh.org.au/education/publications/#position [26 December 2023].

American Conference of Governmental Industrial Hygienists (ACGIH) 2001, *Arsenic and Soluble Inorganic Compounds – Recommended BEI®*, ACGIH®, Cincinnati, OH.

American Conference of Governmental Industrial Hygienists (ACGIH) 2023, *2023 TLVs® and BEIs®*, ACGIH®, Cincinnati, OH.

Australian Pesticides and Veterinary Medicines Authority 2005, *The Reconsideration of Registrations of Arsenic Timber Treatment Products (CCA and Arsenic Trioxide) and Their Associated Labels: Report of Review Findings and Regulatory Outcomes Summary Report*, APVMA, Canberra, apvma.gov.au/sites/default/files/publication/14316-arsenic-summary.pdf [26 December 2023].

Bell, T. 2020, *A Brief History of Lead Properties, Uses and Characteristics*, ThoughtCo, www.thoughtco.com/metal-profile-lead-2340140 [26 December 2023].

GESTIS 2023, *International Limit Values for Chemical Agents (Occupational Exposure Limits, OELs)*, www.dguv.de/ifa/gestis/gestis-internationale-grenzwerte-fuer-chemische-substanzen-limit-values-for-chemical-agents/index-2.jsp [26 December 2023].

Health and Safety Executive (HSE) 2014a, *General Methods for Sampling and Gravimetric Analysis of Respirable, Thoracic and Inhalable Aerosols*, MDHS 14/4, HSE, HSE, London, UK, www.hse.gov.uk/pubns/mdhs/pdfs/mdhs14-4.pdf [26 December 2023].

Health and Safety Executive (HSE) 2014b, *Hexavalent Chromium in Chromium Plating Mists*, MDHS 52/4, HSE, London, UK, www.hse.gov.uk/pubns/mdhs/pdfs/mdhs52-4.pdf [26 December 2023].

Health and Safety Executive (HSE) 2015, *Metals and Metalloids in Air by X-ray Fluorescence Spectrometry*, MDHS 91/2, HSE, London, UK, www.hse.gov.uk/pubns/mdhs/pdfs/mdhs91-2.pdf [26 December 2023].

Health and Safety Executive (HSE) 2020, *Workplace Exposure Limits*, EH40/2005, 4th edn, HSE, London, UK, www.hse.gov.uk/pubns/books/eh40.htm [26 December 2023].

IARC 2023, *Agents Classified by the IARC Monographs, Volumes 1–135*, monographs.iarc.who.int/list-of-classifications [26 December 2023].

International Organization for Standardization 2005, *Workplace Air – Determination of Hexavalent Chromium in Airborne Particulate Matter – Method by Ion Chromatography and Spectrophotometric Measurement Using Diphenyl Carbazide*, ISO 16740:2005, ISO, Geneva, www.iso.org/standard/30432.html [26 December 2023].

International Organization for Standardization 2015, *Workplace Air – Determination of Mercury and Inorganic Mercury Compounds – Method by Cold-Vapour Atomic Absorption Spectrometry or Atomic Fluorescence Spectrometry*, ISO 17733:2015, ISO, Geneva, www.iso.org/standard/67798.html [26 December 2023].

International Organization for Standardization 2020, *Workplace Air – Determination of Metals and Metalloids in Airborne Particulate Matter by Inductively Coupled Plasma Atomic Emission Spectrometry – Part 3: Analysis*, ISO 15202:2020, ISO, Geneva, www.iso.org/standard/74937.html [26 December 2023].

Minamata Convention on Mercury 2023, mercuryconvention.org/en [3 May 2023].

National Institute for Occupational Safety and Health 2020, *NIOSH Manual of Analytical Methods (NMAM®)*, 5th edn, Centers for Disease Control and Prevention, Atlanta, GA, www.cdc.gov/niosh/nmam/default.html [26 December 2023].

Occupational Safety and Health Administration (OSHA) 2021, *Permissible Exposure Limits – Annotated Tables*, OSHA, Washington, DC, www.osha.gov/annotated-pels [26 December 2023].

Safe Work Australia (SWA) 2020, *Health Monitoring Guides*, SWQ, Canberra, www.safeworkaustralia.gov.au/safety-topic/managing-health-and-safety/health-monitoring/resources [26 December 2023].

Safe Work Australia (SWA) 2021, *Hazardous Chemicals Information System (HCIS)*, SWQ, Canberra, hcis.safeworkaustralia.gov.au [26 December 2023].

Safe Work Australia (SWA) 2023, *Model Work Health and Safety Regulations*, SWQ, Canberra, www.safeworkaustralia.gov.au/doc/model-whs-regulations [26 December 2023].

Standards Australia 2009a, *Workplace Atmospheres—Method for Sampling and Gravimetric Determination of Respirable Dust*, AS 2985:2009, SAI Global, Sydney, https://www.standards.org.au/access-standards

Standards Australia 2009b, *Workplace Atmospheres—Method for Sampling and Gravimetric Determination of Inhalable Dust*, AS 3640:2009, SAI Global, Sydney, https://www.standards.org.au/access-standards

Weld Australia, 2021, *Fume Minimisation Guidelines—welding, cutting, brazing and soldering*, Technical Guidance Note TGN-SW01,, https://portal.weldaustralia.com.au/resources/?search=Fume+Minimisation+Guidelines [26 December 2023].

World Health Organization (WHO), 2020, *Brief Guide to Analytical Methods for Measuring Lead in Blood*, 2nd edn, World Health Organization (WHO), Geneva, Switzerland, www.who.int/publications/i/item/9789240009776 [2 January 2024].

10 Gases and Vapours

Aleks Todorovic and Dr Michael Logan

10.1 INTRODUCTION

In occupational work environments, a range of atmospheric contaminants, including gases and vapours, can be found. Although many gases are essential to life, the presence of hazardous gases and vapours in the air poses a significant risk to the health and well-being of workers.

This chapter aims to focus on several common gases and vapours frequently encountered in industrial settings, examining their implications for health and hygiene within the workplace. The toxicological review will be limited to specific gases, while also exploring mechanisms to control these compounds and reduce health risks. However, it is essential to note that numerous other gases and vapours can be harmful to health, and readers are advised to consult more comprehensive resources for a deeper understanding.

Keeping up with the latest developments is crucial for health and safety practitioners, and today's professionals have an assortment of passive, active, and real-time detection devices at their disposal to measure the workplace environment. In this chapter, we will investigate various methods used to monitor gases and vapours, exploring the distinctions between different monitoring techniques and their inherent limitations.

Furthermore, the chapter will discuss potential hazards that confined spaces present, emphasising the importance of comprehending and addressing these risks effectively. Lastly, we will discuss how health and safety practitioners can harness the latest emerging technologies to monitor gases and vapours and analyse data efficiently.

Readers of this chapter will gain valuable insights to make well-informed decisions in handling gases and vapours in their work settings, promoting a safer and healthier work environment for all.

10.2 GASES

Gases ranging from simple gases such as helium to complex hydrocarbon compounds are found in the petrochemical industry. A gas is defined as a substance which at (i) at 50°C has a vapour pressure greater than 300 kPa (absolute) or (ii) completely gaseous at 20°C at a standard atmosphere of 101.3 kPa (ADGC 2022). Gases may be heavier, lighter or have the same density as air and may be colourless and odourless.

Many can be stored as pressurised liquids until released for use. As gases will fill any available volume, if not controlled, they may enter and contaminate the workplace environment. Gases used in workplace processes include:

- chlorine in water treatment plants;
- nitrous oxide (laughing gas) in anaesthesia;
- ammonia in refrigeration plants;
- oxygen from liquid oxygen sources;
- phosphine/methyl bromide in grain fumigation; and
- ethylene oxide in sterilisation machines in hospitals.

DOI: 10.1201/9781032645841-10

Other gaseous contaminants may arise as by-products of industrial processes. Typically, this occurs where some sort of chemical reaction has taken place, or where the gas is produced in the breakdown of a complex chemical—for example:

- carbon monoxide from incomplete burning of natural gas in ovens and kilns;
- oxides of nitrogen from diesel exhausts;
- ozone from photocopying or some electric arc welding machines;
- formaldehyde off-gassing from particle board; and
- hydrogen sulfide from sewers and wastewater treatment plants.

10.3 VAPOURS

A vapour is a gaseous form of a substance or mixture released from its liquid or solid state. In the workplace, thousands of organic chemicals used as solvents, adhesives, paints, chemical reactants, catalysts, sterilants or processing aids produce vapours through natural evaporation, heating or spraying.

Organic chemicals in the form of solvents are by far the largest source of vapours in workplaces. These organic compounds are often chosen as solvents due to two properties—they will dissolve a material and will readily evaporate. This last property leads to many exposure problems. Some typical examples of organic solvents are acetone, chloroform, ethanol, chlorofluorocarbons (CFCs), hexane, perchloroethylene, benzene, toluene and xylene. Organic solvent-based chemicals are found in the factory, the farm, the office and the home.

In industries such as the petrochemical industry, elements such as mercury can produce hazardous vapours while similarly hazardous vapours such as benzene and toluene may also be produced.

A few materials sublime—that is, convert from the solid directly to the vapour. Examples include naphthalene (mothballs) and paradichlorobenzene (deodorant).

Vapour pressure is defined as the saturation pressure above a solid or liquid substance (OECD 104:2006).

Vapour density is the density of a gas/vapour compared to the density of air. Other factors to consider are temperature, pressure and vapour's volume and surface area.

10.4 WARNING SIGNS AND INDICATORS

10.4.1 Odour

Gases and vapours are mostly invisible. Many have strong and characteristic odours, which give warning of their presence in the workplace, but others have no warning odour (e.g., carbon monoxide), and with others, harmful effects may be caused at concentrations well below the odour threshold. With some substances, warning odorants are added to prevent inadvertent exposure and/or to permit detection of the gas's presence. For example, the odorant ethyl mercaptan is used in liquefied petroleum gas (LPG) and natural gas supplies. Lastly, individuals vary greatly in their ability to detect and recognise odours, and even in the same person, this ability may vary from day to day. For example, the range of odour detection of methyl ethyl ketone (MEK) lies between 0.07 and 339 ppm. In other words, while odour can be useful in detection, there are several limitations to its use. Odour is often a good warning indicator for bacterial decay. However, odour cannot easily indicate the degree of exposure; for some gases (e.g., hydrogen sulfide), higher concentrations may cause olfactory fatigue and the contaminant can no longer be smelled even when the concentration of the gas may be at lethal levels. Although odour is often mistakenly associated with harm, it is an **unreliable indicator of harmful conditions** and is not a measure of toxicity. Some gases are deadly yet do not smell, while others are odorous at levels far below those that will cause harm—they just smell unpleasant.

TABLE 10.1
Comparison of Odour Thresholds and WESs for Some Common Workplace Gases and Vapours

Substance	Odour Threshold (ppm)	Australian WES TWA (2022) (ppm)
Acetone	0.4–11 745	500
Ammonia	0.043–60.3	25
Benzene	0.47–313	1
Ethylene oxide (Oxirane)	0.82–690	1
Hydrogen sulfide	0.00004–1.4	10
MEK (methyl ethyl ketone)	0.07–339	150
Ozone	0.0031–0.25	0.1 Peak
Styrene (monomer)	0.0028–61	50
Toluene	0.021–157	50
1,1,1-Trichloroethane	0.97–715	100
Xylene	0.012–316	80

Adapted from: Murnane et al. 2013, pp. 22–45; SWA (2022).

For many volatile materials, the odour threshold may be one-tenth to one-thousandth of the Workplace Exposure Standard (WES). Table 10.1 shows the relation between approximate odour thresholds and WES for several common workplace gases and organic vapours.

10.4.2 Irritation

Some gases and vapours may reveal their presence by various irritating effects:

- respiratory irritation, coughing and asthma;
- lachrymatory action on the eye (tearing);
- cloudy vision (at high concentrations) or other visual disturbances;
- acidic and alkali taste;
- metallic taste (organometallic compounds).

Acidic gases, including chlorine and sulfur dioxide, may be evident because of respiratory irritation rather than smell. In some workers, formaldehyde may cause eye irritation before it can be detected by smell. Investigation of the workplace may reveal a range of processes that produce gases or vapours, and the H&S practitioner will need to distinguish between those which have reliable warning properties and those which do not.

10.5 OCCUPATIONAL HEALTH ASPECTS OF GASES AND VAPOURS

For health and safety purposes, gases and vapours can be classified into three groups:

- irritants;
- asphyxiants; and
- those with miscellaneous effects.

Both irritants and asphyxiants tend to give rise to relatively acute responses, leading to rapidly observed effects. The miscellaneous effects may be acute or chronic.

10.5.1 Irritant Gases and Vapours

Irritants cause inflammation of the tissues exposed to them. Symptoms of exposure can range from mild irritation of the mucous membranes (eyes, nose and throat) to severe damage of the lung (e.g., by ammonia, chlorine, phosgene, formaldehyde and nitrogen dioxide).

10.5.2 Asphyxiants

Asphyxiant gases fall into two groups:

- **Simple asphyxiants**, a vapour or a gas that can cause unconsciousness or death by suffocation (lack of oxygen). Most simple asphyxiants are harmful to the body only when they become so concentrated that they reduce oxygen in the air (normally 21%) to dangerous levels (18% and under). Asphyxiation is one of the principal potential hazards of working in confined spaces.
- **Chemical asphyxiants** interfere with the bodies' ability to take up and transport oxygen such as exposure to low concentrations of carbon monoxide, hydrogen sulfide, or hydrogen cyanide.

10.5.3 Gases and Vapours with Miscellaneous Effects

Many gases and nearly all vapours from solvents fall into this category. Effects include acute effects, for example, on the central nervous system (CNS), and chronic toxic effects on many different organs of the body. Some typical examples of miscellaneous effects caused by gases and vapours that may occur in industrial workplaces are shown in Table 10.2.

TABLE 10.2
Gas or Vapour Hazards Demonstrating Various Health Effects

Gas or Vapour	Health Effect
Benzene	Leukaemia
Carbon disulfide	Cardiac disease
Coal tar pitch volatiles	Skin sensitisation, lung cancer
Ethyl glycol monoethyl ether	Fetotoxic effects
Fluorocarbons	Cardiac arrhythmias
Helium	Vocal changes
Methyl bromide	Cardiac effects
Mercury	Brain damage, kidney damage, gingivitis, tremors and erethism
n-Hexane	Peripheral-nerve neuropathy
Nitroglycerine	Vasodilation and decreasing blood pressure
Nitrous oxide	Analgesia
Oxygen (deficiency)	Brain disturbance and CNS effects
Oxygen (excess)	Pulmonary inflammation and lung oedema
Toluene	Headache, confusion and loss of memory
Trichloroethylene	Psychoactive effects
Vinyl chloride	Angiosarcoma of the liver

Adapted from: Grantham, 1992, p. 159, Table 7.2.

10.6 UNDERSTANDING EXPOSURE MEASUREMENT IN THE WORKPLACE

The H&S practitioner needs to understand what exposure is occurring in the workplace. While the exposure standards are listed as a single value for an eight-hour average and/or a 15-minute average, or a peak limitation, in reality, the exposure changes continually throughout a work shift. Exposures to a particular contaminant are averaged over the period of the shift to give the daily exposure of the worker as a single value, which is then compared with the exposure standard for that contaminant. Figure 10.1 shows the output from a data logger measuring at one-minute intervals the hydrogen sulfide (H_2S) exposure of a worker near the hatch of a tank. This highlights the changing concentrations that can occur over a work shift. Any instantaneous reading from a 'grab sample' will be only one result at one time in a changing environment. (A grab sample is an air sample collected over a short period, usually between one and five minutes.)

In the above case, if no further exposure occurs in the eight-hour shift, then the hydrogen sulfide TWA is only 0.2 ppm, well below the WES of 10 ppm. Yet clearly, during one short period, the instantaneous readings were well more than this standard.

Further information on the variation of exposure with work routine can be compiled using Video Exposure Monitoring (VEM). VEM combines video of a worker with continuous measurements, which are presented in graphic form on screen.

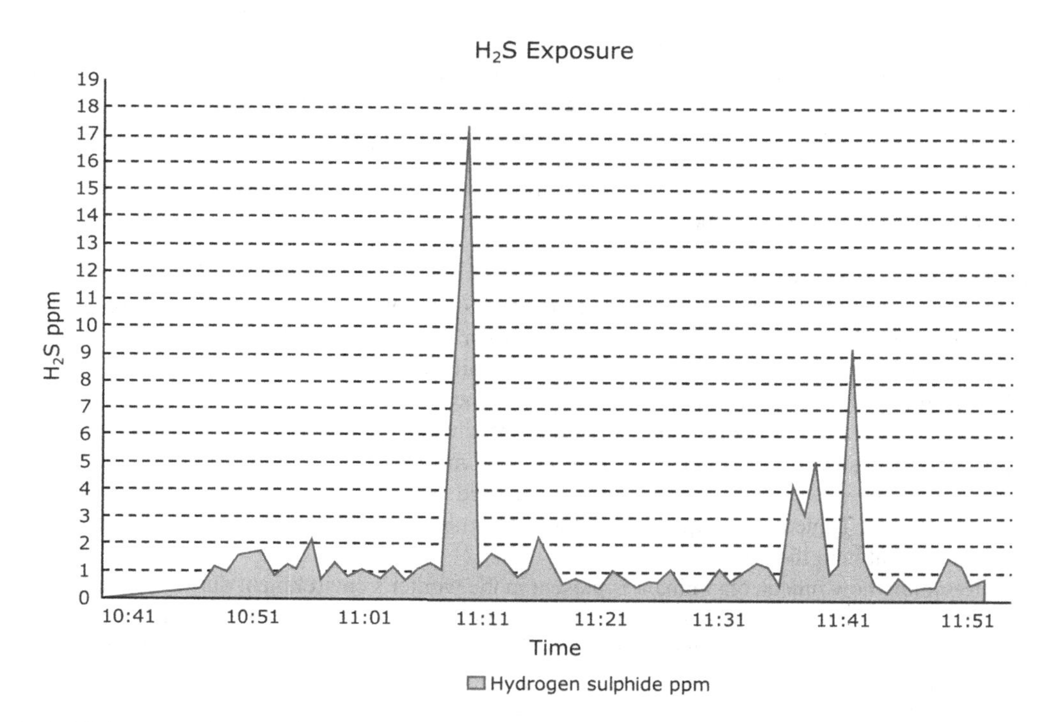

FIGURE 10.1 Hydrogen sulfide levels at a bitumen tank hatch.

Source: A. Todorovic.

10.7 MONITORING OF GASES AND VAPOURS

Many gases and organic vapours occur in various workplaces, and a variety of different techniques are required to assess them. It is essential prior to obtaining a sample the H&S practitioner write a sampling plan. In this context, sampling may include real-time detection and monitoring to a sample collected and analysed at a laboratory. There are many elements to a sampling plan, and they may include:

- sampling objectives;
- hazards, risk and risk control measures specific to that sampling environment;
- gas(es) of interest including likely concentration;
- worker exposure characteristics such as work, location, duration, PPE use and entry route of interest;
- sampling location; and
- sampling measurement method/technique, frequency and duration and suitability.

The information is summarised in a sampling plan and reviewed to ensure that it addresses the sampling objectives, the gases of interest can be sampled, and the plan is implemented safely.

The practitioner should pay particular attention to the nature of the gas or vapour when developing the sampling plan. For example: mists or aerosols change the characteristics from a gas phase to a finely dispersed liquid in the atmosphere.

While some gases and vapours may be dissolved in the moisture of the upper airways, others can reach the alveolar regions of the lung, from where they may pass into the bloodstream. While a considerable proportion of inhaled gas or vapour will be exhaled, exposure measurement is based on the total amount available for inhalation.

In the monitoring of gases and vapours, detection is identifying the presence of something and monitoring is the measurement of something as a function of time: its progression and quantity. Detection can be considered in two parts. They are:

- Identification: Which gas (or gases) or vapours—which contaminants—are present?
- Quantity: How much of the gaseous contaminant is present?

The answers to these two questions are affected by our prior knowledge about the workplace and the workers exposure context. The answer to the first question informs how the second question is answered. By themselves do not quantify the worker's exposure, but together with the workers exposure context then monitoring is applied to determine the workers exposure.

Determining what is in the air (identification) can be achieved by collecting an air sample and analysing it in the laboratory. Alternatively direct-reading instruments such as portable gas chromatographs, or instruments for detecting specific contaminants such as hydrogen sulfide or carbon monoxide can be used. These approaches can be used together to obtain a complete picture of what is in the air. Typically, the quantity of the contaminants is simultaneously determined when identifying what is in the air.

Determining how much contaminant is present in the worker's environment during their exposure requires monitoring the specific contaminant over a certain time. This requires either a direct-reading instrument or collection of the contaminant during the work period and subsequent laboratory analysis.

It is important for the H&S practitioner to also consider the suitability of the sampling equipment when selecting a detection or monitoring instrument. These include:

- measurement task;
- selectivity to the target gas or vapour and sensitivity to interfering gases and vapours;
- measurement range, limit of detection measurement error and measurement frequency;

- susceptibility to environmental factors;
- fitness for purpose, for example, weight, size and durability;
- data storage and analysis;
- training requirements for the reliable operation, maintenance and calibration;
- the total cost of purchase and operation, including calibration and maintenance;
- compliance with the performance requirements of appropriate national or local governmental regulations.

10.8 OPTIONS FOR MONITORING THE WORKPLACE AIR

Several techniques are available for gas and vapour monitoring:

- Conventional **monitoring of the air**. More than 98% of all gases and vapours are investigated in this fashion.
- **Monitoring the worker** for uptake of a gas or vapour by examining a biological index for the substance. This may be done by measuring the excretion of the substance or measuring a metabolite of the inhaled substance. The results are then compared to Biological Exposure Indices (BEIs).
- **Analysing the air** exhaled by the worker at some time after the end of a work shift (e.g., measuring the carbon monoxide in exhaled air).

10.9 MONITORING THE WORKPLACE AIR

For the present, emphasis will be placed on air monitoring because of its simplicity and practical utility. The H&S practitioner will find that many regularly occurring gases and vapours can conveniently be measured with modest equipment. As discussed in Chapter 5, 'Control of workplace health hazards', and Chapter 7, 'Personal protective equipment', air monitoring is a requirement when respiratory protection programs are introduced. Two approaches are widely used:

- direct reading instruments for use in the field; and
- sample collection with subsequent laboratory analysis.

10.10 DIRECT-READING INSTRUMENTS

The ability to examine exposure profiles in real-time using simple handheld devices enables the identification of specific tasks or procedures that require exposure control without the need for laboratory analysis which may take considerable time. The use of direct-reading instruments improves the effectiveness of the H&S practitioner and the safety of workers. The benefits of employing direct reading equipment are both qualitative and quantitative. Other reasons for using direct-reading instruments in the field are:

- Find the sources of emission of hazardous contaminants in rapid time.
- Ascertain if select WES's are being exceeded.
- Check the performance of control equipment.
- Continuously monitor fixed locations.
- Obtain permanent recorded documentation of the concentrations of a contaminant in the atmosphere.
- In legal actions, to inform employees as to their exposure, and for information required for improved design of control measures.
- Allows immediate assessment of undesirable exposures and allows industrial hygienists to make an immediate correction of an operation.

Hundreds of common gas or vapour contaminants require monitoring in the workplace, and modern chemical and electronic technologies have provided a range of instruments for measuring many of them. A good number are well within the operational capability of an appropriately trained H&S practitioner.

Here we examine only the instrument types that find regular use in workplace surveillance. Their use may range from occasional to constant, depending on the nature of the hazard, how often processes change and the effectiveness of the control procedures.

10.11 DIRECT READING: COLOURIMETRIC DETECTOR TUBES

For many gas and vapour monitoring tasks, the most convenient testing instrument is a colourimetric detector tube or detector tubes. Figure 10.2 shows the Dräger Accuro and the RAE piston-type hand-pump.

Detector tubes operate on the principle of a reaction between the test gas (contaminant being measured) and a chemical reagent bound to an inert matrix.

A hand-operated or mechanically operated bellows or a piston pump draws a calibrated volume of contaminated air through a tube containing a reagent chemical. Contaminants in the air will react with the reagent in the tube to produce a progressive colour change along the tube as the air passes through. Stain length in the tube is proportional to the concentration of contaminant present and is read directly from the graduations marked on the tube.

Detector tubes are inexpensive, and tubes are available for a wide range of applications. Special expertise is not required to operate them. They are excellent for monitoring *single* known contaminants. Considerable expertise is required to interpret results. Samples collected by this means are grab samples, representing concentrations or a 'snapshot' at points in time. Potential problems arise when concentrations are potentially subject to rapid change. The user must consider the work task being monitored to account for measurements of contaminants that may be discontinuously present, and how the results should be interpreted in terms of the relevant exposure standard.

H&S practitioners responsible for work areas where gases and vapours constitute a significant proportion of air contaminants should consider obtaining these useful gas-detector tubes. (The kit includes a hand-operated sampling pump and accessories; the appropriate detector

(a)

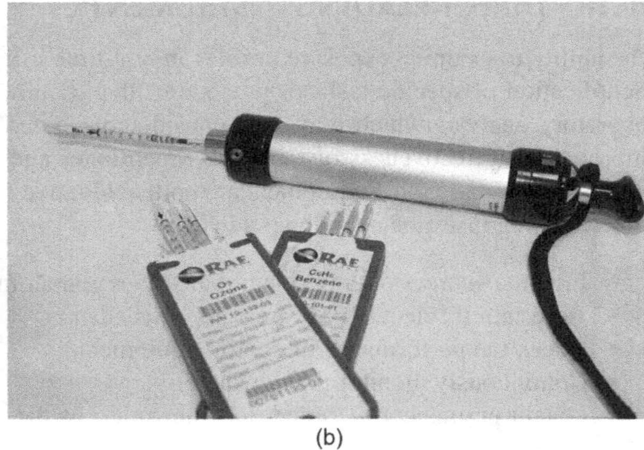

(b)

FIGURE 10.2 A Dräger Accuro in use and an RAE piston-type hand-pump with a selection of detector tubes.

Source: Dräger and A. Todorovic.

tubes need to be specified for the gas/vapour of interest and are purchased separately.) Several brands are available.

10.11.1 Limitations of Colourimetric Tube Systems

Detector tube systems should not be used to establish whether a contaminant is present, but only for measuring already known contaminants. Common problems include:

- Chemical mixtures may negatively influence the results of detector tubes. Cross sensitivity from the presence of interferents other than the target analyte may lead to false positive or false negative results.
- False negatives can arise if the incorrect measuring range is chosen.
- False negatives can arise if the identity of the contaminant is not known, and an incorrect tube is selected.
- Humidity and temperature may affect detector tube performance leading to possible false readings.
- Readings may need to be adjusted to account for variations in atmospheric pressure (e.g., at high altitudes).
- Inability to provide dynamic results for concentrations that vary with time.
- The tubes often have a limited shelf life and are not always available in the range of concentrations required.
- Accuracy varies from tube to tube, range to range and manufacturer to manufacturer.

10.11.2 Practical Instructions

H&S practitioners intending to purchase and use one of these devices should consult the various manufacturers. Dräger, Uniphos, Gastec, Kitagawa and Honeywell produce different designs, including hand bellows, thumb pump and piston types. All manufacturers offer a wide range of tubes, but a measuring range will need to be chosen to suit the workplace and its contaminants. Pumps and tubes from different systems cannot be interchanged, for there is no guarantee that the pumped sample volumes will be identical for each system unless they are confirmed to have the same sampling volume and flow rate. Manufacturers or suppliers usually offer training in the use of this equipment, together with appropriate selection of measuring tubes and as with all monitoring equipment, the manufacturer's instructions should always be read carefully before use.

10.11.3 Passive Dosimeter Tubes

Passive dosimeter tubes closely resemble standard detector tubes. These tubes are placed in the worker's breathing zone. Rather than a piston or a bellows pump drawing the sample into the tube, contaminants passively diffuse into the tube. Stain length is proportional to the concentration of the atmospheric contaminant being sampled. These tubes provide a time-weighted average (TWA) reading rather than instantaneous peak measurement. Users should be aware of limitations when considering using passive dosimeter tubes such as concentration, exposure time, face velocity and orientation.

10.12 DIRECT-READING BADGES

Several specific direct-reading badges are available (e.g., for formaldehyde, isocyanates and ethylene oxide). When exposed to contaminating vapours of a specific kind, the badge colour changes and as the analyte concentration and exposure time increase, the colour change becomes more intense. To use these badges successfully, the H&S practitioner should be certain of the identity of the contaminant to be measured.

10.13 SPECIFIC DIRECT-READING MONITORS

As technology has advanced, so has the range and sophistication of specific direct reading monitors and detectors. Instruments for gases such as carbon monoxide, sulfur dioxide, phosphine, carbon dioxide, hydrogen cyanide, chlorine, nitrogen dioxide, nitric oxide, ammonia, flammable gases, hydrogen sulfide, hydrogen, formaldehyde and chlorofluorocarbons, as well as for oxygen are widely used. These instruments work on various principles of electrochemistry, infrared spectrometry, photoionisation or direct-reading paper tape colourimetry. Figure 10.3 shows a carbon monoxide monitor being used to sample a work area.

Instrument monitors are made for fixed locations as well as for portable use. Fixed-location instruments are typically found in processing plants where fugitive gases and vapours can pose potential hazards to unprotected workers. In oil refineries and petrochemical facilities, fixed location detectors are deployed to detect leaks or emissions from flanges and seals in processing areas, bulk storage areas and plant modifications. These fixed detectors are normally part of an integrated monitoring system whereby readings are relayed to a central monitoring point normally located in a control room or on a centralised status panel. Instruments for personal use are often fitted with visible, audible and vibration alarms set at or just below the appropriate WES.

Modern instruments continuously monitor the surrounding atmosphere and have onboard datalogging capabilities which allow for vast amounts of data to be collected. Data logging permits the calculation of TWA concentrations, identification of any peak exposure and identification of any exposure pattern in the workplace (as illustrated in Figure 10.1). These instruments are easily able to be connected to computers to allow for the data to be downloaded and the data to be interrogated further.

Figure 10.3 shows one of many instruments that are useful for direct field measurements. Modern direct-reading monitors are easy to use, but their application needs to be properly considered. Some designs are primarily used for personal monitoring whereas others are used as survey monitors.

FIGURE 10.3 Carbon monoxide monitor in use.

Source: Watchgas.

10.13.1 Selection of a Direct-Reading Monitor

The decision to invest in direct-reading monitors should ideally be based on being fit for purpose ensuring that their procurement is justified as these monitors may vary considerably in cost.

Many models can detect up to six gases in a single instrument package. For example, the oxygen flammable gas/hydrogen sulfide and carbon monoxide type of meter are indispensable for confined-space entry (e.g., sewers, silos, silage pits and mines) and particularly useful for other confined-space applications (discussed in Section 10.21). Figure 10.4 shows two multi-sensor types. When considering purchase of one of these instruments, the following should be considered:

- The measurement task.
- The limits of operation and sensitivity to environmental changes.
- Sampling site.
- Use in potentially flammable atmospheres.
- Limit of detection.
- Selectivity to the target gases or vapours and sensitivity to interfering gases and vapours (for gas detectors).
- Mains or battery powered.
- Frequency of function checks, maintenance and calibration
- Compliance with national or international performance standards and appropriate national regulations.
- The level of sophistication and training of the workers who will be using the instruments
- The type of DRD required (personal, portable, transportable or fixed).
- Robustness, durability and shock-resistance.
- Visual or audible alarm warning levels.
- Concentration range required.
- Accuracy and precision or overall uncertainty
- Response and recovery time.
- Operating time for battery powered DRD.
- Training requirements for reliable operation, maintenance and calibration.
- The monitoring environment in which the instrument will be used
- The requirements for recordkeeping

Catalytic explosive sensors can be poisoned by sulfide, silicone compounds (e.g., insect repellents), phosphates and lead alkyl compounds.

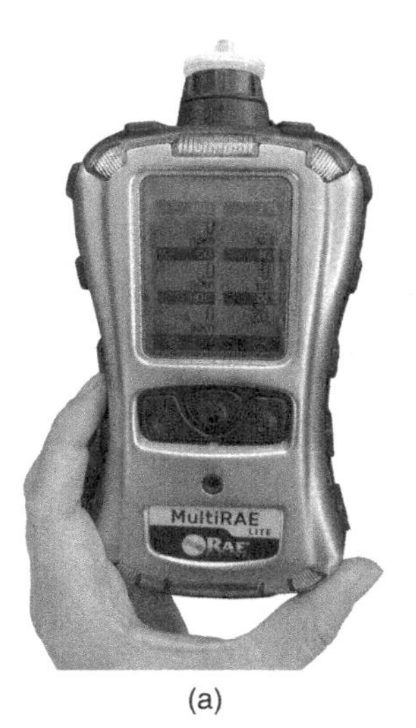

(a)

(b)

FIGURE 10.4 (a) MultiRAE gas monitor **(Source: A. Todorovic).** (b) Dräger X-AM 5800 6 **(Source: Dräger)**.

10.14 GENERAL-PURPOSE DIRECT-READING ANALYSERS

Several devices are now available for general-purpose gas and vapour detection:

- portable gas chromatographs;
- opto-acoustic gas analysers; and
- infrared analysers.

These very versatile instruments provide occupational hygienists with great sensitivity and selectivity in monitoring workplace organic vapours but are not generally recommended for use by H&S practitioners. In a workplace that encounters some dozens of organic vapours (e.g., a chemical manufacturing plant, solvent reprocessing plant and large manufacturing plant), general-purpose devices can provide an efficient and rapid means of surveillance during various processes.

Another kind of general testing instrument, the photoionisation detector (see Figure 10.5), finds use with a wide range of organic vapours. This device cannot distinguish between the different contaminants they are able to detect, rather they provide a single cumulative reading for all the detectable substances present at any moment. However, these instruments can be calibrated for known substances and are useful for single-contaminant situations. It has excellent sensitivity for many compounds encountered in the workplace, particularly solvents such as acetone, MEK and toluene.

Recent advances in photoionisation detection instruments mean they can now be used as personal monitors, for example, the ToxiRAE Pro PID as shown in Figure 10.5.

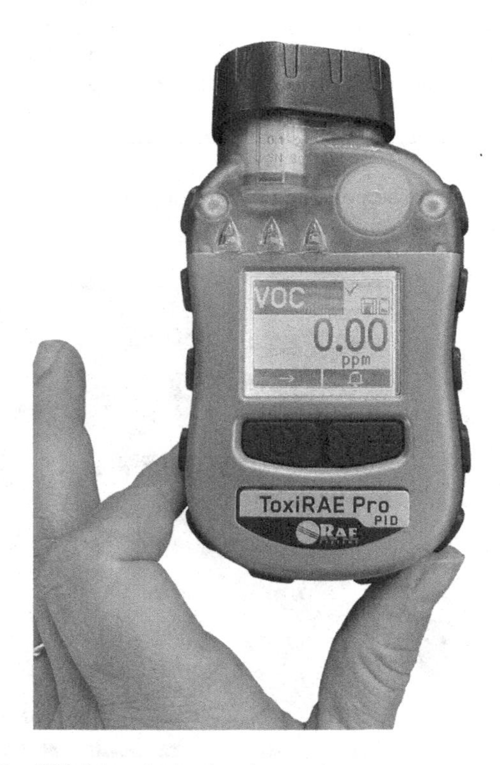

FIGURE 10.5 ToxiRAE Pro PID (photoionisation detector).

Source: A. Todorovic.

10.15 SAMPLE COLLECTION TECHNIQUES

Samples may be collected for analysis either on-site or at a laboratory. Sample collecting techniques may need to be applied where:

- TWA concentrations are required; or
- no direct reading instruments are available.

Chapter 2, on occupational health, indicates that much of our knowledge of the long-term hazards from exposure to hazardous gases and vapours is based on average long-term exposures in the workplace. While a few short-term high or peak exposures may lead to irritation, narcotic effects and feelings of nausea, giddiness and so on, it is often the long-term lower exposures that lead to sensitivity, organ damage, neurotoxic effects and cancer. For this reason, we may need to estimate exposure over longer periods.

Measuring exposure to chemicals implies that there is a WES that can be used to determine compliance or otherwise. While WESs vary slightly from country to country, the method of measurement does not. Reference should be made to the regulated standards in the country in question. If none exists for a particular chemical, then most professionals use either the American Conference of Governmental Industrial Hygienists (ACGIH®) Threshold Limit Values (TLVs®) (ACGIH 2023) or the UK's HSE Workplace Exposure Limits as guidelines.

To comply with these standards, TWA exposure measurements are generally required, principally as personal samples.

Five different techniques are regularly used. In order of practical importance, they are:

- collection of gas or vapour by pumping the contaminated air through a small sampling tube filled with a suitable absorbent;
- passive adsorption of the contaminating gas or vapour onto a badge or tube dosimeter worn on the lapel of the worker;
- passing the gas or vapour through a filter impregnated with a material which reacts with the contaminant;
- collection of the contaminating gas or vapour by bubbling it through a suitable liquid in a liquid impinger (this is less common);
- collection of samples in bags or pumped cylinders.

10.15.1 Tube Sampling

This is the most employed method of reliable vapour monitoring in the workplace. Sampling techniques for monitoring and interpretation are to be found in AS2986.1 Workplace Air Quality—Sampling and Analysis of Volatile Organic Compounds by Solvent Desorption/Gas Chromatography—Pumped Sampling Method (Standards Australia 2003a). Other methods are the US NIOSH Analytical Methods (NIOSH, 2017), US OSHA Methods (OSHA, n.d.) or the UK HSE Methods of Determination of Hazardous Substances (MDHS).

The media inside the tubes are selected to optimise absorbance or adsorbance for a specific contaminant and are normally activated charcoal, silica gel or several special chemicals. A small, calibrated sampling pump is set at a constant flow, suited to the application and draws the contaminated air through the adsorbent. Sampled tubes are usually stable for periods of days or weeks, if properly capped and stored in a refrigerator.

Sampled tubes must be analysed in a chemical laboratory. The contaminant can be stripped from the adsorbent by:

- the action of a powerful solvent such as carbon disulfide; or
- thermal desorption in a hot furnace (this requires special tubes).

The concentration of contaminant in the air is measured by either gas chromatography or high-performance liquid chromatography. These measurement facilities are generally available only in commercial, research or government laboratories. The sampling, however, can be done by any H&S practitioner who has the necessary sampling pumps, and who can apply the appropriate Australian Standard, NIOSH, OSHA or HSE method, and is appropriately trained.

Where a government inspector is auditing risk assessment procedures under the Australian Work Health and Safety Regulations (or their equivalent), the employer or the H&S practitioner may be requested to provide TWA contaminant concentrations for various workers. The use of a pumped sample tube which is analysed by a competent laboratory is generally an acceptable method of meeting this need.

There are some limitations to this method:

- Most permanent gases cannot easily be determined by such methods. (For hydrocarbons, only butane and heavier gases can be measured by this method.) The limitations of the analysis should be discussed with the testing laboratory.
- Breakthrough of a sorbent tube is when the tube becomes saturated and has no further capacity to retain the contaminant on the media. It normally occurs when the sample volumes are too large, or the flow rates are too great. The result of breakthrough is an underestimation of the workers' exposure.

Breakthrough can be determined by using a tube with two segments: a normal trapping segment and a back-up segment. Should contaminant be present in the back-up layer, breakthrough can be assumed to have occurred and the sample should be voided. The total volume of sampled air should be kept low enough that breakthrough is prevented or at least minimised. These sampling tubes again come with a wide variety of adsorbents and with different weights of adsorbent packing. The typical tube is the standard NIOSH tube, which has a 100 mg packing with a 50-mg back-up section. For higher volumes, the 1000-mg 'jumbo' sampling tubes, again with a back-up section, can be used. These sampling tubes are designed for analysis by solvent desorption. Breakthrough cannot be ascertained in tubes which are thermally desorbed, as these samples allow only 'one shot' analysis—that is, repeat testing of the sample is not possible because all the contaminants are destroyed in the analysis.

Figure 10.6 shows the arrangement of a portable low-flow sampling pump and an organic-vapour sampling tube and holder. The sampling tube is placed near the breathing zone of the operator, within about 15–30 cm of the nose.

In petrochemical industries, which handle highly flammable materials, any equipment taken on site, even a small sample pump, must be 'intrinsically safe'—that is, it must not be a source of ignition in these combustible atmospheres. Requirements for intrinsic safety vary from company to company and is commensurate with the expected hazardous area. Approval to use such equipment on site must be obtained before a monitoring survey is undertaken (usually through a daily safe-work permit system). As a guideline, in the United States, monitoring equipment rated intrinsically safe by Underwriters Laboratories Inc. (UL) is accepted; in Europe, equipment meeting the code ATEX II 2 G EEx ib IIC T4 is accepted; and in Australia, instruments with ANZEx or IECEx certification for intrinsic safety are accepted. These ratings are usually displayed on the equipment. Refer to Figure 10.7a–c.

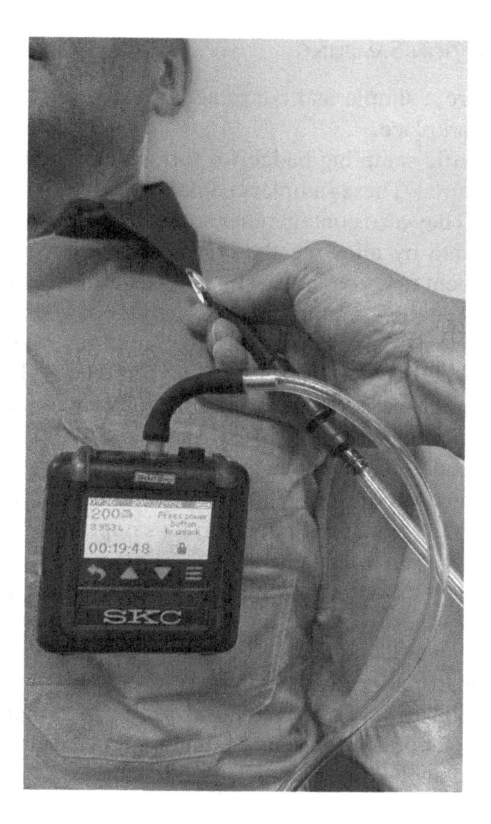

FIGURE 10.6 Worker being fitted with a low-flow sample pump and adsorbent charcoal tube. (Caution: Careful note of sample pump and monitoring equipment requirements for petrochemical industries is required.)

Source: A. Todorovic.

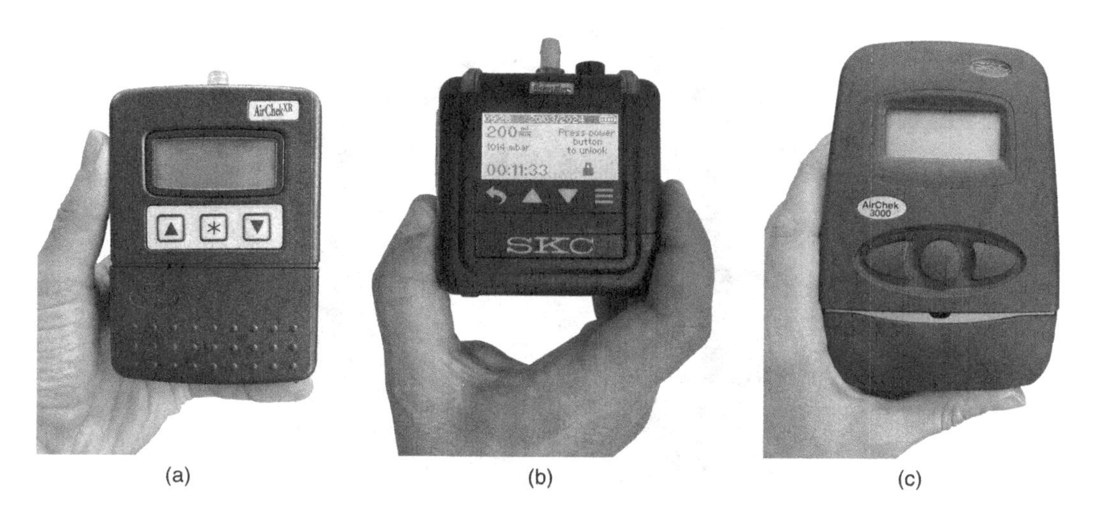

FIGURE 10.7 (a) SKC® Airchek 5000. (b) SKC® Pocket pump. (c) SKC® Airchek 3000.

Source: A. Todorovic.

10.15.2 Passive Adsorption Sampling

Passive samplers or tubes are a simple and convenient method by which to sample for organic vapour contaminants at a workplace.

Figure 10.8 shows a specific sampling badge for formaldehyde. In these passive approaches, sampling pumps are not required. These samplers contain layer/s of material capable of absorbing the contaminant of interest. They also contain other sorbent materials to remove possible interferents. Passive samplers operate by allowing the contaminant molecules to diffuse through the sampler and bond to the sorbent material inside. Techniques for using this method are to be found in AS2986.2 Workplace Air Quality—Sampling and Analysis of Volatile Organic Compounds by Solvent Desorption/Gas chromatography—Diffusive Sampling Method (Standards Australia 2003b), US NIOSH methods, OHSA methods or UK HSE methods (HSE 2000).

For large surveys of organic vapours (e.g., in commercial printing, boat building, spray painting, motor body repairs and petrochemical industry), passive sampling is a very cost-efficient way for the H&S practitioner to undertake vapour sampling. However, before any sampling for organic vapours begins, it is important to contact both the sampling device manufacturer/representative and the analytical laboratory for guidance on the correct sampler for the task and allow the laboratory to arrange proper calibration procedures.

This is expensive laboratory time, for which the H&S practitioner will have to pay. There are several passive adsorption badges available for specific contaminants, but the badges or tubes must be analysed by a competent laboratory.

Once collected and properly capped, the samples are sent to an appropriate laboratory for analysis. As with tube sampling above, the contaminants are desorbed with a specific solvent or by heat. Figure 10.9 shows a laboratory gas chromatograph fitted with an automated thermal desorber suitable for unattended analysis of organic-vapour samples.

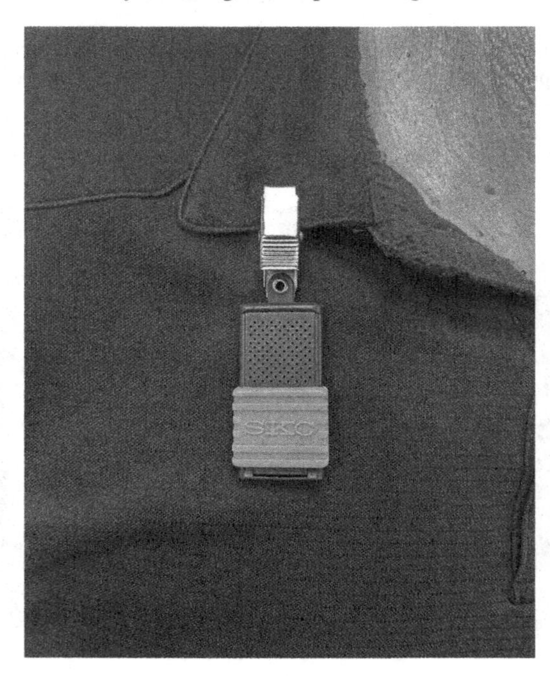

FIGURE 10.8 UMEX® badge.

Source: A. Todorovic.

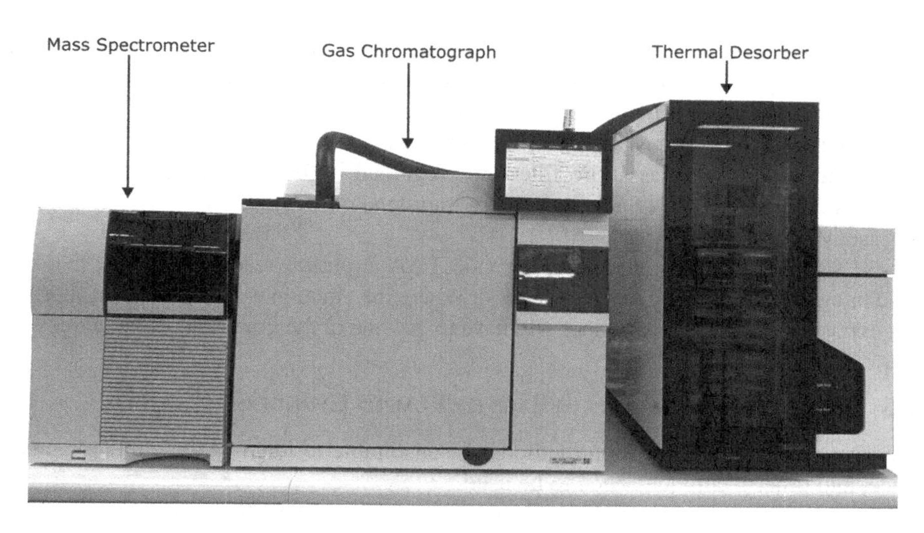

FIGURE 10.9 Laboratory gas chromatograph fitted with an automated thermal desorber.

Source: M. Reed.

10.15.2.1 Limitations of Passive Sampling

As passive samples are designed to work in moving air, it is important to ensure that the worker is moving during sampling. They are generally not designed for static applications to obtain 'workplace environmental' vapour concentrations, unless specifically designated as suitable for this as stagnant air can 'starve' the sampler and yield useless results. Typically, a passive sampler must be worn for more than four hours to obtain sufficient material for the laboratory to analyse.

If the workplace is being monitored for solvent mixtures (e.g., hexane, methylene chloride and xylene), the H&S practitioner will have to provide complete detailed information to the processing laboratory, or if possible, provide a sample of the solvents.

Passive badge or tube samplers are extremely sensitive and can pick up vapour that may leak or diffuse through seals and inappropriate packaging. This will ruin sampling efforts completely.

Ensure the sample is sealed, packaged and transported separately from other items.

10.15.3 Impingement into Liquid Samples

Several gases or vapours contaminating workplaces cannot successfully be trapped on tubes or badges, but they can be trapped by bubbling them through water or some other solvent. These bubblers are called liquid impingers. Contaminated air is drawn by a small pump, at around 1 L/min, through the absorbing liquid. For safety reasons, it is often better to collect area samples rather than personal samples if the absorbing solution contains corrosive substances (acids or alkalis) or hazardous solvents, unless spill-proof impingers can be used. However, in many cases, this sampling method has been superseded by more modern techniques.

10.15.4 Impregnated Filters

Some vapours can successfully be trapped on filters chemically impregnated with a substance that is reactive to the contaminant in question. The method is similar to the vapour tube technique, but more versatile in that it can simultaneously trap particulates. For instance, an aluminium smelter will produce both gaseous fluoride and particulate fluoride, both of which can be trapped on a filter doped with citric acid. This method can likewise trap aerosol and gaseous forms of certain paints, which exist simultaneously in spraying operations.

Other examples are (a) glass-fibre filters impregnated with *2,4*-dinitrophenylhydrazine for sampling of aldehydes and (b) glass-fibre filters impregnated with methoxyphenylpiperazine for sampling isocyanates.

All these sample-collection techniques provide TWA exposure readings because they are collected and integrated over a long sampling period. Where information is also needed on short-term or any peak exposure, the H&S practitioner will have to use one of the grab sample techniques as well.

10.15.5 Collection of Samples in Bags or Pumped Cylinders

In some cases, the above absorption methods are unsuitable to identify the contaminants in the workplace, and other techniques need to be used.

Some samples can be successfully collected in impervious bags (Figure 10.10a and b) or in evacuated pressurised cylinders or canisters (Figure 10.10c and d) and returned to a laboratory for analysis. This technique is required for many of the permanent gases (e.g., oxygen, carbon monoxide, methane and hydrogen), which cannot be trapped by any other means. However, the vessel

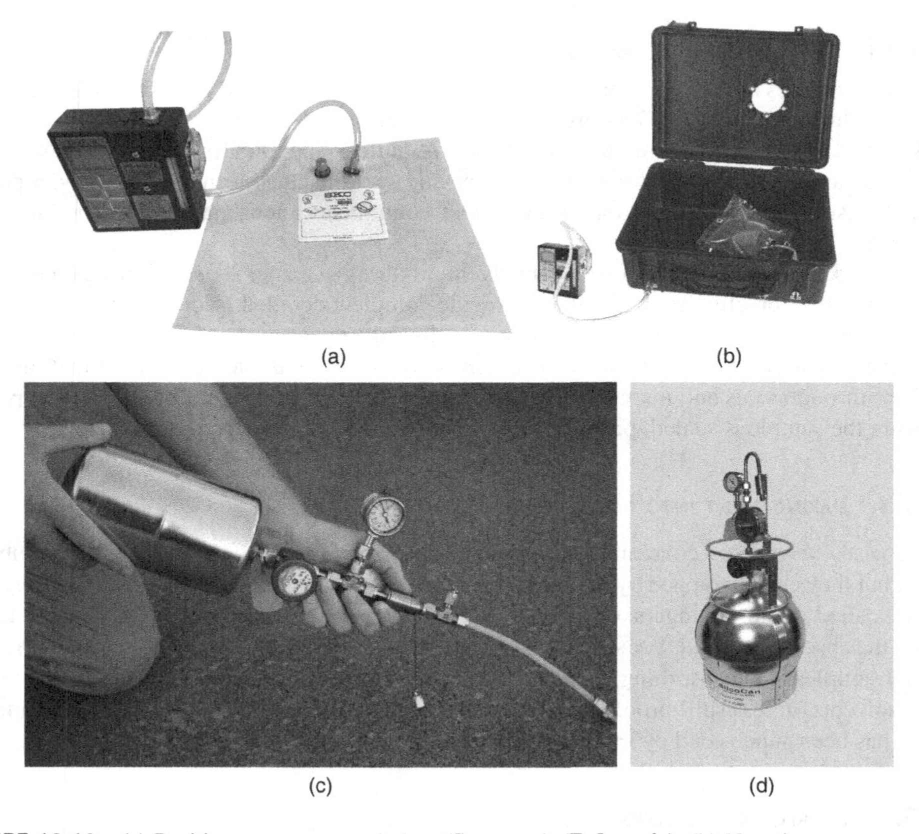

FIGURE 10.10 (a) Positive-pressure sample bag (**Source: A. Todorovic**). (b) Negative-pressure sample bag (**Source: SKC Inc**). (c) Small pressurised canister (**Source: A. Todorovic**). (d) Large pressurised canister (**Source: A. Todorovic**).

must retain the sample adequately, without loss. Some plastics will adsorb organic vapours, and small molecules (particularly hydrogen) will diffuse through plastics. The H&S practitioner must ensure that the plastics are specified as suitable for their intended purpose. As an example, some compounds are photosensitive and hence the sample bag chosen must be able to preserve the integrity of the sample.

To obtain a representative sample, bags are simply inflated and emptied and refilled several times with the gas to be tested. Small sampling pumps with a direct outlet can be used for this purpose. A hand-operated pressure pump can also be used to collect gas analysis samples in bags. Pressurised cylinders are under negative pressure, and the air sample is collected upon opening the valve. Flow restrictors maintain flow stability over the desired sample period from less than one minute to 12 hours. The large cylinder can be fitted with a vacuum gauge to monitor the flow.

This is essentially a grab sample and will provide information only on the contaminants present at the time it was collected; it cannot be used to determine the TWA. It can only be used as a time-weighted sample if the sample has been collected uniformly over the whole shift—that is, if the bag or cylinder was filled at the same flow rate over the shift. It is also indicative only of the general environment where the sample was taken, and not the personal exposure of the worker. It is, however, useful to evaluate whether engineering controls are operating correctly.

10.16 TESTING COMPRESSED BREATHING AIR

Compressed breathing air finds several applications in workplaces, including in underwater breathing apparatus, compressor-fed airline breathing apparatus for abrasive blasting, spray painting in booths and self-contained breathing apparatus (SCBA) breathing sets for firefighting and rescue work.

In all these situations, the air produced by a compressor must be perfectly suitable for breathing under all operating conditions. Compressors can introduce contaminants to the air through incorrect siting of air intakes or through operation of the compressor itself. Oil vapour arises from compressor oil, carbon dioxide arises from contaminated intake air, and carbon monoxide can arise from partial breakdown of the compressor oil as well as from intake air sources. Water vapour will be more concentrated in the compressed air unless it is adequately removed by driers.

Tests need to be conducted on compressed breathing air to ensure its quality (AS/NZS 1715). The instrument most widely used for this purpose is the Dräger Aerotest. Cylinder air or air directly from a compressor receiver (after passage through a conditioner) can be tested. Air compressor systems require regular testing and should be checked thoroughly after initial installation or any servicing.

10.17 TIPS FOR PERSONAL MONITORING

Every personal monitoring sample has a 'story' attached to it, and this should be recorded by answering the following questions:

- Why was there potential for exposure?
- What chemicals and processes were used?
- What did the worker do during the sampling period?
- Did he or she come in contact with these chemicals?
- What was the physical form of the contaminant?
- How was the chemical used? As a liquid, as an aerosol, sprayed?
- Were there any other chemicals used nearby to which this person could inadvertently have been exposed (fugitive emissions)?

- Was respiratory protection used? Was it worn? Was it suitable? Did it work correctly?
- Was other personal protection used? Gloves, apron? (Often, we measure air contaminants, but greater exposure may occur through unprotected skin contact with liquids and vapours.)
- Were engineering controls used? Ventilation/extraction systems?
- Were there any spills, process upsets, known fugitive emissions?

10.18 MONITORING THE WORKER: BIOLOGICAL MONITORING

Monitoring the worker's health directly is another way of checking on the quality of the work environment. This is referred to as **biological monitoring**, and the worker is the 'sampling device', accounting for all exposure routes. Biological tests can provide information on the uptake of a substance by the worker, whereas air testing cannot. For example, blood cyanide tests on workers in the cyanide manufacturing industry can indicate whether they need to be removed from further cyanide exposure.

However, few industries make regular use of biological monitoring, partly because it may involve medical tests and partly because it cannot be done without the complete cooperation of the worker. Confidentiality of medical test results may present another problem. Most H&S practitioners will not become involved in such kinds of sampling. Further detail is found in Chapter 11, 'Biological monitoring of chemical exposure'.

10.19 COMMON GAS OR VAPOUR HAZARDS

This section deals with just a few gases and vapours among the many that occur in workplaces—explosive gases, carbon monoxide, hydrogen sulfide, ammonia, chlorine, nitrous oxides (NOx), benzene and mercury vapour. These substances give rise to typical and recurrent complaints that require investigation by occupational hygienists. Making use of the H&S practitioner's basic recognition skills, basic toxicological information, assessment methods and control procedures, this section examines:

- typical occurrence;
- brief toxicology of the materials;
- WESs and how the risks are assessed;
- possible control procedures that can be recommended.

The six following profiles are very brief and are presented only as basic guidelines. The H&S practitioner who encounters any hazardous gases and vapours in the workplace needs to research each one to be fully informed on safety, health and medical factors, and control processes.

10.19.1 Flammable/Explosive Gases

Many gases encountered in the workplace can be flammable or, in the right mixture with air and with an ignition source, explosive.

10.19.1.1 Lower and Upper Explosive Limits for Flammable Gases and Vapours

Before a fire or explosion can occur, three conditions must be met simultaneously. A fuel (combustible gas) and oxygen (air) must exist in certain proportions, along with an ignition source such as a spark or flame. The ratio of fuel to oxygen required varies with each combustible gas or vapour. The minimum concentration of a particular combustible gas or vapour necessary to

TABLE 10.3
Selected Chemicals and Flammable Range

Gas or Vapour	%LEL	%UEL	Gas or Vapour	%LEL	%UEL
Acetone	2.6	13.0	Hydrogen	4.0	75.0
Acetylene	2.5	100.0	Hydrogen sulfide	4.0	44.0
Ammonia	15.0	28.0	Isobutane	1.8	8.4
Benzene	1.3	7.9	Isobutylene	1.8	9.6
1,3-Butadiene	2.0	12.0	Methane	5.0	15.0
Butane	1.8	8.4	Methanol	6.7	36.0
n-Butanol	1.7	12.0	Methylacetylene	1.7	11.7
Carbon monoxide	12.5	74.0	Methyl bromide	10.0	15.0
Carbonyl sulfide	12.0	29.0	Methyl Cellosolve®	2.5	20.0
Chlorotrifluoroethylene	8.4	38.7	Methyl chloride	7.0	17.4
Cyclopropane	2.4	10.4	Methyl ethyl ketone	1.9	10.0
Dimethyl ether	3.4	27.0	Methyl mercaptan	3.9	21.8
Ethane	3.0	12.4	Pentane	1.4	7.8
Ethanol	3.3	19.0	Propane	2.1	9.5
Ethyl acetate	2.2	11.0	Propylene	2.4	11.0
Ethyl chloride	3.8	15.4	Propylene oxide	2.8	37.0
Ethylene	2.7	36.0	Toluene	1.2	7.1
Ethylene oxide	3.6	100.0	Trichloroethylene	12.0	40.0
Gasoline	1.2	7.1	Xylene	1.1	6.6

support its combustion in air is defined as the gas or vapour's LEL. Below this level, the mixture is too 'lean' to burn. The maximum concentration of a gas or vapour that will burn in air is defined as the UEL. Above this level, the mixture is too 'rich' to burn. The range between the LEL and UEL is known as the *flammable range* for that gas or vapour. Table 10.3 shows the wide flammable range of many gases, particularly hydrogen and acetylene.

The values shown in Table 10.3 are valid only for the conditions under which they were determined (usually room temperature and atmospheric pressure).

Some of these gases will be asphyxiants, others will be toxic or irritants. Each gas should be assessed for its health hazards as well as its flammability.

When assessing the flammability/explosivity of the atmosphere in a workplace, a particular direct-reading instrument is required. Many modern direct-reading monitors include an 'explosimeter' to measure the %LEL in the work environment.

An explosimeter is a device used to measure the amounts of combustible gases in a sample. When a certain percentage of the LEL of an atmosphere is exceeded, an alarm signal on the instrument is activated. The device, also called a combustible gas detector, uses catalytic bead sensors and operates on the principle of resistance being proportional to heat. Other explosive sensors use non-dispersive infrared (NDIR) technology to measure explosive gas concentrations. The explosimeter DOES NOT identify the flammable/explosive gas, but only determines where the work environment falls on the continuum of explosivity. It is essential to measure flammability for any confined space entry (see Section 10.21, 'Confined spaces'). These monitors must be intrinsically safe, or they themselves could spark a fire or explosion.

Note that the detection readings of an explosimeter are accurate only if the gas being sampled has the same characteristics and response as the calibration gas. Most explosimeters are calibrated to methane or pentane.

Many combustible-gas monitors also measure oxygen levels. It should be noted that catalytic explosive sensors require at least 10% oxygen to give an accurate reading. They should therefore not be used for this application if the oxygen level is below 10%, as a false low reading may be obtained, causing underestimation of the risk. In applications where the oxygen concentration is below 10% such as in gas pipeline maintenance use of NDIR, sensor-fitted gas monitors are widely employed.

10.19.2 Ammonia

10.19.2.1 Use and Occurrence

Ammonia (NH_3) is a colourless gas with a strong, pungent smell. It is commonly used in refrigeration, ice making, as a cleaner, as a fertiliser, and in the production of other pharmaceuticals, resins and paper products. Ammonia is usually compressed into liquid form for transport or storage.

Ammonia has a lower flammability limit of 16 % by volume. Although it is difficult to ignite, there have been several explosions and fires caused by the release of ammonia in a confined space.

10.19.2.2 Toxicology

Ammonia is an upper airway irritant. Intense pain in the eyes, mouth and throat is normally a result from single intense exposures. This is due to the formation of ammonium hydroxide when it comes into contact with water. Longer single exposures will produce ulceration of the mouth and nose, cyanosis and pulmonary oedema.

Exposure for 30 minutes at levels higher than 2500 ppm is usually fatal.

10.19.2.3 Standards and Monitoring

The Safe Work Australia (2023) WES for ammonia is 25 ppm TWA and 35 ppm STEL. Ammonia is detectable by the nose at concentrations of 1–5 ppm, but workers who are regularly exposed to ammonia become desensitised and *will lose their ability to smell ammonia* at levels below 50 ppm.

Although detector tubes can be used to measure ammonia, the main means of continuous detection is direct-reading electrochemical instruments, both as fixed installations in plants and as portable personal monitors. PID sensors can also be used to measure high levels of ammonia as most electrochemical sensors do not have the range to measure the concentrations commonly seen during ammonia spills.

10.19.2.4 Controls

Engineering controls should be implemented in the design of industrial plant to reduce the risk of ammonia leaks or spills. A closed handling system is the optimum solution. Where this is not possible, the use of the gas should be restricted to small amounts in well-ventilated areas.

For storage areas or large-scale operations, detectors should be in place in case of leaks and protective equipment of suitable respirators and protective clothing should be donned. Escape breathing apparatus should be available in these working areas.

10.19.3 Benzene

10.19.3.1 Use and Occurrence

Benzene (C_6H_6) is a colourless liquid with a sweet (aromatic) odour. It evaporates very quickly and dissolves to some extent in water. It is highly flammable and vapour/air mixtures are potentially explosive. Benzene's LEL is 1.35%, and its UEL is 6.65%.

Benzene is a major industrial chemical mostly derived from petroleum. It is widely used as an intermediate in the manufacture of products from plastics to pesticides and pharmaceuticals. Key industries involved in the production or use of benzene include:

- petroleum refining;
- coke and coal production;
- rubber tyre manufacturing;
- storage sites (tank farms);
- transportation services (ships, tanker trucks);
- laboratories;
- manufacture of plastics, synthetic rubber, glues, paints, furniture wax, lubricants, dyes, detergents, pesticides and some pharmaceuticals.

Natural sources of benzene include volcanoes and forest fires; it is also present in cigarette smoke.

Benzene is highly mobile in soils. Leaking petrol storage tanks and pipelines have resulted in soil and groundwater contamination, which is of particular concern where town water is drawn from underground sources.

10.19.3.2 Toxicology

Benzene is irritating to the eyes, skin and respiratory tract. If it is swallowed, aspiration into the lungs may result in chemical pneumonitis. It also affects the central nervous system, and exposure to levels far above the occupational WES may lead to unconsciousness and death.

Prolonged exposure dries out and defats the skin and may damage the bone marrow, reducing blood-cell production and impairing the immune system. In sufficient doses, it is carcinogenic.

Benzene can be measured in breath, blood or urine, but only for about 24 hours after exposure, owing to its relatively rapid removal by exhalation or biotransformation. Most people in developed countries have measurable baseline levels of benzene and other aromatic petroleum hydrocarbons in their blood. In the body, benzene is converted by enzyme action to a series of oxidation products, including muconic acid, phenylmercapturic acid, phenol, catechol, hydroquinone and 1,2,4-trihydroxybenzene. Most of these metabolites have some value as biomarkers of human exposure, since they accumulate in the urine in proportion to the extent and duration of exposure, and they may still be present for some days after exposure has ceased. The current ACGIH biological exposure limits for occupational exposure are 500 μg/g creatinine for muconic acid and 25 μg/g creatinine for phenylmercapturic acid in an end-of-shift urine specimen.

10.19.3.3 Standards and Monitoring

The NIOSH puts the IDLH concentration of benzene in air at 500 ppm. Safe Work Australia (2023) sets a WES of 1 ppm TWA and classes benzene as a Category 1 carcinogen. By contrast, the ACGIH recommends a lower occupational exposure limit—TLV: 0.5 ppm as STEL of 2.5 ppm with a skin notation and lists as A1 (confirmed human carcinogen) (ACGIH 2018).

Monitoring can be done by detector tubes and direct-reading instruments; however, for comparison with the WES, a time-weighted sample must be obtained, either by charcoal tube/pump, diffusion badge or direct-reading instruments (specifically modified for benzene) with TWA capabilities.

10.19.3.4 Controls

A closed-loop system should be implemented to prevent release of products containing benzene. Other measures include ventilation extraction systems, explosion-proof electrical equipment and lighting. Open flames, sparks should not be present where benzene may exist. Compressed air should be used for filling, discharging or handling products containing benzene. Hand tools should be non-sparking and the build-up of electrostatic charges prevented (e.g., by grounding or earth-bonding).

10.19.4 Carbon Monoxide

10.19.4.1 Use and Occurrence

Carbon monoxide is a colourless and odourless toxic gas, which burns with a pale blue flame. Prior to the widespread introduction of natural gas as a domestic fuel, town gas supplies contained carbon monoxide. Town gas is still used as an oven gas in coke and steel works. Carbon monoxide is used to render some reactive gases and other materials inert but concerns about it in the workplace arise principally from incomplete combustion of fuels in ovens and furnaces, internal combustion engines (e.g., forklifts used in cold stores) and coal mines. Smouldering materials (e.g., poured metal moulds in foundries and building fires) produce considerable quantities of carbon monoxide. Although workers and vehicle drivers are generally familiar with the dangers, accidental carbon monoxide poisoning is relatively common. Escape from a dangerous situation is often impossible, because the first indication of gross exposure may be collapse.

10.19.4.2 Toxicology

Carbon monoxide (CO) is a chemical asphyxiant and inhibits transport of oxygen in the blood. The affinity of CO to the haemoglobin in blood is over 300 times greater than that of oxygen. Inhaled carbon monoxide is rapidly and extensively absorbed into blood and distributes throughout the body. Short exposures of 2000 ppm results in asphyxia and death, while at 50 ppm discomforting effects are noticeable. When carrying carbon monoxide, arterial blood does not become the bluish colour of venous blood, as in cases of normal asphyxia; the blood and the lips remain a bright cherry red. The tissues of the central nervous system and the brain are the most susceptible. Methylene chloride, a commonly used industrial solvent, metabolises to carbon monoxide, and exposure to high concentrations may produce chemical asphyxia.

Symptoms of carbon monoxide poisoning include, progressively, headache, nausea, drowsiness, fatigue, collapse, unconsciousness and death. These states correspond to increasing percentages of blood carboxyhaemoglobin (COHb), from 10 to 40%. Smokers' blood may contain between 5 and 10% carboxyhaemoglobin.

10.19.4.3 Standards and Monitoring

SafeWork Australia lists a TWA-OES for CO of 30 ppm for an 8-hour day which has a bioequivalent of 5%COHb (carboxyhaemoglobin in blood concentration). The Exposure Standard Working Group believes this value should protect against adverse behavioural effects arising from carbon monoxide exposure as well as minimising the risk to those persons with subclinical cardiovascular disease. The ACGIH BEI (2018) is set at 3.5% COHb as being bioequivalent to the TWA of 25 ppm, not to be exceeded at any time during the work shift to prevent adverse neurobehavioral changes, and to maintain cardiovascular exercise capacity. There are some variations possible for higher short-term exposures for reduced periods, as shown in Table 10.4.

Because of the insidious nature of carbon monoxide, monitoring for it should be carried out in work areas *wherever* there is a possibility that the gas is present. Detector tubes are useful for taking grab samples of potential exposure and provide an instantaneous reading of airborne concentrations. Specific direct-reading monitors are readily available for carbon monoxide from a variety of manufacturers (MSA, Dräger, RAE Systems); many are available as pocket-sized personal gas monitors, with visible and audible alarms that can be set at predetermined levels to alert the user to a potentially hazardous atmosphere.

TABLE 10.4
Australian WESs for Carbon Monoxide

Time	TWA (ppm)	STEL (ppm)
8 hrs	30	No value
1 hr	No value	60
30 min	No value	100
15 min	No value	200
<15 min	No value	400

Adapted from: Safe Work Australia April 2013 p. 20, Table 1.

Portable CO monitors allow for the measurement of a TWA and STEL. Furthermore, the use of portable gas monitors allows the user to track how exposure varies with time by graphing exposure versus time. This helps identify where the main exposure is occurring and can be useful for targeting controls.

10.19.4.4 Controls

Engineering controls are mandatory to prevent build-up of carbon monoxide and sub-sequent exposure of workers. Underground car parks and tunnels must be fitted with extraction systems to remove vehicle exhaust. Ovens, gas furnaces and burners must be correctly vented to carry away both burnt and incompletely burnt combustion products.

Respiratory protective equipment is not recommended for long-term protection against carbon monoxide, since it is more appropriate to use proper ventilation. Emergency escape equipment consists of either self-contained breathing apparatus or a self-rescuing gas converter that turns carbon monoxide into carbon dioxide, with the accompanying production of much heat. Firefighting and rescue equipment usually consists of self-contained breathing apparatus.

10.19.4.5 Training and Education

All workers exposed to carbon monoxide (e.g., from coke oven gases) must be instructed and trained in safe work procedures and in the appropriate use of self-contained breathing apparatus.

10.19.4.6 Medical Requirements

Heavy smokers are at greater risk of ill effects from carbon monoxide exposure because a proportion of their haemoglobin is already incapacitated. Workers with emphysema may also be at greater risk. Because of the possibility of severe health problems in workers with cardiovascular disease, exposure to carbon monoxide should be restricted to ensure that the carboxyhaemoglobin content of all workers' blood is maintained below 5%.

10.19.5 Chlorine

10.19.5.1 Use and Occurrence

Chlorine (Cl_2) is a highly reactive heavy gas which is greenish-yellow in appearance with a characteristic pungent odour. Chlorine solution in water is a strong acid and is corrosive.

The substance is a strong oxidant that reacts violently with bases, combustible substances and reducing agents. It reacts with most organic and inorganic compounds, causing fire and explosion hazard, and attacks metals, some forms of plastic, rubber and various coatings.

Chlorine is supplied in cylinders or in tanks as a liquid under pressure. Its principal applications are in the production of a wide range of industrial and consumer products, such as plastics, solvents for dry cleaning and metal degreasing, textiles, agrochemicals and pharmaceuticals, insecticides, dyestuffs and household cleaning products. The most significant of organic compounds in terms of production volume is 1,2-dichloroethane and vinyl chloride, intermediates in the production of PVC. Other particularly important organochlorines are methyl chloride, methylene chloride, chloroform, vinylidene chloride, trichloroethylene, perchloroethylene, allyl chloride, epichlorohydrin, chlorobenzene, dichlorobenzenes and trichlorobenzenes. The major inorganic compounds include hydrochloric acid, dichlorene oxide, hypochlorous acid, sodium chlorate and chlorinated isocyanurates.

10.19.5.2 Toxicology

Chlorine is corrosive to the eyes, skin and respiratory tract. Rapid evaporation of the liquid may cause frostbite. Inhalation may cause asthma-like reactions and pneumonitis. It may also cause lung oedema, but only after initial corrosive effects on eyes and/or airways have become manifest. The symptoms of lung oedema are often not apparent for several hours and are aggravated by physical effort. Rest and medical observation are therefore essential. Exposure to chlorine in sufficient concentration or over a long period can be lethal.

Chlorine's effects on the respiratory tract and lungs may result in chronic inflammation and impaired function. It may also damage the teeth.

10.19.5.3 Standards and Monitoring

Chlorine is detectable by the nose at 0.08 ppm as a suffocating/sharp/bleach odour. The Safe Work Australia WES for Chlorine is 1 ppm Peak limitation. By contrast, the ACGIH (2023) recommends lower occupational exposure limits—TLV: TLV: 0.1 ppm as TWA; 0.4 ppm as STEL; A4 (not classifiable as a human carcinogen); (ACGIH 2018). Chlorine is immediately dangerous to life and health (IDLH) at 10 ppm.

Although indicator stain tubes can be used to measure chlorine, the main means used for continuous detection are direct-reading, devices as these provide far more information on the work environment.

10.19.5.4 Controls

Engineering controls should be included in the design of new industrial plant to reduce the risk of leaks or spills. A closed handling system is the optimum solution. Where this is not possible, the use of chlorine should be restricted to small amounts in a well-ventilated area.

For storage areas or large-scale operations, detectors should be put in place in case of leaks and protective equipment of suitable respirators and clothing should be worn. Escape breathing apparatus should be available in these working areas.

10.19.6 Formaldehyde

10.19.6.1 Use and Occurrence

Formaldehyde is a colourless gas with a disagreeable sharp odour. It is readily miscible in water and often other additives such as methanol are found in formaldehyde solutions. It is a strong reducing agent.

Formaldehyde is used in many industries ranging from the manufacture of formaldehyde-based resins, medical and research industry, funeral industry, textile and leather industries. It is also widely used in personal care and consumer products. It is also a common combustion by product and readily emitted from furnishings and wood products.

10.19.6.2 Toxicology

Formaldehyde (or its solutions) are irritating to the eyes, skin and respiratory tract. The solution is a strong skin sensitiser. There is little if any evidence reported concerning impacts on the cardiovascular system. It is considered a probable carcinogen. Inhaled formaldehyde at concentrations normally found in workplaces is readily detoxified by metabolism within the body.

10.19.6.3 Standards and Monitoring

The NIOSH puts the IDLH concentration of ozone in air at 20 ppm. Safe Work Australia (2023) sets a WES of 1 ppm TWA. By contrast, the ACGIH recommends a lower occupational exposure limit—TLV Ceiling: 0.3 ppm (ACGIH 2023).

Monitoring can be done by test strips, detector tubes and direct-reading instruments; There are also impinger and similar sampling methods.

10.19.6.4 Controls

Engineering controls should be included in the design of workplaces where formaldehyde is to be used to reduce the risk of leaks or build-up of ozone within enclosed spaces.

Respiratory protection is required when the WES is likely to exceed the action level.

10.19.7 Hydrogen Sulfide

10.19.7.1 Use and Occurrence

Hydrogen sulfide (H_2S) is a colourless gas with a rotten-egg odour that some people can smell at levels of 0.2 ppb. It is produced by anaerobic sulfur fixing bacteria, especially associated with raw sewage. It is also found in crude oil, marine sediments, tanneries, pulp and paper industry. As it is heavier than air, it collects in pits, within protective berms, or in other low-lying areas.

It is a standard recommended gas to test for before entering a confined space.

10.19.7.2 Toxicology

Hydrogen sulfide is a chemical asphyxiant that prevents utilisation of oxygen during cellular respiration, shutting down power source for many cellular processes leading to respiratory paralysis and asphyxia at high concentrations. It also binds to haemoglobin in red blood cells, interfering with oxygen transport. Exposure to hydrogen sulfide occurs primarily by inhalation but can also occur by ingestion (contaminated food) and skin (water and air). Once taken into the body, it is rapidly distributed to various organs, including the central nervous system, lungs, liver, muscle, as well as other organs which may cause muscle cramps, low blood pressure and death at high levels.

Eye, throat and lung irritation, nausea, breathing difficulties, headache and chest pain have been reported in cases of continuous exposure to concentrations above 10 ppm. Though easily smelled in very low concentrations, in high concentrations, hydrogen sulfide may cause a temporary loss of smell termed as 'odour fatigue' (see Section 10.4.1, 'Odour').

10.19.7.3 Standards and Monitoring

The Safe Work Australia WES for hydrogen sulfide is 10 ppm TWA and 15 ppm STEL, while the ACGIH lists a TWA at 1 ppm and a STEL to 5 ppm ACGIH (2023).

Although indicator stain tubes can be used to measure hydrogen sulfide, the main means of continuous detection is direct-reading electrochemical instruments. Detectors for hydrogen sulfide as a single gas or in combination with other gases are now quite cost-effective for long-term monitoring and are standard on most sites where hydrogen sulfide is a common risk.

10.19.7.4 Controls

Attention must be given to the design of industrial plants to reduce the possible production and build-up of hydrogen sulfide. Where it is not possible to significantly reduce or eliminate the gas, fixed-installation detection equipment and personal detection monitors should be used throughout the plant.

Where engineering controls are not feasible, then personal protective equipment of respirators, filters, breathing apparatus and protective clothing must be used. If a gas alarm is sounded, escape filters or breathing apparatus must be worn.

10.19.8 Mercury Vapour

In recent years, a new hazard has emerged in the oil and gas industry. Mercury vapour is contained in certain crude oils and can deposit out in various process units in a petroleum refinery. The hazards of mercury are covered in Chapter 9, 'Metals in the workplace'. This section will cover only monitoring equipment and its applications in the oil industry; the toxicology is covered in Section 10.11.2.

10.19.8.1 Occurrence in the Oil Industry

Mercury (Hg) can be found in trace amounts in some crude oil, and because it can form a vapour, it may be transferred through the production process into the refinery. It can amalgamate with aluminium components in heat exchangers and may cause catastrophic failures in cryogenic units. The most likely locations for mercury deposits are in exchangers, towers, knockout pots and accumulators. The air around a refinery does not usually contain mercury, however. Most mercury is removed following steaming/cleaning.

When refinery equipment is opened during a turnaround or shutdown, it is important to assess whether mercury is present. Refinery exchangers and vessels need to be steamed out before entry. Mercury-vapour monitors can be used to 'sniff' for mercury in and around steam plumes bled off from vessels and equipment. In most cases, the steaming is done into a closed process and the steam can be 'bled' briefly to check for mercury in the air.

Mercury can condense to free liquid and become trapped in the walls of steel vessels and piping. Even after cleaning, some mercury is likely to remain in these walls and may leach out during hot work, because the heat helps liberate it from the steel. This is a health hazard during shutdowns and turnarounds.

Mercury vapour should be measured before any entry into confined spaces suspected of containing it.

10.19.8.2 Standards and Monitoring

A mercury-vapour analyser is used to detect the presence of elemental mercury vapour and measure its concentration in the air. If readings exceed the Australian WES of 0.025 mg/m^3 TWA, then respiratory protection, disposable coveralls, nitrile gloves and rubber/PVC boots are required.

Air monitors may be used in 'sniffer' mode to locate sources of contamination and should be used after cleanups to ensure that removal of mercury was complete.

Several direct-reading instruments are available, including:

- the Ametek Brookfield Jerome range of mercury vapour analysers;
- the Nippon Instruments EMP mercury gas monitor;
- the Ohio-Lumex RA-915+ portable mercury vapour analyser.

There are also colorimetric detector tubes available, but these give only a spot reading and are less accurate than direct-reading devices. A Dräger tube gives readings within ±19% of the actual concentration, while a Kitagawa tube is accurate to ±15%. Tubes are used if mercury meters cannot be used for testing in a hazardous area (since they are not intrinsically safe), or if readings are above 1.0 mg/m^3 (1000 μg/m^3) which exceeds the direct-reading device's range.

The above-mentioned handheld mercury analysers provide a far more accurate means of measuring mercury vapour than a Dräger or Kitagawa tube.

Since none of the mercury meters are intrinsically safe, a Hot Work permit will be required to operate the meter in process areas where flammable vapours may be present.

Measuring devices detect only elemental mercury vapour. To guard against skin contamination by mercury or its compounds, high standards of personal hygiene must always be maintained. If clothing is contaminated, it must be removed, and if skin is accidentally contaminated, the worker must shower, scrubbing thoroughly.

10.19.8.3 Controls

A confined space entry permit requires that the atmosphere inside such a space be tested; results should be entered under 'Other Tests' or as an additional comment. If the mercury vapour concentration is less than 0.012 mg/m^3, no respiratory protection is required. If it exceeds 0.012 mg/m^3 but is less than 0.6 mg/m^3, wear a full-face respirator with a type Hg filter should be worn.

If the concentration is greater than or equal to 0.6 mg/m^3, then the vessel or other space should be further cleaned and/or ventilated, then retested until the levels are acceptable, before entry. Unless a vessel is very well ventilated, mercury vapour concentrations inside it can be expected to correlate with temperature, since mercury will be liberated from the walls of a steel vessel more rapidly as temperature increases.

If hot work is necessary in a confined space, the H&S practitioner should either perform a job safety analysis (JSA) or review it. The results will indicate what PPE is necessary.

Care must be taken when heating steel or other metallic vessels or enclosures, even after they have been decontaminated of mercury, since mercury vapour may still be emitted in high concentrations from below the immediate metal surface of the vessel. Mercury is absorbed into, as well as absorbed onto, metal surfaces when it is cooled, and mercury vapour may be released when the surface is reheated. Mercury inside vessels and exchangers can be seen best at night using a torch: the inside walls of the vessel will 'sparkle'. In some instances, hygienists have noted a foul odour in the air (not H_2S/RSH) when significant levels of mercury are present.

10.19.9 Oxides of Nitrogen (NOx)

10.19.9.1 Use and Occurrence

Oxides of nitrogen that the H&S practitioner needs to consider are mainly nitrogen dioxide (NO_2) and nitric oxide (NO), commonly referred to as NOx. Nitric oxide (NO) is a colourless, odourless gas that is only slightly soluble in water. The main sources of NO are from combustion processes. Nitric oxide is then partly oxidised to form nitrogen dioxide.

NO2 is a red-brown gas with a pungent odour and is produced from diesel powered machinery. Although its acrid, biting odour is detectable by the nose at low concentrations, it can also anaesthetise the nose at continuous low concentrations (4 ppm).

10.19.9.2 Toxicology

Although there is limited evidence that nitric oxide alone is toxic, there is some suggestion that it can react with haemoglobin in the blood to form nitrosyl haemoglobin and methaemoglobin, which reduces the blood's ability to transport oxygen. Exposure to nitric oxide along with carbon monoxide (the other main component of diesel emissions) further increases oxygen starvation of the tissues.

The affinity of NO for haemoglobin is 1400 times that of CO. Inhaled nitric oxide can rapidly react with oxygen in the lung to form nitrogen dioxide, which is a potent pulmonary irritant. NO_2 is a respiratory irritant and occupational exposure is via inhalation where common symptoms of exposure are irregular breathing, reduced lung function and wheezing.

10.19.9.3 Standards and Monitoring

SafeWork Australia (2023) lists a TWA—OES (8hr) for NO_2 of 3 ppm and a STEL of 5ppm which is derived from supporting documentation from ACGIH 1991. However, in 2012, the ACGIH adopted a TWA for NO_2 of 0.2 ppm which was based on controlled human exposure studies to NO_2 where resultant respiratory systems' effects were observed. Safe Work Australia has adopted a TWA-OES (8hr) of 25 ppm for nitric oxide. This recommended exposure limit is designed to reduce the potential for respiratory tract irritation.

Portable instruments and colourimetric tubes are widely used to measure NO and NO_2. The same factors and issues need to be accounted for as in the measurement of CO when using these devices such as cross-sensitivity to other compounds. Care needs to be taken in using NO_2 portable gas meters when considering the adoption of a 0.2 ppm TWA. Although the resolution on the sensor is 0.1 ppm, drift common in many gas monitors may lead to the instrument recording an excursion over the prescribed alarm point.

10.19.9.4 Controls

Exhaust ventilation is the main method used to control acid gases and aerosols within mine tunnels and above acid pickling and electroplating baths. Regular testing of diesel vehicles' exhaust emissions is also important. For example, in underground coal mines, vehicles are tested once a month directly at the exhaust pipe. Results must be within certain levels for the vehicle to remain in use underground. Personal protective equipment must be used to protect against splashes to skin and eyes, and specialised respirator filters are available for use against all acidic gases.

10.20 ATMOSPHERIC HAZARDS ASSOCIATED WITH CONFINED SPACES

Confined spaces present several hazards in the workplace: risk of fire and explosion, engulfment, oxygen deprivation and being overcome by toxic gases and vapours. The potential for the development of these hazardous conditions is affected by:

- the work being performed;
- external sources of contamination;
- products used or produced in conjunction with the space;
- natural processes that occur in the space (such as fermentation and decomposition);
- physical nature of the space.

Adequate oxygen levels are vital, and the risk posed by atmospheric contaminants increases dramatically. In fact, more occupational health-related deaths occur in confined spaces than anywhere else. A common scenario involves a worker who drops a tool in a pit and climbs in to retrieve it. He collapses; his workmate assumes he has had a heart attack and enters the pit to rescue him, and both workers die from lack of oxygen, inhalation of toxic gases or both.

No worker should enter a confined space until the atmosphere has been tested and found to be safe.

Confined space refers to an enclosed or partially enclosed space (not a mine shaft or mining tunnel) that: A 'confined space' means an enclosed or partially enclosed space that:

- is not designed or intended to be occupied by a person;
- is, or is designed or intended to be, at normal atmospheric pressure while any person is in the space; and
- is or is likely to be a risk to health and safety from:
  - an atmosphere that does not have a safe oxygen level;
  - or contaminants, including airborne gases, vapours and dusts, that may cause injury from fire or explosion; or
  - harmful concentrations of any airborne contaminants; or
  - engulfment.

According to the Australian Model Work Health and Safety Regulation 2023a (Part 4.3), a confined space can be inside a vat, tank, pit, pipe, duct flue, oven, chimney, silo, reaction vessel, container, receptacle, underground sewer, shaft, well, trench, tunnel or similar enclosed or partially enclosed structure. The risks associated with confined spaces include:

- loss of consciousness, injury or death owing to the immediate effects of contaminants;
- fire or explosion from the ignition of flammable contaminants;
- asphyxiation resulting from oxygen deficiency when the oxygen level falls below 19.5%;
- enhanced combustibility and spontaneous combustion when the oxygen level exceeds 23.5%;
- asphyxiation resulting from engulfment by 'stored' material—for example, grain, sand, flour, fertiliser.

Working in confined spaces may also greatly increase the risks of injury from the following:

- **Mechanical hazards** associated with plant and equipment, which may result in entanglement, crushing, cutting, piercing or severing of limbs or digits. Sources of mechanical hazards include augers, agitators, blenders, mixers, stirrers and conveyors.
- **Ignition hazards** associated with plant and equipment in or near the confined space. The presence of ignition sources in a flammable atmosphere may result in fire or explosion, and the death or injury of workers. Examples of ignition sources include open flames, sources of heat, static or friction, non-intrinsically safe equipment, welding and oxy-cutting, hot riveting, hot forging, electronic equipment such as cameras, pagers and mobile phones and activities that generate sparks, such as grinding, chipping and sandblasting.
- **Electrical hazards**, which may result in electrocution, electric shock or burns.
- **Uncontrolled substances** such as steam, water or other materials, whose presence or entry may result in drowning, being overcome by fumes, engulfment by earth or rock, and so on.
- **Noise**, whose levels can be greatly increased in confined spaces.
- **Manual handling hazards** arising from work in cramped, confined areas. These may be increased using personal protective equipment such as an airline or harness, which restricts movement, grip and mobility.
- **Radiation hazards** from lasers, welding flash, radio frequency (RF) waves and microwaves, radioactive sources, isotopes and X-rays.
- **Environmental hazards**, including heat or cold stress, which may arise from work or process conditions, wet or damp environments, slips, trips and falls arising from slippery surfaces.

- **Biological hazards**, including infection by microbes and contact with organisms such as fungi, which may cause skin disease or result in respiratory illness, and mites in grain, which may result in dermatitis. Viruses and bacteria may also present a hazard. Exposures to *Leptospira* species and *Escherichia coli* are of particular concern for people who work in sewers. Insects, snakes and vermin are also potential hazards.
- **Traffic hazards**, which may arise where confined-space entry or exit points are located on walkways or roads. Workers entering or exiting the space risk being struck and injured by vehicles such as cars or forklift trucks. The potential for others to fall into the space may also exist.

Several situations may arise during work in confined spaces in which the oxygen level will decrease and/or other gases be generated. Some examples are:

- removal of oxygen by activated carbon, or by some soils undergoing microbiological activity;
- decrease in oxygen level by the oxidation (rusting) of freshly grit-blasted metal surfaces;
- partial oxidation of hydrocarbons, in the presence of catalysts such as alumina, to produce carbon monoxide and decrease oxygen levels, thus generating two hazards;
- decrease in oxygen and the generation of toxic gases from the reaction of pyrophoric materials (pyrophoric iron will self-ignite in the presence of oxygen);
- the presence of hydrocarbons under a 'nitrogen blanket'. Some vessels are purged with nitrogen gas to minimise the risk of flammability, creating a nitrogen-rich atmosphere and all but eliminating oxygen. As catalytic explosive sensors require at least 10% oxygen to give an accurate reading, they should not be used in such conditions, as a false low reading can be obtained, leading to underestimation of the risk. While taking into consideration the suitability of the sampling technique, the best practice is to use an infrared explosive sensor or a thermal conductivity sensor, neither of which needs oxygen to give a measurement.

In summary, confined-space work requires special procedures, training and testing to ensure safe operations and protect the worker. More information can be found in the Safe Work Australia *Confined Spaces Code of Practice* (SWA 2020) and Australian Standard AS 2865–2009 Confined spaces (Standards Australia 2009).

10.21 EMERGING MONITORING TECHNOLOGY

Technological advances over the past decade have seen a progression towards smaller and more functional portable direct reading instruments for monitoring gases and vapours in the workplace and environment. This has led to the opportunity for more workers to be issued a personal monitor and more access for the H&S practitioner to conduct monitoring.

Development and growth of wireless communications combined with advances in cloud-based software programs, storage devices and mobile smartphone capable applications are rapidly becoming important resources for the H&S practitioner.

10.21.1 Wireless Technology and Remote Monitoring

Technology is now available for continuous personal and area monitoring from a central location. Gathering information from multiple points or workers at once, this allows the H&S practitioners to view data remotely in real time and provides a dynamic approach to hazard control. Real-time personal monitoring equipment with integrated Global Positioning System (GPS) is available for monitoring of volatile organic compounds, benzene, and other toxic and hazardous gases. Broadcasting real-time peak level alarms as well as TWA and STEL alarms to a central host computer offers a cost-effective way of increasing safety and minimising exposure risks to employees or contractors. Real-time communications between wirelessly enabled instruments allow the H&S practitioner to *detect change* in a workplace which inherently reduces the risk of exposure to gas and vapour hazards that adversely affect the health and well-being of workers and surrounding communities. Examples are wireless area-monitoring systems that may be configured to warn of potential hazard exposure well before a plume or release enters a working area. Being battery powered, these systems are rapidly deployable from one part of a site to another and are especially useful in areas where access to electricity may be limited.

As the gathered data are stored in the cloud via a central computer, the H&S practitioners may review them at any time from any computer connected to the internet. This is especially useful as a complementary monitoring technique—for example, in the offshore petroleum industry—as it enables the monitoring of unique, uncommon or unplanned maintenance tasks where exposure levels would otherwise be very difficult to capture.

Interoperability of wireless detection and monitoring systems also allows the sharing of data with first responders such as fire or ambulance services, which helps increase safety during emergencies.

Portable wireless enabled gas monitors such as those used in confined space entry have the option to employ 'man-down' technology. Should a worker be incapacitated through a fall, health incident or a sudden catastrophic release of a toxic gas, a remote alarm will be sounded to co-workers either outside the space or close by. This technology reduces the response time of emergency service workers to the incident.

10.21.2 Data Collection and Evaluation

Transmission of occupational exposure data in real time from a monitoring device allows for integration into proprietary software systems. These software solutions have been developed to provide a cost-effective and time-saving mechanism for the analysis of the data for the occupational hygienist. Calculations and analysis which could take hours or days to complete can be performed in minutes.

10.21.3 New and Emerging Resources

There are many online databases that can be referred to for exposure limits and detection options, as well as software programs in which entering a chemical name or CAS number instantly yields basic information on exposure limits, sampling equipment and media, and so on. There are also many instructional videos on YouTube on how to detect gases and use various types of equipment. Most suppliers are using this medium to supply training in the use of their equipment.

REFERENCES

American Conference of Governmental Industrial Hygienists 2018 *Documentation of the Threshold Limit Values for: Chemical Substances, Physical Agents and Biological Exposure*, ACGIH, Cincinnati, OH.

American Conference of Governmental Industrial Hygienists 2023, *Documentation of the Threshold Limit Values for: Chemical Substances, Physical Agents and Biological Exposure*, ACGIH, Cincinnati, OH.

Australian Dangerous Goods Code (ADGC) 2022, *Australian Code for the Transport of Dangerous Goods by Road & Rail (Edition 7.8)*, National Transport Commission, https://www.ntc.gov.au/sites/default/files/assets/files/Australian%20Dangerous%20Goods%20Code%20-%207.8.pdf [9 January 2024].

Grantham, D. 1992, *Occupational Health & Hygiene: Guidebook For The WHSO*, D.L Grantham, Queensland, Australia.

Health and Safety Executive 2000, *Volatile Organic Compounds in Air: Laboratory Method Using Pumped Solid Sorbent Tubes, Solvent Desorption and Gas Chromatography, MDHS 96*, HSE, Norwich, www.hse.gov.uk/pubns/mdhs/pdfs/mdhs96.pdf [10 October 2023].

Murnane, S.S., Lehochy, A. H. & Owens, P.D. (Eds) 2013, *Odor Thresholds for Chemicals with Established Occupational Health Standards* (2nd edn), AIHA, Akron, OH.

National Institute of Occupational Safety and Health 2017, *NIOSH Manual of Analytical Methods (NMAM®)* (5th edn), Centers for Disease Control and Prevention, Atlanta, GA, https://www.cdc.gov/niosh/nmam/5th_edition_web_book.html [10 October 2023].

Occupational Safety and Health Administration n.d., *Sampling and Analytical Methods for Benzene Monitoring and Measurement Procedures, 1910.1028 App D*, US Department of Labor, Washington, DC, https://www.osha.gov/laws-regs/regulations/standardnumber/1910/1910.1028AppD [10th October 2023].

OECD 2006, *OECD Guidelines for the Testing of Chemicals, Section 1 – Test No. 104: Vapour Pressure*, OECD, https://www.oecd-ilibrary.org/environment/test-no-104-vapour-pressure_9789264069565-en [9 January 2024].

Safe Work Australia (SWA) 2013, *Guidance on the Interpretation of Workplace Exposure Standards for Airborne Contaminants*, SWA, Canberra, https://www.safeworkaustralia.gov.au/doc/guidance-interpretation-workplace-exposure-standards-airborne-contaminants [10th October 2023].

Safe Work Australia (SWA) 2020, *Work Health and Safety (Confined Spaces) Code of Practice*, SWA, Canberra, https://www.safeworkaustralia.gov.au/doc/model-code-practice-confined-spaces [10th October 2023].

Safe Work Australia (SWA) 2022, *Workplace Exposure Standards for Airborne Contaminants*, SWA, Canberra, https://www.safeworkaustralia.gov.au/system/files/documents/1804/workplace-exposure-standards-airborne-contaminants-2022_0.pdf [10th October 2023].

Safe Work Australia (SWA) 2023, *Model WHS Regulations*, SWA, Canberra, www.safeworkaustralia.gov.au/doc/model-whs-regulations [9 January 2024].

Standards Australia 2003a, *Workplace Air Quality—sampling and Analysis of Volatile Organic Compounds by Solvent Desorption/Gas Chromatography—Pumped Sampling Method, AS 2986.1:2003*, SWA, Canberra, https://www.standards.org.au/access-standards

Standards Australia 2003b, *Workplace Air Quality—sampling and Analysis of Volatile Organic Compounds by Solvent Desorption/Gas Chromatography—Diffusive Sampling Method, AS 2986.2:2003*, SWA, Canberra, https://www.standards.org.au/access-standards

Standards Australia 2009, *Confined Spaces, AS/NZS 2865:2009*, Standards Association of Australia, Sydney, Standards New Zealand, Wellington, New Zealand, https://www.standards.org.au/access-standards

11 Biological Monitoring of Chemical Exposure

Dr Gregory E. O'Donnell and Dr Martin Mazereeuw

11.1 INTRODUCTION

The role of the occupational hygienist is to assess exposures to hazards and when found to be excessive, to suggest ways to reduce exposures to an acceptable level. Occupational hygienists can assess exposure by performing air monitoring, however, air monitoring may not always be the best option to assess the exposure for several reasons. These can include the nature of the chemical hazard, its chemical properties and the work processes involved and, therefore, may not always give an accurate estimation of the real short- or long-term exposure dose.

Biological monitoring of occupational exposure is defined as the assessment of exposure to a chemical by the measurement of the chemical itself or its metabolites in human samples such as blood, urine or exhaled breath. The measurement result is indicative of the internal dose resulting from an exposure to the chemical. It should be used as an early indicator of exposure before health effects or harm occurs. It is not an indicator of the health status of the worker. Biological monitoring is a measure of the internal dose of a chemical exposure obtained via inhalation, ingestion and dermal absorption. It therefore measures the actual total exposure dose of the individual to the chemical rather than a predicted dose by measuring the chemical in the external breathing zone of the worker. Biological monitoring thus integrates exposures from multiple pathways and sources; however, it cannot easily identify the source or pathway or the relative contributions from the sources. Advances in analytical chemistry and the acknowledgement of the uncertainties involved in external air measurements have led to biological monitoring becoming a well-established method to estimate chemical exposure and manage risk. It is an evolving area with more analytical methods becoming increasingly available.

The theoretical pathway to the development of disease from occupational exposure to a chemical occurs via several pharmacokinetic processes after absorption. Inhalation usually plays a major contribution to the internal dose by breathing in gases, vapours, fumes, dusts or mists. Biological monitoring can be effective in estimating the internal dose; however, other effects can occur in the body which may be a response to the exposure but could also relate to other non-occupational influences. These parameters can be measured by biochemical monitoring which include general health-based markers such as liver function tests and other general pathology tests. Tests that are aimed at effects that have occurred in the body specifically from chemical exposure include the measurement of the activity of the cholinesterase enzyme in whole blood for the exposure to organophosphate insecticides; the level of zinc protoporphyrin for the exposure to lead; chromosomal aberrations, sister chromatid exchange; and other protein, haemoglobin and DNA adducts. These types of measurements are usually called biological effect monitoring.

DOI: 10.1201/9781032645841-11

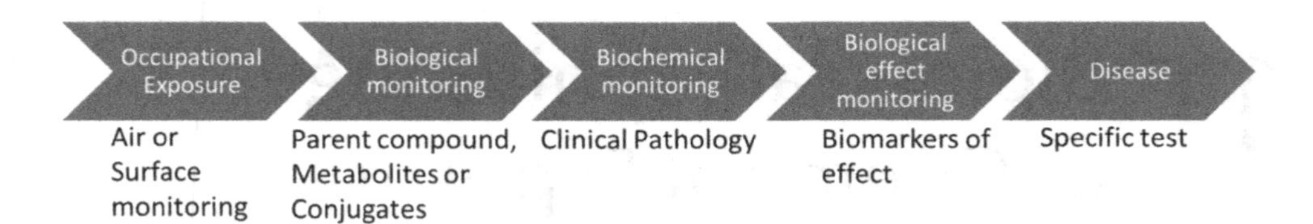

FIGURE 11.1 Schematic illustration of the pathway of occupational exposure to chemicals and the monitoring strategies employed to assess the exposure.

Source: G. E. O'Donnell & M. Mazereeuw.

The final group of tests is used to determine the suggested risk or onset of disease. This process and available monitoring strategies are shown in Figure 11.1.

11.2 USING BIOLOGICAL MONITORING

While the prime objective of biological monitoring is to determine whether individuals are at an increased risk of experiencing future adverse health effects, its use can fulfil other functions such as determining if a particular chemical is above a mandated regulatory limit. Biological monitoring can be useful if the exposure occurs via routes other than inhalation, for example, where skin absorption or inadvertent ingestion may be significant exposure routes. These routes of exposure may not always be obvious to the observer. Biological monitoring can also reveal unanticipated exposures that have occurred outside the workplace.

When controls have been put in place, biological monitoring can be used to monitor if these controls are effective. Personal protective equipment (PPE) such as protective clothing, gloves and respirators are only effective when used in the correct manner, are properly maintained and are free from any previous contamination. Biological monitoring can verify the effectiveness of PPE. Air monitoring cannot confirm if PPE is effective because it is very difficult or even impossible to take an air sample in the breathing zone when a respirator is worn. Biological monitoring can be used to confirm assessments of exposure estimated by air monitoring.

Biological monitoring is also a good technique to determine if an exposure has taken place after an incident has occurred. For instance, if a chemical spill has taken place, it is usually too late to take an air sample; however, a biological monitoring sample can still give some indication of the extent of exposure to the workers.

11.3 HEALTH MONITORING

Health monitoring is the assessment of workers' health status by clinical, biochemical, imaging or instrumental testing to detect any clinically relevant, occupationally dependent change of a single worker's health (Manno, et al. 2010). Health monitoring can only be performed by or under the supervision of a registered medical practitioner experienced in health monitoring. Biological monitoring can be used as part of a health monitoring scheme and can be used to help interpret clinical findings.

11.4 SELECTION OF SAMPLE MATRIX

Biological monitoring can be performed in a variety of matrixes including urine, blood, exhaled breath, oral fluid and hair in order of decreasing significance. The main matrixes used for biological monitoring are the following.

11.4.1 Urine

By far the most frequently used matrix in occupational exposure assessment is urine. It is easy to collect and much less invasive than blood. It can be easily collected before a shift and at the end of a shift to estimate exposure throughout a shift. It can be repeatedly collected throughout the work week to show any accumulation and excretion patterns.

11.4.2 Blood

Most compounds can be measured in blood; however, most organic compounds and their metabolites have short half-lives in blood requiring the sample to be taken within hours of exposure. Also, the sampling of blood is invasive and requires trained medical staff. It is usually performed in a clinical setting. If blood sampling is performed, any individuals handling the samples need to be trained in the risks of infection and should be offered appropriate immunisations to minimise the risk of infection.

11.4.3 Exhaled Air

Exhaled air can be used for some volatile organic compounds. However, if the chemical is quickly metabolised in the blood, then exhaled air is not suitable and urine or blood may be a more appropriate sample.

11.4.4 Oral Fluid

Oral fluid has come to prominence in the area of monitoring for illicit drugs of abuse. Spot test kits using immunoassay analysis are commercially available to perform this analysis and a confirmatory test is performed later in a laboratory. Oral fluid has not as yet been extensively employed in other areas; however, other test kits may become available in the future.

11.4.5 Hair

Hair is not often used in the occupational biological monitoring area. It is increasingly being used in the forensic science area as it is easy to collect, relatively stable, difficult to adulterate or substitute and can give a long chronic estimate of exposure of the previous months or years. However, it is prone to external contamination and pre-analysis washing regimes are yet to be standardised.

11.5 ACCOUNTING FOR DAILY URINE VOLUME IN BIOLOGICAL MONITORING

When two workers are exposed to the same amount of chemical in a work shift, the worker who has consumed more liquid (water, tea or coffee, etc.) in that shift will excrete more urine. Therefore, the amount of chemical excreted will consequently appear to be diluted. Hence, reporting the concentration of the chemical per litre of urine will give a false impression that the hydrated worker has been exposed to a much-reduced amount of the chemical. There are a number of ways to account for the differences in urine volume excreted, and these are listed below. The ideal urine reference parameter should provide little inter-individual variation and should show a constant day-to-day excretion. The parameter should have little influence on other biological factors such as diet or physical activity and should be easy to analyse (Greim & Lehnert 1998). For the most accurate normalisation, the chemical should have come to a steady-state concentration in blood, the time span between the last urination is an important variance and a better correction would be a regression model using an analyte-specific correction (Hertel, et al. 2018). The reference parameters that can be used for biological monitoring are the following.

11.5.1 Volume

It is cheap and easy to collect a 24-hour urine sample. However, the collection is time consuming and requires an amount of discipline by the workers to collect all urinations, making it unreliable. It also presents a danger of sample contamination.

11.5.2 Osmolality

Osmolality is the depression of the freezing point and is independent of diuresis. It is, however, time consuming and influenced by pH and ethanol and has poor international acceptance.

11.5.3 Specific Gravity

Specific gravity is usually performed by reporting the test results to a standard specific gravity of 1.020 g/mL. It is cheap and easy and is relatively independent of the duration of the collection period. However, it is influenced by proteinuria and glucosuria as well as numerous medicines.

11.5.4 Normalisation to Creatinine

The analysis of creatinine in urine is the internationally accepted procedure to account for different urine volumes excreted by individual workers each day. Creatinine is a breakdown product of the muscles and a constant amount is excreted each day by each individual. Therefore, expressing the test result of the analysis to the amount of creatinine instead of per litre of urine nullifies the effect of different volumes of urine. The advantage of creatinine as the reference parameter is that it reduces the influence of diuresis-related fluctuations and has a good correlation with 24 hours urine collection. It is unaffected by proteinuria and glucosuria or by salts. However, it is influenced slightly by age, sex, meat consumption, certain medicines, heavy physical work and long periods of fasting. It is also unsuitable for substances that are not mainly filtered in the glomeruli of the kidneys.

The examples in Table 11.1 illustrate the process of creatinine adjustment. In the case of exposure to benzene, the measurement of the metabolite *S*-phenylmercapturic acid (*S*-PMA) is performed and an example of exposure to mercury is shown.

TABLE 11.1
Examples of Creatinine Adjusted Calculations for *S*-PMA and Mercury

Analyte	Worker 1	Worker 2
S-PMA	10.0 μg/L	10.0 μg/L
Creatinine	0.5 g/L	2.0 g/L
Creatinine adjusted *S*-PMA	20 μg *S*-PMA/g creatinine	5 μg *S*-PMA/g creatinine
Analyte	**Worker 3**	**Worker 4**
Mercury (Hg)	0.05 μmol/L	0.10 μmol/L
Creatinine	0.0020 mol/L	0.0250 mol/L
Creatinine adjusted Mercury	25.0 μmol Hg/mol creatinine	4.0 μmol Hg/mol creatinine

As shown in Table 11.1, Worker 1 has the same concentration of *S*-PMA µg/L in the urine as Worker 2. However, Worker 1 has one-quarter the concentration of creatinine g/L in the urine showing that Worker 1 is well hydrated, and the urine *S*-PMA µg/L concentration is diluted compared to Worker 2. After normalising to the amount of creatinine, Worker 1 shows a higher concentration of *S*-PMA/g creatinine in the urine and therefore a higher dose. In the second example in Table 11.1, Worker 3 has half the concentration of mercury in urine 0.05 µmol/L to Worker 4 of 0.10 µmol/L. After adjusting for the concentration of creatinine, it is shown that Worker 3 has more than six times the estimated dose compared to Worker 4.

11.6 CHEMICAL ELIMINATION FROM THE BODY

Generally, the pharmacokinetics of exposure of a worker to a chemical goes through four main stages: (i) absorption, (ii) distribution, (iii) metabolism and (iv) excretion. The chemical is taken up initially by absorption through the lungs or the skin or by the digestive tract. It is then distributed to different organs and tissues in the body by the biological fluids. As the chemical goes through this process it is absorbed, metabolised and excreted by various organs. Each of these organs can be theoretically described as a compartment with its own excretion rate before the chemical is reabsorbed by the next compartment which again has its own absorption and excretion rates, influenced by the preceding compartment. The true pharmacokinetic model can quickly become very complex and, for practical reasons, is usually treated as one or two compartments showing a one- or two-phase elimination curve. For instance, a chemical being absorbed by inhalation and dermal routes can sometimes appear to have a bimodal elimination curve as dermal excretion usually takes longer. A one-phase elimination curve is illustrated in Figure 11.2. The time that it takes to excrete half of the concentration of the chemical from the body is known as the elimination half-life, $t_{1/2}$. It should not be confused with the stability of the chemical in the biological media after sampling.

When the time it takes to eliminate the chemical from the body is longer than the time between exposures, then an accumulation can result. With repeated day-to-day exposures over the course of the working week this can result in the chemical not being totally eliminated from the body until exposure is discontinued. This scenario is illustrated in Figure 11.3.

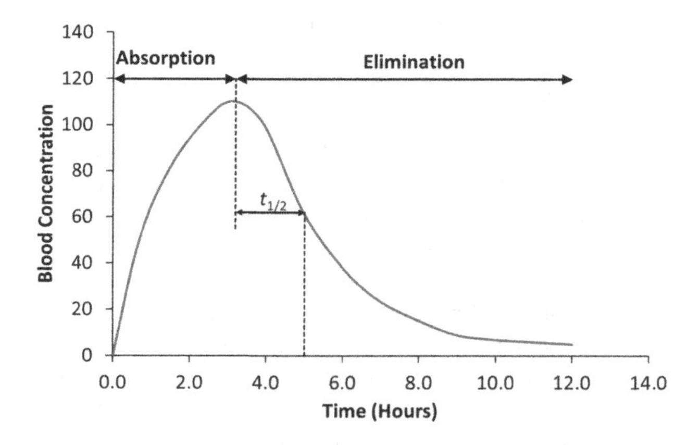

FIGURE 11.2 The absorption and elimination of the concentration of a chemical in blood following the first-order kinetics and the corresponding elimination half-life, $t_{1/2}$.

Source: G. E. O'Donnell & M. Mazereeuw.

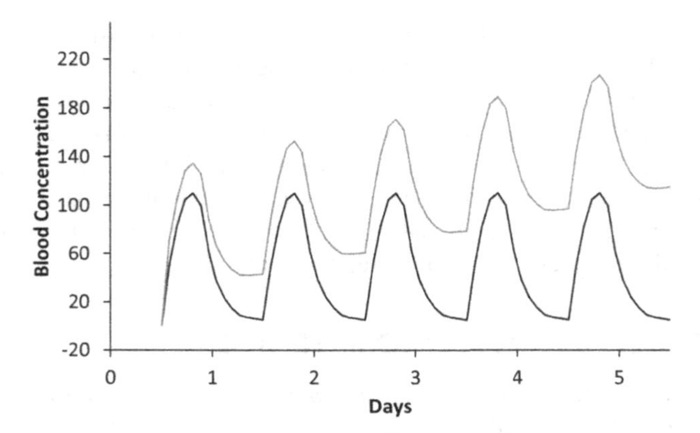

FIGURE 11.3 The absorption and elimination of the chemical over a working week. Black line shows when the elimination is completed before the next exposure. The red line shows when the elimination is longer than the next exposure, showing an accumulation over the working week.

Source: G. E. O'Donnell & M. Mazereeuw.

The elimination half-life is important when estimating the internal dose and is relevant to determine when the most appropriate time is to take a biological monitoring sample. Generally speaking, most polar compounds (parent compounds or metabolites) are excreted fairly quickly within hours and less than one day. These compounds are best sampled at the end of the shift or the following morning. These compounds include: most solvents, organophosphate insecticides, herbicides, isocyanates, 4,4′-methylene-bis-chloroaniline, fluoride, methyl bromide, cyanide and cytotoxic drugs. Compounds with elimination half-lives greater than one day accumulate over the working week and hence are best sampled at the end of shift at the end of the work week. These compounds include: most metals, chlorinated solvents such as trichloroethylene and perchloroethylene, and polycyclic aromatic hydrocarbons. Compounds with longer elimination half-lives include some heavy metals such as mercury, cobalt, cadmium, lead and manganese, and endocrine disruptor chemicals such as polychlorinated biphenyls and organochlorine insecticides. These chemicals are absorbed into body organs and lipophilic tissue and are slowly released into the blood and reach a steady-state concentration which allows sampling at any time.

11.7 SKIN ABSORPTION

Dermal transmission is an important exposure route that can contribute significantly to the total exposure of an individual. The skin is a complex organ which consists of various layers with different functions. Three main regions can be distinguished, namely epidermis, dermis and hypodermis. The primary barrier for dermal exposure is the outermost layer of the epidermis, the stratum corneum, which is typically only 10–20 μm thick. The stratum corneum consists of non-living cells surrounded by a lipid coat. The lipid coating forms a continuous lipid phase and is regarded as the primary route through which chemicals travel through the skin. Hydrophobic components can be absorbed by and diffuse through the lipid layer into the relatively polar epidermis/dermis and eventually enter the bloodstream. Skin penetration varies across the body and is relatively strong at the scalp/forehead, slower at the rest of the body and slowest at the palms and foot soles (Ng & Lau 2015). The diffusion through the skin is determined by the size and weight of the molecules. As a rule of thumb, chemicals with a molar weight of <500 gram, which covers a very large portion of industrial chemicals, diffuse well through the lipid layer. Diffusion is enhanced by an increased temperature and the hydrophobicity of the chemical. Highly polar or hydrophilic components such as water will not penetrate the lipid layer well if at all. Apolar or hydrophobic components will penetrate the lipid layer easily but might only slowly re-enter the polar dermis layer beneath the stratum corneum. In extreme cases of hydrophobicity, the components will not enter the epidermis/dermis. Components with an 'apolar backbone', such a benzene ring or in combination with polar groups, such as –OH (hydroxy), –COOH (carboxy acid), –O– (ether), –C=O (ketone/aldehyde), and $-NH_2$ (amine) groups will readily enter and diffuse through the lipid layer, enter the polar environment of the underlying skin layers and be transported through the body. Typical examples are industrial solvents, pesticides, aromatic amines, phenolic components, etc. Within these groups, the penetration rate depends again on the hydrophobicity. For example, relatively hydrophilic compounds, such as glyphosate, will penetrate the skin but at a lower rate compared to, for example, benzoic acid. When components have been given a biological exposure index skin notation, then skin absorption may be a significant exposure route.

Skin penetration is increased in damaged skin, due to cuts or UV radiation exposure, or skin with a reduced lipid layer, due to the use of lipid-dissolving chemicals, like acetone and degreasers, but also surfactants.

Protective clothing and gloves are essential to manage skin absorption, thus the glove material should be chosen with care depending on the chemical used. Leather gloves are not recommended when working with chemicals as they don't provide an impermeable barrier. Quality glove providers publish the suitability of their glove options to handle different chemicals. As chemicals, just like with skin, can migrate through gloves, regular changes are important and damaged gloves provide no protection. In most cases, damaged gloves or use of gloves after exposure to the hands will promote skin absorption, as part of the chemicals will be trapped close to the warm skin for a prolonged period (Holmgaard & Nielsen 2009; Riviere & Minteiro-Riviere 2010).

11.8 EXPOSURE STANDARDS

Biological monitoring occupational exposure limits have been developed over the years by several prominent bodies from countries such as the United States (US), Germany, the United Kingdom (UK) and the European Union (EU). The values proposed by these bodies have largely been adopted internationally. These bodies evaluate the current research including background population studies, workplace exposure scenarios, chemical incidents, epidemiological studies and health risk assessments. Typically, health risk thresholds are estimated and a limit is set considerably lower after applying an uncertainty factor. The general approach in research is to determine biological equivalents using the forward dosimetry model where an amount of the chemical is measured in the biological media resulting from a known concentration in air (Hays, et al., 2007). Obviously, this approach is only applicable to exposures where the main route of exposure is inhalation. When dermal exposure is predominant other studies such as patch testing are more applicable. A more pragmatic approach has been introduced in the United Kingdom in recent times which equates the 90th percentile of the resulting concentration of the chemical in biological fluid when good occupational hygiene controls are in place. This is particularly useful when limited research data is available and also for carcinogenic chemicals where no threshold limits can be established. The foremost biological occupational exposure standards and their standard setting bodies are listed below:

- Biological Exposure Index (BEI®), set by the American Conference of Governmental Industrial Hygienists (ACGIH®) in the United States (ACGIH 2023).
- Biological occupational tolerance value (Biologischer Arbeitsstoff Toleranz-Wert or BAT), set by the German Research Foundation, Deutsche Forschungsgemeinschaft (DFG) (DFG 2022).

- Biological Limit Values (BLV) from the EU Scientific Committee on Occupational Exposure Limits (SCOEL). Through the publications of the SCOEL, information on occupational exposure values from various EU countries can be found (SCOEL 2023).
- Biological Monitoring Guidance Value (BMGV) from the Health & Safety Executive (UK) (HSE 2023).
- In Australia, SafeWork NSW has set Biological Occupational Exposure Limits (BOEL) which have largely been adopted from the above organisations and are published with the associated guidance notes by Safe Work Australia (Safe Work Australia 2013).

Currently, most biological exposure standards, except for lead in blood, are not cited in regulations in Australia. However, health monitoring guidance published by Safe Work Australia references biological exposure levels at which actions should be taken to improve controls.

Exposure limit values are not values where health effects occur but rather values below which health effects should not occur. They are warning levels that should trigger an investigation into the work environment, processes and exposure controls. Some organisations set their own action limits that are usually below an official exposure limit. Care must be taken when assuming safe levels of chemical exposure, as the impact of exposures will vary per individual, depending on their general health, genetics, age, diet, any personal medication and other environmental factors. A regular exceedance of the exposure limit is liable to affect the long-term health of the individual. However, lower exposures and/or small but extended exposures are not necessarily harmless and should be reduced as far as reasonably practicable.

11.9 BIOLOGICAL MONITORING PROGRAMME

The regulations that apply to health monitoring can be found in the Model Work Health and Safety Regulations, which are available through the Safe Work Australia website (Safe Work Australia 2023). These model regulations have been adopted by most Australian work health and safety jurisdictions; however, duty holders should ensure they are familiar with the specific regulations applying to their state, territory or jurisdiction.

Biological monitoring is essentially a workplace exposure measurement analogous to air or swab samples and therefore can be undertaken by any competent professional. However, when biological monitoring is undertaken as part of health monitoring, it must be conducted or supervised by a medical professional with experience in health monitoring. This could be a general practitioner, however, a medical professional with more specialised workplace exposure skills such as an occupational physician is preferred. It is worthwhile to discuss monitoring strategies with a medical professional in the planning phase of the assessment (AIHA 2004).

The use of urine samples for biological monitoring is considered here and is the usual preferred option and the tests available are usually focused on this media for reasons noted earlier. If, however, the collection of blood is necessary, then a qualified phlebotomist must be involved.

To develop a biological monitoring programme, the following aspects should be addressed.

11.9.1 Programme Manager

A Programme Manager should be appointed and should be a qualified professional with relevant knowledge of biological monitoring and interpretation of the data, as well as the work processes used in the workplace.

11.9.2 Monitoring Process

The following aspects are important when setting up the monitoring process.

11.9.2.1 What, Where and Who to Measure

Before the commencement of a programme, a thorough understanding of the work processes, the use of the chemicals and the workers involved is important for a meaningful review of the occupational exposures. Is biological monitoring the most appropriate method to assess the exposure?

11.9.2.2 Analytics

Is there a laboratory available capable of performing the required analysis? Analytical chemistry has progressed recently with the introduction of very sensitive instruments and is key to biological monitoring. Care must be taken to select the target molecule and the associated analytical method. The target molecule can be the original chemical or its metabolite. When testing for the original chemical, contamination from clothes, hands and hair could occur and invalidate the results. Hence, correct sampling protocols should be sought from the laboratory.

The following aspects are important to discuss with the laboratory:

- **Specificity** – ideally a target molecule is specific for the hazardous chemical and has no or minimal interference from non-occupational influences, such as diet or general metabolism within the body. A typical example is the use of phenol and *S*-phenylmercapturic acid (*S*-PMA) as target molecules for benzene exposure. Although both are metabolites of benzene, a background level of phenol can be found in a large number of urine samples due to diet and other non-occupational sources. As such, phenol is only valuable as a marker for high exposures where the phenol levels are beyond typical background levels. *S*-PMA, on the other hand, is highly specific and there are no known interferences that could influence the reliability of the measurement (Lauwerys & Hoet 2001).
- **Selectivity** – the selectivity of a method is the ability to separate and quantitatively identify the target chemical from the sample matrix. Urine is a complex and variable matrix and highly selective analytical methods are typically necessary. With the advent of mass spectrometric detection systems coupled to gas chromatography instruments, inductively coupled plasma instruments and more recently to liquid chromatography instruments have allowed the analytical power required for selective biological monitoring analysis.
- **Sensitivity** – the analysis should have sufficient response to clearly show a change in the chemical concentration.
- **Limit of quantitation** – the analysis should be able to measure well below the exposure standard or reference value and, ideally, right down to the background environmental levels.
- **Quality assurance** – a laboratory ideally should be accredited to a recognised quality accreditation standard, such as ISO 17025 or ISO 13485 and successfully participate in a relevant interlaboratory proficiency testing scheme.

11.9.2.3 The Location, Timing and Frequency of the Sample Collection

The collection location should be fit for purpose. The timing of the sample collection should be less than three times the elimination half-life of the chemical. The use of pre-shift and post-shift samples will allow you to exclude non-occupational exposures and might reduce the impact of dietary background influences. When the elimination half-life is longer than the interval between exposures pre-shift and post-shift samples can show the accumulation of the chemical over the working week. Increasing the sampling frequency will create a more detailed overview of the investigated exposure. The run time for the entire study should be determined and in line with the aim of the study.

11.9.2.4 Reference Level

The found levels should be compared against a known exposure limit or known environmental background levels (see Section 11.9.2.2). Where no exposure limit is available, biological monitoring may still be very useful provided the testing will give a quantitative result that assists in the management of the exposure and assessing the controls. In this instance, it is important to ensure that the right metrics are available to establish internal reference values.

11.9.2.5 Statistics

Statistics can help to interpret large and complex data sets. Also, it can help with determining the minimum required sample numbers while maintaining a statistically sound study. Advice from a statistician should be considered in the design phase of a monitoring study.

11.9.2.6 Additional Data

Additional information on diet, smoking, start and end times, chemicals and application processes used, episodes of skin contact, Safety Data Sheet information, non-work exposure sources, controls and PPE used, and timing of an exposure incident (as applicable) will be necessary for the interpretation of the results and should be collected at the time of sampling.

11.9.2.7 Sample Volume and Storage

The specimen must be collected in amounts that are sufficient for analysis. This is particularly important for multiple analyses on a single sample and the laboratory can advise on this. It is best practice to store samples as fast as is practical after collection in a separate portable esky, fridge or freezer with a biological hazard label. The storage unit should not be used for food and drinks. Lower storage temperature will prolong the quality of the sample. There are no universally applicable rules for all chemicals, but fridge storage will typically extend the shelf-life of samples to several weeks and when frozen a further extension to 6–12 months can be expected for many chemicals.

11.9.2.8 Sample Transport

Transport to the laboratory should be done by road courier or by air transport companies. Air transportation of biological monitoring samples should comply with the International Air Transport Association (IATA) Dangerous Goods Regulations. Transportation should be done overnight or, if feasible, using a same day delivery to the laboratory. To maintain sample integrity, it is strongly recommended to avoid transportation over the weekend or public holiday periods. Furthermore, while in transit, the samples should be kept under cold conditions until the samples are received by the laboratory. This is usually achieved by transporting the samples in an esky loaded with frozen esky bricks or ice packs. Care should be taken that samples are clearly labelled with indelible ink. Wet ice is not preferred by couriers due to the increased risk of leakage and should only be used when separated from the samples by an impermeable barrier that allows sample labels to remain legible and intact. Dry ice is not necessary for short transport periods. However, for longer, international transport, it might be worth considering. Specialised transport companies provide transportation services around the world, while maintaining a constant specimen temperature. Note that the very low temperature of dry ice (< –65°C) can damage marker pen writing and strongly affects the glue layer of labels, making them drop off the containers. Special labels are available for dry ice transport.

11.10 CONSULTATION

All workers and/or their representatives should be consulted in the planning phase of a biological monitoring programme to discuss the following.

11.10.1 The WHS Regulations

The model WHS Regulations on health monitoring links to specific jurisdictional regulations and health monitoring guidance material that can be found on the Safe Work Australia website and should be explained to the staff involved.

11.10.2 Consent Process and Privacy

Prior to collecting samples, *informed consent* must be obtained from the individual worker. This involves giving a clear overview of the purpose of the monitoring programme, outlining what samples were taken and any risk involved, what analyses are to be performed and who will have access to the test results and how the data will be used. Many workplaces choose to seek consent to biological monitoring on a condition of employment basis, as biological monitoring data can be a key piece of information in managing exposure risks.

There might be concerns over the use of the sample for other analyses, like illicit drugs, medication or alcohol. Strict controls must be in place to ensure that the samples are solely used for workplace exposure monitoring. Testing for drugs and alcohol at work has its own place and should not be entangled with biological monitoring to evaluate exposure to hazardous chemicals.

11.10.3 Refusal

A worker has the choice to refuse to participate in a biological monitoring programme. The benefits of biological monitoring should be fully explained and assurances of strict privacy controls should be clarified to the participants.

11.10.4 Feedback and Actions in Case of Exposure

Prior to the start of the biological monitoring programme, participants should be counselled on the aims of the programme and how a found level of exposure will be handled, the impact on their work and how access to a medical professional can be obtained.

11.10.5 Privacy and Confidentiality

In the United States, OSHA stipulates that a test report from a biological monitoring programme that is undertaken to monitor exposure only is not a medical record. However, a biological monitoring test report that is used in conjunction with health monitoring is a medical record. Nevertheless, a medical practitioner is obliged by law to give the employee the results of the health monitoring and to give the employer the general outcome of the monitoring. Employers who conduct exposure monitoring can remove personal identifiers when privacy and confidentially is of concern.

Employers are required under WHS Regulations to provide health monitoring results to the work health and safety regulator if the results indicate an illness has developed or that controls should be improved. In some circumstances, there is also a requirement for the medical practitioner to notify the work health and safety regulator and/or the health department of the results.

11.10.6 Results and Feedback

All biological monitoring analytical results should be passed on to the worker, as well as being interpreted and explained by a competent person. Where available, the exposure should be compared with an occupational exposure standard. Access to a medical professional should be made available when:

- Required as part of a health monitoring programme under WHS law.
- A medical interpretation of the results is requested by the worker.
- Symptoms are present that could be related to the exposure.
- Exposure levels are above an occupational exposure standard or similar guidance value.

11.11 EVALUATION

Most biological monitoring reports contain guidance on how to interpret the test result. Direct comparison to an exposure standard is the usual approach to determine if excessive exposure has occurred on an individual basis. In the absence of an available exposure standard, information on the population background levels can be found from population studies that have been undertaken. A particularly useful source of information is the NHANES survey that has been undertaken by the Centre for Disease Control and Prevention in the United States for many years (NHANES 2023).

To evaluate if exposure has occurred on a group basis in a particular workplace, pre-shift and post-shift samples are useful. The mean of the difference between the pre-shift and post-shift samples can be assessed by using Student's *t*-test. This can show that an exposure has occurred during that shift but does not compare it to a particular occupational limit.

The accepted procedure to compare the group exposures to an occupational limit is to first take the natural logarithm of the data, as exposure data is usually log normally distributed, transforming it to a normal distribution. A one-sided comparison of the mean to the occupational limit is then performed by Student's *t*-test. When a more comprehensive study is undertaken to establish the compliance of a similarly exposed group (SEG), then the methodology outlined in BS EN 689 should be followed (BS EN 689.2018).

Based on a reliable data set, improvements in the workplace should be made and the effectiveness of the implemented changes evaluated using the same monitoring. The frequency of the testing should be considered and would typically be reduced for work processes that are under control, while in the opposite situation, a continuation or increase of the frequency might be desirable. Upon completion of a monitoring programme, surveillance samples should be planned for ongoing, low-frequency monitoring, if required on a risk basis.

The biological monitoring process and results should be documented with a sufficient level of detail for reporting and interpretation by a person who was not involved in the process. Company policies and regulatory requirements around retaining documentation on biological monitoring should be considered.

11.12 COMMON BIOLOGICAL MONITORING CHEMICALS

Some of the most common chemicals that are monitored by biological monitoring are listed in Table 11.2.

TABLE 11.2

Common Chemicals Exposures Assessed by Biological Monitoring, the Analysis of the Parent or Metabolite Compounds Performed, the Biological Elimination Half-Lives and the Recommended Sampling Time

Chemical Exposure	Analysis	Biological Elimination Half-Life	Recommended Sampling Time
Arsenic	Speciation of arsenic metabolites in urine	1–4 days	End of shift
Benzene	Urinary S-phenylmercapturic acid	9 hours	End of shift
Carbon disulfide	Urinary 2-thiothiazolidine-4-carboxylic acid (TTCA)	9 hours	End of shift
Chlorinated solvents	Urinary trichloroacetic acid	2–4 days	End of shift
Cresol	Urinary cresol	3 hours	Pre- and post-shift
Cyanide	Urinary thiocyanate	Hours	End of shift
Cytotoxic drugs	Urinary cytotoxic drugs screen	3–12 hours	End of shift
Ethyl benzene	Urinary mandelic acid	4 hours	End of shift at end of work week
Fluoride	Urinary fluoride	4–7 hours	Pre- and post-shift
Furfural	Urinary furoic acid	2–2.5 hours	End of shift
Herbicides	Urinary chlorophenoxy acetic acid herbicide screen Urinary glyphosate	<6 hours	End of shift or end of work week
Insecticides	Blood organochlorine screen Urinary organophosphate screen	Persistent 1–2 days	Blood: timing not critical Urine: end of shift
Isocyanates	Urinary isocyanate diamine metabolites	2–4 hours	End of shift
Lead	Lead in blood or free erythrocyte protoporphyrin or zinc erythrocyte protoporphyrin	Triphasic: 6 weeks; 6 months; 20 years	Timing not critical
Mercury	Mercury in blood/urine	40–60 days	Blood: end of shift at end of work weekUrine: pre-shift at end of work week
Metals	Urinary: Be, Bi, Cd, Cr, Co, Cu, Pb, Mn, Ni, Se, Th, U, V Blood: Co, Cd, Pb, Mn	Varies	Varies
Methyl bromide	Blood bromide	9–15 days	End of shift at end of work week
MOCA/MBOCA (4,4-methylene bis (2-chloroaniline))	Urinary MOCA	20 hours	End of shift at end of work week
Phenol	Urinary phenol	1–4 hours	End of shift
Polychlorinated biphenyls (PCBs)	Blood PCB screen	Persistent	Timing not critical
Polycyclic aromatic hydrocarbons (PAHs)	Urinary 1-hydroxypyrene	6–35 hours	End of shift at end of work week
Solvents	Urinary polar solvent screen	Hours	End of shift
Styrene	Urinary mandelic and phenylglyoxilic acid	Biphasic 3–4 hours; 25–40 hours	End of shift

REFERENCES

ACGIH 2023, *TLVs and BEIs Threshold Limit Values for Chemical Substances and Physical Agents & Biological Exposure Indices*, Cincinnati, OH: ACGIH.

AIHA 2004, *Biological Monitoring – A Practical Field Manual*, Fairfax, VA: American Industrial Hygiene Association.

BS EN 689:2018 2018, *Workplace exposure - Measurement of exposure by inhalation to chemical agents – Strategy for testing compliance with occupational exposure limit values*, https://www.en-standard.eu/bs-en-689-2018-workplace-exposure-measurement-of-exposure-by-inhalation-to-chemical-agents-strategy-for-testing-compliance-with-occupational-exposure-limit-values/ [6 Aug 2023].

DFG 2022. *List of MAK and BAT Values*. Report No. 58 ed. Bonn: Wiley-VCH, https://www.dfg.de/en/service/press/press_releases/2022/press_release_no_26/index.html [6 Aug 2023].

Greim, H. & Lehnert, G. 1998, 'Creatinine as a parameter for the concentration of substances in urine', In: D. Henschler, H. Greim & G. Lehnert (Eds). *Biological Exposure Values for Occupational Toxicants and Carcinogens*, Vol 3. Weinheim: Wiley-VCH, pp. 35–44.

Hays, S.M., Becker, R.A., Leung, H.W., Aylward, L.L. & Pyatt, D.W. 2007, 'Biomonitoring equivalents: a screening approach for interpreting biomonitoring results from a public health risk perspective', *Regulatory Toxicology and Pharmacology*, vol. 47, no.1, pp. 96–109. https://doi.org/10.1016/j.yrtph.2006.08.004

Hertel, J., Rotter, M., Frenzel, S., Zacharias, H.U., Krumsiek, J., Rathkolb, B., Hrabe de Angelis, M., Rabstein, S., Pallapies, D., Brüning, T., Grabe, H.J. & Wang-Sattler, R. 2018, 'Dilution correction for dynamically influenced urinary analyte data', *Analytica Chimica Acta*, vol. 1032, pp. 18–31. https://doi.org/10.1016/j.aca.2018.07.068

Holmgaard, R. & Nielsen, J. B. 2009, *Dermal absorption of pesticides – evaluation of variability and prevention*, Danish Ministry of the Environment, https://www2.mst.dk/udgiv/publications/2009/978-87-7052-980-8/pdf/978-87-7052-981-5.pdf [6 Aug 2023].

HSE 2023. *Biological Monitoring Guidance Values*, https://www.hsl.gov.uk/online-ordering/analytical-services-and-assays/biological-monitoring/bm-guidance-values [6 Aug 2023].

Lauwerys, R.R. & Hoet, P. 2001, *Industrial Chemical Exposure: Guidelines for Biological Monitoring* (3rd edn). Boca Raton, FL: Lewis.

Manno, M., Viau, C., Cocker, J., Colosio, C., Lowry, L., Mutti, A., Nordberg, M. & Wang, S. 2010, 'Biomonitoring for occupational health risk assessment (BOHRA)', *Toxicology Letters*, vol. 192, no. 1, pp. 3–16. https://doi.org/10.1016/j.toxlet.2009.05.001

Ng, K.W. & Lau, W.M. 2015, 'Skin deep: The basics of human skin structure and drug penetration', In: N. Dragicevic & H. I. Maibach, eds. *Percutaneous Penetration Enhancers Chemical Methods in Penetration Enhancement*. Berlin-Heidelberg: Springer, pp. 3–11, https://www.researchgate.net/publication/277309261_Skin_Deep_The_Basics_of_Human_Skin_Structure_and_Drug_Penetration [6 Aug 2023].

NHANES 2023. *National Health and Nutrition Examination Survey*, https://www.cdc.gov/nchs/nhanes/index.htm [6 Aug 2023].

Riviere, J.E. & Minteiro-Riviere, N.A. 2010. 'Dermal exposure and absorption of chemicals and nanomaterials'. In: C.A. McQueen, ed. *Comprehensive Toxicology*. Oxford: Academic Press, pp. 111–122.

Safe Work Australia 2013. *Health Monitoring for Exposures to Hazardous Chemicals*, https://www.safeworkaustralia.gov.au/system/files/documents/1702/guide-pcbu-health-monitoring-exposure-hazardous-chemicals.pdf [6 Aug 2023].

Safe Work Australia 2023. *Law and regulation*, https://www.safeworkaustralia.gov.au/law-and-regulation [6 Aug 2023].

SCOEL 2023. *Scientific Committee on Occupational Exposure Limits*, http://ec.europa.eu/social/main.jsp?catId=148&intPageId=684&langId=en [6 Aug 2023].

12 Indoor Air Quality

Michael Shepherd and Dr Claire Bird

12.1 INTRODUCTION

Poor indoor air quality (IAQ) is a significant health issue, given that we spend about 90% of our time indoors (Morawska et al. 2022). It affects health, environment satisfaction, stress, and the economy. The Royal Children's Hospital recently showed that children's screen times exceed recommended levels, encouraging greater online than outdoor play (AIHW 2022) meaning that future generations may spend even more time indoors. Research shows that perceived indoor air quality is a primary discomfort factor indoors (Danza et al. 2020), impacting workplace and educational attendance and performance. With heightened health awareness after COVID-19, IAQ concerns have escalated for governments, employers, and building managers, becoming crucial for workplace health and safety (Lawrence Berkeley 2020). This chapter explores IAQ monitoring and management during this critical period in IAQ history.

IAQ is defined differently across building industry sectors; in this chapter, we treat IAQ as one of several environmental facets impacting the comfort of building occupants. Other indoor environment factors, for example, pathogen transmission, thermal comfort, light, and noise can leave occupants dissatisfied or sick and are discussed in Chapters 13–17.

Spengler et al. (2001) emphasise the vital balance between the built environment and IAQ (Figure 12.1). Disequilibrium can result in specific and non-specific building and human-related illnesses. Escalating infection risks and antimicrobial resistance (Fisher & Denning 2023), including fungal diseases like *Candida auris* and severe Aspergillosis (Barnes & Marret 2006) with high mortality (CDC 2019) may be expedited by, climate change which may be hastened by fungal infection spread. This underscores the interplay between air quality and human health (Mora et al. 2022).

The Department of Climate Change, Energy, the Environment and Water (DCCEEW) defines IAQ based on well-being as well as health. Well-being is individually subjective, making assessing, managing, and monitoring IAQ challenging (Rabone et al. 1994).

Chemical pollutants in built environments stem from furnishings, adhesives, printers, outdoor air, personal products, and occupants. Poor ventilation allows harmful gases and aerosols while cardiovascular risks from exposure indoors mainly from road traffic. Escalating Australian bushfire intensity and increased controlled vegetation burning threatens individuals with bronchopulmonary and cardiovascular conditions (Haikerwal et al. 2015; Morawska et al. 2022).

Sensitivity to airborne pollutants varies widely among individuals. Infants, young children, the elderly, and those with suppressed immunity are more prone to respiratory issues tied to tobacco smoke, wood burner and bush fire smoke, dust mites, and gas combustion products like nitrogen dioxide. Asthma sufferers react to various pollutants, triggering attacks or increasing their frequency (Levik et al. 2005). The issue of multiple chemical sensitivity (refer to Section 12.2.3) and the role of indoor air pollution in its aetiology remains a topic of debate.

Sick building syndrome (SBS) is a World Health Organization (WHO)-defined condition characterised by mild irritation of eyes, nose, throat, headaches, chest tightness, and lethargy. SBS symptoms result from various chemical and/or microbial exposures, primarily but not exclusively

DOI: 10.1201/9781032645841-12

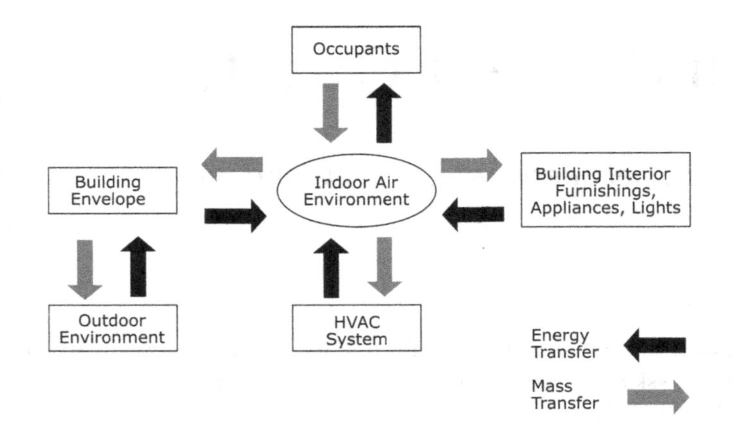

FIGURE 12.1 Indoor air quality model.

Source: Spengler et al. 2001, p. 57.2.

in air-conditioned office buildings (Ganji et al. 2023), with unclear causes (WHO 2010). Aziz et al. (2023) demonstrated that air quality may significantly contribute to SBS and reduced productivity.

Special risk considerations for sensitive groups such as those in residential, health, aged care, and educational settings are not extensively covered in this chapter. It is important to acknowledge that sensitive individuals might use or be in buildings primarily designed for healthy occupants. H&S practitioners should also address IAQ exposure beyond traditional workplaces, as people now work from home and travel via various modes of transport as part of their job.

Inadequate prevention, removal, or dilution of indoor pollutants can result in harmful concentrations. Indoor pollutant levels often surpass outdoor levels, with mixed contaminant groups such as volatile organic compounds (VOCs), microorganisms, and airborne particles (particulate matter or PM) differing in composition indoors and outdoors. Even with compliance to Australian Standard AS-1668.2: 2024 (Standards Australia 2024) and/or the National Construction Code (NCC) (formerly the Building Code of Australia) (ABCB 2022), proximity to a pollutant source can lead to issues, including odour complaints from dry floor traps or exposure to infective pathogens due to close contact.

Potential outdoor pollutant ingress points include windows, ventilation system inlets, and façade cracks. Ingress risk hinges on the orientation and location of air intake points when close to sources like industry, car parks, and roads. Concentrations reduce once indoors due to dilution, surface interactions, or chemical reactions. Carbon dioxide (CO_2) enters from outdoors and is exhaled indoors by humans (Shen et al. 2020a). Common pollutants encompass:

- Outdoors: combustion products – PM (especially $PM_{2.5}$ and $PM_{1.0}$), nitrogen dioxide (NO_2), carbon monoxide (CO), carbon dioxide (CO_2), organic gases (e.g., semi-volatile organic gases from bushfire – SVOC), and bioaerosols.
- In residential and commercial buildings: combustion products (as above) from, for example, gas cooking and unflued gas heaters in residences and schools; formaldehyde, VOCs from fibreboard, furnishings, and adhesives and semi-volatile organic gases (SVOCs) from furnishings and fire retardants.
- Exaggerated in commercial buildings: CO_2 from occupants in densely occupied spaces, for example, classrooms and lecture theatres.

- Exaggerated in residential buildings: combustion products from wood burner smoke and poor housekeeping increase risks of house dust mites, rodents, and cockroaches.
- Geographically dependent pollutants: house dust mite allergen levels in homes, for example, some child-care facilities and schools in our most heavily populated, coastal regions.

Environmental tobacco smoke (ETS) was banned in enclosed public areas in Australia, leading to substantial improvements in IAQ in public spaces (Davidson et al. 2011). Regulations concerning smoking outdoors vary among States and Territories (DHAC 2023). Evidence suggests that second-hand e-cigarette smoking exposes non-smokers to similar levels of nicotine compared to cigarette smoke and that ultrafine PM and heavy metal exposure risks are also elevated in non-smokers (Shearston et al. 2023) resulting in respiratory symptoms (Islam et al. 2022).

12.1.1 The Role of Ventilation in Indoor Air Quality

Heating, ventilation and air conditioning (HVAC) systems were originally designed to enhance thermal comfort and energy efficiency in buildings, curbing expenses and environmental impact. Originally, ventilation aimed to prevent odours and CO_2 buildup from exhalation, but well-designed systems enhance IAQ by supplying fresh air. However, many buildings, including Australian schools, lack mechanical ventilation, relying on enclosed split air conditioning systems that recirculate air often in rooms with closed windows.

Amid the recognition that SARS-CoV-2 (referred to as COVID-19 for the remainder of this chapter) spreads by airborne transmission, ventilation is deemed paramount in curbing its spread along with other respiratory pathogens like influenza and RSV, while concurrently enhancing IAQ. Recent governmental and industry emphasis on modifying ventilation and filtration for infection control caused confusion in construction, building management, and facility sectors regarding their roles in air quality provision. IAQ management now centres on reducing particulate matter since respiratory pathogens are often released in coarse and fine PM via mucous droplets and may persist in the environment in protozoa (Dey et al. 2022). CO_2, released with pathogens during exhalation is used to gauge ventilation efficacy in clearing pollutants and lowering infection risk. However, this approach overlooks the risk of close-range transmission and other sources of indoor pollutants like gases, particles, and microbial aerosols whilst ignoring the impact of air filtration on infection risk. Filtration and technologies like UV light, while not addressing CO_2, are now seen as potential substitutes for fresh air intake, despite potential negative impacts from CO_2 accumulation in densely occupied areas and areas with poor air movement.

The American Society of Heating, Refrigerating and Air-Conditioning Engineers (ASHRAE) Standard 241 (2023) infection control guideline focuses on decreasing airborne pathogen levels by increasing effective ventilation with pathogen-free air through three strategies:

1. Reducing the number of people in the space to reduce the chance of an infected person being present and allow people to remain distant from each other.
2. Bringing in more air from outdoors or from unused building areas.
3. Using supplemental technologies, such as filtration or specific UV irradiation, which have been tested through the standard to show that they filter aerosols or deactivate pathogens; and provide an increase in the "effective" ventilation rate.

The standard addresses a second current disconnect, by provisioning methods for measuring harmful by-products created by air purification devices. However, the testing only measures formaldehyde, ozone, and total PM and does not account for by-products like potential carcinogenic aldehydes, as well as changes in size distributions and surface properties of PM.

Ventilation is crucial for safety, but it can falter in outdoor pollution events like bushfires (Borchers et al. 2020) or if intake air is affected by pollution upwind. First steps in IAQ inquiries should verify that the ventilation rate adheres to the minimum Australian Buildings Code Board and AS 1668.2 standard.

Indoor Environment Quality (IEQ), as defined by the Australian Building Code Board (ABCB) in their Indoor Air Quality Handbook alongside the Australian Institute of Refrigeration, Air Conditioning and Heating (AIRAH), encompasses all indoor environment factors (ABCB 2021a & b). These guidelines, aligned with the NCC (ABCB 2022), maintain "adequate" IAQ. Yet the NCC recognises that some discomfort may persist under adequate conditions. The ABCB also prioritises baseline performance for new construction rather than fostering continuous IAQ improvement of existing buildings.

The ABCB guideline directs performance improvement to the National Australian Built Environment Rating System, NABERS, (DPIE 2021), an IAQ certification program covering thermal comfort, noise, light, and IAQ), managed by the NSW DPIE with fourteen key stakeholder bodies.

12.1.2 Current Incentives and Regulations around IEQ in Buildings

Building certification schemes in Australia rate indoor environment quality based on owner/manager- or tenant-controlled factors or a combination of both. These tools can involve compliance with predetermined conditions, like LEED (USGBC 2024), Green Star Interiors® (GBCA 2020) and Fitwel® (Fitwel 2023) ratings, and IEQ tools. Alternatively, they reward performance compared to other participating buildings, as seen in NABERS (NABERS 2023), WELL® (IWBI 2023) and RESET® (RESET 2018). WELL (IWBI 2023) and Fitwel (2023) focus solely on improving indoor spaces for physical and mental health. NABERS Energy is the only mandatory performance rating in Australia. The occupant satisfaction scoring for its Indoor Environment (IE) score holds a 50% weightage and is determined through the Building Occupant Satisfaction Survey Australia (BOSSA nd – http://www.bossasystem.com).

People's satisfaction with indoor spaces is often influenced by their perception of temperature comfort. Nevertheless, an issue arises when the climate zones specified by building codes do not precisely align with the real climatic conditions of a region (Law 2023). Additionally, the NCC falls short in effectively tackling the difficulties presented by the increasingly frequent occurrences of extreme weather events. IAQ as an accessibility issue is gaining significant traction in Australia on the policy front, on the basis of providing safe shared air for everyone including the large number of people vulnerable to poor air quality and infection.

The importance of managing condensation within buildings as part of IAQ management is highlighted in The Condensation in Buildings Handbook (ABCB 2019). The handbook highlights the relationship of elevated condensation to microbial growth in the context of designing buildings in an environment where extreme weather events are predicted to be more frequent and climatic patterns may shift to create unfamiliar combinations of precipitation, humidity, and seasonal temperatures. Note, the Handbook provides non-mandatory advice and guidance.

12.2 WHAT IS IAQ AND WHO IS RESPONSIBLE?

12.2.1 Definition of IAQ

According to the Dept of Climate Change, Energy the Environment and Water (DCCEEW 2023a) "....*(NHMRC) defines indoor air as air within a building occupied for at least one hour by people of varying states of health...Indoor air quality can be defined as the totality of attributes of indoor air that affect a person's health and well-being*." The definition of an occupant as a person present for as little as 1-hour has been adopted by the ASHRAE 241 in the standard for infection control, in setting of minimum required ventilation rates, which is not adjusted for exposures over a more typical workday.

Different experts such as occupational hygienists, toxicologists, engineers, architects, or laboratories may define satisfactory IAQ differently, so putting IAQ into the right context is important. IEQ factors such as thermal comfort, noise, and lighting can change occupant perception of building health. Human interactions and workplace stress also influence IAQ perception, being governed by social, organisational, and external factors (Derksen et al. 2008; Finell et al. 2018). Linking IAQ to measurable health outcomes is therefore complex and may require input from multiple stakeholders.

Before the COVID-19 pandemic, urban dwellers spent their work time in commercial or industrial settings and the rest in residential buildings. A small percentage of time (5–10%) was spent in travel or vehicles (Klepeis et al. 2001) and outdoors (4%). With the uncertain long term future of working from home and commuting uncertain in Australia, we now need to include pollutant exposure in homes and work-related transport when assessing workplace health risks.

When there are no internal pollutant sources, IAQ is governed by the air coming from the outside. In such cases, outdoor air quality sets the standard for IAQ (Leung 2015). Consequently, the WHO outdoor pollutant targets (WHO 2021) also apply to indoor environments. The speed and effectiveness of external pollutant infiltration into indoor spaces depend on the point of air ingress, effectiveness of air filtration/purification, and the ratio of outdoor (makeup) air to recycled air used by the HVAC system, as well as the surface area size of pollutant-absorbing materials.

IAQ is also governed by pollutants generated within indoor spaces by human activity and the materials and appliances they use. Indoor sources add to pollutants already introduced from outdoor air. Similar types of pollutants can have cumulative effects on health, and on odour, as well as antagonistic effects where impact is reduced.

Industrial settings with well-defined airborne pollutants follow exposure standards (ES) set by Safe Work Australia and state occupational health and safety (OHS)/work health and safety (WHS) regulations. The term IAQ tends to be excluded when discussing typical industrial situations; yet IAQ is increasingly flagged as a workplace risk. There is limited specific workplace guidance on general IAQ (Gunderson & Bobenehausen 2011) except for ventilation rate requirements and in fact, in Australia, few indoor air pollutants have been sufficiently documented in the published literature to establish typical indoor levels (Reed 2009) and IAQ is poorly regulated (Morawska et al. 2022).

12.2.2 Responsibility for Regulating IAQ

In developed countries, environmental or health agencies, and sometimes building authorities, are responsible for regulating IAQ. The US Environmental Protection Agency (US EPA) carries out extensive research and industry/community activities, while in October 2022, the WHO updated its key air quality goals for sulfur dioxide, NO_2, ozone, CO, formaldehyde, and PM while stating clearly that these were relevant to both indoor and outdoor environments (WHO 2021). This may mean that the regulatory framework for outdoor air quality becomes applicable indoors.

IAQ regulation in Australia is the responsibility of individual States and territories (Morawska et al. 2022). The ABCB governs building practice through the NCC and is adopted by most states with minor variations (ABCB 2022). The NCC is performance-based rather than prescriptive with IAQ provisions relating only to ventilation with outdoor air (Section 12.6). Its compliance is mandatory for all building designs and constructions in Australia. Ventilation standards in commercial buildings are AS 1668.2:2024 and AS/NZS 3666.1:2011. Low-rise residential building ventilation requirements follow AS 1668.1:2012. The ABCB has also published a number of Handbooks which provided non-mandatory advice and guidance, for example, the Condensation in Buildings Handbook mentioned in Section 12.1.2.

In Australia, IAQ regulatory processes are influenced by Australian Standards and the ASHRAE guidelines. The recent release of ASHRAE 241 standard focuses on infection control and suggests higher air exchange rates during high infection transmission risk periods which it does not define. The standard does not address IAQ improvements outside those periods nor the potential for public health measures to enhance IAQ. In addition, the efficiency of how the air spreads within the building is a critical factor for IAQ.

Workplace health and safety regulations are administered by States and Territories under a series of Acts, Regulations and Codes of Practice.

In 2022, Safe Work Australia set out its requirements for employers, small businesses, and workers around reducing the transmission of COVID-19, making it clear that '*removal of mandatory isolation does not impact on the duties of an employer to do all that is reasonably practicable to minimise the risks of COVID-19 at the workplace, including asking workers to stay at home when unwell*' (Safe Work Australia 2022).

12.2.3 Building Related Illnesses, Sick Building Syndrome, Biotoxin-related Illness and Multiple Chemical Sensitivities

Health effects from indoor air issues range from severe conditions like asthma, allergies, and cancer to non-specific symptoms that are collectively termed 'building-related illnesses' (United States Environmental Protection Agency 2023a).

To pose a risk, the pollutant must originate from a source, travel through an air pathway, and deleteriously affect the health of an exposed individual. The air pathway may be very short, for example, VOCs from carpet or protracted by complex movement such as movement of odours through a ducted recycled mechanical air handling system.

Many building-related illnesses are those caused by identifiable sources like VOCs, SVOCs, and primary or secondary particulate and gaseous contaminants such as oxidised terpene-based cleaning products, dust mite allergens, or *Legionella pneumophila*.

In contrast, sick building syndrome (SBS) has a less well-established or defined source, pathway and/or receptor sensitivity (ScienceDirect 2022). In 1983, the World Health Organization (WHO) used the term 'Sick Building Syndrome' for the first time to describe situations in which building occupants experience acute health and comfort effects that appear to be linked to the time spent in a building, but no specific illness or cause can be identified (Jafari et al. 2015). Due to this subjectivity, the prevalence of SBS has not been established; however, the WHO indicates it may affect over 30% of occupants in certain buildings. Wolkoff (2013) summarises the symptoms of SBS as follows:

- Mucous membrane irritation:
  - Dry or watering eyes (sometimes described as itching, tiredness, smarting, redness, burning, and difficulty in wearing contact lenses).
  - Runny or blocked nose (sometimes described as congestion, nose bleeds, itchy, or stuffy nose) dry or sore throat (sometimes described as irritation, oropharyngeal symptoms, upper-airway irritation, and difficulty swallowing).
  - Dry or sore throat (sometimes described as irritation, oropharyngeal symptoms, upper-airway irritation, and difficulty swallowing).

- Dermal changes: dry, itching, or irritated skin, occasionally with rash (also described in clinical terms such as erythema, rosacea, urticaria, pruritis, and xerodermia).
- Gastrointestinal disturbance.
- Neurotoxic symptoms: headache, lethargy, irritability and poor concentration, and increased sensitivity to odours.

The causes of SBS, as reviewed by Nag (2019), include poor ventilation, inappropriate humidifiers, odours from damp building materials and bioaerosols, poor thermal comfort, poor lighting, oxidative stress from indoor or outdoor-sourced pollutants (DCCEEW 2023b), VOCs, psychosocial stress, and work-related factors. Recent research in molecular biology has demonstrated that bacteria (in particular, some Gamma-proteobacteria), and specific fungi associated with dampness are linked to SBS (Fu et al. 2021). Previous studies linked endotoxin (from bacteria), and ergosterol and muramic acid from fungi, as well as fungal DNA to SBS symptoms. The main identified building causes of SBS include (Burge 2004; Joshi 2008; United States Environmental Protection Agency 1991):

- **Inadequate fresh air intake**: Insufficient fresh air raises CO_2 levels, increases humidity, and may cause condensation. High humidity can lead to quicker release of surface gases and alter inhalable particle sizes in the air.
- **Poor ventilation systems**: Improperly operated or managed air-conditioning systems are linked to various SBS symptoms. Causes include incorrect air distribution, inadequate maintenance, insufficient clean makeup air, and environments promoting microbial growth.
- **Airborne chemical pollution**: Many items within a building may emit gaseous pollutants associated with SBS; building structure, wall/surface/floor coatings and coverings, office chemicals, building activities, and the occupants themselves.
- **Microorganisms (fungi and bacteria) and microbial particulates**: Poorly maintained air-conditioning systems with a mix of non-organic and organic dust create, respectively, provide a surface for growth and a food source for microorganisms. High humidity and excess moisture, including after leaks and floods, can lead to microbial multiplication and the build-up of particulates on surfaces and furnishings which become airborne upon disturbance.
- **Temperature**: Ideal indoor temperatures vary according to the time of year, type of building (naturally/mechanically ventilated), and location. Thermal discomfort can exacerbate SBS symptoms. If the indoor temperature is too low, water molecules in the air condensate onto cold surfaces, providing a source of moisture for fungal growth.
- **Relative humidity (RH)**: RH represents the amount of water vapour in the air relative to its maximum capacity before condensation at a specific temperature. For thermal comfort, ISO 7730: 2005 suggests maintaining RH between 40% and 65%. RH below 40% may cause dry eyes and skin, while high RH levels can promote SBS via fungal growth and increased emission rates released from surfaces.
- **Lighting**: Certain SBS symptoms may be promoted by poor lighting, the absence of windows, or flicker from fluorescent tubes operated.
- **Personal and organisational factors**: Symptoms of SBS are more common among individuals with higher education levels (Sayan & Dülger 2021), women, workers in routine jobs, those with a history of allergies, computer users, and those who feel they have little control over their indoor environment.

The 2001 inquiry by the NSW Standing Committee on Public Works (NSW Legislative Assembly 2001) revealed that SBS significantly affects building occupant health and productivity. The committee suggested regulations and improved standards for ventilation and building materials, commissioning of new buildings, and setting indoor pollutant criteria. However, these recommendations have still not been implemented in NSW or other jurisdications. Instead, many building owners have voluntarily adopted one or more IAQ rating programs like NABERS or Green Star schemes, which have similar requirements.

Multiple Chemical Sensitivity (MCS), also known as Chemical Sensitivity (CS), Chemical Intolerance (CI), Idiopathic Environmental Illness (IEI), and Toxicant Induced Loss of Tolerance (TILT), is in contrast a chronic, poorly understood health condition, causing symptoms in a few individuals exposed to concentrations of chemicals that would not affect most people (Zucco & Doty 2021). Poor IAQ, cosmetics, food additives, and household products are blamed for MCS (Zucco & Doty 2021). Steinemann (2018) reported that 6.5% of Australians were diagnosed with MCS, and 74.6% had asthma-related symptoms. However, John Hopkins Medicine (2023) and Zucco and Doty (2021) note MCS is under debate in the medical community at this time and there is debate as to whether MCS should be classified and diagnosed as an illness.

12.2.4 Relationship of IAQ to Productivity

Addressing worker and visitor building complaints is costly while enhancing IAQ potentially cuts healthcare and sick leave costs (Fisk & Rosenfeld 1997; Kajander et al. 2014). Modest productivity gains can warrant building upgrades, given staff salaries form a substantial share of organisational expenses (Bell et al. 2003; Fisk & Rosenfeld 1997). The Hawkins et al. (2020) value proposition for IAQ improvements showed:

- Every two-fold increase in ventilation rate above 3 L/s resulted in a 1.1% increase in overall performance (Wargocki et al. 1999).
- School student performance increased when CO_2 levels remained below 1,000 ppm (Bakó-Biró et al. 2012).
- Workers provided with occupant-controlled ventilation at their workstation showed an 11% productivity increase after 16 months (Menzies et al. 1997).
- Inadequate ventilation worsens airborne respiratory pathogen transmission.

As we experience global efforts to adopt and enhance air handling due to personal and economic COVID-19 impacts, the economic and public health benefits of proper ventilation, filtration, and purification technologies cannot be overlooked.

12.3 INDOOR AIR QUALITY GUIDELINES

The WHO has raised concerns about specific indoor criteria pollutants, proposing thresholds, but their adoption and monitoring remain limited (WHO 2010). Harmonised standards are crucial for various indoor settings, but Australia lacks robust IAQ guidelines. Understanding the evolving global landscape is complex as governments across developed countries prioritise IAQ. However, the STC34 IEQ Guidelines by ISIAQ (2023a, b) offer an extensive global IAQ guideline and standard list.

Ventilation codes, pollutant-reducing products, and public education bolster IAQ. However, a more systematic approach is essential for comprehensive IAQ resolution. With Clean Air Campaigns emerging globally, emphasising the role of IAQ pollution reduction but ASHRAE,

ABCA, Standards Australia, and the Centers for Disease Control proposing diverse minimum ventilation rates, the aspiration is for governmental campaigns to establish a unified national IAQ authority or framework (Morawska et al. 2022).

12.4 MAJOR INDOOR AIR POLLUTANTS AND SOURCES

The types of pollutants commonly found in Australian buildings (Table 12.1) include:

- formaldehyde from new building materials in new or renovated buildings (less than six months old) and transportable buildings with high loadings of reconstituted wood-based panels, from oxidised terpene-based cleaning products, and from unflued gas heaters;
- VOCs from new building materials (e.g. paints and varnishes) in new or renovated buildings, (less than six months old);
- allergens in carpets and bedding (except well inland and in Central Australia), from dust mites, microorganisms, and rodents;
- microbial toxins, pathogens, asthmatogens, and inflammation triggers;
- semi-VOCs, including perfluoroalkyl and polyfluoroalkyl substances (PFAS) from fire retardants, combustion products (NO_2, CO, and particles in diameter ranges from nanometre to $PM_{2.5}$) from gas cookers, and unflued gas heaters and wood burning stoves;
- vehicle exhausts (benzene, 1,3-butadiene, $PM_{2.5}$, CO, and diesel particulate matter) from adjacent garages or enclosed car parks,
- asbestos from the disturbance of friable and non-friable building materials, especially insulation products (e.g., pipe lagging, sprayed-on fire-retardant, and service penetration packing material between floor slabs);
- respirable crystalline silica, lead-based paints;
- trichlorethylene and tetrachloroethylene (from subfloor vapour intrusion/industrial sources, and dry-cleaning products, respectively);
- gases from building materials subject to methamphetamine manufacture; and
- polyaromatic hydrocarbons (PAH) including naphthalene from combustion and cooking, bushfire smoke.

TABLE 12.1
Typical Indoor and Outdoor Air Pollutant Levels in Australia

Pollutant	Typical Indoor Air Concentrations (μg/m³)			Typical Outdoor Air (μg/m³)
	New House/Office	Established House	Established Office	
Formaldehyde	100–800	20–120	40–120	10–20
Total VOCs	5000–20000	11–300	100–300	20–100
Nitrogen dioxide				10–50
No unflued gas heater	—	10–35	—	(300 peak)
Unflued gas heater	—	60–1500	—	
Fine particles (PM_{10})				5–70
Smoking	—	>90	100–300	
Non-smoking	40–60	5–40	10–40	
Dust mite allergens (per gram of house dust)	<0.1	10–60 coastal <1 inland	<2 (data limited)	<0.1

Sources: Brown 1996, 1998, 2000, 2002; Brown et al. 2004; Goodman et al. 2017; Lawson et al. 2011; Mannins 2001.

Other pollutants include:

- biological contaminants including bacteria, viruses, animal dander and cat saliva, house dust, mites, cockroaches, and pollen (United States Environmental Protection Agency 2023a);
- particle emissions from laser printers (McGarry et al. 2011; Morawska et al. 2019);
- Ozone from photocopiers, ultraviolet (UV) light sources (ABCB 2021a).

12.4.1 Microorganisms in Indoor Air

Fungi and bacteria are ubiquitous in air and in the general environment. While most are harmless, some can have adverse effects on human health and can create biofilms which impact the integrity and performance of buildings when they proliferate in indoor environments (Ghosh et al. 2015). The impact of helpful building microbiomes on future infrastructure is gaining importance as a crucial consideration (Bruno et al. 2022).

For IAQ health assessment, understanding microorganism viability is vital. Infections require viable microorganisms, yet around 99.9% of viable airborne microbial cells are not culturable on laboratory agar. The dormant state, such as spores or Viable but Not Culturable (VBNC) cells, presents health concerns as current methods struggle to detect them economically.

Microbial cell wall components can trigger non-infection-related illnesses due to the immune system's response to compounds from live, dormant, and dead cells. This can lead to asthma symptoms, including wheezing and peak airflow issues. Mould exposure can worsen asthma symptoms in sensitised children and young adults, increasing respiratory risks. The effectiveness of allergen avoidance in preventing adult asthma remains uncertain, but biochemical detection methods like immuno-based or nucleic acid-based analysis, microscopy, and gene function analysis can better identify asthma triggers than culturing. High levels of inhalable and respirable particulates, like rodent or dust mite proteins, pollen, human skin, dust, and combustion-generated matter, can heighten health risks and are quantifiable.

Outdoor microorganisms thrive indoors in damp spaces. Stagnant water, moist surfaces, and surface dust become microbial reservoirs. Microorganisms can spread through ventilation and contaminated carpets, and even after settling, they can repeatedly settle and resuspend if not filtered or removed from the air.

Most people are exposed to fungi every day, but in low concentrations. A review of the literature points to three key causes of elevated fungal concentration or shift in taxonomic profile to a more harmful microbial community:

- Poor construction techniques.
- Failure to rapidly identify/repair water incursion.
- Failure to correctly operate and maintain air-conditioning systems.

Guidelines (ASTM D7338:14 2023a) stress tracing moisture sources, pathways, and final locations in assessing microbially contaminated buildings, as moisture is essential for microbial growth. Visual dampness, moisture-related staining, or odour is linked to asthma and can be used to quantitatively assess risk (Park & Cox-Ganser 2022). Dampness, linked to chronic inflammation (WHO 2009), prompts testing that can be used to identify altered building microbial ecology due to dampness. The importance of managing condensation as part of building design as part of IAQ management is highlighted in The Condensation in Buildings Handbook (ABCB 2019).

Air sampling for microorganisms is often not successful in identifying the source of contamination and is therefore often regarded as a secondary or last-resort measure. Instead, assessment should focus on identifying historic moisture and growth and understanding microbial ecology, concentrations, and (bioaerosol) particle size distribution.

12.4.2 Dust Mites, Rodents, and Surface Microorganisms

12.4.2.1 Dust Mites

Dust mites are ubiquitous in nature and have been recognised as a cause of allergies for many decades. Their secretions and faeces can cause allergic reactions in susceptible people (American Lung Association 2023). Dust mites are typically 0.25–0.5 mm long (too small to see easily with the naked eye), and their food sources include flakes of shed skin, animal dander, and other organic materials. Therefore, they can be found in most indoor environments, where they favour warm, moist, and dark areas such as upholstered furniture, long-fibred or deep-pile carpets and mattresses.

While it is not possible to remove all dust mites from most office environments, allergic reactions are typically related to the quantity, or dose, of allergen. Mitigating mites involves strategies like HEPA-filtered vacuuming, washing, humidity control, and fabric replacement. Maintaining sub-50% humidity and good ventilation discourages mite growth (American Lung Association 2023).

12.4.2.2 Rodents

One of the primary concerns with rodents is their ability to contaminate the indoor air with allergens and pathogens. Rodent droppings, urine, and dander can contain inhalable allergenic proteins that become airborne when disturbed, which can trigger allergic reactions, exacerbate asthma symptoms, and cause other respiratory issues (United States Environmental Protection Agency 2023b).

Rodents can compromise the integrity of the building's ventilation system. They are notorious for gnawing on materials, including ductwork and insulation. This destructive behaviour can create openings and gaps, providing avenues for outside pollutants and allergens to enter the indoor environment. Damaged ventilation systems can also lead to improper air circulation, causing stagnant air and build-up of pollutants indoors.

Another significant issue with rodents is their attraction to stored food and waste. Rodents not only consume and contaminate stored food but also leave behind droppings and urine that can become airborne and impact the IAQ. The unpleasant odour associated with rodent infestations can also be a concern for IAQ, leading to discomfort and potential productivity loss among occupants.

12.4.2.3 Surface Microorganisms

Microscopic surface debris results from settling particulates, potentially containing microorganisms. Surface microorganisms mirror cleaning efficacy, while airborne microorganisms arise from fresh growth or resuspended particles from surfaces. Surface microorganisms indicate growth or settled particles yet to be stirred while resuspension relies on air movement and ease of aerosolising microbial particles. Microbial sources encompass pets, HVAC, plants, and outdoor air (Prussin & Marr 2015). Airborne pathogen releases depend on respiratory airway properties, viral load, and vocal/airway activity of infected individuals.

Inadequate cleaning or air filtration can allow dust to accumulate on most indoor surfaces. As this dust can provide a food source, controlling the other growth requirement, moisture is crucial to prevent microbial growth, especially in high-risk areas like air-conditioning ductwork, cooling coils, and drip trays. Notable risk areas include uninsulated wet areas with water usage or carpets with gripper tracks under windows subject to condensation.

Microorganisms thrive on surfaces when heat radiates from the opposite side, during a winter night on a south-facing window. Mould can arise when airborne humidity reaches just 60% RH, even without condensation. Cold-exposed surfaces, like ceilings near cooling vents, foster growth due to colder air holding less moisture than warm air, as the water molecules in the air cool and condensate onto surfaces.

Airborne microbial levels change swiftly, and spore counts seldom reflect the concentration or taxonomy of surface materials. To interpret mould samples, it is necessary to grasp taxonomy, aerosol size and its distribution uniformity, spore clustering, and adhesion to surfaces and the impact of air movement. For assessing health risks, it is wise to focus on dampness or infection risk potential pending scientific consensus and agreed safe airborne mould levels in Australia.

12.5 ASSESSMENT OF IAQ

Kirchner et al. (1997) describe an IAQ investigation strategy that includes – interviewing building occupants, a walkthrough survey and details investigations including environmental monitoring. Interviews of occupants and the walkthrough survey should occur before, and inform, any environmental monitoring. Section 12.8 provides further detail on an assessment strategy.

12.5.1 Interviewing Building Occupants

Interviewing building occupants provides information regarding on-site activities that may impact air quality, date of building construction, type of ventilation systems, previous problems, recent renovations, number of occupants, the work environment, type of work, kinds of symptoms, and other environmental elements (such as odours) to help understand contributing factors. Targeted questionnaires are sometimes used.

12.5.2 Walkthrough Survey

It is important not to underestimate the utility of a thorough site survey. A walkthrough survey of the building allows for validating information from interviews, visually examining the site for mould, structural issues such as water leaks, and other contributing factors.

12.5.3 Measurement Approaches (Environmental Monitoring)

The two measurement approaches to assessing indoor air are direct measurement using real-time monitoring devices or by sample collection and analysis of target pollutants. The other approach is to set ventilation rates at a level that prevents build-up of contaminants or transmission of airborne infectious diseases. Indicators of IAQ include the following.

Satisfaction indication: Building Occupants Survey System Australia (BOSSA nd) – occupant satisfaction rating, based on IAQ, thermal, lighting, and noise tolerance.

Comfort indicators:

- Radiant heat, for example, from sunlight through a window: note when air monitoring, it is necessary to avoid surfaces impacted by heat coming through windows.
- Ambient heat: (targets vary to accommodate seasonal indoor clothing choices indoors) usually between 20°C and 25°C.
- Optimal humidity range: 40–65% relative humidity.

Ventilation effectiveness indicators:

- Carbon dioxide is also used as a proxy for ventilation adequacy based on the assumed exhalation concentration by a defined number of individuals in a given indoor space. The NCC Indoor Air Quality Handbook (ABCB 2021a) calls for limits of carbon dioxide of 850 ppm as an 8-hour TWA (i.e., 450 ppm above ambient CO_2 level of 400 ppm and demand control ventilation provisions in AS 1668.2). This represents what is considered to be an adequately ventilated building from an occupant's 'odour amenity' point of view.

- Fresh air infiltration rates: usually determined in Australia under AS 1668.2 (current version is 2024) are currently 10 litres of fresh air/person/second for most buildings. Recommended levels may vary at times of high infection risk (ASHRAE 241:2023) or building uses and activity.

Source indicators:

- **Asbestos fibres**: Refer to applicable codes and regulations for hazard assessment of products.
- **Particulate matter ($PM_{2.5}$)**: Compare to National Environmental Protection Measures (NEPM 2015) for $PM_{2.5}$ and PM_{10}. WHO (2021) has a series of interim and final targets from 5 to 35 μg/m^3 for $PM_{2.5}$ only, specified as both one-year and 24-hour averaging periods (ABCB 2021a).
- **Legionella species**: Use current codes and regulations.
- **House dust mite**: Measure allergens in dust to determine if below the tenth percentile level for a particular geographical area.
- **Microorganisms**: Moist or damp surfaces, with or without visible growths present, are unacceptable; target pathogens or indicators of dampness or sewage should not be present in air or surface samples at levels suggesting more than a negligible association with infection risk, dampness, or contaminated water.
- **Formaldehyde**: Measure in relation to Air Toxics (National Environment Protection Council [NEPC] 2011) goal, 0.04 ppm (24-hour averaging period). The IAQ Handbook (ABCB 2021a) recommends a maximum concentration of 0.1 mg/m^3, averaged over 30 minutes.
- **Volatile organic compounds (VOCs)**: TVOC concentration <500 μg/m^3 averaged over one hour is recommended as maximum concentration (ABCB 2021a) and may signal significant sources, but comparing to background levels is often more valuable.
- **Pesticides**: Measure concentrations if visible residues are found or if building has 'leaky' floor, especially for post-construction application of termiticides.
- **Nitrogen dioxide (NO_2)**: The threshold in relation to the NEPM (NEPC 2021) goal is 80 ppb (based on 1-hour averaging period) may be suitable for identifying sources in most buildings such as dwellings, schools and hospitals when unflued gas appliances and cookers are operating. WHO (2021) has a series of interim and final targets ranging from 10 to 40 μg/m^3. The IAQ Handbook (ABCB 2021a) recommends NO_2 of 0.0987 ppm averaged over 1 hour.
- **Carbon monoxide (CO)**: The threshold in relation to NEPM goal 9.0 ppm (based on an 8-hour averaging period) which may allow tracking of sources, particularly in dwellings, schools, and hospitals while unflued gas appliances are operating (particularly heaters), and in relation to workplace exposure standard in enclosed parking sites. Targets under WHO (2021) are 4–7 μg/m^3 as a 24-hour mean. Maximum concentrations are recommended in the IAQ Handbook (ABCB 2021a) for various classes of buildings.
- **Ozone (O_3)**: The threshold in relation to NEPM goal 0.10 ppm (based on 1-hour averaging period) in dedicated rooms with heavy use of photocopiers, laser printers, and other sources, and at outlets from ozone-based air sterilisers. The target is 100 μg/m^3 over an 8-hour TWA (ABCB 2021a; WHO 2021), 60 μg/m^3 peak season value.
- **Sulfur dioxide (SO_2)**: The target threshold in relation to the WHO (2021) goal is 40 μg/m^3 as a 24-hour mean with some interim targets ranging from 50 to 125 μg/m^3. NEPM (2021) targets are, respectively, over a 1-hour annual TWA period.

12.5.4 Sampling Protocols

Indoor air sampling protocols vary based on their purpose, with standards being provided in Australia and internationally. Sections 5.8 and 5.9 of the Indoor Air Quality Verification Methods Handbook (ABCB 2023) provide a summary of protocols. Sampling of air may be is carried out for various reasons. These include epidemiological study, compliance to exposure thresholds, control monitoring during demolition or construction, building IAQ or IEQ certification, emissions control, remediation scoping/verification/validation, occupant complaints, monitoring, and source identification. Choosing suitable methods is hence vital for accurate assessments.

Sampling duration must match be appropriate to the body's expected response time to pollutants, or the timeframe needed for adverse health effects to occur. Brief sampling (minutes) suits strong irritants, sensitisers, and rapid responses. Longer sampling (hours to days) suits chronic exposure assessment.

When placing air samplers and planning sample strategy, it is important to consider testing at the pollutant's highest concentration point (its source reservoir and/or its sink), and factors impacting its release duration and intensity. Seifert et al. (1989) highlighted the importance of adapting sampling based on long-term (e.g., formaldehyde from particleboard) or short-term (e.g., spray products) sources. For lasting sources, sampling should provide measurements appropriate to both average and extreme conditions, as consideration noted in NABERS Indoor Environment rating rules (ABCB 2023; DPIE 2021).

Generally, most published methods which include sampling strategies for IAQ sampling or testing recommend that the following be considered as a minimum:

- **Building sample location selection** – areas should be 'zoned' according to known sources of indoor air pollutants in order to focus on high-exposure locations, zones of occupancy, and use of the space.
- **Building operation** – the building should be operated in a manner which maximises pollution during the period of pollutant sampling.
- **Sampling period** – over a timescale relevant to the likely biological effect of the pollutant, and the integrity of the contaminant of concern on sample media.
- **Number of buildings** – sufficient for a reliable estimate of exposure for 'at-risk' populations.

Flow rate and air volume collected should relate to the required collection efficiency, especially of airborne particles, and the minimum threshold required to be meaningful when conducting compliance monitoring or risk assessment. Note that if sampling for culturable bioaerosols, survivability may be very low if sampling at flow rates over 100 L/minute or for longer than 2 minutes.

12.5.5 Measurement Methods

Australian Standards exist for only a limited number of indoor air pollutants and have not been reviewed since 2005, although in recent years, many have been published by the International Organization for Standards (ISO), notably under ISO 16000 'Indoor Air' series. The American Society for Testing and Materials (ASTM) also provides global testing standards. A list of ASTM methods is provided under 'Atmospheric Analysis Standards' (ASTM 2023b).

This section addresses current standard methods for measuring indoor air pollutants. Methods for assessing other occupational exposures can be employed if they are sufficiently sensitive. Health and safety practitioners can refer to Chapters 8, 'Aerosols'; Chapter 10, 'Gases and Vapours'; and Chapter 13, 'Biological Hazards' for specific air sampling instrument guidance. The WHO offers an overview of indoor air pollutants sampling and analysis methods (WHO 2020), and ISO presents a standard approach for indoor air assessment and classification under ISO 16000-41 (2023) and pollutant investigation ISO 16000-32 (2014a).

12.5.5.1 Carbon Dioxide (CO_2)

There is no relevant Australian Standard, although CO_2 is commonly measured using instruments with real-time monitors with non-dispersive infrared (NDIR) sensors at ambient air concentrations from approximately 400 ppm to industrial levels of 5,000 ppm, the latter being the current 8-hour TWA exposure limit under the NES. ISO 16000-26 (2012) provides a standard for sampling for CO_2. According to the Indoor Air Quality Verification Methods Handbook (ABCB 2023), the NABERS uses CO_2 as a proxy to measure the VE based on AS 1668.2, where the minimum ventilation rate should be no less than 10 L/s per occupant. Under NABERS, this equates to a difference between measured indoor and outdoor CO_2 that is less than 400 ppm. That is, if outdoor levels of CO_2 are found to be 350 ppm, then indoor levels must remain below 750 ppm. Scores are determined based on the percentage of locations at which these conditions are met.

12.5.5.2 Particulate Matter (PM) Including Ultrafine PM

Two methods are commonly used to measure airborne particulate matter indoors. The first relies on the gravimetric weight per unit air volume (μg/m³) and is suitable for larger coarse particles, such as the larger particles within the PM_{10} range (particles smaller than 10 μm in diameter) which penetrate the upper respiratory tract.

Direct-reading gravimetric tools measure particulate matter, including $PM_{2.5}$, which can penetrate the lower airways. These instruments offer high sensitivity and detect workday changes across various particle sources. However, they are typically calibrated with standard sizes that might differ from those in the monitored. For instance, Kim et al. (2004) reported good agreement between direct-reading and gravimetric $PM_{2.5}$ measurements for welders exposed to fumes, while Chung et al. (2001) found direct-reading instruments consistently overestimated urban air $PM_{2.5}$ by a factor of three compared to ambient air filter-collected PM, which had a more varied particle composition. However, McGarry et al. (2013, 2016) showed that direct reading instruments reliably record elevated particle concentrations above background in relation to challenge aerosols and that the instrumentation was sensitive to characterising incidental sources of particles, independent of the particles used to calibrate the instruments.

To measure ultrafine inhalable and respirable particulate matter, including sub-0.1 μm diameter PM, such as combustion particles with minimal mass, condensation particle counters (CPCs), optical particle counters (OPCs), or light scattering devices are used. CPCs condense submicrometre particles onto a condensation nucleus, creating light-detectable larger droplets. OPCs employ light scattering to quantify particles in real time, analysing their scattering patterns for size and concentration. Heal et al. (2000) employed a particle counter for indoor air monitoring, which overestimated PM_{10} by a factor of 2.0 and $PM_{2.5}$ by 2.2 compared to actual weight. Hence, when collecting and weighing samples to measure PM for compliance or exposure risk assessment gravimetrically, reference to AS/NZS 3580.9.6 (Standards Australia 2003), AS 3580.9.10 (Standards Australia 2017), AS 2985 (Standards Australia 2009) standards is essential. Standard methods for real-time PM_{10} monitoring methods are provided by AS/NZS 3580.9.11 (Standards Australia 2022).

12.5.5.3 Formaldehyde

ASTM D8407 (2021a) provides a standard guide for measurement techniques for formaldehyde while ISO 16000-2 (2022) provides details on forming a sampling strategy and determining formaldehyde. The most common method for sampling formaldehyde is the dinitrophenylhydrazine

(DNPH) impregnated sampler under ISO 16000-4 (2011). This method is based on a reaction between formaldehyde and DNPH, which forms stable derivatives that can be quantified in a laboratory setting. The DNPH impregnated sampler consists of a cartridge or filter coated with DNPH. When air is drawn through the sampler, formaldehyde molecules in the air react with DNPH, forming stable DNPH–formaldehyde derivatives. After the sampling period, the sampler is sent for laboratory analysis, where the derivatives are extracted and quantified using high-performance liquid chromatography (HPLC) or other suitable analytical techniques. The DNPH impregnated sampler is widely used due to its ease of use, cost-effectiveness, and ability to provide precise measurements of formaldehyde levels. Considering the concentration of formaldehyde will be at the limit of quantification of this analytical method, it is important that the maximum sample volume of the method is collected.

Another device for formaldehyde measurement is the active formaldehyde monitor which utilises a pump to actively draw air through a reactive cartridge or filter. As formaldehyde-laden air passes through the cartridge or filter, formaldehyde molecules react with the reagent, generating a colour change or chemical reaction that is proportional to the formaldehyde concentration. Such instruments do not typically have adequate resolution for IAQ assessments.

12.5.5.4 Volatile Organic Compounds (VOCs)

VOCs in industrial workplaces are usually measured in accordance with AS 2986.1 (Standards Australia 2003 – R2016) and AS 2986.2 (Standards Australia 2003), equivalent to ISO Standard 16200 (2000), in which air is sampled on to activated charcoal or other sorbents and subsequently solvent-desorbed for Gas Chromatography (GC) analysis (ISO 16000-6:2021). The method is sensitive typically down to 10 µg/m^3 depending on the individual VOC. Methods in which thermal desorption or cryo-focusing are used to extract VOC are generally up to one-hundred times more sensitive than when extracted using solvent desorption; given that many VOCs have low odour thresholds, thermal desorption may provide greater sensitivity. For general use, where most VOCs have boiling points above 60°C, solvent desorption is adequate. These approaches are detailed in the US EPA TO-17 and in ASTM D6196-23 (ASTM D6196-23 2023c). Similar sensitivities can be achieved using passive sampling tubes as set out in AS 2986.1 (Standards Australia 2003 (-R2016)). Air may otherwise be captured in an evacuated canister for subsequent GC analysis as specified in US EPA TO-14A and ASTM D5466 (2021b).

GC creates a VOC fingerprint with peaks on a graph, identifying compounds using known reference standards. However, some compounds are tentatively identified, and not all peaks match known VOCs. ISO 16000-5 (2007) offers a VOC monitoring strategy.

The use of a sensitive photoionisation detector (PID) for measuring parts-per-billion (ppb) levels of VOC has been suggested by manufacturers as an alternative to the above methods. However, this instrument does not identify the specific VOCs present, has cross-sensitivity to other gases, and responds differently to different VOC mixtures, missing some compounds if they have ionisation potentials that exceed the energy of the ionising lamp. On the other hand, PIDs can track fluctuations in total VOCs in real time and can therefore be used to identify potential sources and chart differences based on location. When used in conjunction with standard sampling methods, a handheld PID it is a powerful tool.

12.5.5.5 Semi-volatile Organic Compounds (SVOCs)

SVOCs exist primarily in solid or particulate states in indoor environments (Lucattini et al. 2018), posing some of the most significant health risks indoors (Ataei et al. 2023, Wechsler and Nazaroff 2008). They may be measured depending on the target compound using impingers, sorbent tubes, or filter system.

A review of SVOCs is provided by the Lawrence Berkeley Lab (Lawrence Berkeley 2020). Compounds included under SVOCs include polyaromatic hydrocarbons (PAH) (ASTM D6209

2021c), polychlorinated biphenyls (PCBs) (ISO 16000-12 2008; US EPA 2023b), polybrominated compounds, perfluoroalkyl acids (PFAAs), pesticides, and microbial SVOCs such as phthalates (ISO 16000-33: 2017).

The US EPA provides information on how to model complex SVOC emission sources and fate for advanced IAQ assessments (US EPA 2023b).

12.5.5.6 Carbon Monoxide (CO)

The two most common sampling devices used to measure carbon monoxide (CO) are real-time instruments using the non-dispersive infrared (NDIR) sensors and electrochemical sensors. NDIR sensors emit infrared light through the air containing CO and measure the reduction in transmitted light to determine CO concentration (ASTM D3162 2021d), AS 2365.2,1993 (R2014)). Electrochemical sensors use a chemical reaction to generate an electric current, providing portable and cost-effective CO readings (ASTM D8406-2022, AS 3580.7.1-2023).

12.5.5.7 Nitrogen Dioxide (NO_2)

To measure NO_2 levels with very low detection limits, Gas Phase Chemiluminescence Analysers and Photometric Analysers are commonly used. These highly sensitive methods can detect NO_2 concentrations down to sub-parts-per-billion (ppb) levels. Gas Phase Chemiluminescence Analysers use a method where the air sample reacts with ozone (O_3) in a reaction chamber, producing light in proportion to the concentration of NO_2 present. Photometric Analysers rely on the absorption of light by NO_2 molecules in the sample air. By analysing the degree of light absorption, the concentration of NO_2 can be determined. Where sample collection is required using filters or passive sampling devices, AS 2365.1.2 (Standards Australia 1990) may be used.

12.5.5.8 Ozone

Ozone levels can be measured using an Ozone Analyzer or a Gas Phase Chemiluminescence Analyser. The Ozone Analyser utilises UV absorption technology, passing UV light through the air sample containing ozone and measuring the reduction in intensity caused by ozone absorption to determine its concentration. Gas Phase Chemiluminescence Analysers detect ozone by measuring the light emitted from the reaction between ozone molecules and a reagent in the air sample. AS 3580.6.1 (Standards Australia 2023) provides the standard for real-time measurement of ozone.

12.5.5.9 Pesticides

Sampling for pesticides in indoor air is undertaken using air sampling pumps attached to tubes and cartridges filled with specific sorbent materials to adsorb and retain pesticides, followed by Gas Chromatography Liquid Chromatography analysis. Reference air sampling methods for pesticide monitoring have been developed by organizations such as the U.S. Environmental Protection Agency (US EPA TO-10A), the Occupational Safety and Health Administration (OSHA), the National Institute for Occupational Safety and Health (NIOSH), and the American Society for Testing and Materials (ASTM).

12.5.5.10 Bioaerosols

Bioaerosol sampling collects microbe-laden particles from the air using a chosen particle collection medium. As particles settle onto surfaces, testing of surfaces provides a longer-term picture of airborne microorganisms which have settled over a long period of time.

Most notably the ASTM D7338-14 (2023a) for assessing fungal conditions in buildings states that testing should only be conducted based on a hypothesis, and this same argument could be applied to any IAQ investigation.

Sample collection from surfaces should follow ASTM D7910-14 (2021e) (sampling using adhesive tape) and D7789-21 (2021f) (sampling using a swab). The NSW food authority provides guidance on collecting environmental surface swabs (NSW 2001).

Bioaerosol sampling techniques include impaction, filtration, impingement, and cyclonic sampling. Standard methods for bioaerosol sampling are provided in the ISO 16000 standards. Mainelis (2020) provides a useful discussion on bioaerosol sampling, including methods and limitations.

12.5.5.11 House Dust Mite Allergen

Dust mites have been proven to be the greatest indoor allergens (Yadav et al. 2022). The indicator for house dust mite (HDM) exposure (CEC 1993) should be the levels of allergen in accumulated dust. Dust collected by vacuum sampling methods from mattresses, carpets, and furniture is analysed using immunochemical assays, preferably enzyme-linked immunosorbent assay for the main house dust mite allergen of concern (CEC 1993).

12.5.5.12 Radon

Australian Standard AS 2365.4 specifies three methods for radon measurement with detection limits of 3–40 Bq/m^3 (AS 2365.4 [Standards Australia 1995 (R2014)]). Other techniques are listed in the National Library of Medicine (Doyle et al. 2012). ISO 11665-6:2 (2020) provides a recent standard for spot measurements.

The WHO (2021) provides a section 'Methodological Guidance for Health Risk Assessment for Air Pollution' in addition to a section on the role of the health and other industry sectors in education and implementation of IAQ objectives, which may be of benefit to the IAQ and the health and safety professional. Radon is only a concern in basements, so is not a significant concern for much of Australia (DCCEEW 2023a).

12.5.5.13 Odour

Odour is highly complex but a common cause of a complaint and consequent investigation. Testing for odour based on using odour thresholds is complex, as mixtures of compounds can result in non-additive effects on thresholds. Further the literature contains a wide range of odour thresholds for the same compounds. The standard method for testing odour is to collect a sample of air and expose to a panel of individuals with calibrated noses. The ISO 16000-30 (2014b) and AS/NZS 4323.3 (Standards Australia 2001) provide details on odour testing.

12.6 BUILDING VENTILATION

Building ventilation can be achieved via three methods: natural ventilation (using openings like windows and doors), mechanical ventilation (mechanical systems drawing in outside air), and mixed mode ventilation (combining recirculated and external air through fans and pumps) (ABCB 2021a). In Australia, homes and many schools typically use natural ventilation and operable windows with recirculated air cooling, while large commercial buildings usually employ mechanical ventilation. Requirement for natural ventilation is set out under AS 1668.4:2024 (Standards Australia 2024b).

In 2022, Rajagopalan et al. demonstrated that the ventilation rates in Victoria schools were insufficient to safeguard students from airborne disease transmission. Similarly, in Queensland schools, Snow et al. (2022) found that the ventilation was inadequate for preventing infection spread, with levels typically above those promoting cognitive performance. The primary reasons for these issues were disabled windows (to discourage student interaction and save energy), inaccessible extractor fan switches, and a lack of awareness.

Ventilation rate recommendations have fluctuated between effective to ineffective since the start of the 20th century (Swegon Air Academy, nd) often influenced by energy reduction efforts.

There is now a sharp industry focus by organisations such as ASRHAE and AIRAH on balancing environmental sustainability and IAQ.

Mechanical ventilation systems require attention during design, operation, and maintenance to maintain good IAQ. Displacement ventilation and directional airflow are being explored to enhance IAQ; displacement ventilation involves introducing outdoor air at lower room heights and extracting it at the ceiling, preventing air recirculation. Effectiveness of displacement ventilation must be maintained by avoiding disturbance of the upward air movement, so layout of the space is important. AS 1668.2 and AS/NZS 3666.1 are Australian Standards concerning mechanical ventilation and microbial control in buildings.

Increasing ventilation with all or mostly outside air may not always be possible or practical. In a multiroom building, increasing the mechanical ventilation rate may spread the virus rapidly from the source room into other rooms at high concentrations. Increasing the amount of outside air delivered by the mechanical ventilation system can lead to increased humidity and mould growth within the indoor space.

12.6.1 HVAC Filtration

Management of IAQ must include a discussion on filtration within HVAC systems. Filters are designed to filter pollutants or contaminants out of the air that passes through them and can help reduce airborne contaminants, including particles containing microorganisms (United States Environmental Protection Agency (2023c). HVAC systems incorporate air filters in accordance with the NCC and are performance rated in accordance with *AS1324.1-2001 Air filters for use in general ventilation and air-conditioning application, performance and construction*. In 2023, Standards Australia started implementation of ISO 16890 (2016), bringing the Australian Standard for air filters used in general ventilation and air conditioning in line with practice in other parts of the world. The measurement of and performance of filtration devices is now defined under ISO16890:2022.

The performance efficiency of these filters to remove aerosols varies depending on the rating of the filter but all filters remove aerosol to some extent. Filters need to be inspected, cleaned, and replaced on a regular basis. Not all HVAC systems are designed to move air through higher rated filters. Many do not have powerful enough fans and cannot be retrofitted. Not all existing equipment can handle these higher rated filters due to the impact of higher pressure drop on system performance (i.e., the system fan may not be able to move enough air through a thicker filter than it was designed for). Therefore, the highest rated filter suitable for the system is used.

12.6.1.1 Portable Air Cleaners

Air cleaning may be useful when used along with source control and ventilation, but it is not a substitute for either method. Source control involves removing or decreasing pollutants such as smoke, formaldehyde, or particles with viruses. The use of air cleaners alone cannot ensure adequate air quality, particularly where significant pollutant sources are present, and ventilation is insufficient (United States Environmental Protection Agency 2023c).

The decision to install a portable air filter must be based upon a risk assessment and informed by the following:

- Model of air cleaner
    - The Clean Air Delivery Rate (CADR) specified for the air cleaner needs to be sufficient for the room volume. The University of Melbourne has published a comparison of a number of HEPA air cleaners available to the Australian consumer market, including information on the selection of an air cleaner relative to the size of a room. This information can be found at https://sgeas.unimelb.edu.au/engage/guide-to-air-cleaner-purchasing.

- Maximum tolerable noise – fans are noisy, and it can sometimes make sense to have two quiet (<40dB) portable air cleaners rather than one large cleaner (>50 dB).
- Cost.
- Replacement of the filters in accordance with manufacturer specifications.
- Avoid
  - Directional fans without any filtration that blow air from person to person could lead to unintended infection transmission. (Note: Air circulation without filtration has been shown to lead to SARS-CoV-2 super-spreader events.)
  - Ionisers, plasma/ozone/photocatalytic oxidation/precipitators, and UV purification or disinfecting add-ons. (Note: These are often unproven/untested technologies, and in some cases dangerous technologies, significantly degrading the air quality by producing ions, ozone and oxidation products. Ozone and ions can trigger asthma, so these technologies should be avoided.)

12.6.2 A Historical Perspective on Ventilation

Ventilation needs evolved over centuries. Initially, the focus was on controlling body odours, but priorities shifted to thermal comfort, energy efficiency, and now IAQ. In 1836, Tredgold estimated a ventilation rate of 14 L/sec/person to combat diseases like tuberculosis. The American Society of Heating, Refrigeration, and Air-conditioning Engineers (ASHRAE) then played a pivotal role in shaping global ventilation standards Persily (2015) offers a comprehensive review of the history and development of ASHRAE Standard 62, detailing how the standard has evolved over time in response to emerging research, shifts in building practices, and advancements in indoor air quality knowledge up to 2015.

Notably, in 1989, ASHRAE (ASHRAE Standard 62 1989) created a mandatory minimum ventilation requirement (ventilation rate requirements – VRP) and an alternative performance-based requirement (Indoor Air Quality Procedure – IAQP). IAQP requires thresholds and control measures to be targeted through ventilation, but only in 2021 did ASHRAE specify design limits for target pollutants many in common with the WHO (2009), criteria pollutants, comprising: $PM_{2.5}$ acetaldehyde, formaldehyde, tri- and tetra-chloroethylene, acetone, benzene, dichloromethane, naphthalene, phenol, BTX (benzene, toluene, and xylene), ozone, carbon monoxide, and ammonia.

In 2012, ASHRAE added ventilation criteria based on flow rate per person to their previous units of only flow rate per unit area. ASHRAE currently has a minimum ventilation rate of 10 L/s/person, which reportedly allows a 20% level of personal dissatisfaction (PD) with IAQ (Khovalyg et al. 2020). The Centers for Disease Control advised that at least 5 Air Changes per Hour are required to prevent transmission of infective diseases (CDC 2023).

The regulation in Australia has been driven by Standards Australia AS 1668, and later by AS 1668.2. AS 1668.2 which governed needs for mechanical ventilation systems. A full history of changes across Australian Standards is shown in Table 12.2.

12.6.3 Performance and Maintenance of Ventilation Systems

A typical HVAC system is shown in Figure 12.2. It consists of:

- A return air duct, grille and register, if air is recirculated to collect air from the conditioned room.
- Return air ductwork to carry the air to an air handling unit.
- An external air intake to bring air from outside (AS 1668.2).
- An air handling unit (AHU) to mix recirculated air from the building with around 5–20% external "makeup" air, push that air across a filter and heater or cooler.

TABLE 12.2
History of Australian Standard Ventilation Rates

Date of Standard	Outdoor Air Intake (Litres Per Person Per Second)	Setting	Reason for Changing
1668.2 – 1976	5	General buildings	Designed to achieve CO_2 levels below 1000 ppm to overcome odour issues
1668.2 – 1980	3.5 2.5		Energy crisis led to a demand for lower energy consumption
1668.2 – 1989	7.5		Complaints around body odour
1668.2 – 1991	10	Applicable as a starting point for most buildings	Increased outdoor airflow but reductions available where gaseous and particulate pollutants are treated in the return air
	2.5	Minimum allowable rate	
	15	Embalming rooms	
	20	Operating, delivery rooms, and air traffic control	
	50	Autopsy rooms	
1668.2 – 2002	10	Typical rate	Rate depends on use, temperature, and type of ventilation system.
	Values from 5 to 25	Health care buildings	Minimum allowable outdoor airflow rate was based on air cleaning and the source of indoor makeup air.
	3.5		Demand-control ventilation based on automated CO_2 and people movement was included. Natural ventilation and car parks moved to 1668:4.
1668.2 – 2002 Supplemental (Rev 2016)	Separate program for areas with tobacco smoking		Smoking banned in most public places.
1668.2 – 2012 (under review at the time of publication)	10	Offices	Design for ventilation and extraction can be prescriptive or follow more complex engineered procedure based on occupant-related, non-occupant-related and building material-related emission factors for pollutants.
	15	Other workplaces listed under 1991 standard (excluding air traffic control rooms)	

- A filter to remove airborne PM of varying efficiencies and energy requirements.
- A cooling coil to chill and dehumidify air and a heating coil – it is common to pre-chill air to remove excess moisture (RH) before heating it to supply the building.
- A condensate tray to collect water running condensing on coils during cooling.
- A supply air system – registers and diffusers to distribute conditioned air into the occupied space.
- A series of fans for supply, make up air, and exhaust of dumped air or locations where only exhaust is needed (e.g., wet areas and bathrooms).

When air is primarily or entirely recirculated, pollutants can accumulate indoors, for example, an HVAC system operated without any outdoor air supply (or split air conditioning systems) as these have no capacity to reduce levels of non-particulate IAQ pollutants. Poor operation and

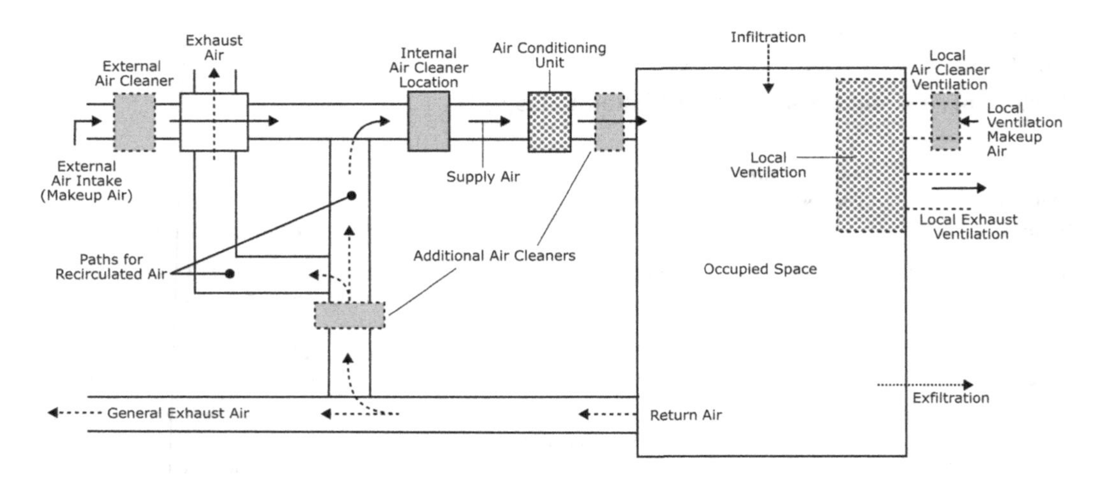

FIGURE 12.2 Schematic diagram of central mechanical ventilation/air-conditioning plant.

Adapted from: ANSI/ASHRAE Standard 62.1-2022.

maintenance of HVAC components can further lead to poor IAQ (Williams 1992) especially where condensate from chiller components reaches materials or surfaces that promote microbial growth.

A written regular maintenance plan should be developed and followed, and maintenance activities documented. Important periodic maintenance actions include:

- checking correct filter and regular inspection and replacement to ensure they are not blocked leading to potential odour or microbial growth;
- cleaning condensate trays and cooling coils (biofilm growth on coils can lead to potential for pathogen transport, antibiotic and antifungal resistance spread, and decreased energy efficiency) (Bakker et al. 2020; Prinzi & Rohde 2023); and
- checking fan and damper operation to ensure appropriate air-flow rates are met.

Measurement and balancing of the ventilation rate from an HVAC system by a specialist mechanical engineering consultant is essential (see Section 12.2) at such times as:

- after significant changes in the building layout or use, HVAC system design or operation, or the occupancy and activity within the building;
- after control settings are readjusted by maintenance personnel; and
- if accurate records of the system's performance are not available.

Cooling coils, positioned after particle filters, may have particles which breached the filter, but containing organic food and skin along with associated viable bacteria and fungi released into the air stream. These particles can impact or settle on chilled surfaces. Over time, dust build-up on duct surfaces, insulation linings, and certain areas after the filter can increase microbial risk. To address this, AIRAH offers a Duct Cleaning guideline (AIRAH 2018, being revised) in accordance with NADCA (2021) primarily focused on removing visible soil from the entire system. Cleaning methods used often leave visible residues, including on chilled surfaces and microbial growth, is not always visible. Consequently, AIRAH is currently re-evaluating its duct hygiene guidelines to better consider recent IAQ insights and the need to separately assess microbial contamination from visible debris.

Duct cleaning in most commercial buildings may not offer significant IAQ benefits unless specific conditions like moisture ingress, visible microbial or biofilm growth, or significant airflow obstruction exist. However, in settings like hospitals and food processing facilities where sterile surfaces are essential, duct cleaning and hygiene testing may be necessary.

AS/NZS 3666.1 (Standards Australia 2011) Air-handling and Water Systems of Buildings – microbial control requires inspection and cleaning where necessary.

Cooling towers are a critical HVAC component that can pose health risks if they become contaminated with *Legionella pneumophila* bacteria, causing Legionnaire's disease and Pontiac fever. Water-cooling towers must be placed away from building air inlets and must comply with state and national regulations for design, operation, antimicrobial treatment, monitoring cleaning, and registration (a legal requirement in some Australian states) as per AS/NZS 3666.1

12.6.4 Relationship between Indoor Air and Outdoor Air

Land animals typically cannot thrive in enclosed spaces due to the lack of external air needed to reduce harmful CO_2 levels. Achieving a balance between airtightness and CO_2 dilution is crucial. External air can infiltrate buildings through cracks and pathways if not properly commissioned. Figure 12.3

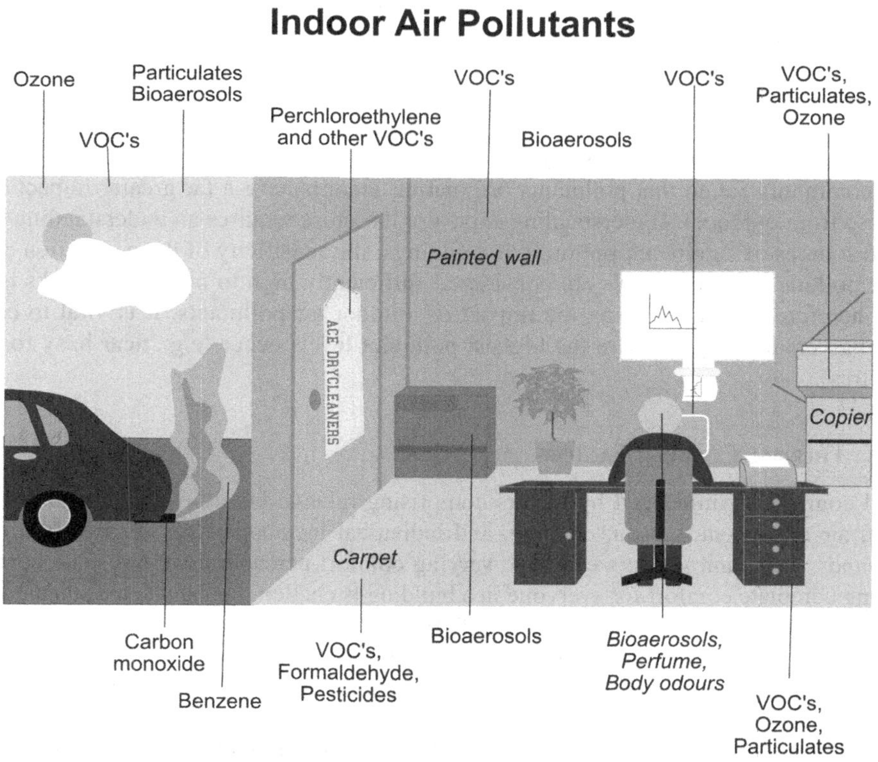

FIGURE 12.3 Primary sources of indoor air pollution (© Copyright CSIRO Australia).

Source: https://www.researchgate.net/figure/Primary-sources-of-indoor-air-pollution_fig1_340930350.

illustrates the intricate sources and flow of air pollutants in a standard office building. Important factors to consider at the indoor-outdoor interface include:

1. Increasing building air tightness leads to increased difficulty in dispersing indoor pollutants in the absence of adequate ventilation.
2. Outdoor air used to ventilate a building may introduce outdoor air pollutants to the indoor space (e.g., combustion products, industrial emissions, and bioaerosols).
3. Without purifying external air, indoor pollutant levels are primarily determined by outdoor conditions. This establishes a baseline level of pollutants that can interact with internally generated pollutants, potentially compounding or conflicting with each other and leading to occupant complaints and health issues. Sources of internal emissions encompass items used in workplace construction, furnishings, and operations, as well as occupant activities like cooking, appliance usage, pesticide application, and the presence of body odours or perfumes.
4. In some cases, the building can become the source of the contamination. SARS (Cov-1) in air exhausted from bathrooms in one multi-storey building infected those in downwind properties in Hong Kong (McKinney et al. 2006). *Legionella pneumophila* (the causative organism for Legionnaires disease) can spread many kilometres in droplets exhausted from a cooling tower.

The total indoor pollutant load is the result of combining both indoor and outdoor source inputs, adjusted for the time the pollutants remain in the air. Residence time is reduced due to dilution, losses to internal surfaces, and reactions with other compounds (e.g., ozone decays rapidly on contact with indoor surfaces, while carbon monoxide does not deplete at all), which are governed by materials and the handling of air between indoors and outdoors (Brown 1996, 1998, 2000, 2002; Brown & Cheng 2003; Brown et al. 2004).

It is commonly found that pollutants originating indoors have a far greater impact on IAQ than those from outdoors. Understanding exposure therefore requires an understanding of relevant end sources of significant pollution in buildings; the sensitivity of the population exposed to those pollutants; and if levels are considered sufficiently high to pose health risks to occupants. Therefore, when assessing the impact of outdoor air pollutants, it is vital to consider those urban environments where the highest pollutant levels occur (e.g., near busy roads and city centres).

12.6.5 Thermal Comfort Factors

Thermal comfort, is influenced by factors comprising relative humidity, air temperature, radiant heat, air movement, season, clothing, and individual metabolic rates. People from diverse backgrounds, ages, and genders can have varying comfort perceptions in the same conditions. Achieving complete comfort for everyone in a building is challenging, and some standards allow for a certain level of personal dissatisfaction (PD). Maintaining acceptable comfort involves controlling indoor temperature and humidity levels within a specified range, regardless of outdoor conditions. The optimal indoor temperature range for comfort identified in building thermal performance ratings is typically between 20°C and 25°C. Temperatures above 26°C may lead to fatigue and headaches, while temperatures below 18°C may cause chills and flu-like symptoms. ANSI/ASHRAE 55 (2020) standardises thermal comfort control.

Maintaining relative humidity within 40–60% is ideal, as low humidity can lead to dry eyes, nose, and throat, while high humidity can cause fatigue and stuffiness. High humidity may also promote the growth of bacteria and mould and increase off-gassing from building materials. Some viruses are immobilised in dry air, while others are negatively affected by very high humidity so applying extremes to combat infection transmission is not a viable approach.

12.7 DYNAMICS OF INDOOR AIR POLLUTANTS

Indoor pollutant concentrations will vary substantially:

- between regions, for example, levels of dust mite allergens and microorganisms will be higher in tropical Queensland than in cold, temperate south-eastern Australia;
- between buildings – depending on building age and type, the materials used, occupancy level and type, and the ventilation system;
- diurnally – for example, CO_2 levels increase in the morning as people arrive and may not reach steady-state levels until the afternoon, while VOCs may build up overnight when ventilation is not operating;
- with building age and timing of renovations – emission rates of VOC from glues, paints, materials, fittings, and furnishings will be at their highest in the first 12 months of use; and
- within a building – proximity of a source such as a copier or a printer to either a receptor or a recycling return air system.

The most significant factors influencing concentrations of indoor pollutants are:

- the sources strength and duration of pollutant emissions;
- the design, maintenance, and operation of the building ventilation system and other processes for pollutant removal (e.g., local air extraction, filtration, cleaning practices, air purification, creation or loss when reacting with surface (heterogenous reactions) or airborne compounds (homogeneous reactions); and
- the level of moisture control in the building.

12.8 CONTROL OF IAQ

Ideally, IAQ control begins at the building **design stage**, with:

- Selecting low-emission building materials, appliances, and furniture.
- Ensuring that ventilation meets NCC and AS requirements.
- Where feasible, deploying best practice such as displacement ventilation, heat recovery, and hybrid ventilation (Spengler et al. 2001).
- Ensuring that persistent condensation and water pooling are prevented on *all* interior surfaces and in all locations, especially in HVAC systems including stagnant water in condensate trays and plumbing dead legs, and ensuring cooling components remain relatively free of soil to prevent microbial proliferation. The importance of managing condensation within buildings as part of the design of the building is highlighted in The Condensation in Buildings Handbook (ABCB 2019).
- Establishing a plan for the operation and maintenance of building services.
- Documenting IAQ control actions for ongoing review and modification.

The health and safety practitioner may be asked to resolve occupant complaints about IAQ where the cause is unknown, few details on the building or mechanical design and construction are available and options to control indoor conditions are limited. A step-by-step assessment plan to address IAQ problems should follow a logical sequence, for example:

Step 1: Liaise with all parties and gather as much information on the history of the event as is available.

Step 2: Inspect the building to ensure that it is functioning as designed and meets accepted general safe practice (e.g. use a checklist of building faults, building contents, and cleaning practices).

Step 3: Inspect operational components of the building for proper function (e.g., ventilation/heating/cooling appliances).

Step 4: Compile a 'standard' indoor air environmental questionnaire from affected occupants to determine personal well-being and environmental comfort. A starting point for such a questionnaire is the CDC Indoor Air Quality Questionnaire (CDC 2008). When investigating an odour complaint, the New Zealand FIDOL assessment is advantageous (NZ Ministry for the Environment 2016) which may also be modified for IAQ investigations.

Step 5: Formulate a hypothesis based on Steps 1–4. Identify potential sources, pathways, sinks, and exposure sites that require sampling. Account for expected concentration fluctuations over time when choosing the sampling duration. Ensure the laboratory's limit of detection and reporting is adequate to detect the contaminant at relevant health or complaint levels or to demonstrate meeting IAQ goals with confidence. Plan for worst-case and typical exposure scenarios when the suspected source is active to cover the entire concentration range. If feasible, obtain background condition samples when the source is not present.

Step 6: Collect sufficient replicate samples to estimate uncertainty in results. Collect at least one field blank sample to prove a positive result was not an outcome of contamination during sampling, handling, or analysis.

Step 7: Record the results of sampling and analysis for comparison with IAQ goals, with baseline results of similar buildings or at times when the suspected source is absent, and with future sampling rounds.

At all steps: Identify areas where action is necessary, take such action as appropriate, and determine its effect on complaints before moving to the next step.

It is recommended that an H&S professional consults with an experienced occupational hygienist with IAQ expertise for suspected IAQ issues.

REFERENCES

ABCB 2019, *Handbook: Condensation in Buildings*. Australian Building Codes Board, Canberra.

ABCB 2021a, *Indoor Air Quality Handbook*. Australian Building Codes Board, Canberra.

ABCB 2021b, *Indoor Air Quality Verification Methods Handbook*. Australian Building Codes Board, Canberra.

ABCB 2022, *National Construction Code (Vols 1, 2 & 3)*. Australian Building Codes Board, Canberra.

ABCB 2023, *Indoor Air Quality Verification Methods Handbook*, Australian Building Codes Board, Canberra.

AIHW 2022, *'Australia's Children – Physical Activity'*, Australian Institute of Health and Welfare, Australian Government, https://www.aihw.gov.au/reports/children-youth/australias-children/contents/health/physical-activity [17 December 2023].

Australian Institute of Refrigeration, Air-conditioning, and Heating (AIRAH) 2004, *AIR-DA26-2004 Indoor Air Quality Design Application Manual*, Melbourne, Australia.

AIRAH 2018, *HVAC Hygiene Best Practice Guideline*. Australian Institute of Refrigeration, Air conditioning and Heating, Commonwealth of Australia and the States and Territories of Australia, Canberra.

American Lung Association 2023, *Dust Mites*. Sourced https://www.lung.org/clean-air/at-home/indoor-air-pollutants/dust-mites [17 September 2023].

American Society of Heating & Refrigerating Airconditioning Engineers (ASHRAE) 1989, *Application of CO2-Based Demand-Controlled Ventilation using ASHRAE Standard 62: Optimising Energy Use and Ventilation, ASHRAE TO-98-21-1*, ASHRAE, Atlanta, GA.

American Society of Heating & Refrigerating Airconditioning Engineers (ASHRAE) 2022, *Ventilation for Acceptable Indoor Air Quality Standard*, 62.1:2022, ASHRAE, Atlanta, GA.

American Society of Heating & Refrigerating Airconditioning Engineers (ASHRAE) 2023, *Control of Infectious Aerosols, ASHRAE Standard 241:2023*, ASHRAE, Atlanta, GA.

American Society of Heating & Refrigerating Airconditioning Engineers (ASHRAE) & American National Standards Institute (ANSI) 2020, *Thermal Environmental Conditions for Human Occupancy*, ANSI/ASHRAE 55:2020, American Society of Heating & Refrigerating Airconditioning Engineers (ASHRAE) Atlanta, GA.

American Society of Testing and Materials (ASTM) 2021a, *Standard Guide for Measurement Techniques for Formaldehyde in Air, D8407-21: 2021*, ASTM International, Philadelphia, PA.

American Society of Testing and Materials (ASTM) 2021b, *Standard Test Method for Determination of Volatile Organic Chemicals in Atmospheres (Canister Sampling Methodology), D5466–21:2021*, ASTM International, Philadelphia, PA.

American Society of Testing and Materials (ASTM) 2021c, *Standard Test Method for Determination of Gaseous and Particulate Polycyclic Aromatic Hydrocarbons in Ambient Air (Collection on Sorbent-Backed Filters with Gas Chromatographic/Mass Spectrometric Analysis), D6209-21:2021*, ASTM International, Philadelphia, PA.

American Society of Testing and Materials (ASTM) 2021d, *Standard Test Method for Carbon Monoxide in the Atmosphere (Continuous measurement by nondispersive Infrared spectrometry), D3162-21:2021*, ASTM International, Philadelphia, PA.

American Society of Testing and Materials (ASTM) 2021e, *Standard Practice for Collection of Fungal Material From Surfaces by Tape Lift, D7910-14(2021e)*, ASTM International, Philadelphia, PA

American Society of Testing and Materials (ASTM) 2021f, *Standard Practice for Collection of Fungal Material from Surfaces by Swab*, D7789-21:2021.

American Society of Testing and Materials (ASTM) 2022, *Standard Practice for Performance Evaluation of Ambient Outdoor Air Quality Sensors and Sensor-based Instruments for Portable and Fixed-point Measurement, D8406-22:2022*, ASTM International, Philadelphia, PA.

American Society of Testing and Materials (ASTM) 2023a, *Standard Guide for Assessment of Fungal Growth in Buildings*, D7338-14 (2023), ASTM International, Philadelphia, PA.

American Society of Testing and Materials (ASTM) 2023b, *Atmospheric Analysis Standards*, ASTM International, Philadelphia, PA, https://www.astm.org/products-services/standards-and-publications/standards/atmospheric-analysis-standards.html [17 December 2023].

American Society of Testing and Materials (ASTM) 2023c, *Standard Practice for Choosing Sorbents, Sampling Parameters, and Thermal Desorption Analytical Conditions for Monitoring Volatile Organic Chemicals in Air*, D6196–23:2023, ASTM International, Philadelphia, PA.

Ataei, Y., Sun, Y., Liu, W.S., Ellie, A., Dong, H. & Ahmad, U.M. 2023, 'Health effects of exposure to indoor semi-Volatile organic compounds in Chinese building environment: A systematic review.' *International Journal of Environmental Research and Public Health*, vol. 20, no.1, pp. 678.

Aziz, N., Adman, M.A., Suhaimi, N.S., Misbari. A.R.A., Aziz, A.A., Lee, L.F. & Khan. M.H. 2023, 'Indoor air quality (IAQ) and related risk factors for sick building syndrome (SBS) at the office and home: A systematic review', *IOP Conference Series: Earth and Environmental Science*, vol. 1140, p. 1. doi: 10.1088/1755-1315/1140/1/012007.

Bakker, A., Siegel, J.A., Mendell, M.J. & Prussin, A.J. 2nd, Marr, L.C. & Peccia J. 2020, 'Bacterial and fungal ecology on air conditioning cooling coils is influenced by climate and building factors', *Indoor Air*, vol. 30, no. 2, pp.326–334.

Bakó-Biró, Z., Clements-Croome, D.J., Kochhar, N., Awbi, H.B. & Williams, M.J. 2012, 'Ventilation rates in schools and pupils' performance', *Building Environment*, vol. 48. pp. 215–223.

Barnes, P.D. & Marr, K.A. 2006, 'Aspergillosis: spectrum of disease, diagnosis, and treatment', *Infectious Diseases Clinical North America*, vol. 20, no. 3, pp. 545–61.

Bell, J., Newton, P., Gilbert, D., Hough, R., Morawska, L. & Demirbilek, N. 2003, 'Indoor environments: Design, productivity and health', *Report No. 2001–005-B, CRC for Construction Innovation*, Brisbane.

Borchers, A. N., Palmer, A.J., Bowman, D.M., Morgan, G.G., Jalaludin, B.B. & Johnston, F.H. 2020, 'Unprecedented smoke-related health burden associated with the 2019–20 bushfires in eastern Australia', *Medical Journal of Australia*, vol. 213, pp. 282–283.

Brown, S.K. 1996, 'Assessment and control of exposure to volatile organic compounds and house dust mites in Australian dwellings', *Proceedings of the 7th International Conference on Indoor Air Quality and Climate, Nagoya, Japan*, vol. 2, pp. 97–102.

Brown, S.K. 1998, 'Case studies of poor indoor air quality in Australian buildings', *Proceedings of the 14th International Clean Air and Environment Conference*, 18–22 October, Melbourne, pp. 205–210.

Brown, S.K. 2000, 'Air toxics in a new Victorian dwelling over an eight-month period', *Proceedings of the 15th International Clean Air and Environment Conference*, 26–30 November, Sydney, pp. 458–463.

Brown, S.K. 2002, 'Volatile organic pollutant concentrations in new and established buildings from Melbourne, Australia', *Indoor Air*, vol. 12, no. 1, pp. 55–63

Brown, S.K. & Cheng, M. 2003, 'Personal exposures of stevedores to VOCs, formaldehyde and isocyanates while loading and unloading new cars on shipping', Paper 306, *21st Annual Conference of the Australian Institute of Occupational Hygiene*, Adelaide.

Brown, S.K., Mahoney, K.J. & Cheng, M. 2004, 'Pollutant emissions from low-emission unflued gas heaters in Australia', *Indoor Air*, vol. 14, suppl. 8, pp. 84–91.

Bruno, A., Fumagalli, S., Ghisleni, G. & Labra, M. 2022, 'The microbiome of the built environment: The nexus for urban regeneration for the cities of tomorrow', *Microorganisms*, vol. 22, no. 10, pp. 2311.

Burge, P.S. 2004. 'Sick Building Syndrome', *Occupational Environmental Medicine*, vol. 61, pp. 185–190. doi: 10.1136/oem.2003.008813

CDC 2008. *Indoor Air Quality Questionnaire*. Industrial Health Program, Office of Health and Safety, Atlanta, GA.

CDC 2019. 'Antimicrobial-resistant *Candida auris*', in: *Antibiotic Resistance Threats in the United States*. Industrial Health Program, Office of Health and Safety, Atlanta, GA.

CDC 2023, *Ventilation in Buildings, Summary of Recent Changes*, updates as of May 12 2023. Industrial Health Program, Office of Health and Safety, Atlanta, GA.

Chung, A., Chang, D.P.Y., Kleeman, M.J., Perry, K.D., Cahill, T.A., Dutcher, D., McDougall, E.M. & Stroud, K. 2001, 'Comparison of real-time instruments used to monitor airborne particulate matter', *Journal of the Air and Waste Management Association*, vol. 51, no. 1, pp. 109–20.

Commission of European Communities (CEC) Working Group 5 1993, '*European collaborative action—Indoor air quality and its impact on man: biological particles in indoor environments*', Report No. 12, CEC, Luxembourg.

Danza, L., Barozzi, B., Bellazzi, A., Belussi, L., Devitofrancesco, A., Ghellere, M., Salamone, F., Scamoni, F. & Scrosati, C. 2020, 'A weighting procedure to analyse the indoor environmental quality of a zero-energy building', *Building and Environment*, vol. 183, 107155.

Davidson, M., Reed, S. & Markham, J. 2011, 'Evaluation of the impact of indoor smoking bans on air quality in Australian licensed clubs', *Air Quality and Climate Change*, vol. 45, no 1, pp. 12–13.

DDCEEW 2022, *Nationwide House Energy Rating Scheme*. Department of Climate Change, Energy, the Environment and Water, Canberra.

DCCEEW 2023a, 'Indoor Air', *Dept of Climate Change, Energy, the Environment and Water*', Department of Climate Change, Energy, the Environment and Water, Canberra.

DCCEEW 2023b, *Total Volatile Organic Compounds*. Dept of Climate Change, Energy, the Environment and Water, Canberra.

Derksen, T., Franchimon, F. & van Bronswijk, J.E.M.H. 2008, 'Impact of management attitudes on perceived thermal comfort', *Scandinavian Journal of Work, Environment and Health*, vol. 4, pp. 43–45.

Dey, R., Dlusskaya, E., & Ashbolt, N. 2022, 'SARS-CoV-2 surrogate (Phi6) environmental persistence within free-living amoebae', *Journal of Water and Health*, vol. 20, no. 1, pp. 83–91.

DHAC 2023. *Smoking and Tobacco Laws in Australia*. Department of Health and Aged Care, Australian Government, Canberra.

Doyle, J., Harper, C., Keith, S., Mumtaz, M.S. & Tarragó, O. 2012. *Toxicological Profile for Radon* United States, Agency for Toxic Substances and Disease Registry, US Department of Health and Human Services.

DPIE 2021, *The Rules – NABERS Indoor Environment for Offices Version 2.0*, New South Wales Department of Planning, Industry and Environment, Australia.

Finell, E., Tolvanen, A., Pekkanen, J., Minkkinen, J., Ståhl, T. & Rimpelä, A. 2018, 'Psychosocial problems, indoor air-related symptoms, and perceived indoor air quality among students in schools without indoor air problems: A longitudinal study', *International Journal of Environment Research and Public Health*, vol. 15, no. 7, pp. 1497.

Fisher, M.C. & Denning, D.W. 2023, 'The WHO fungal priority pathogens list as a game-changer', *Nature Review Microbiology*, vol. 4, pp. 211–212.

Fisk, W.J. & Rosenfeld, A.H. 1997, 'Estimates of improved productivity and health from better indoor environments', *Indoor Air*, vol. 7, no. 3, pp. 158–72.

Fitwel 2023, 'Building health for all', *Center for Active Design*, New York, https://www.fitwel.org/resources/p/fitwel-v21-updates-summary-1 [6 January 2024]

Fu, X., Norbäck, D., Yuan, Q., Li, Y., Zhu, X., Hashim, J.H., Hashim, Z., Ali, F., Hu, Q., Deng, Y. & Sun, Y. 2021, 'Association between indoor microbiome exposure and sick building syndrome (SBS) in junior high schools of Johor Bahru, Malaysia', *Scientific Total Environment*, vol. 20, pp. 753.

Ganji, V., Kalpana, M., Madhusudhan, U., John, N.A. & Taranikanti, M. 2023, 'Impact of air conditioners on Sick Building Syndrome, sickness absenteeism, and lung functions', *Indian Journal of Occupational Environmental Medicine*, vol. 27, no. 1, pp. 26–30.

GBCA, 2020, 'Green Star – Interiors v1.3', *Green Building Council of Australia*, https://new.gbca.org.au/green-star/rating-system/interiors/ [6 January 2024]

Ghosh, B., Lal, H. & Srivastava, A. 2015, 'Review of bioaerosols in indoor environment with special reference to sampling, analysis and control mechanisms', *Environment International*, vol. 85, pp. 254–272.

Goodman, N.B., Steinemann, A., Wheeler, A.J., Paevere, P.J., Cheng, M. & Brown, S. 2017, Volatile organic compounds within indoor environments in Australia, *Building and Environment*, vol. 122, no. 5, pp. 116–125, doi: 10.1016/j.buildenv.2017.05.033.

Gunderson, E.C. & Bobenehausen, C.C. 2011, 'Indoor air quality' in *The Occupational Environment: Its Evaluation, Control and Management* (3rd edn), ed. D.H. Anna, American Industrial Hygiene, Fairfax, VA, pp. 450–500.

Haikerwal, A., Akram, M., Del Monaco, A., Smith, K., Sim, M.R., Meyer, M., Tonkin, A.M., Abramson, M.J. & Dennekamp, M. 2015, *Journal of American Heart Association*, vol. 4, no. 7. https://doi.org/10.1161/JAHA.114.001653

Heal, M.R., Beverland, I.J., McCabe, M., Hepburn, W. & Agius, R.M. 2000, 'Inter-comparison of five PM_{10} monitoring devices and the implications for exposure measurement in epidemiological research', *Journal of Environmental Monitoring*, vol. 2, no. 5, pp. 455–61.

International Organization for Standardization (ISO) 2005, *Ergonomics of the thermal environment – Analytical determination and interpretation of thermal comfort using calculation of the PMV and PPD indices and local thermal comfort criteria*, ISO 7730:2005, ISO Geneva, https://www.standards.org.au/access-standards

International Organization for Standardization (ISO) 2007, *Sampling strategy for volatile organic compounds (VOCs)*, ISO 16000-5:2007, ISO Geneva, https://www.standards.org.au/access-standards

International Organization for Standardization (ISO) 2008, *Sampling strategy for polychlorinated biphenyls (PCBs), polychlorinated dibenzo-p-dioxins (PCDDs), polychlorinated dibenzofurans (PCDFs) and polycyclic aromatic hydrocarbons (PAHs)*, ISO 16000-12:2008, ISO Geneva, https://www.standards.org.au/access-standards

International Organization for Standardization (ISO) 2011, *Determination of formaldehyde – Diffusive sampling method*, ISO 16000-4:2011, ISO Geneva, https://www.standards.org.au/access-standards

International Organization for Standardization (ISO) 2012, *Sampling strategy for carbon dioxide (CO_2)*, ISO 16000-26:2012, ISO Geneva, https://www.standards.org.au/access-standards

International Organization for Standardization (ISO) 2014a, *Investigation of buildings for the occurrence of pollutants*, ISO 16000-32:2014, ISO Geneva, https://www.standards.org.au/access-standards

International Organization for Standardization (ISO) 2014b, *Sensory testing of Indoor air*, ISO 16000-30:2014, ISO Geneva, https://www.standards.org.au/access-standards

International Organization for Standardization (ISO) 2016, *Air filters for general ventilation – Part 1, ISO 16890-1:2016*, ISO Geneva, https://www.standards.org.au/access-standards

International Organization for Standardization (ISO) 2017, *Determination of phthalates with gas chromatography/mass spectrometry (GC/MS)*, ISO 16000-33:2017, ISO Geneva, https://www.standards.org.au/access-standards

International Organization for Standardization (ISO) 2020, *Measurement of radioactivity in the environment Air: radon-222 Part 6: Spot measurement methods of the activity concentration*, ISO 11665-6:2020, ISO Geneva, https://www.standards.org.au/access-standards

International Organization for Standardization (ISO) 2021, *Determination of organic compounds (VVOC, VOC, SVOC) in indoor and test chamber air by active sampling on sorbent tubes, thermal desorption and gas chromatography using MS or MS FID*, ISO 16000-6:2021, ISO Geneva, https://www.standards.org.au/access-standards

International Organization for Standardization (ISO) 2022, *Sampling strategy for formaldehyde*, ISO 16200-2:2022, ISO Geneva https://www.standards.org.au/access-standards

International Organization for Standardization (ISO) 2022, *Air filters for general ventilation*, ISO 16890 Parts 1 and 2:2022, ISO Geneva https://www.standards.org.au/access-standards

International Organization for Standardization (ISO) 2023, *Assessment and classification*, ISO 16000-41: 2023, ISO Geneva, https://www.standards.org.au/access-standards

ISIAQ 2023a, *Indoor Environmental Quality (IEQ) Guidelines Database*. International Society of Indoor Air Quality and Climate.

ISIAQ 2023b, *Indoor Air Quality Guidelines*. International Society of Indoor Air Quality and Climate, Herndon, VA.

Islam, T., Braymiller, J. & Eckel, S.P. 2022, 'Second-hand nicotine vaping at home and respiratory symptoms in young adults', *Thorax*, vol. 77, pp. 663–668.

IWBI 2023, *Air Quality Standards to Ensure a Basic Level of High Indoor Air Quality*. International Well Building Institute.

Jafari, M.J., Khajevandi, A.A., Mousavi Najarkola, S.A., Yekaninejad, M.S., Pourhoseingholi, M.A., Omidi, L. & Kalantary, S. 2015, 'Association of sick building syndrome with indoor air parameters', *Tanaffos*, vol. 14, no. 1, pp. 55–62.

John Hopkins Medicine 2023, Multiple chemical sensitivity,, https://www.hopkinsmedicine.org/health/conditions-and-diseases/multiple-chemical-sensitivity [17 December 2023].

Joshi, S.M. 2008, 'The sick building syndrome', *Indian Journal of Occupational Environmental Medicine*, vol. 12, no.2, pp. 61–64. doi: 10.4103/0019-5278.43262.

Kajander, J.-K.; Sivunen, M. & Junnila, S. 2014, 'Valuing indoor air quality benefits in a healthcare construction project with real option analysis', *Buildings*, *vol.* 4, pp. 785–805. https://doi.org/10.3390/buildings4040785

Khovalyg, D., Kazanci, O.B., Halvorsen, H., Gundlach, I., Bahnfleth, W.P., Toftum, J. & Olesen, B.W. 2020, 'Critical review of standards for indoor thermal environment and air quality', *Energy and Buildings*, vol. 213, 109819.

Kim, J.Y., Magari, S.R., Herrick, R.F., Smith, T.J. & Christiani, D.C. 2004, 'Comparison of fine particle measurements from a direct-reading instrument and a gravimetric sampling method', *Journal of Occupational and Environmental Hygiene*, vol. 1, no. 11, pp. 707–15.

Kirchner, S., Derangere, D., Riberon, J., Skoda-Schmoll, C. & Aubree, D. 1997, 'A method for assessing indoor air quality in office buildings', *Second International Conference Buildings and the Environment*, Paris, https://www.aivc.org/sites/default/files/airbase_10625.pdf

Klepeis, N.E., Nelson, W.C., Ott, W., Robinson, J.P. Tsang, A.M.., Switzer, P., Behar, J.V., Hern, S.C. & Engelmann, W.H. 2001, 'The National Human Activity Pattern Survey (NHAPS): A resource for assessing exposure to environmental pollutants', *Journal of Exposure Analysis and Environmental Epidemiology*, vol. 11, pp. 231–52.

Law, T. 2023, 'Defective non-compliant and unfit for purpose buildings – can HVAC be the solution?', *Australian Institute of Refrigeration, Air Conditioning and Heating (AIRAH)* Indoor Air Quality Conference, Brisbane, 18 July 2023.

Lawrence Berkeley Lab 2020, '*SVOCs and Health*'. Lawrence Berkeley Lab, https://iaqscience.lbl.gov/svocs-and-health#_heading=h.2s8eyo1 [4 July 2023].

Lawson, S.J., Galbally, I.E., Powell, J.C., Keywood, M.D., Molloy, S.B., Cheng, M., & Selleck, P.W. 2011, 'The effect of proximity to major roads on indoor air quality in typical Australian dwellings', *Atmospheric Environment*, vol. 45, no. 13, pp. 2252–2259. https://doi.org/10.1016/j.atmosenv.2011.01.024

Leung, D.Y.C. 2015, 'Outdoor-indoor air pollution in urban environment: challenges and opportunity', *Frontier in Environmental Science*, vol. 2.

Levik, M., Bakke, J.V., Carlsen, K-H., Jensen, J.A., Myhre, K.I., Nafstad, P., Omenaas, E. & Norderhaug, I.N. 2005, 'Indoor exposures and risk of asthma and allergy: a systematic and critical review—preliminary report', *Proceedings of the 10th International Conference on Indoor Air Quality and Climate*, 4–9 September, Beijing, vol. 5, pp. 3576–80.

Lucattini, L., Poma, G., Covaci, A., de Boer, J., Lamoree, M.H. & Leonards, E.G.P. 2018, 'A review of semi-volatile organic compounds (SVOCs) in the indoor environment: occurrence in consumer products, indoor air and dust', *Chemosphere*, vol. 201, pp. 466–482.

Mainelis, G. 2020, 'Bioaerosol sampling: classical approaches, advances, and perspectives', *Aerosol Science and Technology*, vol. 54, no. 5, pp. 496–519. https://doi.org/10.1080/02786826.2019.1671950

Mannins, P. 2001, *Australia State of the Environment Report 2001 (Atmosphere)*, CSIRO Publishing on behalf of Department of Environment and Heritage, Canberra.

McGarry, P., Clifford, S., Knibbs, L., He, C. & Morawska, L. 2016, 'Application of multi-monitor approach to characterisation of particle emissions from nanotechnology and non-nanotechnology sources', *Journal of Environmental and Occupational Hygiene*, vol. 13. No.10, pp. 175–197.

McGarry, P., Morawska, L., He, C, & Jayaratne, R. 2011, 'Exposure to particles from laser printers operating within office workplaces', *Environmental Science and Technology*, vol. 45, no. 15, pp. 6444–52.

McGarry, P., Morawska, L., Knibbs, L. D. & Morris, H. 2013, 'Excursion guidance criteria to guide control of peak emission and exposure to airborne engineered particles', *Journal of Occupational and Environmental Hygiene*, vol. 10, no. 11, pp. 640–651. https://doi.org/10.1080/15459624.2013.831987

McKinney, K.R., Gong, Y.Y. & Lewis, T.G. 2006, 'Environmental transmission of SARS at Amoy Gardens', *Journal of Environmental Health*, vol. 68, no. 9, pp. 26–30.

Menzies, D., Pasztor, J., Nunes, F., Leduc, J. & Chan, C.H. 1997, 'Effect of a new ventilation system on health and well- being of office workers', *Achieves in Environmental Health*, vol. 52, pp. 360–367.

Mora, C., McKenzie, T., Gaw, I.M., Dean, J.M., von Hammerstein, H., Knudson, T.A., Setter, R.O., Smith, C. Z., Webster, K.M., Patz, J.A. & Franklin, E.C. 2022, 'Over half of known human pathogenic diseases can be aggravated by climate change', *Nature Climate Change*, vol. 12, pp. 869–75. https://doi.org/10.1038/s41558-022-01426-1

Morawska, L., Marks, G.B. & Monty. J, 2022, 'Healthy indoor air is our fundamental need: the time to act is now', *Medical Journal of Australia*, vol. 217, no. 11, pp. 578–581. https://doi.org/10.5694/mja2.51768

Morawska, L., Xiu, M., He, C., Buonanno, G., McGarry, P., Maumy, B., Stabile, L. & Thai, P. K. 2019, 'Particle emissions from laser printers: have they decreased?', *Environmental Science & Technology Letters*, vol. 6, no. 5, pp. 300–305. https://doi.org/10.1021/acs.estlett.9b00176

NABERS 2023, *'NABERS Indoor Environment'*, National Australian Built Environment Rating System, NSW Department of Planning and Environment, Sydney.

NADCA 2021, *ARC – The NADCA Standard for the Assessment, Cleaning and Restoration of HVAC Systems*. National American Duct Cleaning Association, Mount Laurel, NJ.

Nag, P.K. 2019, 'Sick building syndrome and other building-related illnesses', *Office Buildings*, vol. 18, pp. 53–103.

NEPC 2011, *National Environment Protection (Air Toxics) Measure*. National Environment Protection Council, Secretariat for Standing Council on Environment and Water, Canberra.

NEPC 2021, '*Variation to the Ambient Air Quality NEPM – ozone, nitrogen dioxide, and sulfur dioxide*'.

NEPM (National Environmental Protection Council). (2015). *National Environment Protection (Air Quality) Measure 2015*. Commonwealth of Australia.

NSW Legislative Assembly 2001, Standing Committee on Public Works Report on Sick Building Syndrome, Report No. 52/07, https://www.parliament.nsw.gov.au/ladocs/inquiries/2181/5207%20Sick%20Building%20Syndrome%20Report.pdf [17 Sept 2023].

NZ Ministry for the Environment 2016, '*Good Practice Guide for Assessing and Managing Odour*', Wellington, Ministry for the Environment, https://environment.govt.nz/assets/Publications/good-practice-guide-odour.pdf [6 January 2024].

Park, J-H. & Cox-Ganser, J.M. 2022, 'NIOSH dampness and mold assessment tool (DMAT): Documentation and data analysis of dampness and mold-related damage in buildings and its application', *Buildings (Basel)*, vol. 8, pp. 1075–1092.

Persily, A. 2015, 'Challenges in developing ventilation and indoor air quality standards: The story of ASHRAE Standard 62', *Building and Environment*, vol. 91, pp. 61–69.

Prinzi, A. & Rohde, R.E. 2023, *The Role of Bacterial Biofilms in Antimicrobial Resistance*, American Society of Microbiology, Bugs and Drugs.

Prussin, A.J. & Marr, L.C. 2015, 'Sources of airborne microorganisms in the built environment', *Microbiome*, vol. 3, no. 1, pp. 1–10. https://doi.org/10.1186/s40168-015-0144-z

Rabone, S., Phoon, W.O., Seneviratne, M., Gutirrez, L., Lynch, B. & Reddy, B. 1994, 'Associations between work related symptoms and recent mental distress, Allowing for work variables and physical environment perceptions in a sick office building', *Indoor Air: An Integrated Approach, Proceedings of the International Workshop on Indoor Air*, ed. Morawska, L. et. al. pp. 243–246.

Rajagopalan, P, Andamon, M.M. & Woo, J. 2022, 'Year long monitoring of indoor air quality and ventilation in school classrooms in Victoria, Australia', *Architectural Science Review*, vol. 65, no. 1, pp. 1–13.

Reed, S. 2009, 'What standards should we use for indoor air quality?', *19th International Clean Air & Environment Conference*, 6–9 September, Perth, Clean Air Society of Australia and New Zealand.

RESET 2018, *'RESET air standard v2.0'*.

Safe Work Australia 2022, *COVID-19 Information for Workplaces*, SafeWork Australia.

Sayan, H.E. & Dülger, S. 2021, 'Evaluation of the relationship between sick building syndrome complaints among hospital employees and indoor environmental quality', *La Medicina del Lavoro*, vol. 112, no. 2, pp. 153–161.

ScienceDirect 2022, *Sick Building Syndrome – An Overview*, https://www.sciencedirect.com/topics/earth-and-planetary-sciences/sick-building-syndrome [17 September 2023].

Seifert, B., Knoppel, H., Lanting, R.W. Person, A., Siskos, P. & Wolkoff, P. 1989, *'Report No. 6: Strategy for Sampling Chemicals Substances in Indoor Air'*, Commission of the European Communities.

Shearston, J.A., Eazor, J., Lee, L., Vilcassim, M.J.R., Reed, T.A., Ort, D., Weitzman, M. & Gordon, T. 2023, 'Effects of electronic cigarettes and hookah (waterpipe) use on home air quality', *Topology Control*, vol, 32, no. 1, pp. 36–41.

Shen, G., Ainiwaer, S., Zhu, Y., Zheng, S., Hou, W., Shen, H., Chen, Y., Wang, X., Cheng, H., & Tao, S., 2020b, 'Quantifying source contributions for indoor CO_2 and gas pollutants based on the highly resolved sensor data', *Environmental Pollution*, vol. 267. https://doi.org/10.1016/j.envpol.2020.115493

Shen, G., Zhu, S.A.Y.Z., Zheng, S., Hou, W., Huizhong Shen, Chen, Y., Wang, X., Hefa Cheng, H. & Tao, S. 2020a, 'Quantifying source contributions for indoor CO_2 and gas pollutants based on the highly resolved sensor data', *Environmental Pollution*, vol. 267, p. 115493.

Snow, S. Danam, R., Leardini, P., Glencross, M., Beeson, B., Ottenhaus, L. & Boden, M. 2022, 'Human factors affecting ventilation in Australian classrooms during the COVID-19 pandemic: Toward insourcing occupants' proficiency in ventilation management', *Frontiers in Computer Science*, vol. 4. https://doi.org/10.3389/fcomp.2022.888688.

Spengler, J.D., Samet, J.M. & McCarthy, J.F. 2001, *Indoor Air Quality Handbook*, McGraw-Hill, New York.

Standards Australia 1990 (R2014) *Methods for the Sampling and Analysis of Indoor Air—Determination of nitrogen dioxide—spectrophotometric method—treated filter/passive badge sampling procedures*, AS 2365.1.2:1990 (R2014). SAI Global, Sydney, https://www.standards.org.au/access-standards

Standards Australia 1993, *Methods for the Sampling and Analysis of Indoor Air—determination of carbon monoxide—direct-reading portable instrument method*, AS 2365.2:1993 (R2014), SAI Global, Sydney, https://www.standards.org.au/access-standards

Standards Australia 1995, *Methods for the Sampling and Analysis of Indoor Air—determination of radon*, AS 2365.4:1995 (R2014), SAI Global, Sydney, https://www.standards.org.au/access-standards

Standards Australia 2003a, *Workplace air quality – Sampling and analysis of volatile organic compounds by solvent desorption/gas chromatography Part 1: Pumped sampling method*, AS 2986.1:2003 (R2016), SAI Global, Sydney, https://www.standards.org.au/access-standards

Standards Australia 2003b, *Workplace Air Quality—sampling and analysis of volatile organic compounds by solvent desorption/gas chromatography, Part 2: Diffusive sampling method*, AS 2986.2:2003, SAI Global, Sydney, https://www.standards.org.au/access-standards

Standards Australia 2009, *Workplace Atmospheres—method for sampling and gravimetric determination of respirable dust*, AS 2985:2009, SAI Global, Sydney, https://www.standards.org.au/access-standards

Standards Australia 2012, *The use of ventilation and air conditioning in buildings – Mechanical ventilation in buildings*, AS 1668.2:2012, SAI Global, Sydney- previously released in 1976, 1980, 1989,1991, 2002, SAI Global, Sydney, https://www.standards.org.au/access-standards

Standards Australia 2017, *Methods for sampling and analysis of ambient air Determination of suspended particulate matter – PM 2.5 low volume sampler – Gravimetric method*, AS 3580.9.10:2017. SAI Global, Sydney, https://www.standards.org.au/access-standards

Standards Australia 2023, *Methods for Sampling and Analysis of Ambient Air – Determination of carbon monoxide – Direct-reading instrumental method*. AS 3580.7.1:2023, SAI Global, Sydney, https://www.standards.org.au/access-standards

Standards Australia and Standards New Zealand 2001a, *Stationary Source Emissions – Determination of Odour Concentration by Dynamic Olfactometry*. AS/NZS 4323.3-2001, SAI Global, Sydney, https://www.standards.org.au/access-standards

Standards Australia and Standards New Zealand 2001b, *Air filters for use in general ventilation and air-conditioning application, performance and construction*, AS/NZS1324.1-2001 SAI Global, Sydney, https://www.standards.org.au/access-standards

Standards Australia and Standards New Zealand 2003, *Methods for Sampling and Analysis of Ambient Air—determination of suspended particulate matter—PM_{10} High Volume sampler with size-selective inlet—gravimetric method*, AS/NZS 3580.9.6:2003, SAI Global, Sydney, https://www.standards.org.au/access-standards

Standards Australia and Standards New Zealand 2011, *Air-handling and Water Systems of Buildings—microbial control*, AS/NZS 3666.1:2011, SAI Global, Sydney, https://www.standards.org.au/access-standards

Standards Australia and Standards New Zealand 2022, *Determination of suspended particulate matter — PM_{10} beta attenuation monitors*, AS/NZS 3580.9.11:2022, SAI Global, Sydney, https://www.standards.org.au/access-standards

Standards Australia and Standards New Zealand 2023, *Methods for the sampling and analysis of ambient air – Determination of ozone—direct-reading instrument method*, AS/NZS 3580.6.1:2023. SAI Global, Sydney, https://www.standards.org.au/access-standards

Standards Australia 2024a, The use of mechanical ventilation and airconditioning in buildings Part 2, Mechanical ventilation in buildings, AS 1668.2-2024.

Standards Australia 2024b, The us of ventilation and airconditioning in buildings, Part 4, Natural ventilation of buildings, AS 1668.4-2024.

Steinemann, A. 2018, 'Prevalence and effects of multiple chemical sensitivities in Australia', *Preventive Medicine Report*. vol. 10, no. 10., pp. 191–194.

Swegon Air Academy n.d., '*The History of Ventilation*', https://www.swegonairacademy.com/good-ventilation/what-is-ventilation/history-of-ventilation/ [6 January 2024].

United States Environmental Protection Agency 1991, *Indoor Air Facts No. 4: Sick Building Syndrome*, US EPA, Cincinnati, OH, https://www.epa.gov/sites/default/files/2014-08/documents/sick_building_factsheet.pdf [17 September 2023].

United States Environmental Protection Agency 2023a. *Biological Pollutants' Impact on Indoor Air Quality*, https://www.epa.gov/indoor-air-quality-iaq/biological-pollutants-impact-indoor-air-quality [17 September 2023].

United States Environmental Protection Agency 2023b, *Indoor semi-volatile organic compounds 9i-SVOC – Version 1.0*, US EPA, Cincinnati, OH.

United States Environmental Protection Agency 2023c. *Air Cleaners, HVAC Filters, and Coronavirus (COVID-19)*. US EPA, Cincinnati, OH, https://www.epa.gov/coronavirus/air-cleaners-hvac-filters-and-coronavirus-covid-19 [17 September 2023].

Wargocki, P., Wyon, D.P., Sundall, J., Clausen, G & Fanger, P.O. 1999, 'The effects of outdoor air supply rate in an office on perceived air quality, sick building syndrome (SBS) symptoms and productivity', *Indoor Air*, vol. 10, pp. 222–236.

Wechsler, C. J., & Nazaroff, W. W. (2008). 'Semivolatile organic compounds in indoor environments', *Atmospheric Environment*, 42, no. 40, pp. 8816–8840.

WHO 2009, *WHO Guidelines for Indoor Air Quality: Dampness and Mould*, World Health Organization, Copenhagen.

WHO 2010, '*WHO Guidelines for indoor air quality: Selected pollutants*'

WHO 2020, '*Methods for sampling and analysis of chemical pollutants in indoor air*'.

WHO 2021, '*WHO global air quality guidelines: particulate matter ($PM_{2.5}$ and PM_{10}), ozone, nitrogen dioxide, sulfur dioxide and carbon monoxide*'.
Williams, P. 1992, *Airconditioning System Faults Affecting Health and Comfort in Melbourne Office Buildings*, Department of Architecture and Building, University of Melbourne, Melbourne.
Wolkoff, P. 2013, 'Indoor air pollutants in office environments: Assessment of comfort, health, and performance', *International Journal of Hygiene and Environmental Health*, vol. 216, no. 4, pp. 371–394.
Yadav, S. K., Mishra, R. K. & Gurjar, B. R. 2022, 'Assessment of the effect of the judicial prohibition on firecracker celebration at the Diwali festival on air quality in Delhi, India'. *Environmental Science and Pollution Research*, vol. 29, no. 57, pp. 86247–86259. https://doi.org/10.1007/s11356-021-17695-w
Zucco, G.M. & Doty, R.L. 2021, 'Multiple chemical sensitivity', *Brain Science*, vol. 12, no. 1, p. 46. https://doi.org/10.3390/brainsci12010046

13 Biological Hazards

Dr Margaret Davidson, Associate Professor Ryan Kift, Ken Martinez and Dr Joshua Schaeffer

13.1 INTRODUCTION

From The Bible to Bernadino Ramzinni's 18th-century masterpiece *De Morbis Artificum Diatriba*, biological hazards, their associated diseases, and their control have been the subject of scientific discussion and debate for millennia. In the past five years, there has been somewhat of a renaissance in the scientific research of, as well as public and governmental focus on, infectious (communicable) diseases and other biological hazards such as biotoxins and allergens. The resurgence was largely driven by the COVID-19 pandemic, bringing a paradigm shift in airborne virus transmission and renewed interest in indoor air quality and ventilation (Morawska et al. 2023). Other important events leading to this resurgence included the Australian House of Representative Inquiries into *biotoxin-related illness* (2018), *allergies and anaphylaxis* (2020), and reprisal of the seminal texts *Bioaerosols: Assessment and Control* and the AIHA *Mold in Buildings* (Hung et al. 2020). Advances in next-generation sequencing (NGS) technology continued the further understanding of how human bodies respond to infectious agents, toxins and allergens, along with the relationship between the human microbiome, physical and psychological health. However, work is still needed in the continued develop and review of standard methods for the investigation and evaluation of biological hazards. Standardised approaches are critical in data collection for the evaluation and risk management of specific biological hazards such as mould, viruses, bacteria, endotoxin, pollens in occupational environments, of which the mechanisms of disease are still poorly understood. The COVID-19 pandemic has underscored the critical need for preparedness to address biological hazards exacerbated by climate change, geopolitical tensions, and technological advancements.

13.2 BIOLOGICAL AGENTS

Biological hazards vary widely and include potentially infectious bacteria, archaea, viruses, fungi, protozoa, and parasitic organisms, as well as the non-living structural components and products of living organisms such as microbes, plants, algae, animals, and insects. Examples of non-viable agents include bacterial and fungal toxins and cell wall, prions (infectious proteins), animal and insect dander, dust mites, plant fibres, pollen, as well as grain, cotton, and other organic dusts to name but a few (Macher 1998). Zoonosis is a term used to describe infectious diseases transmitted between animals and humans.

Biological hazards are ubiquitous and are present in all workplaces. They may be of human or environmental origin (soil, air, water, animals, plant, etc.). Exposure may occur through multiple pathways including ingestion (toxins and infectious agents), inhalation (allergens, SARS-CoV-2 virus, *Mycobacterium tuberculosis*, and *Bacillus anthracis*), puncture/injection (spider venom and *Hepatitis B*), vector transfer by mosquitoes and ticks (specific viruses and bacteria) and

DOI: 10.1201/9781032645841-13

dermal contact (saps, resins, and plants). The following sections provide a brief introduction to the following biological hazards and associated terminology including bacteria, fungi, allergens, viruses, protozoans, and prions.

13.2.1 Bacteria

Bacteria are single-celled organisms with no nucleus, or membrane-bound organelles. Bacteria may have structure that includes flagella (motility), glycocalyces (protective outer coating), pili (reproduction) and fimbriae (attachment and protective biofilm/slime layer). Bacteria typically reproduce asexually through division (binary fission) (Tortora et al. 2020). Some bacteria, such as *B. anthracis* (anthrax), can produce hardy endospores enabling them to survive in inhospitable environments and making them difficult to eliminate and control. In 2016, an anthrax outbreak occurred when endospores were released from buried reindeer in the arctic permafrost, which is also known to contain bodies contaminated with 1918 influenza and smallpox, as well as streptomycin and tetracycline-resistant *Psychrobacter psychrophilus*, and a giant viruses known as *Mollivirius sibericum* and *Pithovirus sibericum* (Alempic et al. 2023).

Bacteria can be categorised into Gram-negative or Gram-positive, based on a Gram stain that targets the cell wall which identifies the presence of lipopolysaccharide (LPS) or peptidoglycan (PG) in their outer cell wall (Bauman 2012). Gram-positive bacteria typically stain purple and Gram-negative bacteria, with LPS in their outer wall, stain pink. Some bacteria such as *Neisseria* species and environmental isolates can produce a Gram-variable stain that is both pink and purple (Tortora et al. 2020). Discussion of Gram-positive and Gram-negative bacteria may arise in relation to organic dust and bioaerosol exposures, particularly in relation to the respiratory conditions associated with Lipopolysaccharide (LPS) exposure. LPS, also commonly called endotoxin, is a proinflammatory microbial cell-wall constituent that can cause lung inflammation and reduced function, and as a possible cause of organic toxic dust syndrome (OTDS) (Liebers et al. 2020; Pickering, 2020). It has also been used as a marker for bioaerosol exposure with its own exposure standard of 90 EU/m^3 (Dutch Expert Committee on Occupational Safety [DECOS] 2010) proposed algorithm for estimating exposures (Friesen et al. 2023), and standardised occupational endotoxin sampling methods (ASTM E 2144:2021; BS EN 14042:2003). Industries with the potential for high endotoxin exposures include agriculture, forestry and fisheries; laboratory workers, health care, metal working (oil fluids), waste management and recycling, as well as textile and food production (Liebers et al. 2020; Pickering 2020). Conversely, endotoxin exposure has also been linked to reduced risk of occupational cancers (Khedher et al. 2017).

Bacterial exotoxins are proteins that may be secreted by specific Gram-negative and -positive bacteria or released upon cell lysis (Sheehan et al. 2022). There are three main subtypes based on mode of action.

- **Type 1** surface active superantigens that can cause toxic shock syndromes, for example, *Staphylococcus* sp. and *Streptococcus* sp.).
- **Type II** membrane disrupting toxins that cause cell death by interfering with pores, for example, *Staphylococcal* haemolysin-α (Hla or α-toxin), haemolysin-β, and phenol-soluble modulinos (PSMs).
- **Type III** intracellular A (active) and B (binding) toxins and other toxins that disrupt cell function, for example, *Clostridium tetani* (tetanus) and *Clostridium botulinum* (botulism).

Cyanobacteria (blue-green algae) are among the Earth's oldest microbes (Yates et al. 2016), they can become problematic, and their accelerated growth causes 'blooms' that can produce toxic cyanotoxins. To date, no human deaths have been directly associated with cyanotoxins, but their ingestion may cause gastroenteritis, and direct contact can cause skin rashes and eye irritation (NHMRC 2022).

13.2.2 Fungi

Fungi are single or multicellular organisms. Their cell walls are predominantly composed of chitin. There are over 81,000 known species which vary in size from microscopic to macroscopic and also in shape. Fungi are immobile, non-photosynthetic organisms capable of both sexual and asexual reproduction. They are typically classified as yeasts (single cells that reproduce by budding or spores) or moulds (unicellular/multicellular that reproduce by spores or gamete formation) (Tortora et al. 2020). Some fungi are edible, but others are poisonous and difficult to differentiate from harmless varieties. Eating a single mushroom of some species can be fatal.

Mycoses are direct fungal growths on human and animal hosts, and mycotoxicoses are fungal diseases that can be caused by the ingestion, dermal contact, and inhalation of some toxic fungi (Bennett & Klich 2003). Fungal exposure can cause serious occupational diseases such as infectious aspergillosis, lung diseases such as hypersensitivity pneumonitis (HP) and OTDS, as well as allergic rhinitis, sinusitis, and occupational asthma (Miller 2023; Pickering 2020). High-level exposures may occur in agriculture (hay, grain, and animal facilities), forestry, wood chipping, waste recycling, composting, cannabis, and other fibre processing industries (Donham & Thelin, 2016; Miller 2023;). However, moulds and yeasts can present a significant risk for workers in poorly maintained and/or ventilated buildings promoting their proliferation. Mould species of health significance include species of the genera *Aspergillus*, *Penicillium*, *Cladosporium*, *Fusarium*, *Alternaria*, *Stachybotrys*, and *Wallemia sebi* (Cherrie et al. 2021; Dutkiewicz et al. 2011; Miller 2023). ISO 16000-19:2012 Annex A identifies mould species that may be indicative of excessive moisture. These include: *Acremonium* spp. *Aspergillus penicillioides*, *A. restrictus*, *A. versicolor*, *Chaetomium spp. Cladosporium sphaerospermum*, *Engyodontium album*, *Penicillium chrysogenum*, *Pithophora* spp. *Scopulariopsis brevicaulis*, S. *fusca*, *Stachybotrys chartarum*, and *Trichoderma* spp. Some species can produce mycotoxins such as aflatoxin, ochratoxin, fumonsins, citrinin, and trichothecenes (Samson 2015), which can pose significant health and economic risks in agriculture and food production sectors. However, the "agricultural relevant" mycotoxins are typically not considered a significant health risk in relation to indoor relevant mould species (Miller 2023; WHO 2009)

Other fungal components of immunological interest are fungal spores, mycelia, hyphae fragments, cell wall β-(1,3)-D-glucans (also in plants and bacteria), and *Candida albicans* and *Malassezia furfur* mannams (sugars). More than 180 fungal species produce allergenic proteins, with *Ascomycota* and *Basidiomycota* genus species of particular importance (Stewart et al. 2014). There is conflicting published evidence on the health effects of the various β-1,3-D-glucan forms, although triple-helical β-1,3-D-glucan associated with damp building materials has been identified as a strong inflammation causing agent (Miller 2023). Further research is required to understand the health risks associated with β-1,3-D-glucans, including any potential dose–response relationships.

13.2.3 Allergens

Occupational allergens encompass a wide range of agents, mostly protein complexes, that stimulate immune-mediated responses (Stewart et al. 2014). Allergens may also be metals, agrichemicals, adhesives, etc. discussed in earlier chapters. The focus here will be on allergens from living and previously living organisms. A curated database of officially recognised allergens can be accessed at WHO site http://allergen.org.

Allergen-related illnesses such as work-related asthma (WRA) can significantly impact on workers physically, financially, and socially (Rui et al. 2022), even occasioning death as recently seen in the recreational cannabis industry (OSHA 2022). While occupational allergens may not pose a significant risk for most workers, they can be potentially life-threatening or fatal for atopic (sensitised) workers. Allergens trigger an exaggerated immune response, with localised and/or systemic effects, varying from mild (sneezing, itching, and watery eyes) to severe symptoms such as anaphylaxis, a life-threatening response requiring immediate hospitalisation (Stewart et al. 2014). In addition, repeated exposures by inhalation, contact and ingestion, or single high-level exposure such as handling mouldy hay or grain, can trigger sensitisation and development of conditions such as HP, contact dermatitis, asthma, rhinosinusitis, and rhinitis (Chu et al. 2020; Donham and Thelin 2016;).

Other occupational allergens include flour, egg powder, hydrolytic enzymes in bakeries and mills, washing powder enzymes, latex, laboratory animal proteins, wood dust, animal and insect dander (hair, urine, faeces, saliva, skin cells, etc.), tick bites, plant materials including pollen and resins, rubber, latex, fungal spores and hyphae, bacteria including toxins, protozoa, bird droppings, feathers, and cyanotoxin (Dobashi et al. 2014; Donham & Thelin 2016; Hoy & Brims 2017; Stewart et al. 2014). Allergic disease and anaphylaxis cases continue to rise in Australia, with one in five Australians thought to be affected (Commonwealth of Australia 2020).

13.2.4 Viruses

Viruses are found in all environments, in the home and the occupational environments, and they form part of our personal microflora, referred to as the human virome. Viruses are composed of protein and nucleic acid, either deoxyribonucleic acid (DNA) or ribonucleic acid (RNA). They are extremely small (20–450 nm) and can only replicate inside other living cells (Cowan 2012). Interest in airborne exposure to viruses has increased significantly with the COVID-19 pandemic, creating greater understanding of virus transmission and control. Monitoring for some viruses in Australia can be conducted for sampling of the vectors (mosquitoes) and serological testing of sentinel chicken flocks (Knope et al. 2016). Human cases of selected viral diseases, especially those that are bloodborne, are monitored through the National Notifiable Disease Surveillance System (NNDSS). Current limitations to public health virus and virome research include limited diagnostic markers, absence of 'universal' viral sequence similar to the 16s region used in bacterial metagenomics, bias towards DNA sampling strategies over RNA which is the structural component of many viruses, limited genomic databases for matching molecular sequences, obtaining sufficient virus sample for identification, and the need for greater development of culture systems for propagating viruses and experimental animal-infection models (Wang 2020; Zou et al. 2016).

13.2.5 Other Biological Hazards

Other biological hazards include protozoan pathogens, heterotrophic single-celled organisms capable of independent movement including:

- *Giardia lamblia* (G. *intestinalis*, G. *duodenalis*), *Cryptosporidium parvum*, and *C. Hominis* which cause gastroenteritis when infectious cysts are ingested.
- The malaria parasite *Plasmodium ovale*, *P. malariae*, *P. knowlesi*, *P. vivax*, and *P. falciparum* which are a concern in deployed defence personnel and overseas travellers (Heymann et al. 2015). Australia was declared as malaria free in 1981, and the greatest risk for people is contracting the disease while travelling overseas.
- Amoeba, such as *Entamoeba histolytica*, a human parasite which is transmitted through ingestion of contaminated food or water, as well as contact with contaminated surfaces.

Algae, photosynthetic eukaryotic organisms, include aquatic diatoms, and dinoflagellates both of which can produce neurotoxins that humans may be exposed to through the consumption of contaminated fish and shellfish (Tortora et al. 2020). This must not be confused with bacterial 'blue-green' algae.

Prions are microscopic self-replicating proteins that cause neurodegenerative diseases such as classical Creutzfeldt–Jakob Disease (CJD) and variant CJD (vCJD) in humans, and bovine spongiform encephalopathy (BSE) in animals. These diseases are always fatal. Prions are infectious proteins that replicate by acting as a template that causes normal cellular proteins to reshape into prion proteins (Heymann 2022). Spasmodic cases of Classical CJD have been recorded in Australia since 1970; however, there have been no reported cases of vCJD. Australia is considered a negligible-risk country for BSE (NSW Health 2019).

Finally, there are other macroscopic hazards such as spiders, snakes, predatory and stinging aquatic organisms, and even farm animals which can cause crush injuries and even occasion death. There are also parasitic worms (helminths), tapeworms, nematodes, liver flukes, hookworm, which can be contracted working with animals, handling soil and waste products, and working in outdoor environments.

13.3 OCCUPATIONAL DISEASE

Occupational diseases associated with biological hazards can be grouped as infectious diseases, respiratory diseases (allergic and irritant), and cancers. This chapter does not provide an exhaustive coverage of all diseases associated with biological exposures, it focusses on those most likely to occur in the occupational environment. It will not cover toxin-mediated disease, foodborne illness, or dermatoses (allergic and irritant).

Occupations with an increased susceptibility to occupational diseases linked to biological agents encompass a wide variety of sectors including health care, social assistance, veterinary medicine, waste management, biomedical research, biosecurity, charitable aid, defence, agriculture, forestry, and fisheries (Edmonds et al. 2016; Rapp et al. 2014; SWA 2011; Viegas et al. 2017). Notably, unlike many chemical and physical agents of occupational significance, biological agents are everywhere, adding to the burden of workplace exposures, especially allergens, for example, grass pollen, fungi, insect, and animal dander.

13.3.1 Infectious Disease

Infectious diseases, also called communicable diseases, can be passed between humans and their environment (abiotic and biotic), including zoonotic diseases transmitted between humans and animals. Transmission may be by direct contact, accidental autoinoculation (needlestick), inhalation of contaminated aerosols, contact with contaminated objects (fomites), or ingestion of contaminated substances. Australia employs a mandatory reporting system for 70 plus notifiable infectious diseases of public health significance. These are grouped as bloodborne (Hepatitis B, C, and D), gastrointestinal (listeria and Hepatitis A), sexually transmissible infections (chlamydia), vaccine preventable diseases (measles and tetanus), respiratory diseases (influenza and SARs-COV-2), vector-borne diseases (malaria and Dengue), zoonoses (anthrax, Q fever, and tularaemia), listed human diseases (plague and COVID-19), and other notifiable diseases (leprosy and Group A Streptococcal disease).

Depending on the workplace and worker activities, there is potential for any of the notifiable diseases to be the result of an occupationally acquired infection. However, some will inevitably pose a greater risk to workers based on the proximity to the disease vectors, hosts and reservoirs.

In Australia, "Deemed Diseases" of occupational significance have been published. Infectious diseases included are anthrax, avian influenza, brucellosis, COVID-19, Hepatitis A, B and C, HIV/AIDS, Influenza A (H1N1), leptospirosis, Middle East Respiratory Syndrome (MERS), Orf, Psittacosis, Q fever, and tuberculosis (Driscoll 2021). Additional infectious diseases may be listed as 'notifiable' or 'deemed' in individual States and Territories, and people should check with their public health and/or workers compensation agencies for the latest information (Driscoll 2021). Occupations/activities most frequently identified in the Deemed Diseases documentation include those handling domestic and/or wild animals, including pelts, meat, and/or regulatory work accessing these facilities; frontline healthcare work with direct patient contact, childcare workers, carers, pathologists, laboratory workers, and waste management (all types) (Driscoll 2021). Sex workers are a glaring omission from vulnerable occupations and industries (infectious diseases) and the list should be updated to be inclusive of this profession.

In Australia, cases of workplace acquired COVID-19, brucellosis, HIV, leptospirosis, mycosis, tuberculosis, measles, legionellosis, Q fever, Hendra virus, cellulitis (bacterial skin infection), and viral hepatitis have all been reported over the last two decades (SWA 2006, 2011, 2020). However, published data on the true occupational burden of infectious diseases in Australia is limited, and likely underreported and not adequately addressed.

Outdoor workers in Australia face a notable health risk from zoonotic and arboviral diseases transmitted by mosquitoes and ticks. Australia has many endemic vector-borne viruses, including Ross River virus (RRv), Murray Valley Encephalitis (MVE), and Barmah Forest virus (BFv), along with other bacterial pathogens, which can lead to severe illness or even fatalities (Hime et al. 2022). Additionally, sporadic tickborne diseases including viral Flinders Island and Australian spotted fevers, as well as bacterial melioidosis caused by *Burkholderia pseudomallei* (Dehhaghi et al. 2019) have been reported. Tick bites can also trigger allergic responses such as mammalian meat allergy, paralysis, and anaphylaxis (Pek et al. 2016; van Nunen 2018). Sporadic cases of locally acquired exotic diseases like malaria, dengue, and Japanese encephalitis have been reported in Northern Australia (Ong et al. 2021; Russell & Doggett 2012).

Legionnaires' disease is a pneumonia caused by the *Legionella* species. Workers at risk include those exposed to aerosols from cooling towers, decorative fountains, water sprayed to cool people, or water sprays being used to control dust levels, or for cleaning, which potentially contain *Legionella pneumophila*. Smokers and the immuno-compromised are at higher risk of contracting *Legionella* illness (Heymann 2022).

Q fever is one of the most reported zoonotic diseases in Australia, with most cases likely being of occupational origin (Driscoll 2021). This flu-like illness is caused by the organism *Coxiella burnetii*. The disease presents a significant economic burden on Australian agricultures. Between 2002 and 2012, 177 workers compensation claims were made for Q fever in NSW, totalling around $3.5 million in claims. Workers may be infected through handling animal placental material, inhalation of contaminated materials and drinking unpasteurised milk from infected cows. At risk, workers include veterinary staff, farm, dairy, stockyard and abattoir and rendering plant workers. Council workers involved in animal carcass recovery and landscapers are also at risk, along with zoo staff and wildlife carers/volunteers for high-risk animals (Heymann 2022; SafeWork NSW 2023). Q fever can be effectively controlled with a vaccine and is mandatory for some occupations. In 2019, a Queensland (QLD) Museum taxidermist contracted Q fever while prepping exhibits, and a co-worker was prosecuted in 2023 for failing to provide a safe workplace under the QLD Workplace Health and Safety legislation (AIHS 2023).

For further information on notification and control of notifiable conditions, consult the NNDSS documentation at the Department of Health (2023), or your local health department.

13.3.2 Respiratory Disease

Respiratory diseases are complex responses by the body to various agents that involve allergic (immune-mediated) and non-allergic reactions, which can cause significant discomfort, impairment, and death. Biological agents with allergic and non-allergic actions are presented in Table 13.1, and several are also on the Deemed Diseases list (Driscoll 2021). It is important to distinguish between allergic and non-allergic respiratory diseases as their treatment approaches and management strategies differ for both the patient's health, return to work and management of the agent in the workplace. For example, for some allergic individuals where life-threatening anaphylaxis is a risk, exclusion from the environment is the only control, while for irritant asthma, other controls could be explored. Respiratory diseases are diagnosed through careful evaluation of the patient's medical history, symptoms, physical examination, and diagnostic tests. Proper diagnosis and targeted interventions are crucial for effectively managing and treating these conditions.

13.3.2.1 Allergic Diseases

An allergic response occurs when the immune system is triggered by the presence of a foreign substance (antigen) that triggers the antigen-binding antibodies to eliminate the threat. However, on occasion the body overreacts (hypersensitivity response) causing a range of symptoms from mild to severe, dependent on the antigen, dose, and other intrinsic (internal) and extrinsic (external) variables such as genetics, smoking status, childhood exposures, diet, and living on farm (Reynolds et al. 2013). Historically, there were four hypersensitivity classifications (Gell & Coombs), now expanded to seven mechanisms [Sells classification] (Tuano & Chinen 2020). Type I and Type IV hypersensitivity are of key significance in occupational settings.

Type I is a rapid immunoglobulin E (IgE) mediated reaction (15–30 minutes) to allergens. Examples include occupational asthma, eczema, conjunctivitis, allergic rhinitis and in severe cases anaphylaxis. *Allergic Rhinitis* (hay fever) is a common type I occupational disease, along

TABLE 13.1
Respiratory Diseases from Exposure to Non-communicable Biological Hazards

Respiratory Disease	Agent	High-risk Occupations
Non-allergic		
Non-allergic asthma, non-allergic rhinitis/mucous membrane irritations (MMI), chronic bronchitis, chronic airflow obstruction, and ODTS	Fungi, bacteria actinomycetes, endotoxin, ß-(1,3)-D-glucans, peptidoglycan, and mycotoxin	Agriculture, forestry and fishing workers, waste treatment workers, composting and recycling workers, textile and food production workers, horticulturalists, metal workers and machinists, veterinarians, zookeepers, laboratory workers, construction workers, archaeologists, and biofuel production workers
Allergic		
Allergic asthma, allergic rhinitis, and hypersensitivity pneumonitis (allergic alveolitis/farmer's lung)	Fungi, thermophilic mycelial bacteria, *Mycobacterium immunogenum* microbial volatile organic compounds (MVOCs), spores, allergens (dust mite, plant, insect and animal proteins, pollens, etc.), and endotoxin, ß-(1,3)-glucans	Waste treatment, composting and recycling workers, biomedical researchers, enzyme production workers and lab animal tenders, healthcare workers, bakers, industrial (detergent and biopesticide manufacturing) workers, pet shop owners, agricultural, forestry and fishing workers, wood processors/ furniture makers, and horticulturalists

Adapted from: Liebers et al. 2008; Donham & Thelin 2016.

with *Occupational* (new onset) and *Work-aggravated* (pre-existing allergies) *Asthma*. Triggers include pollen, fungi, dust mite, pet dander, cockroaches, detergents, biological enzymes, and flour, among many others (Alif et al. 2020; Dao & Bernstein 2018, Hoy et al. 2020). Symptoms of OA may occur outside of work hours making it harder to identify the cause and can have significant health, social, and financial impacts for workers (Hoy et al. 2020).

Type IV reactions are delayed 'sensitising' responses, typically1–3 days post allergen exposure (latex, poison ivy). Examples include contact dermatitis, psoriasis, rheumatoid arthritis, and other autoimmune diseases. Hypersensitivity pneumonitis (HP) is a debilitating occupational respiratory disease that involves both Type III and Type IV hypersensitivity reactions and has over 300 suspected allergens. This condition is commonly referred to by the occupation or allergen involved, for example, Farmer's lung and mushroom workers lung (Chandra & Cherian 2023).

13.3.2.2 Non-allergic Response

Non-allergic responses may occur from mechanical blockage, irritation, or scaring of tissues, as through innate (passive) immune-system responses (Donham & Thelin 2016). Symptoms can include irritation of eyes, nose, and throat (mucous membrane irritation), airway blockages, and skin irritation. Mycobacterial and fungal infections may cause the rare lung disease *Alveolar Proteinosis* [over accumulation of alveoli surfactant and lipids] (Alif et al. 2020). *Non-Allergic Asthma* may occur with excessive bioaerosol exposures, causing big cross-shift declines in lung function. Although it presents like occupational asthma, it is not a Type 1 (IgE) allergy (Linaker & Smedley 2002).

Inhalation fevers (flu-like illnesses) occur from inhalation of bioaerosols such as agricultural dust, air humidifier, air conditioner, and cooling tower emissions. They include:

- **Humidifier fever** (humidifiers and air-conditioning). Symptoms appear within 4–12 hours of exposure, and the illness is self-limiting, with recovery happening within days.
- **Pontiac fever** from exposure to *Legionella pneumophila* and other *Legionella* species. A legionellosis variant that is mild, self-limiting, and does not result in pneumonia (Tortora et al. 2020).

Organic toxic dust syndrome (ODTS) is a toxic pneumonitis (alveolitis) caused by high-level organic dust and endotoxin exposures (Donham & Thelin 2016) and fungal spores and fragments, including mouldy hay, silage, and corn (Madelin & Madelin 1995). Symptoms are typically self-limiting (24 hours) but may persist for up to 7 days. They include fever, shivering, dry cough, chest tightness, dyspnoea, headache, muscular and joint pain, fatigue, nausea, and general malaise (Douwes et al. 2008).

13.3.3 Cancer

Specific microorganisms or their components are carcinogenic to humans. These include Hepatitis C, Hepatitis B, Epstein Barr virus, *Helicobacter pylori*, *Aspergillus* aflatoxins, *Plasmodium* sp. and *Schistosoma haematobium*. Some wood and leather-based dusts are also group 1 carcinogens (IARC, 2023).

13.4 REGULATIONS AND STANDARDS PERTAINING TO BIOLOGICAL HAZARDS

There are workplaces where biological hazards are either regulated or subjected to Australian and/or International Standards, as well as general duties of care under applicable OHS/WHS legislation.

13.4.1 Security-Sensitive Biological Agents (SSBAs)

The Department of Health and Ageing administers the legislation relating to SSBAs. The *National Health Security Act 2007* (as amended 2016) and the associated *National Health Security Regulations 2008* (as amended 2013) describe regulated biological agents based on their potential for use as biological weapons.

13.4.2 Genetically Modified Organisms (GMOs)

The Australian Office of the Gene Technology Regulator (OGTR) administers the legislation relating to genetically modified biological materials. The *Gene Technology Act 2000* (as amended 2016) and the *Gene Technology Regulations 2001* (as amended 2020) aim to 'protect the health and safety of people, and the environment, by identifying risks linked to gene technology, and by managing those risks through regulating certain dealings with genetically modified organisms' (GMOs) (Ley 2015, p. 1). As well as regulating GMOs, the OGTR also sets requirements for laboratory construction and procedures to be used when handling genetically modified biological materials (OTGR, 2023).

13.4.3 Department of Agriculture and Water Resources (DAWR)

DAWR regulates the importation of biological materials from overseas. It also authorises and inspects Quarantine Approved Premises (QAP) within Australia. QAP may be used to hold or work on imported biological materials, including plants, animals, or other products. The DAWR (previously DAFF and AQIS) administers the *Biosecurity Act 2015*, *Export Control Act 2020*, and the *Imported Food Control Act 1992*.

13.4.4 Australian Standard/New Zealand Standard 2243.3: Safety in Laboratories

The recently updated AS/NZS 2243.3:2022 is relevant to the operation of biological laboratories in Australia and New Zealand, including those used for research, teaching, and pathology among others – particularly laboratories that are not regulated by other means (e.g., by the OGTR). AS/NZ 2243.3 sets out facility construction and practices' requirements for laboratory, animal, plant, and invertebrate containment facilities. This Standard classifies microorganisms into risk groups (RGs) according to their pathogenicity (Table 13.2). Laboratories are divided into four levels of containment depending on the work to be conducted (Table 13.3). The majority of Australian biological laboratories are classified as physical containment (PC) level 2.

TABLE 13.2
Definition of Microorganism Risk Groups

Risk Group Level	Work Undertaken
RG1	Unlikely to cause human or animal disease
RG2	May infect and individual worker, but treatment is likely to be available and spread to the wider community is unlikely
RG3	May pose a significant risk to an individual worker, but further spread to the community is most likely able to be controlled
RG4	May cause life-threatening human or animal disease and may spread in the community or environment. Treatment may not be available

Adapted from: p. 17 from AS/NZS 2243.3:2022.

TABLE 13.3
Physical Containment Levels for Laboratories

Physical Containment Level	Work Undertaken
PC1	Low hazard, risk group (RG) 1 microorganisms, for example, undergraduate teaching laboratory
PC2	Research or diagnostic laboratory, RG2 microorganisms
PC3	Research or diagnostic laboratory, RG3 microorganisms
PC4	High-risk specialist work with RG4 microorganisms

Adapted from p. 34 & 35 from AS/NZS 2243.3:2022.

Risk groupings of organisms relate directly to the containment levels of the associated laboratories. In other words, RG1 organisms should be handled in a PC1 laboratory, and so on. AS 2243.3 gives other useful information for laboratories, such as standards for spill clean-up, transportation of biological materials, use of specialist equipment (centrifuges, biological safety cabinets, etc.), useful disinfectants for various organisms, and waste disposal.

13.4.5 Notifiable Disease Surveillance

The NNDS diseases discussed in previous sections can be accessed in the *National Health Security (National Notifiable Disease List) Instrument 2023*.

13.5 OCCUPATIONAL EXPOSURE STANDARDS

The number of occupational exposure standards (OES) for biological hazards is extremely limited, and all are related to chemical, non-viable, constituents such as wood dust, bacterial endotoxin, or gaseous emissions rather than living microorganisms. The Senate Biotoxin Inquiry has called for further research on the development of standards and/or accreditation requirements for the mould testing and remediation, which may see a push towards the development of Australian guidelines and benchmarks for assessing and evaluating damp and mould-affecting buildings, or the adoption of international standards for bioaerosol sampling and analysis such as ASTM D7338-14 (2023) *Standard Guide for Assessment of Fungal Growth in Buildings* or ISO 16000-19:2012 *Indoor Air – Part 19: Sampling strategy for moulds*.

The establishment of quantitative viable bioaerosol OES is complicated by the highly variable nature of exposures, inherent variability in peoples' immunological response, and other factors. For this reason, quantitative limits are not recommended because their application could potentially cause over or underestimation of potential health risks (Miller 2023). Viable bioaerosol concentrations can vary by species and in orders of magnitude depending on ventilation of the building, its occupancy, activities, and the sampling strategy applied. Other factors include season, climate, construction, and age of the building and outdoor air intake. The measurement of viable airborne microorganisms is further complicated by the limited sampling duration used to prevent media overload (30 seconds to 3 minutes) and can easily miss significant shifts in airborne microbiota.

The following should always be considered when evaluating biological hazards:

- The infectious dose (viable cells) can vary greatly between workers and is dependent on worker health, behaviours, and genetics.

- There is very limited data on the dose–response relationship for specific microorganisms, especially in workplaces, as well as for microbial components, for example, toxins, cell-wall components, spores, seeds, and hyphae.
- For some agents, there is no pathological difference between an occupationally and domestically acquired infectious disease.
- The lack of standardisation in sampling and analytical methods makes it difficult for critical review of historical studies and meta-analysis (Gorny 2007; Walser et al. 2015).

The use of biological markers to estimate microbial exposure, such as cell-wall components, has gained popularity. The Dutch Expert Committee on Occupational Safety (2010) has recommended an exposure limit of 90 EU/m^3 for bacterial endotoxins, and researchers are undertaking epidemiological studies on workplace exposures to other microbial constituents such as muramic acid (Gram-positive bacteria), ergosterol (fungi), and 3-OHFA (Gram-negative bacteria). For biological agents other than bacteria and fungi, occupational exposure limits have been established for:

- nuisance dusts, wood, textile, cellulose, and grain dusts (GESTIS 2023; SWA 2023);
- volatile compounds such as ammonia and hydrogen sulfide, which can be produced by biological organisms (SWA 2023); and
- subtilisin and enzymes obtained from *Bacillus subtilis* (ACGIH® 2023).

The GESTIS International Limit Values database is a valuable resource for OES (GESTIS 2023), and safety data sheets (SDS) are available internationally for microorganisms (Canada Health). It is hard to identify the point at which exposure to bioaerosols becomes a health hazard, and, for the majority, their role in the initiation and development of many occupationally acquired illnesses are still poorly understood. Host health status, gene–environment previous sensitisation, interactions also play a role in the induction and severity of diseases (ACNEM 2023; Kwo & Christiani 2017).

13.6 SAMPLING AND ANALYSIS OF BIOLOGICAL HAZARDS

There is no perfect method for the investigation of biological hazards, with hygienists and researchers using combinations of traditional and novel sampling and analytical approaches. Investigation of biological agents, bioaerosols, and aerobiology encompasses many disciplines including occupational hygiene, microbiology, mycology, aerobiology, food technology, veterinary, medical, environmental science, and engineering to name but a few. The hygienist should always consult with their laboratory before designing their sampling plan and depending on the complexity/purpose of the work they may also want to consult other specialists such as building engineers, medical, and allied health practitioners.

A comprehensive discussion of sampling and analytical approaches is beyond the scope of this chapter, and readers are encouraged to consult key texts by professional bodies, such as:

- Hung et al.'s (2020) *Recognition, Evaluation and Control of Indoor Mold*, 2nd Edition. This text has comprehensive sections on sampling design, methodology (bulk, air, surface), analysis, reporting and remediation for mould investigations.
- Bioaerosols: Assessment and Control, 2nd edition (refer to AIHA website for more information). The revised text of this seminal textbook addresses a variety of bioaerosols including fungi, amoebae, bacteria, dust mite, endotoxin, fungal toxins, MVOCs as well as investigation and remediation.

- Flannigan et al.'s (2011) Microorganisms in Home and Indoor Work Environments: Diversity, Health Impacts, Investigation and Control, 2nd edition.
- *NIOSH Manual of Analytical Methods; Sampling and Characterization of Bioaerosols* (Lindsley et al. 2017) is free to download from the NIOSH website.
- *Field Guide for Determination of Biological Contaminants in Environmental Samples*, 2nd edition (Hung et al. 2005).
- *Manual of Environmental Microbiology* (Yates et al. 2016).
- Relevant *ASTM, ANSI, ISO, Australian and CEN Standards.*

13.6.1 Sampling Strategy and Study Design

The reasons for sampling biological agents may vary widely and may involve:

- Regulatory and compliance investigations for notifiable agents, for example, *Legionella*, Q Fever, HIV, COVID-19, or for non-infectious biological hazards with OES such as flour, wood, or cotton dusts.
- Complaints, source identification, staff reassurance, exposure documentation, for example, public health complaints, workers compensation investigation, and building certification (Green Star, etc.).
- Epidemiological or toxicological research, dose–response, development of OES, etc.
- Remediation assessments; determining the adequacy of remediation work such as flood damage and the potential for mould contamination.
- Assessment of exposure controls such as the decontamination of HAZMAT gear used in emergency response to bioterrorism events, or filtration efficiency.

The purpose of the investigation will always govern the design. This includes the sample types and devices used (bulk, aerosol, surface, swab, and viable/non-viable), how samples will be analysed (culture-based, gravimetric, microscopy, biochemical, metagenomic, etc.), as well as sample number, duration, frequency, etc. This is especially so where an OES/WES is being applied because the publisher will define what approach must be used to ensure the collection of appropriate data. For infectious notifiable disease outbreaks, the lead government agency will apply their state or territory guidelines that are based on the Communicable Diseases Network Australia (CDNA) *Series of National Guidelines* (SoNGs) for notifiable diseases (Dept of Health 2023). Other key design considerations include:

- Observation of any preliminary data (reports, complaints, building plans, and past sampling) that may give insight into where issues may be occurring and why.
- Accessibility of sampling equipment and/or analytical technology. For example, rental options for cotton dust elutriators are limited and there are no Australian manufacturers.
- The physiochemistry of the biological agent and its potential concentration. Over and underestimation of biological agents is problematic due to rapid airborne changes in, cell desiccation during sampling creating viable but non-culturable (VBNC) organisms, and short sampling periods required by many devices (grab sampling rather than full shift).
- The environmental matrix (soil, water, construction material, air, etc.) which can result in sample losses on filters and other media, as well as potential inhibition or amplification during analysis, particularly with biochemical assays like Limulus Amoebocyte Lysate (LAL) assay for endotoxin or molecular sequencing (soil chemicals can inhibit amplification).
- Finances can significantly impact, culture-based analysis can be expensive, requiring multiple replicates and controls to address issues such as high uncertainty and random error associated with bioaerosol sampling.

- Transportation is another key issue, particularly culture-based analysis where samples need to be transferred to the laboratory within specific times and kept under strict temperature control.
- Cross-contamination, sterility of consumables, and cleaning of sampling devices are also very important considerations due to the ubiquity of microorganisms and allergens, as well as their components like endotoxin, muramic acid, and the fact that humans have their own personal dust cloud and skin microflora that can potentially be transferred to samples.

Data analysis is also not always straightforward for biological hazards because there are often no quantitative limits and the data is often qualitative in nature (swabs, tapes, dip slides, etc.) and may have large uncertainty due to very short sampling periods (30 s–30 minutes) (Hung et al. 2020; Flannigan et al. 2016). Differential statistical analysis of microbial diversity and taxa abundance may be required to compare controls (outdoor air, pre-treatment/remediation, and sterile products) and samples. This approach is more typical for studies incorporating molecular sequencing technologies analysing arrays of DNA, RNA, proteins, and lipids in various samples (air, soil, water, biological tissues, etc.) to identify biomarkers of disease susceptibility or resilience, study changes in microbial communities, evaluation of the effectiveness of controls, development and testing of new therapeutic drugs, and many other reasons. Mould and damp building evaluations also compare microbial abundance and diversity to look for shifts that may indicate a potential problem area (Hung et al. 2020).

13.6.2 Sample Collection

Sampling methods may include the collection of airborne contaminants, surface (swabs, RODAC plates, tape lifts, and dip slides), bulk materials (soil, house dust, wash water, etc.), or biological samples (blood, urine, and nasal secretions). Australian Standards are available for bulk materials (water and food), along with sampling of respirable and inhalable particulates for dust and other bioaerosols. For microorganisms and microbial constituents, international standards are available for fungi and bacteria (ASTM and ISO 16000s including impactors, impingers, spore traps, swabs, and tape lifts), viruses (VDI 4251 BLATT 1:2019-09), endotoxin (BS EN 14031:2021; ASTM E2144-21), MVOCs (VDI 4254 BLATT 1:2018-06), and pollen (BS EN 16868:2019). The hygienist may also want to consult their laboratory as they may have in-house methods for the sampling and analysis of specific agents. Table 13.4 presents some common commercially available devices which may or may not be incorporated in published standard methods.

13.6.2.1 Impactors

Impactors are commonly used for indoor air-quality assessments, clean-room testing, and industrial operations (Haig et al. 2016). Commercially available impactors range from slit to single-stage or multistage (cascade) impactors, which enable size-selective sampling of viable and culturable microorganisms in different aerosol fractions (Lindsley et al. 2017).

Slit impactors direct the air stream through a rectangular slit on to agar, which undergoes incubation for culturable organisms (slit-to-agar sampler), or on to glass slides or moving tapes for microscopic analysis of pollen and spore traps (Ruzer & Harley 2013). Commercial devices include the BioSlide direct-to-slide sampler and the Hirst volumetric spore sampler.

The single-stage and multistage cascade impactors work by directing an air stream on to a plate containing agar, which is subsequently incubated, and the cultivated microorganisms counted, or on to a filter which can be analysed by a variety of methods including direct microscopy analysis (counts) and the extraction and biochemical/chemical analysis of

TABLE 13.4
Sampling Devices for Biological Hazards

Device	Media	Method	Aerosol	Comments
Impactors (ASTM D7788-14, ASTM D8068-19, ISO 16000-18:2011, ASTM D7391-20)				
Single-stage	Agar	Area	Viable microorganisms or spores (agar, tape, and gelatin filters)	Relatively inexpensive, and equipment and analytical facilities readily available for viable microorganism analysis in Australia
Cascade	Agar or filters	Area and personal		Slit impactors can give real-time indication of temporal variation in bioaerosols
Slit	Agar or tape strip		Allergens, dusts, and microbial cell-wall components (filters)	Collection and impaction on agar and gelatin filters can stress microorganisms, making them viable but non-culturable, selective for more resilient organisms. May require short sampling periods of 1–3 minutes in heavily contaminated atmospheres. Cascade samplers enable size-selective measurement of biological agents, and both personal and area samplers are commercially available. Gelatin filters have short sampling periods of 30 minutes, after which filters become brittle. Microorganisms may become viable and non-culturable through sampling process. Use of different filter media alters recovery of cell-wall components like endotoxin
Impingers – BS EN 17359:2020				
AS PCR	Liquid	Area	Viable and countable microorganisms	Samples can undergo a wide variety of culture- and non-culture-based analyses
Rotating cup	Liquid	Area		
AGI–30	Liquid	Area	Cell-wall components, MVOCs, and toxins	Subject to inlet sampling lossesRequire short sampling periods of less than 30 minutes owing to sample media evaporation
BioSampler	Liquid	Area		Glass presents personal injury risk with midget impingers
Midget	Liquid	Personal	Viable and countable microorganisms	Bulky and cumbersome to wear
Filtration – ISO 16000-16:2008, ISO 16000-20:2014, AS 2985-2009, AS 3640-2009, BS EN 16868:2019				
IOM	Filters	Area Personal	Dust, microbial and cell-wall components, spores, MVOC, toxins, and allergens	Can be adapted for full work-shift sampling Best suited for microbial indicators such as cell wall components (endotoxin, muramic acid, etc.) or direct counts owing to sampling stress such as desiccation on viable organisms
Button				Gelatin filters for viable microorganisms have short sampling periods, maximum 30 minutes
Cyclone Versa Trap Microspore cassettes	Custom cassettes			Particle removal efficiency and biological recovery efficiency will vary widely for various filter materials
Other				
NIOSH personal bioaerosol sampler	Filter and centrifuge tubes	Personal	Inhalable Dust	Highly suited to culture-independent analytical approaches

(Continued)

TABLE 13.4 (CONTINUED)

Device	Media	Method	Aerosol	Comments
Vertical elutriators	Liquid	Area	Organic dust (<15 μm diameter) and endotoxin	Vertical elutriators are standard methods in cotton and textile industry in relation to byssinosis. BS 3406-3:1963, OSHA 1910.1043
Electrostatic precipitators	Liquid	Area	Vegetative cells	Experimental technology
Wetted-wall cyclone	Liquid	Area	Vegetative cells	Experimental technology may not be efficient for hydrophobic bacteria or fungal spores. Subject to impaction and rehydration stresses
Settling plates	Agar	Area	Vegetative cells and spores	Unsuitable for spore sampling. Highly subject to environmental conditions and low repeatability
Surface vacuuming	Certified micro-vacuum cassettes or HEPA vacuum	Surface	Viable cells, dust, cell-wall components, toxins, etc.	Not representative of airborne exposures. ASTM D7144-21 and ASTM D5438-17
Tape, wipes, and swab samples	Sticky tape, biotapes, wipes, and swabs	Surface	Viable cells, dust, cell-wall components, toxins, adenosine triphosphate (ATP), etc.	Not representative of airborne exposures, qualitative in nature and subject to sampling error. Standard methods ASTM D7658-17-21, ASTM D7789-21, ASTM D7910-14, ISO 18593:2018, ASTM E3226-19, ASTM D6602-22, ASTM D7144-21, ASTM D7789-21, ASTM D7910-14, ASTM E1216-21, and ASTM E2458-17

Adapted from: Broadwater et al. 2022; Cage et al. 1996; Engelhart et al. 2007; Görner et al. 2006; Hung et al. 2020; Jensen et al. 1992; Kulkarni et al. 2011, Lindsley et al. 2017; Madelin & Madelin 1995; Reponen et al. 2011; Spurgeon 2007; Yao & Mainelis 2007; Yamamoto et al. 2011.

cell-wall constituents. Particles in the airstream collect on the agar based on their velocity, with heavier/large particles with greater momentum collected on the top stages and smaller particles on the lower stages.

Commonly used agars include non-selective nutrient or tryptic soy agar (TSA) for bacteria and malt extract (MEA), Sabouraud dextrose agar (SDA), cellulose extract, or DG-18 agars for fungi. ISO 16000-17 specifies the types of, and preparation of agars for culture-based sampling, and some commercial laboratories will provide prepared plates as part of their service. Other selective agars such as eosin methylene blue (EMB), and MacConkey agars may also be used but advice should be sought from the laboratory. Differential media such as blood agar can be used to help presumptively identify bacteria through their growth morphology and ability to lyse red blood cells. Some agars are classified as both differential (microbes of interest from colonies with distinctive features for identification) and selective (inhibits growth of microbes that may obscure species of interest), examples include EMB agar, *Escherichia coli* develop a gold sheen, and chromogenic agars, which are used to select for, and presumptive identification of specific microorganisms such as *Salmonella* colonies that form mauve colonies when cultured on CHROMagar™. However, chromogenic agars can be expensive in comparison and prone to false positive results. Presumptive samples must be verified by more advanced identification methods.

The advantages of using impactors such as the Andersen microbial sampler, SAS microbial air sampler, or BioSampler are that they are readily accessible, easy to operate and calibrate, and the instruments, media, and sample analysis are all relatively inexpensive compared with other methods, such as endotoxin or MVOC analysis. The disadvantages of the impactors are that the results may not be truly representative of the microbial community because the impaction of microorganisms on to solid surfaces can render them viable but non-culturable and various species may go undetected. Sampling times need to be very short (30 seconds to three minutes) in heavily contaminated environments because the agar plates quickly become overloaded. Many impactors are not designed for personal sampling and hence are unsuitable for this purpose. Due to loading bioaerosol concentrations may be underestimated because they are based on plate counts of individual colonies, which may grow from an aggregate of cells or spores, rather than the individual cells. It is estimated that only a fraction of the microbial organisms can be cultured in the laboratory (Schloss & Handelsman 2005), and this is why the advent of culture-independent approaches enables better understanding of complex microbial communities.

13.6.2.2 Impingers

Impingers, historically, have been used to sample a variety of biological aerosols, including viable and countable bacteria, fungi, viruses, pollens and cell-wall components. They collect air contaminants by drawing an air stream through a thin glass tube and aspirating it into a liquid medium which separates out large particles. The liquid medium can be plated directly on to agar, examined by microscope, and analysed either biochemically or using molecular techniques. The action of the air stream impinging into the liquid aids in the separation of aggregates, which, when plated on agar, can provide a more representative count than impactors can. Most impingers, including the SKC BioSampler and the AGI-30, are designed for area sampling and have large, cumbersome pumps. Midget impingers designed for personal sampling have been around since the 1930s but could be considered a hazard due to their glass construction. Multistage impingers are also available, but their glass construction makes it difficult to accurately predict the mass median diameters collected by each stage (Vincent 2007). Limitations of impingers for biological hazards monitoring include short sampling time owing to rapid evaporation of media (maximum 30 minutes), the hydrophobic particles will bounce and escape the system (Hung et al. 2005), and the effects of moderate changes in flow rate and liquid quantity on sample collection (Dart & Thornburg 2008). Impingers may be subject to inlet and internal losses as well as re-aerosolisation of organisms at lower concentrations (Han & Mainelis 2012; Kesavan et al. 2010). ViaTrap mineral oil can be used as a collection liquid to reduce evaporation, allowing for eight-hour sampling periods, but such oils do not plate well and are better suited for microscopic analysis.

13.6.2.3 Filtration

Biological aerosols can be collected using the methods for particulate sampling: AS 2985:2009 (respirable dust) and AS 3640:2009 (inhalable dust). These methods are suitable for the collection of non-viable bioaerosol samples such as pollen, allergens, spores, endotoxins, and DNA/RNA, but not necessarily viable cells to undergo culturing in the laboratory. This is because microbial cells can be sub-lethally injured during sampling, making them viable but non-culturable and causing both underestimation of bacterial loadings and sampling bias for hardier organisms such as the endospore-forming *Bacillus* species. Choice of collection media is also critical. Gelatine filters have been recommended for collection of viable microorganisms, but they become brittle and fragile during sampling and have a maximum sampling period of 30 minutes, depending on the environmental conditions. Filter materials must be tested for background endotoxin and genetic materials, as well as for recovery efficiency of various bioaerosols such as endotoxin or β-D-glucans (Davidson et al. 2004; Hung et al. 2005). Metal sampling heads, such as IOM cartridges or button samplers, are ideal because they can be heat treated to remove background contamination. This is important when sampling for endotoxin and genomic DNA material, as they may not be fully removed from devices using standard cleaning methods such as autoclaving and/or alcohol rinsing (Davidson et al. 2005). There are now some single-use filter sampling devices, such as spore traps, which are commercially available for collection of fungal spores, and reduce the risk of sample cross-contamination.

13.6.2.4 Other Samplers

Other commercially available samplers include wetted-wall, centrifugal, and electrostatic samplers. Wetted-wall cyclones operate on the same principle as particle-sampling cyclones except that the internal surfaces are coated with a suitable liquid to wash particles into the collection chamber (Reponen et al. 2011; Vincent 2007). The hand-held, battery-operated Reuter Centrifugal Sampler (RCS) has been commercially available for several years and has been adopted in Canada as a reference sampling method for fungal contamination assessments (Vincent 2007). The operating principle is a centrifugal fan (impeller) which draws air into the sampling head and directs the stream on to an agar strip fitted to the internal walls. The strip is then incubated, and the number of colonies is counted to estimate the bioaerosol loading (Cherrie et al. 2021).

Real-time monitors that operate by measuring the fluorescence emitted by particles are commercially available. Real-time monitors can be linked to existing alarm systems as engineering controls for biohazard containment or clean-room efficiency (Fennelly et al. 2017). A personal polymerase chain reaction (PCR)-based detection unit has also been developed to assess bioaerosol exposures in real-time (Agranovski et al. 2017), as well as wearable sensors to detect biological threat agents (Ozanich 2018).

13.6.3 Sample Storage and Transportation

Before sampling commences, the technician must consult with their analytical laboratory regarding special storage and transport requirements which will change depending on sample type and endpoint analysis. The basic principles for sample storage and transport are to protect and stabilise samples from environmental disturbances (sunshine, humidity, desiccation, heat, dust, etc.). Consideration needs to be given to temperature fluctuations and storage time, or cell growth in favourable conditions. Freeze/thaw cycling of endotoxin causes sample losses and should be avoided (Hung et al. 2005), while collection and storage methods for metagenome studies must be optimised based on the sample type to avoid confounding relative abundance estimates and community profiles (Quince et al. 2017). Workplace atmosphere guidelines for the measurement of airborne microorganisms and endotoxin, as well as local standard methods published by governmental and other professional agencies, should be consulted.

13.6.4 Sample Analysis

The selection of analytical methods will be primarily driven by the agent of interest, regulatory requirements, for example, *Legionella* sp. in cooling towers or bloodborne viruses, budget, and

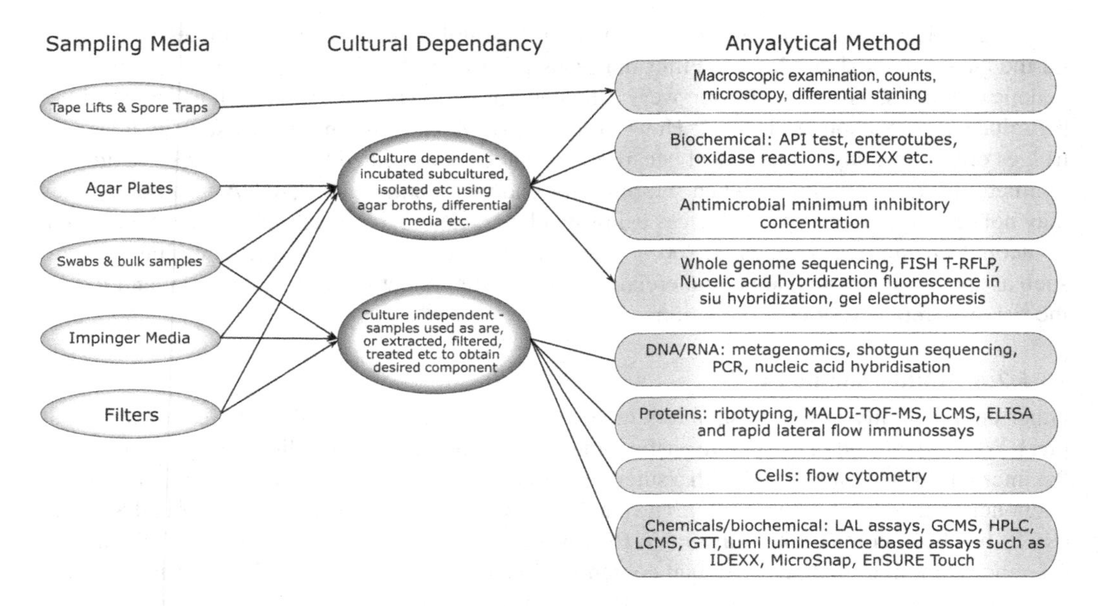

FIGURE 13.1 Sampling and analysis options for fungal samples

Source: M. Davidson.

access to analytical services. There are a wide variety of biological and chemical-based methods that can be harnessed depending on the sample matrix and target organism as illustrated in Figure 13.1 showing the options for microbial analysis of air samples.

Over the last decade, there has been a shift from culture-based analysis to culture-independent approaches in the study of microbial diversity. Some of these culture-independent approaches include metagenomics, whole genome sequencing, real-time, and quantitative PCR (Mbareche et al. 2017). Culture-based approaches and spore identification remain more accessible and affordable for practitioners and have internationally recognised standardised sampling and analytical methods. Further work on standardised sampling protocols for microbial investigations in occupational environments using Next-Generation Sequencing (NGS) technologies and bioinformatics is needed and could be based on environmental microbiology protocols, for example, Earth Microbiome Project.

Genomic sequencing does provide greater insight into the complexity of microbial communities that inhabit the body and environment, including non-culturable microorganisms, such as archaea (prokaryotic bacteria with no peptidoglycan in cell wall) and viruses (Blais-Lecours et al. 2015). Given the flexibility for sampling methodologies that can be harnessed with culture-independent analysis, it is anticipated that these methods will continue to increase in popularity as observed in the investigation of SARs-COV2 virus transmission (Lednicky et al. 2020) and adoption for foodborne disease investigations, notifiable diseases, and biosecurity surveillance (Gerner-Smidt et al. 2019).

It is possible to distinguish live from dead microorganisms using specialised dyes, ribose nucleic acid (RNA), and protein-based methods (Emerson et al. 2017). Table 13.5 outlines the various methodologies that can be used for analysis of bioaerosols, but for more comprehensive overviews, consult your local analytical laboratory.

TABLE 13.5
Analytical Methodologies for Biological Agents

Category	Methods	Endpoint	Comments
Culture-based	Plate counts, most probable number, membrane filtration, turbidity, and IDEXX	Viable bacteria, fungi, and viruses	Most common method in Australia at present. Accredited labs readily accessible in all states and territories. Most inexpensive method. Australian Standard methods available
Countable	Microscopic counts and flow cytometry	Total bacteria, fungi, spores, and pollen	Accredited labs readily available for microscopy work. Flow cytometry primarily a research method
Chemical	GC/MS & GC/ MSMS, LC/MSMS, HPLC, TLC, and MALDI-TOF	MVOCs, lipopolysaccharide (Gram-negative), muramic acid (Gram-positive), ergosterol (fungi), and 16s rDNA	Limited number of labs in Australia perform these analyses, which are mainly used for epidemiological and exposure research at present
Molecular	Next Generation Sequencing.single cell sequencingReal-time PCR, Western blot, RibotypingPulsed-field gel electrophoresis	RNA, DNA, viruses, fungi, bacteria, proteins, cytokines, and allergens	Expensive and laboratories offering these services commercially are limited in Australia, but steadily becoming more accessible Analysis of next-generation sequencing data can be complex and requires specialised bioinformatics software
Biochemical	Bacteria and fungi ID kits (API, Enterotube, IDEXX, bioluminescence assays, and rapid lateral flow tests)	Bacteria and fungi	Commercial labs offer bacteria, fungi and protein identification, ELISA, and other standardised testing. For in-field/rapid testing a variety of commercial lateral flow kits and luminometry based detection devices are available, mostly developed for food and beverage and health care sector.
	ELISA and Western blot	Proteins/allergens	
	LAL and rFC assay	β-D-glucans, fungal extracellular polysaccharides, endotoxin, and peptidoglycan	Endotoxin and β-D-glucans analysis of air samples not readily accessible in Australia and mainly used for research at present. Kits are available, but many variables must be addressed when using. Refer to BS EN 14031:2021; ASTM E2144-21 for more information

13.7 CONTROL

For the general principles of control, refer to Chapter 5. The legislative and other documentation are outlined in this chapter. The principles of this chapter are applied with specific applications and examples for biological hazards below.

13.7.1 Focus on Infectious/Communicable Diseases

For infectious/communicable diseases, the intent is to break the chain of infection by disrupting the agent–host–environment interaction, often referred to as the epidemiological triad. The *agent* is the infectious organism, which is required for disease to occur, the *host* is the human who contracts the disease, and the *environment* includes the external/extrinsic biological, physical, and socioeconomic factors that can influence the host and agent interaction.

Control of or preventing the spread of infectious disease can be achieved by targeting the host (immunisation and diet), agent/pathogen (genetic modification), environment (sanitation and removal of water sources), or the vector (sterilisation of organisms and retractable sharps for injections). Another model that aligns to the epidemiological triad but is better suited to the occupational safety and health perspective was proposed by Sietsema et al. (2019). The authors use the concepts of source (agent), pathway (environment), and receiver (host). The Sietsema model emphasises the application of a hierarchy of controls that focusses first on source, followed by pathway, and finally receptor control.

In response to the COVID-19 pandemic, AIHA (2021) released the second edition of the Role of the Industrial Hygienist in a pandemic with the purpose to enhance and expand the resources, information, and tools, the industrial hygienist needs to protect the working public from pandemic risks. For other communicable diseases, the American Public Health Association's Control of Communicable Diseases Manual is a handy guide that sets out methods of control for significant public health pathogens (Heymann 2022). For healthcare settings, in Australia, the NHMRC (National Health and Medical Research Council) has published a comprehensive guideline on the prevention and control of infection in health care (NHMRC 2022). Other key guidelines for control of infectious microorganisms such Legionellosis include the enHealth *Guidelines for Legionella Control* and AS3666: Air-handling and water systems of buildings – Microbial control.

13.7.2 Non-infectious Biological Hazards

When controlling biological hazards, the standard principles of the hierarchy of control, beginning with elimination as part of a risk management approach, are prescribed in AS/NZS ISO45001:2018. Administrative controls and personal protective equipment (PPE) are considered the lowest order controls and should not be used alone, rather they should be combined with administrative and engineering controls such as time limits on work shifts to prevent fatigue or use of positive/negative pressure environments depending on the situation/agent. Table 13.6 provides examples of the different types of controls that may be applied in various industries.

13.8 LOCAL AND GLOBAL DISEASE ISSUES

The precise health burden of biological hazards through infectious and other diseases on Australian workers is unknown. Reports and workers compensation data are available for a limited set of conditions such as respiratory disease and notifiable conditions including COVID-19 (Alif et al. 2020; SWA 2011; SWA 2020). Issues that obscure the true picture of what is happening include underreporting of cases, non-notification, asymptomatic cases, non-notifiable conditions, and the fact that disease/health conditions that are not well captured or defined by current workers compensation statistics. The last comprehensive publications of work-related infectious and parasitic diseases were released over a decade ago, and lung infections were outside the scope of the most recent Australian respiratory disease report (Alif et al. 2020).

Recent pandemics signify how biological hazards can have significant consequences for workers (physical and psychological), in addition to their social and economic impacts. With climate

TABLE 13.6
Potential Controls for Biological Hazards

Hierarchy of Control	Examples
Elimination	Elimination of contaminated water sources. Elimination of vectors, such as mosquitoes.
Substitution	Substituting high-dust animal feed with a low-dust variety. Substituting a high-risk bacterium with a low-risk organism in a laboratory. Substitute latex gloves with nitrile. Undertaking in vitro experiments instead of using an animal model.
Engineering	Use of biological safety cabinets when handling hazardous microorganisms in a laboratory. Use of fully enclosed and air-conditioned tractor cabs to protect the operator from outside dusts. Use of approved sharps containers to dispose of needles or other sharp items. Using single-use equipment to remove the risk of transferring an infectious disease in medical centres, beauty salons, or tattoo parlours. Maintenance of negative or positive air pressure, via ventilation, in spaces where biological hazards may be handled or naturally present.
Administrative	Development of a vaccination program to reduce the risk of infection from an identified pathogen, for example, Hepatitis A or B. Use of a documented biological waste segregation and disposal system, for example, in a hospital, laboratory or home healthcare setting. Scheduling work to limit exposure during high exposure activities. Implementation of an induction or training activities and appropriate supervision. Use of biohazard signage and barricading. Frequent changing of machine oil in sumps to prevent bacterial growth. Instigation of routine maintenance schedule for mechanical ventilation systems to ensure biological material does not accumulate.
PPE	Appropriate approved respiratory protection. Safety glasses, goggles, or face shield as appropriate. Lab coat, gown, overalls, or Hazmat suite. Gloves.

change, habitat destruction, and globalisation, the geospatial boundaries of infectious agents, their hosts, and reservoirs are expanding, with no country or workforce immune to this shift. Increased global temperatures have already released some (anthrax) and may potentially release other pathogenic microorganisms from stasis in glaciers and permafrost such as smallpox and the 1918 influenza pandemic (Sköld 2017; Stevens et al. 2004). Changing temperatures can also promote survival and proliferation of pathogens in their environmental niches, thereby increasing their outbreak potential (Watts et al. 2015).

Antimicrobial resistance (AMR) is a significant public health concern, especially for occupationally acquired infections in healthcare services. AMR is the ability of microorganisms to develop immunity to antimicrobial agents such as antibiotics, antivirals, and antimalarials. This immunity renders standard medical treatments ineffective and results in persistent infections in hosts. The rise in AMR has been attributed to over-prescribing of antibiotics, patients not finishing courses of treatments, poor infection control practices, lack of hygiene and sanitation, limited research into new treatments, and the overuse of antibiotics in agriculture both for preventing infection and the use as growth promoters in animal production. (Australian Government 2023; DoH 2023; National Academies of Sciences 2017; WHO 2021). The prevalence of AMR is increasing both in Australia and internationally at a pace exceeding the development of new antimicrobials. Controlling the risk of workplace-acquired infections of AMR organisms must include the addition of higher-order infection controls (isolation, engineering, and administrative) in addition to therapeutic drugs and PPE, especially when protecting frontline health workers and other at-risk professions (veterinary, agricultural, sex workers, water treatment, etc.). These include the use of isolation to limit spread within facilities; implementation of cleaning strategies using appropriate chemicals; hand hygiene and other hygiene strategies and surveillance for AMR organisms where required (NHMRC 2022).

REFERENCES

ACGIH 2023, *Threshold Limit Values for Chemical Substances and Physical Agents and Biological Exposure Indices*, ACGIH®, Cincinnati, OH.

ACNEM 2023, *Biotoxins (indoor damp and mould) Clinical Pathway*, https://www.acnem.org/about/advocacy/ [21 September 2023].

Agranovski, I.E., Usachev, E.V., Agranovski, E. & Usacheva, O.. 2017, 'Miniature PCR based portable bioaerosol monitor development', *Journal of Applied Microbiology*, vol. 122, no. 1, pp.129–138. https://doi.org/10.1111/jam.13318

AIHA 2021, *The Role of the Industrial Hygienist in a Pandemic*, 2nd edn, AIHA, Cincinnati, OH, https://online-ams.aiha.org/amsssa/ecssashop.show_product_detail?p_mode=detail&p_product_serno=2522 [21 September 2023].

AIHS 2023, *QLD: Worker Sentenced over Q Fever Case*, https://www.aihs.org.au/news-and-publications/news/qld-worker-sentenced-over-q-fever-case [8 August 2023].

Alif, S., Glass, D., Abramson, M., Hoy, R. & Sim, M. 2020, *Occupational Lung Disease in Australia 2006-2019*, Safe Work Australia, https://www.safeworkaustralia.gov.au/sites/default/files/2020-08/Occupational%20lung%20diseases%20in%20Australia%202006-2019.pdf [21 September 2023].

Australian Government 2023, *Antimicrobial Resistance*, https://www.amr.gov.au/ [21 September 2023].

Bauman, R.W. 2012, *Microbiology with Diseases by Body System*, Benjamin Cummings, Boston, MA.

Bennett, J.W. & Klich, M. 2003, 'Mycotoxins', *Clinical Microbiology Reviews*, vol. 16, no. 3, pp. 497–516.

Blais-Lecours, P., Perrot, P. & Duchaine C. 2015, 'Nonculturable bioaerosols in indoor settings: Impact on health and molecular approaches for detection', *Atmospheric Environment*, vol. 110, pp. 45–53. https://doi.org/10.1016/j.atmosenv.2015.03.039

Broadwater, K. Ashley, K. & Andrews, R. 2022, 'Surface sampling guidance, considerations, and methods in occupational hygiene' in *NIOSH Manual of Analytical Methods (NMAM)* (5th edn), https://www.cdc.gov/niosh/nmam/5th_edition_web_book.html [29 November 2023].

Cage, B.R., Schreiber, K., Portnoy, J. & Barnes, C. 1996, 'Evaluation of four bioaerosol samplers in the outdoor environment', *Annals of Allergy, Asthma and Immunology*, vol. 77, no. 5, pp. 401–6.

Chandra, D. & Cherian, S.V. 2023, *Hypersensitivity Pneumonitis*. StatPearls, https://www.ncbi.nlm.nih.gov/books/NBK499918/ [21 September 2023].

Cherrie, J.W., Semple, S. & Coggins, M.A. 2021, 'Chapter 16: Bioaerosols', in *Monitoring for Health Hazards at Work* (5th edn), John Wiley & Sons.

Chu, C., Marks, J.G. & Flamm, A. 2020, 'Occupational contact dermatitis: common occupational allergens', *Dermatologic Clinics*, vol. 38, no. 3, pp. 339–349.

Commonwealth of Australia 2020, *Walking the allergy tightrope: Addressing the rise of allergies and anaphylaxis in Australia*, https://www.aph.gov.au/Parliamentary_Business/Committees/House/Health_Aged_Care_and_Sport/Allergiesandanaphylaxis/Report [14 October 2023].

Cowan, M.K. 2012, *Microbiology: A Systems Approach* (3rd edn), McGraw-Hill, New York

Dao, A. & Bernstein, D.I. 2018, 'Occupational exposure and asthma', *Annals of Allergy, Asthma & Immunology*, vol. 120, no. 5, pp. 468–475. https://doi.org/10.1016/j.anai.2018.03.026

Dart, A. & Thornburg, J. 2008, 'Collection efficiencies of bioaerosol impingers for virus-containing aerosols', *Atmospheric Environment*, vol. 42, no. 4, pp. 828–32.

Davidson, M., Reed, S., Markham, J. & Kift, R. 2004, 'Filter performance in endotoxin sampling and quantification in the Australian agricultural environment', *AIOH 22nd Annual Conference Proceedings*, Fremantle WA, Australian Institute of Occupational Hygienists.

Davidson, M., Reed, S., Markham, J. & Kift, R. 2005, 'Guide to Environmental Endotoxin Sampling', *AIOH 23rd Annual Conference Proceedings*, Terrigal NSW, Australian Institute of Occupational Hygienists.

Dept Health 2023, *CDNA Series of National Guidelines* (SoNGs), https://www.health.gov.au/resources/collections/cdna-series-of-national-guidelines-songs [21 September 2023].

Dobashi, K., Akiyama, K., Usami, A., Yokozeki, H., Ikezawa, Z., Tsurikisawa, N., & Okumura, J. 2014, 'Japanese Guideline For Occupational Allergic Diseases 2014'. *Allergology International*, vol. 3, no. 63, pp. 421–442. https://doi.org/10.2332/allergolint.14-rai-0771.

DoH 2023, *FINAL PROGRESS REPORT Australia's First National Antimicrobial Resistance Strategy 2015–2019*, https://www.amr.gov.au/sites/default/files/2022-10/final-progress-report-australia-s-first-national-antimicrobial-resistance-strategy-2015-2019.pdf [21 September 2023].

Donham, K.J. & Thelin, A. 2016, *Agricultural Medicine: Rural Occupational and Environmental Health, Safety, and Prevention*, 2nd edn, John Wiley & Sons, New Jersey.

Douwes, J., Eduard, P. & Thorne, P.S. 2008, 'Bioaerosols', in K. Heggenhougen and S. Quah (Eds) *International Encyclopaedia of Public Health*, Elsevier Science/Academic Press, Oxford.

Driscoll, T. 2021, *Deemed Diseases in Australia. Safe Work Australia.*, https://www.safeworkaustralia.gov.au/doc/revised-list-deemed-diseases-australia [21 September 2023].

Dutch Expert Committee on Occupational Safety. 2010, *Endotoxins: Health-based recommended occupational exposure limit*, Health Council of the Netherlands, The Hague, https://www.gezondheidsraad.nl/sites/default/files/201004OSH.pdf

Dutkiewicz, J., Cisak, E., Sroka, J., Wojcik-Fatla, A. & Zajac, V. 2011, 'Biological agents as occupational hazards—selected issues', *Annals of Agricultural and Environmental Medicine*, vol. 18, no. 2, pp. 286–93.

Edmonds, J., Lindquist, H. D. A., Sabol, J., Martinez, K., Shadomy, S., Cymet, T., Emanuel, P. & Schuch, R. 2016, 'Multigeneration cross-contamination of mail with bacillus anthracis spores', *Plos One*, vol. 11, no. 4. https://doi.org/10.1371/journal.pone.0152225

Emerson, J.B., Adams, R.I., Román, C.M.B., Brooks, B., Coil, D.A., Dahlhausen, K., Rothschild, L.J. 2017, 'Schrödinger's microbes: Tools for distinguishing the living from the dead in microbial ecosystems', *Microbiome*, vol. 5, no, 1, pp. 86. https://doi.org/10.1186/s40168-017-0285-3

Engelhart, S., Glasmacher, A., Simon, A. & Exner, M. 2007, 'Air sampling of Aspergillus fumigatus and other thermotolerant fungi: comparative performance of the Sartorius MD8 airport and the Merck MAS–100 portable bioaerosol sampler', *International Journal of Hygiene and Environmental Health*, vol. 210, no. 6, pp. 733–9. https://doi.org/10.1016/j.ijheh.2006.10.001

Fennelly, M., Sewell, G., Prentice, M., O'Connor, D. & Sodeau, J. 2017, 'Review: The use of real-time fluorescence instrumentation to monitor ambient primary biological aerosol particles (PBAP)', *Atmosphere*, vol. 9, no. 1, pp. 1–1. https://doi.org/10.3390/atmos9010001

Flannigan, B., Samson, R. A., & Miller, J. D. 2011, *Microorganisms in Home and Indoor Work Environment: Diversity, Health Impacts, Investigation and Control.* (2nd edn) CRC Press. http://www.crcnetbase.com/isbn/9781420093353

Friesen, M.C., Xie, S., Viet, S.M., Josse, P.R., Locke, S.J., Hung, F., Andreotti, G., Thorne, P.S., Hofmann, J.N. & Beane Freeman, L.E. 2023, 'An algorithm for quantitatively estimating occupational endotoxin exposure in the biomarkers of exposure and effect in agriculture (BEEA) study: I. development of task-specific exposure levels from published data', *American Journal of Industrial Medicine*, vol. 66, no. 7, pp. 561–572. https://doi.org/10.1002/ajim.23486

Gerner-Smidt, P., Besser, J., Concepción-Acevedo, J., Folster, J.P., Huffman, J., Joseph, L.A., Kucerova, Z., Nichols, M.C., Schwensohn, C.A. & Tolar, B. 2019, 'Whole genome sequencing: Bridging one-health surveillance of foodborne diseases', *Frontiers in Public Health*, vol. 7, pp. 172. https://doi.org/10.3389/fpubh.2019.00172

GESTIS 2023, *GESTIS International Limit Values.* DGUV, Germany, http://limitvalue.ifa.dguv.de/ [21 September 2023].

Görner, P., Fabries, J. F., Duquenne, P., Witschger, O. & Wrobel, R. 2006, 'Bioaerosol sampling by a personal rotating cup sampler CIP 10-M', *Journal of Environmental Monitoring*, vol. 8, no. 1, pp. 43–8.

Gorny, R. 2007, '*Biological agents; need for occupational exposure limits (OELs) and feasibility of OEL Setting*, Depart of Biohazards, WHO Collaborating Center; Poland, https://osha.europa.eu/sites/default/files/seminars/documents/en/seminars/occupational-risks-from-biological-agents-facing-up-the-challenges/speech-venues/speeches/biological-agents-need-for-occupational-exposure-limits-oels-and-feasibility-of-oel-setting/Presentation%20by%20Mr%20Gorny.ppt [8 August 2023].

Haig, C.W., MacKay, W.G., Walker, J.T., & Williams, C. 2016. "Bioaerosol sampling: sampling mechanisms, bioefficiency and field studies." *Journal of Hospital Infection*, vol. 93, no. 3, pp. 242–255.

Han, T. & Mainelis, G. 2012, 'Investigation of inherent and latent internal losses in liquid-based bioaerosol samplers', *Journal of Aerosol Science*, vol. 45, pp. 58–68. https://doi.org/10.1016/j.jaerosci.2011.11.001.

Heymann, D.L. 2022, *Control of Communicable Diseases Manual* (21st edn). APHA Press, an imprint of American Public Health Association.

Heymann, D.L., Chen, L., Takemi, K., Fidler, D.P., Tappero, J.W., Thomas, M.J., Kenyon, T.A., Frieden, T.R., Yach, D., Nishtar, S., Kalache, A., Olliaro, P.L., Horby, P., Torreele, E., Gostin, L. O., Ndomondo-Sigonda, M., Carpenter, D., Rushton, S., Lillywhite, L. & Rannan-Eliya, R.P. 2015, 'Global health security: The wider lessons from the west African Ebola virus disease epidemic', *The Lancet*, vol. 385, no. 9980, pp. 1884–1901. https://doi.org/10.1016/S0140-6736(15)60858-3

Hime, N.J., Wickens, M., Doggett, S.L., Rahman, K., Toi, C., Webb, C. & Lachireddy, K. 2022, 'Weather extremes associated with increased Ross River virus and Barmah Forest virus notifications in NSW: Learnings for public health response'. *Australian and New Zealand Journal of Public Health*, vol.46, no. 6, pp. 842–849.

Hoy, R. & Brims, F. 2017, 'Occupational lung diseases in Australia', *The Medical Journal of Australia*, vol. 207, no. 10, pp. 443–448.

Hoy, R., Burdon, J., Chen, L., Miles, S., Perret, J. L., Prasad, S., Radhakrishna, N., Rimmer, J., Sim, M. R., Yates, D., & Zosky G. 2020, Work-related asthma: a position paper from the Thoracic Society of Australia and New Zealand and the National Asthma Council Australia. *Respirology*, Vol. 25, no. 11, pp. 1183–1192. https://doi.org/10.1111/resp.13951.

Hung, L.-L., Caufield, S.M. & Miller, J.D. 2020, *Recognition, Evaluation, and Control of Indoor Mold* (2nd edn), American Industrial Hygiene Association.

Hung, L.L., Miller, J.D. & Dillon, K.H. 2005, *Field Guide for the Determination of Biological Contaminants in Environmental Samples*, (2nd edn), American Industrial Hygiene Association, Fairfax, VA.

IARC 2023, *IARC MONOGRAPH*, Lyon, France, https://monographs.iarc.fr [7 August 2023].

Jensen, P.A., Todd, W.F., Davis, G.N. & Scarpino, P.V. 1992, 'Evaluation of eight bioaerosol samplers challenged with aerosols of free bacteria', *American Industrial Hygiene Association Journal*, vol. 53, no. 10, pp. 660–7.

Kesavan, J., Schepers, D. & McFarland, A.R. 2010, 'Sampling and retention efficiencies of batch-type liquid-based bioaerosol samplers', *Aerosol Science and Technology*, vol. 44, no. 10, pp. 817–29.

Khedher, B.S., Guida, N.M., Matrat, F., Cenée, M., Sanchez, S., Menvielle, M., Molinié, G., Luce, F.D. & Stücker, I. 2017, 'Occupational exposure to endotoxins and lung cancer risk: Results of the ICARE Study', *Occupational and Environmental Medicine*, vol. 74, no. 9, pp. 667–79.

Knope, K.E., Kurucz, N., Doggett, S. L., Muller, M., Johansen, C.A., Feldman, R., Hobby, M., Bennett, S., Sly, A., Lynch, S., Currie, B.J. & Nicholson, J. 2016, 'Arboviral diseases and malaria in Australia, 2012–13: annual report of the national arbovirus and malaria advisory committee'. *Communicable Diseases Intelligence Quarterly Report*, vol. 40, no. 1, pp. 17–47.

Kulkarni, P., Baron, P.A. & Willeke, K. 2011, *Aerosol Measurement: Principles, Techniques, and Applications* (3rd edn), Wiley.

Kwo, E. & Christiani, D. 2017, 'The role of gene–environment interplay in occupational and environmental diseases: current concepts and knowledge gaps', *Current Opinion in Pulmonary Medicine*, vol. 23, no. 2, pp. 173–176. https://doi.org/10.1097/MCP.0000000000000364

Lednicky, J.A., Shankar, S.N., Elbadry, M.A., Gibson, J.C., Alam, M.M., Stephenson, C.J., Eiguren-Fernandez, A., Morris, J.G., Mavian, C.N., Salemi, M., Clugston, J.R. & Wu, C.Y. 2020, 'Collection of SARS-CoV-2 virus from the air of a clinic within a university student health care center and analyses of the viral genomic sequence' *Aerosol and Air Quality Research*, vol. 20, no. 6, pp. 1167–1171. https://doi.org/10.4209/aaqr.2020.02.0202

Ley, S. 2015, 'Gene Technology Amendment Bill', The Parliament of the Commonwealth of Australia. Canberra.

Liebers, V., Brüning, T. & Raulf, M. 2020, 'Occupational endotoxin exposure and health effects', *Archives of Toxicology*, vol. 94, pp. 3629–3644.

Liebers, V., Raulf-Heimsoth, M. & Brüning, T. 2008, 'Health effects due to endotoxin inhalation (review)', *Archives of Toxicology*, vol. 82, no. 4, pp. 203–10.

Linaker, C. & Smedley, J. 2002, 'Respiratory illness in agricultural workers', *Occupational Medicine*, vol. 52, no. 8, pp. 451–9.

Lindsley, W.G., Green, B.J., Blachere, F.M., Law, B.F., Jensen, P.A. & Schafer, M.P. 2017, 'Sampling and characterisation of bioaerosols', in *NIOSH Manual of Analytical Methods (NMAM)* (5th edn), NIOSH, Cincinnati, OH, https://www.cdc.gov/niosh/nmam/pdf/Chapter-BA.pdf [14 October 2023].

Macher, J. 1998, *Bioaerosols: Assessment and Control, ACGIH®*, Cincinnati, OH.

Madelin, T.M. & Madelin, M.F. 1995, 'Biological analysis of fungi and associated molds', in *Bioaerosols Handbook*, eds C.S. Cox and C.M. Wathes, Lewis Publishers, Boca Raton, FL.

Mbareche, H., Brisebois, E., Veillette, M. & Duchaine, C. 2017, 'Bioaerosol sampling and detection methods based on molecular approaches: No pain no gain', *Science of The Total Environment*, vol. 599–600, pp. 2095–104.

Miller, J. 2023, 'Fungal bioaerosols as an occupational hazard', *Current Opinion in Allergy and Clinical Immunology*, vol. 23, no. 2, pp. 92–97. https://doi.org/10.1097/ACI.0000000000000886.

Morawska, L., Bahnfleth, W., Bluyssen, P.M., Boerstra, A., Buonanno, G., Dancer, S.J. & Wierzbicka, A. 2023, 'Coronavirus disease 2019 and airborne transmission: Science rejected, lives lost. Can society do better?', *Clin Infectious Diseases*, vol. 76, no. 10, pp.1854–1859.

National Academies Sciences 2017, *Forum on Microbial Threats. Combating Antimicrobial Resistance: A One Health Approach to a Global Threat: Proceedings of a Workshop*. Washington (DC), https://www.ncbi.nlm.nih.gov/books/NBK469957/ [21 September 2023].

NHMRC 2022, *Australian Guidelines for the Prevention and Control of Infection in Healthcare*, Department of Health, Canberra, https://www.nhmrc.gov.au/about-us/publications/australian-guidelines-prevention-and-control-infection-healthcare-2019 [21 September 2023].

NSW Health 2019, *Creutzfeldt-Jakob Disease (CJD)*, NSW Government, Sydney, https://www.health.nsw.gov.au/Infectious/factsheets/Pages/creutzfeldt-jakob-disease.aspx [7 August 2023].

Ong, O.T.W., Skinner, E.B., Johnson, B.J. & Old, J.M. 2021, 'Mosquito-borne viruses and non-human vertebrates in Australia: A review', *Viruses*, vol. 13, no. 2, pp. 265. https://doi.org/10.3390/v13020265

OSHA 2022, *Inspection: 1572011.015 - Trulieve Holyoke Holdings Llc.*, https://www.osha.gov/ords/imis/establishment.inspection_detail?id=1572011.015 [7 August 2023].

OTGR 2023, *'Office of the Gene Technology Regulator'*, Australian Government, Canberra, http://www.ogtr.gov.au/ [9 August 2023].

Ozanich, R. 2018, 'Chem/bio wearable sensors: current and future direction', *Pure and Applied Chemistry*, vol. 90, no. 10, pp. 1605–1613. https://doi.org/10.1515/pac-2018-0105

Pek, C.H., Cheong, C.S.J., Yap, Y.L., Doggett, S., Lim, T.C., Ong, W.C. & Lim, J. 2016, 'Rare cause of facial palsy: Case report of tick paralysis by ixodes holocyclus imported by a patient travelling into Singapore from Australia', *The Journal of Emergency Medicine*, vol. 51, no. 5, pp. e109–e114. https://doi.org/10.1016/j.jemermed.2016.02.031

Pickering, A. 2020, 'Subacute inhalation injssuries (inhalation fevers)', In A. N. Taylor, P. Cullinan, P. Blanc & A. Pickering (Eds.), *Parkes' Occupational Lung Disorders* (4th edn, pp. 381–392). CRC Press. https://doi.org/10.4324/9781315381848

Quince, C., Walker, A.W., Simpson, J.T., Loman, N.J. & Segata, N. 2017, 'Shotgun metagenomics, from sampling to analysis', *Nature Biotechnology*, vol. 35, no. 9, pp. 833–844. https://doi.org/10.1038/nbt.3935.

Rapp, C., Aoun, O., Ficko, C., Andriamanantena, D. & Flateau, C. 2014, 'Infectious diseases related aeromedical evacuation of French soldiers in a level 4 military treatment facility: A ten year retrospective analysis', *Travel Medicine and Infectious Disease*, vol. 12, no. 4, pp. 355–9. https://doi.org/10.1016/j.tmaid.2014.03.005

Reponen, T., Willeke, K., Grinshpun, S. & Nevalainen, A, 2011, 'Biological particle sampling', in P. Kulkarni, P. A. Baron and K. Willeke (Eds*), Aerosol Measurement: Principles, Techniques, and Applications* (3rd edn, pp. 549–570. Wiley.

Reynolds, S.J., Nonnenmann, M.W., Basinas, I., Davidson, M., Elfman, L., Gordon, J., Kirychuck, S., Reed, S., Schaeffer, J.W., Schenker, M.B., Schlünssen, V. & Sigsgaard, T. 2013, 'Systematic review of respiratory health among dairy workers', *Journal of Agromedicine*, vol. 18, no. 3, pp. 219–243. https://doi.org/10.1080/1059924X.2013.797374

Rui, F., Otelea, M.R., Fell, A.K.M., Stoleski, S., Mijakoski, D., Holm, M., Schlünssen, V. & Larese Filon, F. 2022, 'Occupational asthma: The knowledge needs for a better management', *Annals of Work Exposures and Health*, vol. 66, no. 3, pp. 287–290. https://doi.org/10.1093/annweh/wxab113

Russell, R.C. & Doggett, S.L. 2012, *Vector Borne Diseases*, Department of Medical Entomology, University of Sydney, Sydney, http://medent.usyd.edu.au/fact/fact.htm

Ruzer, L.S. & Harley, N.H., 2013, *Aerosols Handbook: Measurement, Dosimetry, and Health Effects* (2nd edn). Taylor & Francis.

Safe Work Australia (SWA) 2006, *Work Related Infectious and Parasitic Diseases*, SWA, Canberra, https://www.safeworkaustralia.gov.au/system/files/documents/1702/workrelated_infectious_parastitic_disease_australia.pdf [14 October 2023].

Safe Work Australia (SWA) 2011, *National Hazard Exposure Worker Surveillance: Exposure to Biological Hazards and the Provision of Controls Against Biological Hazards in Australian Workplaces*, SWA, Canberra, https://www.safeworkaustralia.gov.au/system/files/documents/1702/nhews_biologicalmaterials.pdf [07 August 2023].

Safe Work Australia (SWA) 2020, *COVID-19 Related Workers' Compensation Claims*, SWA, Canberra, https://www.safeworkaustralia.gov.au/resources-and-publications/c19-workers-compensation-claims [7 August 2023].

Safe Work Australia (SWA) 2023, *Workplace Exposure Standards*, SWA, Canberra, https://hcis.safeworkaustralia.gov.au/ [7 August 2023].

Samson, R.A. 2015, 'Cellular constitution, water and nutritional needs, and secondary metabolites', in Carla Viegas s (ed.), *Environmental Mycology in Public Health: An Overview Fungi and Mycotoxins Risk Assessment and Management*, Elsevier, pp. 5–13.

Schloss, P.D. & Handelsman, J. 2005, 'Metagenomics for studying unculturable microorganisms: Cutting the gordian knot', *Genome Biology*, vol. 6, no. 8, pp. 229–229.

Sheehan, J.R., Sadlier, C. & O'Brien, B. 2022, 'Bacterial endotoxins and exotoxins in intensive care medicine'. *BJA education*, vol. 22, no. 6, pp. 224–230. https://doi.org/10.1016/j.bjae.2022.01.003

Sietsema, M., Radonovich. L., Hearl, F., Fisher, E., Brosseau, L., Schaffer, R. & Koonin, L. 2019, 'A control banding framework for protecting the US workforce from aerosol transmissible infectious disease outbreaks with high public health consequences', *Health Security*, vol. 17, no. 2, pp. 124–132. https://doi.org/10.1089/hs.2018.0103

Sköld, P. 2017, 'The Health Transition: A Challenge to Indigenous Peoples in the Arctic', in K Latola & H Savela (eds), *The Interconnected Arctic – UArctic Congress 2016*, Springer International Publishing, Cham, pp. 107–13.

Spurgeon, J. 2007, 'A comparison of replicate field samples collected with the Bi-Air, Air-O-Cell, and Graesby-Andersen N6 bioaerosol samplers', *Aerosol Science and Technology*, vol. 41, no. 8, pp. 761–9.

Standards Australia 2022, *Safety in laboratories, Part 3: Microbiological Safety and Containment, AS/NZS 2243.3:2022*, SAI Global, Sydney, https://www.standards.org.au/access-standards

Stevens, J., Corper, A.L., Basler, C.F., Taubenberger, J.K., Palese, P. & Wilson, I.A. 2004, 'Structure of the uncleaved human H1 hemagglutinin from the extinct 1918 influenza virus', *Science*, vol. 303, no. 5665, pp. 1866–70.

Stewart, G. A., Richardson, J. P., Zhang, J. & Robinson, C. 2014, 'The structure and function of allergens', in N. F. Adkinson Jr, W. W. Busse, B. S. Bochner, A. W. Burks, & S. T. Holgate (Eds.), *Middleton's Allergy: Principles and Practice* (Ch 26, pp. 387–427.e1).

Tortora, G.J., Funke, B.R. & Case, L. 2020, *Microbiology: An Introduction*, 13th edn, Pearson, London

Tuano, K.S. & Chinen, J. 2020, 'Adaptive immunity', in N.F. Adkinson Jr., W.W. Busse, B.S. Bochner, S.T. Holgate, R.F. Lemanske Jr., & F.E.R. Simons (Eds.), *Middleton's Allergy: Principles and Practice* (pp. 20–29), Elsevier.

van Nunen, S.A. 2018, 'Tick-induced allergies: Mammalian meat allergy and tick anaphylaxis', *Medical Journal of Australia*, vol. 208, no. 7, pp. 316–321.

Viegas, C., Viegas, S., Gomes, A., Täubel, M. & Sabino, R. 2017, *Exposure to Microbiological Agents in Indoor and Occupational Environments*, Springer International Publishing.

Vincent, J.H. 2007, *Aerosol Sampling: Science, Standards, Instruments and Applications*, John Wiley and Sons, West Sussex, UK.

Walser, S.M., Gerstner, D.G., Brenner, B., Bünger, J., Eikmann, T., Janssen, B., Herr, C. E. W., 2015, 'Evaluation of exposure–response relationships for health effects of microbial bioaerosols – A systematic review', *International Journal of Hygiene and Environmental Health*, vol. 218, no. 7, pp. 577–589. https://doi.org/10.1016/j.ijheh.2015.07.004

Wang, D. 2020, '5 challenges in understanding the role of the virome in health and disease', *PLoS Pathogens*, vol. 16, no. 3, pp. e1008318. https://doi.org/10.1371/journal.ppat.1008318

Watts, N., Adger, W. N., Agnolucci, P., Blackstock, J., Byass, P., Cai, W., Costello, A. 2015, 'Health and climate change: Policy responses to protect public health', *The Lancet*, vol. 386, pp. 1861–1914. http://dx.doi.org/10.1016/S0140-6736(15)60854-6

WHO 2009, *WHO Guidelines for Indoor Air Quality: Dampness and Mould*. World Health Organisation, Geneva., https://www.who.int/publications/i/item/9789289041683 [21 September 2023].

WHO 2021, *Antimicrobial Resistance*, World Health Organisation, Geneva, https://www.who.int/newsroom/fact-sheets/detail/antimicrobial-resistance [21 September 2023].

Yamamoto, N., Schmechel, D., Chen, B.T., Lindsley, W.G. & Peccia, J. 2011, 'Comparison of quantitative airborne fungi measurements by active and passive sampling methods', *Journal of Aerosol Science*, vol. 42, no. 8, pp. 499–507.

Yao, M. & Mainelis, G. 2007, 'Analysis of portable impactor performance for enumeration of viable bioaerosols', *Journal of Occupational and Environmental Hygiene*, vol. 4, no. 7, pp. 514–24.

Yates, M.V. Nakatsu, C.H., Miller, R. & Pillai, S.D. 2016, '3.1.2.1 Approaches to Cyanohab Monitoring', In M. V. Yates, C. H. Nakatsu, R. V. Miller, & S. D. Pillai, (Eds.), *Manual of Environmental Microbiology* (4th edn). ASM Press.

Zou, S., Colombini-Hatch, S., Srinivas, P., Glynn, S. & Caler, L. 2016, 'Research on the human virome: where are we and what is next', *Microbiome*, vol. 4, no. 1, pp. 1–4. https://doi.org/10.1186/s40168-016-0177-y

14 Noise and Vibration

Beno Groothoff and Dr Jane Whitelaw

14.1 INTRODUCTION

Noise is one of the most pervasive health hazards that the Health and Safety (H&S) practitioner has to deal with. Noise can be found almost everywhere, industrial settings, offices and places of entertainment, traffic and transport vehicles of all kinds. Apart from libraries, few workplaces escape without the intrusion of noise, which is distracting, annoying, or hazardous to hearing and health. Mechanisation and modern lifestyles have not alleviated the problem; in fact, the opposite has happened, with widespread and sustained noise exposures occurring in an increasingly noisy world. Complaints from the public about increasingly noisy urban environments and its associated health problems are on the rise, in particular relating to not only road traffic but also air traffic, trains, and construction work, which are often exacerbated by poor sound insulated buildings.

Occupational exposure to noise, its physiological and psychological consequences, and methods of control are all very large areas of study. Noise-induced hearing loss (NIHL) is a well-documented, major occupational disease and accounts for a large percentage of workers' compensation claims each year.

Safe Work Australia's Report 'Occupational Noise Induced Hearing Loss' of August 2010 states in its Summary on page XI:

> 'Occupational noise-induced hearing loss (ONIHL) is a significant health and economic problem in Australia. Between July 2002 and June 2007 there were about 16 500 successful workers' compensation claims for industrial deafness, or ONIHL, involving permanent impairment due to noise. Access Economics in its 2006 report'.

Listen, Hear, the economic impact and cost of hearing loss in Australia, 'estimates that 37 % of hearing loss is caused by noise, which amounts to about $4.3 billion annually'. The economic burden of ONIHL is borne by workers and their families, business owners and managers, and the wider society (Safe Work Australia 2010, p. XI). The report further estimates that about 20% of adult onset of hearing loss is caused by exposure to loud noises from all sources. The Deloitte Access Economics report 'Social and Economic Costs of Hearing Loss in Australia' commissioned by the Hearing Care Industry Association estimates that 37% of adult hearing loss is noise-induced and demonstrates a deterioration since its 2006 'Listen, Hear' report with a new annual estimated cost to the nation of about $5.9 billion, up from $4.3 billion (Deloitte Access Economics 2017).

In its July 2014 'Occupational Disease Indicators' report, Safe Work Australia states that: 'Among those who experience noise-induced hearing loss, 20% or more also suffer from tinnitus' (Safe Work Australia 2014, p. 7). The Australian Government Department of Health and Aged Care (2022) also estimated that approximately 3.6 million people have some level of hearing loss and one in three Australians experience hearing impairment from noise-related exposure.

The effects certain industrial chemicals have on hearing, either alone or in combination with noise, have become better understood. These chemicals are known as ototoxins, and a list of recognised ototoxic chemicals/substances is detailed in Table 14.6. Many workers are frequently

 DOI: 10.1201/9781032645841-14

exposed to both noise and chemicals at work; hence, the National Code of Practice for Managing Noise and Prevention of Hearing Loss at Work (Safe Work Australia 2020) recommends that hearing conservation programs account for the effects of ototoxic chemicals where:

- workers are potentially exposed to airborne concentrations (without regard to respiratory protection worn) greater than 50 per cent of the national exposure standard for the substance, regardless of the noise level, and/or
- ototoxic substances at any level and noise with LAeq,8 h greater than 80 dB(A) or LC, peak greater than 135 dB(C), and/or
- hand-arm vibration at any level and noise with LAeq,8 h greater than 80 dB(A) or LC, peak greater than 135 dB(C).

The effects of hand-arm and whole-body vibration on workers have also gained more attention as the effects of vibrating plants and equipment are better understood. In Australia, legal precedents have been set with rather large compensation payouts, for example, in 2021 a central Queensland mine worker was awarded almost $1.5 million in damages by Supreme Court Queensland judge, Justice Graeme Crow, after suffering white finger syndrome due to his employer's negligence (Tyndall v Kestrel Coal Pty Ltd (No 3) [2021] QSC 119). The effects of whole-body vibration on drivers of vehicles such as trucks and earth movers as well as rigid hull inflatable boats (RHIBs) are also being studied more extensively to avoid or minimise adverse health effects.

In this chapter, the recognition, evaluation, and control of noise and vibration are examined in general terms. The H&S practitioner is referred to the reference list for more comprehensive information. The chapter also introduces phenomena of acoustic shock and ototoxicity. Acoustic shock is a particular concern for workers in the call-centre industry and emergency services control rooms and can have severe health consequences if no adequate protective measures are implemented.

14.2 REGULATORY NOISE LIMITS

To prevent the onset and development of ONIHL, Chapter 4, 'Hazardous Work', of the model harmonised Work Health and Safety Regulation 2023 (Safe Work Australia 2023), sets as the exposure standard for occupational noise at a limit of $L_{Aeq,8h}$ 85 dB(A) (i.e. an average of 85 dB(A) over an eight-hour exposure), or a C-weighted peak sound pressure level of $L_{C,peak}$ 140 dB(C), with both limits referenced to 20 μPa.

This does not mean that lower noise levels should be tolerated in all occupations nor guarantees that adverse health effects won't develop. For example, workers in an open-plan office or a call centre would not be able to work effectively with the distraction presented by noise levels of $L_{Aeq,8h}$ 85 dB(A) or even $L_{Aeq,8h}$ 75 dB(A). Effective communication and mental concentration are impaired by high levels of noise and may in turn affect health and other aspects of safe work. The Work Health and Safety legislation requires obligation holders to conduct risk assessments to prevent or minimise risk. This applies equally to low-level and high-level noises, both of which cause communication and other problems that can affect health. To minimise health effects occurring in these low-level noise work environments, the Code of Practice for Managing Noise and Preventing Hearing Loss at Work recommends that total noise levels, including noise from other work being carried out within these workplaces, be kept below:

- 50 dB(A) where work is being carried out that requires high concentration or effortless conversation.
- 70 dB(A) where more routine work is being carried out that requires speed or attentiveness or where it is important to carry on conversations.

14.3 PHYSICAL CHARACTERISTICS OF SOUND AND NOISE

Sound originates when a vibrating source causes variations in atmospheric pressure that are detected by the ear and interpreted by the brain. Sound may convey useful information, but when sound is unintelligible, unwanted, or may cause damage to hearing, it is referred to as **noise**. A person perceives, evaluates, and identifies the uniqueness of a sound by virtue of only three features: **combinations of frequencies**, their **relative intensities**, and their rate of **onset and decay**. Distinguishing between sounds (e.g. a bird call, a violin, or a chainsaw) can all be ascribed to differences in these three features.

As sound is transmitted through an elastic medium, in this case air, the air is compressed and rarefied to form a pressure wave (called a longitudinal wave), similar to the ripples that appear on a pond when a pebble is thrown in the water. The number of pressure variations per second is called the frequency of sound and is measured in Hertz (Hz) (Figure 14.1). The wavelength is the distance between two similar points on the sine (curved, peak-and-trough) wave. The velocity of sound (wavelength × frequency) depends on the mass and elasticity of the conducting medium. In air, sound propagates at about 344 m/s at 20°C. In water, sound propagates at about 1500 m/s, and through steel it propagates at about 6000 m/s. The spectrum of good human hearing ranges between about 20 and 20000 Hz, and everyday sounds contain a wide mixture of frequencies. Speech communications rely on sound in the frequencies range between 100 and 5000 Hz. Audible sound pressure variations are superimposed on the atmospheric air pressure (about 100000 Pa) and normally range between 20 μPa and 100 Pa.

14.4 SOUND PRESSURE, DECIBELS, LOUDNESS, AND THE L_{EQ}

The quietest sound that can be detected by a young person with healthy ears is 20 millionths of a Pascal (20 μPa) at 1000 Hz. This has been standardised as the threshold of hearing (0 dB) for the purpose of sound-level measurements and is used as a reference level (P_0). A sound pressure of 100 Pa (130 dB) is so loud it causes physical pain to the average person and is therefore called the threshold of pain.

Sound pressure variations of a pure tone fluctuate so that for half the duration of a complete cycle, they are above (+) and the other half they are below (–) atmospheric pressure. The average pressure fluctuation is zero since there are as many positive changes as negative ones. To overcome this, the average pressure is squared over many cycles, and the square root taken. The resulting value, the root mean square (rms), is used in sound level measurements, as it is proportional to the energy content of the sound wave. For pure tones, the rms is equal to 0.707 times the amplitude (volume) of the sine wave.

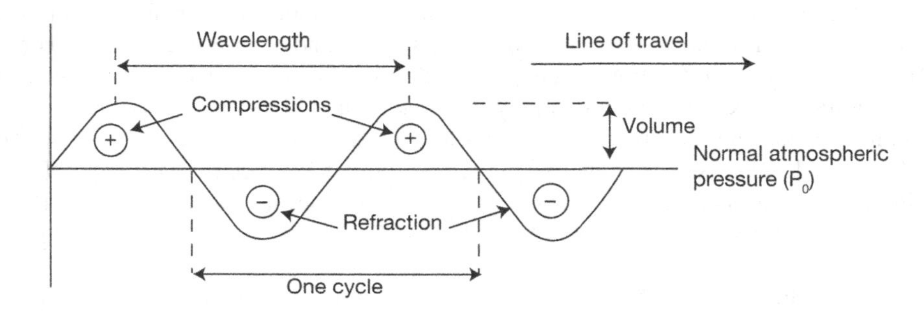

FIGURE 14.1 Sound as represented by an advancing pressure wave in air.

Source: B. Groothoff.

In terms of sound pressure detected by the ear, the loudest sounds heard at the threshold of pain are up to 10 million times greater than the softest ones. If calculations were made using a linear scale, the resulting numbers would be unwieldy. Additionally, the human ear does not respond linearly but rather logarithmically to sound stimuli. It is therefore more practicable to express acoustic parameters as a logarithmic ratio of the measured value to the reference value (20 μPa). This produces a more manageable basis for comparisons of sound pressures. The resulting unit, the Bel (after Alexander Graham Bell 1874–1922), is defined as the logarithm to the base 10 of the ratio between two acoustical powers or intensities. This unit is still large, so the **decibel** (one-tenth of a Bel) is generally used. The decibel is defined as:

$$L = 20 \log \frac{P}{P_0} [dB] \tag{14.1}$$

where

L = root mean square (rms) sound pressure level, re 20 μPa
P = the rms sound pressure in Pascal, and
P_0 = reference sound pressure (20 μPa)

For a doubling of sound energy, the logarithmic scale increases by 3 dB, and a 100-fold increase in noise results in a 20 dB increase in scale. Some examples of indicative noise levels in decibels are shown in Table 14.1.

Sound pressure levels represent only part of the picture. The apparent loudness of a sound depends very much on frequency, especially since the ear detects sound intensity with different sensitivities, depending on the frequency composition of the sound. The ear is most sensitive to frequencies between 1000 and 4000 Hz, with sensitivity falling off at low and very high frequencies. High-frequency sounds are heard much better than low-frequency ones of equal intensity. Further, an individual's interpretation and experience of loudness are subjective, depending on the harshness or intrusiveness of the noise (e.g. bagpipes compared with a flute), These factors are not reflected by the values (in dB) measured with a sound-level meter.

To relate measured sound levels to the response of the human ear, measuring instruments are fitted with sound-level weighting filters. The internationally adopted A-weighting is most commonly

TABLE 14.1
Indicative Noise Levels of Various Sources

Sound	Sound Pressure Level dB (re: 20 μPa)
Firearm	155–165
Jet engines	140–150
Threshold of pain	130
Jackhammer	110
Nightclub	100
Noisy factory	90–95
Passing heavy traffic	80–90
Office environment	45–70
Speech	60–65
Inside a home	40
Whisper	20
Threshold of hearing	0

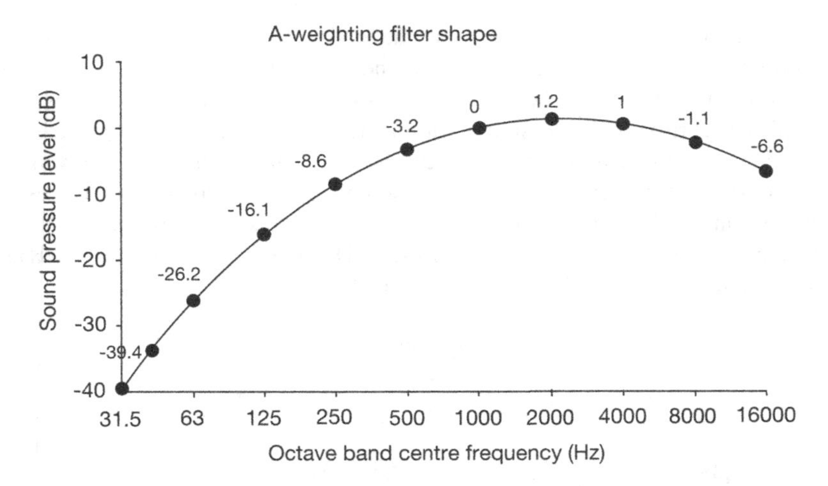

FIGURE 14.2 Frequency–response attenuation characteristics for the A-weighted network.

Source: B. Groothoff.

used, as it best corresponds to the way in which the human ear responds to sound. Sound pressure levels measured with the A-weighting filter are denoted as dB(A). This A-weighting response is illustrated in Figure 14.2, which shows clearly how the low-frequency response is weighted downwards—that is, a sound level meter using the A-weighting reads lower than the actual (linear) response at frequencies below 1000 Hz.

The A-weighted scale has been generally adopted because it attenuates broadband frequencies in a way that reflects their association with NIHL. It is thought that the higher frequencies, considered to be above 1000 Hz, give rise to most hearing damage. The eight-hour noise-exposure limits of the exposure standard are based on A-weighted measurements (Safe Work Australia 2010).

Few workplaces have constant noise levels such as that produced by dust extraction cyclones or electrical generators. Generally, noise tends to be variable and intermittent, like the clanking of machinery, the roar of passing traffic, or the mechanical percussion of hammers, chippers, or grinders. WHS legislation accommodates for this by requiring the average A-weighted noise level over an eight-hour work period ($L_{Aeq,8h}$) to be determined and compared with the legislated exposure limit. To obtain an 'average' value of the noise, an integrated value called the equivalent continuous sound (energy) level (L_{Aeq}) over a measurement period of time (T), reported as $L_{Aeq,T}$. The equivalent continuous sound (energy) level of say, 60 seconds ($L_{Aeq,60}$), provides a measure of the acoustic energy of the fluctuating noise during the measurement period. The $L_{Aeq,60}$ is one parameter that finds convenient use in surveys of occupational noise exposure levels.

14.5 EFFECTS ON HEARING

14.5.1 Hearing Loss

The principal health-related effect of noise exposure is **hearing loss**. Excessive noise can destroy the ear's ability to hear properly, and the damage is irreversible; hence, the H&S practitioner should put great emphasis on prevention. Damage to hearing depends on the loudness of the noise as well as the length of the exposure. Other health effects of noise exposure include cardiovascular effects such as increased blood pressure and heart rate, psychosocial stress, excretion of stress hormones, and decrease of peripheral blood flow due to an increase in vasoconstriction.

Noise can also impair communication, making it harder to do a job or hear emergency signals. In the social sphere, it spoils much of our enjoyment of life, leading to social isolation, loneliness, and potentially depression. Hearing normally deteriorates with age, known as age-related hearing loss or presbycusis. ONIHL can accelerate age-related hearing loss and drastically affect the quality of life.

Hearing-related effects from noise exposure include the following.

14.5.1.1 Temporary Threshold Shift

Temporary threshold shift (TTS) in hearing occurs during or immediately after exposure to significant loud sounds. Quiet sounds can no longer be heard, and the condition may last for periods ranging from minutes to hours and days. This occurs when the hair cells in the hearing organ (cochlea) become fatigued and reversibly desensitised. Repeated TTS may lead to a permanent loss of hearing.

14.5.1.2 Permanent Threshold Shift

Permanent threshold shift (PTS) usually results from long-term regular exposure to loud noise but can sometimes occur from a single or a few extremely loud impact or impulsive noise. The ear canal acts as a resonating tube and amplifies sounds at between 3000 and 4000 Hz. This amplification increases the sensitivity of a persons' hearing at these frequencies and adds to the susceptibility of damage. Loud sounds, irrespective of their frequency composition, cause hearing damage, typically around the 4000–6000 Hz range. ONIHL is irreversible as it arises from the destruction of the hair cells in the cochlea, the human ear does not have the ability to repair or regrow hair cells, and no medical treatment can restore them. The left-hand photograph in Figure 14.3 shows healthy hair cells—a single row of inner hair cells and three rows of outer hair cells. The right-hand photograph shows a large number of destroyed hair cells.

14.5.1.3 Acoustic Trauma

Acoustic trauma normally results from high-intensity explosive- or loud impact-type impulsive noise, which can destroy the hair cells and other ear mechanisms after one or relatively few exposures.

14.5.1.4 Tinnitus

Tinnitus consists of the perception of sounds in the ears or head, which do not originate from any external source (Jastreboff & Hazell 2004). Tinnitus sounds may be experienced as a tone,

FIGURE 14.3 Destruction of hair cells resulting from noise-induced hearing loss.

Source: CDC 2020.

buzzing, hissing, whistling, ringing, cicadas, musical sounds, pulsating, or a 'flickering' sound. These sounds are variable and are often complex and can range from being annoying to causing severe distress. Tinnitus is permanent with currently no cure available, can have a devastating impact on sufferers, and may even lead to suicide. Tinnitus is most likely experienced after exposure to loud noise that causes a temporary threshold shift, for example, the sound of a jackhammer or a firearm used without hearing protectors, or after a rock concert or drag racing. Loud noise through headphones is another important cause. Other causes for tinnitus to occur include:

- conductive hearing loss such as from wax in the ear;
- Meniere's disease;
- Flu and head colds;
- excessive alcohol consumption;
- stress;
- ototoxic drugs such as used in chemotherapy;
- ototoxic chemical exposure in work situations;
- head injury, including from head banging during rock concerts;
- whiplash (from a car accident); and
- tumour at the vestibulocochlear nerves.

There are three types of tinnitus people experience. These are Objective tinnitus, Subjective tinnitus, and Phantom tinnitus.

Objective tinnitus is relatively uncommon and results from measurable bodily noise within the area of the ear such as from blood flow through either normal vessels but with increased flow such as caused by atherosclerosis, or abnormal vessels such as with tumours or malformed vessels, near the middle ear. There is no measurable defect in hearing ability.

Subjective tinnitus is the most common tinnitus experienced. The brain receives signals that are perceived as noise but are not caused by sound. It is considered to originate from abnormal neuronal activity in the auditory cortex due to damage in the auditory pathway chain, most commonly hair cell damage. It is thought that the damage may cause a loss of suppression of intrinsic cortical activity and possibly the creation of new neural connections.

Phantom tinnitus is when the vestibulocochlear nerves are damaged and hearing loss has occurred. The surviving nerve tissue compensates for the lack of information by spontaneously creating a phantom noise which the brain interprets as sound.

14.5.1.5 Treatment

There is currently no cure for tinnitus and treatment must consist therefore of managing the presence of tinnitus. The management of tinnitus consists mainly of relaxation therapy and sound therapy. Relaxation therapy is probably the most difficult one as it depends on the commitment of the person to want to relax. In a busy world, most people do not know how to relax and must learn and practise to achieve it. During relaxation, the heart rate and breathing slows down and brain activity decreases. This causes the body to feel mentally and physically refreshed afterwards and may help to reduce the intensity of the tinnitus to manageable levels. Sound therapy consists of using devices that introduce low-level sounds in various forms to mask the tinnitus. It aims to enable the brain to 'filter out' the tinnitus sounds. A second component of sound therapy consists of creating a positive understanding about tinnitus management and the emotional stress often associated with tinnitus. Sound therapy can be used as a self-help tool or as part of a management program delivered by qualified audiologists or specialty clinics. The sounds that are often used include environmental sounds, radio, and smartphone apps providing broadband noise or gentle relaxing music and wearable sound generators producing constant 'white noise'. The latter should be fitted by a tinnitus specialist as part of a tinnitus management program. It is important that the generated sound is pleasant to listen to but basically blends into the background meaning the sufferer is aware of the sound, but it does not demand too much attention. It is important to avoid complete silence with severe tinnitus.

Non-hearing-related effects from noise exposure include cardiovascular effects and cognitive decline.

14.5.2 Cardiovascular Effects

Several recent systematic reviews found evidence of a correlation between noise exposure and cardiac health effects. There is a dose–response relationship between occupational noise exposure >80 dB and hypertension (Pretzsch et al. 2021), and a study conducted by the WHO and ILO found an increased risk of ischemic heart disease at >85 dB (Teixeira et al. 2021).

14.5.3 Cognitive Decline

NIHL has also been suggested as a risk factor in cognitive decline and neurodegenerative diseases such as Alzheimer's and dementia (Bernabei, et al. 2014; Huang et al. 2021).

14.6 ASSESSMENT OF NOISE IN THE WORKPLACE

Two practical approaches can be employed by the H&S practitioner in assessing noise in the workplace. Both require sophisticated equipment and reasonable levels of competency for proper measurement and interpretation. Provided appropriate equipment is available and the H&S practitioner is adequately trained, the noise assessments could be carried out in-house. In complex situations or where either equipment or competency cannot be secured acoustic consultants should be sought to advise the H&S practitioner.

14.6.1 The Sound Level Meter

AS/NZS 1269.1 Occupational Noise Management—Part 1: Measurement and assessment of noise immission and exposure states that an integrating sound level meter of at least Class 2 should be used when undertaking occupational noise assessments (Standards Australia 2005a). A Class 1 integrating sound level meter is the preferred instrument for measuring sound pressure levels. Such instruments, similar to that shown in Figure 14.4, provide simultaneous noise measurements in a range of different parameters, such as rms sound level $L_{Aeq,T}$, peak sound levels ($L_{C,peak}$), A- and C-weighting, and linear (Z) dB measurements.

Figure 14.2, in Section 14.4, shows the A-weighting response curve (matching the human ear's sensitivity) of the sound level meter. This curve follows different values at different frequencies from the horizontal (0 dB) line, representing dBZ. The A-weighting response curve values are used in noise control where the results of octave band analysis, measured linearly, have to be expressed as an A-weighted value. Another weighting often used in noise control is the C-weighting. This weighting follows a different curve which is essentially horizontal (flat) between 80 and 3150 Hz and tapers off only outside this frequency range. The C-weighting is predominantly used when measuring peak sound pressure levels and determining the suitability of hearing protectors in a particular work environment.

Modern sound level meters can, depending on the software installed, provide 'octave', or 'one-third octave' (frequency) analysis at the same time as other parameters are measured. This enables

FIGURE 14.4 Sound level meter monitoring noise exposure of polishing wheel operator.

Source: M. Reed.

measurement of the sound spectrum in a range of frequencies determined by the meter's capabilities. Frequency analysis is necessary for the proper analysis of noise in the design of noise-control measures, or provision of appropriate hearing protectors in a particular work environment. The use of frequency analysis is examined further in Section 14.8.

The sound level meter is used in work locations principally for measuring noise emissions from machinery, equipment and processes, or the noise exposure of workers (noise immission).

There are many cheap and easy-to-operate sound level meters available that may be useful to the H&S practitioner who is inexperienced in conducting sound level measurements. There are also numerous sound level meter Apps available for smartphones. If such an App is intended to be used, the H&S practitioner should be careful to select one with a sound intensity exchange rate of 3 dB as many have a 5 dB exchange rate, which are based on American legislative requirements, and can therefore not be used. These cheap meters and phone Apps do *not* meet the requirements of AS/NZS 1269 and should *never* be used to demonstrate compliance with WHS legislation. They may, however, be useful for indicative purposes, for example, in 'preliminary surveys' or 'walk through surveys' to determine if the workplace or a work area is likely to contain excessive noise. Where this is found to be the case, more sophisticated instruments complying with the relevant Standards should then be employed to conduct more detailed noise surveys using Class 1 instruments, and to identify options for noise control.

14.6.2 Personal Noise Dose Meters

Many workers move in and out of noisy areas when doing their job, thus sustaining different noise exposures. Their total noise exposure is therefore the sum of many different partial exposures. The personal noise dose meter, also known as dosimeter (Figure 14.5), provides an integrated measure of noise exposure over a known period, usually the work shift.

Modern noise dose meters provide data logging with computer-readable output in noise dose, $L_{Aeq,8h}$ and A-weighted noise exposure, $E_{A,T}$, expressed in Pascal squared hours (Pa^2h), and C-weighted peak readings; they also warn when daily exposure limits have been exceeded. Noise dose meters may have the capability for results to be downloaded and its data stored on a computer using appropriate software (Figure 14.6a). Often workers' exposures are reported as a histogram of the worker's exposure over the duration of the shift (Figure 14.6b). In recent years, noise dose badges have appeared and are frequently used in favour of the larger traditional noise dose meters.

These noise dose badges (Figure 14.7) perform similar functions as the traditional meters but are much smaller with built-in microphone and thus do not have microphone cables attached to them which may get in the way of the worker performing tasks. Feedback from workers wearing noise dose badges indicates that they are perceived as less intrusive. This has had the positive

FIGURE 14.5 Personal noise dose meter: correct microphone fitting relative to the position of the ear.

Source: Photo courtesy of TSI Incorporated.

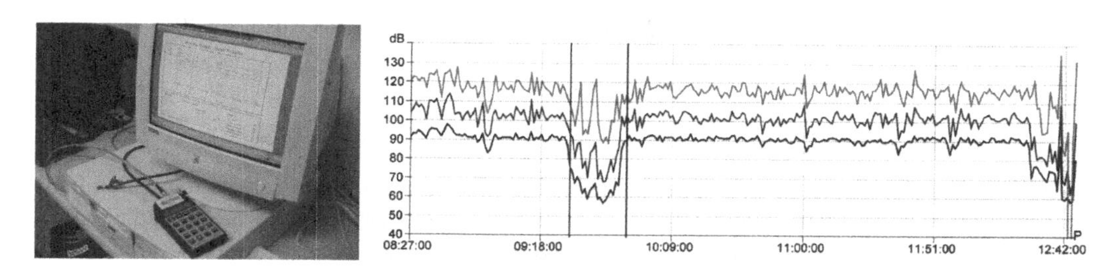

FIGURE 14.6 Downloading a histogram of a worker's exposure.

Source: B. Groothoff.

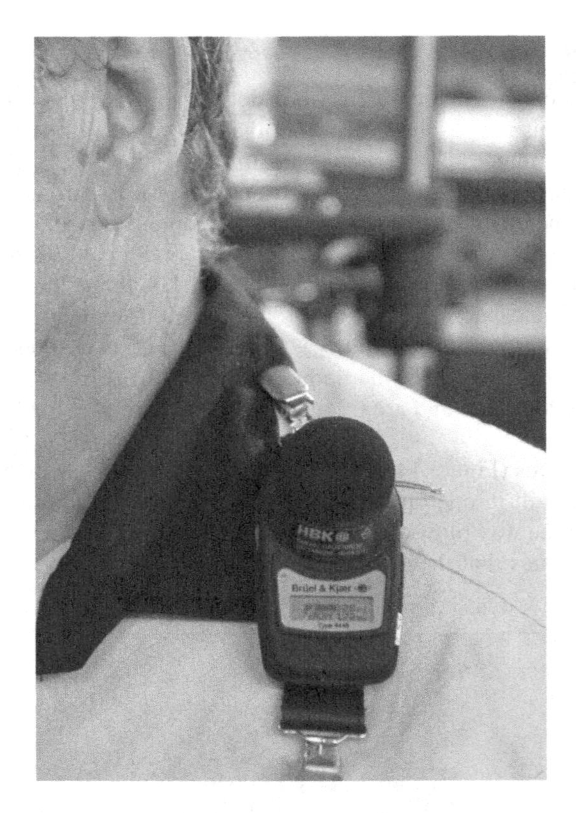

FIGURE 14.7 Noise dose badge.

Source: M. Reed.

effect that the worker, in the course of conducting work, basically forgets wearing it and is therefore less likely to interfere with the badge being worn, leading to more reliable measurement results. Technology is currently advancing at a rapid pace, and so the functionality and capabilities of the noise dose badges. Some include Bluetooth capability, remote control, and pre-set alarms to warn workers when they are exposed to high noise levels or approaching their daily dose limit.

All measurements must be made in accordance with measuring protocols recommended by the manufacturer and as required by the relevant Australian Standards, that is, AS/NZS 1269 for occupational noise (Standards Australia 2005a) and AS 1055 (Standards Australia 2018) for environmental noise.

14.6.3 Calibration

Both the sound level meter and the personal noise dose meter will require calibration with a field calibrator immediately before and after a noise survey in accordance with the manufacturer's instructions. Most calibrators generate a pure tone of 1000 Hz and typically produce a sound pressure level of 94 dB, which equates to a sound pressure of 1 Pascal. These parameters are chosen so that the calibrator can be used with the different weightings, such as the A-weighting, C-weighting, and linear, indicated as Z-weighting, which all intersect at 1000 Hz (refer Figure 14.2).

If the calibrator uses a different frequency, for example, such as 250 Hz, the A-weighted response of the sound level meter must be adjusted. For instance, if a sound level meter set to A-weighting is calibrated at 250 Hz and 114 dB, the sound level meter indicator must read 105.4 dB(A) as the A-weighted response at 250 Hz is −8.6 dB relative to the reference level at 1000 Hz.

In addition to the above-mentioned field calibration procedures, AS/NZS 1269.1 requires in clause 7.10 that complete measuring systems to be used for detailed or follow-up assessments be calibrated in accordance with the relevant standards at regular intervals not exceeding two years. Such calibrations shall be performed by a suitably equipped and independently audited laboratory with full traceability to national measurement standards (Standards Australia 2005a). The two-yearly certification of sound level measuring instruments would therefore normally be carried out by a NATA registered laboratory. The certification of field calibrators should be carried out annually by a NATA registered laboratory to ensure their accuracy (Groothoff 2015).

14.7 ADDING OF SOUND LEVELS

A sound level meter allows for the actual sound pressure level to be measured and confirmed when the sources are generating noise simultaneously.

For example, if two sources were measured at 90 dB and 93 dB, their combined sound pressure level can be found by taking the difference between 90 and 93, which is 3, looking up this value in the left-hand column of Table 14.2, and finding the corresponding correction value in the right-hand column, which in this case is 1.8. Because the two sources operate side by side, their combined value must be higher than the highest individual value (93), and the outcome is therefore 93 + 1.8 = 94.8 dB, which can be rounded to 95 dB. Where the difference between two sources is 10 or more, it is generally considered that the lower of the two does not contribute to the higher value. For instance, one source at 85 dB(A) and one at 95 dB(A) will have a combined sound level of about 95 dB(A). Where there are more than two sources operating with a difference of 10 dB or more, their combined level may be affected and needs to be determined by using Table 14.2.

TABLE 14.2
Logarithmic Addition of Noise Levels

Difference in dB Value	Add to the Higher dB Value
0	3.0
1	2.6
2	2.1
3	1.8
4	1.4
5	1.2
6	1.0
7	0.8
8	0.6
9	0.5
10.0	0.4
11.0	0.3
12.0	0.2
13.0 or more	0

This may be the case with items of plant that start and stop intermittently, but which on occasion may all run at the same time and may cause exposure limits to be exceeded.

Because workers are usually exposed to different noise levels from different tasks, each lasting for a different period, it is helpful to have a means of integrating all these exposures into a single, useful noise exposure value. AS/NZS 1269.1 uses the 'eight-hour equivalent continuous A-weighted sound pressure level' ($L_{Aeq,8h}$) and the 'A-weighted noise exposure' ($E_{A,T}$) in Pa^2h (Pascal squared hours) to express a worker's exposure to noise in a single numerical value. Modern WHS legislation limits a worker's eight-hour unprotected exposure to $L_{Aeq,8h}$ 85 dB(A). Traditionally the eight-hour allowable exposure has been expressed as a daily noise dose (DND) with a value of 1.0; in other words, a DND of 1.0 is equal to $L_{Aeq,8h}$ 85 dB(A).

As discussed earlier, when sound propagates from its source, it causes small disturbances in atmospheric pressure. The static atmospheric pressure is about 10^5 Pa, and audible pressure variations are in the order of 20 μPa (20×10^{-6} Pa) to 100 Pa. As we saw in Section 14.4, 20 μPa corresponds to the threshold of hearing and 100 Pa to the threshold of pain. The magnitude of the disturbance is measured with a sound level meter as sound pressure level. The sound level meter measures the square of the relative value of the sound pressure in Pascals over a time period, T. The result is expressed in Pa^2 and if present for eight hours is expressed as Pa^2h. A sound pressure level of 85 dB(A) corresponds to a sound pressure of 0.126 Pa^2. For an eight-hour work period, the Pa^2h value would be $8 \times 0.126 = 1.008$ Pa^2h. For practical purposes, the $E_{A,T}$ value of 1 Pa^2h corresponds to $L_{Aeq,8h}$ 85 dB(A). It follows that:

$$1\,Pa^2h = L_{Aeq,8h}\,85\,dB(A) = DND\,1 \quad (14.2)$$

As the noise exposure level increases, the permitted time of exposure for unprotected ears decreases. This is shown in Table 14.3 as eight hours of continuous exposure at 85 dB(A), or four hours at 88 dB(A), and so on. The 'doubling of sound intensity exchange rate' used by Australian authorities is 3 dB.

Both the daily noise exposure and Pascal squared hours methods provide for the integration of partial noise exposures over a given period and enable normalisation of partial noise exposures to

TABLE 14.3
Period of Maximum Exposure for Various Equivalent Continuous Noise Levels

Limiting dB(A)	Maximum Exposure Period Allowed to Stay within $L_{Aeq,8h}$ 85 dB(A)
82	12 h*
85	8 h
88	4 h
91	2 h
94	1 h
97	30 min
100	15 min

* Extended shifts pose a degree of risk that is greater than from a normal 8 hour shift due to additional damaging exposure once maximum Temporary Threshold Shift is reached after 10 hours and a shorter recuperation time between shifts. AS/NZS 1269.1 states that organisations should add an adjustment as listed in Table 14.2 of the Standard to the normalised noise exposure level before comparing that level to the noise exposure criterion. In the case of a 12 h shift, that adjustment would be 1 dB on top of the measurement result after having been normalised to an 8 hour exposure ($L_{Aeq,8h}$).

TABLE 14.4
Correction Table for Extended Shifts

Shift Length, h	Adjustment to $L_{Aeq,8h}$, dB
<10	+0
>10 to <14	+1
>14 to <20	+2
>20 to 24	+3

Source: Reproduced Table 2 from AS/NZS 2169.1:2005, p. 18.

a five-day working week where the worker may work more or less than five 8-hour days. Modern integrating noise dose meters will have built-in capabilities to calculate the daily noise dose (DND), $L_{Aeq,8h}$dB(A), $E_{A,T}$ and Pa^2h (Pascal squared hours) and methods of integrating the noise exposure. AS/NZS 1269.1 (Standards Australia 2005a) provides tables to convert noise level measurements to Pascal squared values, and relatively simple tabular methods of converting the summed Pascal squared values to an equivalent $L_{Aeq,8h}$.

Be mindful that exposures are determined for workers who are considered not to be wearing hearing protection devices. A worker wearing hearing protectors in a noisy workplace is in a situation of protected exposure provided the hearing protectors are worn and worn correctly and consistently. However, this cannot always be guaranteed, and that protected exposure does not mean no exposure, as the workplace noise is still present.

Shifts exceeding 10 hours present a greater risk of hearing damage than eight-hour ones. This is partly due to the additional damaging effect of continued exposure once the maximum temporary threshold shift is reached after about 10 hours of exposure, and partly because of the reduced recovery time between shifts. Corrections must therefore be made to the noise exposure levels to accommodate the greater risk. AS/NZS 1269.1 incorporates in Clause 9.4 a correction table, reproduced in Table 14.4, which indicates the adjustments to be made to the normalised exposure level with extended shifts.

Suppose a worker works a 14-hour shift and is exposed during this shift to $L_{Aeq,14h}$ 89 dB(A). The normalised total daily noise exposure is therefore:

$$L_{Aeq,8h} = 89 + 10_{\log 10}\left(14/8\right) = 91.43\,\text{dB}\left(\text{A}\right) \quad (14.3)$$

For a shift of 14 hours, 1 dB must be added. The adjusted $L_{Aeq,8h}$ therefore becomes 91.43 + 1 = 92.43 dB(A). It is this value that must be used for noise management and control.

14.8 NOISE CONTROL STRATEGIES

Where the results of a noise survey identify the likelihood of exposure to excessive noise (>85dB(A)), further investigation is necessary to identify noise sources and prioritise control strategies to reduce the hazard. Workplace noise control strategies fall into three categories, ranged here from most to least effective:

a) **Control at source**: elimination or modification of noise source or process.
b) **Control of transmission path**: enclosures, barriers, sound-proofed control rooms.

c) **Control of noise at the receiver**: enclosures, barriers, sound-proofed control room, training and the use of personal hearing protectors. Hearing protectors should not be seen as noise-control devices but rather as a temporary way of minimising noise exposure by reducing the noise that enters the ear canal (Figure 14.8).

Occasionally, the H&S practitioner may need to use two or even all three approaches to noise control. Attacking the source of the noise or preventing its transmission is always preferable to requiring workers to wear hearing protection. Hearing protection should be used only if alternatives are not feasible, or reasonably practicable or economical, or as an intermediate phase until more permanent (higher order) noise-reduction measures are in place. In practice, it is often difficult to enforce such requirements, so reliance on the use of hearing protectors by workers will in the long run cost more than implementing higher-order controls. Many (engineering) noise-reduction programs, although perhaps initially expensive, can be very cost-effective over the long term and will remain consistently effective. Some excellent practical examples can be found in the references in Section 14.14.

14.8.1 Control at Source

Many noise hazards arise because little thought, if any, has been given to the acoustic qualities of the building, for example, modern concrete tilt-up buildings are popular for industrial areas as they are quick to erect, but they consist of acoustically highly reflective concrete floors and walls and are often covered with a metal roofing. This combination of hard reflective surfaces actually increases the noise production once the workplace is in operation. There is often little or no thought given to the processes or the correct design or installation of equipment. High levels of continuous and percussive noise are often traded off for immediate ease of operation—for example, bending metal by hammering on it rather than using a machine or lever. Noisy motors, compressors, power saws and grinders can all be found in many workplaces, located near workstations with little regard to their impact on the worker.

The following are examples of ways to minimise noise or prevent its generation.

14.8.1.1 Substitution of Processes

- Use welding instead of riveting.
- Use hot working of metals instead of cold forging.
- Use impact-absorbing materials (plastic, rubber and nylon) rather than metal.
- Use procedures such as lowering rather than dropping.

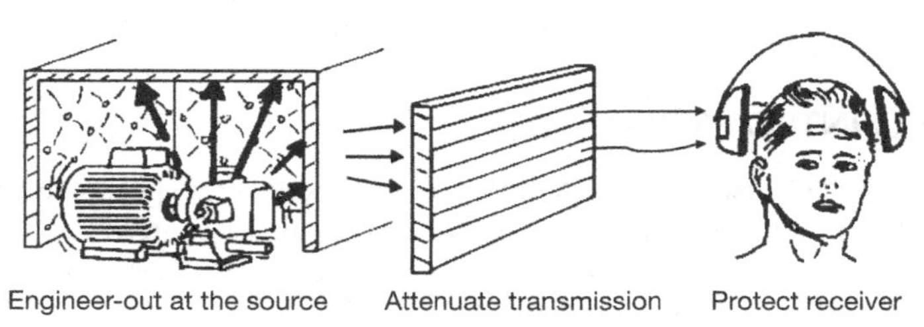

FIGURE 14.8 Essential elements to be addressed in noise control strategies.

Source: Grantham 1992, p. 193, Figure 8.7.

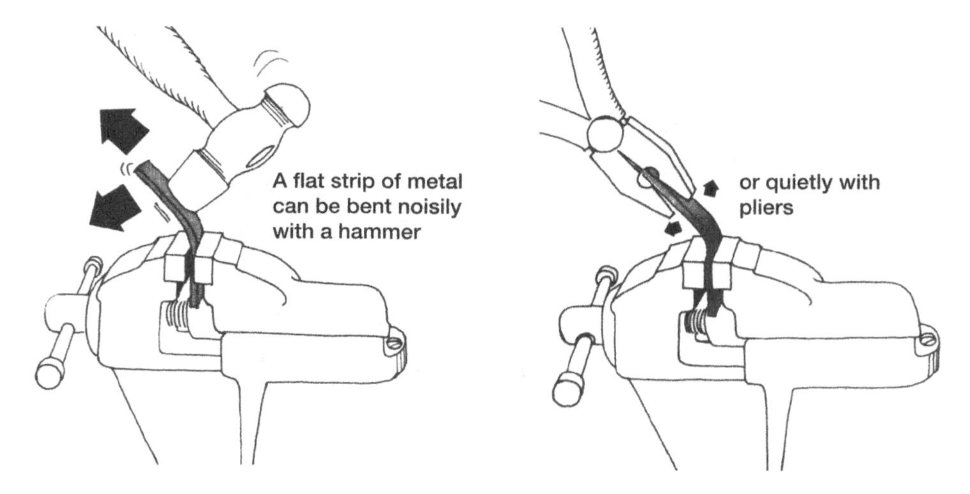

FIGURE 14.9 Changing the force, speed or pressure to reduce noise.

Source: OSHA, 1980, p. 12.

14.8.1.2 Minimising Changes in Force and Pressure That Produce Noise

- Eliminate impact noises, for example, use compression riveters rather than impact riveters.
- Replace hammering with slow application of force (Figure 14.9).
- Use hydraulic presses rather than mechanical-impact presses.

14.8.1.3 Reducing the Speed

The higher the speed, the higher the frequency, and so the louder (as perceived by the ear) and potentially more damaging the noise.

To alleviate this:

- Use larger, slower machines rather than small, fast ones.
- Run machines at lower speeds, but with higher torque.
- Air guns with air channels around the central channel reduce noise and pitch by not having all the air concentrated at high speed through the central air channel but spreading the flow of air through several channels around the central orifice at lower speed while not reducing effectiveness. The air gun on the left in Figure 14.10 has air channels around the central orifice. This lowers the speed of the air and changes the pitch of the sound. This type of air gun can be up to 7 dB quieter than the 'traditional' air gun shown on the right.

14.8.1.4 Prevent Mechanical Vibration from Being Converted into Sound-Generating Sources

- Isolate the vibrating unit to prevent it from transmitting noise.
- Optimise rotational speed (usually decreasing speed) to prevent oscillating vibrations.
- Alter size or mass to change resonant (i.e. natural) frequencies. Depending on their composition, size and shape, all objects tend to vibrate more freely at a particular frequency, called the resonant frequency. A machine transfers the maximum energy to an object when the machine vibrates at the object's resonant frequency.

FIGURE 14.10 The airgun on the left spreads the airflow around the central orifice and is quieter than the traditional airgun on the right.

Source: B. Groothoff.

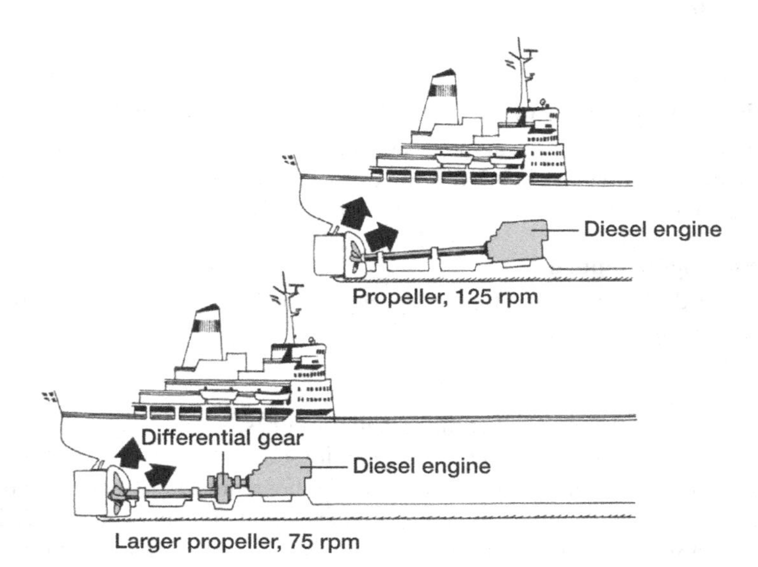

FIGURE 14.11 Dampening of resonant surfaces to reduce noise generation.

Source: OSHA 1980, p. 27.

- Dampen the resonance. A ship's engine running at 125 rpm and directly connected to the propeller shaft is shown in Figure 14.11. This speed coincides with the resonant frequency of the propeller shaft and causes disturbing noise and vibration. When a reduction gear is added (Figure 14.11), the propeller shaft now runs at 75 rpm, shifting the vibration to a lower frequency and thus reducing noise and vibration.
- Use extra support or stiffeners to withstand vibrations. The circular-saw sharpening machine in Figure 14.12 causes an intense resonant (ringing) noise. Clamping the rubber sheet against the saw blade greatly reduces the resonance and thus the ringing noise.

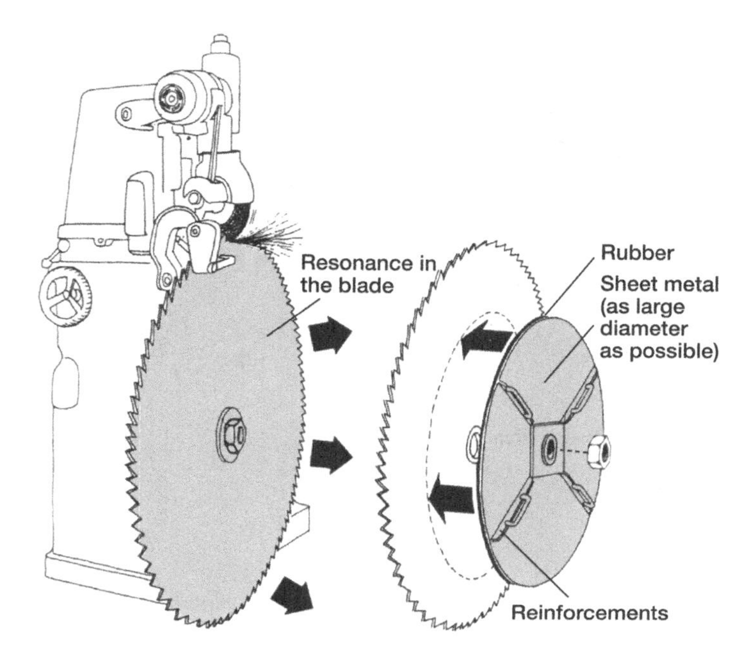

FIGURE 14.12 Stiffening of resonant surfaces to reduce noise generation.

Source: OSHA 1980, p. 41.

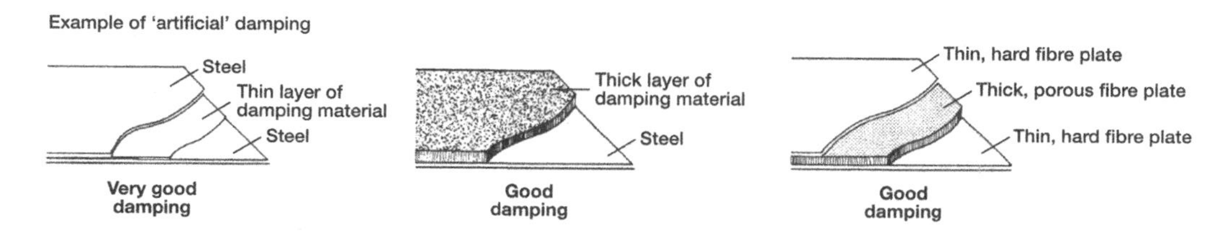

FIGURE 14.13 Various constructions for vibration dampening.

Source: OSHA, 1980, p. 38.

- Use vibration-damping surfaces (Figure 14.13). Steel plates have very poor internal vibration damping. Adding coatings of intermediate layers with better damping properties can reduce vibration and thus noise.
- Reduce radiating area (Figure 14.14). The control panel on the left is mounted on the pump system and vibrates with the system. Isolating the control panel (as shown on the right) reduces the vibrating surface of the system and therefore the noise level.
- Use perforated non-resonant surfaces (Figure 14.15). On the left, the guard over the flywheel and belt consists of a solid metal cover which causes the large surface area of the guard to vibrate and act as a sounding board for noise. Replacing the guard with a perforated one (as shown at right) reduces the vibration and in turn the noise.
- Use active cancellation (artificial noise created 180 degrees out of phase) to negate the effect of the original source.

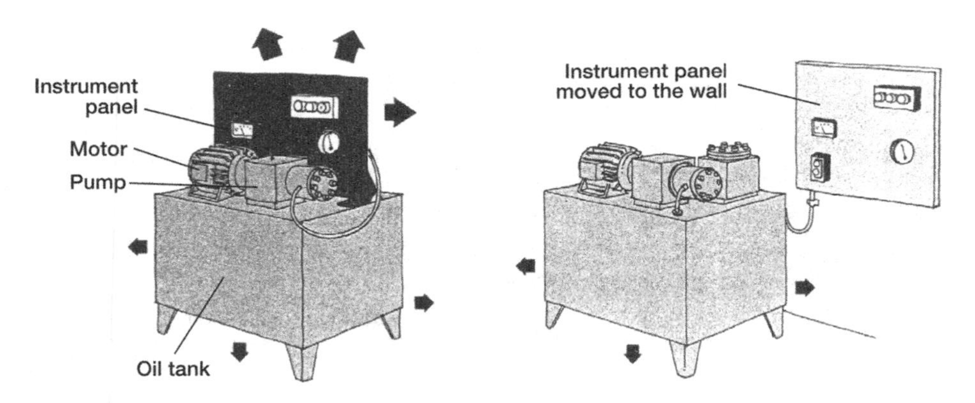

FIGURE 14.14 Reducing radiating area.

Source: OSHA, 1980, p. 29.

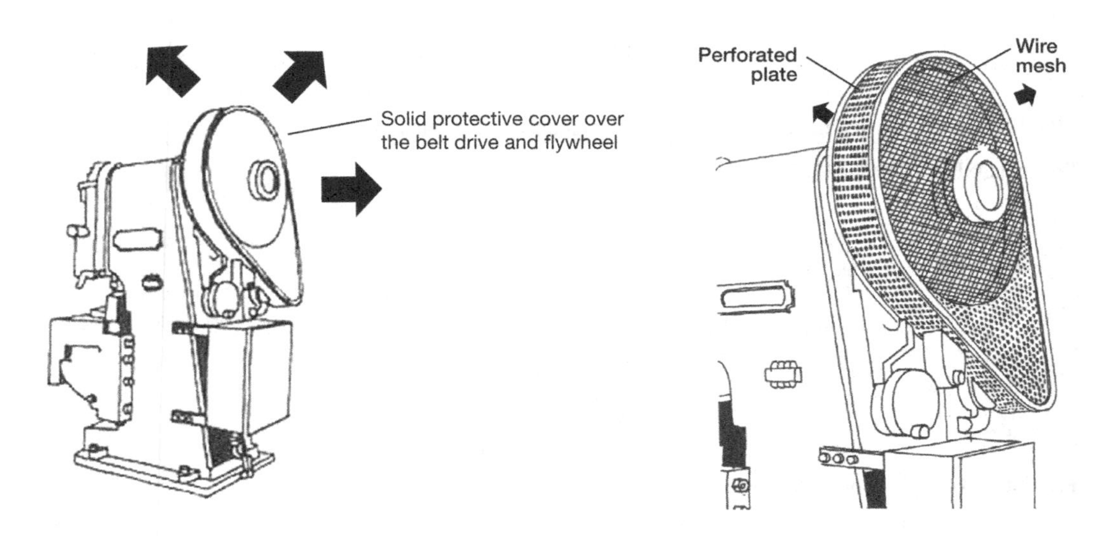

FIGURE 14.15 Use of non-resonant surfaces.

Source: OSHA, 1980, p. 3.

14.8.1.5 Reduce Transmission Possibilities

Fluid pumping systems can be very noisy due to the intense pressure shocks created in the liquid by the compressors driving the systems. If such systems are mounted rigidly to a building structure, noise can travel through the building and cause problems for the occupants. To avoid this issue compressors and pipe systems should be isolated from the building, as shown in Figure 14.16:

- use springs, dampers, flexible couplings and mountings;
- ensure that ducts cannot carry sound.

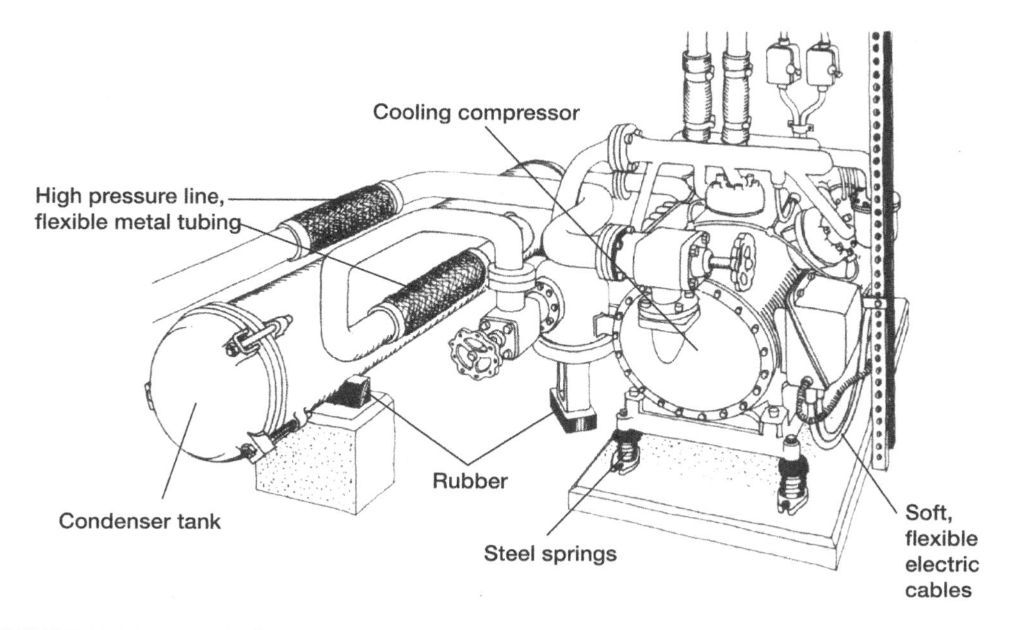

FIGURE 14.16 Use of a flexible connector to decouple noise source.

Source: OSHA, 1980, p. 93.

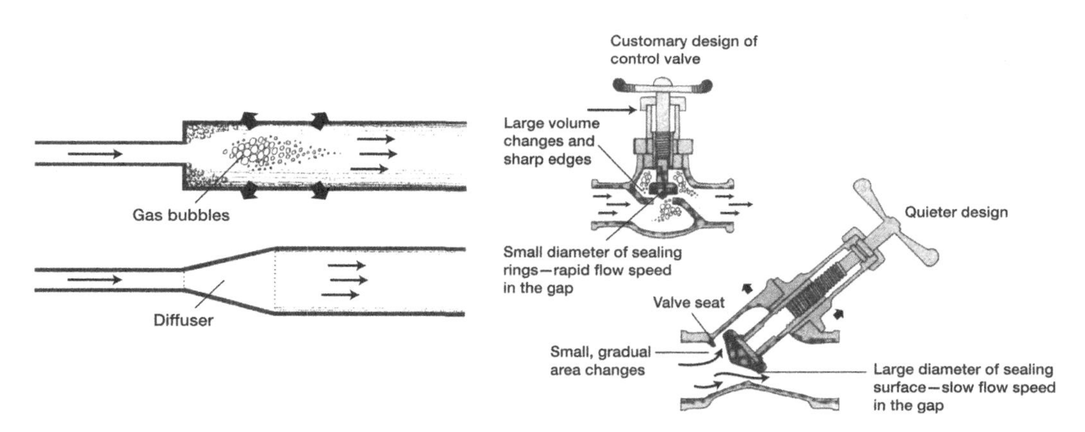

FIGURE 14.17 Slowly reducing pressure reduces noise transmission in liquids.

Source: OSHA, 1980, pp. 58 & 59.

14.8.1.6 Reduce the Likelihood of Noise Being Generated in Air or Fluid Flow

- Use fit for purpose, appropriately designed fans to reduce air turbulence. The air reaching a fan's rotor must be as unobstructed as possible to achieve quiet operation.
- Slowly reduce speed of air or fluid flow to avoid turbulence from sudden change in volume and pressure drop.
- Use reduced pressures and velocities to minimise noise generated by air or fluid flow (Figure 14.17).

- Avoid rapid pressure changes, which can produce 'cavitation' sound (left side of Figure 14.18) for example rapid pressure drops near control valves, propellers and pumps. To prevent cavitation noise, inserts can be strategically placed in the fluid line, to prevent greater pressure drops than required (see right side of Figure 14.18).
- Reduce turbulence on air exits (Figure 14.19). When high-speed exhaust air escaping from the grinder's handle mixes with the relatively still outside air, the resulting turbulence creates noise. Inserting a silencer made of porous sound-absorbing material tames the turbulence and reduces the exhaust noise.

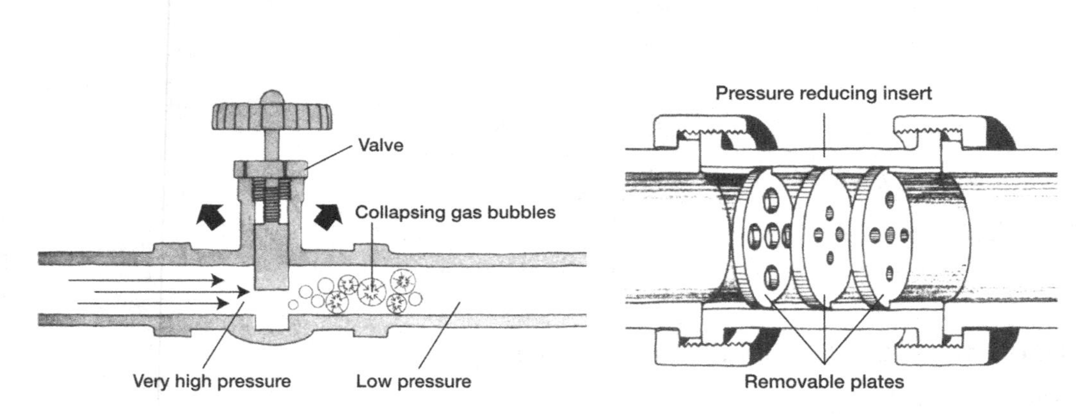

FIGURE 14.18 Use of pressure reducers to reduce 'cavitation' sound.

Adapted from: OSHA, 1980, p. 60.

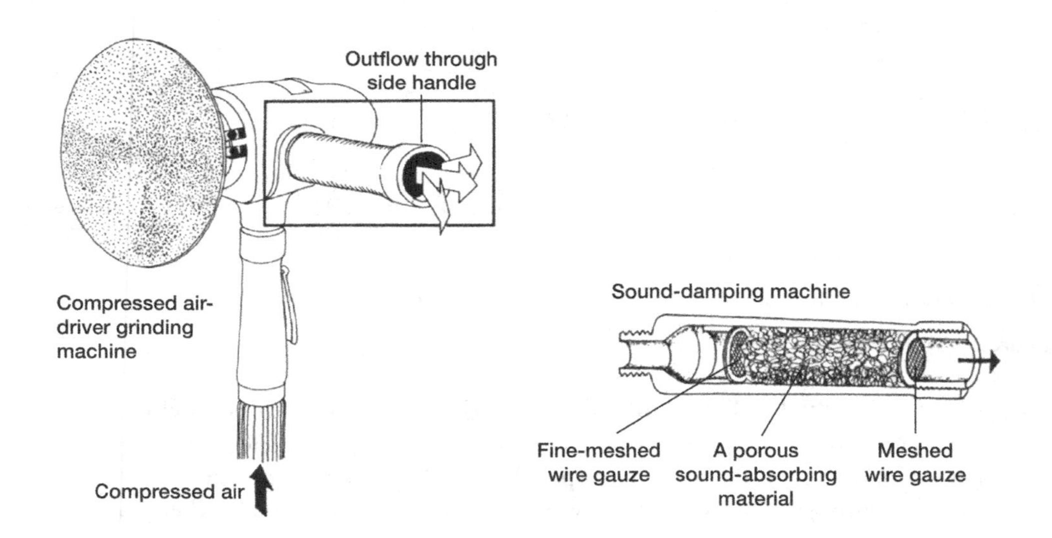

FIGURE 14.19 Minimising turbulence at air exits to reduce noise.

Source: OSHA 1980, p. 51.

14.8.1.7 Buy Quiet

Noise control techniques should be incorporated in the design and installation of equipment since the cost of doing so is minimal compared to that of after-market design and installation. However, there is rarely an off-the-shelf solution for the suppression or control of noise, and it is necessary for the H&S practitioner to assess each workplace situation separately.

- While a piece of equipment may correctly claim to produce sound levels less than 85 dB(A), two or more operating side by side may still produce combined sound levels in excess of the regulatory noise level.
- Companies should be advised to include noise limits in purchasing specifications, keeping in mind the additive effects of more than one piece of equipment. The basic information to be requested from potential suppliers should include the A-weighted sound power level (L_{Wa}). The sound power of a piece of equipment is independent of its environment. The sound power level can be converted to a sound pressure level. If enough H&S practitioners and employers insist on quiet machinery, manufacturers will begin to provide it.

A good example of 'buying quiet' is the German Blue Angel program, under which manufacturers of construction industry machines and equipment can have their products tested against environmental noise criteria. If their products meet the criteria, they are entitled to carry the Blue Angel logo. From the manufacturer's perspective, this is a selling point, because German (and other European) building sites must comply with strict (local) government regulations, and quiet plant and equipment are often demanded in requests for tender.

14.8.2 Control of Noise Transmission Path

The location of a machine or process is often a significant factor in the noise it transmits to the workplace. Several methods of noise attenuation focus on this point.

14.8.2.1 Isolating the Noise Source from the Worker

Where noise generation cannot be prevented, the following options can be considered:

- Placement of the noise source at a distance from the workplace: for example, pumps, compressors, generators and the like can be installed outside the building (provided this does not lead to environmental noise complaints).
- Use noise-absorbing enclosures around the noise source (see Figure 14.20).
- Confine sources to a noise-insulated room, using double walls with insulating material, double-glazed windows and solid core doors.
- Use an isolating enclosure around the worker such as.

14.8.2.2 Preventing the Noise from Reaching the Worker

- Place noise sources away from natural reflectors, for example, in corners.
- Use sound absorbers on ceilings and walls to avoid reflection and amplification of noise from standing waves (see Figure 14.21).
- Use noise baffles or deflectors to direct noise away from workers.

The efficacy of this method depends on the frequency of the sound produced. Low-frequency noise tends to pass through openings and around objects (see Figure 14.21), while high-frequency noise is more easily deflected (see Figures 14.22 and 14.23).

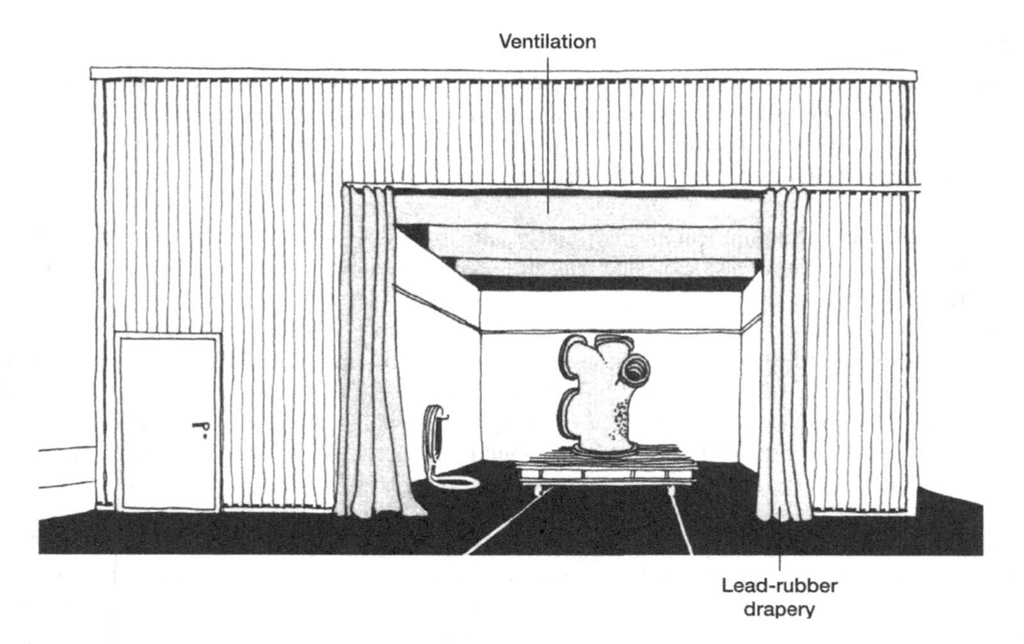

FIGURE 14.20 Use of an isolated room for noisy operations.

Source: OSHA 1980, p. 95.

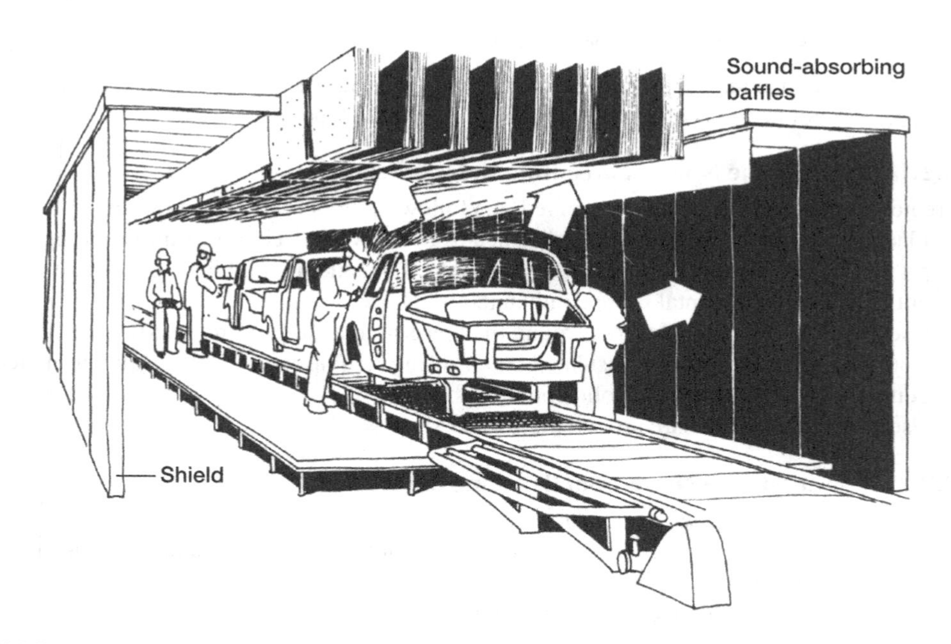

FIGURE 14.21 Sound-absorbing baffles can minimise sound movement indoors.

Source: OSHA 1980, p. 71.

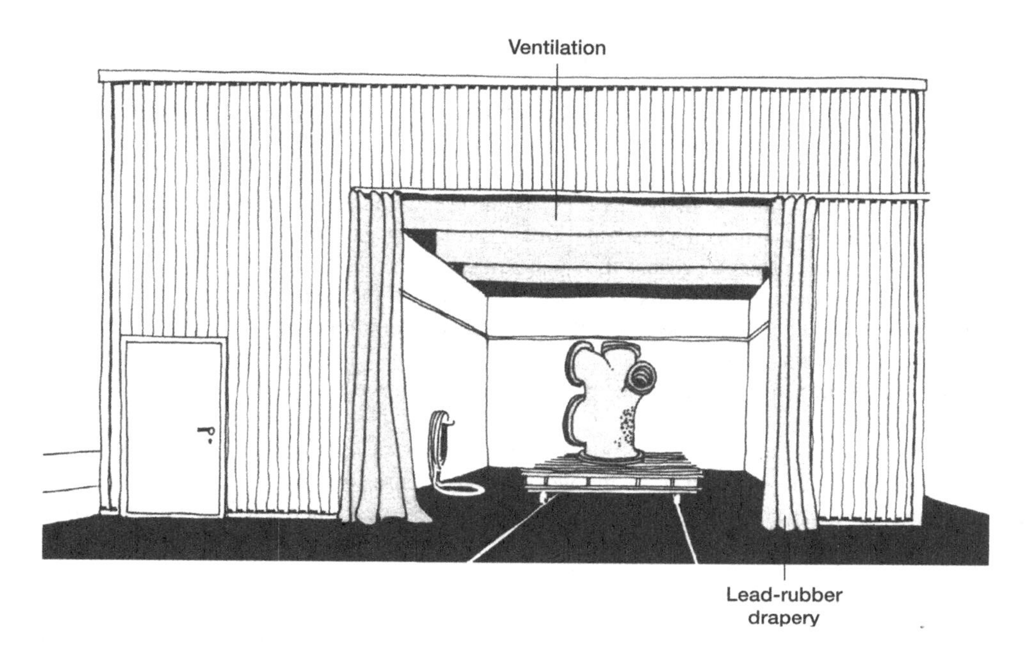

FIGURE 14.22 Use of sound absorbent to reduce sound transmission.

Source: OSHA 1980, p. 23.

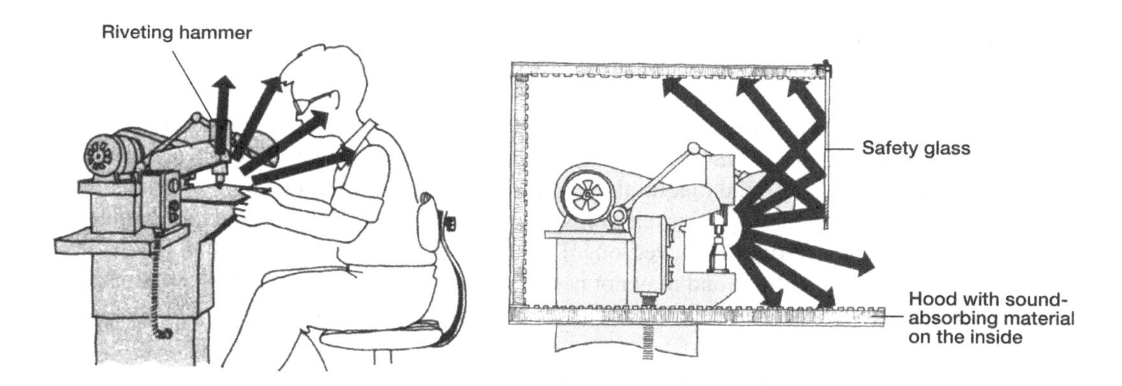

FIGURE 14.23 High-frequency noise can easily be deflected away from the worker.

Source: OSHA 1980, p. 21.

14.9 PERSONAL HEARING PROTECTION

Where it has been confirmed that workplace noise levels are in excess of regulatory noise limits, and the noise is unable to be controlled at the source or along the transmission path, personal hearing protective devices (HPD) must be used until it can be demonstrated that the noise exposure is reduced below the applicable regulatory limit.

The personal hearing protection program must include, at minimum:

- a noise-control policy;
- a system for conducting noise level and noise level exposure assessments on a regular basis;

- a program for the planning and implementation of engineering and administrative noise control, where possible;
- a suitable hearing protection training program, including regular refresher training, and records of training topics and attendance;
- frequent supervision and implementation of corrective actions by line management to ensure correct fitting and use of earplugs to ensure attenuation is maximised;
- selection and provision of personal hearing protectors and documented reasons why the selected hearing protectors were the most suitable;
- provision for the use, maintenance, care and storage of the hearing protectors;
- provision for audiological assessment of workers on commencement of employment (or the hearing protection program) and regularly thereafter;
- ongoing monitoring and review of the effectiveness of the program.

To have any chance of success, a hearing protection program requires the full involvement and cooperation of management and from the workers. To secure such involvement, the H&S practitioner must demonstrate that a risk-management process has been carried out. Workers and management should be given training on:

- how the ear works, including how hearing loss occurs;
- reasons why hearing protection is required—that is, importance of preservation of hearing and health; legislative requirements, limitations and other control options;
- selection of personal hearing protectors;
- fitting and use of hearing protectors, and the importance of good fit and comfort;
- good and bad habits when wearing hearing protectors;
- the requirement to wear the hearing protectors *all the time* when exposed to noise;
- correct use and maintenance of hearing protection.

AS/NZS 1269.3 (Standards Australia 2005b) provides further guidance on the elements and training requirements for the management of workplace noise through hearing protector programs.

The H&S practitioner will find that implementing effective hearing protection programs is far more complicated than simply buying earplugs or earmuffs for workers. No given earmuff or earplug will fit or be comfortable for everyone. Different types of hearing protectors have different noise attenuation capabilities and may not necessarily be compatible with all work situations. The choice of hearing protectors depends very much on factors, such as:

- the noise levels measured in the workplace.
- the frequency spectrum of the noise.
- the attenuation required to achieve compliance with regulations.
- the worker's hearing ability, for example needing to wear hearing aids.
- the worker's acceptance of and the degree of good fit and comfort of the hearing protector. An earplug should be inserted for at least three-quarters of its length to achieve the rated attenuation.
- compatibility with other personal protective equipment.
- the need to communicate with others and to hear important signals and sounds.
- cost.

Regardless of the type of hearing protector used, its effectiveness is drastically reduced when it is not always worn correctly and consistently in noisy environments. This is illustrated in Figure 14.24, which shows the reduction in effectiveness of hearing protectors with stated attenuation values of SLC_{80} 28 dB(A) and 20 dB(A) respectively.

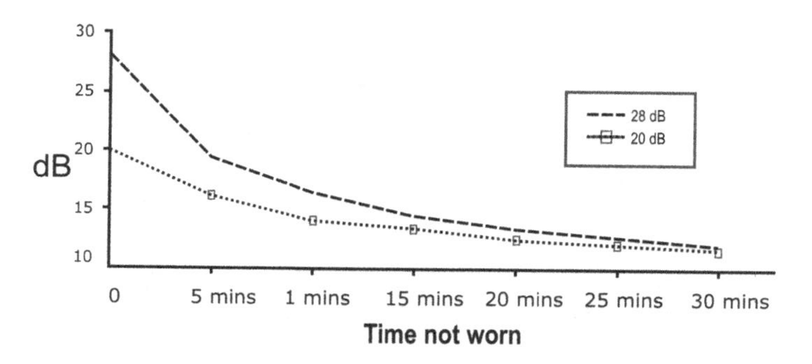

FIGURE 14.24 Reduction of effectiveness of a hearing protector with time not worn.

Source: Workplace Health and Safety Queensland 2019, p. 2, Figure 3.

If the hearing protector with a stated reduction of 28 dB(A) is removed for only five minutes during an eight-hour exposure time, the effective attenuation is reduced to only 20 dB(A); if it is removed for 10 minutes, the effective attenuation is reduced to about 16 dB(A). The longer the hearing protector is not worn, the less effective it becomes.

To further illustrate the importance of the correct wearing of hearing protectors, there is documented evidence to suggest that 'real-world' attenuation may be up to 6 dB less than in laboratory tests for earmuffs and up to 9 dB for earplugs (Berger 2000; Eden & Piesse 1991).

14.9.1 Types of Hearing Protector

There are four basic types of hearing protector, of which only two are in widespread use.

Acoustic helmets cover a large part of the head and the outer ear and resemble those worn by helicopter and fighter pilots. These helmets are bulky, expensive and not suited for general industrial use. Noise attenuation may be as much as 50 dB at the lower frequencies and 35 dB at 250 Hz; they may also diminish noise conduction through bone.

Ear-canal caps are made from a light rubber or PVC type material and seal off the entrance to the ear canal without entering it as an earplug does (Figure 14.25). Ear-canal caps are held in place by a spring headband. They are gaining popularity on construction sites because they can easily be placed over the ears when a noise—usually of short duration—starts suddenly and unannounced.

Earplugs are widely used, both as single-use disposable types and as reusable types (Figure 14.26). They are typically manufactured from foam rubber, plastic or silicone and may come user-formable, pre-moulded or custom-moulded. All these types can provide good noise attenuation, but attention must be paid to correct fitting. All earplugs are designed to prevent noise reaching the inner ear by filling and blocking the ear canal. The formable types are designed to fit all ears; they are rolled between the thumb and finger into a cylindrical shape prior to insertion. Once the earplug is rolled tightly with one hand, insertion in the ear is done by raising the other arm over the head, grabbing the pinna and pulling it up and towards the back of the head so as to straighten the ear canal and inserting the plug in the ear and holding it in place with a finger or thumb on it until the plug expands again and block the ear canal. The earplug should be inserted sufficiently deep so that when looking front on the plug should be hidden behind the Tragus. When removing the earplug, it would be good practice to grab the plug again between the finger and thumb and give it a quarter turn to break the seal and allow air to enter the ear canal before pulling the earplug out completely. Reusable earplugs (Figure 14.27) should be able to be inserted into the ear canal without having to be squashed to fit. Some pre-moulded earplugs have an acoustic resonator

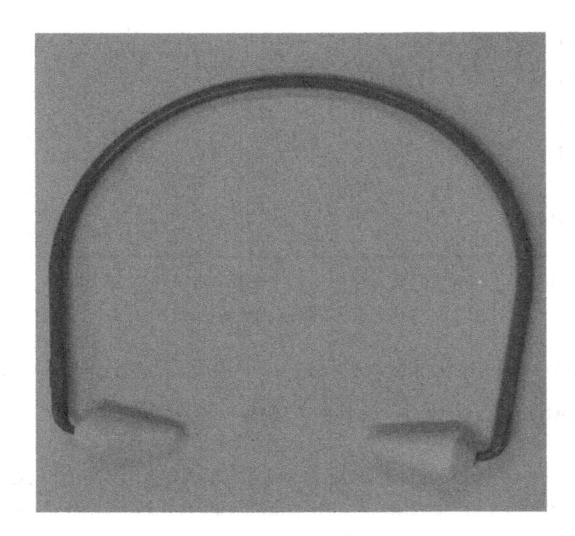

FIGURE 14.25 Types of ear-canal caps.

Source: J. Whitelaw.

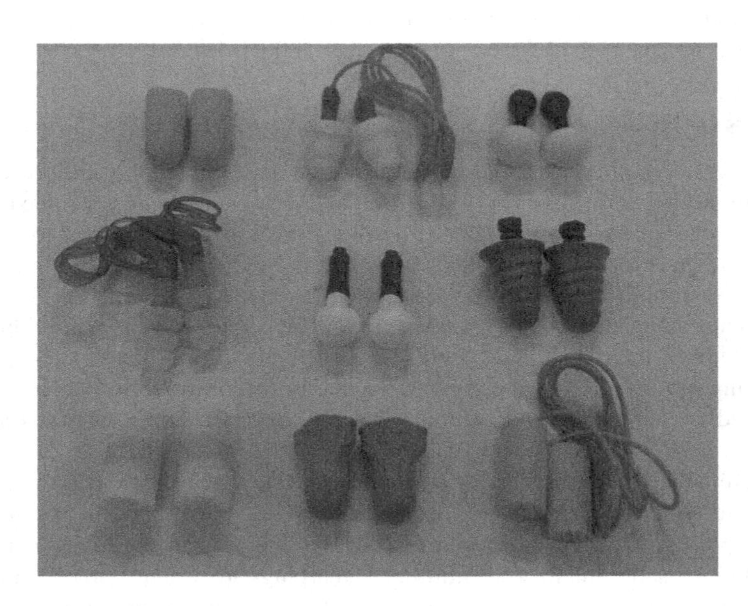

FIGURE 14.26 Selection of earplugs.

Source: J. Whitelaw.

chamber included to provide an almost flat attenuation across the frequency range. This type of earplug may be beneficial for workers who have a higher need for communicating verbally amidst noise, or who have a mild hearing loss, as the attenuation at 2 and 4 kHz (4 kHz is where NIHL typically manifests itself) is not as great as with normal industrial-type earplugs.

Individually moulded earplugs are made from an acrylic or a silicone mould of the wearer's ear canal. This means they can be worn only in the ear canal from which the mould was made. Individually moulded earplugs can also be fitted with acoustic resonator chambers to provide an

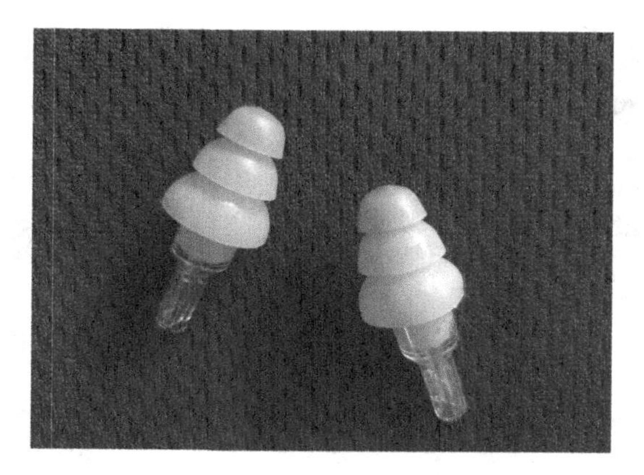

FIGURE 14.27 Example of a reusable flat attenuating earplug.

Source: B. Groothoff.

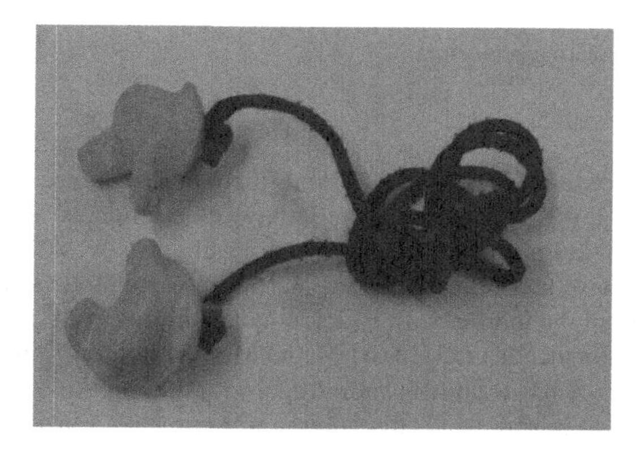

FIGURE 14.28 Individually moulded earplugs with resonator chambers fitted.

Source: J. Whitelaw.

almost flat attenuation over the frequency range and can therefore assist in communication by offering less distortion than ordinary individually moulded earplugs. Individually moulded earplugs must be expertly fitted, since performance and comfort may be poor if they are incorrectly shaped or sized. Flat attenuating or individually moulded earplugs with built in acoustic resonators are sometimes also called musicians earplugs and may be beneficial to musicians as, compared to normal industrial earplugs, the music is far less distorted enabling their wearing during the making of music and preserve their hearing and livelihood. The attenuation provided is affected by the expertise of the person doing the mould as well as the quality control of the process. They are not currently Standards Australia approved. An example of individually moulded earplugs is shown in Figure 14.28.

Earmuffs consist of two padded, internally insulated domes which cover the entire ear (Figure 14.29). A spring-torsioned headband holds the padded cups to the sides of the head at a clamping force to provide the attenuation desired. When selecting earmuffs, ensure that the cup is just large enough to clear the ear lobes. It is important that the cushions attached to the cups be soft and not cracked, as they are essential to provide a proper seal. They should be cleaned after use with warm soapy water and a wet cloth (no solvents or harsh chemicals) and their condition regularly checked. When hard or cracked, the cushions can easily be replaced.

FIGURE 14.29 Earmuff hearing protector.

Source: B. Groothoff.

In situations where communication is important, the so-called noise cancelling earmuffs can be used. These types of level dependent earmuffs have electronically activated shut-off valves, which enable the muffs to admit noise up to a certain level—usually 82 dB(A); above this level, the earmuff blocks the noise and acts as a normal earmuff. Some modern and relatively inexpensive earmuffs have entered the market with Bluetooth capabilities enabling workers to communicate, make and receive phone calls or listen to FM and MP3 without having to remove the earmuff.

Another type of earmuff has a built-in radio receiver. Although this may seem attractive for workers doing mundane or boring work, the volume of radio sound inside the earmuff cup combined with the level of workplace noise entering the cup may cause the eight-hour exposure limit to be exceeded. An employer providing this kind of earmuff to workers must ensure that the workers' in-ear noise level does not exceed the exposure limits imposed by WHS legislation.

14.9.2 Frequency Attenuation

Sound can be transmitted to the inner ear via conduction through bone, leaks in the ear canal seal, oscillation of an earplug, transmission of sound through an earmuff, or vibration through a sealing cushion of an earmuff. For these reasons, attenuation of more than 50 dB is not possible. Some types of hearing protectors cannot attenuate low frequencies as well as others. Figure 14.30 compares the typical attenuations of an earplug and an earmuff.

When selecting hearing protection, it is important to know which frequencies need to be attenuated. For example, if the noise is predominantly low frequency, a hearing protector that offers poor protection at low frequencies would not be an appropriate choice. On the other hand, a worker might need to hear some of the lower frequencies of speech. Overall, the aim is to ensure that the daily noise exposure of the hearing-protected worker is still less than $L_{Aeq,8h}$ 85 dB(A), irrespective of the frequencies making up the noise in the workplace.

There are several possible approaches to selecting hearing protective devices (HPDs), ranging from a simple ad hoc assessment of the noise to a sophisticated assessment that takes account of all the frequencies involved. All approaches require some workplace noise measurements to be made. Failure to measure the workplace noise could result in either under- or over-attenuation,

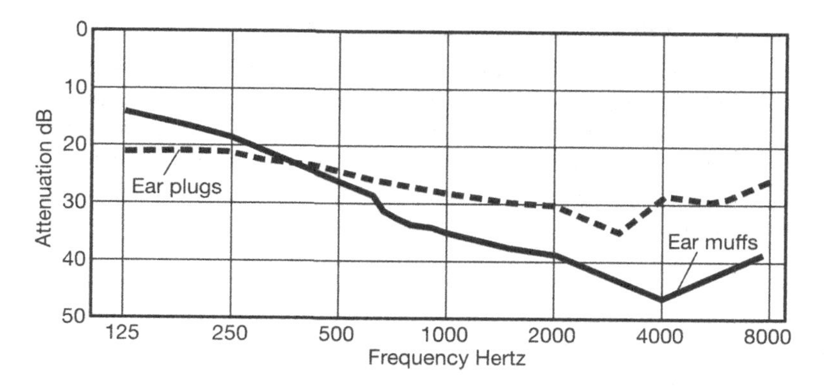

FIGURE 14.30 Noise attenuation of earplugs and ear muffs.

Source: Grantham 1992, p. 203, Figure 8.24.

which can jeopardise a worker's hearing and safety respectively. There are several different methods of making the assessment, with each offering a different level of precision and useful for different purposes.

14.9.2.1 Classification Method

The simplest way of selecting an HPD is to use the New Zealand classification method, as referenced in AS/NZS 1269.3 (Standards Australia 2005b). Selection depends on the extent to which the $L_{Aeq,8h}$ is above the critical level of $L_{Aeq,8h}$ 85 dB(A). For the classification method, a noise dose meter or sound level meter is used to measure the A-weighted average exposure in the workplace (if no noise dose meter is available, some equivalent measure such as $L_{Aeq,T}$ can be used, from which, based on a knowledge of workplace activities and work histories, an eight-hour exposure can be calculated). Selection of the appropriate class of hearing protector is then made from Table 14.5.

This method suits those H&S practitioners who can measure only an A-weighted workplace noise exposure. It does not consider the actual frequency of the workplace noise being measured and can lead to inappropriate choices if the noise contains predominantly very high or very low frequencies.

TABLE 14.5
Relationship between Measured $L_{Aeq,8h}$, Class Method for Rating Hearing Protection and SLC_{80}

Measure Exposure $L_{Aeq,8h}$ dB(A)	Class	SLC_{80} Range
Less than 90	1	10–13
90 to less than 95	2	14–17
95 to less than 100	3	18–21
100 to less than 105	4	22–25
105 to less than 110	5	26 or greater

Adapted from: Table A1 & Table E1 from AS/NZS 1269.3:2005, pp. 20 & 29.

14.9.2.2 Sound-Level Conversion Method

The sound-level conversion (SLC_{80}) method was originally developed as a simplified version of the octave band method. It uses the difference between the C-weighted L_{eq} and the A-weighted L_{eq} reaching the wearer's ear, and therefore accounts better for low-frequency components (<80 Hz) of the workplace noise as well as high-frequency ones (>3150 Hz). The SLC_{80} technique is a simple method for determining the HPD best suited for a hazardous noise area. Its name refers to a rating given to hearing protectors that provide the stated attenuation expected for 80 per cent of wearers.

The first step is to measure the workplace noise level in dB(C) at the worker's ear (i.e. using C-weighting rather than A-weighting). The second step is to calculate the minimum SLC_{80} the HPD should have to provide adequate hearing protection for the wearer. This is determined by simply subtracting the 'desired' A-weighted in-ear noise level of the wearer from the C-weighted noise level measurement.

The 'desired' wearer's ear noise level is typically chosen to be between 75 and 80 dB(A), a level that presents minimal risk of NIHL. To enable communication in a noisy environment, the wearer must be able to hear and understand speech.

The HPD must therefore provide sufficient noise attenuation without blocking too much noise. This means the in-ear noise level should at least meet legal requirements, but preferably meet the target level of between 80 and 75 dB(A). The 'bigger is better' syndrome definitely does not apply here; over-protection should be avoided as much as under-protection. Over-protection leads to feelings of acoustic isolation and prompts workers to remove their hearing protector when communication is required.

All HPDs should, as a minimum, provide information of an SLC_{80} rating, a class rating and an octave band rating. Sometimes the manufacturer will provide an example of how the rating is applied. For example, if the work environment has a C-weighted noise level of L_{Ceq} 105 dB(C), wearing a hearing protective device with an SLC_{80} rating of 27 dB should provide an in-ear sound level of 105–27 = 78 dB(A). An HPD with an SLC_{80} of only 15 dB would permit in-ear noise levels of up to 90 dB(A), inadequate for use in that noise environment. Table 14.5 shows the relationship between class and SLC_{80} range for use in determining suitable hearing protection.

14.9.2.3 Octave Band Method

The third method of measuring frequency attenuation is the more sophisticated octave band method. This method is the most suitable where special noise characteristics in the workplace need to be addressed (e.g. high-pitched noise, or a low-frequency component) and must also be applied where the $L_{Aeq,8h}$ exceeds 110 dB(A). It involves conducting a frequency analysis of the noise in the workplace where the hearing protector is to be used. The octave band attenuation of the hearing protector is then subtracted from the measured octave band results in the frequency range between 125 and 8000 Hz. Then the A-weighted sound level, as presented to the ear, is calculated for the resulting attenuated spectrum. Application of the octave band method requires a precision sound level meter designed for octave band analysis because of the need to measure unweighted (i.e. linear) sound pressure levels.

To identify the degree of attenuation that the hearing protector needs to provide, two sets of data are required:

- the measured octave band analysis of the noise in the workplace;
- the frequency attenuation of the hearing protector, provided by the manufacturer.

Calculation of octave band attenuation

Step 1: Carry out a sound-level survey of the workplace, using octave band analysis at the frequencies shown in the following example. Use unweighted sound pressure levels, that is, linear (dB(Z)), never use A- or C-weighted levels for this type of assessment. This is done by using a sound level meter which has the capacity to record the unweighted sound pressure levels of at least seven different frequency bands centred on the following frequencies: 125 Hz, 250 Hz, 500 Hz, 1 kHz, 2 kHz, 4 kHz and 8 kHz.

	Frequency (Hz)						
	125	**250**	**500**	**1k**	**2k**	**4k**	**8k**
Measured octave band level (dB re 20 μPa)	87	88	90	94	97	106	104

Step 2: Obtain the frequency attenuation of the hearing protector. In each octave, subtract the mean-minus-standard-deviation octave band attenuation of the hearing protector from the measured octave band results.

	Frequency (Hz)						
	125	**250**	**500**	**1k**	**2k**	**4k**	**8k**
Attenuation of earmuff (mean-minus-standard-deviation) (dB*)	13	13	14	18	25	35	28

* AS/NZS 1270-2002 Table A2.

Step 3: Calculate attenuated sound levels.

	Frequency (Hz)						
	125	**250**	**500**	**1k**	**2k**	**4k**	**8k**
Unweighted sound pressure level* (from noise survey)	87	88	90	94	97	106	104
Mean-minus-standard-deviation (dB)	13	13	14	18	25	35	28
Attenuated sound level (dB)	74	75	76	76	72	71	76

Step 4: Apply A-weighting corrections and calculate A-weighted attenuated sound levels. Make the A-weighting correction by adding the A-weighting corrections at the same frequencies. (The origin of the A-weighting correction values was explained in Section 14.4.)

	Frequency (Hz)						
	125	**250**	**500**	**1k**	**2k**	**4k**	**8k**
Attenuated sound level (dB)	74	75	76	76	72	71	76
A-weighting correction (dB)	–16	–9	–3	0	+1	+1	–1
A-weighted attenuated levels	58	66	73	76	73	72	75

Step 5: Rearrange the A-weighted attenuated sound levels from lowest to the highest—in this case: **58, 66, 72, 73, 73, 75** and **76**. Now add the levels two at a time, using the procedure in Table 14.2, until all are used, as follows:

The difference between 58 and 66 is 8. Looking up the difference in Table 14.2 shows a correction factor of 0.6 to be added to the higher level. Combining 58 and 66 is 66.6.

The next level is 72. The difference between 66.6 and 72 is 5.4. Looking up Table 14.2, the closest correction factor is 1.2. Adding 1.2 to 72 is 73.2.

The next level is 73. The difference between 73 and 73.2 is 0.2. Looking up Table 14.2, the closest correction factor is 3.0. Adding 3.0 to 73.2 is 76.2.

The next level is 73. The difference between 76.2 and 73 is 3.2. Looking up Table 14.2, the closest correction factor is 1.8. Adding 1.8 to 76.2 is 78.

The next level is 75. The difference between 78 and 75 is 3. Looking up Table 14.2, the correction factor is 1.8. Adding 1.8 to 78 is 79.8.

The next level is 76. The difference between 79.8 and 76 is 3.8. Looking up Table 14.2, the closest correction factor is 1.4. Adding 1.4 to 79.8 is 81.2.

The attenuated level under the earmuff is therefore about 81 dB(A).

This is within the daily noise dose, $L_{Aeq,8h}$ 85 dB(A), but 1 dB over the target level of 80 dB(A), provided the worker always wears the hearing protector in this environment while exposed to noise.

To combine the attenuated A-weighted values of the hearing protector, we can use this formula instead of the correction table:

$$L = 10_{\log 10}\left(10^{L1/10} + 10^{L2/10} + 10^{L3/10} + 10^{Ln/10}\right) \tag{14.4}$$

Applying the formula, we should get:

$$L = 10_{\log 10}\left(10^{58/10} +^{1066/10} +^{1073/10} +^{1076/10} +^{1073/10} +^{1072/10} +10^{75/10}\right) L = 81\,dB(A)$$

Reputable manufacturers of HPDs will provide full details on attenuation at the various frequencies on the packaging of their products as well as on their websites. Alternatively, when the HPD has been tested in accordance with AS/NZS 1270 (Standards Australia 2002), the frequency attenuation of many HPDs can be found in the New Zealand Ministry of Business, Innovation and Employment booklet "Classified Hearing Protectors" of March 2013 which is freely available for download from https://docplayer.net/11727956-Classified-hearing-protectors.html.

Select a hearing protector that provides adequate protection for the determined workplace noise, taking into consideration cost and wearer compatibility and acceptability.

H&S practitioners who believe that hearing protectors may be required in the workplace but do not have sound level measuring equipment can ask a supplier or an acoustic consultant or Occupational Hygienist to conduct octave band noise analysis in each noisy location. The expected sound levels for the various hearing protectors can then be calculated from the manufacturer's published information and a decision can be made about which protector is suitable.

Other information on noise measurement can be found in:

- AS IEC 61672.1 Electroacoustics—sound level meters, part 1: Specifications (Standards Australia 2004)
- AS/NZS 2399 Acoustics—Specifications for Personal Sound Exposure Meters. (Standards Australia 1988)
- AS/NZS 1269 Parts 0–4—Occupational Noise Management (Standards Australia 2005a, 2005b, 2005c, 2005d, 2014)

14.9.2.4 NRR Method

The NRR method is an American system of determining the expected attenuation provided by a hearing protector. This method cannot be used in Australia or compared to, say, SLC_{80} values, because of the different laboratory test methods used in the United States and the different exchange rates for doubling of sound intensity. The United States (OSHA 1980) uses a 5 dB exchange rate, whereas Australia and most other countries use the natural 3 dB rate.

14.10 OCCUPATIONAL AUDIOMETRY AND THE HEARING CONSERVATION PROGRAM

Occupational audiometry, the testing of workers' hearing acuity, may well identify hearing disabilities. Such tests may not necessarily indicate, however, that hearing loss is the result of current noise exposure.

Audiometry is not generally a task for the H&S practitioner. It requires well-trained, qualified, experienced audiometrist or audiologist with properly calibrated test equipment and a specially constructed soundproof booth. The current Standard for audiometric testing is AS/NZS 1269.4: 2014. The main changes from the 2005 edition of the Standard are replacement of the table of maximum allowable ambient noise levels for particular makes and models of earphones/enclosure combinations with a method to calculate the maximum permissible ambient noise level for any earphone/enclosure combination (Appendix C) (Standards Australia 2014). It is important that for conducting air conduction audiometry, the maximum permissible ambient noise levels stated in Table C1 of Appendix C are not exceeded. If these ambient noise levels are exceeded, then the audiometric test results are likely to be unreliable due to masking of the test signals. The H&S practitioner is therefore referred to Appendix C of the Standard to ensure the test environment does not exceed the required ambient noise levels before audiometric tests are carried out in that environment. To prove useful, audiometric testing must be part of an ongoing program in which hearing is regularly tested, preferably from the time the worker joins the workplace, to check whether there is any discernible deterioration above that attributable to ageing (presbycusis).

A health practitioner trained in conducting hearing tests (an audiometrist or audiologist) will test the worker's baseline auditory threshold for both ears. The frequencies used for both reference and monitoring audiometry are 500 Hz, 1000 Hz, 1500 Hz, 2000 Hz, 3000 Hz, 4000 Hz, 6000 Hz and 8000 Hz.

Reference audiometry must be conducted within three months after a worker starts employment, but should ideally be conducted before exposure to a noisy workplace occurs, e.g. as part of a pre-employment assessment regime. Reference audiometric testing must be conducted immediately after a period of not less than 16 hours of quiet—i.e. noise exposures below 75 dB(A), which are unlikely to produce a temporary threshold shift.

The reference audiogram, also known as a baseline audiogram, shall be updated whenever a significant threshold shift has occurred or at least every 10 years, whichever occurs sooner. Future monitoring audiograms shall then be compared with the most recent reference audiogram and records of previous reference audiograms shall be retained (Standards Australia 2014).

Monitoring audiometry should be performed well into or at the end of the work shift and carried out within 12 months of initial reference audiometry and ongoing at least every two years, as prescribed by WHS legislation. The results should be compared with the reference audiometry test results to see whether the hearing threshold has changed. If it has, this will indicate inadequacies in the use of personal hearing protectors or changes in workplace noise exposure levels. For workers exposed to high exposure levels, >100 dB(A), more frequent audiometric testing—for example, every six months—may be required.

Where significant hearing impairment is detected at the initial reference audiometric test, the person should undergo a medical examination to see whether a repeat audiometric test, conducted on another day, confirms the original finding.

Where the monitoring audiometry results, when compared with the results of the reference audiometry, show:

- a shift in average threshold at 3000, 4000 and 6000 Hz greater than or equal to 5 dB; or
- a shift in mean threshold ≥ 10 dB at 3000 and 4000 Hz; or
- a change in mean threshold ≥ 15 dB at 6000 Hz; or
- a threshold shift ≥ 15 dB at 500, 1000, 1500 or 2000 Hz; or
- a threshold shift ≥ 20 dB at 8000 Hz.

The worker shall be requested to have a second confirmatory audiometric test on another day, again after 16 hours in quiet conditions. If the threshold shift is confirmed during this second test, the person should be referred for a medical opinion. If the shift in threshold is confirmed as noise-induced, the worker should be advised of this in writing. Figure 14.31 shows the various degrees of hearing loss and their relationship to everyday sounds via the scale on the right.

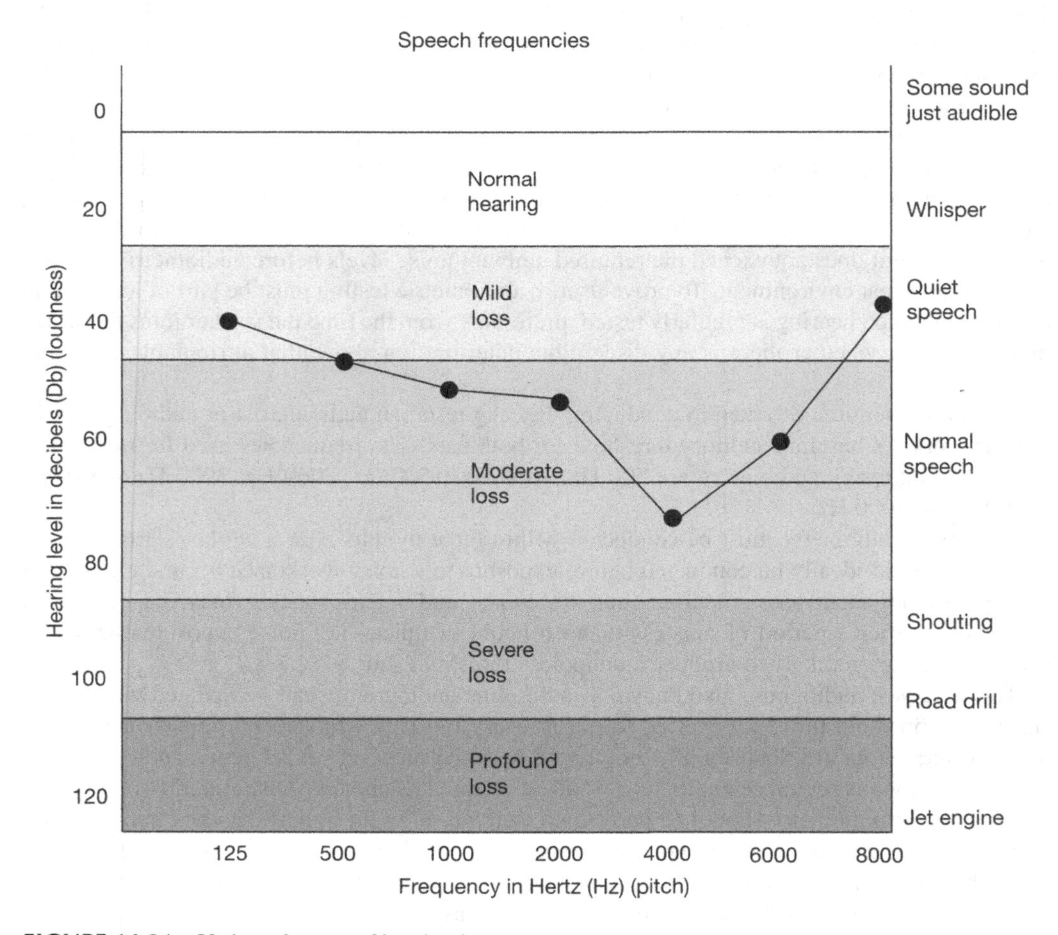

FIGURE 14.31 Various degrees of hearing loss.

Source: B. Groothoff.

Audiometric testing can be of benefit to both employers and workers in excessively noisy industries (foundries, canneries, construction, metal industries, transport and storage), but only if it is an integral part of a rigorous noise management program. Audiometric testing in isolation from other elements of a noise management program serves only to record the deterioration in hearing. In combination with other elements, it can detect the early onset of NIHL and enable countermeasures to be put in place. The tests may also provide useful information for assisting in workers' compensation claims. For instance, an employer who knows the hearing ability of workers at the start of their employment and can demonstrate adherence to a robust noise management program may well have a defence against future claims for NIHL compensation.

Hearing Conservation Programs are designed to prevent permanent hearing impairment, principally by maintaining noise exposure within the limits required by the legislation. Any of the technical control processes discussed above might be used in such a program. Certainly, the best are noise-reduction programs. However, if noise sources cannot be quietened sufficiently through higher-order noise-control measures, the worker's hearing must be protected by using an HPD. Management must then also provide training and instruction in, and supervision of, the efficient use of HPDs. Management's commitment to the use and benefits of HPDs must be matched by the rigid enforcement of the noise management program.

14.11 ACOUSTIC SHOCK

14.11.1 Introduction

ACIF G616 *Industry Guideline—Acoustic Safety for Telephone Equipment* (Australian Communications Industry Forum [ACIF] 2013), defines an **acoustic incident** as:

> The receipt by a telephone user of any unexpected sound that has acoustic characteristics that may cause an adverse reaction in some telephone users. Depending on the characteristics of the sound and of the user, an acoustic shock may result from the incident.
>
> (*ACIF 2013, p. 3*)

An acoustic incident is typically a high-intensity, high-frequency monaural squawk, screech, shriek or howling tone that occurs without warning. This type of noise is typically produced by such things as misdirected fax calls, feedback between the microphone and speaker of the telephone, or mobile phone signal interference.

Exposure to an acoustic incident usually does not last more than a few seconds, as the operator generally tears off the headset almost as soon as the signal is heard. If it is severe enough, however, an acoustic incident may lead to acoustic shock.

Acoustic shock is defined by the European Telecommunications Standards Institute (ETSI 2000) and ACIF G616 as:

> Any temporary or permanent disturbance of the functioning of the ear, or the nervous system, which may be caused to the user of a telephone earphone by a sudden sharp rise in the acoustic pressure produced by it.
>
> (*ACIF 2013, p. 3*)

Tests have shown that the signals in question are less than 120 dB sound pressure level at the eardrum because of the electronic limiters in the headset systems. Measurements by the Health and Safety Laboratory of the UK Health and Safety Executive (2001) indicate that fax tones produce on average 83 dB(A), holding tones 88 dB(A) and carrier tones 95 dB(A) when measured at maximum volume.

14.11.2 How Is an Acoustic Shock Experienced?

When an acoustic incident occurs, telephone workers may initially feel startled or shocked. In extreme cases, they may quickly experience a varied combination of vertigo, nausea, vomiting, stabbing pain in the ear, feeling of fullness in the ear and/or facial numbness. Tingling, tenderness or soreness around the ear and neck and arms, and tinnitus, may also be experienced. There is seldom loss of hearing, and symptoms tend to disappear over time.

Secondary and tertiary symptoms may develop consistent with stress from trauma, including headache, fatigue, feelings of vulnerability, anger, hypervigilance and hyper-sensitivity to loud sound, depression, substance abuse and anxiety, especially about returning to work.

14.11.3 Why Call Centres?

An explanation for the symptoms experienced by a call-centre telephone operator after an acoustic incident must be sought in both the psychosomatic and physiological areas. The main psychosomatic factors are the stress of handling large volumes of telephone calls within certain time limits, the performance monitoring systems applied and targets to be achieved, the company's management culture, the operator's perceived lack of control over the stressful work situation and the inability to anticipate an acoustic incident. Furthermore, the work environment, with often inadequate office equipment and acoustics; lack of proper training; mistakes by incoming callers or clients such as sending a fax over a telephone line; interference from mobile phone signals; and loud noises deliberately made by irate clients may all contribute. These issues combine to cause stronger physiological responses (e.g. a startle reflex) when an acoustic incident does take place than if the worker were exposed to a similar sound in an industry where the sound was anticipated.

The main physiological response to an acoustic shock occurs in the middle ear. The middle ear contains two muscles, the stapedius and the tensor tympani, which are attached to the ossicles (the middle ear bones, the malleus, incus and stapes). The tensor tympani muscle is attached to the malleus, and the stapedius muscle to the stapes. When loud sounds occur, both muscles react, causing what is known as the 'aural reflex'. The stapedius muscle reacts to loud sounds, causing an increased pressure on the oval window membrane of the cochlea and the movements of the stapes. The tensor tympani muscle causes a startle reflex. If triggered, it restricts the movements of the malleus and incus and is capable of placing large forces on the alignment of the eardrum. The tensor tympani muscle's reflex threshold can be 'reprogrammed' to react at much lower sound levels.

With acoustic shock incidents, the loud noises occur with rise times of between 0 and 20 milliseconds. The time for the middle ear muscles to react by contracting is about 25 milliseconds. When they do contract, they do so with added force because of the combination of loud noise and the startle reflex. In extreme cases, this may lead to a tearing of the oval or round window membrane and subsequent leaking of fluid from the cochlea.

14.11.4 What Is the Evidence for the Existence of Acoustic Shock?

The available evidence indicates that acoustic shock and its effects constitute a real health issue because of the specific characteristics of call centres. In Australia alone, there have been several hundred reported cases of acoustic shock in call-centre telephone operators. A minority of affected individuals are unable to continue working as telephone operators. Other countries report similar acoustic incidents with remarkably similar symptoms. The chance that all these cases are due to malingering is non-existent.

14.11.5 How Can Acoustic Shock Be Managed?

Management of acoustic incidents must focus on two main areas:

- prevention of sudden loud noises over the telephone through the incorporation of appropriate telephone equipment and electronic sound-limiting devices;
- the reduction of stress levels in workers, through appropriate workplace design, management systems and worker education.

These must be implemented in combination, as either one will fail without the other.

14.11.6 Prevention of Loud Noises over Telephone Lines

Telephone systems should comply with AS/CA S004 Voice Performance Requirements for Customer Equipment (Standards Australia 2013a) and limited to 120 dB for handsets and 118 dB for headsets. However, this level does not prevent acoustic incidents. To prevent loud noises from entering an operator's headset, a so-called volume-limiting amplifier must be incorporated between the telephone and the headset. Such a device searches for high-frequency acoustic tones and eliminates them within about 20 milliseconds. If the device is working correctly, the telephone operator is often not even aware that a loud tone has occurred.

14.11.7 Reduction of Stress Levels in Call Centre Workers

The two main areas for reducing negative stress are the work environment (including management systems) and the workplace design (including construction, layout and equipment).

WHS legislation is quite specific about the need to conduct risk assessments for different tasks and systems of work in order to ensure a safe and healthy work environment. It is less specific about workplace design. The acoustics in a given call centre will limit the number of operators because having too many close together leads to mutual speech interference. Proper acoustic design is therefore beneficial to owners because it optimises the number of telephone operators who can occupy a given space without speech interference. It is also beneficial to telephone operators because, with better ability to communicate, their stress levels will decline to acceptable levels.

14.12 OTOTOXINS

14.12.1 Introduction

A wide variety of chemicals and medication may, alone or in concert with noise, result in hearing loss. These substances, known as ototoxins (*oto* = ear, *toxin* = poison), affect the hair cells and/or the auditory neurological pathways. Inhalation or absorption of certain chemicals through the skin may cause hearing loss independent of noise exposure, while other chemicals may have an additive or synergistic effect. Some chemical toxicants that do not cause permanent hearing loss themselves may, in combination with noise exposure, cause permanent hearing loss. Hearing damage is more likely to occur if the ear is exposed to a combination of substances or to the combination of a substance and noise. Because of the different mechanisms of interaction between noise and ototoxins, there are difficulties for both risk assessment and standard setting in the industrial environment. These are compounded by the use of different 'languages' by different professionals when measuring chemical agents and physical stressors, and the tendency in industry to attribute NIHL in noisy environments to noise alone.

Ototoxins can be divided into two general classes: workplace chemicals and medications. In this section, medications will be mentioned only briefly. Of the workplace chemicals, three major classes have been identified: solvents, heavy metals and chemical asphyxiants.

14.12.2 Workplace Chemicals

In most industrial cases, ototoxic hearing loss is caused by aromatic and aliphatic hydrocarbon solvents. These solvents are well recognised for their neurotoxicity to the central and peripheral nervous systems. The most common ototoxic solvents are alcohol, toluene, ethyl benzene, styrene, n-hexane, carbon tetrachloride, carbon disulfide, trichloroethylene, perchloroethylene and acrylonitrile. Other known ototoxic substances in the workplace include mercury, manganese, lead, arsenic and cobalt, and asphyxiants such as hydrogen cyanide and carbon monoxide. Carbon monoxide is also released as a metabolic by-product of the paint stripper methylene chloride.

The 2020 Safe Work Australia Work Health and Safety Model Code of Practice, Managing Noise and Preventing Hearing Loss at Work recommends that workers should have regular audiometric testing if they are exposed to any of the ototoxic substances listed in Appendix B, where the airborne exposure (without regard to respiratory protection worn) is greater than 50 per cent of the national exposure standard for the substance regardless of noise level, or if they are exposed to ototoxic substances at any level *and* noise with $L_{Aeq,8h}$ greater than 80 dB(A) or $L_{C,peak}$ greater than 135 dB(C).

The substances listed in Appendix B of the 2020 Model Code of Practice are shown in Table 14.6, occupations where noise and ototoxins often combine.

Occupations where noise and ototoxic substances are most commonly found together include:

- printing;
- painting;
- construction;
- boat building;
- furniture making;
- petroleum product refinery/manufacture;
- vehicle/aircraft fuelling;
- firefighting;
- weapons firing (armed forces and shooting clubs);
- rural/agriculture.

Research is ongoing to establish the effects of human exposure to workplace ototoxins. One remaining problem is to relate the results of animal studies to humans, as few studies have been conducted with workers to date. The occupational exposure studies conducted so far seem to confirm laboratory studies and suggest that simultaneous exposure to noise and chemicals produces a significantly greater hearing loss than exposure to either agent alone. In the Code of Practice of the Model Work Health and Safety 2023 legislation, *Managing Noise and Prevention of Hearing Loss at Work* (Safe Work Australia 2020), the effects of known ototoxic chemicals and noise have been considered for the first time. The Code of Practice states that where workers are exposed to ototoxic substances at any level and noise with $L_{Aeq,8h}$ greater than 80 dB(A) or $L_{C,peak}$ greater than 135 dB(C), monitoring with regular health monitoring is recommended. It is expected that in the near future it will be possible to predict the effect on hearing of chemical/noise combination. It is also expected that the individual effects of substances on the auditory system will be better understood and can be used in noise management programs.

TABLE 14.6
Some Common Ototoxic Substances

Type	Name	Skin Absorption
Solvents	Butanol	√
	Carbon disulfide	√
	Ethanol	
	Ethyl benzene	
	n-heptane	
	n-hexane	
	Perchloroethylene	
	Solvent mixtures and fuels or Stoddard solvent (white spirits)	√
	Styrene	
	Toluene	√
	Trichloroethylene	√
	Xylenes	
Metals	Arsenic	
	Lead	
	Manganese	
	Mercury	√
	Organic tin	√
Others	Acrylonitrile	√
	Carbon monoxide	
	Hydrogen cyanide	√
	Organophosphates	√
	Paraquat	

Agents that are considered synergistic with noise exposure (i.e. the interaction of the agents produces a threshold shift which is greater than the sum of the effects of the individual agents), include carbon disulfide, carbon tetrachloride, carbon monoxide, hydrogen cyanide, styrene, methyl ethyl ketone and methyl isobutyl ketone.

Agents that are considered to have either an additive or synergistic effect include toluene, ethyl benzene, styrene, carbon monoxide and hydrogen cyanide.

Agents that are known to cause auditory system impairment by themselves include:

- organic solvents such as toluene, styrene, xylene and trichloroethylene;
- metals such as cobalt, mercury, lead and trimethyltin;
- asphyxiants such as carbon monoxide and hydrogen cyanide.

As stated earlier, it is recommended that workers in any of the occupations listed above be included in audiometric testing programs. Reviewers of audiometric test results should be alert to the possible additive or synergistic effects between exposure to noise and ototoxins. Where necessary, they should suggest reducing exposure to one or both agents. Employers must therefore ensure that they know the airborne concentrations of the ototoxic substances to which their workers are exposed and take the actions required to minimise these concentrations. Employers should also ensure that information on ototoxins and their effects is included in training sessions. It is further recommended that at least biannual audiograms be taken of employees whose airborne exposures to known ototoxic substances (without regard to respiratory protection worn) are at 50 per cent or more of Safe Work Australia's workplace exposure standard for the substances in question and the lower noise levels of the Model Code of Practice (Safe Work 2020).

14.12.3 Medication

Some medications have been identified as ototoxins, including antibiotics such as streptomycin; quinine; salicylates such as aspirin when used over long periods of time; anti-inflammatory, anti-thrombosis and anti-rheumatic agents; and loop diuretics.

Employees who have any concerns about the ototoxic effects of medication should be encouraged to discuss their concerns with their doctor or pharmacist.

14.13 VIBRATION

Exposure to vibration is widespread in modern industries. Many tools, machines and vehicles such as chainsaws, jackhammers, chipping tools, tractors and earth-moving vehicles vibrate. Vibration occurs as a side effect of industrial activities or may be deliberately introduced—for example, in concrete pours, where vibration is used to shake the wet concrete into place.

Prolonged exposure to vibration causes health effects, disorders and/or disease. According to Safe Work Australia during the period 2001–2015, there were 5260 worker's compensation claims for injuries or illness attributed to exposure to vibration costing $134 million (Safe Work Australia 2015). The risk depends on the characteristics of the vibration, the part(s) of the body exposed and the duration of exposure.

14.13.1 Human Exposure to Vibration

There are basically three kinds of vibration to which workers are exposed:

- vibrations transmitted to the whole-body surface or substantial parts of it when the body is immersed in a vibrating medium such as air or water through which high-intensity sound is travelling;
- vibrations transmitted to the body as a whole through the supporting surface—for example, in vehicles, on drill platforms or in the vicinity of working machinery, where vibration is transmitted through the feet, the buttocks or the supporting area of a reclining person; and
- vibrations applied to particular parts of the body, such as the head or limbs, by vibrating handles, pedals or headrests or by hand-held power tools and appliances.

Vibrations of specific interest in the occupational environment are normally classified as either:

- whole-body vibrations, in the range of 1 to 80 Hz; or
- segmental vibrations, for example, hand-arm vibrations, in the range of 8 Hz–1 kHz.

Because the body acts as a mechanical system, its various parts resonate at various frequencies. When vibration occurs at or near any of the body's resonant frequencies, the effect on the body is greatly increased. The smaller the body part or limb, the faster it can vibrate and the higher its resonant frequency will be. For example, the head and shoulders resonate at a frequency of about 5–25 Hz, and the eyeball resonates in the range of about 30–80 Hz (Figure 14.32).

Exposure to vibration causes a number of physiological and psychological responses, which are outlined below.

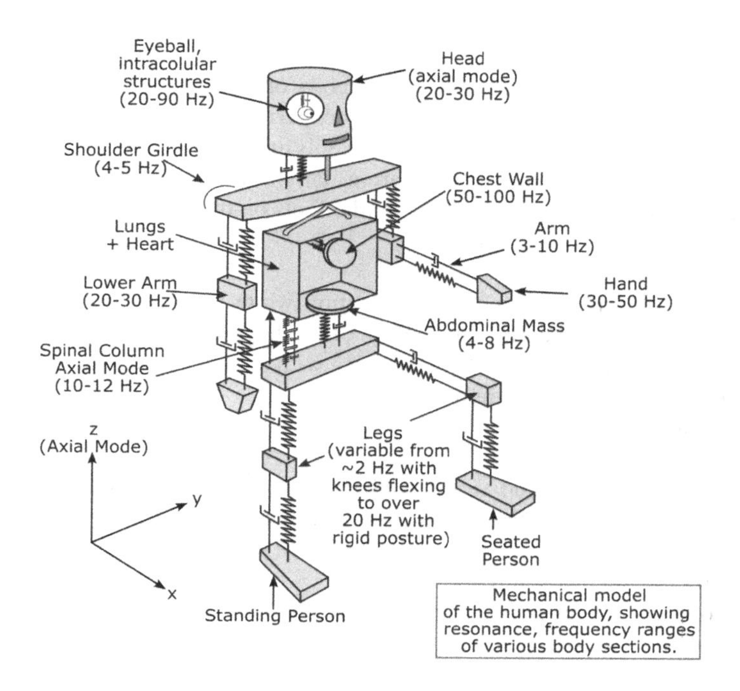

FIGURE 14.32 Frequency response ranges of different parts of the body.

Image: Copyright © Hottinger, Brüel & Kjaer, adapted from: Brüel and Kjaer, 1989, p. 8.

14.13.1.1 Whole-Body Vibration

Research has shown that the human body is most sensitive to vertical vibrations in the range 4–10 Hz. Studies by Kroemer and Grandjean (1997) show that vibrations between 2.5 and 5 Hz generate strong resonance in the vertebrae of the neck and the lumbar region. Vibrations between 4 and 6 Hz cause resonances in the trunk, shoulders and neck, and vibrations between 20 and 30 Hz set up the strongest resonances between the head and shoulders of seated persons. Other health effects and some associated frequency ranges include:

- motion sickness in the range 0.2–0.7 Hz, with the greatest effect at 0.3 Hz;
- faults in the vestibular system of the ear caused by disturbance to the inner ear's balancing system;
- visual impairment affecting the efficiency of drivers of tractors and heavy vehicles and increasing the risk of accidents, in the range 10–30 Hz;
- damage to bones and joints at frequencies below 40 Hz, especially in the lower spine—for example, ischaemic lumbago;
- problems in the digestive system;
- variations in blood pressure that may lead to cardiovascular problems;
- disorders of menstruation, internal inflammation and abnormal childbirth in women exposed to 40–45 Hz vibrations;
- increase in foetal heart rate when vibration (120 Hz) is applied to the mother's abdomen;
- fatigue, loss of appetite, irritability and headache;
- general reduced efficiency, which may lead to errors and/or accidents.

14.13.1.2 Hand-Arm Vibration

Segmental vibration affects a specific part of the body or an organ. The most prominent type of segmental vibration is hand-arm vibration, mainly experienced by operators of hand-held power tools such as jackhammers, pneumatically driven tools and chainsaws.

Vibrations caused by these types of tools are usually found in the higher frequencies, such as 40–300 Hz. In this range, vibrations may have ill effects on the blood vessels and nerve endings and blood circulation in the hands, causing:

- Raynaud's syndrome, also known as vibration white finger (VWF), or dead finger syndrome. Initially the whitening (blanching) of the fingers is localised on the tips of the fingers most exposed to the vibrating source, but with continued exposure it spreads to involve all fingers and the tips of the thumbs;
- nerve and blood vessel degeneration, resulting in loss of the sense of touch, heat perception and grip strength;
- endothelial damage to blood vessel wall elasticity resulting in fibrosis thus reducing the internal vessel diameter with subsequent restriction of blood flow;
- pain and cold sensations between attacks of VWF;
- muscle atrophy and tenosynovitis;
- damage to joints and muscles in wrists and/or elbows;
- decalcification of the carpal tunnel;
- bone cysts in fingers and wrists.

The symptoms and effects of VWF are aggravated when the hands are exposed to cold and/or the operator is a smoker.

The first symptoms are relatively mild (e.g. a tingling sensation in the fingertips) and tend to disappear when exposure ends. With continued exposure, however, symptoms become progressively more severe and eventually irreversible. Prevention is therefore a high priority.

14.13.2 Exposure Guidelines

The Model Work Health and Safety Regulation, in Part 4.2, 'Hazardous manual tasks', requires that risks to health and safety from manual tasks, including vibration, be controlled. The regulation does not give limits for vibration exposure. The Safe Work Australia 2020 Model Code of Practice on Managing Noise and Preventing Hearing Loss does state a limit, but this is primarily aimed at minimising hearing loss by providing audiometric testing of workers exposed to hand-arm vibration at any level and to noise with $L_{Aeq,8h} > 80$ dB(A) or $L_{C,peak} > 135$ dB(C). Australian Standards for the assessment and evaluation of whole-body and hand-arm vibration also do not specify limits for exposure, but they do provide information and guidelines regarding the risks of exposure level and duration. Although no vibration-specific exposure standard exists under WHS legislation, the general requirement to ensure health and safety at work also applies to vibration. Following the Code of Practice on Hazardous Manual Tasks (Safe Work Australia 2015) goes a long way towards demonstrating that proper diligence has been applied in a particular situation. In October 2015, Safe Work Australia released a series of information material on both hand-arm and whole-body vibration. The series include guides to managing risks of exposure to vibration in workplaces as well as guides to measuring and assessing workplace exposure to hand-arm and whole-body vibration. Following the guides will further assist workplaces with demonstrating compliance with their duty of care. The guidance material can be accessed and downloaded from Safe Work Australia's website. The current WHS legislation obliges manufacturers, suppliers and installers of plant and equipment to provide information that workers need to use the equipment without endangering their health and safety. Applying the control hierarchy is extremely important, as there are basically no effective options for minimising vibration exposure through PPE. An effective way of exercising proper diligence is to use international exposure guidelines, such as the European *Directive 2002/44/EC* on vibration, which provides 'daily exposure action' and 'daily exposure limit' values for both whole-body vibration and hand-arm vibration (European Union 2002). In addition, H&S practitioners should prepare a documented health and safety program, which puts it into practice and regularly reviews its effectiveness. Such a program should incorporate risk assessment and use of the relevant Australian Standards. Those applicable to human vibration are:

- AS 2670.1 Evaluation of Human Exposure to Whole Body Vibration, Part 1: General requirements (Standards Australia 2001).
- AS ISO 5349.1-2013 Mechanical vibration – Measurement and evaluation of human exposure to hand-transmitted vibration; Part 1 General requirements.
- AS ISO 5349.2-2013 Mechanical vibration – Measurement and evaluation of human exposure to hand-transmitted vibration; Part 2 Practical guidance for measurement at the workplace.

In AS ISO 5349.1:2013, vibration exposure evaluation is based on the measurement of vibration magnitude at the grip zones or tool handles and exposure times and expressed as the 'Vibration Total Value' (VTV). Measurements of the VTV have values that are greater than measurements in a single axis of up to 1.7 times (typically between 1.2 and 1.5 times) the magnitude of the greatest component. The daily vibration exposure is based on the 8 h energy equivalent acceleration value 'A8' (in AS 2763-1988—this was based on 4 h exposures). Conversion of 4 h exposures to 8 h can be done by multiplying the four-hour exposure by 0.7 to obtain the 8-hour exposure.

The Standard provides useful information in a series of appendices, called annexes.

- Annex A gives definitions for frequency weighting W_h and band limiting filters to be applied with measurements.
- Annex B gives information on health effects.
- Annex C gives guidance to competent authorities for setting exposure limits or action values.
- Annex D gives information on human response to vibration.
- Annex E gives guidance on preventative measures.
- Annex F gives guidance on uniform methods for measuring and reporting exposure of human beings to HAV.

Standards for human vibration in the workplace are expressed in terms of **acceleration** (m/s^2) and **duration of exposure** and take into account the frequency of the vibration.

The European *Directive 2002/44/EC* (the 'Vibration Directive') was published on 6 July 2002, and is the most used in Australian workplace situations to determine the acceptability or otherwise of human vibration exposure. In Safe Work Australia's guidance material, the 'Vibration Directive' is referred to as containing the most widely used and accepted exposure action value and exposure limit value. The Vibration Directive requires employers to:

- assess the risk and exposure;
- plan and implement control measures;
- provide and maintain a suitable work environment;
- provide training and information on vibration risks and their control to workers;
- monitor and review the effectiveness of the risk-control program.

Where a risk has been identified, workers have a right to health surveillance. The 2015 Safe Work Australia guides for hand-arm vibration do not mention or refer to health surveillance. As a guide, the provisions in Appendix B of the now superseded Australian Standard 2763:1988 could be consulted (Standards Australia 1988). Appendix B states:

> Segmental workers shall be medically examined.* The examination should be conducted by a medical practitioner or delegate (such as a qualified nurse who has received suitable instruction in the detection of vibration disease).
>
> * A segmental vibration worker is any person who is expected to encounter exposure of the upper limbs to vibratory motion within the frequency range of 5 Hz to 1500 Hz, at suspected weighted levels at or above 2.9 m/s^2.
>
> (*Standards Australia 1988, p 13*)

The Vibration Directive provides for 'daily exposure action' (EAV) and 'daily exposure limit' (ELV) values, which are specified as eight-hour energy equivalent frequency-weighted acceleration values and expressed as A(8). For whole-body vibration, the directive also gives 'vibration dose value' (VDV) as an alternative. Exposure values are:

For hand-arm vibration

- EAV: 2.5 m/s^2 A(8)
- ELV: 5 m/s^2 A(8)

For whole-body vibration

- EAV: 0.5 m/s^2 A(8) or VDV 9.1 m/s$^{1.75}$
- ELV: 1.15 m/s^2 A(8) or VDV 21 m/s$^{1.75}$

The above exposure values are derived from ISO 5349–1:2001 (ISO 2001) for hand-arm vibration and ISO 2631–1:1997 (ISO 1997) for whole-body vibration.

Most exposures to vibration are for less than eight hours and because the VDV is more sensitive than the root mean square acceleration to shocks with high peak accelerations, the VDV options for the EAV and ELV will often be more protective than their A(8) counterparts (Nelson & Brereton 2005).

In industries such as mining, employees commonly work 12-hour shifts. This increases the risk for workers exposed to vibration, and the increased risk needs to be accounted for. One way of doing this is by reducing the allowable exposure to vibration. For a 12-hour shift, the allowable hand-arm action value should be reduced to 2.0 m/s^2 and the exposure limit to 4.0 m/s^2. However, it is highly unlikely that a worker's 'trigger time' would last the entire 12-hour shift. For whole-body vibration, the allowable exposure action value should be reduced to 0.41 m/s^2 and the exposure limit to 0.94 m/s^2. Exposure to whole-body vibration may well persist for the whole of the 12-hour shift in some industries.

The H&S practitioner should, however, use the accepted practice of working with normalised eight-hour shifts in order to demonstrate compliance or otherwise with legislative and Australian Standard requirements.

European *Directive 92/85/EEC* provides guidance on limits for workers who are pregnant or have recently given birth or are breastfeeding. With respect to vibration exposure, the Directive states that:

> ...Work shall be organised in such a way that pregnant workers and those who have recently given birth are not exposed to work entailing any risk arising from unpleasant vibration of the entire body, particularly at low frequencies; micro-traumas, shaking, and shocks; or situations where jolts or blows are delivered to the lower body....
>
> (*European Union 2000, p 19*)

14.13.3 Vibration Measurement

The vibration must be measured in three planes—that is, Z (up-down), X (back-forward) and Y (side-to-side)—and, for this purpose, a triaxial accelerometer is normally used. One plane of vibration is usually dominant.

Common types of transducers fit onto the seat of a vehicle (whole-body vibration) or via an adaptor on the back of the hand (segmental vibration), as shown in Figure 14.33.

Hand-arm vibration is measured in the three orthogonal vibration directions, X, Y and Z, and the root sum of the squares of the three single axes is taken. Whole-body vibration is determined separately for each of the three orthogonal vibration directions at the point where vibration enters the body—for example, the buttocks or feet (see Figure 14.34).

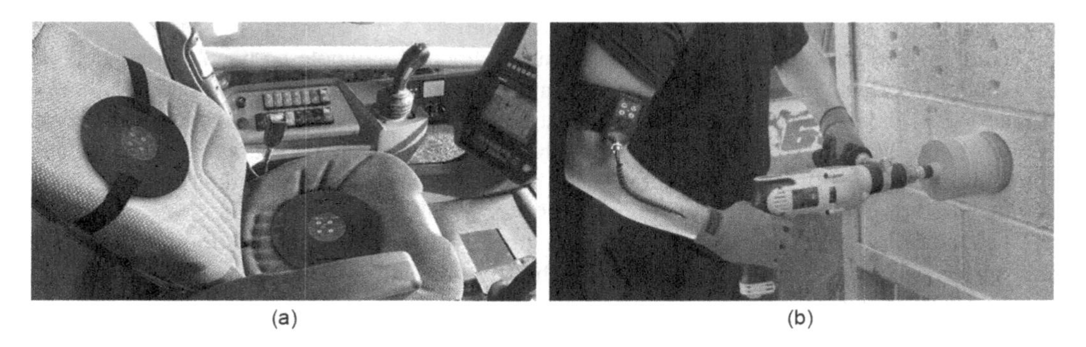

FIGURE 14.33 (a and b) Measurement of whole-body and hand-arm vibration.

Source: SVANTEK.

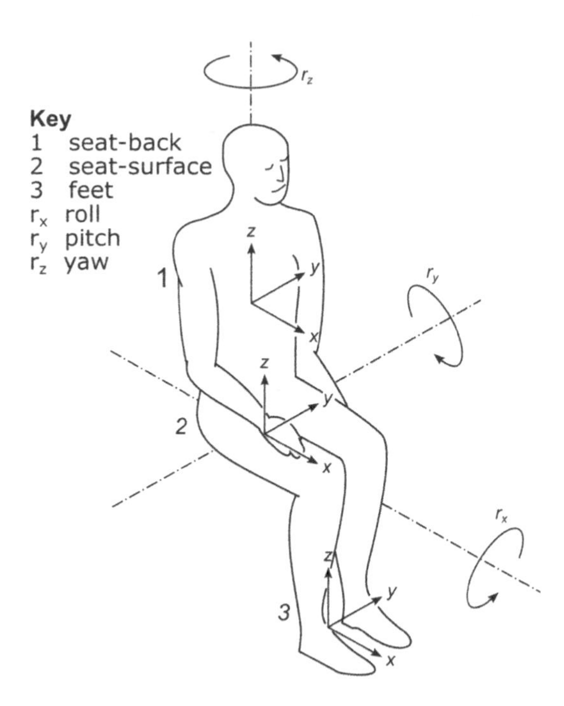

FIGURE 14.34 Whole-body orthogonal coordinate system.

Source: Reproduced Figure 1a from AS 2670.1-2001, p. 33.

14.13.3.1 Measurement of Whole-Body Vibration

ISO Standard 2631.1 (ISO 1997) and AS 2670.1 (Standards Australia 2001) indicate three criteria for the assessment of whole-body vibration in different situations in the workplace. These are:

- the effects on human health and comfort;
- the probability of vibration perception; and
- the incidence of motion sickness.

The Standard does not specify or recommend limits of exposure, but the annexes provide information on the possible effects of vibration on health, comfort and perception.

Annex B is the more important one for the assessment of whole-body vibration with respect to human health. It applies to rectilinear vibration along the X, Y and Z axes of the body for people in normal health who are regularly exposed to vibration.

Figure 14.35 shows the health guidance caution zones for whole-body vibration. For the assessment of effects on health, two relationships can be used: the average acceleration value (equation B1) and the vibration dose value (equation B2). Exposures below the zones are believed unlikely to pose a threat to health. For the zones between the dotted lines, caution is indicated, and above the zone's health risks are likely. These recommendations are, according to AS 2670.1, mainly based on exposures of four to eight hours.

The caution zone should be viewed as an 'action level' at which intervention to control the exposure is necessary. Exposures in the 'likely health risk zone' would be considered unacceptable under WHS legislation.

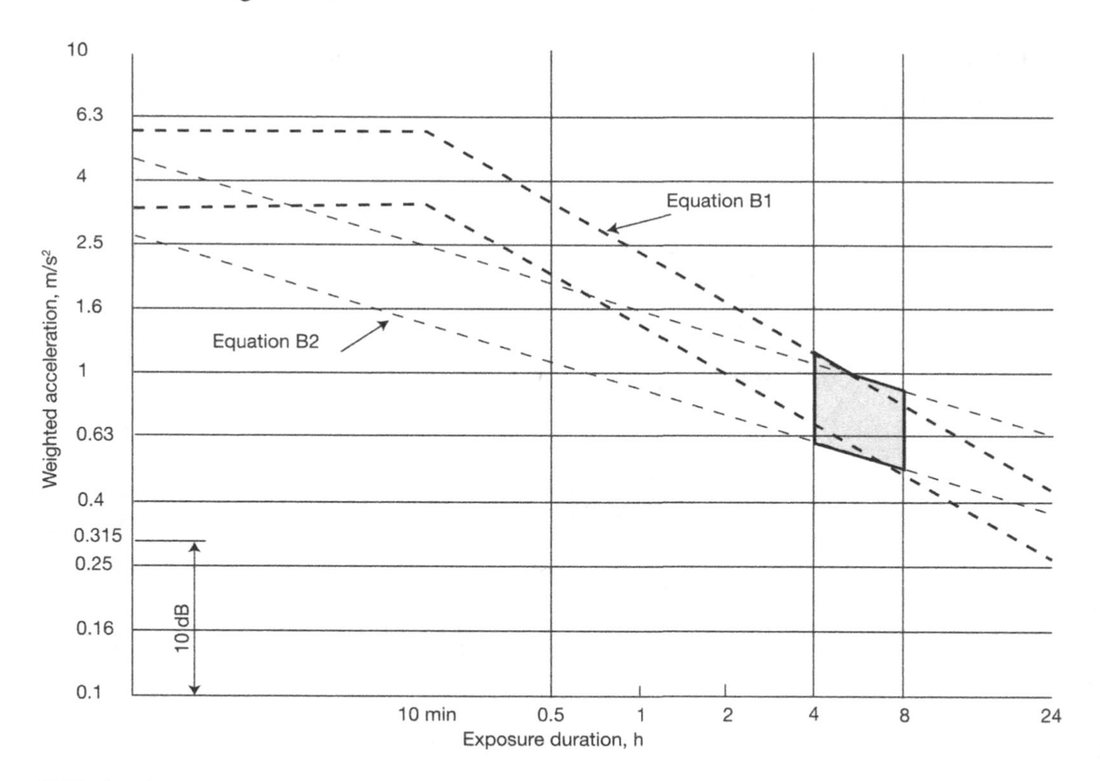

FIGURE 14.35 Health guidance caution zones.

Source: Reproduced Figure B.1 from AS 2670.1-2001, p. 36.

For vibration in more than one direction, as is typically the case with whole-body vibration, AS 2670.1 (Standards Australia 2001) suggests that the effects can be calculated by taking the vector sum, a, of the three weighted acceleration values, a_x and a_y and a_z, as follows:

$$a = \sqrt{(1.4 \times a_x)^2 + (1.4 \times a_y)^2 + (a_z)^2} \quad (14.5)$$

The actual exposure time, expressed as a percentage of the total allowed exposure time, is known as the equivalent exposure percentage.

14.13.3.2 Measurement of Hand-Arm Vibration

AS ISO 5349.1-2013 and AS ISO 5349.2-2013 (Standards Australia 2013b & c) do not provide limits for safe exposure to hand-arm vibration, but they do provide guidelines for its assessment. These Standards suggest that the directions of vibrations be measured and reported using two orthogonal coordinate systems: the basic entre system and the biodynamic system. The basic entre system refers to the tool, and the biodynamic system refers to the hand.

AS ISO 5349.1-2013 and AS ISO 5349.2-2013 (SAI Global 2013) provide a chart allowing one to predict when the first signs of VWF in workers will occur on the basis of exposure levels and duration. An example from this chart is shown in Figure 14.36.

The probability line of Figure 14.36 can be used to assess the long-term effects of eight hour-per-day exposures to hand-arm vibration. For example, the 10 per cent line predicts that exposure to 9 m/s^2 vibration acceleration will cause 10 per cent of exposed workers to reach stage 1 of VWF (blanching of and tingling sensation in fingertips that tends to disappear with cessation of exposure) in about three years. At a vibration acceleration of 2.5 m/s^2, it would take about 18 years for the same percentage of workers to reach stage 1.

A worker's daily exposure to vibration can be calculated from the intensity (magnitude) and duration of the exposure using the equation below:

$$A(8) = a_{hv}\sqrt{\frac{T}{T_0}} \quad (14.6)$$

where:

a_{hv} = vibration magnitude in m/s^2
T = the actual duration (trigger time) of the exposure
T_0 = the reference duration i.e. the normalised eight hour period

If the operation is such that the total daily exposure is made up from several exposures, then the total frequency-weighted acceleration may be calculated using the equation:

$$A(8) = \sqrt{A_1(8)^2 + A_2(8)^2 + \ldots} \quad (14.7)$$

where:

$A_1(8)$, $A_2(8)$, etc. are the partial vibration exposures from the different tools used.

The advantage of the above method is that effects from longer shifts are already taken into account with the calculation of vibration exposure to a normalised eight-hour shift period.

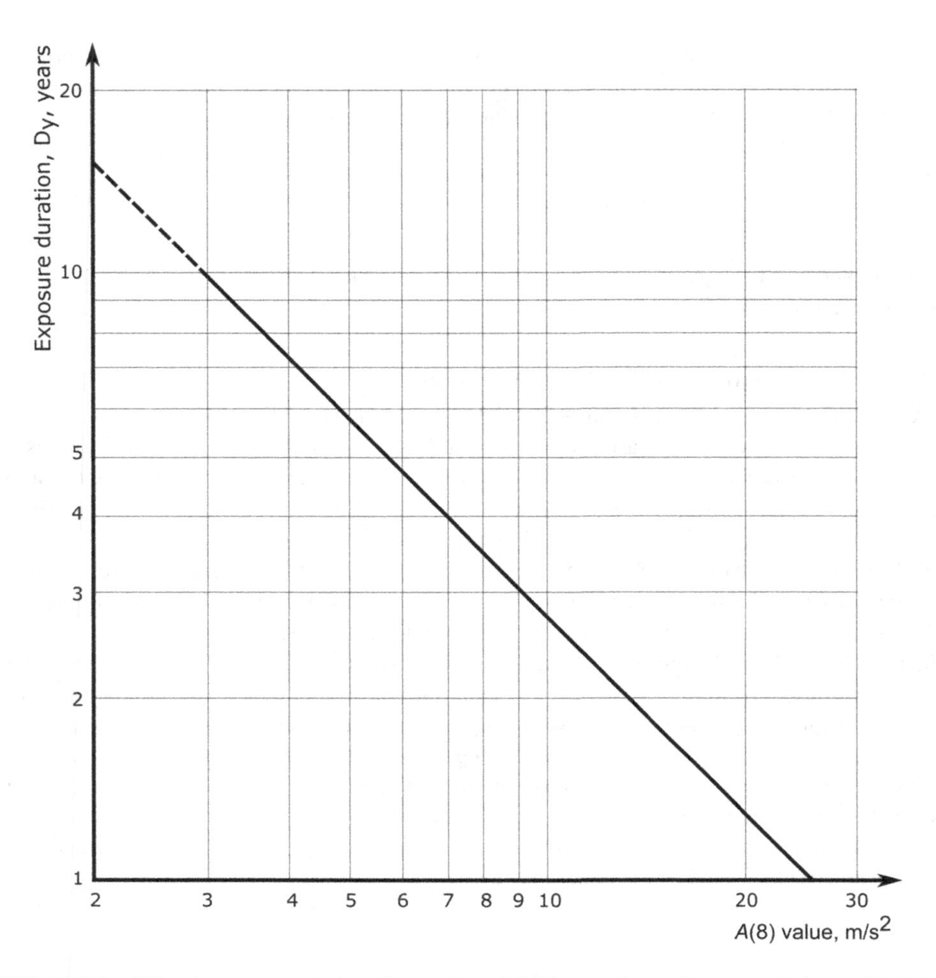

FIGURE 14.36 Vibration exposure time for projected 10% prevalence in a population suffering mild effects on the tip of the finger (white finger).

Source: Reproduced Figure C.1 from AS ISO 5349.1-2013.

14.13.3.3 Motion Sickness

One other kind of vibration, which can lead to the special problem of motion sickness, is vibration in the frequency range of 0.1–0.63 Hz in the vertical Z axis. In this range, a significant proportion of unaccustomed people will experience discomfort, depending on frequency and exposure times.

14.13.4 Control Measures

Control of vibration usually requires a combination of appropriate tool selection, good work practices and education programs, as well as medical surveillance. It also requires identification of the hazards through vibration measurements and reduction of vibration at the source or transmission. Where vibration is a problem, workers should be warned of the hazards of vibrating tools, and medical supervision should be employed to identify those workers showing early signs of adverse health effects or reversible VWF.

Some considerations for control measures for whole-body vibration include:

- ensuring that all on-site road and work-area surfaces are well maintained to minimise rough rides;
- ensuring that a traffic management system that incorporates speed limits operates on-site;
- ensuring that drivers of vehicles are familiar with the on-site road conditions;
- insulation of seat and head-rest vibration through springs and dampers;
- installation of vibration-dampening seats (suspension seats) that allow correct adjustment for the handling of the vehicle and do not interfere with visibility from the cab;
- implementing a seat maintenance system to ensure that suspension seats are regularly checked and maintained in a serviceable condition;
- ensuring that only vehicles suitable for the job, and which provide driver comfort, are used;
- ensuring that cab layouts are such that drivers do not have to adopt awkward and potentially damaging postures;
- ensuring that regular mini breaks are incorporated in shifts;
- where required, and practicable, providing special boots with vibration-absorbing soles to protect against vibration through the floor;
- limiting the time spent by workers on vibrating surfaces;
- ensuring that plant and equipment are well maintained;
- mounting machines and plant on vibration-isolating mounting pads;
- some control measures for hand-arm vibration include:
  - isolating the vibrations—for example, by using special mountings and/or adjusting the centre of gravity as low as possible; and
  - damping of the vibrations, for example, by wearing padded gloves, provided this does not lead to an increase in clamping force (which results in increased transmission of vibrations).

The adverse effects can also be minimised if the operator's hands are kept warm or the handles of the vibrating tools are warmed in cold work situations.

REFERENCES

Access Economics 2006, *Listen hear! the economic impact and cost of hearing loss in Australia*, Sydney.

Australian Communications Industry Forum 2013, *Industry Guideline—acoustic safety for telephone equipment*, ACIF G616, ACIF, Sydney, https://www.commsalliance.com.au/__data/assets/pdf_file/0004/38956/G616_2013.pdf [30 December 2023].

Australian Government Department of Health and Aged Care 2022, *About ear health*, https://www.health.gov.au/topics/ear-health/about#ear-health-in-australia [30 December 2023].

Berger, E., 2000, *The Noise Manual*, 5th edn, American Industrial Hygiene Association, Fairfax, VA

Bernabei, R., Bonuccelli, U., Maggi, S., Marengoni, A., Martini, A., Memo, M., Pecorelli, S., Peracino, A. P., Quaranta, N., Stella, R., & Lin, F. R., 2014, 'Hearing loss and cognitive decline in older adults: Questions and answers.' *Aging Clinical and Experimental Research*, vol. 26, no. 6, pp. 567–573. https://doi.org/10.1007/s40520-014-0266-3

Brüel, P. & Kjaer, V. 1989, *Human Vibration*, BR 0456–12, Brüel & Kjaer, Nærum, Denmark, https://www.scribd.com/document/138618880/Bruell-Kjaer-Human-Vibration-br056 [5th October 2023].

Centres for Disease Control and Prevention 2020, *How Does Loud Noise Cause Hearing Loss*, https://www.cdc.gov/nceh/hearing_loss/how_does_loud_noise_cause_hearing_loss.html [30 December 2023]

Deloitte Access Economics 2017, *Social and Economic Costs of Hearing Health in Australia, Canberra*, https://www.hcia.com.au/hcia-wp/wp-content/uploads/2017/08/Social-and-Economic-Cost-of-Hearing-Health-in-Australia_June-2017.pdf [30 December 2023].

Eden, D. & Piesse, R. 1991, 'Real World Attenuation of Hearing Protectors', *Proceedings of the 4th Western Pacific Regional Acoustic Conference*, Brisbane, pp. 508–13

European Telecommunications Standards Institute 2000, *Acoustic Shock from Terminal Equipment (TE): An investigation on standards and approval documents*, ETSI TR 101 800 V1.1.1 (2000–07), ETSI, Valbonne, France, www.etsi.org/deliver/etsi_tr/101800_101899/101800/01.01.01_60/tr_101800v010101p.pdf [26 September, 2023].

European Union 2000, *Directive 92/85/EEC—Pregnant workers*, European Agency for Safety and Health at Work, Brussels, https://osha.europa.eu/en/legislation/directives/sector-specific-and-worker-related-provisions/osh-directives/10 [26 September, 2023].

European Union 2002, *Directive 2002/44/EC—Vibration*, European Agency for Safety and Health at Work, Brussels, https://osha.europa.eu/en/legislation/directives/exposure-to-physical-hazards/osh-directives/19 [30 December 2023].

Grantham, D. 1992, *Occupational Health & Hygiene: Guidebook for the WHSO*, D.L Grantham.

Groothoff, B. 2015, *Occupational Noise Management*, Brüel & Kjaer, Sydney.

Huang, L., Zhang, Y., Wang, Y., & Lan, Y., 2021, 'Relationship between chronic noise exposure, cognitive impairment, and degenerative dementia: Update on the experimental and epidemiological evidence and prospects for further research', *Journal of Alzheimer's Diseases.*, vol. 79, no. 4, pp. 1409–1427. doi: 10.3233/JAD-201037.

International Organization for Standardization, 1997, *Mechanical Vibration and Shock—evaluation of human exposure to whole body vibration—Part 1: general requirements, ISO 2631-1:1997*, ISO, Geneva, https://www.standards.org.au/access-standards

International Organization for Standardization, 2001, *Mechanical Vibration: Guidelines for the movement and the assessment of human exposure to hand-transmitted vibration*, ISO 5349:2001, ISO, Geneva, https://www.standards.org.au/access-standards

Jastreboff, P.J. & Hazell, J.W.P. 2004, *Tinnitus Retraining Therapy. Implementing the Neurophysiological Model.* Cambridge University Press, https://assets.cambridge.org/97805215/92567/frontmatter/9780521592567_frontmatter.pdf [30 December 2023].

Kroemer, K.H.E. & Grandjean, E. 1997, *Fitting the Task to the Human: A Textbook for Occupational Ergonomics*, Francis & Taylor, London.

Ministry of Business, Innovation and Employment (MBIE) 2013, Classified Hearing Protectors, NZ Ministry of Business, Innovation and Employment, https://healthandsafetybydesign.co.nz/wp-content/uploads/2019/02/classified_hearing_protectors.pdf [30 December 2023].

Nelson, C.M. & Brereton, P.F. 2005, 'The European vibration directive', *Industrial Health*, vol. 43, pp. 472–9

Occupational Health and Safety Administration (OSHA) 1980, *Noise Control: A guide for workers and employers*, U.S. Department of Labor, Washington, DC, https://www.nonoise.org/hearing/noisecon/noisecon.htm [26 September, 2023].

Pretzsch, A., Seidler, A. & Hegewald, J., 2021, 'Health effects of occupational noise', *Current Pollution Report*, vol.7, pp. 344–358. https://doi.org/10.1007/s40726-021-00194-4

Safe Work Australia (SWA) 2010, *Occupational Noise-Induced Hearing Loss in Australia: Overcoming Barriers to Effective Noise Control and Hearing Loss Prevention*, SWA, Canberra, https://www.safeworkaustralia.gov.au/system/files/documents/1702/occupational_noiseinduced_hearing_loss_australia_2010.pdf [30 December 2023].

Safe Work Australia (SWA) 2014, *Occupational Disease Indicators*, SWA, Canberra, https://www.safeworkaustralia.gov.au/system/files/documents/1702/occupational-disease-indicators-2014.pdf [30 December 2023].

Safe Work Australia (SWA) 2015, *Media Release – Safer Work Australia Releases Workplace Vibration Guidance Material, Friday 2 October 2015*, SWA, Canberra.

Safe Work Australia (SWA) 2020, *Model Code of Practice-Managing Noise and Preventing Hearing Loss at Work*, SWA, Canberra, https://www.safeworkaustralia.gov.au/doc/model-code-practice-managing-noise-and-preventing-hearing-loss-work [30 December 2023].

Safe Work Australia (SWA) 2023, *Law and Regulation*, SWA, Canberra, https://www.safeworkaustralia.gov.au/law-and-regulation [30 December 2023].

Standards Australia 1988, *Vibration and Shock-hand-Transmitted Vibration-Guidelines for the Measurement and Assessment of Human Exposure*, AS 2763:1988, SAI Global, Sydney, https://www.standards.org.au/access-standards

Standards Australia 2001, *Evaluation of Human Exposure to Whole Body Vibration, Part 1: General Requirements, AS 2670.1:2001*, SAI Global, Sydney, https://www.standards.org.au/access-standards

Standards Australia 2002, *Acoustics—Hearing Protectors*, AS/NZS 1270:2002, SAI Global, Sydney, https://www.standards.org.au/access-standards

Standards Australia 2004, *Electroacoustics—Sound Level Meters, Part 1: Specifications*, AS IEC 61672.1:2004, SAI Global, Sydney, https://www.standards.org.au/access-standards

Standards Australia 2013a, *Voice Performance Requirements for Customer Equipment*, AS/CA S004:2013, SAI Global, Sydney, https://www.standards.org.au/access-standards

Standards Australia 2013b, *Mechanical Vibration – Measurement and Evaluation of Human Exposure to Hand-Transmitted Vibration; Part 1 General Requirements*, AS ISO 5349.1-2013, SAI Global, Sydney, https://www.standards.org.au/access-standards

Standards Australia 2013c, *Mechanical Vibration – Measurement and Evaluation of Human Exposure to Hand-transmitted Vibration; Part 2 Practical Guidance for Measurement at the workplace*, AS ISO 5349.2-2013, SAI Global, Sydney, https://www.standards.org.au/access-standards

Standards Australia 2018, *Acoustics—description and Measurement of Environmental Noise*, AS 1055:2018, SAI Global, Sydney, https://www.standards.org.au/access-standards

Standards Australia and New Zealand Standards 2005a, *Occupational Noise Management, Part 1: Measurement and Assessment of Noise Immission and Exposure*, AS/NZS 1269.1:2005, SAI Global, Sydney, https://www.standards.org.au/access-standards

Standards Australia and New Zealand Standards 2005b, *Occupational Noise Management, Part 3: Hearing Protector Program*, AS/NZS 1269.3:2005, SAI Global, Sydney, https://www.standards.org.au/access-standards

Standards Australia and New Zealand Standards 2005c, *Occupational Noise Management, Part 0: Overview*, AS/NZS 1269.0:2005, SAI Global, Sydney, https://www.standards.org.au/access-standards

Standards Australia and New Zealand Standards 2005d, *Occupational Noise Management, Part 2: Noise Control Management*, AS/NZS 1269.2:2005, SAI Global, Sydney, https://www.standards.org.au/access-standards

Standards Australia and New Zealand Standards 2014, *Occupational Noise Management, Part 4: Auditory Assessment*, AS/NZS 1269.4:2014, SAI Global, Sydney, https://www.standards.org.au/access-standards

Teixeira, L. R., Pega, F., Dzhambov, A. M., Bortkiewicz, A., da Silva, D. T. C., de Andrade, C. A. F., Gadzicka, E., Hadkhale, K., Iavicoli, S., Martínez-Silveira, M. S., Pawlaczyk-Łuszczyńska, M., Rondinone, B. M., Siedlecka, J., Valenti, A., & Gagliardi, D., 2021, 'The effect of occupational exposure to noise on ischaemic heart disease, stroke and hypertension: a systematic review and meta-analysis from the who/ilo joint estimates of the work-related burden of disease and injury,' *Environment International*, vol. 154. https://doi.org/10.1016/j.envint.2021.106387

Workplace Health and Safety Queensland, 2019, *Personal hearing protectors – protecting your hearing*. https://www.worksafe.qld.gov.au/__data/assets/pdf_file/0022/15880/perhearpro_hear.pdf [25 September 2024]

15 Radiation-Ionising and Non-ionising

Dr Geza Benke and Dr Martin Ralph

15.1 INTRODUCTION

Radiation is energy travelling in the form of either electromagnetic waves or particulates (high-speed particles). Radiation is ionising when it has sufficient energy to remove an electron from an atom of a molecule (Table 15.1).

Electromagnetic waves travel in air and vacuum at the speed (v) of light, 3.0×10^8 metres per second (m/s), and are characterised by their wavelength (l, or λ, in metres) and frequency (f, in cycles per second, or hertz (Hz)), as shown in Equation 15.1. Wavelength and frequency are inversely proportional:

$$f = \frac{v}{\lambda} \tag{15.1}$$

Radiation made up of high-speed particles travels at a range of speeds and there are a range of different particle types (e.g. alpha radiation travels at approximately one-twentieth the speed of light and beta radiation travels at close to the speed of light).

The first half of this chapter discusses ionising radiation, its common sources, and its quantification, health effects, measurement and controls. The second half discusses non-ionising radiation, which includes ultraviolet, visible, infrared and radiofrequency (RF) radiation and extremely low-frequency and static electric and magnetic fields. Lasers are also included as a special application of visible, infrared and ultraviolet radiations.

TABLE 15.1
Radiation

	Electromagnetic Wave	Particle
Ionising	Gamma and X-ray	Alpha, beta and neutrons
Non-ionising	Ultraviolet, visible, infrared, radiofrequency radiation; and extremely low-frequency and static electric and magnetic fields	

DOI: 10.1201/9781032645841-15

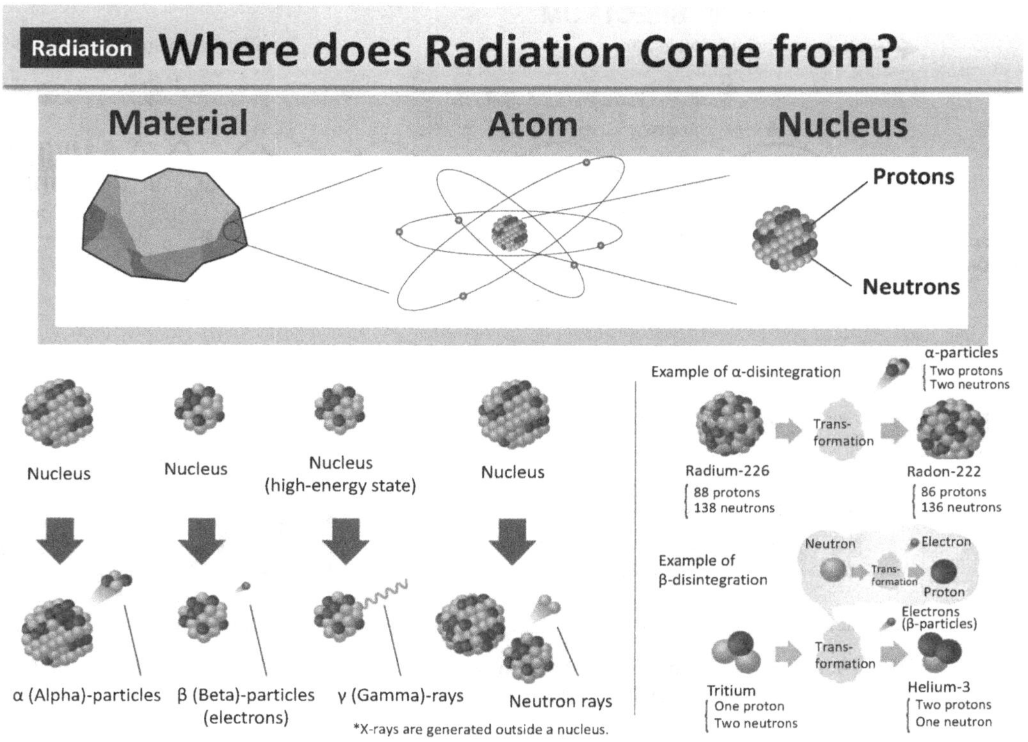

FIGURE 15.1 Types of ionising radiation.

Source: Ministry of the Environment website 2022, p.13, https://www.env.go.jp/en/chemi/rhm/basic-info/1st/pdf/basic-1st-vol1.pdf.

15.2 IONISING RADIATION

There are two types of ionising radiation: electromagnetic waves or particles. Figure 15.1 shows these.

Ionising radiation occurs naturally from primordial radionuclides such as uranium-238 (^{238}U), thorium-232 (^{232}Th) or potassium-40 (^{40}K) and also arises from human-made sources. It has many practical uses. Sealed radiation sources and X-rays are used in industrial and medical applications, and unsealed sources are used in medical, research, and radiation monitoring applications. Overexposure to ionising radiation can cause a range of adverse health effects, including increasing the risk of cancer.

15.2.1 Electromagnetic Ionising Radiation

X-rays and gamma rays are electromagnetic radiation and can be thought of as photons. These types of radiation have the highest energy in the electromagnetic spectrum (see Figure 15.2). Unlike the lower-energy electromagnetic radiations, such as visible light, they have enough energy to cause ionisation of stable atoms.

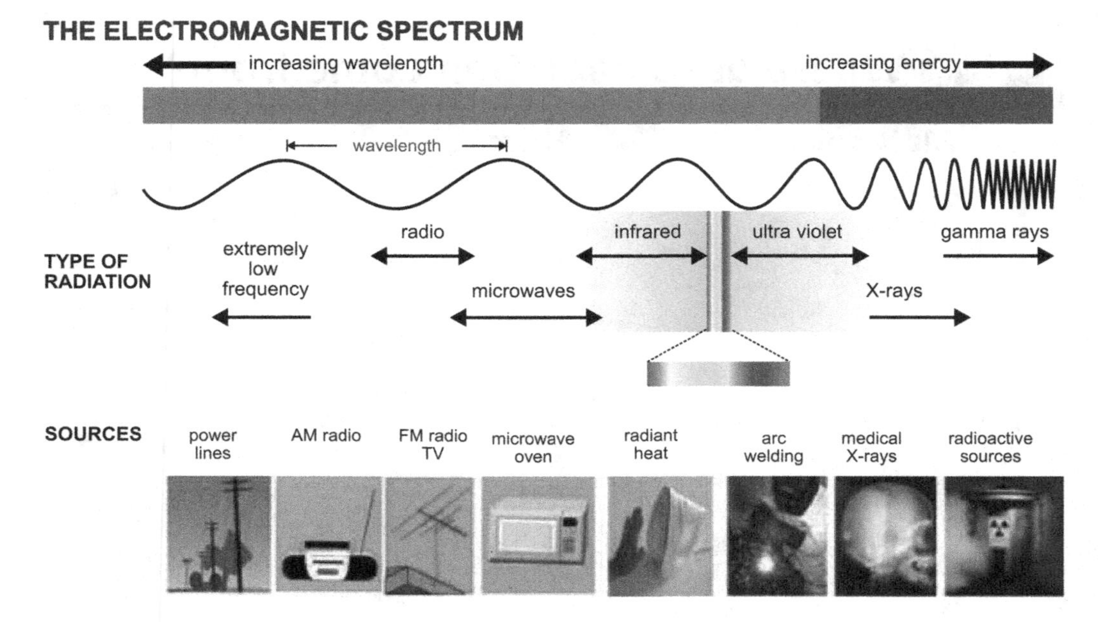

FIGURE 15.2 Electromagnetic spectrum.

Source: ARPANSA, n.d.-a, b; © Commonwealth of Australia as represented by the Australian Radiation Protection and Nuclear Safety Agency (ARPANSA).

15.2.2 Particulate Ionising Radiation

There are also highly energetic particulate forms of ionising radiation. The most common types used in industrial and biomedical research applications are alpha and beta particles. Neutron radiation, another form of particulate radiation, is typically used in specialist applications such as physics research, neutron imaging and environmental monitoring. The most common industrial application is moisture-density gauges.

15.2.3 International and National Legislative Framework

The International System of Radiological Protection has been developed by the International Commission on Radiological Protection (ICRP), which publishes dose limits and many other ionising-radiation safety publications. Many international standards on ionising radiation, such as the Basic Safety Series, are also developed by the United Nations' International Atomic Energy Agency (IAEA).

The International System of Radiological Protection defines three principles:

- **The justification of practice**: No practice involving exposure to ionising radiation should be adopted unless the benefit outweighs the potential or actual harm. In practical terms, when considering the use of ionising radiation in a procedure or activity, the benefits must outweigh the risks.
- **The optimisation of protection** (the ALARA principle): The size of individual doses, the number of people exposed and the likelihood of incurring exposure should be kept as low as reasonably achievable (ALARA), with economic and social factors being taken into account. ICRP has defined dose constraints and reference levels to assist in the application of this principle.

- **Dose limitations**: The total dose to any individual from regulated sources in planned exposure situations other than medical exposure of patients should not exceed the recommended dose limits (ICRP 2007). Planned exposure situations involve the deliberate introduction and operation of radiation sources.

In Australia, the Australian Radiation Protection and Nuclear Safety Agency (ARPANSA) provides guidance material such as radiation protection advice and guidelines. ARPANSA also regulates Commonwealth entities such as the Australian Nuclear Science and Technology Organisation (ANSTO) and the Commonwealth Scientific and Industrial Research Organisation (CSIRO). State and territory governments regulate non-Commonwealth entities such as hospitals, universities and industry, based on model regulations and recommendations from ARPANSA. Other countries have National Regulators. Harmonisation across jurisdictions in Australia is achieved via all states and territories being signatories to the National Directory for Radiation Protection (2nd Edition, 2021).

ARPANSA bases its regulations and recommendations on the ICRP's International System of Radiological Protection, as well as on documentation and data from the IAEA (to which Australia is a signatory). Australia's current dose limits are based on the dose limits recommended by the ICRP.

15.2.4 Sources of Ionising Radiation

Elements are characterised by the number of protons in the nuclei of their atoms. Atomic nuclei also contain neutrons. When two atoms of the same element contain different numbers of neutrons, the atoms are called isotopes. Examples of naturally occurring isotopes are: carbon-12 (6 protons and 6 neutrons), carbon-13 (6 protons and 7 neutrons) and carbon-14 (6 protons and 8 neutrons). Some isotopes are stable and some are unstable—for example, carbon-12 and carbon-13 are stable and carbon-14 is unstable. An atom is unstable when there is an imbalance of energy or mass (e.g. when it gains an extra neutron and does not have enough energy to bind its neutrons together). Radioactive decay occurs when an unstable atom emits radiation. For example, carbon-14, which has a half-life of 5,700 years (refer to Section 15.2.8.2) emits a beta particle accompanied by a 156.5 keV gamma ray as it decays. Unstable atoms are commonly called radioisotopes or radionuclides.

Often when radioisotopes decay, they do so in multiple stages, with a number of radioactive progeny products arising before a stable end-form is reached. This is called a decay chain.

Radioisotopes can occur naturally in the environment or can be created using equipment such as an X-ray tube, a cyclotron or a nuclear reactor.

15.2.4.1 Background Radiation from Natural Sources

All human beings are exposed to natural background radiation.

- **Cosmic radiation** comes from outer space and the sun, but the earth's atmosphere acts as a shield, so that at sea level the cosmic radiation level is very low.
- **Naturally occurring radioactive material** (NORM) is distributed in the soil, air and water. It ends up in human beings, in food and in building materials and other objects. For example, potassium-40 (gamma emitter) is naturally elevated in bananas because they have a high total potassium level. Uranium-238 and thorium-232 (mainly gamma and alpha emitters) are found in varying concentrations in soil and rocks.

Another example of elevated NORM levels is seen in the Darling Scarp, in south-western Western Australia. The geology is particularly rich in the alpha-radiation emitters uranium-238 and thorium-232, which results in an average background radiation dose from terrestrial sources that is twice as high as in other areas of Australia.

Some other areas of the world have high natural levels of NORM. Brazil's black beach sands contain a range of naturally occurring radioisotopes. These can result in an average background radiation dose from terrestrial sources up to 100 times higher than elsewhere (Gonzalez & Anderer 1989).

The naturally occurring radioisotope radon-222 gas (alpha emitter) and the short-lived radioactive products of its decay (called radon progeny) are generated when radium-226, within building materials and soil, undergoes radioactive decay. Radium-226 is a member of the uranium-238 decay chain, which can be found in trace concentrations in most building materials. Radon gas can accumulate inside poorly ventilated buildings, especially in areas below the ground (basements, cellars or underground mines) and can be readily breathed in and its progeny particulates deposited within the body.

Radon and radon progeny are recognised by the International Agency for Research on Cancer (IARC 2012) as a cause of lung cancer. In 2017, ICRP re-evaluated its estimates of lung cancer risk for radon and its progeny – almost doubling the risk. Radon effective dose coefficients (the dose per unit of intake) are now calculated using biokinetic and dosimetry models with specific radiation and tissue weighting factors. As a result, radon is now a significant contributor to our natural background radiation dose (ICRP 2014).

Although radon levels are generally much lower in Australia than in many other countries for a range of reasons, including geology, climate (no snow cover) and building styles (fewer residential basements). In Australia, there are only a few areas where radon levels are likely to build up. These include underground and uranium mines, and some caves in Victoria, Tasmania and New South Wales that are popular tourist sites. These do not significantly increase the dose for casual visitors but could significantly raise doses for cave-tour guides (Solomon et al. 1996; ARPANSA 2018). An Australian action level for residential indoor radon is 200 Bq/m^3, corresponding to an effective dose of approximately 10 mSv per year (ARPANSA 2018).

15.2.4.2 Background Radiation from Human-Made Sources

A human-made source of ionising radiation can be generated either by concentrating naturally occurring radioisotopes or by using special apparatus such as an X-ray tube or a cyclotron.

Large-scale releases of ionising radiation such as occurred in the Chernobyl nuclear accident, the atomic bombing of Japan and various nuclear tests have added to the background radiation to which all humans are exposed, though the additional exposure from these sources is small (United Nations Scientific Committee on the Effects of Ionising Radiation 2008).

NORM may be found in minerals, oil and gas. Although it may initially occur at low levels, the refined product, the process line and waste (especially corrosion scale in steel pipes and vessels) may contain higher levels of radioactivity. Examples include thorium-232 and uranium-238 in the mining and processing of mineral sands, and radium-226 (and associated radon progeny) in oil and gas production (Cooper 2005). Contamination of equipment used to process NORM can become a problem for disposal and personal exposure during its operation and at demolition.

15.2.4.3 Categories of Human-Made Radiation Sources

There are three physical forms of human-made ionising radiation:

- **Radiation apparatus, or a radiation generator**, is equipment that generates radiation. Examples include X-ray machines, cyclotrons and neutron generators (ICRP now uses the term radiation generator).
- **Sealed sources and sealed-source apparatus** refer to any quantity of radioisotope whose physical form is enclosed as to prevent the escape of any of the radioisotope (but not the radiation). Some examples are:
  - electron capture detectors in gas chromatographs, which may use nickel-63 (beta emitter);
  - density, level and thickness gauges, which use a variety of radioactive sources such as americium-241 (alpha emitter), caesium-137 and cobalt-60 (both gamma emitters);

- bore-hole loggers which use americium-241 in beryllium to cause fission, producing neutrons which are then used to measure moisture;
- smoke detectors, which may use foil containing americium-241 (alpha emitter).
- **Unsealed sources** are usually in a liquid or powder form and may readily escape into the environment if they are not carefully contained. They are most often used for nuclear medicine and scientific research. Some examples are:
  - iodine-125 (gamma emitter) for labelling peptides in research;
  - sulfur-35 (beta emitter) for labelling steroids in research;
  - technetium-99m (gamma emitter) for imaging studies in nuclear medicine;
  - phosphorus-32 (high-energy beta emitter) and phosphorus-33 (low-energy beta emitter) for labelling DNA in research.

It is useful to classify sources encountered within the workplace based on their form, because these forms influence the type of hazard presented by the source.

15.2.5 Exposure Categories

Everybody is exposed to background radiation, but ionising radiation is widely used in medical applications and research, as well as industry. For this reason, the ICRP defines three exposure categories: public, medical and occupational.

Public exposure is primarily from background sources such as those discussed above and varies based on the local environment and dietary habits. The annual average dose from public exposure to background radiation in Australia (excluding medical and work-related sources) is 1500 microsieverts per year (μSv/year – units are discussed in Section 15.2.9.2) plus an increase in radon exposure (ARPANSA 2012c, 2015, 2018).

Medical exposure is experienced by all persons who are treated using ionising radiation. Radioisotopes may be used for treating cancer, because at high doses, their ionising radiation is effective at destroying rapidly dividing cells, and because it can be directed to selectively irradiate specific tissues or organs. For example, a beam of external gamma rays emitted by a cobalt-60 source can be aimed at the site of an internal tumour, or a small iodine-125 source can be implanted in the tongue to treat tongue cancer. Radioisotopes and X-ray generators are used in a wide range of diagnostic and treatment procedures, such as the injection of technicium-99m (gamma emitter) for brain scans or iodine-131 (gamma emitter) for thyroid scans.

Occupational exposure occurs among people who are required to work with sources of ionising radiation. Work-related doses vary depending on the work. Table 15.2 shows a comparison of work-related annual doses for workers in various occupations in Australia. The majority of these occupational doses are significantly below the annual average dose from public exposure in Australia.

Aircrew are considered to be occupationally exposed to ionising radiation because their cosmic radiation exposure increases as the atmospheric density decreases at altitude. Aircrew thus have a significantly higher exposure over the course of a year than their grounded counterparts. Crews who routinely fly domestic routes receive about double the annual average background dose of the general public, while crews on international routes, who fly higher and stay in the air longer, receive more than three times this dose (ARPANSA 2012b).

15.2.6 Properties of Ionising Radiation

15.2.6.1 Interaction with Matter

Alpha and beta particles, neutrons, gamma rays and X-rays are known as ionising radiation because they have sufficient energy to ionise the medium through which they travel. These types of radiation remove electrons from atoms to produce positive ions.

TABLE 15.2
Typical Average Work-related Radiation Doses for Workers in Various Occupations in Australia in 2004

Occupation	Average Whole-body Dose (μSv/year)
Radiologists in large hospitals	108
Dentists in private practice	12
Uranium miners	1,125
Users of radioactive tracers in research	31
Undergraduate students in tertiary education	19
Note: the occupational exposures are in addition to background radiation	
Background	
Australia – average natural radiation (incorporating an increase in Radon exposure of 300 μSv/year)	1,800
World – natural radiation	1,000–13,000

Adapted from: ARPANSA 2015, 2018; Morris et al. 2004.

The charge carried by **alpha** and **beta** particles allows them to interact with target atoms. Alpha particles have a double positive charge and beta particles have a negative charge. When charged particles come close to one another, they are either repulsed if they have the same type of charge or attracted if they have opposite charges. Whenever these attractions or repulsions occur, energy is transferred from alpha and beta radiations to target atoms, causing ionisation.

Gamma rays and **X-rays** are electromagnetic waves and do not have any mass or charge, so they cannot directly ionise other atoms via interaction with charged fields. However, through various complex interactions, they can transfer their energy to an atom's orbital electrons and thereby ionise the atom.

Neutrons have mass but they have no charge. Like gamma and X-rays, they do not cause ionisation by direct interaction with an atom's charged electron fields. Instead, they cause ionisation indirectly, by interacting with the nucleus of the atom. By this process, neutrons can cause stable materials to become radioactive. Neutrons are the only type of ionising radiation able to do this.

15.2.6.2 Penetrating vs. Ionising Ability

The penetrating and ionising abilities of the different types of ionising radiation vary. Penetration ability refers to the distance a type of radiation can travel, and ionising ability refers to the degree of ionisation it can cause as it travels. The greater the penetrating ability, the greater the risk of interaction with the human body, and as ionising ability increases, so does biological damage.

Alpha particles (helium nuclei) consist of two protons and two neutrons tightly bound together. They are heavy, slow-moving (around one-twentieth the speed of light) and have a positive charge of 2. They cause a lot of ionisation over a short distance and usually travel only a few centimetres in air. Their large mass and charge means that alpha particles interact easily with the matter through which they pass, but each interaction reduces the particles' energy, until they no longer have enough energy to cause ionisation.

Beta particles are electrons that travel at close to the speed of light. They have a very low mass and a negative charge of 1. Because they have half the charge and much less mass than alpha particles, they cause less ionisation over the same travel distance, and typically travel further before they use up their energy. Beta particles are emitted from the nucleus during radioactive

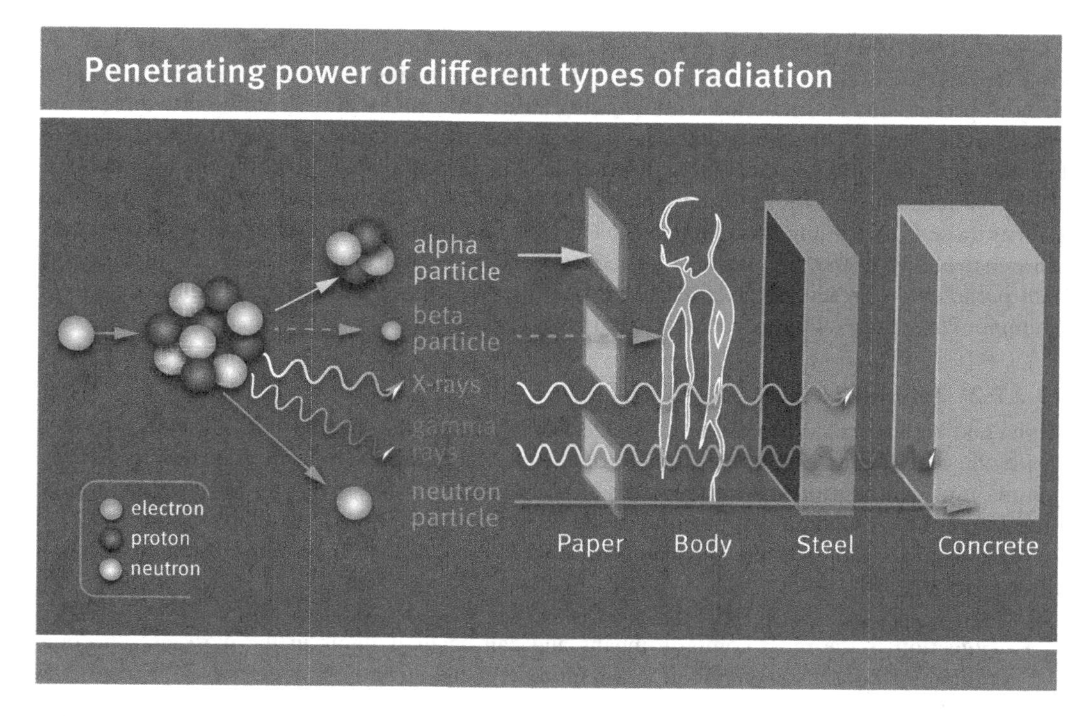

FIGURE 15.3 Penetrating and ionising ability of ionising radiation.

Source: UNEP 2016, p9.

decay with a spectrum of energies ranging from close to zero to a maximum energy that is characteristic of the radioisotope. Low-energy beta radiation such as that from tritium (hydrogen-3) travels less distance in air than most alpha particles. Medium-energy beta particles typically travel up to 3 metres in air, but the distance varies according to their energy. Phosphorus-32 is a high-energy beta emitter, and the highest-energy beta particles it emits may travel up to 7 metres in air.

Neutron particles are fast-moving neutrons with a spectrum of energies and speed. They have a variable ionising capability, which is dependent on their energy. They may be more ionising than alpha and beta particles, gamma rays or X-rays. They may travel many metres in air before stopping completely. They are more penetrating than alphas and betas, as they do not interact with the charged fields of atoms as they pass through matter.

X-rays and gamma rays are the least ionising of all types of ionising radiations. They are generally the most penetrating because they do not interact with the charged fields of atoms as they pass through matter. They will travel very large distances through matter—they typically travel many metres in air.

Penetrating and ionising abilities for all types of ionising radiation are summarised in Figure 15.3.

15.2.7 External and Internal Hazard

Unlike many other hazards, ionising radiation can present a risk when it is inside the body, when it is outside the body, and even if it is at some distance from the body. The degree of internal and/or external hazard is different for the different types of radiation because it depends on their penetrating and ionising ability.

15.2.7.1 External Hazard

An external radiation hazard is present when a source of ionising radiation is located outside the body. A person may be irradiated without being aware of it, and without coming into direct contact with the source. The greatest external radiation hazards are X-rays, gamma rays and neutrons because of their ability to travel large distances and penetrate matter, including the human body. Alpha particles, such as those from uranium-238 and thorium-232, and low-energy beta particles, such as those from tritium, do not represent a significant external radiation hazard, even if they are close to the body, as they are unable to penetrate the outer layer of skin. Medium- and high-energy beta particles do present an external hazard, as they are more energetic and penetrating—for example, Phosphorus-32 may cause skin burns and eye damage.

15.2.7.2 Internal Hazard

Alpha and beta radiation represent the greatest internal radiation hazards when they are emitted inside the body because they are capable of causing intense ionisation in a local area. X-rays, gamma rays and neutrons cause less ionisation, but because they are more penetrating, a large proportion of the radiation will pass out of the body without causing ionisation.

Radioisotopes can enter the body and cause an internal radiation exposure by three different exposure pathways:

- **Ingestion** is the most common means and can occur if items in the work area become contaminated. This contamination can ultimately end up being transferred to the hands and then to items put into the mouth.
- **Inhalation** of radioactive vapours, gases, aerosols or dusts can occur depending on the chemical form of the radioisotopes or the processes in which the radioisotopes are used. For example, unbound radioactive iodine presents a particular danger owing to its volatility, while some equipment and processes, such as centrifuges, can generate a radioactive mist. In the mining industry, inhalation of dusts containing naturally occurring radioisotopes, radon and radon progeny is a significant contributor to worker doses.
- **Skin uptake** occurs when radioactive materials penetrate intact skin under favourable conditions. Skin uptake rates are determined by contact time, the chemical form of the radioisotope and whether the radioisotope is mixed with other chemicals, such as organic solvents. Where there is a wound, the uptake rate is higher.

15.2.8 Quantifying Ionising Radiation

To quantify ionising radiation, it is necessary to know the:

- activity and activity concentration;
- radiological half-life;
- energy.

15.2.8.1 Activity and Activity Concentration

The activity tells us how much radiation is being emitted. The international standard (SI) unit for activity is the becquerel (Bq), which is equivalent to one disintegration, or nuclear transformation, per second. The non-SI unit of activity is the curie (Ci), which is still in active use. The relationship between the becquerel and the curie is shown in Equation 15.2:

$$1\,\text{Ci} = 3.7 \times 10^{10}\,\text{Bq} \tag{15.2}$$

Activity concentration describes the amount of radiation emitted per unit mass (e.g. Bq per kilogram) or volume (e.g. Bq per litre). Although activity concentration is a derived quantity, it is nonetheless important as many licensing and regulatory conditions are based on the limits of activity per unit mass or volume, for example, the Australian action level for residential indoor radon of 200 Bq/m^3, as cited in Section 15.2.4.1.

15.2.8.2 Radiological Half-Life

The radiological half-life of a radioisotope describes its rate of decay. As radiation is emitted, the number of unstable atoms decreases exponentially, so the activity of a set amount of radioactive material decreases with time. This pattern can be used to predict how long the decay process might take, as shown in Figure 15.4.

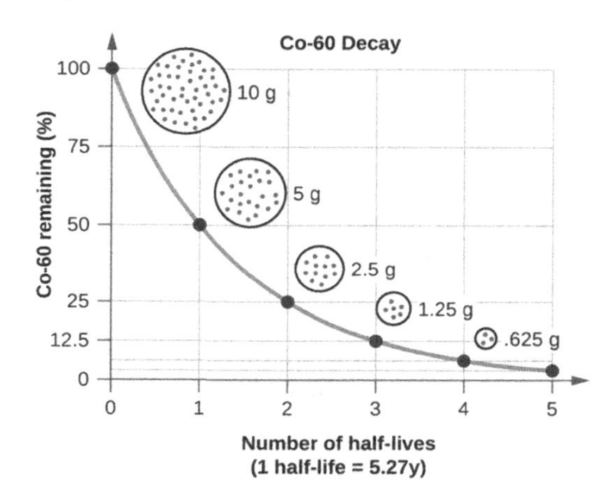

FIGURE 15.4 The pattern of radioactive decay.

Boise State University, n.d., 116 19.3 Radioactive Decay, Figure 6, https://boisestate.pressbooks.pub/chemistry/chapter/20-3-radioactive-decay/.

Radiological half-life ($T_{1/2}$) is the time taken, on average, for half of a given amount of radioisotope to undergo radioactive decay. The relationship between activity and radiological half-life is defined by Equations 15.3 and 15.4:

$$A_t = A_0 e^{-\lambda t} \text{ or } A_t = \frac{A_0}{2^n} \tag{15.3}$$

where

A_t = activity at time elapsed (t),
A_0 = original activity at t = 0,
λ = decay constant
n = number of half-lives $\left(n = {}^{t}/_{T_{1/2}}\right)$
t = time elapsed (secs)
$T_{1/2}$ = radiological half-life

$$T_{1/2} = \frac{\ln}{\lambda} \text{ or } = \frac{0.693}{\lambda} \tag{15.4}$$

If the starting activity of the radioisotope (A_0) is known, the half-life can be used to determine how much radiation will still be present after a given time—that is, for how long, the radioisotope will continue emitting radiation. Half-life varies depending on the type of radioisotope. For example, phosphorus-32 has a half-life of 14.29 days, while radium-226 has a half-life of 1600 years. A rule of thumb is that after 10 half-lives the radiation emitted by the material will be about one thousand times lower than it was at the start.

15.2.8.3 Specific Activity

The specific activity (SA) of a material is the activity per unit mass of a radionuclide and is measured in Bq/g. The SA is a specific property of a radionuclide and is calculated from Equation (15.5)

$$SA = \frac{4.17 \times 10^{23}\left[\text{mol}^{-1}\right]}{T_{\frac{1}{2}}\left[\text{s}\right] \times M\left[\text{g/mol}\right]} \tag{15.5}$$

where
- SA = specific activity (Bq/g),
- M = mass number of radionuclide (g/mol)
- $T_{1/2}$ = radiological half-life

For radiation protection purposes, it is important to note that the SA is inversely proportional to half-life, the inference being that long-lived radionuclides have lower SA than their shorter-lived counterparts.

15.2.8.4 Energy

The SI unit for energy is the electron volt (eV). All types of ionising radiation have either a specific energy or a spectrum of energies. The amount of energy affects the relative penetrating and ionising ability of each ionising radiation type. Typical energies for various types of ionising radiation are shown in Table 15.3.

TABLE 15.3
Energies of Ionising Radiation

Type of Radiation	Typical Energy Range
Alpha	3–9 MeV
Beta*	0–3 MeV*
Gamma**	10 keV and 10 MeV
X-rays***	A few eV to several MeV
Neutrons	0–10 MeV

* Each radioisotope emits beta radiation with a spectrum of energies. The most probable energy is approximately one-third of the maximum energy.
** Have a characteristic energy.
*** Have a spectrum of energy.
Adapted from: Simon 2007.

15.2.9 Quantifying Exposure and Dose

To measure the amount of ionising radiation absorbed and the amount of harm it is likely to inflict, it is necessary to understand the following concepts:

- absorbed dose and exposure;
- equivalent dose and radiation weighting factor;
- effective dose and tissue weighting factor;
- biological half-life and effective half-life;
- committed dose.

15.2.9.1 Absorbed Dose and Exposure

Absorbed dose is the primary quantity used to define the energy deposited in matter due to ionising radiation exposure. Historically, exposure was defined as the amount of ionisation that gamma or X-rays produce in air. Exposure was formerly measured in roentgen (R); now the SI unit coulomb per kilogram of air (C/kg) is used. However, in the latest version of the Basic Safety Series, exposure is defined as the state of being subject to irradiation (International Atomic Energy Agency 2011).

Absorbed dose (D) is a measure of the energy absorbed per unit mass of *any matter*. For a given type of ionising radiation of defined energy, the absorbed dose increases with the density of the absorbing matter. The absorbed dose is measured in the SI units of joule per kilogram (J/kg) or gray (Gy). The gray is a large unit, so normally milligray (mGy), microgray (μGy) and nanogray (nGy) are used.

The rate of absorbed dose from electromagnetic radiation can be quantified by modern instrumentation and used for purposes of controlling exposure. The derived absorbed dose rate is most often indicated in units of milli or micro grays per hour (mGy/h or μGy/h).

Absorbed dose indicates how much energy was absorbed, but not how much ionisation has occurred. Knowing the amount of ionisation, however, is crucial to quantifying the biological damage caused by the radiation exposure.

15.2.9.2 Equivalent Dose and Radiation Weighting Factor

The equivalent dose is an average measure of the dose or doses received by a particular mass of tissue from all types of radiation.

Different types of radiation have different relative biological effectiveness owing to their different ionising abilities. The ICRP defines radiation weighting factors (w_R) to account for these differences (see Table 15.4). Equivalent dose (H_T) is the sum of the absorbed dose times the radiation weighting factor as shown in Equation 15.6:

$$H_T = \Sigma\left(W_R \times D\right) \quad (15.6)$$

where

H_T = equivalent dose
W_R = radiation weighting factor
D = absorbed dose

The radiation weighting factor is a dimensionless number (i.e. it has no unit), and it is based on experimental data. Like the absorbed dose, the equivalent dose is measured in J/kg, but to distinguish it from absorbed dose its unit is given the name sievert (Sv), rather than gray.

TABLE 15.4
Radiation Weighting Factors*

Radiation Type	Radiation Weighting Factor, w_R
Gamma and X-ray	1
Beta	1
Alpha	20
Neutrons	A continuous curve as a function of neutron energy*

Based on ICRP (2007). For a full list of radiation weighting factors and the energy curve for neutrons, refer to ICRP (2007).

15.2.9.3 Effective Dose and Tissue Weighting Factor

Equivalent dose limits apply to a tissue or an organ. When considering the risk of health effects such as cancer, it is important to know that some tissues and organs are more likely to be damaged by ionising radiation than others. Tissue weighting factors (w_T) represent the relative amount of damage likely to be caused to various types of tissue based on their sensitivity (Table 15.5). Effective dose (ED) is the sum of the equivalent doses multiplied by the tissue weighting factors and is also expressed in sievert (Sv), as shown in Equation 15.7:

$$ED = \Sigma\left(H_T \times W_T\right) \tag{15.7}$$

where
ED = effective dose
H_T = equivalent doses
W_T = tissue weighting factor

Tissue weighting factors are defined by the ICRP after examination and modelling of data from many different human epidemiological studies as well as other research, such as animal studies. It is important to stay abreast of the current ICRP recommendations, as the tissue weighting factors can change, based on new evidence.

TABLE 15.5
Tissue Weighting Factors*

Organ/Tissue	Number of Tissues	Tissue Weighting Factor, w_T	Total Contribution
Lung, stomach, colon, bone marrow, breast and remainder tissues**	6	0.12	0.72
Gonads (testes and ovaries)	1	0.08	0.08
Thyroid, oesophagus, bladder and liver	4	0.04	0.16
Bone surface, skin, brain and salivary glands	4	0.01	0.04

* Based on ICRP (2007)

** Remainder tissues are: adrenals, extrathoracic tissue, gall bladder, heart, kidneys, lymphatic nodes, muscle, oral mucosa, pancreas, prostate, small intestine, spleen, thymus and uterus/cervix.

15.2.9.4 Biological Half-Life and Effective Half-Life

The biological half-life of a radioisotope is the time it takes the body to eliminate half of an intake of radioisotope by natural biological processes such as urination. The effective half-life combines the radiological half-life and the biological half-life. The effective half-life is the time taken for half of the radioisotope to be removed from the body by both radioactive decay and biological processes.

15.2.9.5 Committed Dose, Dose Coefficient and Limits of Intake

The committed dose is the exposure that will be received from a radioisotope that has entered the body. Once a radioisotope is inside the body, it will irradiate the tissues until it either decays completely (radiological half-life) or is excreted (biological half-life). To allow calculation of the committed dose, the ICRP has recommended dose coefficients or doses per unit of intake of a radioactive substance. Different dose coefficients exist for different pathways of exposure, and they take into account both the physical half-life of the radioisotope (radiological half-life) and its biological retention rates (biological half-life). The committed dose is calculated for a 50-year period for adults and a 70-year period for children, and then—per ICRP recommendations—assigned to the year in which the intake occurred. Both committed equivalent doses and committed effective doses can be calculated.

ICRP dose coefficients can also be used to derive the secondary operational limit called the annual limit of intake (ALI) for inhalation or for ingestion for occupational exposure (e.g. ICRP 2017, 2019). The ALI is the amount of a radioisotope which, when taken into the body, would produce a committed effective dose equal to the annual effective dose limit (see Section 15.2.11). As ALIs are calculated limits, it is important to understand the underlying assumptions when using an ALI from another source such as a website or an older national standard. The ALI must be based on the appropriate dose limit and the latest ICRP publications.

15.2.10 Health Effects of Ionising Radiation

The health effects of ionising radiation fall into two main categories:

- **Harmful tissue reactions** (also called deterministic effects) occur when enough cells have been killed or damaged to affect the function of an organ or tissue. These reactions are a consequence of the death or malfunction of cells following high doses. In its 2007 recommendations, the ICRP noted that no tissues show loss of function when exposed to either a single acute dose or a fractionated annual dose in the absorbed dose range up to 100 mGy (ICRP 2007).
- **Stochastic effects** may or may not occur, but their likelihood increases with the exposure. These effects are mainly due to damage to the DNA, and the resulting effects are primarily cancer and hereditary disorders.

Additional effects that need to be considered are the risk to the unborn child and chronic diseases other than cancer.

15.2.10.1 Radiation Sickness and Death

A short-term dose of approximately 1 Gy may result in radiation sickness (ICRP 1990). This manifests as symptoms of nausea, vomiting, rapid pulse and fever within a few hours of the exposure. Doses of this magnitude are not common, and historically have usually been associated with exposure due to accidents.

Death from a short-term dose of ionising radiation occurs only at very high whole-body doses. For example, a short-term dose of 3–5 Gy to the whole body results in damage to the bone marrow and death within a few months. Doses of 5–15 Gy to the whole body can cause damage to the gastrointestinal tract and lungs, and death within a few weeks. Short-term whole-body doses exceeding 15 Gy damage the central nervous system and cause death within a few days (ICRP 1990).

15.2.10.2 Skin Effects

Skin effects of radiation exposure include hair loss, reddening of the skin, burns and cancer. Reddening and damage to the outer layer of skin, called erythema, occurs with short-term absorbed dose to the skin of around 3–5 Gy. The initial reddening lasts only a few hours but may be followed several weeks later by a wave of deeper and more prolonged reddening (ICRP 1990). Erythema is unlikely in modern workplace settings, where ionising radiation is well controlled.

15.2.10.3 Effects on the Eyes

Cataracts can form on the lens of the eye when ionising radiation breaks down the dividing cells in the lens epithelium, although this is unlikely where ionising radiation exposure is well controlled. Cataracts may be induced by a single, relatively large short-term dose, or by a series of smaller doses, particularly of beta radiation. The threshold is now thought to be 0.5 Gy, as opposed to the 5 Gy that was defined as the threshold in the past (ARPANSA 2011).

15.2.10.4 Reproductive System and Hereditary Effects

Ionising radiation can seriously affect the reproductive system in both sexes, as the reproductive cells are sensitive to radiation. In males, a single acute dose of 0.15 Sv can cause temporary sterility, and a single acute dose of 3.5–6 Sv may cause permanent sterility. In females, a single acute dose of 2.5–6 Sv may cause permanent sterility (ICRP 2007). Doses high enough to result in sterility would also cause other health effects such as radiation sickness but are unlikely in settings where exposure to ionising radiation is well controlled.

In addition to direct effects such as sterility, there has been concern about whether radiation can cause hereditary damage in the form of genetic disorders being passed to future generations. The summary of the current state of knowledge is that there is no direct evidence of heritable disease having been caused by exposure of human parents to radiation, but there is compelling evidence of this from experimental studies of animals (ICRP 2007).

15.2.10.5 Effects on the Unborn Child

The unborn child (embryo and foetus) is more susceptible to ionising radiation than an adult. The effect of such radiation depends entirely on the stage of development. Early in a pregnancy, a high short-term dose may cause death of the foetus and miscarriage but absorbed doses of less than 100 mGy are very unlikely to cause any damage. When the organs are being formed, exposure to ionising radiation may cause deformities in that organ, though the current evidence appears to indicate that the threshold for malformations is about 100 mGy (ICRP 2007). In the later stages of pregnancy, the cognitive ability of the unborn child may be adversely affected by exposure to high doses of ionising radiation; doses below 300 mGy pose a very low risk of damage (ICRP 2007). Ionising radiation exposure must be carefully controlled, even in the most modern facilities, to ensure that unborn children are not affected. Cancer risk from exposure in utero is similar to the risk of exposure in early childhood—that is, three times the population risk.

15.2.10.6 Cancer

The association between radiation exposure and cancer is mostly based on data from situations where relatively high exposures have occurred (e.g. in Japan, among atomic bomb survivors). The association between low doses of radiation and cancer is complicated and difficult to prove because of the relatively high rate of cancer within the general population. This is a stochastic effect, where the risk of cancer increases as the radiation dose increases (what is commonly referred to as the 'linear-no-threshold' effect of ionising radiation exposure). Some cancers known to be induced by exposure to ionising radiation include:

- leukaemia in atomic bomb survivors;
- thyroid cancer due to exposure to radioactive iodine;

- lung cancer in uranium miners as a result of inhaling radioactive dust;
- bone sarcoma in workers who applied radium-containing paint to the faces of luminous clocks and dials. The workers used their lips to get a fine point on their brush and ingested radium in the process.

15.2.10.7 Importance of Rate of Exposure to Cancer Risk

Epidemiological research has been unable to establish unequivocally that there are effects of statistical significance at acute doses below a few tens of millisieverts. However, there are useful risk estimates for acute exposures that result in doses in the range of 50–100 millisieverts. The risk factor averaged over all ages and cancer types is about 1 in 10,000 per millisievert (ARPANSA 2019).

The risk from chronic exposures to ionising radiation is less well quantified; however, the international radiation protection community has adopted a reduction of the dose risk factor by a factor of two for chronic exposures. For radiation protection purposes, the risk of radiation-induced fatal cancer is taken to be about 1 in 20,000 per millisievert of dose for the population as a whole (ARPANSA 2019).

15.2.10.8 Diseases Other than Cancer

There has been some evidence in highly exposed groups (e.g. Japanese atomic bomb survivors) that ionising radiation may be a risk factor for non-cancer diseases such as heart disease, stroke, digestive disorders and respiratory disease. However, the data have been judged insufficient to extrapolate any of these effects to lower dose exposures—that is, below 100 mSv (ICRP 2007).

15.2.11 Dose Limits (Exposure Standards)

Exposure standards that apply to exposure to ionising radiation from external and internal sources are called dose limits and are defined in Table 15.6. Dose limits for use in Australia are established by ARPANSA based on international recommendations and are adopted into legislation by each

TABLE 15.6
Dose Limits for Planned Exposure to Ionising Radiation

	Dose Limit	
Application	**Occupational**	**Public**
Effective dose	20 mSv per year, averaged over five consecutive years with no more than 50 mSv in any single year*	1 mSv in a year** The limit for the unborn child is 1 mSv from the declaration of pregnancy
Annual equivalent dose:		
Lens of the eye***	20 mSv***	15 mSv
Skin (averaged over 1 cm^2 of skin)	500 mSv	500 mSv
Hands and feet	500 mSv	—

* Refer to the ARPANSA publication for all of the provisos.

** A higher value is allowed in special circumstances, provided the average over five years does not exceed 1 mSv.

*** The dose limit for the lens of the eye has been significantly reduced by the ICRP based on epidemiological data. The current dose limit in Australia is still 150 mSv, but ARPANSA has advised that a new limit of 20 mSv will be introduced. *Note*: No statement has been made about the public dose limit for the lens of the eye.

(*Sources:* ARPANSA 2011).

state or territory (see Section 15.2.3). Dose limits are defined for occupational exposure as well as for members of the public. An arbitrary safety factor is applied to the public dose limits to take account of the sick, the elderly and the very young. These groups are thought to be more sensitive to radiation than workers exposed to it on the job.

The effective dose limit has been set to minimise the risk of cancer and other effects. The additional equivalent dose limits, for the lens of the eye, the skin and the hands and feet, have been set to ensure that individual tissue thresholds are not exceeded.

15.2.12 Detecting Ionising Radiation

Ionising radiation monitors can be categorised according to:

- what they measure
  - count rate
  - dose rate
  - dose
- the principle of detection
  - gas-filled detectors
  - scintillation detectors
  - semiconductor detectors
  - thermoluminescent detectors

15.2.12.1 Types of Measurement

Instruments for monitoring ionising radiation provide three main measurements:

- **Count rate** (counts per second). This does not mean disintegrations per second (Bq) unless the device has had appropriate calibration, so count rate monitors typically provide a relative measure of the radiological activity that is present. They are particularly useful for conducting contamination surveys and finding sources.
- **Dose rate** (dose per unit time). Users need to know for which type of radiation dose the meter is calibrated. Dose-rate monitors are particularly useful for estimating potential exposures for different exposure scenarios. They are also useful for checking surface dose rates for radioisotope waste disposal and for labelling packages for transport.
- **Equivalent and effective doses**. These measures are provided by dosimeters.

Some monitors are multi-functional—for example, they provide both a count rate and a dose rate—while many digital dosimeters measure dose rate.

Some examples are shown in Figure 15.5.

15.2.12.2 Types of Instruments

Instruments can be classified based on the principle of detection that they utilise:

- **Gas-filled detectors** have a gas-filled chamber of known volume with conducting electrodes connected to an electronic circuit to collect the charges created by ionisation events that occur within the chamber. The three main types in common use are Geiger–Müller (GM) detectors, ionisation chambers and proportional counters. In an ionisation chamber, one ionisation event gives rise to one pulse that can be detected. In a GM detector, each ionisation event is multiplied, which causes a cascade of pulses. The proportional counter has some multiplication for each event but does not produce a cascade. This makes the

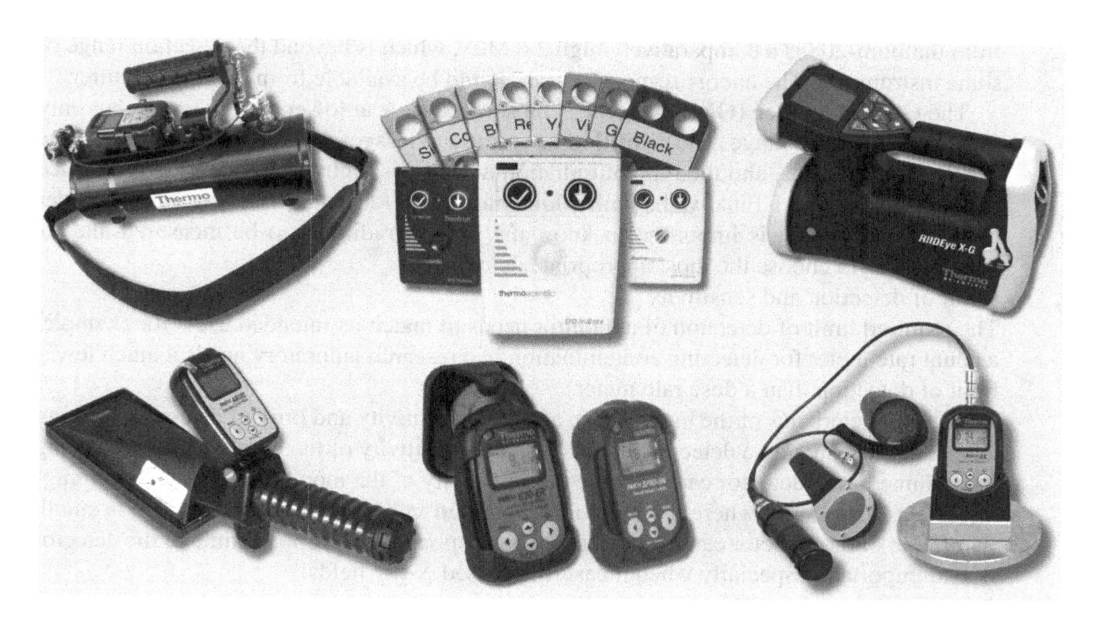

FIGURE 15.5 Instruments for measuring ionising radiation

Courtesy of ADM systems.

GM more sensitive in general than both the ionisation chamber and the proportional counter. The downside of the GM detection method is that the electronics can detect only one cascade at a time, which can result in overload or dead time in high-radiation areas. The proportional counter can be set up to distinguish between ionisation events caused by alpha and beta particles.

- **Scintillation detectors** contain a material that emits light when it absorbs ionising radiation energy. This light can then be detected using an appropriate electronic device, such as a photomultiplier tube. The scintillant material is often a crystal such as sodium iodide or a powder such as zinc sulfide. The type of scintillant material has to be matched to the type of radiation that needs to be detected. A variant of the scintillation detector is a laboratory-based device in which the scintillant is a liquid that is physically mixed with a sample.
- **Semiconductor detectors** use a semiconductor material that generates a charge when exposed to gamma radiation. Coupled with the correct electronics, this type of detector can be used to analyse the energy spectrum of the radiation.
- **Thermoluminescent (TLD) or Optically Stimulated Luminescent (OSL) detectors** use polycrystal materials that change to a metastable excited state when exposed to ionising radiation. When TLD material is warmed, or OSL material is optically stimulated, the material releases detectable light, which is correlated to the amount of radiation incident upon the detector.

15.2.12.3 Choice of a Monitoring Instrument

- Energy response curves

  No single instrument is suitable for measuring all types and all energies of ionising radiation. Hence, it is important to consider the type of radiation and the energy response curve for any detector in relation to the energy of the radiation that needs to be measured. This is important when measuring the gamma emissions from the thorium-232 decay chain, as one emission, from thallium-208 is a comparatively high 2.6 MeV, which is beyond the detection range of some instruments. The energy response curve should be available from the manufacturer.

The **Geiger–Müller (GM) detector** is most responsive at lower energy levels, but only for a limited energy range. The **scintillation counter** is less responsive, but to a wider range of radiation energies, and the type of radiation detected is highly dependent on the choice of scintillant material. Finally, the **ionisation chamber** is fairly insensitive to an even wider range of energies. It is important to know the type of radiation to be measured and its energy, so as to choose the most appropriate detector.

- Limit of detection and sensitivity
  The required limit of detection of a monitor needs to match its intended use—for example, a count rate meter for detecting contamination in a research laboratory needs a much lower limit of detection than a dose rate meter.

  The characteristics of the monitor can affect its sensitivity and limit of detection. Having a large surface area on a detector can increase the sensitivity of the monitor. But increasing the volume of the detector can decrease the sensitivity of the monitor as a result of averaging effects, especially where the radiation is emitted in beams or the source size is small. Saturation of the detector can lead to a false low response. The response time of the detector is also important, especially when measuring pulsed X-ray fields.

15.2.12.4 Calibration and Performance Checking

Performance checking should be undertaken every time a radiation measuring instrument is used. The battery should be checked for sufficient energy and the detector should be inspected for damage. The detector should be checked against a suitable source (preferably of the same radiation being monitored), and its response noted. This should be logged and referred to over time as a benchmark for any deterioration in detector performance.

Calibration is a more in-depth examination of the performance of the radiation monitor and requires the services of a specialist calibration service such as the one provided by ARPANSA. It should be undertaken at regular intervals, and any time the radiation monitor is repaired, knocked, damaged or otherwise altered.

15.2.12.5 Monitoring Techniques

The technique used for monitoring ionising radiation depends on the aspect of ionising radiation being examined. The main techniques are:

- area surveys with direct read-out monitor—contamination survey and area survey;
- area monitoring with wipe testing;
- external exposure assessment—passive and active dosimetry;
- internal exposure assessment—personal contamination surveys, air monitoring, external monitoring and bioassay.

15.2.12.5.1 Area Surveys with Direct Read-Out Monitor

Area monitoring is necessary to ensure that appropriate control strategies are implemented to minimise exposure. Area monitoring for ionising radiation must be done at various heights over a wide area because ionising radiation can diffract and scatter. An instrument measuring count rate or dose rate is most commonly used. The area to be monitored should be mapped out on a grid pattern and measurements taken at each grid point. For the sake of personal safety, choose an instrument with an adequate response time and commence monitoring at the farthest point from the ionising radiation source. Using this method of monitoring, it is possible to quickly establish areas where no person should enter or where exposure time should be limited.

Surfaces should be checked for contamination before and after any procedure that uses unsealed sources of ionising radiation, or if the integrity of a sealed source is questioned.

15.2.12.5.2 Area Monitoring with Wipe Testing

Wipe testing is used to identify surfaces contaminated with radioisotopes. It is particularly useful when:

- the radioisotope has very low energy, such as tritium, as wipe testing is the only effective way to detect surface contamination;
- the contamination is very dilute on the surface;
- there is a high radiation background that may interfere with the survey instrument—for example, where the integrity of a sealed source is being checked or surfaces in a radiation—source storage area are being tested for contamination.

Wipe testing is done by wiping a moistened filter paper across the surface to be tested. This concentrates any radioactive contamination on the filter paper. The filter paper can then be monitored using a survey instrument (usually a calibrated count rate meter) or counted using laboratory instrumentation such as a liquid scintillation counter. Wipe testing can also be done to determine whether radioactive contamination is fixed or removable and will provide information about whether clean-up or shielding is required.

15.2.12.5.3 External Exposure Assessment

Dosimetry is the measurement of the radiation dose to either the whole body or certain parts of the body.

Passive dosimetry can be performed by a number of monitors. Typical passive dosimeters are thermoluminescent detectors (TLD badges) or optically stimulated luminescence (OSL) monitors. Other older forms of passive dosimeters that utilise other detection methods, such as a charged quartz fibre or film badge dosimeters, are also available. A TLD badge can be used to monitor for beta, gamma and x-ray radiation, while the OSL monitor (ARPANSA 2018) is suitable for gamma and x-rays.

The whole-body dose is assessed by using a single badge at the point of expected highest exposure (usually the waist or the chest) for a period of 4–12 weeks. At the end of this period, the badge is sent away for analysis. If a lead apron is worn, then the badge should be positioned beneath it at the waist or chest to ensure that it represents the actual dose. Passive dosimeters and analytical services are available from ARPANSA and several other suppliers.

Equivalent doses to parts of the body can be measured using small TLD badges. For example, the dose to the hands can be measured using a TLD that is wrapped around the fingers. A person's radiation dosimetry results should be recorded and provided to the regulatory authority. In Australia, ARPANSA maintains a dose register of personal annual radiation doses called the Australian National Radiation Dose Register.

Passive monitoring results indicate what dose of radiation has been received but do not provide any real-time data to help choose interventions or controls.

Active electronic dosimeters can be used to measure dose or dose rate and can provide immediate feedback on dose rates associated with specific tasks. Dosimeters of this type often contain very small Geiger-Müller (GM) tubes. Practitioners planning to use an active electronic dosimeter should ensure that it will detect the types of radiation that are present and that it complies with the standards specified by the International Electrotechnical Commission (Voytchev et al. 2011).

To avoid spurious results electronic dosimeters should be kept away from other electronic devices like mobile phones.

15.2.12.5.4 Internal Exposure Assessment

- **Personal contamination surveys**: Whenever radioisotopes are handled, it is necessary to routinely monitor the hands, clothes and body to check for contamination. This type of contamination can lead to intake of the radioisotope and receipt of a committed dose.

  Personal contamination surveys are usually done using a direct read-out instrument of the kind used for an area contamination survey. Where contamination is likely, a clean person (someone who has not recently had direct contact with radioisotopes) should survey a person who is potentially contaminated. The surfaces of laboratory coats, cuffs, hands and feet are the usual sites of personal external contamination. Many common laboratory contamination monitors have a stand or clip that allows researchers to use them in hands-free mode to check the hands during experiments.
- **Air monitoring**: Monitoring for airborne contamination should be undertaken where there is a risk of inhaling the radioisotope, such as a radioactive dust, radon and radon progeny or radioactive iodine (because it can sublime—that is, change directly to vapour). Monitoring for airborne radioactive material requires specialist methods and equipment, and the H&S practitioner will probably need to seek expert advice.
- **External monitoring**: Exposure to gamma- or X-ray-emitting radioisotopes can often be estimated directly on the basis of external monitoring of either the whole body, or specific organs or tissues. For example, when using iodine radioisotopes, the thyroid gland can be monitored with a sensitive scintillation detector to ascertain the dose of radioactive iodine. Whole-body monitoring is usually done only when someone suspects that they have received a large dose of radioisotope internally. It is performed using a similar monitor to that used for thyroid monitoring, and detects only gamma and X-ray radiation. Only a few facilities in Australia operate instruments for whole-body monitoring. ARPANSA, ANSTO or local health authorities would be a useful first point of contact for locating such an instrument.
- **Bioassay**: Indirect radioisotope intake can be gauged indirectly by measuring the amount in a biological tissue or product, such as urine, blood, faeces, hair, sweat or exhaled breath. Common bioassay methods involve monitoring urine for radioisotopes such as tritium (beta emitter), sulfur-35 (beta emitter) and sometimes carbon-14 (beta emitter). Urine samples can be analysed using an instrument such as a laboratory liquid scintillation counter. If workers routinely use radioisotopes that can be assessed via urine monitoring, it is useful to establish a baseline count before work using the radioisotope commences.

15.2.13 Controlling Radiation Hazards

Anyone working with ionising radiation should aim to minimise the potential for exposure. This should always be done in the planning stages of the work, in line with the three principles defined in the International Radiological System of Protection (see Section 15.2.3). Before a decision to use ionising radiation is made, the benefits need to be carefully weighed against the potential harm. If the use is justified, then appropriate controls should be used.

15.2.13.1 Minimising External Exposure

The primary methods for controlling or minimising external exposure to radiation are:

- **shielding**: using a barrier to absorb the radiation;
- **distance**: increasing the distance between the person and the radiation source;
- **time**: handling the radiation source as briefly as possible.

15.2.13.1.1 Shielding

Shielding is the most important control measure to reduce or eliminate external exposure to ionising radiation. The shielding required for each type of radiation depends on its properties:

- Alpha radiation
  - does not penetrate far in air, so does not require a specific shield (a sheet of paper is thick enough to stop alpha particles).
- Beta radiation
  - can penetrate a reasonable distance in air;
  - needs low-density types of shielding such as Perspex (approx. 1 centimetre thick) or aluminium (several millimetres thick);
  - needs to be shielded with the correct material. Do NOT use high-density shielding such as lead. X-rays known as Bremsstrahlung radiation are generated when beta particles are slowed down too quickly in high-density materials.
- Gamma and X-rays
  - can penetrate a long distance in air;
  - are best controlled with high-density shielding. Relatively effective shields can also be made from thicker slices of less dense materials such as brick, concrete, sand or gravel;
  - need thicker shielding as their energy increases.
- Neutrons
  - can penetrate moderate to long distances in air;
  - need shielding with specific elements such as boron and cadmium, or with materials containing high levels of hydrogen—for example, paraffin wax, water and concrete.

15.2.13.1.2 Distance

If it is not practical to shield a radiation source, or if the shielding does not stop all the radiation, then increasing the worker's distance from the source should be considered. This can be done by using remote handling devices such as tongs or mechanised devices. The intensity of ionising radiation is inversely proportional to the square of the distance from the source, so a worker twice as far from the source receives one-quarter of the radiation exposure.

15.2.13.1.3 Time

If shielding and distance do not completely stop ionising radiation, then reducing the exposure time should be considered.

15.2.13.2 Minimising Internal Exposure

The main methods for minimising internal exposure are:

- choice of work practices;
- use of contamination controls;
- multiple layers of containment; and
- managing the risk of inhalation.

15.2.13.2.1 Choice of Work Practices

Work practices should be chosen that minimise the number of steps in the radioisotope handling process. Complex processes provide more opportunities for contamination to occur: the chances for contamination with a multi-step chemical process are much higher than for a simple dilution.

15.2.13.2.2 Use of Contamination Controls

Minimising the amount of contamination reduces the potential for a radioisotope to enter the body. Cleanliness is very important in avoiding the internal uptake of radioisotopes. This includes measures such as changing gloves frequently, wearing gloves that do not absorb radioactive materials and, if using surgical gloves, wearing two pairs.

Workers carrying out experiments involving radiation should frequently monitor their clothing, hands and workspace, and any spills should be cleaned up without delay. Routine area contamination surveys should be done and the results recorded to ensure that the general work environment remains uncontaminated.

15.2.13.2.3 Multiple Layers of Containment

Multiple layers of containment should be used to minimise the risk and the consequences of spillages. This should be done for both storage and handling. An example of good practice is to work within a tray lined with absorbent material to contain a spill.

15.2.13.2.4 Managing the Risk of Inhalation

The risk of inhaling radioactive material can be managed by installation of local exhaust ventilation, choosing work practices that minimise aerosol production, minimise use of volatile chemical forms of a radioisotope and avoid the use of dry powder forms.

15.3 NON-IONISING RADIATION

Non-ionising radiation is electromagnetic radiation of a wavelength greater than 100 nm that does not have sufficient energy to ionise the matter with which it interacts. It includes ultraviolet, visible, infrared, radiofrequency and extremely low-frequency radiation. With such a wide range of wavelengths, frequencies and therefore energies, non-ionising radiation has the potential to cause various adverse health effects in humans in certain body locations.

15.3.1 Ultraviolet Radiation

15.3.1.1 Properties

Ultraviolet radiation is the highest-energy form of non-ionising radiation and exists in three bands, from the highest to the lowest energy:

- **Far (frequency), short (wavelength) or UV-C**—wavelengths 100–280 nm and frequencies around 10^{16} Hz. Wavelengths below 180 nm are absorbed by air and are therefore of little biological significance. This is why the wavelength range of UV-C is often listed as 180–280 nm.
- **Middle, erythemal or UV-B**—wavelengths 280–315 nm and frequencies around 10^{15} Hz.
- **Near, long or UV-A**—wavelengths 315–400 nm and frequencies around 10^{14} Hz.

15.3.1.2 Sources

The most common source of ultraviolet radiation is the sun. Other sources capable of providing significant exposure in the occupational setting include arc sources and specialised lamps.

Sometimes ultraviolet radiation is produced as an unwanted side-effect, as in plasma torches, gas and electric arc welding. Ultraviolet lamps are used for applications such as:

- tanning and dermatology (UV-A and UV-B);
- black lights for non-destructive testing (UV-A);
- scientific research (all UV);
- photocuring of inks and plastic (UV-A and UV-B);
- photoresist processes (all UV);
- germicidal uses (UV-C and UV-B).

Some lasers also emit ultraviolet radiation. Lasers are discussed further in Section 15.3.3.

Ultraviolet light sources with emissions below 250 nm can interact with the workplace atmosphere to produce ozone, oxides of nitrogen and phosgene.

15.3.1.3 Health Effects and Quantification

The skin and eyes may be affected by exposure to ultraviolet radiation. At one time, wavelengths below 315 nm were collectively known as 'actinic radiation'—that is, radiation that can induce biological effects, but those health effects are also observed in UV-A at substantially higher doses. Health effects can result from occupational exposures in the absence of effective controls.

- **Reddening (or erythema)** results from overexposure of skin in the middle and far ultraviolet range (200–315 nm), with the greatest sensitivity occurring at 295 nm. Exposure to near ultraviolet alone requires far higher levels to induce erythema; however, exposure to near and middle ultraviolet together intensifies the response.
- Chronic exposure to ultraviolet light, especially middle ultraviolet, increases **skin ageing, skin pigmentation changes** and the **risk of developing skin cancer**. IARC classifies UV radiation as a Group 1 carcinogen (IARC 2012).
- The cornea and conjunctiva of the eye strongly absorb middle and far ultraviolet radiation. The resulting condition, photo-keratoconjunctivitis, is generally known as **'welder's flash'** because it often occurs after welding. Wavelengths above 295 nm penetrate the cornea and are absorbed by the lens, increasing the risk of cataracts.

Individuals who have had an eye lens replaced or exposed to photosensitising agents are at additional risk from ultraviolet exposure. Exposure to ultraviolet radiation incident on the skin or eyes is measured in joules per square metre (J/m^2). The maximum allowable exposure varies according to the wavelength of the radiation—that is, its relative spectral effectiveness—with the most harmful wavelength being 270 nm. Where a person is exposed to a range of ultraviolet wavelengths simultaneously (in light from a broad-spectrum or incoherent source), the spectrally weighted effective irradiance (E_{eff}) is measured or calculated to determine the exposure and compare with exposure limits. Effective irradiance (E_{eff}) is the surface exposure dose rate and has units of watt per square metre (W/m^2). Since 1 W equals 1 J/s, effective irradiance can be used to calculate the time taken to reach a certain maximum permissible exposure to ultraviolet radiation.

Measuring ultraviolet radiation is a task usually undertaken by a professional with specific expertise in the area (see Section 15.3.1), using the correct equipment and considering the geometry of exposure. ARPANSA provides a Solar UV index on its website that is updated hourly and can be accessed for use by outdoor workers.

15.3.1.4 Control Methods

Exposure to ultraviolet radiation can be reduced by shielding, distance and time:

- **Hats, eye protection, clothing and sunshades serve as shields** against sunlight. Clothing and hats chosen should have a formal rating for ultraviolet protection (ARPANSA nd). Wide-brimmed hats and high-protection-factor sunscreens are also useful but may not provide sufficient protection.

  Protective glasses for outdoor or general use, or face shields and glasses for welding, should meet the relevant International or Australian Standards. Where shields are required for industrial sources, polycarbonate or methyl methacrylate plastics strongly absorb most ultraviolet radiation. However, where a high-intensity source exists, the radiation may not be absorbed fully. Ultraviolet radiation can also reflect from shiny surfaces.

  Exposures to artificial or industrial sources of ultraviolet radiation should be controlled by engineering solutions such as light-tight cabinets and enclosures, and ultraviolet-light-absorbing glass and plastic shielding. Shields, curtains and barriers are used to shield against UV light from welding processes.
- **Distance** from the source should be maximised, since the intensity of non-ionising radiation falls off rapidly as this distance increases.
- The **time of exposure** to the source should be limited wherever possible. For outdoor workers, avoiding working outside without proper protective equipment during the middle of the day can also help reduce exposure.

Control measures can be found in ARPANSA RPS-12, which provides best practice for workers (ARPANSA 2006). In Australia, the Cancer Council and Safe Work Australia both provide practical information and advice on implementing good sun protection policies and practices in the workplace. They provide a comprehensive sun protection program that describes various sun protection control measures in the workplace.

15.3.2 Visible and Infrared Radiation

15.3.2.1 Properties

Visible and infrared radiations are less energetic forms of non-ionising radiation than ultraviolet. They can be divided into bands from the highest to the lowest energy:

- **visible light**—wavelengths 400–780 nm and frequencies around 10^{14} Hz;
- **near infrared or IR-A**—wavelengths 780–1400 nm and frequencies around 10^{14} Hz;
- **middle infrared or IR-B**—wavelengths 1400–3000 nm and frequencies around 10^{14} Hz;
- **far infrared or IR-C**—wavelengths 3000 nm–1 mm and frequencies around 10^{11}–10^{14} Hz.

15.3.2.2 Sources

Visible light comes mainly from the sun but lacks the intensity to harm the eyes unless a person stares directly at the sun for a sustained period. There are also human-made visible light sources such as incandescent lamps, lasers and gas discharge sources that produce very intense visible light. Intense visible light may also be a by-product of industrial processes such as welding. Most artificial sources used in industry, consumer, scientific and medical applications emit visible radiation that is not dangerous. Usually, the eye is protected by the natural tendency to blink or avert the gaze in response to intense visible light.

Infrared radiation from the sun is felt as warmth. Transfer of infrared radiation also occurs from any object that is at a higher temperature than another receiving object. Intense sources of infrared radiation in the workplace and the home are many and include heating devices such as furnaces, ovens, infrared lamps and some lasers. In general, most high-intensity broadband sources, such as incandescent lamps, produce negligible levels of IR-C compared with emissions of visible light and other infrared bands. Where substantial IR-C exposure is present (e.g. steel furnaces), it may contribute significantly to heat stress (International Commission on Non-Ionizing Radiation Protection [ICNIRP 2006]).

15.3.2.3 Health Effects and Quantification

The eye is the most vulnerable organ to visible and near-infrared radiation because the wavelengths from these sources are focused by the lens on to the retina, where thermal or photochemical damage may result. Photochemical injury to the retina is most likely due to long exposures to visible light, peaking at 440 nm. Thermal injury to the retina is more common with short exposures to near-infrared radiation. The lens is most vulnerable to middle infrared, and the formation of cataracts is common in glass-blowers ('glass-blower's cataract') and furnace workers unless eye protection is worn. Far infrared is absorbed at the surface of the eye and may cause superficial burns to the cornea or skin. However, skin is not usually at risk unless the source is very intense and pulsed, because the pain reflex limits the duration of exposure. There is no link to cancer for exposure to IR; however, an emerging issue is blue light, that is, shorter-wavelength of visible radiation. There are suggestions that this accelerates retinal aging and suppresses the secretion of melatonin, which is involved in the regulation of sleep and wake cycles (ICNIRP 2020a).

To injure the eye or skin, visible and infrared radiation must be transmitted to and absorbed by the tissues. Control therefore requires knowledge of the spectral distribution of radiation from the source and the total irradiance, measured at the eye or on the skin. At least five types of injury may be caused by visible and infrared radiation:

- thermal injury of the retina (UV-A, visible, IR-A), which requires very high-intensity sources;
- photochemical injury of the retina (UV-A, visible (blue light));
- thermal injury of the lens (IR-A, IR-B);
- thermal injury (burns) to the skin and cornea (visible, IR-C);
- whole body heat stress;
- photosensitisation of the skin, including side effects of medication.

The amount of photochemical injury depends on dose and exposure duration. An extremely bright light viewed for a short time will have the same effect as a bright light viewed for a longer period. These photochemical exposures are accumulative over a working day. However, the amount of thermal injury to a body tissue will depend on whether the heat absorbed can be conducted away from the exposed site quickly enough. If not, the temperature rise will damage the tissue. Sub-threshold thermal exposures are not cumulative over a working day.

Assessing exposure for small sources of visible-light and near-infrared radiation requires knowledge of the spectrally weighted total irradiance (E, W/m^2), measured at the location of the eye. Irradiance measurements are made using a spectrally weighted radiometer matching the applicable action spectrum. Once total irradiance is known, the appropriate exposure standard can be applied, and a maximum permissible duration of exposure can be established.

Assessing exposure for large sources of visible-light and near-infrared radiation requires information about the radiance (or 'brightness') of the source and not the irradiance at the eye. Radiance is measured with a radiometer and is independent of distance from the source. A large source creates an image on the retina that changes in proportion to the irradiated power on the retina. This means the power level of incident radiation and risk of injury remain in proportion. Allowed exposures to large optical sources are expressed as radiance of the source in watts per square metre per steradian—that is, $W/(m^2.sr)$.

The measurement of infrared radiation is a task usually undertaken by a professional with specific expertise in the area (see Section 15.4). Particular consideration should be given to evaluating intense infrared sources that are not visibly bright, as these sources could be easily gazed upon, putting the eye at risk of thermal injury.

H&S practitioners often measure visible light to determine whether a workplace has sufficient illumination. A photometer is used to measure quantities such as luminance (brightness as perceived by a standard human observer) or illuminance (the light falling on a surface) in lux (lx) weighted to the response of the human eye (see Chapter 17).

15.3.2.4 Control Methods

Exposure to visible and infrared radiation can be controlled by shielding, distance and time:

- Certain materials can be used to **shield** against or attenuate both visible and infrared radiation. Protective glasses for outdoor or general use, or face shields and glasses for welding, should meet the relevant International or Australian Standards for eye and face protection. Glasses or goggles are used to protect the eye and are typically made from plastics such as polycarbonate. Sources of infrared radiation should be shielded close up, using reflective materials such as aluminium. Shields, curtains and barriers are used to protect people from welding processes.
- **Distance** from the source should be maximised, since the intensity of non-ionising radiation falls off rapidly with distance.
- Limiting the **time of exposure** to the source will reduce photochemical injury.

Further details regarding controls can be found in the ICNIRP Guidelines (ICNIRP, 2013).

15.3.3 Lasers

15.3.3.1 Properties

LASER is an acronym for Light Amplification by Stimulated Emission of Radiation. Lasers currently produce light in the ultraviolet, visible or infrared region of the electromagnetic spectrum. Laser light differs from ordinary electromagnetic radiation in that the light beam is coherent in space (all the waves are in phase) and time (all the waves are of the same frequency). As a result, the beam has very little divergence and is able to be transmitted large distances while retaining a relatively high level of energy per unit area. In other words, a laser beam remains very intense over long distances. This is in contrast to ordinary light, whose intensity falls off with the square of distance. Lasers produce a single or narrow-wavelength band of light.

15.3.3.2 Sources

Lasers have a wide range of industrial and commercial applications, including for alignment, range-finding in construction, printing, drilling, welding, cutting, advertising and entertainment. Lasers are also used for surgery, ophthalmic and cosmetic procedures, and as pointers in lectures and business presentations.

15.3.3.3 Health Effects

The eyes and skin are the body parts most susceptible to damage from lasers. The intensity of the beam influences the extent of damage, while the emitted wavelength determines the site of damage. A powerful laser, that is not appropriately controlled, may cause serious injury before the body's natural aversion response (blinking or turning away) is triggered. Reflections from mirrors, equipment and other surfaces, including watch faces, are also a significant hazard. A laser may cause photochemical effects (chemical reactions, after-images), thermal effects (burns) and nonlinear effects.

Nonlinear effects are caused by a rapid heating and thermal expansion of biological tissue. The high irradiance produced by some lasers delivers a lot of energy to the biological target in a very short time. The target tissues experience such a rapid rise in temperature that the liquid components of their cells are converted to gas and the cells rupture.

15.3.3.4 Control Methods

The laser classification system has been standardised internationally and indicates a laser's level of risk along with the measures needed for safe use (Table 15.7). Products that contain (embedded) lasers are also classified according to the risk they present. Such 'laser products' should be handled and dismantled with caution, as the lasers within may be more hazardous (i.e. a higher

TABLE 15.7

Summary of Laser Classes, Characteristics and Control Measures

Laser Class	Risk	Characteristics	Specific Control Measures
1	No risk to eyes or skin	Safe under most circumstances including when viewed with optical instruments	None
1M	No risk to naked eyes No risk to skin	These lasers emit in the range 302.5–4000 nm but may be hazardous if optics are used in the beam	i–iii
2	No risk to eyes or skin for short exposure times, including viewing through optical instruments	These lasers must emit visible radiation 400–700 nm as eye and skin protection is afforded by normal aversion responses (e.g. blinking) including where optical instruments are used for intra-beam viewing	None
2M	No risk to eyes or skin for short exposure times	As for class 2, except that viewing may be more hazardous if optics are used in the beam	i–iii
3R	A risk to eyes No risk to skin	These low-power lasers (<5 mW) emit in the range 302.5–10^6 nm, where direct intra-beam viewing is potentially hazardous	i–iii
3B	Medium to high risk to eyes Low risk to skin	As for class 3R, except that direct intra-beam viewing is hazardous. Viewing diffuse surface reflections is normally safe	i–vii
4	High risk to eyes High risk to skin Fire hazard	These lasers have power of 500 mW and above. They cause eye injury and are capable of producing hazardous diffuse reflections, skin injuries and may also be a fire hazard	i–viii

Explanation of control measures:

i. Training
ii. Beam stop or attenuator
iii. Avoid specular reflections
iv. Protective eyewear
v. Warning signs on enclosures
vi. Remote interlock to laser
vii. Key control of laser
viii. Protective clothing

Adapted from AS/NZS IEC 60825-1. Standards Australia 2014.

class) when exposed than when properly encased. It is a legal requirement in Australia for lasers to be labelled according to their particular hazard.

Laser pointers are increasing in efficiency and power, with many capable of dazzling and causing an after-image (flash blindness) if the beam enters the eye. This dazzling may have catastrophic consequences for operators of machinery, vehicles or aircraft.

Laser pointers of power above 1 mW are prohibited from public use in some countries. They are prohibited imports into Australia (Australian Customs and Border Protection Service 2020) and controlled or prohibited as potential weapons in many jurisdictions (e.g. NSW Police Force 2009). Therefore, teachers and public speakers should not exceed a class 1 or 2 laser pointer.

15.3.4 Radiofrequency Radiation

15.3.4.1 Properties

Radiofrequency (RF) radiation covers the frequency range from 1001 kHz to 300 GHz and has corresponding wavelengths from approximately 1000 km to 1 mm (ARPANSA 2021a).

15.3.4.2 Sources

Microwaves and communications devices are the major sources of RF radiation. There are also natural sources, such as the earth, and the cosmos, whose RF radiation covers a range of frequencies but is of very low intensity. Human-made sources, many of which may be found in workplaces, include:

- radar;
- RF induction heaters;
- Wood-glueing and plastic welding;
- electronic article-surveillance systems, RF identification systems;
- microwave ovens;
- mobile telephones, Wi-Fi;
- television transmitters;
- FM and AM radio transmitters.

Exposure to sources such as radar devices would only be high if a person is very close to them. Medical uses of RF radiation include diathermy, microwave treatment and magnetic resonance imaging (MRI). These sources are in the MHz and GHz range and can potentially expose patients to high-intensity levels.

15.3.4.3 Health Effects and Quantification

The human body responds differently to different electromagnetic field frequencies, resulting in different health effects and safety limits. Health effects from RF radiation may be broadly grouped as either thermal or non-thermal. Thermal effects are well established and occur when human tissue is irradiated especially with frequencies between 1 and 10,000 MHz. Thermal effects include both localised heating which can cause burns and tissue damage and whole-body heating which can cause hyperthermia and heat stress. In much the same way, as water absorbs heat in a microwave oven, body tissues absorb and are heated by RF radiation. Different water content means that tissues are not heated uniformly, and some are more sensitive to heat than others (e.g. the brain, lens, testes and the unborn child). The interaction of RF with biological material is dependent on a number of factors including the frequency, the intensity and the duration of the exposure, as well as the size and shape of the receiving material and its composition in terms of its susceptibility to EMF (often called dielectric characteristics). For example, localised warming occurs in

TABLE 15.8
Basic Restrictions Criteria for Radiofrequency Radiation

Frequency Range	Basic Restriction	Health Effects
100 kHz–10 MHz	Induced electric field, E_{ind}	To prevent electrostimulation of excitable tissue
100 kHz–300 GHz	Whole body average specific absorption rate, SAR	To prevent whole-body heat stress
100 kHz–6 GHz	Local SAR (head, torso and limbs)	To prevent excessive localised temperature rise in tissue
400 MHz–6 GHz	Local specific energy absorption, SA	To prevent rapid temperature elevation
6 GHz–300 GHz	Local absorbed power density, S_{ab}	To prevent excessive heating in tissue at or near the body surface
6 GHz–300 GHz	Local absorbed energy density, U_{ab}	To prevent rapid temperature elevation

Adapted from ARPANSA (2021a).

the ankles at 30–80 MHz. Above 10 GHz, radiation is absorbed at the skin surface, increasing the likelihood of cataracts and skin burns.

Non-thermal effects include the induction of electrical currents in tissues, shocks and burns due to contact currents. Shocks and burns can be caused by contact with a source of RF radiation between 100kHz and 110 MHz range. Interference with implanted medical devices such as pacemakers can also occur and it has been suggested that there is no substantiated evidence of a causal relationship between RF radiation and cancer (ICNIRP 2020b).

In order to limit exposure to excessive RF radiation, there are guidelines by ICNIRP which are designed to prevent the established health effects mentioned earlier (ICNIRP 2020a). In Australia, there is an RF exposure standard by ARPANSA (2021a) which has adopted the ICNIRP exposure limits. The exposure limits for RF radiation are defined as 'basic restrictions' and take different forms depending on the band of the RF spectrum under consideration. Basic restrictions are quantities that are impractical to measure. Thus, standards also refer to 'reference levels' which are quantities that can be measured to assess compliance with basic restrictions. The basic restrictions and reference levels have criteria for both occupational and general public scenarios.

The basic restrictions criteria for various frequency ranges are provided in Table 15.8. A description of their derivation is provided in the ICNIRP guidelines (2020).

Measuring non-ionising radiation is often complex, and it is easy to obtain false readings. It is best to engage a health physicist or occupational hygienist with specific expertise in type of radiation with access to the appropriate instrumentation.

15.3.4.4 Control Methods

Exposure to RF fields is best controlled by avoidance (including distance) and shielding at the source.

Leakage of RF radiation should be limited at the source, using shielding materials to absorb or contain the radiation. While electric fields are easier to shield, the magnetic field component penetrates most materials. Choosing such shielding is a specialist task, as inappropriate shielding can enhance the RF field and lead to higher exposures (see Section 15.4).

In the case of communications technology such as mobile phone antennas and radar devices, the RF radiation cannot be suppressed. Care must be taken to prevent contact between the human body and RF fields by limiting access to transmitters. Conductive suits can partially shield the user if fully enclosed. Such suits have the disadvantage of high thermal load and limited visibility. In addition, any opening in the suit will result in an enhanced local field.

Care should also be taken to prevent access to objects like tools that may cause high levels of contact current and thus pose a risk of shocks. Ideally, these objects should be electrically grounded. Using electrically insulating gloves when handling metal tools can prevent 'startle' currents.

More detailed discussions regarding controls to RF field exposure can be found in the current national standard RPS S-1 (ARPANSA 2021b).

15.3.5 Extremely Low-Frequency Electric and Magnetic Fields

15.3.5.1 Properties

Extremely low-frequency (ELF) electric and magnetic fields have frequencies up to 100 kHz and wavelengths above 1000 km.

15.3.5.2 Sources

The ELF region of the electromagnetic spectrum is associated with the generation, distribution and use of electricity. Electricity supply in Australia is at 50 Hz (6 km). Exposure to relatively high fields can occur in non-electrical occupations, for example, welders, railway drivers and sewing machine operators. The presence of electric charges gives rise to electric fields, which are measured in volts per metre (V/m). The motion of the electric charges (the current) gives rise to magnetic fields, measured in tesla (T), millitesla (mT) or gauss (G; 10000 G = 1 T). Electric fields are easily shielded with common materials, but magnetic fields pass through such materials. Both types of fields are strongest close to the source and diminish with distance.

15.3.5.3 Quantification and Health Effects

Health effects of ELF radiation depend on the component electric and magnetic fields. A WHO task group (WHO 2007) concluded that there are no substantive health issues related to ELF electric fields at levels generally encountered by members of the public. External ELF magnetic fields induce electric fields and currents in the body which, at very high field strengths (well above the 200 μT general public exposure limit), can cause nerve and muscle stimulation and changes in nerve–cell excitability in the central nervous system (ICNIRP 2010).

Both electric and magnetic fields are known to interfere with cardiac pacemakers at levels of exposure that may not otherwise cause adverse health effects. There is conjecture surrounding the possible carcinogenic effects of extremely low-frequency radiation and static electric and magnetic fields, but this link remains unproven (WHO 2007).

Measurement of ELF radiation is a task for a radiation specialist (see Section 15.4).

15.3.5.4 Control Methods

Control of ELF radiation also requires specialist expertise (see Section 15.4). In power-transmission applications, it is impractical to shield ELF radiation. Instead, attempts are made to control the electric field strength at ground level by locating high-voltage transmission lines high in the air and with a corridor of land around them.

15.3.6 Static Fields

15.3.6.1 Properties

Static magnetic and electric fields occur where the frequency is 0 Hz and are characterised by magnetic and electric field strengths that do not change over time.

15.3.6.2 Sources

The earth has a static magnetic field generated by the electric current flowing in its core. Human-made static magnetic fields are usually stronger and occur around direct current (DC) sources, DC transmission lines, electric trains, nuclear magnetic resonance spectrometers, aluminium production plants, powder coating, galvanisation and in some welding. Medical resonance imaging can expose the patient from 0.2 to 7 T as well as medical staff performing MRI procedures.

Static electric fields (also known as electrostatic fields) occur where there are charged bodies. Friction can separate charges and generate strong static electric fields that create a spark on discharge.

15.3.6.3 Health Effects and Quantification

The health effects of static electric fields are poorly understood. Known effects are currently limited to discomfort from spark discharges. However, it is recommended that electric field strength (E) be controlled in order to limit both currents on the body surface and induced internal currents. Field strength should also be controlled in order to prevent safety hazards such as spark discharge and high contact currents on metal objects (ICNIRP 2009).

Magnetic field strength (H), measured in amperes per metre (A/m), and magnetic flux density (B), measured in tesla (T), are used to quantify a magnetic field. Static magnetic fields are likely to cause health effects only when there is movement within the field. A person moving within a field above 2 T can experience sensations of vertigo, nausea, a metallic taste and perceptions of light flashes (ICNIRP 2014). Static magnetic fields also affect implanted metallic devices such as cardiac pacemakers.

15.3.6.4 Control Methods

Protection from static electric fields is similar to ELF and entails grounding objects carrying current and prescribing protective suits similar to those used in the presence of RF radiation.

Static electric fields in isolated applications may be shielded using an earthed conducting enclosure (Faraday cage) and static magnetic fields may be shielded using magnetic shielding. Protection is also gained by maximising distance from the source and minimising time spent in close proximity to it. Recent controls include equipment design which allows for cancellation of fields (due to opposing fields) as a mitigation method. Small metal objects need to be kept away from strong static magnetic fields, which can turn them into missiles.

15.4 THE ROLE OF THE H&S PRACTITIONER WITH REGARD TO RADIATION HAZARDS

15.4.1 Ionising Radiation

H&S practitioners may become involved with ionising radiation when a source is present in their workplace. Their role is generally one of identifying sources of ionising radiation, assisting in complying with any routine requirements such as registration, licensing and managing radiation dose badges for radiation workers within their organisations. They may undertake basic surveys for contamination, but other monitoring methods and interpretation of the results can be complex and can require specialist expertise.

If a source has very high activity or high energy, or generates beams of ionising radiation, a specialist will be needed to advise on correct placement of shielding and recommend other appropriate controls.

15.4.2 Non-ionising Radiation

H&S practitioners may become involved with non-ionising radiation when a source is present in their workplace. Their role is generally one of the identifying and characterising sources of non-ionising radiation, assisting in complying with any routine requirements such as registration, licensing and managing radiation dose badges for radiation workers within their organisations. They may also be involved in

- reviewing new workplace plans, equipment or arrangements;
- identifying potential hazards associated with equipment, processes or environments, like a new high-intensity radiation source, ultraviolet lamp or laser;
- dealing with concerns of the workforce;
- arranging the appropriate specialist to assess and address issues related to non-ionising radiation;
- prescribing simple control measures, such as hats and sunscreen for protection from ultraviolet radiation outdoors. Some controls should be carefully chosen to avoid increasing the risk of exposure to non-ionising radiation;
- ensuring radiation workers undertake the appropriate training;
- reviewing procedures and processes.

15.4.3 Specialist Expertise

Specialists such as health physicists and occupational/industrial hygienists are often engaged to measure and prescribe controls for both ionising and non-ionising radiation. These specialists should be members of relevant National or International professional societies. In part, specialists have access to special equipment required for surveys that determine the risk to workers.

It should be remembered that the electromagnetic compatibility (EMC), that gives rise to the CE Mark for immunity from electromagnetic interference of equipment, does not indicate direct health effects on people, but indirect health such as interference with a pacemaker could lead to a health consequence. Human exposure should be determined by appropriate radiation surveys and personal monitors.

REFERENCES

Australian Customs and Border Protection Service 2020, *Can You Bring It in?* https://www.abf.gov.au/entering-and-leaving-australia/can-you-bring-it-in/list-of-items# [12 December 2023].

Australian Radiation Protection n.d.-a, *Ultraviolet Radiation Services: Testing and Services Guide*. ARPANSA, Melbourne, https://www.arpansa.gov.au/sites/default/files/uv_catalogue.pdf [12 December 2023].

Australian Radiation Protection n.d.-b, *What Is Radiation? What Is the Electromagnetic Spectrum*, ARPANSA, Melbourne, https://www.arpansa.gov.au/understanding-radiation/what-is-radiation#whatistheelectromagneticspectrum [12 December 2023].

Australian Radiation Protection and Nuclear Safety Agency (ARPANSA) 2006, *Radiation Protection Standard for Occupational Exposure to Ultraviolet Radiation*, Radiation Protection Series. No. 12, ARPANSA, Melbourne, https://www.arpansa.gov.au/sites/default/files/legacy/pubs/rps/rps12.pdf [12 December 2023].

Australian Radiation Protection and Nuclear Safety Agency (ARPANSA) 2011, *Monitoring, Assessing and Recording Occupational Radiation Doses on Mining and Mineral Processing*, Radiation Protection Series No. 9.1, ARPANSA, Melbourne, www.arpansa.gov.au/pubs/rps/rps9_1.pdf [28 September 2023].

Australian Radiation Protection and Nuclear Safety Agency (ARPANSA), 2012b, *Cosmic Radiation Exposure When Flying*, Fact Sheet 27, ARPANSA, Melbourne, www.arpansa.gov.au/radiationprotection/Factsheets/is_cosmic.cfm [28 September 2023].

Australian Radiation Protection and Nuclear Safety Agency (ARPANSA) 2012c, *What Is Radiation*, Melbourne, www.arpansa.gov.au/Radiation, ARPANSA, Melbourne, Protection/basics/understand.cfm [28 September 2023].

Australian Radiation Protection and Nuclear Safety Agency (ARPANSA) 2015, *Ionising Radiation and Health*, ARPANSA, Melbourne, www.arpansa.gov.au/understanding-radiation/radiation-sources/more-radiation-sources/ionising-radiation-and-health [28 September 2023].

Australian Radiation Protection and Nuclear Safety Agency (ARPANSA) 2018, *ARPANSA Advisory Note, New Dose Coefficients for Radon Progeny: Impact on Workers and the Public*, ARPANSA, Melbourne, https://www.arpansa.gov.au/understanding-radiation/sources-radiation/radon/new-dose-coefficients-radon-progeny-impact-workers [12 December 2023].

Australian Radiation Protection and Nuclear Safety Agency (ARPANSA) 2019, *ARPANSA Health Effects of Ionising Radiation*, ARPANSA, Melbourne, https://www.arpansa.gov.au/understanding-radiation/what-is-radiation/ionising-radiation/health-effects [28 September 2023].

Australian Radiation Protection and Nuclear Safety Agency (ARPANSA) 2021a, *Radiation Protection Standard—for Limiting Exposure to Radiofrequency Fields—100 kHz to 300 GHz*, Radiation Protection Series S-1, ARPANSA, Melbourne.

Australian Radiation Protection and Nuclear Safety Agency (ARPANSA) 2021b, National Directory for Radiation Protection National Directory for Radiation Protection (2nd Edition, 2021) (NDRP 2nd edition, 2021). ARPANSA, Melbourne, https://www.arpansa.gov.au/regulation-and-licensing/regulatory-publications/national-directory-for-radiation-protection [12 December 2023].

Boise State University, n.d., *116 19.3 Radioactive Decay, Figure 6*, https://boisestate.pressbooks.pub/chemistry/chapter/20-3-radioactive-decay/ [12 December 2023].

Cooper, M.B. 2005, *Naturally Occurring Radioactive Materials (NORM) in Australian Industries—Review of Current Inventories and Future Generation*, Radiation Health and Safety Advisory Council, ARPANSA, Melbourne.

Gonzalez, G.A. and Anderer, J. 1989, 'Radiation versus radiation: nuclear energy in perspective—a comparative analysis of radiation in the living environment', *IAEA Bulletin*, vol. 31, no. 2, pp. 21–31.

International Agency for Research on Cancer (IARC) 2012, Radiation. Volume 100D. *IARC Monographs on the Evaluation of Carcinogenic Risks to Humans*. IARC, Lyon.

International Atomic Energy Agency 2011, *Radiation Protection and Safety of Radiation Sources: International Basic Safety Standards: Interim edition No. GSR Part 3* (Interim), www-pub.iaea.org/MTCD/publications/PDF/p1531interim_web.pdf [28 September 2023].

International Commission Non-Ionizing Radiation Protection 2006, 'ICNIRP statement on far infrared radiation exposure', *Health Physics*, vol. 91, no. 6.

International Commission Non-Ionizing Radiation Protection 2009, 'ICNIRP Guidelines on limits of exposure to static magnetic fields'. *Health Physics*, vol. 96, no. 4, pp. 504–14.

International Commission Non-Ionizing Radiation Protection 2010, 'ICNIRP Guidelines for limiting exposure to time-varying electric and magnetic fields (1 Hz to 100 kHz)', *Health Physics*, vol. 99, no. 4, pp. 813–836.

International Commission Non-Ionizing Radiation Protection 2013, 'ICNIRP guidelines on limits of exposure to incoherent visible and infrared radiation', *Health Physics*, vol. 105, no. 1, pp. 74–96.

International Commission Non-Ionizing Radiation Protection 2014, 'ICNIRP Guidelines for limiting exposure to electric fields induced by movement of the human body in a static magnetic field and by time-varying magnetic fields below 1 Hz', *Health Physics*, vol. 106, no. 3, pp. 418–425.

International Commission Non-Ionizing Radiation Protection 2020a, 'ICNIRP light-emitting diodes (LEDS): Implications for safety'. *Health Physics*, vol. 118, no. 5, pp. 549–561.

International Commission Non-Ionizing Radiation Protection 2020b, 'ICNIRP Guidelines for Limiting Exposure to Electromagnetic Fields (100 kHz to 300 GHz)'. *Health Physics*, vol. 118, no. 5, pp. 483–524.

International Commission on Radiological Protection 1990, '1990 recommendations of the International Commission on Radiological Protection, ICRP publication 60', *Annals of the ICRP*, vol. 31, no. 1–3, pp. 100–5.

International Commission on Radiological Protection 2007, 'The 2007 recommendations of the International Commission on Radiological Protection, ICRP publication 103', *Annals of the ICRP*, vol. 37, pp. 2–4.

International Commission on Radiological Protection 2017. 'Occupational Intakes of Radionuclides: Part 3. ICRP Publication 137', *Annals of the ICRP*, vol. 46, no. 3–4.

International Commission on Radiological Protection 2019. 'Occupational Intake of Radionuclides: Part 4. ICTP Publication 141', *Annals of the ICRP*, vol. 48, no. 2–3.

Ministry of the Environment, Government of Japan & National Institutes for Quantum Science and Technology 2022, *Basic Knowledge and Health Effects of Radiation*, basic-1st-vol1.pdf (env.go.jp) [12 December 2023].

Morris, N.D., Thomas, P.D. & Rafferty, K.P., 2004, *Personal Radiation Monitoring Service and Assessment of Doses Received (2004)*, Technical report series no. 139, ARPANSA, Melbourne.

NSW Police Force 2009, *Laser Pointers—Questions and answers*, NSW Government, Sydney, https://www.police.nsw.gov.au/online_services/firearms/laser_pointers/laser_pointers_-_questions_and_answers [28 September 2023].

Simon, S.L. 2007, 'Introduction to radiation physics and dosimetry', *Radiation Epidemiology Course*, U.S. National Cancer Institute, Bethesda, Maryland.

Solomon, S.B., Langroo, R., Peggie, J.R., Lyons, R.G. & James, J.M. 1996, *Occupational Exposure to Radon in Australian Tourist Caves: An Australia-wide study of radon levels*, Australian Radiation Laboratory, Melbourne.

Standards Australia 2014, *Safety of Laser Products, Part 1: Equipment Classification, Requirements and User's Guide, AS/NZS IEC 60825.1*, Standards Australia, Sydney, pp. 13–14, 100. https://www.standards.org.au/access-standards

UNEP, 2016, Radiation: Effects and Sources, United Nations Environment Programme, https://www.unep.org/resources/report/radiation-effects-and-sources [12 December 2023].

United Nations Scientific Committee on the Effects of Ionising Radiation 2008, *Sources and Effects of Ionizing Radiation*, Vol. 1, United Nations, New York.

Voytchev, M., Ambrosi, P., Behrens, R. and Chiaro, P. 2011, 'IEC standards for individual monitoring of ionising radiation', *Radiation Protection Dosimetry*, vol. 144, no. 1–4, pp. 33–6.

World Health Organisation 2007, *Electromagnetic Fields and Public Health—exposure to extremely low frequency fields*, Fact sheet no. 322, WHO, Geneva.

16 The Thermal Environment

Dr Ross Di Corleto and Jodie Britton

16.1 INTRODUCTION

Workplaces are seldom an ideal environment and can often expose the individual to physical extremes of temperature. This chapter looks at the impact of these parameters on the individual and outlines assessment protocols and guidelines for their management.

The iron ore mines in the Pilbara region of northern Western Australia and the diamond mines of the Canadian Northwest Territories have one thing in common, extremes of temperature and what can often be an inhospitable working environment. The predicted influence of global climate change will no doubt further exacerbate working in these extreme environments with all regions of the planet expected to experience warming and extreme weather events in the future. The forecasted increase in environmental heat is likely to affect large populations of people, causing both acute and chronic health impacts with the average global temperature expected to increase between 1.8 and 4.9°C by the year 2100. (Kjellstrom et al. 2010, Raftery et al. 2017). The human body functions best in a moderate climate, and variations that can result in a lowering or increase of the core body temperature can lead to serious physiological injury or illness. While the identification of such conditions is often obvious, their control and successful correction are not always straightforward, owing to the myriad variations that can be involved. The three key areas where variation can occur include the environment, task and the individual. By taking a systematic approach to the investigation of heat strain, such confusion can often be overcome, and practical controls identified and implemented.

16.2 WORK IN HOT ENVIRONMENTS

Workers in hot environments, around furnaces, smelters, boilers, confined spaces or out in the sun, can be subjected to considerable thermal stress. Because of natural climatic conditions, outdoor lifestyles and work styles, many environments have a high potential for heat-related illnesses. The health and safety practitioner should be able to recognise the physical factors contributing to heat stress and how the body responds to them and be familiar with control procedures for these adverse factors. This section provides a basic introduction to the concepts of heat stress and its management. More comprehensive discussion in this area may be found in specialist texts and guides such as the Australian Institute of Occupational Hygienists' *A guide to managing heat stress: Developed for use in the Australian environment* (Di Corleto et al. 2013), from which sections of this text have been adapted.

16.2.1 Heat Stress and Heat Strain

It is important at the outset to define two key terms associated with work in the thermal environment. The combined effect of external thermal environment and internal metabolic heat production constitutes the **thermal stress** on the body. The response to the thermal stress from bodily systems such as the cardiovascular, thermoregulatory, respiratory, renal and endocrine systems constitute the **thermal strain**. Thus, environmental conditions, metabolic workload and clothing,

DOI: 10.1201/9781032645841-16

individually or combined, can create **heat stress** for the worker. The body's physiological response to that stress—for example, sweating, increased heart rate and elevated core temperature—is the **heat strain** (Di Corleto et al. 2013).

16.2.2 The Heat Balance Equation

In order for the body to maintain thermal equilibrium and avoid illness or injury, a thermal balance must be maintained. This is represented by the equation:

$$S = M \pm C \pm R \pm K - E \tag{16.1}$$

where

S = net heat accumulation by the body
M = metabolic heat output
C = convective heat input or loss (can be positive or negative)
R = radiant heat input or loss (can be positive or negative)
K = conductive input or loss (can be positive or negative)
E = evaporative cooling by sweating (can only be negative).

Environmental or personal work factors can prevent the body from maintaining heat balance for a number of reasons, these can be:

- the air temperature is too high;
- humidity is too high;
- there is a high radiant heat load;
- the worker is constricted by insulating clothing;
- the worker's personal factors (age, health, acclimatisation and drugs (illicit or prescribed); or
- the level of hydration.

During any activity, the body automatically attempts to maintain a constant core body temperature range from 36.8°C to 37.4°C by balancing out the heat gain and heat loss. Working creates metabolic heat and that heat is carried by the blood to the surface of the skin. The work causes the heart to pump faster and so carries the blood faster to the surface. The body dissipates heat through the skin via vasodilation. Heat is transferred from the blood to the air surrounding the skin surface against a temperature gradient via radiation, convection and the cooling mechanism provided by evaporation of sweat off the skin.

16.2.3 Acclimatisation

Workers exposed to repetitive bouts of work in hot environments (either artificial or natural) eventually become acclimatised, which ultimately reduces heat strain. Key physiological responses to acclimatisation are a lower heart rate, an earlier onset of sweating and dilute sweat content. There are different rates of physiological change and adaptation in the acclimatisation process with some occurring more rapidly than others.

- The first stage of the acclimatisation process usually involves the cardiovascular processes of the body such as heart rate decreases which can occur in the first four to five days in physically fit individuals. Plasma volume expansion of up to 16% can occur over the first three to five days (Pryor et al. 2018).

- The intermediate phase occurs when cardiovascular stability has been assured, and surface and internal body temperatures are lower. Usually, 75–80 per cent of optimum can be achieved by day 8 (Pandolf 1998; Roussey et al. 2021), although some research carried out in northern Australia (Brake & Bates 2001) suggests that about 70–80 per cent will be achieved after about 7–10 days Pryor et al. (2018) suggest approximately 14 days.
- The third phase (>15 days) sees a decrease in the salt content of sweat and urine, and other compensations to conserve body fluids and restore electrolyte balances. Usually, 93 per cent of optimum is achieved by day 18 and 99 per cent by day 21 (ACGIH 2000).

It is very important to note that employees who have been on extended leave, new employees and contract workers from a cooler climatic location will not be acclimatised, and this must be taken into consideration when scheduling work in a hot environment. Generally, new workers in hot environments must be permitted time to acclimatise. While programs exist which are designed to achieve faster acclimatisation (Wardenaar et al. 2021) these are not generally employed in the workplace. Additionally, as acclimatisation is obtained to the level of the heat exposure present, a person will not be able to fully acclimatise to a sudden higher level of exposure such as those experienced during a heat wave.

The rate of decay of this heat acclimatisation has been suggested to occur such that, one day of acclimatisation is lost for every two days spent without working in the heat (Pryor et al. 2018). The first heat adaptations to decline are usually those that were first to have been achieved such as the cardiovascular adaptations (Garrett et al. 2011).

Recent studies have identified a potential gender difference in acclimatisation and modified heat adaptation regimes have been trialled to improve performance in the heat (Kelly et al. 2023; Kirby et al. 2019).

16.2.4 The Basic Forms of Heat Illness

The body experiences physiological heat strain, with a range of different symptoms and illnesses which are dependent on the degree of heat stress. The conditions of medical importance, ranging from least to most hazardous, are:

- **Behavioural disorders**: Chronic or transient simple physical heat fatigue often occurs in workers from colder climates who are unacclimatised to continuously hot weather. This can often manifest itself as a change in demeanour, irritability, tiredness and lethargy, impaired judgement, loss of cognitive function and poor concentration. The relationship between performing manual work in heat and a subsequent reduction in cognitive function has been shown to have a significant link to workplace safety performance (Ganio et al. 2011; Knapik et al. 2002, Xiang et al. 2014, Yuan et al. 2022). Lifestyle changes (suitable clothing, mid-day resting), avoiding strenuous work during the heat of the day and acclimatisation, are appropriate to alleviate most symptoms.
- **Heat rash (or prickly heat)**: This usually occurs as a result of continued exposure to humid heat during which the skin remains continuously wet from unevaporated sweat. This can often result in blocked glands, itchy skin and reduced sweating. In some cases, prickly heat can lead to lengthy periods of disablement (Donoghue & Sinclair 2000). Where conditions encourage the occurrence of prickly heat (e.g., damp situations in tropical environments or deep underground mines), control measures may be important to prevent onset. Keeping the skin clean, cool and as dry as possible to allow recovery is generally the most successful approach.

- **Heat cramps**: These are characterised by painful spasms in one or more muscles. Heat cramps may occur in persons who sweat profusely in heat without replacing their salt losses, or unacclimatised personnel with higher levels of salt in their sweat. Resting in a cool place and oral replacement of electrolytes will rapidly alleviate cramps (Casa et al. 2008; Casa 2018). The use of salt tablets is undesirable. Counselling by medical staff or a health-care practitioner should be sought to ensure workers maintain a balanced intake of electrolytes, with meals if required. Note that heat cramps not only occur most commonly during heat exposure but can also occur sometime after heat exposure (Di Corleto et al. 2013).
- **Heat oedema**: Swelling of the limbs caused by vasodilation usually in the in the hands, ankles and feet in response to heat exposure. It is regarded as one of the mildest heat-related illnesses. It does not directly contribute to an increased risk of developing more serious heat-related conditions. Treatment is by removing the individual from heat and elevating the affected limbs (Adams & Jardine 2019; Sorensen & Hess 2022).
- **Fainting** (or **heat syncope**): Exposure of fluid-deficient persons to hot environmental conditions can cause a major shift in the body's remaining blood supply to the skin vessels, in an attempt to dissipate heat. This ultimately results in an inadequate supply of blood being delivered to the brain, one consequence of which is fainting. Fainting may also occur without a significant reduction in blood volume as a result of wearing restrictive or confining clothing, or postural restrictions and changes.
- **Heat exhaustion**: Although serious, heat exhaustion is initially a less medically severe heat injury than heat stroke, it can become a precursor to heat stroke. Heat exhaustion is generally characterised by clammy, moist skin; weakness or extreme fatigue; nausea; headache; no excessive increase in body temperature; and low blood pressure with a weak pulse. Without prompt treatment, collapse is inevitable. Heat exhaustion occurs most often in persons whose total blood volume has been reduced by dehydration (i.e., depletion of body water as a consequence of deficient water intake) but can also be associated with inadequate salt intake even when fluid intake is adequate. Individuals who have a low level of cardiovascular fitness and/or are not acclimatised to heat have a greater potential to suffer heat exhaustion, sometimes recurrently. This is particularly important where self-pacing of work is not practised. Note that in workplaces where self-pacing is practised, both fit and unfit workers tend to have a similar frequency of heat exhaustion (Brake & Bates 2001). Lying down in a cool place and drinking an electrolyte supplement will usually result in rapid recovery of the victim of heat exhaustion, but a physician should be consulted prior to resumption of work. Heat exhaustion from salt depletion may require further medical treatment under supervision (Glazer 2005).
- **Heat stroke**: This is a state of thermoregulatory failure and is the most serious of the heat illnesses. There are two forms of heat stroke, classic and exertional. Classic heat stroke is most often observed in the very young or elderly who may be immunocompromised and induced by passive heat exposure (Laitano et al. 2019). This form of heat stroke is usually (but not always) characterised by hot, dry skin (anhydrosis); rising body temperature; collapse; loss of consciousness; and convulsions. Exertional heat stroke can occur with young physically fit workers or athletes who collapse during physical activity or exercise in the heat. This form of heat stroke does not require extreme temperature as the thermal loading may have been exacerbated by high physical demand or high levels of personal protective equipment (PPE). Unlike classical heat stroke, it is not unusual to find patients exhibiting heavy sweating during these events (Leon 2015). Without prompt and appropriate medical attention, including removal of the victim to a cool area and applying a suitable method for reduction of the rapidly increasing body temperature (exceeding 40°C), heat stroke

can be fatal. Immediate cooling is necessary to reduce the body's core temperature. Rapid and effective body cooling is a key initial treatment for heat stroke victims. It has been recommended that whole-body immersion in an ice bath be the method of choice when a quick reduction of core temperature is needed. (Casa et al. 2007). Alternatively, rotating ice water-soaked towels with ice packs over the major vessels in the neck, axilla and groin may be used (Laitano et al. 2019). Caution needs to be taken to ensure that over-cooling of hyperthermic individuals does not occur (Gagnon et al. 2010). A heat-stroke victim is a medical emergency and needs immediate and experienced medical attention.
- **Chronic illness**: While the acute illnesses associated with heat exposure are well known, there is increasing evidence that chronic exposure to heat can exacerbate existing chronic diseases and result in other illnesses in the long term. Studies have shown potential increase in susceptibility to kidney stones (Atan et al. 2005; Borghi et al. 1993) and kidney disease (Jimenez et al. 2014; Wesseling et al. 2020). Chronic dehydration has been linked with cardiopulmonary disorders, gastrointestinal dysfunction (El-Sharkawy et al. 2015) and bladder cancer (Jones & Ross 1999).

16.2.5 Factors Influencing Heat Stress and Strain

The working body gains heat from several sources:

- **Muscular activity** from the work. As the muscles of the body undertake work and oxygen is consumed, heat is released, which increases the core temperature.
- **Conductive and convective heat** from working in hot environmental conditions. In some cases, heat is transferred to the body when hot objects are handled. Cool air can cool the body directly. If air temperature is hotter than body temperature, heat flows from the hotter air to the cooler skin surface. Air speed is also very important in workplace cooling because it influences evaporation rate and convective cooling. The humidity of the air is also able to effect the evaporation rate. High humidity has less capacity to absorb moisture hence poorer evaporation of sweat off the skin and less consequent cooling.
- **Radiant heat** from nearby or distant hot bodies. These radiate heat in the infrared region, which passes through air (or vacuum) unobstructed. The infrared energy is absorbed by the body of the worker, equipment in the workplace and surrounding materials. This can be a major factor contributing to heat stress.

It is important to note that in addition to environmental factors and metabolic workload, there are several personal factors that may exacerbate a worker's physiological response to working in heat, these include age, health, diet, hydration, medication and acclimatisation, further detail is provided in Section 16.2.15.

16.2.6 Fluid Intake and Thirst

The importance of adequate fluid intake and the maintenance of correct bodily electrolyte balance cannot be overemphasised. Maintaining heat balance in conditions of heat stress demands the production and evaporation of enough sweat to cool the body and assist with the balance of the heat gain from the environment and metabolism. Continuous production of sweat is influenced by the upper limit of fluid absorption from the digestive tract. Fluid absorption is dependent on gastric emptying and intestinal absorption. Gastric emptying is the process in which the fluids in the stomach pass into the small intestine, where they are absorbed into the bloodstream. These processes are affected by a number of factors, which include the volume, temperature, calorie content and osmolality of the fluid, as well as exercise intensity (Casa et al. 2000). Bariatric surgery is another factor that must be taken into consideration when assessing adequate fluid absorption and subsequent hydration in an industrial setting with the challenges around sufficient fluid intake for those post bariatric surgery often posing a challenge. Physiological effects from dehydration may commence at 1.5–2.0% change in total body weight (Casa et al. 2000; Hunt et al. 2009), while a net fluid loss of 5% or more in an occupational setting is considered severe dehydration.

Urine-specific gravity (U_{sg}) is being used as a measure of dehydration in sporting and industrial settings. The methodology usually employed utilises refractometers, and self-testing via hydration-specific urine test strips is now becoming popular. Urine results can involve a lag factor though this has not been considered as a major issue if samples are collected in the first void in the morning. It is important to note that spot samples taken when there have been acute changes in water flux within the body, that is, high sweating or large intake of water can result in false negatives and positives (Cheuvront & Zambraski 2015). Also, reagent strips can be easily misused due to the wide variety of manufacturers, each containing a specific set of instructions for immersion timing and result reading (Trabelsi et al. 2018). The relationship between urine-specific gravity and hydration level is illustrated in Table 16.1. The dehydration level is described from an athletic performance perspective. It should be noted that if taken from a clinical perspective, the levels are higher, that is, mild is 1–5%, moderate is 5–10% and severe >10% (McDermott et al. 2017).

Saliva has been identified as another method for testing for dehydration; however, it has had less acceptance. Munoz et al. (2013) found that while it could be used for moderate-to-severe dehydration, based on body weight loss, it was not effective for assessing mild dehydration. In a study undertaken by Owen et al. (2019), they concluded that while saliva osmolality showed adequate diagnostic accuracy to identify mild intra-cellular dehydration it did not identify extracellular dehydration with adequate diagnostic accuracy. The reason proposed by the authors was that these markers compared with previous studies (Ely et al. 2014; Oliver et al. 2008) 'may relate to the smaller fluid-deficit and osmotic, volume and autonomic nervous system alterations'. A key benefit of saliva testing is that it is less invasive than other methods and more readily accepted by the workforce. Recently modern instruments have made this and other methods of testing and predicting hydration more reliable and easier. Providing the instructions for use and limitations are understood, they can be useful tools.

The US National Athletic Trainers' Association (NATA) recommends '....to maintain hydration and not allow more than a 2% body mass loss' (McDermott et al. 2017). Research has shown that body weight loss levels of 2% or more can be regarded as indicating that an individual is in the early stages of dehydration (Casa et al. 2010; Cheuvront & Sawka 2005; Ganio et al. 2007; Sawka et al. 2007). In its guideline, the American College of Sports Medicine recommends drinking 0.4–0.8 L/h of fluid during exercise, depending on the size of the individual and the level of work/exercise being undertaken (McDermott et al. 2017; Sawka et al. 2007).

TABLE 16.1
National Athletic Trainers Association Index of Athletic Hydration Status

	Body Weight Loss (%)	Urine-Specific Gravity (USG)
Hydrated	<2	<1.015
Mild-to-moderate dehydration	2–5	1.015–1.030
Severe dehydration	>5	>1.030

Adapted from: McDermott et al. 2017.

While drinking to thirst may be sufficient to offset fluid losses generally in less aggressive climates and during low-intensity exercise of shorter duration (<90 min) (Kenefick 2018), thirst is not a reliable indicator of the need for fluid replacement (Williams 2018) in an occupational environment. The sensation of thirst lags behind the loss of fluid, and most individuals are dehydrated to some degree. Workers should be encouraged to drink small amounts of water frequently rather than larger quantities infrequently. The water should be cool (10–15°C) and be available close to the workplace. In some cases, it may be desirable to flavour the water to make it more palatable, in these cases, low-sugar flavouring should be used. It should be noted that high solute levels of fluids reduce the rate of water absorption in the gastrointestinal system. Alcohol is a diuretic, and at levels above 2% has been found to reduce fluid retention during rehydration and above 4% increases urine output (McDermott et al. 2017). While caffeine can increase non-sweat body fluid losses (urine production) in some individuals, recent studies (Armstrong et al. 2007; Killer et al. 2014; Roti et al. 2006; Seal et al. 2017) have shown that moderate caffeine intake does not have the diuretic effect first thought and can add to the overall fluid intake of an individual. In most situations, the diet provides sufficient salt to maintain electrolyte requirements for acclimatised individuals; however, in situations of severe fluid loss, electrolyte replacement may also be required. With unacclimatised workers in a high heat-stress scenario, a deficiency in electrolyte can occur even when large volumes of fluid are consumed. In such situations commercially available electrolyte replacement drinks can be used sparingly, salt tablets should not be used to fulfil this role. Table 16.2 lists the advantages and disadvantages of a number of drinks commonly used for fluid replacement.

16.2.7 Education

Education and training are key components in any health-management program. In relation to heat stress, it should be conducted for all personnel likely to be involved with:

- hot environments;
- physically demanding work at elevated temperatures; and
- the use of impermeable protective clothing.

Any combination of the above conditions will further increase the risk of a heat-related illness. The education and training of workers and supervisors should encompass the following:

- mechanisms of heat exposure;
- potential heat exposure situations;
- recognition of predisposing factors;
- the importance of fluid intake and adherence to a liquid replacement schedule;
- the nature of acclimatisation;
- effects of alcohol and drug (illicit and prescription) use in hot environments;
- early recognition of symptoms of heat illness;
- prevention of heat illness;
- first aid treatment of heat-related illnesses;
- self-assessment;
- management and control; and
- medical surveillance programs and the advantages of employee participation in such programs.

Training of all personnel in heat-stress management should be recorded on their personal training record.

TABLE 16.2
Analysis of Fluid Replacement

Beverage Type	Uses	Advantages	Disadvantages
Tea/coffee	Before, during and after work	Provide energy Palatable	Excessive quantities, which result in high levels of caffeine my result in nervousness, insomnia and gastrointestinal upset
Sports drinks	Before, during and after work	Provide energy Aid electrolyte replacement Palatable	May not be correct mix Excessive use may exceed salt replacement requirement levels Low pH levels may affect teeth Elevated carbohydrate content may contribute to excess calorie consumption and weight gain
Fruit juices	Recovery	Provide energy Palatable Low in sodium	Not absorbed rapidly
Carbonated drinks	Recovery	Palatable Variety of flavours Provide potassium Low sodium Quick 'fillingness'	Belching 'Diet' drinks supply no energy Risk of dental cavities Some may contain caffeine and excessive intake is undesirable Elevated carbohydrate content may contribute to excess calorie consumption and weight gain
Water and mineral water	Before, during and after exercise	Palatable Most obvious fluid Readily available Low sodium	Not as good for high output events of 60 minutes + No energy provided
Milk	Before and after recovery	Contains sodium Provides some energy Use with fruit and cereal	Slowly absorbed Has fat Not suitable during an event

Adapted from: Pearce 1996.

16.2.8 Self-Assessment

Self-assessment is a key element in the training of workers potentially exposed to heat stress. With the correct knowledge in relation to signs and symptoms, individuals will be in a position to identify the onset of a heat illness in the earliest stages and take appropriate actions. This may simply involve taking a short break and a drink of water, which in most cases should take only a matter of minutes. This brief intervention can help significantly in preventing the onset of more serious heat-related illnesses, particularly when workers are also allowed to carry out tasks at their own pace.

16.2.9 Assessment of the Hot Thermal Environment

Numerous factors can affect the heat stress associated with a particular task or environment, and no single factor can be assessed in isolation. A structured assessment protocol is the suggested approach, with the flexibility to address a variety of situations.

The use of a heat-stress index alone to determine heat stress and the resultant heat strain is not recommended. Each situation requires an assessment that will incorporate the many parameters that may impact an individual working in hot conditions. In effect, a risk assessment must be carried out which includes additional observations such as workload, worker characteristics and PPE, as well as measurement and calculation of the thermal environmental conditions. This process may involve a variety of heat-stress indices, including but not limited to wet bulb globe temperature (WBGT), basic effective temperature (BET), Apparent Temperature (AT), predicted heat strain and thermal work limit (TWL).

WBGT uses air temperature (T_a), globe temperature (T_g) and a natural wet bulb temperature (T_{nwb}) (Figure 16.1). These parameters are incorporated into one of the two formulae for either indoor or outdoor measurements:

$$\text{Indoor}: \text{WBGT} = 0.7\,T_{nwb} + 0.3\,T_g \tag{16.2}$$

$$\text{Outdoor}: \text{WBGT} = 0.7\,T_{nwb} + 0.2\,T_g + 0.1\,T_a \tag{16.3}$$

The BET is predominantly used in the underground coalmining industry and combines dry bulb and aspirated wet bulb temperatures with air velocity. It should be noted that caution should be exercised when utilising BET in situations where values above 31°C may be encountered, since the index is less accurate above this value (Hanson et al. 2000).

Predicted heat strain (PHS) is a rational index (i.e., it is an index based on the heat balance equation—see Section 16.2.2). It estimates the required sweat rate and the maximal evaporation rate, utilising the ratio of the two as an initial measure of 'required wettedness'. This required wettedness is the fraction of the skin surface that would have to be covered by sweat in order for the required evaporation rate to occur. The evaporation rate required to maintain a heat balance is then calculated (Di Corleto et al. 2013).

The TWL was developed in Australia in the underground mining industry by Brake and Bates (2002). TWL is defined as the limiting (or maximum) sustainable metabolic rate that hydrated and

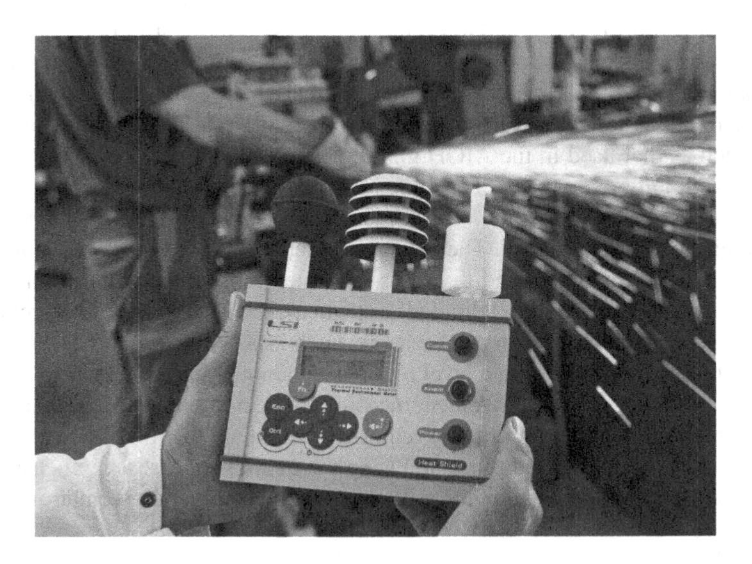

FIGURE 16.1 A modern wet bulb globe temperature (WBGT) instrument.

Source: Active Environmental Solutions.

acclimatised individuals can maintain in a specific thermal environment, within a safe deep body core temperature (<38.2°C) and sweat rate (<1.2 kg/h).

16.2.10 Three-Stage Assessment Protocol

A recommended method of assessment is as follows:

- **Stage 1**: Conduct a basic heat-stress risk assessment incorporating a simple index, and/or a basic review of the task and the environment.
- **Stage 2**: If a potential problem is indicated from Stage 1, then progress to a second level rational index (e.g., PHS and TWL) to make a more comprehensive investigation of the situation and general environment. Ensure that factors such as temperature, radiant heat load, air velocity, humidity, clothing, metabolic load, posture and acclimatisation are taken into account.
- **Stage 3**: Where the calculated allowable exposure time is less than 30 minutes, or there is an involvement of high-level PPE, then employ some form of physiological monitoring.

The first two stages involve measuring the environmental parameters, measuring metabolic work rate factors and/or estimating workload factors from task observation. The third stage looks at the individual's physiological response to the exposure. Technological methods for measuring deep core temperature include swallowing a continuous radio-transmitting temperature sensor. However, for the average H&S practitioner, such techniques or other physiological measurements have limitations. This is often in relation to resistance of participants to the invasive nature of some of the methodology or privacy issues.

16.2.11 Stage 1: Basic Risk Assessment

The first level of assessment utilises a basic observational risk assessment in conjunction with a simple index such as AT or WBGT. It is important that the initial assessment involve a review of the work conditions, the task and the personnel involved. Risk assessments may be carried out using checklists or proformas designed to prompt the assessor to identify potential problem areas. The method may range from a short checklist to a more comprehensive calculation matrix that will produce a numerical result for comparative or priority listing. Also now available are mobile phone and tablet applications to aid in conducting a basic thermal risk assessment (BTRA).

A BTRA, such as that used in the AIOH guidelines (Di Corleto et al. 2013) and illustrated in Table 16.3, is a simple first approach. The table incorporates a number of heat-stress-related factors that could influence an individual. These factors are given a numerical value and weighted according to their potential influence. The values are then used in a simple calculation that yields a numerical value which may be used to assess the potential risk according to a predetermined scale. This approach encourages the individual or team assessing the situation to review a number of parameters and not focus solely on one measure, such as air temperature. The simple AT index assists with the final result by adding a level of environmental measurement and objectivity.

The AT index provides a basic, convenient measure for heat-stress evaluations as in its simplest form only air temperature and humidity are required to estimate the final temperature. The original simple version of the AT is used to try and keep the qualitative assessment easy to use.

Two simple temperature measurements will provide the necessary information:

1. Measure the dry air temperature in the workplace with a thermometer (e.g., mercury/alcohol in glass or thermocouple junction), shielded from radiant energy. A sling psychrometer which is composed of a dry bulb and wet bulb thermometer is often used for this purpose as it will also allow the calculation of humidity.

TABLE 16.3
Example of Basic Thermal Risk Assessment

Hazard Type	Assessment Point Value 0	1	2	3
Sun exposure	Indoors ❑	Shade ❑	Part Shade ❑	No Shade ❑
Hot surfaces	Neutral ❑	Warm on Contact ❑	Hot on contact ❑	Burn on contact ❑
Exposure period	< 30 min ❑	30 min–1 hour ❑	1 hour–2 hours ❑	>2 h ❑
Confined space	No ❑			Yes ❑
Task complexity		Simple ❑	Moderate ❑	Complex ❑
Climbing, up/downstairs or ladders	None ❑	One level ❑	Two levels ❑	> Two levels ❑
Distance from cool rest area	<10 metres ❑	<50 metres ❑	50–100 metres ❑	>100 metres ❑
Distance from drinking water	<10 metres ❑	<30 metres ❑	30–50 metres ❑	>50 metres ❑
Clothing (permeable)		Single layer (light) ❑	Single layer (mod) ❑	Multiple layer ❑
Understanding of heat strain risk	Training given ❑			No training given ❑
Air movement	Strong wind ❑	Moderate wind ❑	Light wind ❑	No wind ❑
Resp. protection (−ve pressure)	None ❑	Disposable half-face ❑	Rubber half-face ❑	Full face ❑
Acclimatisation	Acclimatised ❑			Unacclimatised ❑
Sub-total A				☐
		2	**4**	**6**
Metabolic work rate*		Light ❑	Moderate ❑	Heavy ❑
Sub-total B				☐
		1	**2**	**3** / **4**
Apparent temperature		<27°C ❑	>27°C ≤ 33°C ❑	>33°C ≤ 41°C ❑ / > 41°C ❑
Sub-total C				☐
Total = A plus B	Multiplied by C		=	☐

2. Measure the humidity at the same point in the workplace, which can be readily accomplished with the psychrometer or a hygrometer. The wet bulb is a standard thermometer in which the temperature-sensing element is enclosed in a white cotton wick that is wetted by immersion in distilled water. This temperature reading will take into account the ability of the air to cool by evaporation and the air movement that aids cooling by sweat evaporation from the skin.
3. The dry bulb temperature is aligned with the corresponding relative humidity to determine the apparent temperature in Table 16.4. Numbers in () refer to skin humidity's above 90% and are only approximate.

TABLE 16.4
Apparent Temperature: Temperature—Humidity Scale

Dry Bulb Temperature	Relative Humidity (%)										
(°C)	0	10	20	30	40	50	60	70	80	90	100
20	16	17	17	18	19	19	20	20	21	21	21
21	18	18	19	19	20	20	21	21	22	22	23
22	19	19	20	20	21	21	22	22	23	23	24
23	20	20	21	22	22	23	23	24	24	24	25
24	21	22	22	23	23	24	24	25	25	26	26
25	22	23	24	24	24	25	25	26	27	27	28
26	24	24	25	25	26	26	27	27	28	29	30
27	25	25	26	26	27	27	28	29	30	31	33
28	26	26	27	27	28	29	29	31	32	34	(36)
29	26	27	27	28	29	30	30	33	35	37	(40)
30	27	28	28	29	30	31	33	35	37	(40)	(45)
31	28	29	29	30	31	33	35	37	40	(45)	
32	29	29	30	31	33	35	37	40	44	(51)	
33	29	30	31	33	34	36	39	43	(49)		
34	30	31	32	34	36	38	42	(47)			
35	31	32	33	35	37	40	(45)	(51)			
36	32	33	35	37	39	43	(49)				
37	32	34	36	38	41	46					
38	33	35	37	40	44	(49)					
39	34	36	38	41	46						
40	35	37	40	43	49						
41	35	38	41	45							
42	36	39	42	47							
43	37	40	44	49							
44	38	41	45	52							
45	38	42	47								
46	39	43	49								
47	40	44	51								
48	41	45	53								
49	42	47									
50	42	48									

Source: Steadman, 1979, Table 2, p. 862.

16.2.12 Worked Example of Basic Thermal Risk Assessment

An example of the application of the BTRA would be as follows:

A fitter is working on a pump out in the plant at ground level that has been taken out of service the previous day. The task involves removing bolts and a casing to check the impellers for wear, approximately 2 h of work. The pump is situated approximately 25 metres from the workshop. The fitter is acclimatised, has attended a training session, is wearing a standard single layer long shirt and trousers and is carrying a water bottle, and a respirator is not required. The work rate is light, there is a light breeze, and the air temperature has been measured at 30°C, and the relative humidity at 70%. This equates to an apparent temperature of 35°C (see Table 16.3).

An example of the application of the BTRA using this information is given in Table 16.5.

Examples of Work Rate

Light work: Sitting or standing to control machines; hand and arm work assembly or sorting of light materials.
Moderate work: Sustained hand and arm work such as hammering, and handling of moderately heavy materials.
Heavy work: Pick and shovel work, continuous axe work and carrying loads upstairs.

Subtotal A = 9 This is a general measure of the working environment (excluding temperature) and some key influencing factors associated with heat stress.
Subtotal B = 2 This evaluates the metabolic load on the individual (i.e., the intensity of work being undertaken in the task).
Subtotal C = 3 Here the assessment incorporates the actual environmental temperatures using the apparent temperature index. As this is an important factor in the scenario, it is given added weight by using a multiplication factor rather than addition.

The formula used to arrive at the numerical assessment of risk is (A + B) x C. Hence, the total = (9 + 2) x 3 = 33.

- If the total is **less than 28**, then the risk from thermal conditions is low to moderate.
- If the total is **28 to 60**, there is a risk of heat-induced illnesses occurring if the conditions are not addressed. Further analysis of heat-stress risk is required.
- If the total **exceeds 60**, then the onset of a heat-induced illness is very likely and action should be taken as soon as possible to implement controls.

As the total lies between 28 and 60, there is a risk of heat-induced illness if the conditions are not addressed, and more comprehensive analysis of heat-stress risk or implementation of controls is required.

16.2.13 Stage 2 of the Assessment

When stage 1 of an assessment indicates that the conditions may be unacceptable, as in the above example, relatively simple and practical control measures should be considered based on the outcomes identified in the BTRA. Where these are unavailable, a more detailed assessment is required. This stage usually involves a more extensive measurement survey of the environment, including humidity, air velocity, clothing, posture, globe temperature and metabolic load. These additional data would then be used in a higher-level heat-stress index such as the rational index ISO 7933:2023 *Ergonomics of the thermal environment – Analytical determination and interpretation of heat stress using calculation of the Predicted Heat Strain* (ISO 2023), or other rational indices.

TABLE 16.5
Worked Example of Basic Thermal Risk Assessment

	Assessment Point Value				
Hazard Type	**0**	**1**	**2**	**3**	
Sun exposure	Indoors ❑	Shade ☑	Part shade ❑	No shade ❑	
Hot surfaces	Neutral ☑	Warm on contact ❑	Hot on contact ❑	Burn on contact ❑	
Exposure period	< 30 min ❑	30 min–1 hour ❑	1 hour–2 hours ☑	>2 h ❑	
Confined space	No ☑			Yes ❑	
Task complexity		Simple ❑	Moderate ☑	Complex ❑	
Climbing, up/downstairs or ladders	None ☑	One level ❑	Two levels ❑	> Two levels ❑	
Distance from cool rest area	<10 metres ❑	<50 metres ☑	50–100 metres ❑	>100 Metres ❑	
Distance from drinking water	<10 metres ☑	<30 metres ❑	30–50 metres ❑	>50 Metres ❑	
Clothing (permeable)		Single layer (light) ☑	Single layer (mod) ❑	Multiple layer ❑	
Understanding of heat strain risk	Training given ☑			No training given ❑	
Air movement	Strong wind ❑	Moderate wind ❑	Light wind ☑	No wind ❑	
Resp. protection (-ve pressure)	None ☑	Disposable half face ❑	Rubber half face ❑	Full face ❑	
Acclimatisation	Acclimatised ☑			Unacclimatised ❑	
		3	6	0	
Sub-total A				9	
		2	**4**	**6**	
Metabolic work rate*		Light ☑	Moderate ❑	Heavy ❑	
Sub-total B				2	
		1	**2**	**3**	**4**
Apparent temperature		< 27°C ❑	>27°C ≤ 33°C ❑	>33°C ≤ 41°C ☑	> 41°C ❑
Sub-total C				3	
Total = A plus B	Multiplied by	C	=	33	

Assessments at this level require technical interpretation, and assistance may be required from a suitably qualified specialist such as an occupational hygienist. The number of calculations required with rational indices necessitates the use of pre-programmed instrumentation, a computer program or a calculation spreadsheet. Annex E2 (International Standards Organisation 7933:2023, p 23) details an example of code that may be used to develop a computer program for performing predicted heat-strain model computations. There are also mobile phone and tablet applications available, and some equipment providers include software to calculate PHS with their instruments. This allows the calculation of predicted body core temperatures for specific environmental conditions and workloads for a number of tasks or work phases and may be used as a guide in the development of heat-stress controls. It should always be taken into consideration that the aforementioned tools are guides only and should not be used for the development of absolute safe/unsafe limits.

A heat-stress risk assessment checklist should also be used as part of the stage 2 process to ensure a comprehensive assessment is undertaken. An example of a suitable checklist is presented in Table 16.6.

Figure 16.2 shows a simple, practical portable device for WBGT measurements being used to measure temperatures in an outdoor environment at a mine site.

TABLE 16.6
Heat-Stress Risk Assessment Impacts

Assessment Parameter	Impact
Dry bulb temperature	Elevated temperatures will add to the overall heat burden
Globe temperature	Will give some indication of the radiant heat load
Air movement—wind speed	Poor air movement will reduce the effectiveness of sweat evaporation; high air movements at high temperatures (>42°C) will add to the heat load
Humidity	High humidity is also detrimental to sweat evaporation
Hot surfaces	Can produce radiant heat as well as result in contact burns
Metabolic work rate	Elevated work rates can potentially increase internal core body temperatures
Exposure period	Extended periods of exposure can increase heat stress
Confined space	Normally results in poor air movement and increased temperatures
Task complexity	Will require more concentration and manipulation
Climbing, ascending, descending—work rate increase	Can increase metabolic load on the body
Distance from cool rest area	Long distances may be disincentive to leave hot work area or seen as time wasting
Distance from drinking water	Prevents adequate rehydration
Employee condition	
Medications	Diuretics, some antidepressants and anticholinergics may affect the body's ability to manage heat
Chronic conditions, that is, heart or circulatory	May result in poor blood circulation and reduced body cooling
Acute infections, that is, colds, influenza and fevers	Will affect the way the body handles heat stress, that is, thermoregulation
Acclimatisation	Poor acclimatisation will result in poorer tolerance of heat, that is, less sweating and more salt loss
Obesity	Excessive weight will increase the risk of a heat illness
Age	Older individuals (>50) may cope less well with the heat
Fitness	A low level of fitness reduces cardiovascular and aerobic capacity

(Continued)

TABLE 16.6 (CONTINUED)

Assessment Parameter	
Alcohol in the last 24 hours	Will increase the likelihood of dehydration
Chemical exposure factors	
Gases, vapours and dusts soluble in sweat	May result in chemical irritation/burns and dermatitis
Impermeable clothing	Significantly affects the body's ability to cool
Respiratory protection (negative pressure)	Will affect the breathing rate and add additional stress on the worker
Increased workload owing to personal protective equipment (PPE)	Items such as self-contained breathing apparatus SCBA will add weight and increase metabolic load
Restricted mobility	Will affect posture and positioning of employee

FIGURE 16.2 Use of a portable WBGT in an outdoor industrial setting.

Source: R. Di Corleto.

16.2.14 Stage 3: Individual Heat Stress Monitoring

In some circumstances, rational indices cannot provide the necessary information to guide the assessment of the exposed work group, and the use of individual physiological monitoring may be required. This may include situations of high heat-stress risk or where the rational index indicates that the exposure time is limited to less than 30 minutes or where the individual's working environment cannot be accurately assessed. A common example is work involving the use of encapsulating suits or within enclosed/confined spaces.

Instruments for personal heat-stress monitoring do not measure the environmental conditions leading to heat stress; rather, they monitor the physiological indicators of heat strain—usually body temperature and/or heart rate. Temperature may be measured by a number of routes, including oral, rectal, aural (ear canal or tympanic), oesophageal, skin and internal telemetry. Each of these methods (measuring heart rate and body temperature) has advantages and drawbacks as demonstrated in Table 16.7. The method chosen must not only provide the required data but also be acceptable to the individual being monitored.

TABLE 16.7
Core Temperature Measurement Techniques

Measurement Type	Characteristics	Limitations
Skin (forehead) infrared radiation, electronic thermistor and liquid crystal strip	Rapid, non-invasive, convenient, low-cost and hygienic measurement	• Can be several degrees lower than core temperature • Highly influenced by ambient temperature
Skin (upper arm or axillary). Core temperature algorithms	Rapid, non-invasive, convenient and hygienic measurement	• Accuracy dependent on the algorithm used • Can cause skin irritations • Lower accuracy at lower core temperatures
Oral (electronic device or glass thermometer)	Rapid, non-invasive, hygienic and convenient measurement	• Influenced by positioning and breathing with open mouth • Not accurate in hot and cold environments • Low accuracy • Influenced by head temperature
Axillary (electronic device or glass thermometer)	Rapid, non-invasive, hygienic and convenient measurement	• Strongly affected by ambient temperature and positioning • Reading is lower than in other locations • Significant time lag during cooling or rewarming • Low accuracy
Tympanic (infrared radiation)	Rapid, non-invasive, hygienic and convenient measurement	• Inaccurate in hot environment, if incorrectly positioned • If the tympanic membrane is obstructed by blockages of the external auditory canal by: o ear wax o water o foreign objects • Low accuracy
Oesophageal (electronic thermistor)	Lower third of the oesophagus (approx. 40 cm insertion depth from the incisors) Good correlation with arterial blood temperature, especially in steady state	• Very invasive • Insertion of probe may provoke vomiting, aspiration, nasal bleeding, cardiac arrhythmias, and cardiac arrest. • Can be misplaced in the trachea. • Relatively contraindicated in patients with unsecured airways.
Rectal (electronic thermistor or glass thermometer)	Close correlation with arterial blood temperature in steady state	• Significant lag time during cooling and rewarming • Very invasive and may be embarrassing for the patient • Non-hygienic • Possible perforation of the rectum
Gastrointestinal temperature (telemetry temperature sensor)	Higher validity compared to rectal measurement	• Slower response to changes than oesophageal measurement • Unpredictable location • Must be ingested 4–8 h before use • Invasive

Adapted from: Paal et al. 2022; Moyen et al. 2021.

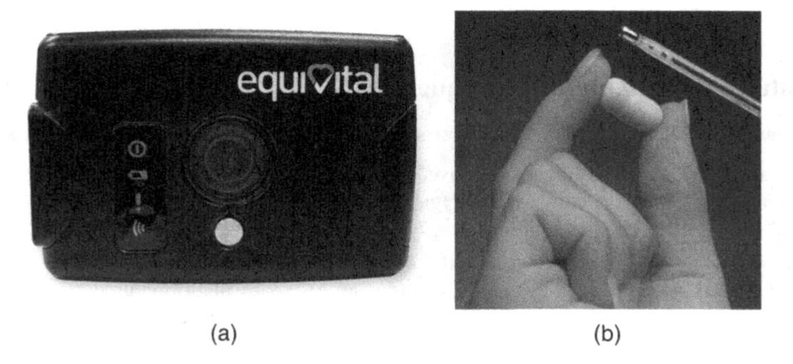

(a) (b)

FIGURE 16.3 Personal heat-stress monitor: (a) data-logging sensor; (b) ingestible capsule.

Source: Active Environmental Solutions.

Figure 16.3 shows a personal heat-stress monitor that employs an ingestible core temperature capsule, which transmits physiological parameters to an external data-logging sensor or laptop computer. It is small, lightweight can be slipped into a shirt pocket or worn on a belt and is fitted with an audible indicator to warn of stressing conditions. Thresholds are not fixed for heart rate or temperature but are established as the physiological measurements are taken.

Such instruments provide a logged time-history graph that may be downloaded to a computer for further analysis. Several of these also have the ability to be able to connect to a mobile phone or tablet via an application with the data being able to be viewed real time as the task at hand is being performed. Depending on the circumstances, physiological monitoring can be involved and complex. When a risk assessment deems it necessary, a competent person with proven technical skills and experience in heat stress and/or human physiology must undertake the assessment. H&S practitioners involved in industries where heat stress is a continuing problem may find this approach very useful. A good example is the stripping of asbestos while dressed in impermeable plastic clothing, a practice not at all suitable for tropical and subtropical climates.

16.2.15 Pre-placement Health Assessment

Pre-placement health assessment screening should be considered for identifying those susceptible to heat injury, or for tasks involving high heat-stress exposures. The standard ISO 12894 *Ergonomics of the Thermal Environment – Medical supervision of individuals exposed to extreme hot or cold environments* (ISO 2001) provides guidance for medical supervision of individuals exposed to extreme heat. Health assessment screening should consider the worker's physiological and biomedical aspects and provide an interpretation of job fitness for the tasks to be performed. Specific indicators of heat intolerance should be included.

Some workers may be more vulnerable to heat stress than others. They include but are not limited to individuals who:

- are dehydrated;
- are unacclimatised to workplace heat levels;
- are physically unfit;
- have low aerobic capacity, as measured by maximal oxygen consumption;
- carry excessive weight;

- are more than 50 years old;
- suffer from diabetes;
- suffer from hypertension;
- suffer from heart, circulatory or skin disorders;
- suffer from thyroid disease;
- are anaemic;
- use medications that impair temperature regulation or perspiration by disrupting hypothalamic thermoregulation, blocking sweat gland excretion or cause narrowing of the blood vessels in the skin, for example, antipsychotics, tricyclic antidepressants, antihistamines, phenothiazines and epinephrine (Stadnyk & Glezos 1983).

Workers with a history of renal, neuromuscular or respiratory disorders, previous head injury or fainting spells, or previous susceptibility to heat illness, may also be at risk (Brake et al. 1998; Hanson & Graveling 1997). In a meta-analysis, Westwood et al. (2021) identified numerous potential contributing factors. They concluded that risk factors tend to be closely related and there is greater risk as these accumulate rather than on single predisposing factor (Westwood et al. 2021). Those individuals who are at greater risk may be excluded from certain work conditions or have more frequent medical checks.

Workers with short-term disorders or illnesses, such as colds or flu, diarrhoea, vomiting, lack of sleep and hangover, should also be considered at risk. These acute disorders will limit their ability to tolerate heat stress and, hence, make them more susceptible to heat illness.

16.2.16 Setting Limits for Heat Exposure

As can be seen from the preceding section, a number of variables associated with individual well-being can affect workers' response to heat. When climatic variables such as humidity and air speed, the clothing/PPE required for the job and the workload itself are taken into account, the equation for estimating heat loads becomes extremely complex. For this reason, the practice of selecting a single parameter such as air temperature as a work/no-work limit is dangerous and should be discouraged. It is impossible to select any one temperature that is suitable for work for all individuals and tasks. Heavy work at temperatures not normally classed as elevated can produce a significant heat stress risk equal to low workloads at higher temperatures. A temperature of 35°C at 15 per cent humidity is more tolerable than one of 30°C at 95 per cent humidity. Considering individual health and fitness variations reveals how difficult it would be to select a single valid temperature for limiting work. Using the 'one temperature' principle, it is quite possible to overprotect some workers and under protect others within the same group. Management of heat stress requires investigating all variables in the workplace rather than only one easily measurable parameter.

16.2.17 Control Measures Against Heat Stress

A number of controls may be utilised in the workplace to modify environmental and task factors and therefore reduce heat stress. Ideally, controls should be prioritised based on the hierarchy of controls. Controls may be determined from the data collected in the risk assessment and measurement stages. Some examples of these potential controls are listed in Table 16.8.

Application of these guidelines should ensure that workplaces are free of heat stress-inducing conditions. Illness from heat stress is totally preventable.

TABLE 16.8
Heat Exposure Control Examples

Elimination	
	Utilise remote-controlled equipment such as mini-diggers and automated hydroblasting equipment
	Conducting the task at cooler times of the day
	Allow hot equipment or vessels to cool before working on or nearby
Engineering controls	
	Erect shade or barriers when radiant heat sources are involved
	Insulating and/or clad equipment
	Improve ventilation by using force draft fans at air temperatures below 4°C or with chiller units at temperatures above
	Use cranes, forklifts and manual handling aids to reduce metabolic work rate
	Establish cooled rest areas in close vicinity to the task
	Changing emissivity of the hot surface
	Dehumidifying air to increase evaporative cooling from sweating
	Eliminating sources of water vapour from leaks in steam lines or standing water evaporating from floors
Administrative	
	Using extra manpower or mechanisation to reduce exposure time for each worker
	Establish appropriate work/rest regimes
	Simplify tasks into smaller components
	Pre-planning of tasks to include heat as a risk
	Ensuring cool palatable water is readily available close by
	Supply of electrolyte replacement where appropriate
Personal protective equipment	
	Providing specialised vortex air-cooled or ice/phase change vests for some continuous-demand tasks

16.3 WORK IN COLD CLIMATES

Concerns relating to work in cold conditions occur predominantly in countries or regions where snow falls during winter and outdoor work must continue. Cold is also relevant to work in freezer plants and cold-storage facilities, and for a few outdoor occupations in winter. Efforts in the prevention of cold-related health problems are directed towards the maintenance of body heat.

During exposure to cold environments, the body responds by constricting blood vessels in the skin. When heat is continually lost and body temperature decreases, muscles begin to involuntarily contract (shiver) to produce heat. The body's 'thermometer' is a pea-sized region of the brain called the hypothalamus. This area regulates heat loss and heat gain mechanisms such as changes in skin blood flow and initiation of sweating and shivering. As a result of severe cold, nerve reactions become impeded and fingers and hands lose dexterity, thus presenting additional safety hazards.

16.3.1 Health Effects of Exposure to Extremes of Cold

Local injuries are generally the most common form of cold injuries. Non-freezing cold injuries (NFCI) are characterized by damage to the soft tissues, nerves and vessels of the hands and feet due to long stay (usually greater than two days) in wet, cold (but not freezing: usually from 0 to 15 °C) conditions (Kravets et al. 2022). These conditions include Trenchfoot (immersion foot), cold immersion injury frostbite and chilblains, which result when insufficient blood reaches the extremities and the fluids around cells freeze. This causes tissue damage and can occur on any superficial tissues in the body. Frostbite can occur when very cold objects are handled or when cold air is passed over the skin for a period of time. The initial stages are sometimes also referred to as 'frostnip'. A sensation of cold is followed by numbness. Frostbite occurs in three degrees: freezing; freezing with blistering or peeling; and freezing with tissue death. These injuries are particularly prevalent for those sleeping rough, or generally exposed to cold wet environments (Tipton & Eglin 2023) as in the military or occupational fields as well as those participating in cold water swimming (Tipton et al. 2017).

Actual freezing is not always necessary to cause serious tissue injury. Injury can also occur from prolonged local cooling at temperatures well above freezing. Trench foot was common in the First World War and immersion foot was described in the Second World War. Both conditions can occur as a result of prolonged exposure in damp or wet conditions from 0°C to 10°C. The feet become cold, swollen and may be painful, itchy or numb. Trenchfoot is usually classified into three forms, a mild form with loss of tactile sensation with pain, a medium to severe form which is characterised by the appearance of blisters and a severe form in which serious infection develops which can lead to necrosis and gangrene (Burton & Edholm 1955; Kravets et al. 2022).

Generalised effects of exposure to severe cold include uncontrollable shivering accompanied by slowing heart rate and a decrease in blood pressure. Similarly, to heat stress, exposure to extreme cold is known to have an impact on mood (tension, depression and anger) as well as fatigue and cognitive performance. Cold-induced cognitive decline has also been linked to workplace safety (Muller et al. 2012). Speech may slur and become incoherent, with drowsiness, irregular breathing and cool skin noted as core body temperature decreases to around 30°C. Serious problems occur at a body temperature below 30°C. Thermogenesis ceases, heat loss becomes pronounced and the respiratory rate decreases markedly. The stages of hypothermia are detailed in Table 16.9.

Popular remedies such as drinking alcohol to keep warm can be dangerous because they exacerbate heat loss by dilating surface blood vessels.

Acclimatisation in cold environments is less well understood and less significant than its counterpart in the heat and there have only been limited studies which have quantified the effect (Blondin et al. 2017; Davis 1961). Humans have learnt to behave in cold environments such that they can survive and keep warm. Physiological acclimatisation is difficult to demonstrate and the evidence for such processes is inconclusive. Recent research by Blondin et al. (2019) demonstrated that a period of consecutive days of cold water immersion can decrease shivering intensity providing some form of acclimatisation.

There is also evidence of local acclimatisation to cold of the fingers and hands. It is often observed that people whose hands are regularly exposed to cold (fishermen and Eskimos)

TABLE 16.9
Stages of Hypothermia

Stage	Symptoms	Estimated Core Temperature (°C)
Hypothermia I (mild)	Conscious and shivering	35–32°C
Hypothermia II (moderate)	Impaired consciousness	<32–28°C
Hypothermia III (severe)	Unconscious	<28°C
Hypothermia IV (very severe)	Apparent death: vital signs absent	Classically <24°C

Source: Lott et al. 2021.

maintain hand temperature. This may be caused by less vasoconstriction and more cold-induced vasodilatation.

Similarly, studies that have shown some cold adaptive traits such as cold tolerance on the extremities (i.e., face and hands) of Korean breath-hold women divers known as haenyeo. Notably, their ability to tolerate cold temperatures while only wearing thin cotton bathing suits has significantly diminished since they have begun to wear wetsuits (Lee et al. 2017).

16.3.2 Assessment of the Cold Thermal Environment

Assessment of cold environments can be undertaken using a similar approach to that of hot environments; however, the effects of cold should be evaluated from two different perspectives:

1. The impact of local cooling on the extremities of the body (e.g. hands, feet and face), employing approaches such as the Wind Chill Index (WCI) (Equation 16.4).
2. The impact of the cooling effect on the body overall, utilising thermal indices based on the heat balance equations, such as the Required Clothing Index (IREQ). The method for calculation of IREQ is defined in terms of the heat balance equation in ISO document ISO 11079 (2007).

The main factors contributing to cold injuries are humidity, wind contact with cold bodies, improper clothing and general state of health. Assessment of localised cooling also requires the consideration of parameters such as convective cooling (wind chill), conductive cooling, extremity cooling and airway cooling.

Wind chill, in which the wind blows away insulating layers of air near the body, can make cold conditions feel bitterly cold. The wind-chill factor is significant for individuals working in cold because threshold limit values (TLVs®) (Table 16.10) are based on workload and wind speed. Not all countries have standards for cold exposure, but the TLV® table, developed in North America, can be applied.

The WCI can be calculated via an equation which was derived to estimate the rate of cooling of exposed skin, which is commonly expressed in SI units as:

$$\mathrm{WCI} = 1.16 \times \left(10 \times \sqrt{\mathrm{v}} + 10.45 - \mathrm{v}\right) \times \left(33 - \mathrm{t_a}\right) \tag{16.4}$$

Where

WCI = Wind Chill Index in W/m^2

v = air velocity in m/s[1]

t_a = temperature of the atmosphere in °C

The IREQ (Required Clothing Insulation) index is used in the assessment of general body cooling. It is defined as 'the resultant clothing insulation required to maintain the body in thermal equilibrium under steady state conditions when sweating is absent and peripheral vasoconstriction is present' (BOHS 1996, p. 56). It incorporates the effects of elements of the heat balance equation, which include:

- air temperature;
- mean radiant temperature;
- relative humidity;
- air velocity; and
- metabolic rate.

TABLE 16.10
Threshold Limit Values as Work/Warm-Up Schedule for a Four-Hour Shift

Air Temp °C (Approx.), Sunny Sky	No Noticeable Wind		8 km/h Wind Speed		15 km/h Wind Speed		25 km/h Wind Speed		35 km/h Wind Speed	
	Max Work (mins)	No. of Breaks	Max Work (mins)	No. of Breaks	Max Work (mins)	No. of Breaks	Max Work (mins)	No. of Breaks	Max Work (mins)	No. of Breaks
–26 to –28	(Norm. breaks)	1	(Norm. breaks)	1	75	2	55	3	40	4
–29 to –31	(Norm. breaks)	1	75	2	55	3	40	4	30	5
–32 to –34	75	2	55	3	40	4	30	5	EWO*	
–35 to –37	55	3	40	4	30	5	EWO*			
–38 to –39	40	4	30	5	EWO*					
–40 to –42	30	5	EWO*							
–43 and below	EWO*									

Source: Adapted from Occupational Health and Safety Division, Saskatchewan Department of Labor, Canada. Reproduced by permission of the American Conference of Governmental Industrial Hygienists.

Notes: Schedule applies to any four-hour work period with moderate-to-heavy work activity, with warm-up periods of 10 minutes in a warm location, and with an extended break (e.g., lunch) at the end of the four-hour work period in a warm location.

Examples of wind movement are: 8 km/h: light flag moves; 15 km/h: light flag is fully extended; 25 km/h: newspaper sheet is lifted; 35 km/h: snow blows and drifts.

TLVs® are applied only for workers who are dressed in dry clothing.

*EWO is emergency work only.

An important aspect of the IREQ index is that it can also be utilised to identify and evaluate controls and improvements in work planning for tasks under cold environmental conditions. As with the heat rational indices such as PHS, any of the parameters of the heat balance equation may be changed, and the relative impact on the IREQ can then be assessed.

ISO 11079:2007 provides more comprehensive detail on the principles and application of the index (ISO 2007). The risk assessment and management of work in cold environments can follow similar principles as that employed for heat stress. In both cases, the strategy of a three-stage protocol is readily applied:

- **Observation**: basic thermal risk assessment or checklist as per annex A in International ISO 15743:2008. ISO Geneva. Organization for Standardization 2008, *Ergonomics of the Thermal Environment—cold workplaces—risk assessment and management*
- **Analysis**: determination of wind cooling effects or incorporating a cold-stress index such as IREQ (ISO 11079:2007, ISO, Geneva).
- **Expertise**: utilisation of individuals with specific competencies such as occupational health-care professionals and occupational hygienists and specialised monitoring equipment (ISO 15743:2008).

16.3.3 Prevention and Control Measures

In formulating measures to reduce the risk of hypothermia and cold injury, factors of relevance are:

- Skin exposure at equivalent chill temperatures of –32 °C or less is to be prevented.
- Wet clothing should be changed for exposures at temperatures below –2 °C.
- Special care needs to be taken if working with evaporative liquids such as alcohol, petrol or solvents that may spill on the hands (Parsons 2014).
- Acclimatisation to cold conditions can have a beneficial, though small, effect.
- Cold air is often very dry, and insidious dehydration via water loss through the skin must be prevented by the consumption of warm, non-alcoholic drinks.
- Salt balance is best controlled by normal dietary means.
- Engineering controls can include:
  - provision of windshields outdoors, or against circulated air indoors in freezer rooms;
  - provision of local heating, hot air jets, radiant heating if bare hands have to be used;
  - avoiding metal tools;
  - avoiding seats for low-temperature work (< –1 °C);
  - use of powered equipment to reduce physical workload;
  - heated shelters for recovery or, if possible, for working in;
- Administrative controls can include:
  - staying within the TLVs® for cold work (this involves measuring air temperature and wind speed and minimising work in cold);
  - maintaining a schedule of rest and liquid refreshment;
  - having an adequate workforce;
  - using the least cold part of the day for the coldest work (i.e., work with the highest exposure potential);
  - instructing workers to recognise and act on adverse effects of cold;
  - avoiding long shifts or excessive overtime in the cold.

PPE against cold primarily takes the form of adequate clothing. Working in cold environments is one of the few work situations where PPE is the first line of defence. Layers of clothing provide a number of insulating air layers. The clothing must be permeable to sweat—inner layers of cotton are ideal. Particular attention must be paid to the hands, the feet and the head—a large heat emitter.

REFERENCES

Adams, W.M. & Jardine, J.F. 2019, 'Minor heat illnesses'. In *Exertional Heat Illness: A Clinical and Evidence Based Guide* (pp. 137–147). Springer International Publishing AG.

El-Sharkawy, A.M., Sahota O. & Lobo D.N. 2015, 'Acute and chronic effects of hydration status on health', *Nutrition Reviews*, vol. 73, no. suppl_2, pp. 97–109.

American Conference of Governmental Industrial Hygienists 2000, *Threshold Limit Values for Chemical Substances and Physical Agents and Biological Exposure Indices*, 6th edn, ACGIH®, Cincinnati, OH.

Armstrong, L.E., Casa, D.J., Maresh, C.M. & Ganio, M.S. 2007, 'Caffeine, fluid-electrolyte balance, temperature regulation, and exercise-heat tolerance', *Exercise in Sport Science Review*, vol. 35, no. 3 pp 135–140.

Atan, L., Andreoni, C, Ortiz, V, Silva, E. K., Pitta, R., Atan, F. & Sougi, M. 2005, 'High kidney stone risk in men working in steel industry at hot temperatures', *Urology*, vol. 65, no. 5, pp 858–861.

Borghi, L., Meshi, T., Amato, F., Novarini, A., Romanelli, A. & Cigala, F. 1993, 'Hot occupation and nephrolithiasis', *Journal of Urology*, vol. 150, no. 6, pp 1757–1760.

Brake, D.J. & Bates, G.P. 2001, personal communication with the author.

Brake, D.J. & Bates, G.P. 2002, 'Limiting metabolic rate (thermal work limit) as an index of thermal stress', *Applied Occupational and Environmental Hygiene*, vol. 17, no. 3, pp. 176–86.

Brake, D.J., Donoghue, A.M. & Bates, G.P. 1998, 'A new generation of health and safety protocols for working in heat', in *Proceedings of Queensland Mining Industry Health and Safety Conference: New Opportunities*, 30 August–2 September, Yeppoon, Queensland, pp. 91–100.

British Occupational Hygiene Society (BOHS) 1996, *The Thermal Environment*, Technical Guide no. 12, 2nd edn, H and H Scientific Consultants, Leeds, UK.

Burton, A.C. & Edholm, O.G. 1955, *Man in a Cold Environment—Physiological and Pathological Effects of Exposure to Low Temperatures*, Edward Arnold, London.

Blondin, D.A., Taylor, T., Tingelstad, H.C., Bézaire, V., Richard, D., Carpentier, A.C., Taylor, A.W., Harper, M., Aguer, C. & Haman, F. 2017, 'Four-week cold acclimation in adult humans shifts uncoupling thermogenesis from skeletal muscles to brown adipose tissue', *The Journal of Physiology*, vol. 595, no. 6, pp. 2099–2113.

Blondin, G.D.P., Friesen, B.J., Tingelstad, H.C., Kenny, G.P. & Haman, F. 2019, 'Seven days of cold acclimation substantially reduces shivering intensity and increases nonshivering thermogenesis in adult humans', *Journal of Applied Physiology*, vol. 126, no. 6, pp. 1598–1606.

Casa, D.J., Armstrong, L.E., Hillman, S.K., Montain, S.J., Reiff, R.V., Rich, B.S., Roberts, W.O. & Stone, J.A. 2000, 'National Athletic Trainers Association position statement: fluid replacement for athletes', *Journal of Athletic Training*, vol. 35, no. 2, pp. 212–24.

Casa, D.J., Ganio, M.S., Lopez, R.M., McDermott, B.P., Armstrong, L.E. & Maresh, C.M. 2008, 'Intravenous versus oral rehydration: physiological, performance, and legal considerations', *Current Sports Medicine Reports*, vol. 7, no. 4, pp. S41–S49

Casa, D.J., McDermott, B.P., Lee, E.C., Yeargin, S.W., Armstrong, L.E., & Maresh, C.M. 2007, 'Cold water immersion: The gold standard for exertional heatstroke treatment', *Exercise and Sport Science Reviews*, vol. 35, no. 3, pp. 141–9.

Casa, Stearns, et al. 2010, 'Influence of hydration on physiological function and performance during trail running in the heat', *Journal of Athletic Training*, vol. 45,no. 2, pp. 147–156.

Casa, D.J. 2018. *Sport and Physical Activity in the Heat.* Springer International Publishing, pp. 42–43.

Cheuvront, S.N. & Sawka, M.N. 2005, 'Hydration assessment of athletes', *Sports Science Exchange*, vol. 18, no. 2, pp. 1–8.

Cheuvront, Kenefick, & Zambraski, E.J. 2015, 'Spot urine concentrations should not be used for hydration assessment: A methodology review', *International Journal of Sport Nutrition and Exercise Metabolism*, vol. 25, no. 3, pp. 293–297.

Davis, T.R.A. 1961, 'Chamber cold acclimatization in man', *Journal of Applied Physiology*, vol. 16, pp. 1011–1015.

Di Corleto, R, Firth, I., & Maté, J. 2013, *A Guide to Managing Heat Stress: Developed for Use in the Australian Environment*. Australian Institute of Occupational Hygienists (AIOH).

Donoghue, A.M., & Sinclair, M.J. 2000, 'Miliaria rubra of the lower limbs in underground miners', *Occupational Medicine*, vol. 50, no. 6, pp. 430–3.

Ely, B.R., Cheuvront, S.N., Kenefick, R.W., Spitz, M.G., Heavens, K.R., Walsh, N.P. & Sawka, M.N. 2014, 'Assessment of extracellular dehydration using saliva osmolality', *European Journal of Applied Physiology*, vol. 114, no. 1, pp. 85–92. https://doi.org/10.1007/s00421-013-2747-z

Gagnon, D., Lemire, B.B., Casa, D.J. & Kenny, G.P. 2010, 'Cold-water immersion and the treatment of hyperthermia: using 38.6°C as a safe rectal temperature cooling limit', *Journal of Athletic Training*, vol. 4, no. 5, pp. 439–44.

Ganio, M.S., Casa, D.J., Armstrong, L.E. & Maresh, C.M. 2007, 'Evidence based approach to lingering hydration questions', *Clinics in Sports Medicine*, vol. 26, no. 1, pp. 1–16.

Ganio, M.S., Armstrong, L.E., Casa, D.J., McDermott, B.P., Lee, E.C., Yamamoto, L.M., Marzano, S., Lopez, R.M., Jimenez, L., Le Bellego, L., Chevillotte, E. & Lieberman, H.R. 2011, 'Mild dehydration impairs cognitive performance and mood of men', *The British Journal of Nutrition*, vol. 106, no. 10, pp. 1535–43. https://doi.org/10.1017/S0007114511002005

Garrett, A.T., Rehrer, N.J. & Patterson, M.J. 2011, 'Induction and decay of short-term heat acclimation in moderately and highly trained athletes', *Sports Medicine*, vol. *41*, no. 9, pp. 757–771. https://doi.org/10.2165/11587320-000000000-00000

Glazer, J. L. 2005, 'Management of heatstroke and heat exhaustion', *American Family Physician*, vol. 71, no. 11, pp. 2133–40.

Hanson, M.A. & Graveling, R.A. 1997, *Development of a Code of Practice for Work in Hot and Humid Conditions in Coal Mines*, IOM Report TM/97/06, Institute of Occupational Medicine, Edinburgh.

Hanson, M.A., Cowie, H.A., George, J.P.K., Graham, M.K., Graveling, R.A. & Hutchison, P.A. 2000, *Physiological Monitoring of Heat Stress in UK Coal Mines*, IOM Report TM/00/05, Institute of Occupational Medicine, Edinburgh.

Hawkins, V.R., Marcham, C.L., Springston, J.P., Miller, J., Braybrooke, G., Maunder, C., Feng, L., & Kollmeyer, B. 2020, *The Value of IAQ: A Review of the Scientific Evidence Supporting the Benefits of Investing in Better Indoor Air Quality*. AIHA, Retrieved from https://commons.erau.edu/publication/1500

Hunt, A.P., Stewart, I.B. & Parker, T.W. 2009, 'Dehydration is a health & safety concern for surface mine workers', in *International Conference on Environmental Ergonomics*, 2–7 August, Boston, International Society for Environmental Ergonomics, UK.

International Organization for Standardization 2001, *Ergonomics of the Thermal Environment—medical supervision of individuals exposed to extreme hot or cold environments*, ISO 12894:2001, ISO, Geneva.

International Organization for Standardization 2007, *Ergonomics of the Thermal Environment—Determination and Interpretation of Cold Stress When Using Required Clothing Insulation (IREQ) and Local Cooling Effects*, ISO 11079:2007, ISO, Geneva.

International Organization for Standardization 2008, *Ergonomics of the Thermal Environment—Cold Workplaces—Risk Assessment and Management*, ISO 15743:2008, ISO, Geneva.

International Organization for Standardization 2023, *Ergonomics of the Thermal Environment: Analytical Determination and Interpretation of Heat Stress Using Calculation of the Predicted Heat Strain*, ISO 7933:2023, ISO, Geneva.

Jimenez, C.A.R., Ishimoto, T., Lanaspa, M.A., Rivard, C.J., Nakagawa, T., Ejaz, A. A., Cicerchi, C., Inaba, S., Le, M. P., Miyazaki, M., Glaser, J., Correa-Rotter, R., González, M. A., Aragón, A., Wesseling, C., Sánchez-Lozada, L. G. & Johnson, R. J. 2014, 'Fructokinase activity mediates dehydration-induced renal injury', *Kidney International*, vol. 86, no. 2, pp. 294–302. https://doi.org/10.1038/ki.2013.492

Jones PA. & Ross, RK. 1999, 'Prevention of bladder cancer', *New England Journal of Medicine.*, vol. 340, pp. 1424–1426.

Kelly, M.K., Bowe, S.J., Jardine, W.T., Condo, D., Guy, J.H., Snow, R.J. & Carr, A.J., 2023, 'Heat adaptation for females: A systematic review and meta-analysis of physiological adaptations and exercise performance in the heat', *Sports Medicine*, vol. 53, no. 7, pp. 1395–1421.

Kenefick, R.W. 2018, 'Drinking strategies: planned drinking versus drinking to thirst', *Sports Medicine (Auckland, N.Z.)*, vol. 48, no. Suppl 1, pp. 31–37. https://doi.org/10.1007/s40279-017-0844-6

Killer, S.C., Blannin, A.K. & Jeukendrup, A.E. 2014, 'No evidence of dehydration with moderate daily coffee intake: A counterbalanced cross-over study in a free-living population', *PloS One*, vol. 9, no. 1. https://doi.org/10.1371/journal.pone.0084154

Kirby, N.V., Lucas, S.J. & Lucas, R.A. 2019, 'Nine-, but not four-days heat acclimation improves self-paced endurance performance in females', *Frontiers in Physiology*, vol. 10, p. 539.

Kjellstrom, T., Butler, A. J., Lucas, R. M. & Bonita, R. 2010, 'Public health impact of global heating due to climate change: potential effects on chronic non-communicable diseases', *International Journal of Public Health*, vol. 55, pp. 97–103.

Knapik, J. J., Canham-Chervak, M., Hauret, K., Laurin, M., Hoedebecke, E., Craig, S. & Montain, S. J., 2002, 'Seasonal variations in Injury rates during US Army basic combat training', *Annals of Occupational Hygiene*, vol. 46, no. 1, pp. 15–23.

Kravets, O.V., Yekhalov, V. V., Trofimov, N. V., Sedinkin, V. A., & Martynenko, D. A. 2022, 'Trench foot and other non-freezing cold injuries (literature review)', *Emergency Medicine*, vol. 18, no. 8, pp. 7–13. https://doi.org/10.22141/2224-0586.18.8.2022.1538

Laitano, O., Leon, L.R., Roberts, W.O. & Sawka, M.N. 2019, 'Controversies in exertional heat stroke diagnosis, prevention, and treatment', *Journal of Applied Physiology*, vol. 127, no. 5, pp. 1338–1348.

Lee, J.-Y., Park, J., & Kim, S. 2017, 'Cold adaptation, aging, and Korean women divers haenyeo', *Journal of Physiological Anthropology*, vol. 36, no. 1, pp. 1–13. https://doi.org/10.1186/s40101-017-0146-6

Leon, L.R. 2015, *Pathophysiology of Heat Stroke*, Morgan & Claypool Life Science Publishers, https://doi.org/10.4199/C00128ED1V01Y201503ISP060

Lott, C., Khalifa, G.E.A., Truhlář, A., Alfonzo, A., Barelli, A., González-Salvado, V., Hinkelbein, J., Nolan, J.P., Paal, P., Perkins, G.D., Thies, K.-C., Yeung, J., Zideman, D.A., Soar, J., Álvarez, E., Barelli, R., Bierens, J.J.L.M., Boettiger, B., & Brattebø, G. 2021, 'European resuscitation council guidelines 2021: cardiac arrest in special circumstances,' *Resuscitation*, vol. 161, pp. 152–219. https://doi.org/10.1016/j.resuscitation.2021.02.011

McDermott, B.P., Anderson, S.A., Armstrong, L.E., Casa, D.J., Cheuvront, S.N., Cooper, L., Kenney, W.L., O'Connor, F.G. & Roberts, W.O., 2017. 'National athletic trainers' association position statement: fluid replacement for the physically active', *Journal of Athletic Training*, vol. *52*, no. 9, pp. 877–895. https://doi.org/10.4085/1062-6050-52.9.02

Moyen, N.E., Bapat, R.C., Tan, B., Hunt, L.A., Jay, O. & Mündel, T. 2021, 'Accuracy of algorithm to non-invasively predict core body temperature using the kenzen wearable device', *International Journal of Environmental Research and Public Health*, vol. 18, no. 24, pp. 13126–13126. https://doi.org/10.3390/ijerph182413126

Munoz, C.X., Johnson, E.C., DeMartini, J.K., Huggins, R.A., McKenzie, A.L., Casa, D.J., Maresh, C.M., & Armstrong, L.E. 2013, 'Assessment of hydration biomarkers including salivary osmolality during passive and active dehydration', *European Journal of Clinical Nutrition*, vol. 67, no. 12, pp. 1257–1263. https://doi.org/10.1038/ejcn.2013.195

Muller, D.M., Gunstad, J., Alosco, M.L., Miller, L.A., Updegraff, J., Spitznagel, M.B. & Glickman, E.L. 2012, 'Acute cold exposure and cognitive function: evidence for sustained impairment', *Ergonomics*, vol. 55, no. 7, pp. 792–798. https://doi.org/10.1080/00140139.2012.665497

Oliver, S.J., Laing, S.J., Wilson, S., Bilzon, J.L.J. & Walsh, N.P. 2008, 'Saliva indices track hypohydration during 48h of fluid restriction or combined fluid and energy restriction', *Archives of Oral Biology*, vol. 53, no. 10, pp. 975–980. https://doi.org/10.1016/j.archoralbio.2008.05.002

Owen, J.A., Fortes, M.B., Ur Rahman, S., Jibani, M., Walsh, N.P. & Oliver, S.J. 2019, 'Hydration marker diagnostic accuracy to identify mild intracellular and extracellular dehydration', *International Journal of Sport Nutrition and Exercise Metabolism*, vol. 29, no. 6, pp. 604–611. doi:10.1123/ijsnem.2019-0022

Paal, P., Pasquier, M., Darocha, T., Lechner, R., Kosinski, S., Wallner, B., Zafren, K. & Brugger, H. 2022, 'Accidental hypothermia: 2021 update', *International Journal of Environmental Research and Public Health*, vol. 19, no. 1, pp. 501–501. https://doi.org/10.3390/ijerph19010501

Pandolf, K.B. 1998, 'Time course of heat acclimation and its decay', *International Journal of Sports Medicine*, vol 19(Suppl2) pp. S157–60. https://doi.org/10.1055/s-2007-971985

Parsons, K. 2014, *Human Thermal Environments*, 3rd edn, CRC Press, London.

Pearce, J. 1996, 'Nutritional analysis of fluid replacement beverages', *Australian Journal of Nutrition and Dietetics*, vol. 43, pp. 535–42.

Pryor, J.L., Minson, C.T. & Ferrara, M.S. 2018, 'Heat acclimation, Chapter 3', in D.J. Casa (Ed) *Sport and Physical Activity in the Heat*. Springer International Publishing. https://doi.org/10.1007/978-3-319-70217-9

Raftery, A.E., Zimmer, A., Frierson, D.M.W., Startz, R. & Liu, P. 2017, 'Less than 2°C warming by 2100 unlikely', *Nature Climate Change*, vol. 7, pp. 637–641. https://doi.org/10.1038/nclimate3352

Roti, M.W., Casa, D.J., Pumerantz, A.C., Watson, G, Judelson, D.Q., Dias, J.C., Ruffin, K., & Armstrong, L.E. 2006, 'Thermoregulatory responses to exercise in the heat: Chronic caffeine intake has no effect', *Aviation, Space & Environment Medicine*, vol. 77, no. 2, pp. 124–129.

Roussey, G., Bernard, T., Fontanari, P. & Louis, J. 2021, 'Heat acclimation training with intermittent and self-regulated intensity may be used as an alternative to traditional steady state and power-regulated intensity in endurance cyclists', *Journal of Thermal Biology*, vol. 98. https://doi.org/10.1016/j.jtherbio.2021.102935

Sawka, M.N., Burke, L.M., Eichner, E.R., Maughan, R.J., Montain, S.J. & Stachenfeld, N.S. 2007, 'ACSM position stand: Exercise and fluid replacement', *Medicine & Science in Sports & Exercise*, vol. 39, no. 2, pp. 377–90.

Seal, A.D., Bardis, C.N., Gavrieli, A., Grigorakis, P., Adams, J.D., Arnaoutis, G., Yannakoulia, M. & Kavouras, S.A. 2017. 'Coffee with high but not low caffeine content augments fluid and electrolyte excretion at rest', *Frontiers in Nutrition*, vol 4, Article. 40. https://doi.org/10.3389/fnut.2017.00040

Sorensen, C. & Hess, J. 2022, 'Treatment and prevention of heat-related illness', *New England Journal of Medicine*, vol. 387, no. 15, pp. 1404–1413. https://doi.org/10.1056/NEJMcp2210623

Stadnyk, A. N. & Glezos, J.D. 1983, 'Drug induced heat stroke', *Canadian Medical Association Journal*, vol. 128, no. 8, pp. 957–9.

Steadman, R.G. 1979, 'The assessment of sultriness. Part I: A temperature-humidity index based on human physiology and clothing science', *Journal of Applied Meteorology (1962–1982)*, vol. 18, no. 7, pp. 861–873.

Tipton, M.J., Collier, N., Massey, H., Corbett, J. & Harper, M. 2017, 'Cold water immersion: Kill or cure?', *Experimental Physiology*, vol. 102, no. 11, pp. 1335–1355. https://doi.org/10.1113/EP086283

Tipton, M. & Eglin, C. 2023, 'Non-freezing cold injury: A little-known big problem', *Experimental Physiology*, vol. 108, no. 3, pp. 329–330. https://doi.org/10.1113/EP091139

Trabelsi, K., Stannard, S.R., Chtourou, H., Moalla, W., Ghozzi, H., Jamoussi, K. & Hakim, A., 2018, 'Monitoring athletes' hydration status and sleep patterns during Ramadan observance: methodological and practical considerations', *Biological Rhythm Research*, vol. 49, no. 3, pp. 337–365. https://doi.org/10.1080/09291016.2017.1368214

Wardenaar, F.C., Ortega-Santos, C.P., Vento, K.A., Beaumont, J.S., Griffin, S.C., Johnston, C. & Kavouras, S.A., 2021. 'A 5-day heat acclimation program improves heat stress indicators while maintaining exercise capacity', *The Journal of Strength & Conditioning Research*, vol. 35, no. 5, pp. 1279–1286. https://doi.org/10.1519/JSC.0000000000003970

Wesseling, C., Glaser, J., Rodríguez-Guzmán, J., Weiss, I., Lucas, R., Peraza, S., da Silva, A.S., Hansson, E., Johnson, R.J., Hogstedt, C. & Wegman, D.H. 2020, 'Chronic kidney disease of non-traditional origin in Mesoamerica: A disease primarily driven by occupational heat stress', *Revista Panamericana de Salud Pública*, vol. 44, no. 1 pp. 1–13. https://doi.org/10.26633/RPSP.2020.15

Westwood, C.S., Fallowfield, J.L., Delves, S.K., Nunns, M., Ogden, H.B. & Layden, J.D. 2021, 'Individual risk factors associated with exertional heat illness: A systematic review', *Experimental Physiology*, vol. 106, no. 1, pp. 191–199. https://doi.org/10.1113/EP088458

Williams, W.J. 2018, 'Behavioural and Technological Adaptation. Chapter 7', in Y. Hosokawa (Ed) *Human health and physical activity during heat exposure*, Springer International Publishing. https://doi.org/10.1007/978-3-319-75889-3

Xiang, J., Peng, B.I., Pisaniello, D. & Hansen, A. 2014, 'Health impact of workplace heat exposure: An epidemiological review', *Industrial Health*, vol. 52, no. 2, pp. 91–101.

Yuan, W., He, B.J., Yang, L., Liu, X. & Yan, L. 2022, 'Heat-induced health impacts and the drivers: Implications on accurate heat-health plans and guidelines', *Environmental Science and Pollution Research*, vol. 29, no 58, pp. 88193–88212. https://doi.org/10.1007/s11356-022-21839-x

17 Lighting

Dr SoYoung Lee, Professor Dino Pisaniello and Professor Bruno Piccoli

17.1 INTRODUCTION

Humans are light-dependent beings and while all our senses are important, our sense of vision is recognised as being our most important link with our surroundings. Approximately 80 per cent of all information to be processed by the human brain is received by way of the eyes, and this has a major influence on the way we function and how well we function.

Good lighting in the design of a workplace refers to the quality of light as well as the quantity. In this regard, the ergonomics of lighting are briefly described to show how workers' operating performance levels improve in a well-lit environment.

This chapter discusses the physiological effects of light and the impact of light on the health and well-being of individuals. Since the early 1980s, a considerable volume of literature has been published on the response of organs such as the pineal gland to light and its impact on the neuro-endocrine system. Light has been shown to be a major factor in entraining the body's circadian rhythms and is also a factor in some disorders, as well as being a therapeutic agent for others. These considerations are all relevant to the design of a well-lit working environment and investigation of complaints about lighting.

17.2 LIGHT

Visible radiation forms the part of the electromagnetic spectrum. 'Light' spans wavelengths of approximately 380–780 nm, with violet light having the shortest and red the longest wavelengths. Adjacent to the visible part of the spectrum, and at shorter wavelengths, is the ultraviolet region, while the adjacent infrared region has longer wavelengths. Light may be natural, from sources such as the sun, or it may be artificial.

17.2.1 Why Is Lighting Important?

The interaction of light with the human body entails multiple effects, both essential and potentially harmful. The principal mechanisms are sensorineural, neuro-endocrine and neuro-psychological. In addition, high-power laser (coherent) light can cause immediate physical damage to exposed tissue and eyes, while exposure to high-intensity incoherent light in the blue region can cause photoreceptor degeneration in the eye. This chapter does not discuss lasers.

The sensorineural response to light enables vision via a complex apparatus associated with the eye. The apparatus, also termed the visual system, is composed of dual and symmetric peripheral organs (eye globes), one signal transmission system to the brain (optical nerve) and specific brain areas (calcarine fissure) for processing and cognitive interpretation of the signal, as an image.

The provision of lighting for vision can be considered a fundamental occupational health and safety requirement, and this can be achieved using natural and/or artificial light sources.

To undertake visually demanding tasks in modern workplaces, workers may suffer overloading of the visual apparatus and adopt unnatural postures leading to pain, fatigue and musculoskeletal

DOI: 10.1201/9781032645841-17

disorders. That is, the 'eye leads the body' since the priority task requires good vision, particularly binocular vision. To complicate matters, only a minority of workers have what is considered to be a 'normal' visual apparatus. For example, refractive corrections to vision may be required to deal with short or long sightedness. Older workers have less light transmission through the eye and will need more light to maintain comfortable vision. The ophthalmological status of individuals in the working population cannot be understated but is unfortunately 'normalised' in lighting standards. That is, the lighting criteria are predicated on normal vision (AS/NZS 1680 2006). Despite the best efforts of lighting designers, complaints of visual disturbance may occur even with lighting optimised for the task and the average worker.

Finally, lighting is only one component of the 'work and vision framework', promulgated by the International Commission on Occupational Health (ICOH) Scientific Committee on Work and Vision (Piccoli et al. 2003). Figure 17.1 indicates the diversity of risk factors for adverse effects on vision, which is medically described as asthenopia.

Lighting of an appropriate quality and adequate quantity contributes to good vision. At the simplest level, it enables workers to see moving machinery and other safety hazards. It also reduces the chance of accidents and injuries from 'momentary blindness' (termed 'adaption') while the eyes adjust to brighter or darker surroundings. The time taken for adaption to occur is much slower than the forming and interpreting of a visual image.

The ability to 'see' at work depends not only on lighting but also on other factors such as:

- proper refraction and ocular motility (abnormal control of eye movement or alignment), and adnexa (e.g., poor lacrimal film or blinking);
- indoor air quality and microclimate;
- the time to focus on an object: fast-moving objects are hard to see;
- the size of an object: very small objects are hard to see;
- the location of the object in the visual field: objects are more difficult to see if too close or at the periphery of the visual field;
- brightness: too much or too little reflected light makes objects hard to see;
- contrast between an object and its immediate background: too little contrast makes it hard to distinguish the two.

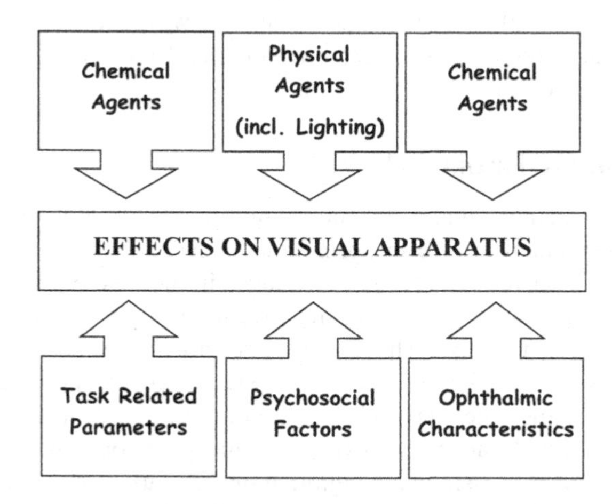

FIGURE 17.1 A conceptual framework for the relationship between individual and work-related influences on the visual apparatus.

Source: Piccoli et al. 2003, p. 401.

17.3 THE EYE AND VISION

Light possesses wave-like properties which enable it to be focused, reflected and refracted. These properties are intrinsic to the sense of vision, as most light is *reflected* off surfaces, *refracted* by the lens of the eye and *focused* to form an image on the retina.

Humans have evolved a complex visual apparatus with two eyes, enabling binocular vision and variable apertures (pupil size).

In relation to vision, the two main functions of the eye are:

- acting as an optical instrument to collect light waves from the environment and project them as images onto the retina;
- functioning as a sensory receptor which responds to the images formed on the retina by sending information by way of the optic nerve to the visual areas of the brain.

In addition, the eye also sends information from light falling on the retina to non-visual parts of the brain, to synchronise neuro-endocrine activity.

The anatomical components of the eye are shown in Figure 17.2.

17.3.1 Vision

Light rays striking the eye first pass through the cornea, which has a high degree of curvature and a refractive index of 1.3765, compared with 1.00 for air. These two factors in combination cause the light rays to bend as they enter the eye. The rays then pass through the lens of the eye, which has a refractive index of 1.40–1.42, causing further bending of the rays. As a result of this process, the light rays for normal sighted persons (emmetropic) are brought to a focus on the retina at the back of the eye.

The lens can change its curvature by the contraction of the muscle. This process is known as accommodation. The ciliary muscles relax when the eye is focused on distant objects, that is, greater than 3 m (with normal vision and eye movement/alignment). Prolonged muscular effort in activating accommodation due to near work (typically < 1 m) leads to overloading and fatigue and may be associated, in the long term (tens of years), to the development of myopia (near-sightedness) (Dutheil et al. 2023). This is why people working with computer screens are advised to take regular breaks and relax their eyes, for example, by gazing out the window at a far distance, that is, more than 3 m.

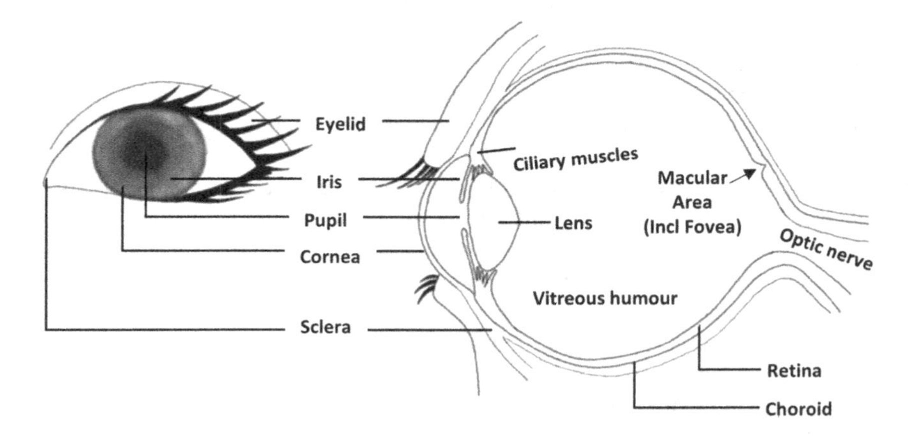

FIGURE 17.2 Schematic diagram of the human eye.

Adapted from: US National Eye Institute 2022.

The retina contains two sets of visual photoreceptor cells, the rods and the cones, the so-called because of their shape. The cones are concerned with colour vision and fine detail at high levels of illumination (photopic vision); the rods are much more sensitive to low levels of illumination than the cones but have less visual acuity, namely clarity and sharpness. The rods are effective for dim light or the so-called scotopic (and mesopic, i.e., twilight) vision.

The human eye has three types of cones to sense light in three respective bands of colour. The biological pigments of the cones have maximum absorption values at wavelengths of about 420 nm (blue), 534 nm (bluish-green) and 564 nm (yellowish-green). Their sensitivity ranges overlap to provide vision throughout the visible spectrum, with maximum efficacy at 555 nm (yellow-green). The fovea is the part of the retina located in the centre of the macula lutea of the retina and is a region where only cones are found and is only about 5 mm in diameter. The cones are particularly densely packed in the fovea, and particularly in the foveola where there is greater ratio of neurons to photoreceptors, hence, it is the region of the highest detailed vision—about 40 times as sharp as that at the retinal border. The photoreceptor cells convert light energy into nerve impulses that travel via the optic nerve to the visual areas of the brain where the image is formed.

Rod cells are most sensitive to wavelengths of light around 498 nm (green-blue) and are insensitive to wavelengths longer than about 640 nm (red). Most people have experienced the Purkinje shift or effect when the eyes transition from predominantly photopic (cone-based) to scotopic (rod-based) vision. The Purkinje shift usually occurs around dusk. As this happens, the maximum sensitivity of the eye shifts towards the blue end of the spectrum at very low illumination levels. The transition vision is termed mesopic vision, where colour appears to vanish from the scene and all objects appear to become progressively more black and white. This has a practical application in the workplace. Under conditions where both the photopic and scotopic systems need to be active, red lighting was thought to provide a solution. For example, submarines and flight decks on aircraft carriers are dimly lit to preserve the night vision of the crew members working there, but the control room must be lit to allow crew members to read instrument panels or maps. Using red lights allows the cones to receive enough light to provide photopic vision (for reading or viewing the control panel). Because the rods are not saturated by bright light and are not sensitive to long-wavelength red light, however, the crew remain dark-adapted, ready and able to view the external environment at night even though some discomfort might be present.

More recently, a third set of photoreceptors has been found in the eye—a small population of intrinsically photosensitive retinal ganglion cells (ipRGCs). These cells contain a photopigment called melanopsin and are connected to the non-visual parts of the brain (Hattar et al. 2002). They show peak response to blue light with a wavelength of 460–480 nm. This action spectrum of ipRGCs is referred to as melanopic. It appears that melanopsin cells play a vital role in the entrainment of the body's circadian rhythms by providing light input to the suprachiasmatic nucleus, which is located in the hypothalamus. The neurons in the hypothalamus control the body's circadian rhythms.

Figure 17.3 shows the spectral sensitivity functions for melanopic and photopic response.

The discovery of ipRGCs is significant to all considerations of the lit environment. Traditionally, lighting industry guidelines followed several scientific principles for efficacy (energy efficiency), light quantity (illumination levels), light quality (colour temperature, colour rendition glare, etc.) and lighting uses (ambient, task or accent lighting). The discovery of ipRGCs introduced a new dimension of considerations for lighting or display designs: that is, how to minimise the adverse effect of artificial lights, via ipRGC photo-transduction (response), on mental and physical health while maximising visual functions and energy efficiency (Cao & Barrionuevo 2015).

The spectral responsivity of ipRGCs is of particular interest to lighting designers. Since the discovery of the ipRGC, the lighting industry has responded by producing design guidelines for

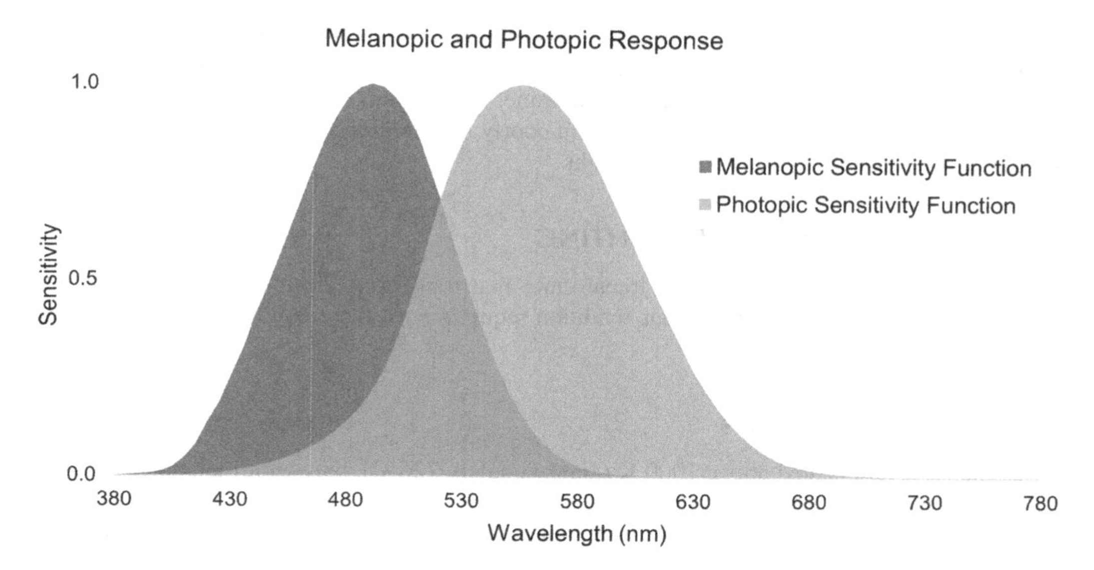

FIGURE 17.3 Melanopic and photopic (visual) response

Adapted from: International Well Building Institute (IWBI) 2022.

creating melanopic (biological) lighting environments. Until recently, lighting designers had no real idea of how to design artificial lighting schemes that would not only provide effective lighting to support visual tasks but also provide non-visual biologically effective illumination. These are now considered in European standards (EN 12464-1 2021) and health building standards, such as WELL (2022) and UL Design Guidelines 24480 (2020).

Relatedly, lighting manufacturers are looking at ways of optimising the spectral power distributions of their products to produce biologically effective white light; that is, white light with an abundance of blue light centred on the melanopic peak wavelength. This is termed human-centric lighting.

17.3.2 The Eye's Response to Differing Light Levels

The level of illumination has a critical effect on the ability of the eye to focus. At low light levels, contrast and image sharpness are diminished, making it more difficult to see. Focusing requires more effort when illumination is poor, especially in older workers, because the transmission of light through the eye is reduced with age.

The size of the pupils is controlled by the muscles of the iris, enlarging or reducing the opening to regulate the brightness of the image projected on the retina. Effort is needed to change the size of the opening, but not to maintain it: when the pupil reaches the required diameter, the muscles of the iris relax. For this reason, repeated viewing of objects that have different brightness levels is more tiring than viewing objects of uniform brightness. Note that light impacting the fovea is not affected by pupil size.

In workplaces, a crucial aspect of safety is avoiding sudden changes from very bright to dimly lit areas or vice versa. Immediately after a person enters a dark room, the eye initially is unable to see anything clearly, particularly small details. Objects become more discernible, usually within a few minutes. This is due to the initial rapid response by the cones in the retina, followed by the slower response of the rods. The rods adapt to a greater degree, although this process may take as long as 25–30 minutes and continue slowly for several hours. When a person moves from dark to light, the reverse process occurs, with a temporary loss of vision occurring until the eyes adapt to the sudden increase in the level of illumination. A typical bright sunny day in summer outside can be 1000 times brighter than an office, which in turn can be 1000 times brighter than a full moon.

Historically, occupational conditions such as miner's nystagmus (rapid, involuntary eye movement), were thought to be caused by working in poorly lit conditions or, more accurately, by trying to maintain foveal fixation under poor light.

17.4 TYPES OF LAMPS AND LIGHTING

An electric lamp is a type of energy transformer that transforms electrical energy into light. According to energy saving and colour rendition requirements, there is a range of lamps commonly used in workplaces.

17.4.1 Incandescent Lamps

When materials are heated above 1000 K (approximately 726°C), they emit radiation as visible light. This property is termed incandescence. A hot object emits a broad spectrum of light. The wavelength at which most of the light is emitted depends on the temperature of the object. The hotter the object, the shorter the wavelength of the energy emitted.

This principle is used in incandescent light bulbs, where a fine filament—typically tungsten wire—is heated by passing an electrical current through it.

Figure 17.4 shows the spectral distribution of different light sources. This diagram compares the spectral distribution of an incandescent lamp, halogen lamp, a compact fluorescent lamp (CFL), a cool white and a warm white LED and a metal halide lamp. Note that the incandescent lamp's curve lies predominantly in the infrared (non-visible) region of the spectrum.

Incandescent lamps are extremely energy inefficient—only around 10 per cent of the energy is emitted as visible light. In quartz halogen lamps, halogen gases such as iodine are used to fill the lamp. This permits the tungsten filament to operate at a higher temperature so that more of its energy is available as visible light.

17.4.2 Fluorescent Lamps

Fluorescent lamps are low-pressure mercury lamps, available as hot cathode and cold cathode versions. The hot cathode type is the type normally found in offices and workplaces, while the cold cathode type is used mainly for signage and advertising. The central element in a fluorescent lamp is a sealed glass tube typically containing a small quantity of mercury and an inert gas, typically argon, under very low pressure. The glass tube also contains a phosphor powder coating on the inside of the glass. Different phosphors are selected by the manufacturer, depending on the specific frequencies of the light they emit. When a current is passed through the tube, the mercury atoms become excited and emit UV radiation, which strikes the phosphors on the inside of the glass tube. The phosphors absorb the radiation and then re-emit it at a different wavelength within the visible spectrum. Fluorescent lamps are much more efficient converters of electrical energy than incandescent lamps.

Claims have arisen that the flickering sometimes perceived with fluorescent lighting is linked to health concerns such as photosensitive epilepsy and migraine (Yoshimoto et al. 2017).

17.4.3 Electric Discharge Lamps

An alternative and more efficient form of lighting is achieved by passing an electric current through a gas. This excites the atoms and molecules of the gas, causing it to emit radiation of

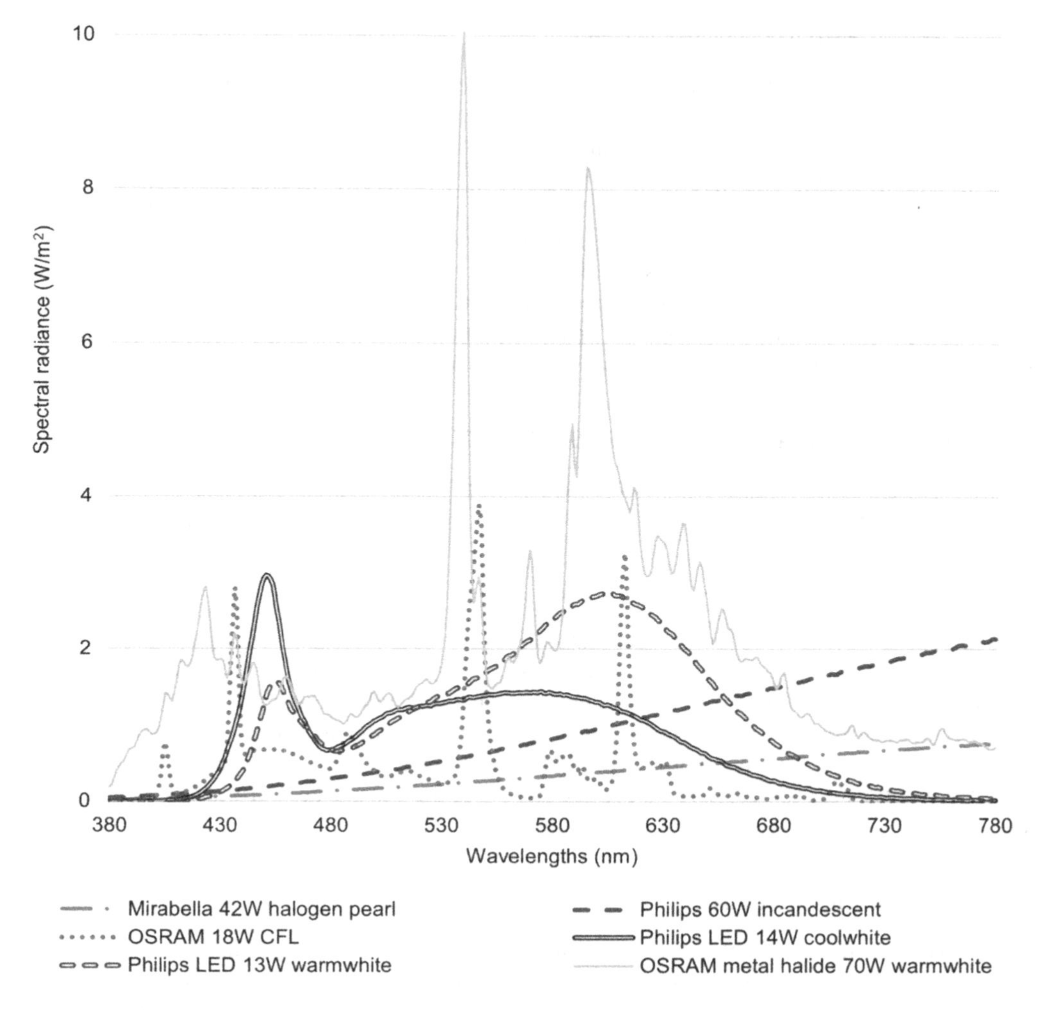

FIGURE 17.4 Spectral distribution of light sources (1 m distance, using Specbos 1211UV spectroradiometer, OD2.5 filter).

Source: Lee, 2020, p. 43.

which the spectral distribution is characteristic of the specific gas used. The most commonly used metals are mercury and sodium vapour, as they emit useful visible radiation. Discharge lamps are often classified as either high- or low-pressure lamps. In a low-pressure lamp with mercury or sodium vapour as an active ingredient, the metal vapour is mixed with an inert gas, often neon or argon. High-pressure lamps are referred to as high-intensity discharge lamps, or HID lamps. They include mercury vapour, metal halide, high-pressure sodium and xenon short-arc lamps. Compared with fluorescent and incandescent lamps, HID lamps are highly efficient in that they produce a large quantity of light in a small package. They generate light by striking an electrical arc across tungsten electrodes housed inside a specially designed inner fused quartz or fused alumina tube. This tube is filled with both gas and metal halides. The gas aids in the starting of the lamps; the metals produce the light once they are heated to a point of evaporation.

17.4.4 Light-Emitting Diodes (LEDs)

A diode is a semiconductor device that allows an electric current to flow in one direction only. LEDs are diodes that emit light when current passes through. The efficiency (measured in lumens per watt) and effectiveness (as shown by their relative brightness) of LEDs have resulted in extensive use in commercial and household settings.

The colour of the LED light is a function of the materials and processes used in making the chip. Red and green LEDs have been around for several decades but were initially used only in electronic equipment such as watches or calculators. It was only in the 1990s that blue LEDs were invented. This advance finally enabled manufacturers to create white LED light. White LED light allows companies to create smartphone and computer screens, as well as light bulbs that last longer and use less electricity than any lamp invented before.

17.5 TERMS AND UNITS OF MEASUREMENT

The following terms are taken from Australian Standard AS 3665 *Simplified definitions of lighting terms and quantities* (Standards Australia 1989).

17.5.1 Luminous Flux

Luminous flux—symbol Φ, unit is the lumen (lm)—the quantity of light energy emitted per second in all directions. One lumen is the luminous flux of a uniform point light source with luminous intensity of 1 candela and contained in 1 unit of solid angle (or one steradian). A solid angle can be regarded as a cone and is measured in steradians. It is the area of the segment of a sphere subtended, or covered, by the light from a point source at the centre of the sphere. This concept is shown in Figure 17.5 for a sphere of 1 m radius. Since the surface area of the sphere is $4\pi r^2$, the luminous flux of the point light source is 4π lumens.

17.5.2 Luminous Intensity

Luminous intensity—symbol I, unit candela (cd)—the concentration of luminous flux emitted in a specified direction. Luminous intensity is the ability to emit light in a given direction, or the luminous flux that is radiated by the light source in a given direction within a given spatial angle (measured in degrees). If the point light source emits (Φ) lumens into a small spatial angle (ß), the luminous intensity is:

$$I = \frac{\Phi}{ß} \tag{17.1}$$

1 candela = 1 lumen per steradian

17.5.3 Illuminance

Illuminance—symbol E, unit lux (lx)—the luminous flux density at a surface (luminous flux incident per unit area, lumens/m^2). It describes the amount of light that reaches a given surface. If Φ is the luminous flux and S is the area of the surface, then the illuminance E is determined by:

$$E = \frac{\Phi}{S} \tag{17.2}$$

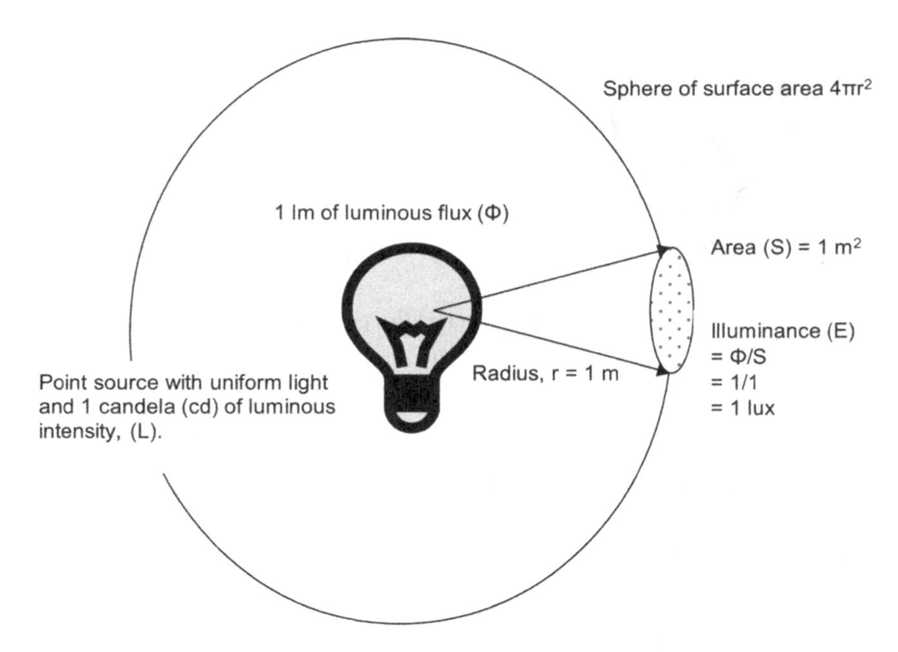

FIGURE 17.5 Relationship between luminous intensity, luminous flux and illuminance.

Source: SY Lee.

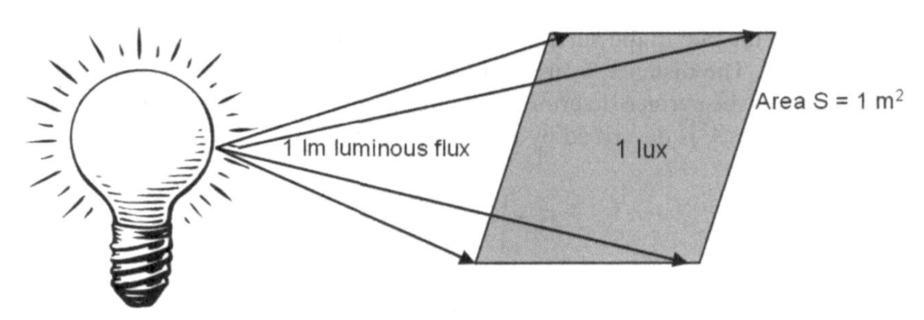

FIGURE 17.6 The relationship between lumen and lux.

Source: SY Lee.

One lux is the illuminance of 1 m^2 surface area uniformly illuminated by 1 lm of luminous flux. Figure 17.6 illustrates this definition.

17.5.4 Luminance

Luminance—symbol L—is the luminous intensity of 1 m^2 of the surface area of a given light source. Mathematically:

$$L = \frac{I}{S} \tag{17.3}$$

where I is the luminous intensity and S is the area of the light source surface perpendicular to the given direction. The unit of luminance is cd/m^2 (Figure 17.7).

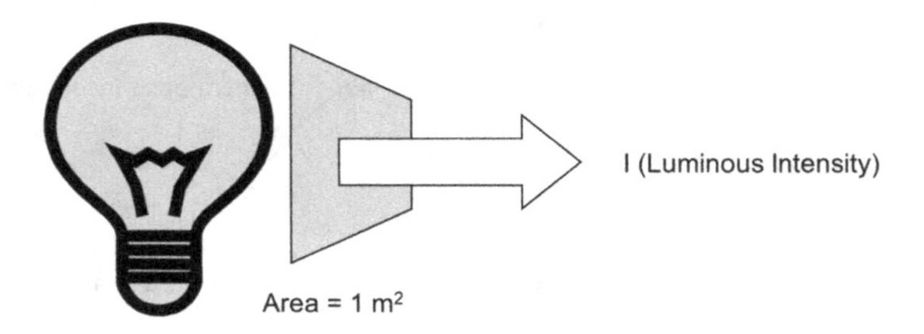

FIGURE 17.7 Luminance, the intensity of light emitted in a given direction by unit area.

Source: SY Lee.

The relationships between luminous intensity, I, luminous flux, Φ, and illuminance, E, are shown in Figure 17.5.

The lumen (lm) is the photometric equivalent of the watt, weighted to match the eye response of the 'standard observer'. Yellowish-green light (555 nm) receives the greatest weight because it stimulates the eye more than blue (450 nm) or red light (660 nm) of equal radiometric power. One watt at 555 nm is equal to 683.0 lumens.

To put this into perspective: the human photoreceptors can detect a flux of about 10 photons per second at a wavelength of 555 nm.

Lighting design and evaluation are concerned with the light entering the eye, the surface being viewed (e.g., the work) and the source of the light. This is known as the lighting triangle and is shown in Figure 17.8.

The distance from the eye to the surface and the distance from the light source to the surface are the critical factors. The distance from the light source to the surface (d) and the angle (θ) at which the light reaches the surface determines the illuminance, E, or the amount of light received by the surface. Illuminance is described by Equation (17.4):

$$E = \frac{(I \times \cos\theta)}{d^2} \tag{17.4}$$

The smaller the angle (θ) for any given light source of luminous intensity (I), and the smaller the distance (d), the higher the illuminance and the better the visibility of the task.

So far, the measurement of light has been described in terms of photometric units. However, the measurement can be in radiometric units. Radiance is analogous to luminance, and irradiance is analogous to illuminance. However, there is no simple relationship between photometric and radiometric units.

17.5.5 Spectral Radiance

Spectral radiance—symbol L_λ—at a particular wavelength is the radiance of a source in watts per square metre per steradian taking into account the spatial angle of light. The unit of spectral radiance is $W/m^2 \cdot sr \cdot nm$. The value of integrated spectral radiance can be used to measure the effective radiance of the light source, especially a blue light source.

17.5.6 Spectral Irradiance

Spectral radiance—symbol E_λ—is the irradiance of a source in watts per square metre taking into account the wavelength (in nm) of light. The unit of spectral irradiance is $W/m^2 \cdot nm$. The value of

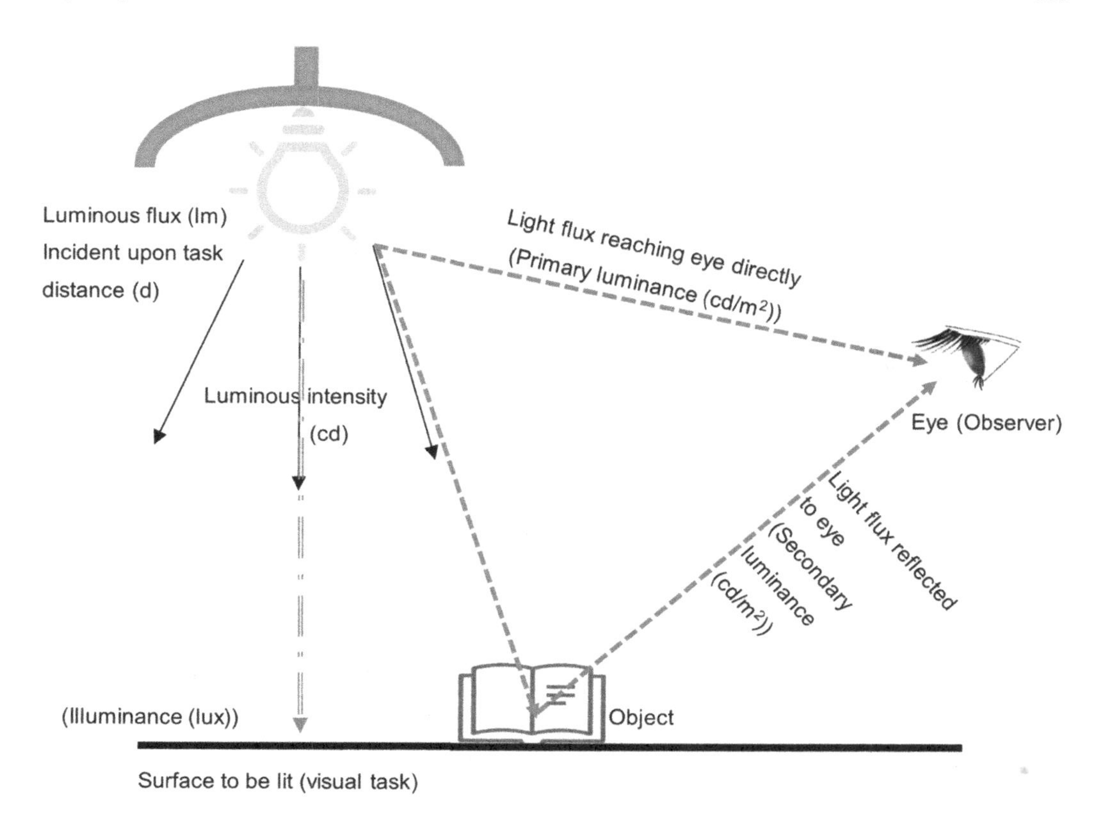

FIGURE 17.8 The lighting triangle.

Source: SY Lee.

integrated spectral irradiance can be used to measure the effective irradiance from a light source, especially a blue light source.

Figure 17.9 illustrates the relationships between key photometric and radiometric terms.

17.6 MEASUREMENT OF LIGHT

17.6.1 Photometers and Radiometers

Photometers are instruments that use special optic filters to reconstruct the exact response of the human eye to light intensity. A light-measuring cell/sensor in the meter (photodiode) converts incident light into an electronic signal that is read and displayed on the meter. The two common types of photometers are the lux (illuminance) meter and the luminance meter (see Figure 17.10).

Ideally, the lux meter should have a measurement range of 0.001–100000 lux, although a range of 0.1–20000 lux is acceptable for most indoor workplace applications. The photo sensor should not be shadowed by the user. This can be achieved with an extension cable. Instruments fitted with colour (spectral response) correction filters to limit the sensitivity to ultraviolet (UV) and infrared (IR) radiation are preferred because their spectral response has been matched to that of the human eye. If the photo sensor is not colour corrected, the appropriate correction factor (usually supplied by the manufacturer) should be applied. The photocell should also be cosine corrected to take account of the effects of light falling on it at oblique angles.

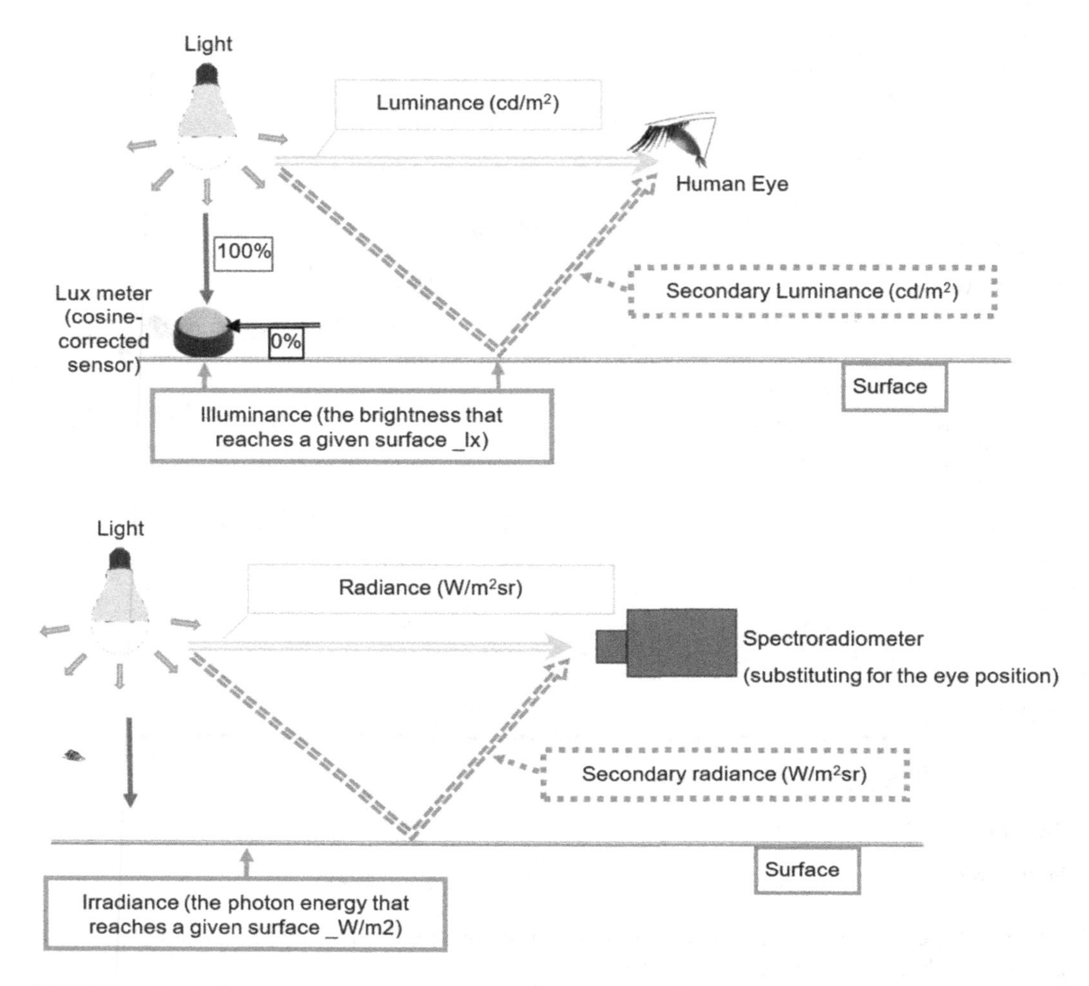

FIGURE 17.9 Luminance and illuminance (top) and radiance and irradiance (bottom).

Source: Lee 2020, p. 53.

The simplest luminance meters resemble a handle held video camera. The image and the associated luminance value in the view finder indicate the brightness as experienced by the observer. These meters should have a wide range, up to 100,000 cd/m². As the measurement is directional, it relates to a particular direction and solid angle. Accordingly, luminance should be assessed in the visual field of the worker, that is, the occupational visual field (OVF) described below.

Radiometers can measure integrated irradiance at a particular position. However, spectroradiometers are capable of providing a spectral distribution of the irradiance or radiance, that is, measurements across the wavelengths of the spectrum. An example of a spectroradiometer is provided in Figure 17.10. Again, spectral radiance measurements should be in the OVF. Spectroradiometers are ideal for determining the spectral distribution of light sources as in Figure 17.4. Note that a light source might appear to be white but may contain light with variable wavelengths and amplitudes. A spectroradiometer can be used to rapidly characterise the emissions of light sources where manufacturer specifications are unavailable.

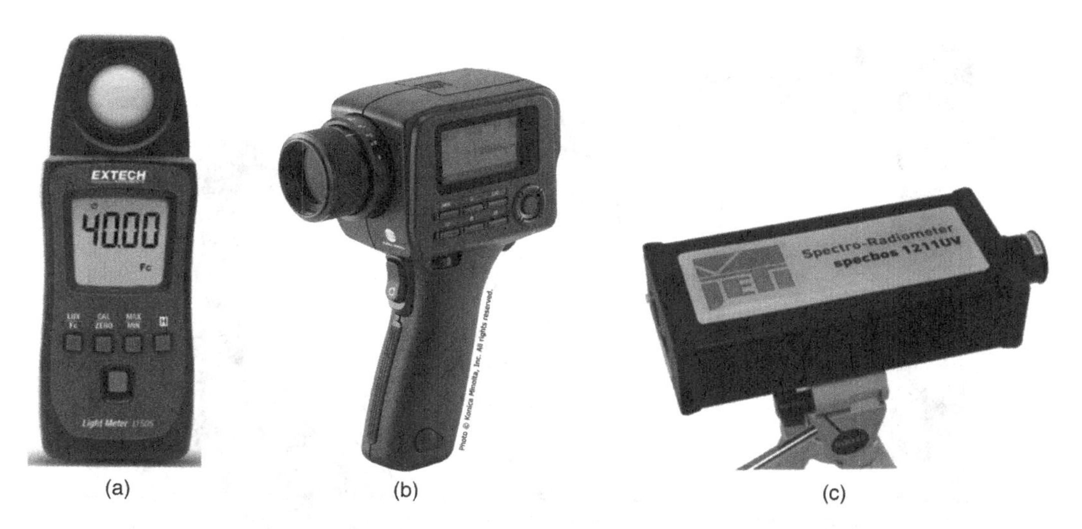

FIGURE 17.10 Illuminance meters (a: Extech LT505, **Courtesy of FLIR**), luminance meter (b: Konica Minolta LS150, **Photo courtesy of Konica Minolta, Inc. All rights reserved**) and spectroradiometer (c: Specbos 1211UV, **Courtesy of Technische Instrumente GmbH**).

All meter types should be regularly calibrated in accordance with manufacturer's recommendations. As mentioned, there is no simple relationship between photometric and radiometric units, but the key consideration for risk assessment is determining what actually enters to eye.

Figure 17.10 illustrates a lux meter, luminance meter and a spectroradiometer.

A prototype wearable spectrophotometer has recently been described for assessing personal photopic and melanopic illuminance (Cain et al. 2020). The small wearable device clipped to clothing records light exposure near eye level.

17.6.2 Occupational Visual Field (OVF)

This has been defined based on the basic anatomy of the eye and its periphery (Piccoli et al. 2004). The most relevant impact of light occurs in the macula, and in particular the fovea where there is a high concentration of colour-sensitive photoreceptors. A photoreceptor may ultimately be serviced by one neuron, enabling the highest image-forming resolution. In order to carry out visually demanding tasks, there is a deliberate geometric alignment of object and fovea. Figure 17.11 illustrates the occupational fixation axes for work involving three visual demand areas (screen, document holder and keyboard). The fixation axis is the theoretical segment extending from the centre of the macular region to the centre of each occupational target. The external projection of the macula is approximately 40 degrees (the theoretical value is 36.8°; i.e., 2 × 18.4) and defines the occupational fixation zone. Hence, the task for the office worker in Figure 17.11 has three visual fields, with an overlap region as the worker shifts gaze from one target to the other. Measurements of luminance and radiance should be conducted in the overall OVF, that is, bounded by the thick solid line in Figure 17.11.

Figure 17.12a illustrates the OVF reticle, including concentric circles of the foveolar (inner fovea) projection, foveal projection, macular and 2 x macular projection. Also indicated is the angle of 110 milliradians (6.3 degrees) used for blue light assessment. Figure 17.12b illustrates the OVF reticle for an office worker in a fixed gaze. Note that the windows occupy a part of the OVF in this case.

Figure 17.12b illustrates the OVF reticle for an office worker in a fixed gaze. Note that the windows occupy a part of the OVF in this case.

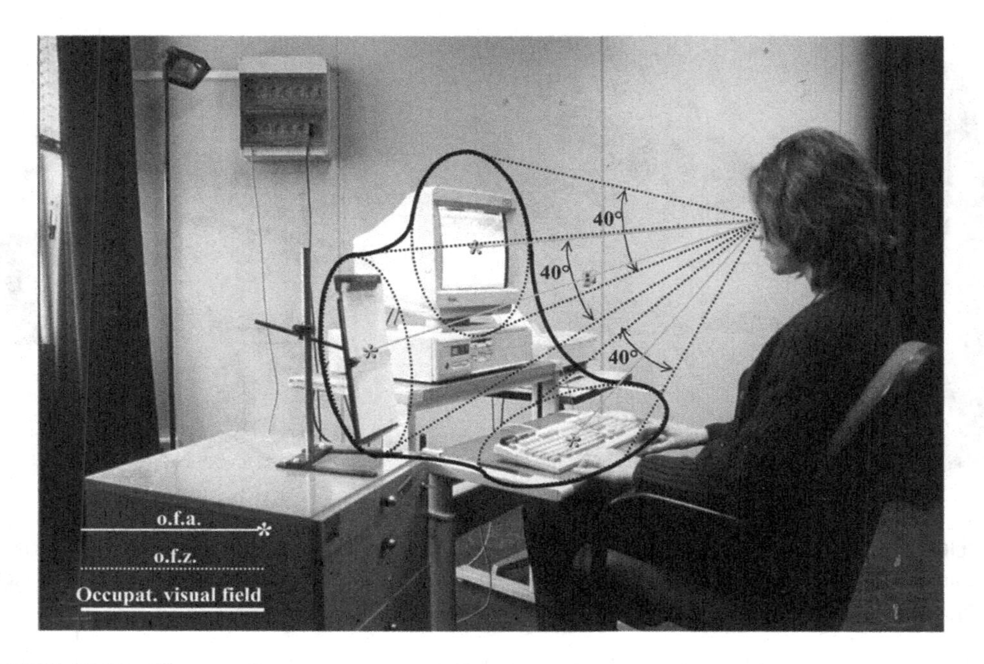

FIGURE 17.11 The overall occupational visual field.

Source: Piccoli et al. 2004, p. 33.

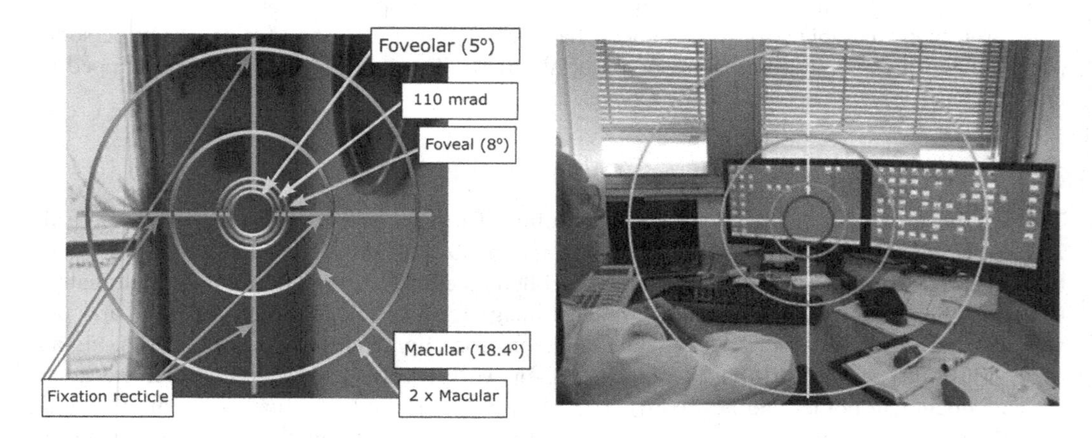

FIGURE 17.12 (a) An occupational visual field reticle (Source: B Piccoli). (b) Occupational visual field reticle for an office worker in a fixed gaze.

Source: B. Piccoli.

A prototype smartphone-based device for recording images, luminance and weighted blue light exposure in the OVF has a patent pending. (European Patent 2023).

17.6.3 Photometric Evaluation

The most common method of measuring light (photometry) is to use a lux meter to measure the level of illuminance. Indeed, standards related to light in the workplace such as AS/NZS 1680.1

(2006) refer to levels of illuminance. However, illuminance alone is not the only factor that should be measured when evaluating lighting in the workplace. Luminance assessment can reveal large light contrast ratios that can lead to visual disturbance and reduced visual acuity. AS/NZ 1680.1 (2006) refers to a maximum luminance ratio of 10:1. Overall, luminance is preferred over illuminance for vision work (Nilsson 2009). To illustrate the value of luminance assessment, consider the same worker and OVF as in Figure 17.11. Figures 17.13 and 17.14 show the luminance and illuminance values of this workstation with curtains closed (Figure 17.13) and curtains open (Figure 17.14). The values of illuminance on the keyboard lit by overhead ceiling lamps and sunlight from the window are similar (490 lx in Figure 17.13 and 500 lx in Figure 17.14); however,

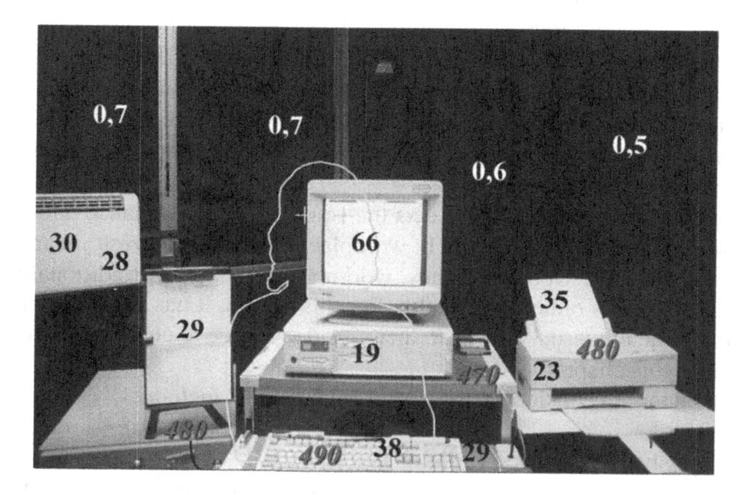

FIGURE 17.13 Laboratory photometry evaluation of a workstation (luminances are in bold letters and illuminances in italics).

Source: Piccoli et al. 2004, p. 32.

FIGURE 17.14 Laboratory photometry evaluation of a workstation with open windows (luminances are in bold letters and illuminances in italics).

Source: Piccoli et al. 2004, p. 32.

the values of luminance in Figure 17.14 are around 100–1000 times higher than in Figure 17.13 (Piccoli et al. 2004). The situation in Figure 17.14 results in reduced visual acuity and potentially visual discomfort (sometimes referred to as discomfort glare).

Thus, photometric evaluation by hygienists must consider multiple factors such as types of light sources and light characteristics in the occupational visual field (OVF).

17.6.4 Lighting Assessment in the Workplace

The visual demands of the different tasks in a given work area will determine the lighting required. AS/NZS 1680.1 (2006) and ISO 8995.1—2002 (ISO 2002) set out general principles and recommendations for the lighting of building interiors for performance and comfort. They apply primarily to interiors in which specific visual tasks are undertaken and takes into account both electric lighting and daylight. The recommendations aim to achieve a visual environment in which essential task details are easy to see and factors that may cause visual discomfort are either excluded or appropriately controlled. It should be noted that the recommendations do not apply to lighting for physiological purposes.

Workplace lighting, or illumination, affects the safety, task performance and the visual environment of the workers. Although a poorly lit workplace can result in eyestrain and/or headaches, as well as in diminishing the work efficiency, workplace lighting has traditionally been considered a matter of ergonomics rather than health. For this reason, it has not traditionally been considered to be a major health issue. However, there is no evidence to challenge this view. This will be discussed later in this chapter.

Other relevant factors to consider in the visual environment include:

- avoiding excessive illuminance variations,
- absence of direct glare from lamps, luminaires (complete lighting unit) or windows,
- an appropriate luminance distribution on interior surfaces,
- use of suitable colours on the main interior surfaces,
- control of flickering lights and
- use of light sources with suitable colour characteristics.

17.6.5 General Considerations

Before measurements are taken, the light meter should be calibrated by covering the light sensor and recording the measured value, which should be zero. Measurement of illuminance of a workplace lit by artificial lighting is made either after dark or with daylight excluded from the interior. Some light sources, such as gas-discharge lamps (e.g., mercury or sodium vapour), should be switched on at least 30 minutes before the measurement to allow the light output to stabilise.

17.6.6 Measurement of Task Illuminance

To measure illuminance at a workstation or equipment operator's position, the following procedure should be followed:

- Use at least four measurement points at each sampling location to obtain a representative average illuminance value.
- Take measurements on the plane (horizontal, vertical or inclined) in which the visual task is performed while the worker is in their normal position, even if this results in shadow on the sensor. Calculate the average illuminance at the task position, as the arithmetic mean and compare the average illuminance to the recommended level for the task, as in AS/NZS

1680.1 (2006) and AS/NZS 1680.2 series (2008a, 2008b, 2017, 2018). Note that recommended maintained illuminances for screen-based tasks may be lower than tasks such as reading and writing, since the screen provides its own luminance. A value of about 300 lx may be appropriate for common office-based tasks involving computer screens.

17.6.7 Measurement of the General Illuminance of an Interior

The average illuminance in a space is determined by measuring the illuminance at each of a series of points, set out in a regular pattern, then calculating the arithmetic mean of those values. Reference should be made to Appendix B of AS/NZS 1680.1 (2006). Accuracy is dependent on the number and spacing of measurement points and the relative location of the points with respect to the luminaires. The following criteria should be applied:

- The selected measurement areas chosen should collectively represent all areas of both the lighting layout and the physical environment, but do not necessarily need to cover the whole space to be measured.
- Where there is a regular array of lighting, the measurement area should cover an entire pattern.
- The measurement area should be divided into a number of squares.
- The illuminance should be measured at the centre of each square, at the height of the working plane. Note: the use of a tripod to support the meter at the correct height and in a horizontal position may be useful.
- The maximum distance between measurement points is 1 m.
- No measurement point should be closer than 1 m from a wall.
- For areas with a floor area less than 25 m^2, the minimum number of measurement points is nine, and the whole area should be measured.
- For areas with a floor area greater than 25 m^2, the minimum area of measurement is 9 m^2.
- The minimum number of luminaires to be included in the measurement area is four.
- For spaces with non-regular lighting layouts or different room conditions, several measurement areas to be selected to give a representative set of measurements.

17.6.8 Uniformity of Illuminance

Where a general lighting system is used, a fairly high degree of uniformity of illuminance in the working plane is required, as the aim of such systems is to allow a particular task or series of tasks to be performed anywhere within the space. Therefore, for general lighting systems, the uniformity of illuminance (the ratio of the minimum illuminance to the average illuminance) should be ≥ 0.5 over the space covered or included by the general lighting system. The uniformity of illuminance should be ≥ 0.7 over the task area.

17.6.9 The Blue Light Hazard and the Measurement of Blue Light

Ocular exposure to intense light sources can cause photochemical photoreceptor degradation, notably when the light is in the wavelength range of 400–500 nm (violet, blue and blue-green). This is termed the blue light hazard (Jennings 2016). The hazard weighting function (wavelength band) is narrow with maximum damage occurring at about 440 nm. Looking directly at the sun is the most obvious example, with many cases of photoreceptor damage being recorded during a solar eclipse. To assess the blue light hazard, it is necessary to characterise the radiant exposure of workers, by exposure level (how much) and duration (how long) workers are exposed to blue light in their workplaces. The exposure assessment should focus on the visual field of a worker. In the work environment, tasks vary and quantification of light exposure in the visual field of all tasks is difficult. Thus, simulations and worst-case scenarios are often used (Lee et al. 2022).

Significant sources of blue light include blue LED arrays, intense white light sources (such as projection lamps, floodlights, video production lights and welding arcs), direct and reflected sunlight (ACGIH 2024). Occupations at risk include stage lighting and video production technicians, where metal halide lamps are common, dentists using handheld curing lamps and, possibly, nail technicians using curing lamps.

Blue light is measured differently from other types of light, in that radiometric measurements are required. Lee et al. (2016) have described the application for a simple LED spot source and explored the exposure methodology for two commercially available nail curing lamps used in beauty salons.

The assessment of the blue light hazard requires the use of a radiometer, or preferably a spectroradiometer, rather than a photometer (lux or luminance meter). The spectral radiance is used to evaluate the retinal photochemical damage. There is no simple or reliable conversion factor between photometric and radiometric quantities.

Both ACGIH and ICNIRP (2013) provide equations for measuring effective radiance ($W/m^2 \cdot sr$) and radiance dose ($J/m^2 \cdot sr$) for blue light exposure. For viewing durations less than 10,000 seconds (2.8 h) in a day, the effective radiance dose must not exceed 10^6 J/m^2, and for periods greater than 10,000 seconds, the radiance limit is 100 W/m^2 sr. (ACGIH 2023; Söderberg et al. 2013). The expressions used for blue light are provided in Table 17.1.

There are nuances in these limits, and aspects of practical measurement are not immediately obvious for those seeking to assess workplace risk according to the guidelines. Specialist assistance may be required if the hygienist is not familiar with blue light radiometry.

Measurements should be conducted in the OVF (Piccoli et al. 2004), with special consideration of the averaging angle of acceptance. According to ICNIRP guidelines, the acceptance averaging angle is time-dependent (from 0.01 to 0.1 radians; 0.6 to 6 degrees) (Söderberg, et al. 2013). However, depending on the actual task and behaviour, a larger averaging angle can be used, provided that any part of the retina, particularly the macula, is not exposed beyond the radiance dose limit.

Damage to the peripheral retina has relatively little adverse impact on visual function, as evidenced by laser treatment for diabetic retinopathy (Palanker et al. 2011). However, if the macula

TABLE 17.1
Equations for Blue Light Exposure to Non-aphakic (Eyes with Natural Lens) Persons, Based on ICNIRP Guidelines

Formulae	Details
$L_B = \sum_{300}^{700} L_\lambda \cdot B(\lambda) \cdot \Delta\lambda$	L_B: effective radiance of blue light ($W/m^2 \cdot sr$)
$D_B = L_B \cdot t = \sum_{300}^{700} L_\lambda \cdot t \cdot B(\lambda) \cdot \Delta\lambda$	D_B: effective blue light radiance dose ($J/m^2 \cdot sr$)
$D^{EL}_B = 10^6$ J/m^2sr	L_λ: spectral radiance ($W/m^2 \cdot sr \cdot nm$)
$L^{EL}_B = 100$ W/m^2sr	$B(\lambda)$: blue light hazard function
	$\Delta\lambda$: wavelength interval (nm)
	t: exposure duration (seconds)
	D^{EL}_B: exposure limit of the radiance dose (0.25 s $\leq t <$ 10,000 s)
	L^{EL}_B: exposure limit of the radiance ($t >$ 10,000 s)

Adapted from: ICNIRP, 2013.

is exposed beyond the radiance dose limit, then visual acuity may be adversely affected, or may contribute to macular degeneration. This means that the risk should be assessed for any intense blue light source within the visual field that may be imaged on the macula. The product of this particular radiance and the viewing time limits the risk for short exposures (up to 2.8 h). This would apply to most real-life exposure scenarios. If it were certain that the worker was constantly exposed to a blue light source within the acceptance angle for more than 2.8 h, the radiance limit would apply. In any case, the averaging is assumed to be an arithmetic mean. There is epidemiological evidence of the association of sunlight exposure and age-related macular degeneration (Sui et al. 2013).

It should be noted that there is no occupational exposure standard for blue light in Australia. The ACGIH and ICNIRP values are based on short-term animal experiments and there is uncertainty about long-term exposures. The European standard is Directive 2006/25/EC (2006).

17.7 LIGHT SOURCE COLOUR PROPERTIES

Colour is not an inherent property of objects but more a human perception facilitated by light. Light sources have two colour-related properties: the apparent colour of the light they emit (colour appearance) and the effect the light has on the colours of surfaces (colour rendering). An application of these properties with which most people will be familiar is the use of lighting in a supermarket to enhance the red colour of meat in the meat produce section.

17.7.1 Colour Appearance

The colour appearance of near-white-light sources is normally defined in terms of their correlated colour temperature (CCT), expressed in Kelvin (K). The higher the CCT, the cooler the appearance of the source. For example, the reddish-yellow flame of a candle has a CCT of about 1900 K, the ordinary incandescent lamp about 2800 K, and cool bluish-white southern-sky daylight has a CCT of over 6500 K (AS/NZS 1680.1 2006, s. 7.2). The CCTs of lamp types have been grouped into three classes as shown in Table 17.2.

17.7.2 Colour Rendering

Colour rendering is the ability of a light source to render the colours of an object as similarly as possible to the way they appear in an ideal or natural light source such as daylight. The colour-rendering properties of a light source can be measured and described according to the CIE Colour Rendering Index (CRI) system, on a scale of 0–100. The higher the CRI, the more natural the reflected light colour.

TABLE 17.2
Lamp Colour Appearance Groups

Colour Appearance Group	Colour Appearance	CCT (K)
1	Warm	<3300
2	Intermediate	$3300 \leq 5300$
3	Cool	>5300

Source: Reproduced Table 7.1 from AS/NZS 1680.1:2006, p. 39.

17.8 ERGONOMICS OF LIGHTING

The objective of good lighting design is to apply visual ergonomics to optimise the perception of visual information, provide conditions conducive to task performance, maintain safety and provide an acceptable level of visual comfort. The International Ergonomics Association defines visual ergonomics as a science which aims to achieve a good balance between what a person can see and the visual demands of a task, but also taking into consideration musculoskeletal issues that may arise (Long & Richter 2014). Ergonomically designed lighting takes human capability into account and, generally, results in improvements in productivity.

There are four parameters that influence the nature of visual information processed and, therefore, a worker's visual performance (Clarke 1989):

- task characteristics, including size of object, distance, texture, colour, contrast, motion and time factors;
- the worker's own visual ability, which depends on ophthalmic condition, age, adaptation, depth and colour perception;
- lighting characteristics, including illuminance, uniformity, glare and flicker;
- workspace factors such as postural constraints, safety requirements, other physical constraints and psychological factors.

These relate to elements of the work and vision framework in Figure 17.1. It is often possible to compensate for a deficit in one or more of these factors by enhancing one or more of the others. An obvious consequence of this is that the application of visual ergonomics can increase the number of options available for providing an acceptable visual environment.

In recent times, there have been a number of building design and verification standards based on ergonomics principles and aligned with green building standards. These address the above-mentioned workspace factors and lighting characteristics. The WELL building standard, primarily for new or renovated buildings, is a rating system, with 11 lighting-related components that include low glare workstation design colour quality, solar glare control, visual lighting design, automated shading and dimming controls and daylighting fenestration (International WELL building Institute 2022).

17.9 BIOLOGICAL EFFECTS OF LIGHT

Light can impact human health via the visual and non-visual systems originating in the retina of the eye or as optical radiation falling on the eye or exposed body surfaces (Boyce 2022). In normal artificial indoor lighting arrangements, photothermal effects on the skin and eye are insignificant. However, sunlight, which has infrared and UV radiation components, may damage skin and conjunctiva with prolonged exposure. Sunlight impinging on windows, especially open windows, may be a source of UV exposure and infrared skin heating and may induce asymmetric thermal discomfort for those near windows. On the other hand, daylight may provide some psychological benefits and modify the potential association between reported neurological symptoms and flickering fluorescent lights or pixel jitter (Watson et al. 2012; Yoshimoto et al. 2017). However, care is required when judging the veracity and relevance of broad assertions about the benefits of lighting for human health (Boyce 2022).

Each of the effects of light upon mammalian tissue may be classified as direct or indirect, depending on whether its immediate cause is a photochemical reaction occurring within that tissue or a neuroendocrine signal generated by a photoreceptor.

17.9.1 The Indirect Effects of Exposure to Light

Light exerts an indirect effect on various metabolic, hormonal and organic functions by way of the eyes. Circadian rhythms are entrained by the light–dark cycle. Free-running circadian rhythms—in which the sleep cycle is not synchronised to environmental cues and usually oscillates with a period slightly more than 24 hours—seem to be influenced by light intensity. Most totally blind people have free-running circadian rhythms. This condition causes recurrent insomnia and daytime sleepiness when the rhythms drift out of phase with the normal 24-hour cycle (Sack et al. 2000).

Budnick et al. (1995) demonstrated that exposure to high levels of bright light (i.e., 6,000–12,000 lux) on at least half of a worker's night shifts over three months was effective in altering the worker's circadian rhythm. In the systematic review and meta-analysis by Jeon et al. (2023), the most effective intervention for sleep outcomes amongst rotating night shift workers was light therapy.

17.9.2 Implications for Occupational Health

From the above, it can be seen that workers such as shift-workers or those whose health and well-being are adversely affected by travelling over several time zones, for example, travellers or commercial aircrew, may benefit from exposure to high levels of light to re-establish their circadian rhythms.

It was predicted that women working a non-day shift would have a higher risk of developing breast cancer than their day-working counterparts. Studies such as those of Davis et al. (2001) suggest that exposure to light at night may increase the risk of breast cancer by suppressing the normal nocturnal production of melatonin by the pineal gland, which in turn could increase the release of oestrogen by the ovaries. This hypothesis has been extended more recently, to include prostate cancer. On the basis of limited human evidence and sufficient evidence in experimental animals, in 2007 the International Agency for Research on Cancer classified 'shift work that involves circadian disruption' as a probable human carcinogen, group 2A (Stevens et al. 2011). In relation to light exposure for shift workers, Dun et al. (2020) found that cancer risk was not significantly elevated with the increased light exposure of night-shift work.

In addition to the diurnal (day/night) circadian rhythms of melatonin, there are also circannual (yearly) rhythms, and melatonin production seems to follow seasonal variations. Moreover, this has been linked to clinical variables in patients with depression (Wetterburg 1990). Exposure to bright light can offset the negative effects of a type of depression called seasonal affective disorder (SAD), which appears to affect a large number of adults who become depressed during the dark days of winter and start to feel better in the spring. The exact cause of SAD is unknown but is thought to be due to circadian misalignment (Boyce 2022) However, SAD is likely to be an occupational health problem only for workers who are 'light deprived', such as those working in polar regions (Arendt 2012).

In a statement on the circadian, neuro-endocrine and neuro-behavioural effects of light, ACGIH states it does not consider it practical to develop TLVs® to protect against light induced changes in circadian rhythms. Instead, ACGIH recommends (ACGIH 2024, Appendix A):

1. Shift work is best addressed by optimal planning of work schedules rather than exposure limits such as TLVs®.
2. Adjusting the colour palette of computer displays.

The lighting conditions in occupational settings should provide the safest and most alerting environment possible, while maintaining typical visual function. Work environments should therefore incorporate high-intensity, blue-enriched (high melanopic) light during both the day and especially at night given the high risk of sleepiness-related accidents and injuries. In occupational settings, there are potentially conflicting needs, such as in a hospital during the night when patients sleep but staff are awake, the patient bedroom or ward environment should be optimised for sleep with low-intensity, blue depleted (low melanopic) light while the staff environment (nursing station, breakrooms) should enhance alertness with high intensity, blue-enriched (high melanopic) light. These more complex environments need careful consideration of the spectrum, location and use of the light but are likely to be solved through the lighting design process. Designs, however, should be consistent with ACGIH and ICNIRP guidelines on the blue light hazard.

From the above, it will be evident that there is a coherent relationship between light and health in humans (Boyce 2022). For many workers, the majority of waking hours are spent in artificial lighting, which provides much less illumination than daylight. Furthermore, the spectral distribution of this illumination is very different from that of natural light. For a number of reasons, such as worker preference and circadian entrainment, natural light is desirable for indoor environments. From an occupational health perspective, the following points should be considered in relation to the design of a workplace:

- buildings should be built so as to maximise the use of natural light;
- individuals should be allowed to control their lighting environment to meet their own requirements;
- all workers should have access to natural light at their workstation;
- windows should allow full transmission of light, but they must be equipped with effective light control systems;
- workers should be encouraged to take 'light breaks' and get out of artificially lit environments during their lunch periods.

17.9.3 Making Full Use of Daylight

Maximising the use of daylight improves morale and reduces energy costs. Examine the workplace layout, material flow and workers' needs, and then consider the following options for making effective use of daylight:

- provide skylights—for example, by replacing roof panels with translucent ones;
- equip the workplace with additional windows;
- place machines or equipment near windows;
- move work requiring more light nearer to windows.

Before planning and installing windows and skylights:

- consider the height, width and position needed for windows or skylights. More light is available when the window is placed high on a wall;
- install shades, screens, louvres, canopies or curtains on the windows and skylights to protect the workplace from external heat and cold while taking advantage of the natural light;
- orient skylights and windows away from direct sunlight to obtain constant but less bright light;
- direct skylights and windows towards the sun if variations in levels of brightness throughout the day do not disturb workers.

Traditionally, most H&S practitioners have confined their efforts with respect to lighting, to ensure that workplaces have adequate illumination for work to be done safely and efficiently. This chapter has demonstrated to the reader that there is much more to the lighting environment than simply measuring light levels. H&S practitioners now need to be aware that the lighting environment directly affects the health not only of workers but that of all members of the community. The onus is on occupational hygienists in particular, to ensure that the impact of the lit environment is considered as an integral part of the health risk assessment and design of any facility, work regime or work environment.

REFERENCES

ACGIH 2023, 'Documentation for Light and Near Infra-red Radiation', *Threshold Limit Values for Chemical Substances and Physical Agents and Biological Exposure Indices*, ACGIH, Cincinnati, OH.

ACGIH 2024, 'Appendix A: Statement on the Occupational Health Aspects of New Lighting Technologies – Circadian, Neuroendocrine and Neurobehavioural Effects of Light', *Threshold Limit Values for Chemical Substances and Physical Agents and Biological Exposure Indices*, pp. 242–244, ACGIH, Cincinnati, OH.

Arendt, J. 2012, 'Biological Rhythms During Residence in Polar Regions', *Chronobiology International*, vol. 29, no. 4, pp. 379–394.

Boyce, P. 2022, 'Light, Lighting and Human Health'. *Lighting Research & Technology.* vol. 24. no.2, pp. 101–144. doi:10.1177/14771535211010267

Budnick, L.D., Lerman, S.E. & Nicolich, M.J. 1995, 'An Evaluation of Scheduled Bright Light and Darkness on Rotating Shift Workers: Trials and Limitations', *American Journal of Industrial Medicine*, vol. 27, no. 6, pp. 771–82. https://doi.org/10.1002/ajim.4700270602

Cain, S.W., McGlashan, E.M., Vidafar, P. Mustafovska, J., Curran, S.P.N., Wang, X., Mohamed, A., Kalavally, V. & Phillips, A.J.K. 2020, 'Evening home lighting adversely impacts the circadian system and sleep', *Scientific Report*, vol. 10, no. 1. doi: https://doi.org/10.1038/s41598-020-75622-4

Cao, D. & Barrionuevo, P.A. 2015, 'The Importance of Intrinsically Photosensitive Retinal Ganglion Cells and Implications for Lighting Design', *Journal of Solid State Lighting*, vol. 2, no.1, pp. 1–8. doi: 10.1186/s40539-015-0030-0

Clarke, G. 1989, 'Lighting the Workplace for People'. *Proceedings of the 25th Annual Conference of the Ergonomics Society of Australia, 26–29 November 1989*, Ergonomics Society of Australia, Canberra

Davis, S., Mirick D.K. & Stevens R.G. 2001, 'Night Shift Work, Light at Night, and Risk of Breast Cancer', *Journal of the National Cancer Institute*, vol. 93, no. 20, pp. 1557–62

Dun A, Zhao X, Jin X, Wei T, Gao X, Wang Y, Hou H. 2020, 'Association Between Night-Shift Work and Cancer Risk: Updated Systematic Review and Meta-Analysis', *Frontier in Oncology.* vol. 10, pp. 1006–1006. doi: https://doi.org/10.3389/fonc.2020.01006

Dutheil, F., Oueslati, T., Delamarre, L., Castanon, J., Maurin, C., Chiambaretta, F., Baker, J. S., Ugbolue, U. C., Zak, M., Lakbar, I., Pereira, B., & Navel, V. 2023, 'Myopia and Near Work: A Systematic Review and Meta-Analysis', *International Journal of Environmental Research and Public Health* vol. 20, no. 1, pp. 875–875. https://doi.org/10.3390/ijerph20010875

European Patent 2023, https://www.knowledge-share.eu/en/proprietario/universita-degli-studi-di-roma-tor-vergata/?_sf_s=Piccoli [28 December 2023].

European Standard (CEN) 2021, *Light and lighting – Lighting of workplaces – Part 1: Indoor work places.* EN 12464-1:2021 European Committee for Standardization. Brussels.

European Union 2006, *Directive 2006/25/EC of the European Parliament and of the Council of 5 April 2006*, http://data.europa.eu/eli/dir/2006/25/2019-07-26 [1 February 2024].

Hattar, S., Liao, H.W., Takao, M., Berson, D. M. & Yau, K.W. 2002, 'Melanopsin-Containing Retinal Ganglion Cells: Architecture, Projections, and Intrinsic Photosensitivity', *Science*, vol. 295, no. 5557, pp. 1065–1070.

ICNIRP (2013). *'ICNIRP Guidelines on Limits of Exposure to Incoherent Visible and Infrared Radiation'*, https://www.icnirp.org/cms/upload/publications/ICNIRPVisible_Infrared2013.pdf [26 February 2024].

International Organization for Standardization (ISO) 2002, *Lighting of indoor workplaces*, ISO 8995.1:2002 (reviewed 2018), International Organization for Standardization, Geneva.

International WELL Building Institute 2022. *WELL Building Standard™ version 2* (WELL v2™). New York, https://v2.wellcertified.com/en/wellv2/overview [29 January 2024].

Jennings, A.M. 2016, 'What Is the Blue Light Hazard? A Review of Recent Developments', *Proceedings of AIOH 34th Annual Conference, 3–7 December 2016*, Gold Coast Australia.

Jeon, B.M., Kim, S.H. & Shin, S.H. 2023, 'Effectiveness of Sleep Interventions for Rotating Night Shift Workers: A Systematic Review and Meta-analysis', *Frontier in Public Health*, vol. 11, pp. 1 – 10. doi: 10.3389/fpubh.2023.1187382.

Lee, S.Y. 2020, *Assessment of Blue Light Exposure in the Occupational Visual Field* (Doctoral dissertation), The University of Adelaide., https://digital.library.adelaide.edu.au/dspace/handle/2440/129181 [3 February 2024].

Lee, S.Y., Gaskin, S., Piccoli, B. & Pisaniello, D. 2022, 'Blue Light Exposure in the Workplace: A Case Study of Nail Salons', *Archives of Environmental & Occupational Health*, vol. 77, no. 5, pp. 351–355. doi: https://doi.org/10.1080/19338244.2021.1924604.

Lee, S.Y., Pisaniello, D., Gaskin, S. & Piccoli, B. 2016, 'Characterising Blue Light Exposure: Methodological Considerations and Preliminary Results', *Proceedings of the 34th Annual Conference of the Annual Institute of Occupational Hygienists Inc*. Gold Coast Queensland, 3–7 December 2016.

Long, J. & Richter, H. 2014, 'Visual Ergonomics at Work and Leisure', *Work (Reading, Mass.)*, vol. 47, no. 3, pp. 419–20. https://doi.org/10.3233/WOR-141820

Nilsson, T.H. 2009, 'Photometric Specification of Images', *Journal of Modern Optics*, vol. 56, no. 13, pp. 1523–1535. doi: https://doi.org/10.1080/09500340902999263

Palanker, D., Lavinsky, D., Blumenkranz, M.S. & Marcellino, G. 2011, 'The Impact of Pulse Duration and Burn Grade on Size of Retinal Photocoagulation Lesion: Implications for Pattern Density', *Retina*, vol. 31, no. 8, pp. 1664–9. doi: https://doi.org/10.1097/IAE.0b013e3182115679

Piccoli, B. & ICOH Scientific Committee 2003, 'A Critical Appraisal of Current Knowledge and Future Directions of Ergophthalmology: Consensus Document of the ICOH Committee on "work and Vision"', *Ergonomics*, vol. 46, no. 4, pp. 348–406. doi: https://doi.org/10.1080/0014013031000067473

Piccoli, B., Soci, G., Zambelli, P.L. & Pisaniello, D. 2004, 'Photometry in the Workplace: The Rationale for a New Method', *Annals of Occupational Hygiene*, vol. 48, no. 1, pp. 29–38.

Sack, R.L., Brandes, R.W., Kendall, A.R. & Lewy, A.J. 2000, 'Entrainment of Free-running Circadian Rhythms by Melatonin in Blind People', *The New England Journal of Medicine*, vol. 343, no. 15, pp. 1070–1077.

Söderberg, P., Matthes, R., Feychting, M., Ahlbom, A., Breitbart, E., Croft, R.J., de Gruijl, F.R., Green, A.C., Hietanen, M., Jokela, K., Lin, J.C., Marino, C., Peralta, A.P., Saunders, R.D., Schulmeister, K., Stuck, B.E., Swerdlow, A.J., Sienkiewicz, Z., Taki, M. & International Commission on Non-Ionizing Radiation Protection (ICNIRP) (2013). 'ICNIRP Guidelines on Limits of Exposure to Incoherent Visible and Infrared Radiation'. *Health Physics*, vol. 105, no. 1, pp. 74–96. https://doi.org/10.1097/HP.0b013e318289a611

Standards Australia 1989, *Simplified Definitions of Lighting Terms and Quantities, AS 3665:1989*, SAI Global, Sydney, https://www.standards.org.au/access-standards

Standards Australia 2006, *Interior and Workplace Lighting Part 1: General principles and recommendations, AS/NZS 1680.1:2006*, SAI Global, Sydney, https://www.standards.org.au/access-standards

Standards Australia 2008a, *Interior and workplace lighting, Part 2.2: Specific applications – Office and screen-based tasks, AS/NZS 1680.2.2:2008*, SAI Global, Sydney, https://www.standards.org.au/access-standards

Standards Australia 2008b, *Interior and workplace lighting, Part 2.1: Specific applications – Circulation spaces and other general area, AS/NZS 1680.2.1:2008*, SAI Global, Sydney, https://www.standards.org.au/access-standards

Standards Australia 2017, *Interior and workplace lighting, Part 2.4: Industrial tasks and processes, AS/NZS 1680.2.4:2017*, SAI Global, Sydney, https://www.standards.org.au/access-standards

Standards Australia 2018, *Interior and workplace lighting, Part 2.5: Hospital and medical tasks, AS/NZS 1680.2.5:2018*, SAI Global, Sydney, https://www.standards.org.au/access-standards

Stevens, R.G., Hansen, J., Costa, G., Haus, E., Kauppinen, T., Aronson, K.J., Castaño-Vinyals, G., Davis, S., Frings-Dresen, M.H.W., Fritschi, L., Kogevinas, M., Kogi, K., Lie, J.A., Lowden, A., Peplonska, B., Pesch, B., Pukkala, E., Schernhammer, E., Travis, R.C. & Straif, K. 2011. 'Considerations of Circadian Impact for Defining "shift work" in Cancer Studies: IARC Working Group Report', *Occupational and Environmental Medicine*, vol. 68, no. 2, pp.154–154. https://doi.org/10.1136/oem.2009.053512

Sui, G.Y., Liu, G.C., Liu, G.Y., Gao, Y.Y., Deng, Y., Wang, W.Y., Tong, S.H. & Wang. L. 2013, 'Is Sunlight Exposure a Risk Factor for Age-related Macular Degeneration? A Systematic Review and Meta-analysis', *British Journal of Ophthalmology*, vol. 97, no. 4, pp. 389–94. doi: https://doi.org/10.1136/bjophthalmol-2012-302281

UL 2020, *UL Publishes Lighting Design Guideline for Circadian Entrainment*, DG 24480, UL Solutions, https://www.ul.com/news/ul-publishes-lighting-design-guideline-circadian-entrainment [29 January 2024].

US National Eye Institute 2022, *How Eyes Work*, National Eye Institute, https://www.nei.nih.gov/learn-about-eye-health/healthy-vision/how-eyes-work [31 January 2024].

Watson, L.M., Strang, N.C., Scobie, F., Love, G.D., Seidel, D. & Manahilov, V. 2012, 'Image Jitter Enhances Visual Performance When Spatial Resolution Is Impaired', *Investigative Ophthalmology & Visual Science*, vol. 53, no. 10, pp. 6004–10. doi: https://doi.org/10.1167/iovs.11-9157

Wetterburg, L. 1990, 'Lighting: Nonvisual Effects', *Scandinavian Journal of Work Environmental Health*, vol. 16, suppl. 1, pp. 26–28.

Yoshimoto, S., Garcia, J., Jiang, F., Wilkins, A.J., Takeuchi, T. & Webster, M.A. 2017, 'Visual Discomfort and Flicker', *Vision Research*, vol. 138, pp. 18–28. doi: https://doi.org/10.1016/j.visres.2017.05.015

18 Occupational Hygiene Tools and Sources

Dr Sue Reed, Linda Apthorpe, Dr Adélle Liebenberg and Ian Firth

18.1 INTRODUCTION TO OCCUPATIONAL HYGIENE TOOLS AND SOURCES

Access to information is much easier compared to earlier years, particularly for H&S practitioners. Most information sources can now be accessed for free via the internet or through membership with relevant associations. Some occupational hygiene tools may also incur a licence fee before a full copy of the tool can be accessed. Note that many of the resources listed in this chapter do not necessarily have the amount of detail or level of precision for formal occupational hygiene measurements, and some require additional interpretation of the results obtained.

Besides written literature, other resources include smartphone applications (apps) that may only work on a specific platform (for example Iphone or Android phones). Most of the relevant apps have been indicated in the respective chapters.

In addition to the topic-specific resources cited in the chapters, the following resources are useful to any H&S practitioner.

Note that the web links were correct and apps current at the time of printing.

18.2 GENERAL SOURCES AND KNOWLEDGE COLLECTION

AIOH Conference Proceedings	https://www.aioh.org.au/annual-conference/
AIOH Publications & Papers	https://www.aioh.org.au/education/publications/
ILO, *Encyclopaedia of Occupational Health and Safety*	https://www.iloencyclopaedia.org/
NSW Centre for WHS	https://www.centreforwhs.nsw.gov.au/
Occupational Hygiene Training Association (OHTA)	http://www.ohlearning.com
Safe Work Australia *Model Code of Practice: How to manage work health and safety risks*	https://www.safeworkaustralia.gov.au/
Silverstein, B.D. 2008, *AIHA Value Strategy Manual*	https://online-ams.aiha.org/amsssa/ecssashop.show_product_detail?p_mode=detail&p_product_serno=1084
The OHS Body of Knowledge	https://www.ohsbok.org.au/

DOI: 10.1201/9781032645841-18

18.3 EXPOSURE STANDARDS

EH40/2005 Workplace exposure limits	https://www.hse.gov.uk/pubns/priced/eh40.pdf
GESTIS International Limit Values	https://limitvalue.ifa.dguv.de/
OSHA Permissible Exposure Limits – Annotated Tables	https://www.osha.gov/annotated-pels
Safe Work Australia – Workplace exposure standards for airborne contaminants (2024)	https://www.safeworkaustralia.gov.au/doc/workplace-exposure-standards-airborne-contaminants-2024

18.4 OCCUPATIONAL HYGIENE STATISTICS

Advanced Reach Tools (ART) 1.5: 'Advanced Reach Tool' https://www.advancedreachtool.com/

This tool incorporates a mechanistic model of inhalation exposure and a statistical facility to update the estimates with measurements selected from an in-built exposure database or the user's own data.

BWStat https://www.bsoh.be/nl/bwstat

This Excel tool allows the estimation of, from three or more measurements, group and individual compliance with the OEL, taking into account the intra- and inter-individual variations in repeated measurements from the same individuals, and measurements in different individuals from a similar exposure group.

Expostats https://expostats.ca/site/en/

Expostats is a Bayesian calculation engine allowing estimation of parameters of the distribution of exposure for a worker or group of workers. The user enters measurement data, and the website performs calculations and returns risk metric estimates as well as uncertainty estimates.

IHSTAT https://aiha-assets.sfo2.digitaloceanspaces.com/AIHA/resources/StrategyBook4/Appendix-IV/EXPASSVG-IHSTATmacrofree.xls

IHSTAT macro-free version is an Excel application that calculates various exposure statistics, performs goodness of fit tests and graphs exposure data.

18.5 CONTROL

Breathe Freely Australia	https://www.breathefreelyaustralia.org.au/
HSE – *Local exhaust ventilation (LEV): workplace fume and dust extraction*	https://www.hse.gov.uk/lev/index.htm
NIOSH – *Control Banding*	https://www.cdc.gov/niosh/topics/ctrlbanding/
NIOSH – *Directory of Engineering Controls*	https://www.cdc.gov/niosh/engcontrols/
RESP-FIT	https://respfit.org.au/

18.6 CHEMICAL HAZARDS

Asbestos and Silica Safety and Eradication Agency	https://www.asbestossafety.gov.au
COSHH Regulations	https://books.hse.gov.uk/COSHH
European Centre for Ecotoxicology and Toxicology of Chemicals (ECETOC)	https://www.ecetoc.org
IFA – GESTIS Substance Database	https://www.dguv.de/ifa/gestis/gestis-stoffdatenbank/index-2.jsp>https://gestis-database.dguv.de/
IFA – GESTIS-DUST-EX – - Database Combustion and explosion characteristics of dusts	https://www.dguv.de/ifa/gestis/gestis-staub-ex/index-2.jsp
IFA – GESTIS – Analytical Methods for Chemical	https://www.dguv.de/ifa%3B/gestis/gestis-analysenverfahren-fuer-chemische-stoffe/index-2.jsp
NIOSH – *Chemicals: Managing Chemical Safety in the Workplace*	https://www.c-dc.gov/niosh/chemicals/
NIOSH – *Manual of Analytical Methods (NMAM)*, 5th Edition	https://www.cdc.gov/niosh/nmam/default.html
NIOSH – *Pocket Guide to Chemical Hazards*	https://www.cdc.gov/niosh/npg/
Safe Work Australia – *Chemicals*	https://www.safeworkaustralia.gov.au/safety-topic/hazards/chemicals
UK Health and Safety Executive – Methods for the Determination of Hazardous Substances (MDHS) guidance	https://www.hse.gov.uk/pubns/mdhs/

18.7 PHYSICAL HAZARDS

Australian Radiation Protection and Nuclear Safety Agency (ARPANSA)	https://www.arpansa.gov.au/
Safe Work Australia – *Noise*	https://www.safeworkaustralia.gov.au/safety-topic/hazards/noise
Safe Work Australia – *Laser classifications*	https://www.safeworkaustralia.gov.au/safety-topic/hazards/laser-classifications
Safe Work Australia – *Working in heat*	https://www.safeworkaustralia.gov.au/safety-topic/hazards/working-heat

18.8 BIOLOGICAL HAZARDS

Safe Work Australia – *National Hazard Exposure Worker Surveillance: Exposure to biological hazards and the provision of controls against biological hazards in Australian workplaces*	https://www.safeworkaustralia.gov.au/resources-and-publications/reports/national-hazard-exposure-worker-surveillance-exposure-biological-hazards-and-provision-controls-against-biological-hazards-australian-workplaces
GESTIS *Biological Agents Database*	https://www.dguv.de/ifa/gestis/gestis-biostoffdatenbank/index-2.jsp
HSE – *Biosafety and microbiological containment*	https://www.hse.gov.uk/biosafety/index.htm
HSE – *The Approved List of biological agents Advisory Committee on Dangerous Pathogens*	https://www.hse.gov.uk/pubns/misc208.pdf

18.9 RELEVANT OCCUPATIONAL HEALTH AND HYGIENE ORGANISATIONS AND PROFESSIONAL BODIES

18.9.1 Australian Societies

Australian Institute of Occupational Hygienists	https://www.aioh.org.au/
Australasian Faculty of Occupational Medicine	http://www.racp.edu.au/
Australian Institute of Health & Safety (AIHS)	https://www.aihs.org.au/
Australian Radiation Protection Accreditation Board	http://www.arpab.net.au/
Australian Radiation Protection Society	http://www.arps.org.au/
Australian Welding Institute (AWI)	https://welding.org.au/
Clean Air Society of Australia & New Zealand	http://www.casanz.org.au/home/Default.aspx
Human Factors & Ergonomics Society of Australia	http://www.ergonomics.org.au/
Risk Management Institution of Australasia	http://www.rmia.org.au/
The Australasian College of Toxicology & Risk Assessment (ACTRA)	https://actra.org.au/

18.9.2 International Societies

American Conference of Governmental Industrial Hygienists	http://www.acgih.org/
American Industrial Hygiene Association	https://www.aiha.org/Pages/default.aspx
Asian Network Occupational Hygiene (ANOH)	https://www.anoh.net/html/
Board for Global EHS Credentialing (BGC) (previously the American Board of Industrial Hygiene)	https://gobgc.org/healthandsafety/
British Occupational Hygiene Society (BOHS)	http://www.bohs.org/
Canadian Registration Board of Occupational Hygiene	http://www.crboh.ca/
International Commission on Occupational Health (ICOH)	https://www.icohweb.org/
International Ergonomics Association (IEA)	https://iea.cc/
International Occupational Hygiene Association (IOHA)	http://www.ioha.net/
International Radiation Protection Association (IRPA)	https://www.irpa.net/
International Society for Respiratory Protection (ISRP)	https://www.isrp.com/
New Zealand Occupational Hygiene Society	http://www.nzohs.org.nz/
Occupational Hygiene Training Association	http://www.ohlearning.com/
South African Institute of Occupational Safety and Health	http://www.saiosh.co.za/
Workplace Health Without Borders	https://whwb.org/

18.10 RELEVANT GOVERNMENT DEPARTMENTS ON OCCUPATIONAL HEALTH AND HYGIENE

18.10.1 Australian Government Sites

18.10.1.1 National

Australian Radiation Protection and Nuclear Safety Agency (ARPANSA)	https://www.arpansa.gov.au/
National Industrial Chemicals Notification & Assessment Scheme (NICNAS)	http://www.nicnas.gov.au/
Safe Work Australia	http://www.safeworkaustralia.gov.au/

18.10.1.2 Australian Capital Territory

WorkSafe ACT	http://www.worksafe.act.gov.au/health_safety

18.10.1.3 New South Wales

SafeWork New South Wales	https://www.safework.nsw.gov.au/
TestSafe Australia – *Chemical analysis branch handbook* (9th Edn)	https://www.nsw.gov.au/sites/default/files/2022-02/TestSafe-Chemical-Analysis-Branch-Handbook-9th-edition-TS033.pdf

18.10.1.4 Northern Territory

NT WorkSafe	http://www.worksafe.nt.gov.au/

18.10.1.5 Queensland

WorkSafe Queensland	https://www.worksafe.qld.gov.au/
Workplace Health and Safety Queensland	https://www.worksafe.qld.gov.au/about/who-we-are/workplace-health-and-safety-queensland
WorkCover Queensland	https://www.worksafe.qld.gov.au/about/who-we-are/workcover-queensland

18.10.1.6 South Australia

SafeWork South Australia	https://www.safework.sa.gov.au/

18.10.1.7 Tasmania

WorkCover Tasmania	http://www.workcover.tas.gov.au/

18.10.1.8 Victoria

WorkSafe Victoria	http://www.worksafe.vic.gov.au/

18.10.1.9 Western Australia

Department of Energy, Mines, Industry Regulation and Safety (DEMIRS)	https://www.dmirs.wa.gov.au/
WorkCover Western Australia	https://www.workcover.wa.gov.au/
WorkSafe Western Australia	https://www.commerce.wa.gov.au/worksafe

18.10.2 International Government Sites

Canada:	
Canadian Centre for Occupational Health & Safety	http://www.ccohs.ca/
Institut de recherche Robert-Sauvé en santé et en sécurité du travail (IRSST)	https://www.irsst.qc.ca/en/
European:	
European Agency for Safety & Health at Work (EU-OSHA)	http://osha.europa.eu/
New Zealand:	
WorkSafe	https://www.worksafe.govt.nz/
United Kingdom:	
Health & Safety Executive	http://www.hse.gov.uk/
Health & Safety Laboratory	http://www.hsl.gov.uk/
United States:	
Centers for Disease Control and Prevention	http://www.cdc.gov/
Mine Safety and Health Administration (MSHA)	http://www.msha.gov/
National Institute for Occupational Safety & Health	http://www.cdc.gov/niosh
Occupational Safety & Health Administration	https://www.osha.gov/

18.10.3 Non-governmental Organisations (NGO)

International Agency for Research on Cancer (IARC)	http://www.iarc.fr/
International Labour Organisation	http://www.ilo.org/public/english/protection/safework/
World Health Organisation	http://www.who.int/occupational_health/en/
National Association of Testing Authorities – Australia	http://www.nata.asn.au/

18.11 OCCUPATIONAL HYGIENE TOOLS

AIHA Apps and eTools Resource Center	https://www.aiha.org/public-resources/consumer-resources/apps-and-tools-resource-center
AIOH Tools & Calculators	https://www.aioh.org.au/resources/tools/
HSE Publications and products	https://www.hsl.gov.uk/publications-and-products
Advanced Reach Tools (ART) 1.5: '*Advanced Reach Tool*'	https://www.advancedreachtool.com/
Stoffenmanager®	https://stoffenmanager.com/en/

18.12 SEARCHING FOR SOURCES

The easiest way to access articles is to find 'Open Access' articles which can be found at Google Scholar (https://scholar.google.com.au/. You need to type the keywords of the topic you are interested in. Links to articles that match your keyword search will be displayed, and those with a link to a pdf file are usually articles available through 'Open Access', are often shown with the symbol 🔓 or the words '*open access*'.

If you have graduated from a university, they may offer lifetime free (or for a small annual fee) access to online databases, e-Journals and on campus libraries. Contact your Institution for further information.

Some journals that provide limited access to Open Access papers and others may charge a fee per article to download.

Open Access (free to download journals) in the field of Occupational Health and Hygiene include:

Environmental and Occupational Health Practice	https://eohp-journal.jp/
Environmental Health	https://ehjournal.biomedcentral.com/
International Journal of Environmental Research and Public Health	https://www.mdpi.com/journal/IJERPH
International Journal of Occupational Hygiene	https://ijoh.tums.ac.ir/index.php/ijoh
International Journal of Occupational Medicine and Environmental Health	http://ijomeh.eu/
International Journal of Occupational Safety and Health	https://www.nepjol.info/index.php/IJOSH
Journal of Environmental and Occupational Medicine	https://journals.lww.com/joem/pages/default.aspx
Journal of Occupational Health	Issues up to end of 2023: https://onlinelibrary.wiley.com/journal/13489585 Issues from beginning of 2024: https://academic.oup.com/joh?login=false
Journal of Occupational Medicine and Toxicology	https://occup-med.biomedcentral.com/
Occupational Medicine	https://academic.oup.com/occmed
Safety and Health at Work	https://www.sciencedirect.com/journal/safety-and-health-at-work
Scandinavian Journal of Work, Environment & Health	https://www.sjweh.fi/

Index

C

H

O

P

Q

R

S

T

U

V

Printed in the United States
by Baker & Taylor Publisher Services